Structural Chemistry across the Periodic Table

Structural Chemistry across the Periodic Table

Thomas Chung Wai Mak

Emeritus Professor and Wei Lun Research Professor,
Department of Chemistry,
The Chinese University of Hong Kong.

Yu-San Cheung

Senior Lecturer,
Department of Chemistry,
The Chinese University of Hong Kong.

Gong-Du Zhou

Professor, College of Chemistry and Molecular Engineering,
Peking University.

Ying-Xia Wang

Professor, College of Chemistry and Molecular Engineering,
Peking University.

OXFORD
UNIVERSITY PRESS

OXFORD
UNIVERSITY PRESS

Great Clarendon Street, Oxford, OX2 6DP,
United Kingdom

Oxford University Press is a department of the University of Oxford.
It furthers the University's objective of excellence in research, scholarship,
and education by publishing worldwide. Oxford is a registered trade mark of
Oxford University Press in the UK and in certain other countries

Published in the United States of America by Oxford University Press
198 Madison Avenue, New York, NY 10016, United States of America

British Library Cataloguing in Publication Data
Data available

Library of Congress Control Number: 2023931642

ISBN 978–0–19–887295–5

DOI: 10.1093/oso/9780198872955.001.0001

Printed and bound by
CPI Group (UK) Ltd, Croydon, CR0 4YY

Preface

Our book *Advanced Structural Inorganic Chemistry* (abbreviated as *ASIC*, Oxford University Press, 2008) matches the Chinese Second Edition (Peking University Press, 2006), which went through five reprints since the appearance of the Chinese First Edition in 2001. The Chinese Third Edition was printed in 2021.

In view of the rapid advances in bonding theory and synthetic inorganic chemistry in recent years, we consider it worthwhile to prepare a new book that is an expanded version of Part III: Structural Chemistry of Selected Elements in *ASIC* that includes updated material for the benefit of upper-class undergraduates and graduate students who aspire to pursue academic and industrial research careers. Selected material from significant advances in the past decade has been added, with particular emphasis on compounds that exemplify new types of bonding such as σ-hole, triel bond, tetrel bond, pnictogen bond, chalcogen bond, halogen bond, halogen–halogen interaction, aerogen bond, as well as quintuple and sextuple metal–metal bonds. Other new topics include an update on novel noble-gas compounds, a new half-chapter on the structural chemistry of actinide compounds that includes new types of metal-ligand δ and φ bonding, as well as sections on metallophilicity, heterometallic macrocycles and cages, com- and dis-proportionation reactions, hydrogen-bonded organic frameworks (HOFs), halogen-bonded organic frameworks, halogen–halogen interactions in supramolecular frameworks, covalent organic frameworks (COFs), and metal-organic frameworks (MOFs). The cut-off date for literature updating was set at the second half of 2021.

Sadly and regrettably, our *ASIC* co-author Professor Wai-Kee Li passed away in January 2016.* As both of us had become members of the Octogenarian Club several years ago, we managed to recruit Dr Yu-San Cheung (Professor Li's MPhil student and now Senior Lecturer at the Chinese University of Hong Kong) and Professor Ying-Xia Wang of Peking University as co-authors in the preparation of this book manuscript.

Broadly speaking, this updated volume faithfully reflects the evolutionary development of undergraduate and postgraduate curricula in General Chemistry, Chemical Bonding, Inorganic Chemistry, and X-Ray Crystallography at the Chinese University of Hong Kong and Peking University in the past half-century.

Thomas Chung Wai Mak*
The Chinese University of Hong Kong

Gong-Du Zhou
Peking University
September 2022

I should like to dedicate this volume to the memory of a brotherly colleague of over 45 years, who had left his indelible mark as a two-time winner of the "University Teaching Award" in the Faculty of Science at the Chinese University of Hong Kong.

Contents

Chapter 1

Structural Chemistry of Hydrogen

1.1 The Bonding Types of Hydrogen

Hydrogen is the third most abundant element (after oxygen and silicon in terms of the number of atoms) in the earth's crust and oceans. Hydrogen atom exists as three isotopes: ^{1}H (ordinary hydrogen, protium, H); ^{2}H (deuterium, D), and ^{3}H (tritium, T). Some of the important physical properties of the isotopes of hydrogen are listed in Table 1.1.1. Hydrogen forms more compounds than any other element, including carbon. This fact is a consequence of the structure of the hydrogen atom.

Table 1.1.1 Properties of hydrogen, deuterium, and tritium

Property	H	D	T
Abundance (%)	99.985	0.015	$\sim 10^{-16}$
Relative atomic mass	1.007825	2.014102	3.016049
Nuclear spin	½	1	½
Nuclear magnetic moment/μ_N	2.79270	0.85738	2.9788
NMR frequency (at 2.35 T)/MHz	100.56	15.360	104.68
Properties of diatomic molecules	H_2	D_2	T_2
mp/K	13.957	18.73	20.62
bp/K	20.30	23.67	25.04
T_c/K	33.19	38.35	40.6 (calc.)
Enthalpy of dissociation/kJ mol^{-1}	435.88	443.35	446.9
Zero point energy/kJ mol^{-1}	25.9	18.5	15.1

Hydrogen is the smallest atom with only one electron and one valence orbital, and its ground configuration is $1s^1$. Removal of the lone electron from the neutral atom produces the H^+ ion, which bears some similarity with an alkali metal cation, while adding one electron produces the H^- ion, which is analogous to a halide anion. Hydrogen may thus be placed logically at the top of either Group 1 or Group 17 in the Periodic Table. The hydrogen atom forms many bonding types in a wide variety of compounds, as described below.

(1) Covalent bond

The hydrogen atom can use its $1s$ orbital to overlap with a valence orbital of another atom to form a covalent bond, as in the molecules listed in Table 1.1.2. The covalent radius of hydrogen is 37 pm.

Structural Chemistry across the Periodic Table. Thomas Chung Wai Mak, Yu-San Cheung, Gong-Du Zhou, and Ying-Xia Wang. Oxford University Press. © Thomas Chung Wai Mak, Yu-San Cheung, Gong-Du Zhou, and Ying-Xia Wang (2023). DOI: 10.1093/oso/9780198872955.003.0001

Table 1.1.2 Molecules containing covalently bonded hydrogen

Molecules	H_2	HCl	H_2O	NH_3	CH_4
Bond	H–H	H–Cl	H–O	H–N	H–C
Bond lengths/pm	74.14	147.44	95.72	101.7	109.1

(2) Ionic bond

(a) Hydrogen present as H^-

The hydrogen atom can gain an electron to form the hydride ion H^- with the helium electronic configuration:

$$H + e^- \longrightarrow H^- \quad \Delta H = 72.8 \, kJ \, mol^{-1}$$

With the exception of beryllium, all the elements of Group 1 and 2 react spontaneously when heated in hydrogen gas to give white solid hydrides M^IH or $M^{II}H_2$. All MH hydrides have the sodium chloride structure, while MgH_2 has the rutile structure and CaH_2, SrH_2, and BaH_2 have the structure of distorted $PbCl_2$. The chemical and physical properties of these solid hydrides indicate that they are ionic compounds.

The hydride anion H^- is expected to be very polarizable, and its size changes with the partner in the MH or MH_2 compounds. The experimental values of the radius of H^- cover a rather wide range.

Compounds	LiH	NaH	KH	RbH	CsH	MgH_2
$r(H^-)$/pm	137	142	152	154	152	130

From NaH, which has the NaCl structure with cubic unit-cell parameter $a = 488$ pm, the radius of H^- has been calculated as 142 pm.

(b) Hydrogen present as H^+

The loss of an electron from a H atom to form H^+ is an endothermic process:

$$H(g) \longrightarrow H^+(g) + e^- \quad \Delta H = 1312.0 \, kJ \, mol^{-1} = 13.59 \, eV$$

In reality H^+, a bare proton, cannot exist alone except in isolation inside a high vacuum. The radius of H^+ is about 1.5×10^{-15} m, which is 10^5 times smaller than other atoms. When H^+ approaches another atom or molecule, it can distort the electron cloud of the latter. Therefore, other than in a gaseous ionic beam, H^+ must be attached to another atom or molecule that possesses a lone pair of electrons. The proton as an acceptor can be stabilized as in pyramidal hydronium (or hydroxonium) ion H_3O^+, tetrahedral NH_4^+, and linear H_2F^+. These cations generally combine with various anions through ionic bonding to form salts.

The total hydration enthalpy of H^+ is highly exothermic and much larger than those of other simple charged cations:

$$H^+(g) + n \, H_2O(\ell) \longrightarrow H_3O^+(aq) \quad \Delta H = -1090 \, kJ \, mol^{-1}$$

(3) Metallic bond

At ultrahigh pressure and low temperature, such as 250 GPa and 77 K, solid molecular hydrogen transforms to a metallic phase, in which the atoms are held together by the metallic bond, which arises from a band-overlap mechanism. Under such extreme conditions, the H_2 molecules are converted into a linear chain of hydrogen atoms (or a three-dimensional network). This polymeric H_n structure with a partially filled band (conduction band) is expected to exhibit metallic behavior. Schematically, the band-overlap mechanism may be represented in the following manner:

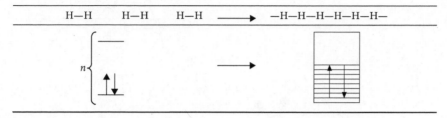

The physical properties of H_n are most interesting. This material has a nearly opaque appearance and exhibits metallic conduction, and has been suggested to be present in several planets.

(4) The hydrogen bond

The hydrogen bond is usually represented as X–H⋯Y, where X and Y are highly electronegative atoms such as F, O, N, Cl, and C. Note that the description of "donor" and "acceptor" in a hydrogen bond refers to the proton; in the hydrogen bond X–H⋯Y, the X–H group acts as a hydrogen donor to acceptor atom Y. In contrast, in a coordination bond the donor atom is the one that bears the electron pair.

Hydrogen bonding can be either intermolecular or intramolecular. For an intramolecular X–H⋯Y hydrogen bond to occur, X and Y must be in favorable spatial configuration in the same molecule. This type of hydrogen bond will be further elaborated in Section 1.2.

(5) Multi-center hydrogen bridged bonds

(a) B–H–B bridged bond

Boranes, carboranes, and metallocarboranes are electron-deficient compounds in which B–H–B three-center two-electron (3c-2e) bridged bonds are found. The B–H–B bond results from the overlap of two boron sp^3 hybrid orbitals and the $1s$ orbital of the hydrogen atom, which will be discussed in Chapter 3.

(b) M–H–M and M–H–B bridged bonds

The 3c-2e hydrogen bridged bonds of type M–H–M and M–H–B can be formed with main-group metals (such as Be, Mg, and Al) and transition metals (such as Cr, W, Fe, Ta, and Zr). Some examples are shown in Fig. 1.1.1.

(c) Triply-bridged (μ_3-H)M$_3$ bond

In this bond type, a H atom is covalently bonded simultaneously to three metal atoms, as shown in Table 1.1.1(d).

(6) Hydride coordinate bond

The hydride ion H^- as a ligand can donate a pair of electrons to a metal to form a metal hydride. The crystal structures of many metal hydride complexes, such as Mg_2NiH_4, Mg_2FeH_6 and K_2ReH_9, have been determined. In these compounds, the M–H bonds are covalent σ coordinate bonds.

(7) Molecular hydrogen coordinate bond

Molecular hydrogen can coordinate to a transition metal as an intact ligand. The bonding of the H_2 molecule to the transition metal atom appears to involve the transfer of σ bonding electrons of H_2 to a

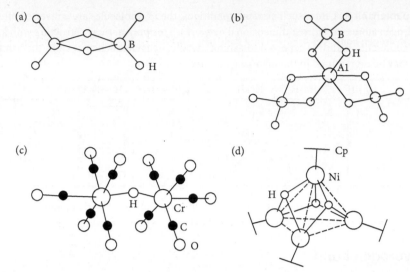

Fig. 1.1.1 Structures containing multiple hydrogen bridged bonds:(a) B_2H_6, (b) $Al(BH_4)_3$, (c) $[(CO)_5Cr–H–Cr(CO)_5]^-$, (d) $H_3Ni_4Cp_4$.

vacant metal d orbital, coupled with synergistic back donation of metal d electrons to the vacant σ^* antibonding orbital of the H_2 molecule. This type of coordinate bond formation weakens the H–H covalent bond of the H_2 ligand and, in the limit, leads to its cleavage to two H atoms.

(8) Agostic bond

The existence of the agostic bond C–H→M has been firmly established by X-ray and neutron diffraction methods. The symbolic representation C–H→M indicates formal donor interaction of a C–H bond with an electron-deficient metal atom M. As in all 3c-2e bridging systems involving only three valence orbitals, the bonded C–H→M fragment is bent. The agostic bond will be further discussed in section 1.5.3.

1.2 Hydrogen Bond

1.2.1 Nature and geometry of the hydrogen bond

In a typical hydrogen bonding system, X–H⋯Y, the electron cloud is greatly distorted toward the highly electronegative X atom, so that the thinly shielded, positively charged hydrogen nucleus becomes strongly attracted by the electronegative atom Y. The double- and triple-bonded carbon atoms, which have high electronegativity, can also form hydrogen bonds. Some examples are given in Fig. 1.2.1.

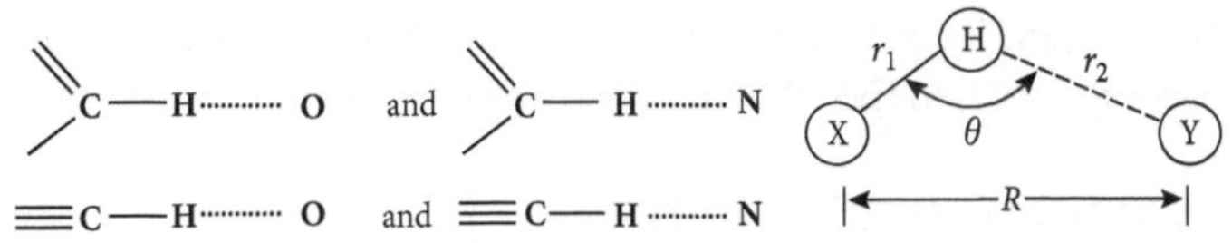

Fig. 1.2.1 Geometry of the hydrogen bond.

Table 1.2.1 Experimental hydrogen bond lengths and calculated values

Hydrogen bond	Experimental bond length/pm	Calculated value/pm
F–H·F	240	360
O–H·F	270	360
O–H·O	270	369
O–H·N	278	375
O–H·Cl	310	405
N–H·F	280	360
N–H·O	290	370
N–H·N	300	375
N–H·Cl	320	410
C–H·O	320	340

The geometry of the hydrogen bond can be described by the parameters R, r_1, r_2, and θ, as shown in Table 1.2.1. Numerous experimental studies have established the following generalizations about the geometry of hydrogen bonds:

(a) Most hydrogen bonds X–H···Y are unsymmetric; that is, the hydrogen atom is much closer to X than to Y. A typical example of hydrogen bonding is the interaction between H_2O molecules in ice-I_h. The data shown below are derived from a neutron diffraction study of deuterated ice-I_h at 100 K.

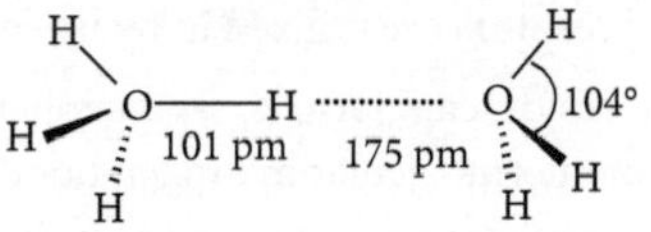

(b) Hydrogen bonds X–H···Y may be linear or bent, though the linear form is energetically more favorable. However, in the crystalline state the packing of molecules is the deciding factor.

(c) The distance between atoms X and Y is taken as the bond length of the hydrogen bond X–H···Y. Similar to all other chemical bonds, the shorter the bond length the stronger the hydrogen bond. As the bond length X···Y shortens, the X–H distance is lengthened. In the limit there is a symmetrical hydrogen bond, in which the H atom lies at the mid-point of the X···Y line. This is the strongest type of hydrogen bond and occurs only for both X and Y equal to F or O (see section 1.2.3).

(d) The experimental hydrogen bond length is in general much shorter than the sum of the X–H covalent bond length and van der Waals radii of H and Y atoms. For example, the average O–H···O hydrogen bond length is 270 pm, which is shorter than the sum (369 pm) of the O–H covalent bond length (109 pm) and van der Waals contact distance of H···O (120 pm + 140 pm). Table 1.2.1 compares the experimental X–H···Y hydrogen bond lengths and the calculated values by summing the X–H covalent bond length and van der Waals radii of H and Y atoms.

(e) The valence angle α formed between the H···Y line and the Y–R bond usually varies between 100° and 140°.

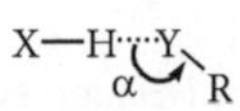

(f) Normally the H atom in a hydrogen bond is two-coordinate, but there are a fair number of examples of hydrogen bonds with three-coordinate and four-coordinate H atoms. From a survey of 1509 NH···O=C hydrogen bonds observed by X-ray and neutron diffraction in 889 organic crystal structures, 304 (about 20%) are found to be three-coordinate and only six are four-coordinate.

Three-coordinate
hydrogen bond

Four-coordinate
hydrogen bond

(g) In most hydrogen bonds only one hydrogen atom is directed toward a lone pair of Y, but there are many exceptions. For example, in crystalline ammonia each N lone pair accepts three hydrogen atoms, as shown in Fig. 1.2.2(a). The carbonyl O atom in the tetragonal phase of urea forms four acceptor hydrogen bonds. In the inclusion compound $[(C_2H_5)_4N^+]_2 \cdot CO_3^{2-} \cdot 7(NH_2)_2CS$, the carbonate ion proves to be the most prolific hydrogen-bond acceptor, being surrounded by 12 convergent NH donor groups from six thiourea molecules to form a hydrogen-bonded aggregate shaped like two concave three-leaved propellers sharing a common core, as illustrated in Fig. 1.2.2(b).

Organic compounds generally conform to the following generalized rules in regard to hydrogen bonding:

(a) All good proton donors and acceptors are utilized in hydrogen bonding.

(b) If intramolecular hydrogen bonds can form a six-membered ring, they will usually do so in preference to the formation of intermolecular hydrogen bonds.

(c) After the formation of intramolecular hydrogen bonds, the remaining set of best proton donors and acceptors tend to form an intermolecular hydrogen bond with one another.

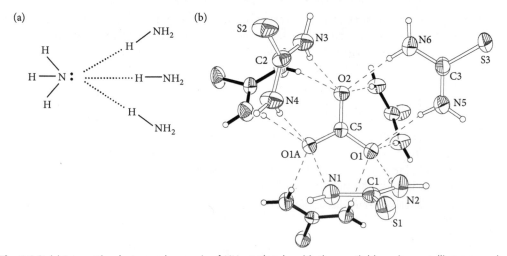

Fig. 1.2.2 (a) Interaction between lone pair of NH_3 molecule with three neighbors in crystalline ammonia. (b) Carbonate ion forming 12 acceptor hydrogen bonds with six thiourea molecules.

1.2.2 **IUPAC definition of the hydrogen bond**

The term "hydrogen bond" has been used in the scientific literature for nearly a century by physicists, chemists, biologists, and material scientists. In 2011 a new definition, *based on direct experimental evidence for a partial covalent nature and taking the debate on the observation of a blue shift in stretching frequency following X–H···Y hydrogen bond formation (X–H being the donor and Y being the acceptor) into account*, was proposed by an IUPAC Task Group on Categorizing Hydrogen Bonding and Other Intermolecular Interactions, *as follows*:

> The hydrogen bond is an attractive interaction between a hydrogen atom from a molecule or a molecular fragment X–H in which X is more electronegative than H, and an atom or group of atoms in the same or a different molecule, in which there is evidence of bond formation.

The Task Group adapted six criteria based on experimental evidence for hydrogen bond formation. For a hydrogen bond X–H···Y–Z:

(E1) The forces involved in the formation of a hydrogen bond include those of an electrostatic origin, those arising from charge transfer between the donor and acceptor leading to partial covalent bond formation between H and Y, and those originating from dispersion.

(E2) The atoms X and H are covalently bonded to one another and the X–H bond is polarized, the H···Y bond strength increasing with the increase in electronegativity of X.

(E3) The X–H···Y angle is usually linear (180°) and the closer the angle is to 180°, the stronger is the hydrogen bond and the shorter is the H···Y distance.

(E4) The length of the X–H bond usually increases on hydrogen bond formation, leading to a red shift in the infrared X–H stretching frequency and an increase in the infrared absorption cross-section for the X–H stretching vibration. The greater the lengthening of the X–H bond in X–H···Y, the stronger is the H···Y bond. Simultaneously, new vibrational modes associated with the formation of the H···Y bond are generated.

(E5) The X–H···Y–Z hydrogen bond leads to characteristic NMR signatures that typically include pronounced proton deshielding for H in X–H, through hydrogen bond spin–spin couplings between X and Y, and nuclear Overhauser enhancements.

(E6) The Gibbs energy of formation for the hydrogen bond should be greater than the thermal energy of the system for the hydrogen bond to be detected experimentally.

One criterion focuses on the geometrical requirement that the three atoms X–H···Y usually tend toward linearity. Two criteria relate to the physical forces involved: dispersion and electrostatic (X is required to be more electronegative than H). Two criteria are based on spectroscopy: IR (red shift in X–H vibrational frequency) and NMR (deshielding of H in XH). In addition, Gibbs energy is mentioned explicitly as both enthalpy and entropy changes are involved in hydrogen bond formation.

References

E. Arunan, G. R. Desiraju, R. A. Klein, J. Sadlej, S. Scheiner, I. Alkorta, D. C. Clary, R. H. Crabtree, J. J. Denenberg, P. Hobza, H. G. Kjaergaard, A. C. Legon, B. Mennucci, and D. J. Nesbitt, Defining the hydrogen bond: an account. *Pure Appl. Chem.* **83**, No. 8, 1619–36 (2011).

E. Arunan, G. R. Desiraju, R. A. Klein, J. Sadlej, S. Scheiner, I. Alkorta, D. C. Clary, R. H. Crabtree, J. J. Denenberg, P. Hobza, H. G. Kjaergaard, A. C. Legon, B. Mennucci, and D. J. Nesbitt, Definition of the hydrogen bond. *Pure Appl. Chem.* 83, No. 8, 1637–41 (2011).

M. Goswami and E. Arunan, The hydrogen bond: a molecular beam microwave spectroscopist's view with a universal appeal. *Phys. Chem. Chem. Phys.* **11**, 8974–83 (2009).

1.2.3 **The strength of hydrogen bonds**

The strongest hydrogen bonds resemble covalent bonds, the weakest ones are like van der Waals interactions, and the majority have energies lying between these two extremes. The strength of a hydrogen bond corresponds to the enthalpy of dissociation of the reaction:

$$X\text{–}H \cdots Y \longrightarrow X\text{–}H + Y$$

Strong and weak hydrogen bonds obviously have very different properties. Table 1.2.2 lists the properties observed for different types of hydrogen bonds.

In ice-I_h, the O–H⋯O bond energy is 25 kJ mol^{-1}, which results from the following interactions:

(a) Electrostatic attraction: this effect reduces the distance between the atoms of H⋯O.

$$O^{\delta-}\text{–}H^{\delta+} \cdots\cdots O^{\delta-}$$

(b) Delocalization or covalent bonding: the valence orbitals of H and O atoms overlap with each other, so that the bonding effect involves all three atoms.

(c) Electron cloud repulsion: the sum of the van der Waals radii of hydrogen and oxygen is 260 pm, and in a hydrogen bond the H⋯O distance often approaches to within 180 pm. Thus the normal electron–electron repulsive forces will occur.

(d) Van der Waals forces: as in all intermolecular interactions these forces contribute to the bonding, but their combined effect is relatively small.

The results of a molecular orbital calculation of the energies involved in the O–H⋯O system are tabulated in Table 1.2.3.

Table 1.2.2 Properties of very strong, strong, and weak hydrogen bonds

Property	very strong	strong	weak
X—H⋯Y interaction	mostly covalent	mostly electrostatic	electrostatic
Bond lengths	X–H ≈ H–Y	X–H < H⋯Y	X–H ≪ H⋯Y
H⋯Y/ pm	120–150	150–220	220–320
X⋯Y/ pm	220–250	250–320	320–400
Bond angles	175° ~ 180°	130° ~ 180°	90°–150°
Bond energy/kJ mol^{-1}	> 50	15–50	< 15
Relative IR ν_s vibration shift/cm^{-1} *	> 25%	10 ~ 25%	< 10%
^{1}H NMR chemical shift downfield/ppm	14 ~ 22	< 14	–
Examples	acid salts, acids, proton sponges, HF complexes	acids, alcohols, hydrates, phenols, biological molecules	weak base, basic salts, C–H⋯O/N O/N–H⋯π

* Observed ν_s relative to ν_s for a non-hydrogen bonded X–H.

Table 1.2.3 The energy contributions in a O–H···O hydrogen bond

Type of energy contribution	Energy/kJ mol^{-1}
(a) Electrostatic	−33.4
(b) Delocalization	−34.1
(c) Repulsion	41.2
(d) van der Waals energy	−1.0
Total energy	−27.3
Experimental	−25.0

1.2.4 Symmetrical hydrogen bond

The strongest hydrogen bond occurs in symmetrical O–H–O and F–H–F systems. The linear HF_2^- ion has the H atom located midway between the two F atoms:

$$\text{113 pm} \quad \text{113 pm}$$
$$\text{F} \underline{\quad\quad} \text{H} \underline{\quad\quad} \text{F}$$

This symmetrical hydrogen bond is highly covalent, which may be viewed as a 3c-4e system. If the molecular axis is taken along the z direction, the $1s$ orbital of the H atom overlaps with the two $2p_z$ orbitals of the F atoms (A and B), as shown in Fig. 1.2.3(a), to form three molecular orbitals:

$$\psi_1\,(\sigma) = N_1\,[2p_z\,(A) + 2p_z\,(B) \;+\; c1s]$$
$$\psi_2\,(n) = N_2[2p_z\,(A) - 2p_z\,(B)]$$
$$\psi_3\,(\sigma^*) = N_3[2p_z\,(A) + 2p_z\,(B) - c1s]$$

where c is a weighting coefficient and N_1, N_2, N_3 are normalization constants. The ordering of the molecular orbitals is shown qualitatively in Fig. 1.2.3(b). Since there are four valence electrons, the bonding (ψ_1) and non-bonding (ψ_2) molecular orbitals are both occupied to yield a 3c-4e bond. The bond order of each F–H link in HF_2^- is 0.5, which can be compared with that in the HF molecule:

Molecule	Bond order	d/pm	k/N m^{-1}
HF	1	93	890
HF$_2^-$	0.5	113	230

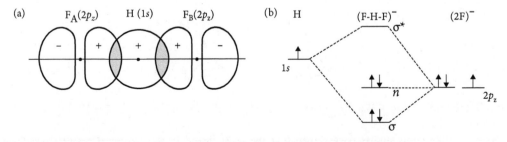

Fig. 1.2.3 Bonding in HF$_2^-$: (a) orbital overlap; (b) qualitative MO energy level diagram.

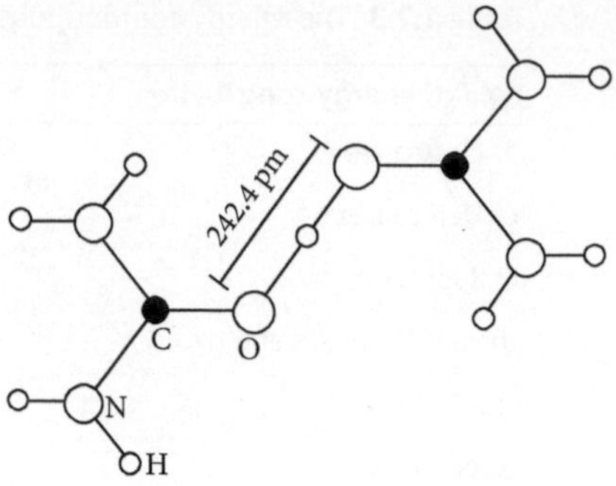

Fig. 1.2.4 Structure of $\{[(NH_2)_2CO]_2H\}^+$.

In the $\{[(NH_2)_2CO]_2H\}(SiF_6)$ crystal, there are symmetrical hydrogen bonds of the type O–H–O, with bond lengths 242.4 and 244.3 pm in two independent $\{[(NH_2)_2CO]_2H\}^+$ cations. Fig. 1.2.4 shows the structure of the cation.

1.2.5 **Hydrogen bonds in organometallic compounds**

In transition metal carbonyls, the M–C=O group acts as a proton acceptor, which interacts with appropriate donor groups to form one or more C=O⋯H–X (X is O or N) hydrogen bonds. The CO ligand can function in different μ_1, μ_2, and μ_3 modes, corresponding to a formal C–O bond order of 3, 2, and 1, respectively. Examples of hydrogen bonding formed by metal carbonyl complexes are shown below:

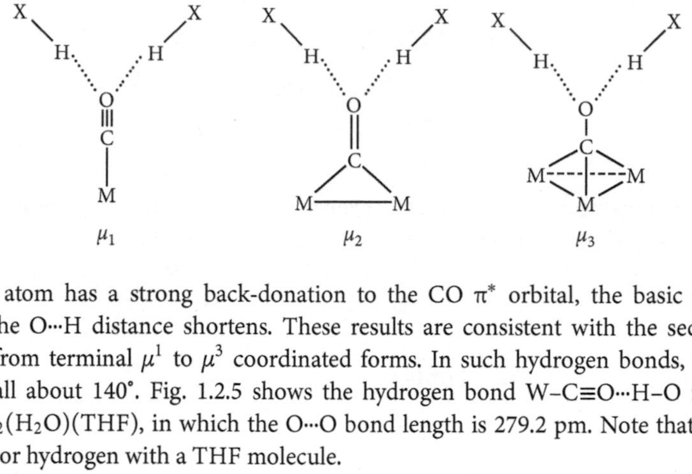

$$\mu_1 \qquad \mu_2 \qquad \mu_3$$

When the M atom has a strong back-donation to the CO π^* orbital, the basic property of CO increases, and the O⋯H distance shortens. These results are consistent with the sequence of shortened distances from terminal μ^1 to μ^3 coordinated forms. In such hydrogen bonds, the bond angles of C–O⋯X are all about 140°. Fig. 1.2.5 shows the hydrogen bond W–C≡O⋯H–O in the crystal of $W(CO)_3(P^iPr_3)_2(H_2O)(THF)$, in which the O⋯O bond length is 279.2 pm. Note that the aqua ligand also forms a donor hydrogen with a THF molecule.

In organometallic compounds, the μ_3-CH and μ_2-CH$_2$ ligands can act as proton donors to form hydrogen bonds, as shown below:

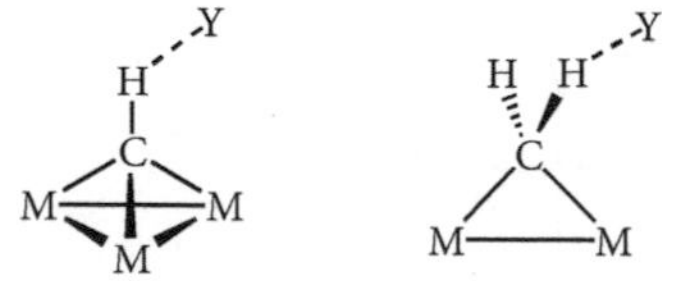

Since the acidity of μ_3-CH is stronger than that of μ_2-CH$_2$, the length of a μ_3-CH hydrogen bond is shorter than that of μ_2-CH$_2$.

Neutron diffraction studies have shown that the –CH ligand in the cluster $[Co(CO)_3]_3(\mu_3\text{-CH})$ forms three hydrogen bonds, with H⋯O distances of 250, 253, and 262 pm. In the compound $[CpMn(CO)_2]_2(\mu_2\text{-CH}_2)$, the μ_2-CH$_2$ group forms hydrogen bonds with the O atoms of Mn–CO groups, as shown in Fig. 1.2.6.

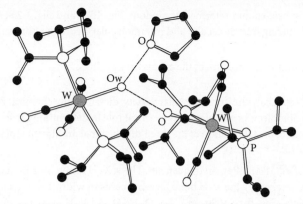

Fig. 1.2.5 The hydrogen bond W–C≡O···H–O in crystalline W(CO)$_3$(P^{i}Pr$_3$)$_2$(H$_2$O)(THF).

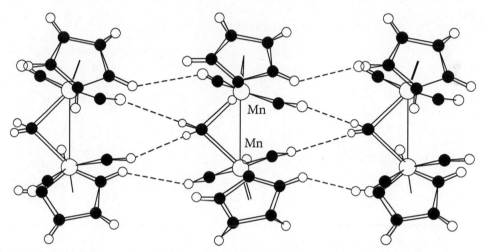

Fig. 1.2.6 C–H···O hydrogen bonds between μ_2-CH$_2$ and Mn–CO groups in [CpMn(CO)$_2$]$_2$(μ_2-CH$_2$).

1.2.6 **The universality and importance of hydrogen bonds**

Hydrogen bonds exist in numerous compounds. The reasons for its universal appearance are as follows:

(a) The abundance of H, O, N, C, and halogen elements. Many compounds are composed of H, O, N, C, and halogens, such as water, HX, oxyacids, and organic compounds. These compounds generally contain functional group such as –OH, –NH$_2$, and >C=O, which readily form hydrogen bonds.

(b) Geometrical requirement of hydrogen bonding. A hydrogen bond does not require rigorous conditions for its formation as in the case for covalent bonding. The bond lengths and group orientations allow for more flexibility and adaptability.

(c) Small bond energy. The hydrogen bond is intermediate in strength between the covalent bond and van der Waals interaction. The small bond energy of the hydrogen bond requires low activation energy in its formation and cleavage. Its relative weakness permits reversibility in reactions involving its formation and a greater subtlety of interaction than is possible with normal covalent bonds.

(d) Intermolecular and intramolecular bonding modes. Hydrogen bonds can form between molecules, within the same molecule, or in a combination of both varieties. In liquids, hydrogen bonds are continuously being broken and reformed at random. Fig. 1.2.7 and Fig. 1.2.8, respectively, display some structures featuring intermolecular and intramolecular hydrogen bonds.

Hydrogen bonds are important because of the effects they produce:

(a) Hydrogen bonds, especially the intramolecular variety, dictate many chemical properties, influence the conformation of molecules and often play a critical role in determining reaction rates. Hydrogen bonds are responsible for stabilization of the three-dimensional architecture of proteins and nucleic acids.

(b) Hydrogen bonds affect IR and Raman frequencies. The $\nu(X–H)$ stretching frequency shifts to a lower energy (caused by weakening the X–H bond) but increases in width and intensity. For N–H$\cdots$F, the change of frequency is less than 1000 cm^{-1}. For O–H$\cdots$O and F–H$\cdots$F, the change of frequency is in the range 1500–2000 cm^{-1}. The $\nu(X–H)$ bending frequency shifts to higher wave numbers.

(c) Intermolecular hydrogen bonding in a compound raises the boiling point and frequently the melting point.

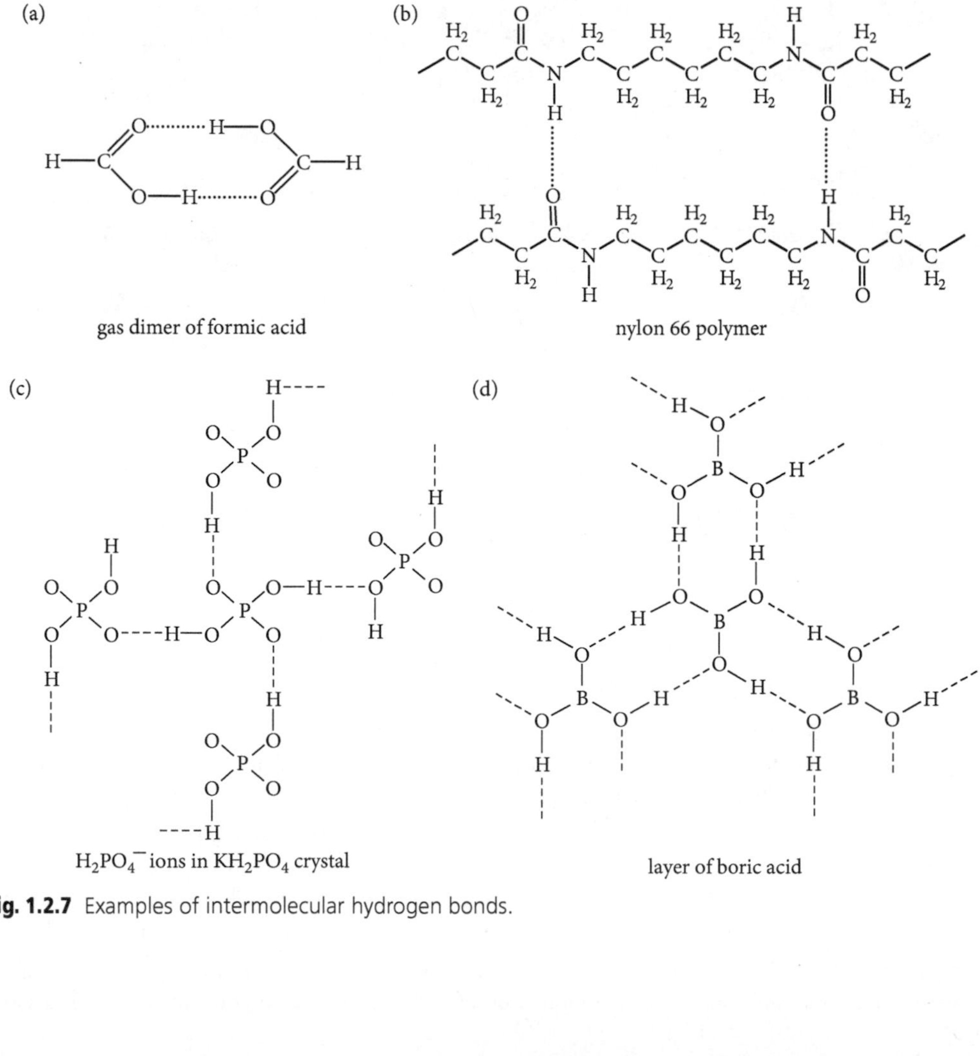

Fig. 1.2.7 Examples of intermolecular hydrogen bonds.

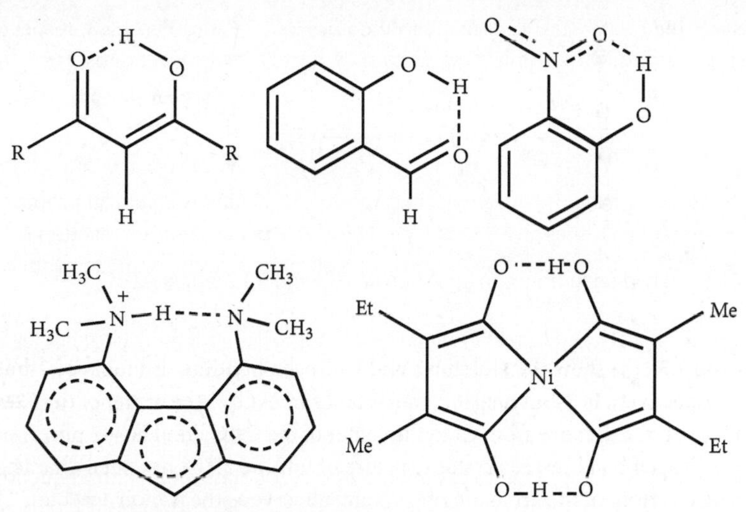

Fig. 1.2.8 Examples of intramolecular hydrogen bonds.

(d) If hydrogen bonding is possible between solute and solvent, solubility is greatly increased and often results in infinite solubility. The complete miscibility of two liquids, for example water and ethanol, can be attributed to intermolecular hydrogen bonding.

1.3 Non-conventional Hydrogen Bonds

The conventional hydrogen bond X–H⋯Y is formed by the proton donor X–H with a proton acceptor Y which is an atom with a lone pair (X and Y are all highly electronegative atoms such as F, O, N, and Cl). Some non-conventional hydrogen bonds that do not conform to this condition are discussed below:

1.3.1 X–H⋯π hydrogen bond

In a X–H⋯π hydrogen bond, π bonding electrons interact with the proton to form a weakly bonded system. The phenyl ring and delocalized π system as proton acceptors interact with X–H to form X–H⋯π hydrogen bonds. The phenyl group is by far the most important among π-acceptors, and the X–H⋯Ph hydrogen bond is termed an "aromatic hydrogen bond." The N–H and phenyl groups together form aromatic hydrogen bonds which stabilize the conformation of polypeptide chains. Calculations show that the bond energy of a N–H⋯Ph bond is about 12 kJ mol^{-1}. Two major types of N–H⋯Ph hydrogen bonds generally occur in the polypeptide chains of biomolecules:

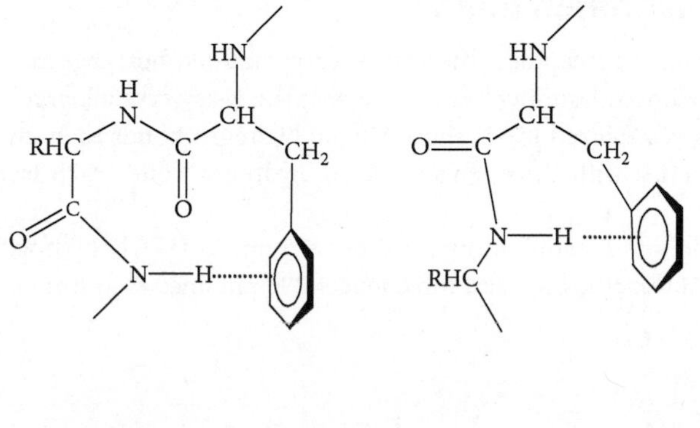

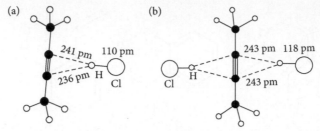

Fig. 1.3.1 The Cl–H⋯π hydrogen bonds in (a) 2-butyne·HCl and (b) 2-butyne·2HCl.

Figs. 1.3.1(a) and 1.3.1(b) show the structures and hydrogen bonding distances of 2-butyne·HCl and 2-butyne·2HCl, respectively. In 2-butyne·HCl, the distance from Cl to the center of the C≡C bond is 340 pm. In 2-butyne·2HCl, the distance from Cl to the center of the C≡C bond is 347 pm.

The Cl–H⋯π hydrogen bond in the crystal structure of toluene·2HCl has been characterized. In this case the π bonding electrons of the aromatic ring of toluene serve as the proton acceptor. The distance of the H atom to the ring center is 232 pm, as shown in Fig. 1.3.2.

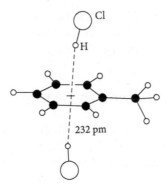

Fig. 1.3.2 Cl–H⋯π hydrogen bonds in the crystal of toluene·2HCl.

In addition to N–H⋯π and Cl–H⋯π hydrogen bonds, there are also O–H⋯π and C–H⋯π hydrogen bonds in many compounds. Two examples are shown in Fig. 1.3.3(a) and (b). A neutron diffraction study of the crystal structure of 2-acetylenyl-2-hydroxyl adamantane (Fig. 1.3.3(c)) has shown that intermolecular O–H⋯O and C–H⋯O hydrogen bonds coexist [Fig. 1.3.3(d)] with the O–H⋯π hydrogen bonds (Fig. 1.3.3(e)).

1.3.2 **C–H⋯Cl⁻ hydrogen bond**

A chloride-selective synthetic receptor in the form of a cryptand-like host cage molecule (Fig. 1.3.4) using multiple C–H⋯Cl⁻ hydrogen bonding has been assembled. X-ray crystallography established that the chloride guest species is stabilized by six short 270 pm hydrogen bonds each involving a 1,2,3-triazole CH donor group, together with three longer 290 pm hydrogen bonds each from a phenylene group (Fig. 1.3.5).

The central chloride ion is stabilized by six short 270 pm C–H⋯Cl⁻ hydrogen bonds each involving a 1,2,3-triazole CH donor group, and three longer 290 pm ones each from a phenylene CH donor group.

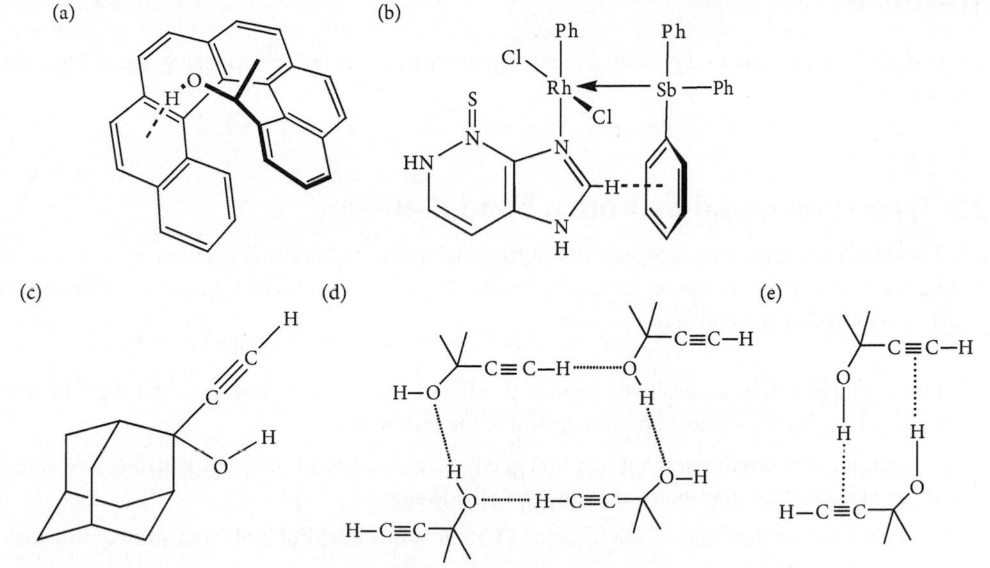

(a)

(b)

(c)

(d)

(e)

Fig. 1.3.3 Some compounds containing O–H···π and C–H···π hydrogen bonds.

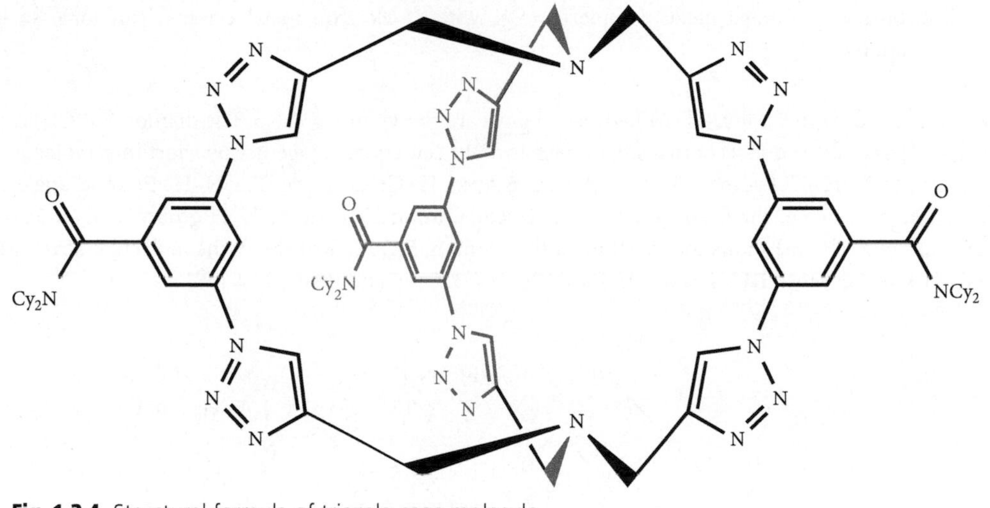

Fig. 1.3.4 Structural formula of triazolo cage molecule.

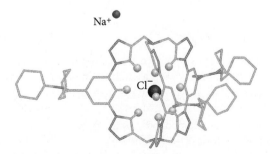

Fig. 1.3.5 Structure of triazolo cage·NaCl

Reference

Y. Liu, W. Zhao, C.-H. Chen, and A. H. Flood, Chloride capture using a C–H hydrogen bonding cage, *Science*. **365**, 159–61 (2019).

1.3.3 Transition metal hydrogen bond X–H···M

The X–H···M hydrogen bond is analogous to a conventional hydrogen bond and involves an electron-rich transition metal M as the proton acceptor in a 3c-4e interaction. Several criteria that serve to characterize a 3c-4e X–H···M hydrogen bond are:

(a) The bridging hydrogen is covalently bonded to a highly electronegative atom X and is protonic in nature, enhancing the electrostatic component of the interaction.

(b) The metal atom involved is electron rich, that is, typically a late transition metal, with filled d-orbitals that can facilitate the 3c-4e interaction involving the H atom.

(c) The ^{1}H NMR chemical shift of the bridging H atom is downfield of TMS and shifted downfield relative to the free ligand.

(d) Intermolecular X–H···M interactions have an approximately linear geometry.

(e) Electronically saturated metal complexes (e.g., with 18-electron metal centers) can form such interactions.

Two compounds with 3c-4e X–H···M hydrogen bonds are shown in Fig. 1.3.6. The dianion {(PtCl$_4$)·cis-[PtCl$_2$(NH$_2$Me)$_2$]}$^{2-}$ consists of two square-planar d^8-Pt centers held together by short intermolecular N–H···Pt and N–H···Cl hydrogen bonds: H···Pt 226.2 pm, H···Cl 231.8 pm. The N–H···Pt bond angle is 167.1°. The presence of the filled Pt $d_z{}^2$ orbital oriented toward the amine N–H group favors a 3c-4e interaction. Fig. 1.3.6(a) shows the structure of this dianion. Fig. 1.3.6(b) shows the molecular structure of PtBr(1-C$_{10}$H$_6$NMe$_2$)(1-C$_{10}$H$_5$NHMe$_2$) with Pt···N = 328 pm and Pt···H–N = 168°.

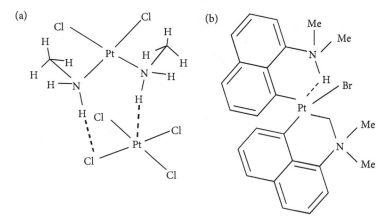

Fig. 1.3.6 Molecular structures of compounds with X–H···M hydrogen bond:(a) {(PtCl$_4$)·cis-[PtCl$_2$(NH$_2$Me)$_2$]}$^{2-}$, (b) PtBr(1-C$_{10}$H$_6$NMe$_2$)(1-C$_{10}$H$_5$NHMe$_2$).

1.3.4 **Dihydrogen bond X–H⋯H–E**

Conventional hydrogen bonds are formed between a proton donor, such as an O–H or N–H group and a proton acceptor, such as oxygen or nitrogen lone pair. In all such cases a non-bonding electron pair acts as the weak base component.

A wide variety of E–H σ bonds (E = boron or transition metal) act unexpectedly as efficient hydrogen bond acceptors toward conventional proton donors, such as O–H and N–H groups. The resulting X–H⋯H–E systems have close H⋯H contacts (175–190 pm) and are termed "dihydrogen bonds." Their enthalpies of formation are substantial, 13–30 kJ mol^{-1}, and lie in the range for conventional H bonds. Some examples are listed below:

(1) **N–H⋯H–B bond**

A comparison of the melting points of the isoelectronic species H_3C–CH_3 (–181 °C), H_3C–F (–141 °C) and H_3N–BH_3 (104 °C) suggests the possibility that unusually strong intermolecular interactions are present in H_3N–BH_3. As compared to the H_3C–F molecule, the less polar H_3N–BH_3 molecule has a smaller dipole–dipole interaction, and both slack a lone pair to form a conventional hydrogen bond. The abnormally high meting point of H_3N–BH_3 originates from the presence of N–H⋯H–B bonding in the crystalline state.

Many close intermolecular N–H⋯H–B contacts in the range 170–220 pm have been found in various crystal structures. In these dihydrogen bonds, the (NH)⋯H–B angle is strongly bent, falling in the range 95–120°. Fig. 1.3.7(a) shows the (NH)⋯H–B bond between H_3N–BH_3 molecules.

(2) **X–H⋯H–M bonds**

X-ray and neutron diffraction studies have established the presence of transition metal N–H⋯H–M and O–H⋯H–M dihydrogen bonds, in which the hydride ligand acts as a proton acceptor. Fig. 1.3.7(b) shows the dihydrogen bond N–H⋯H–Re in the crystal structure of $ReH_5(PPh_3)_3$·indole·C_6H_6. The H⋯H distances are 173.4 and 221.2 pm. Fig. 1.3.7(c) shows a similar dihydrogen bond with measured H⋯H distances of 168 and 199 pm in the crystal structure of $ReH_5(PPh_3)_2$(imidazole).

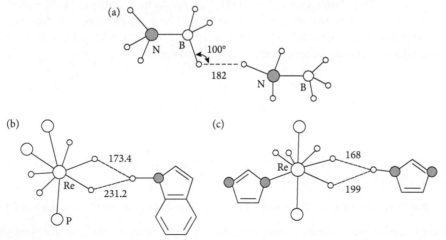

Fig. **1.3.7** Structure of compounds containing X–H⋯H–E bonds: (a) H_3N–BH_3⋯H_3N–BH_3, (b) $ReH_5(PPh_3)_3$·indole, (c) $ReH_5(PPh_3)_2$(imidazole).

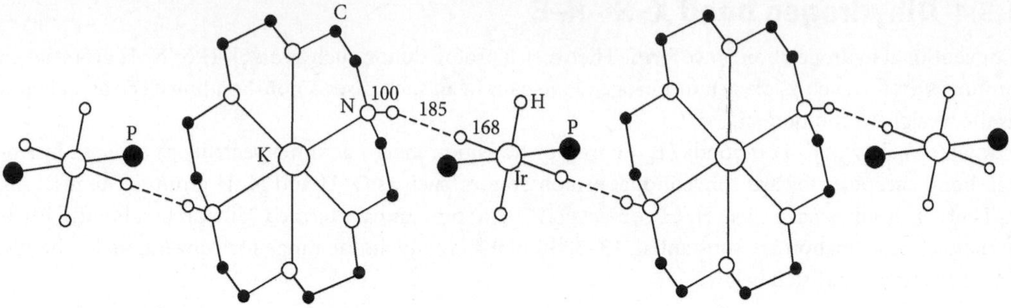

Fig. 1.3.8 Chain structure of $[K(1,10\text{-diaza-18-crown-6})][IrH_4(P^iPr_3)_2]$ and relevant dihydrogen bond lengths in pm. P^iPr_3 C and H atoms and non-essential crown H atoms have been omitted for clarity.

In the crystal structure of $[K(1,10\text{-diaza-18-crown-6})][IrH_4(P^iPr_3)_2]$, the two kinds of ions form an infinite chain held together by N–H···H–Ir bonds, as shown schematically in Fig. 1.3.8. The observed distance of H···H is 207 pm, and the observed N–H bond length of 77 pm is likely to be less than the true value. The corrected N–H and H···H distances are 100 and 185 pm, respectively.

The X–H···H–M systems generally tend to lose H_2 readily, and indeed such H···H bonded intermediates may be involved whenever a hydride undergoes protonation.

1.3.5 **Inverse hydrogen bond**

In the normal hydrogen bond X–H···Y, the H atom plays the role of electron acceptor, while the Y atom is the electron donor. The interaction is of the kind $\text{X—H}\overset{e}{\curvearrowleft}\text{Y}$. In the inverse hydrogen bond, the H atom plays the role of electron donor, while the Y atom becomes the electron acceptor, and the interaction is of the kind $\text{X—H}\overset{e}{\curvearrowright}\text{Y}$. Some examples of inverse hydrogen bonds are presented below:

(1) The so-called lithium hydrogen bond, Li–H···Li–H, occurs in the hypothetical linear $(LiH)_2$ dimer. The inner Li atom is electron deficient, and the inner H atom is sufficiently electron rich to act as a donor in the formation of an inverse hydrogen bond. The calculated bond lengths (in pm) and electron donor-acceptor relationship are illustrated below:

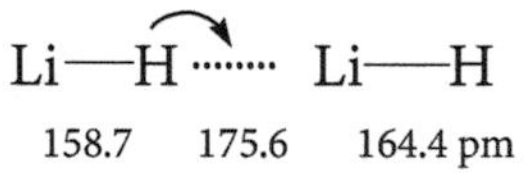

$$\text{Li—H}\overset{e}{\xrightarrow{\hspace{1cm}}}\text{Li—H}$$

158.7   175.6   164.4 pm

The distance H···Li is shorter than the sum of the atomic van der Waals radii of H and Li, and the linkage of Li–H···Li is almost linear.

(2) The dihydrogen bond X–H···H–M may be formally regarded as a normal hydrogen bond $\text{X—H}\overset{e}{\curvearrowleft}\text{Y}$ where Y stands for the hydridic H atom as the electron donor, or an inverse hydrogen bond $\text{M—H}\overset{e}{\curvearrowright}\text{Y}$ where Y is the H atom bonded to X.

(3) The inverse hydrogen bond B^-–H···Na^+ is found in the adduct of $Na(Et_3BH)$ and $Nb_2(hpp)_4$, where hpp is the anion of 1,3,4,6,7,8-hexahydro-2H-pyrimido-[1,2-a]-pyrimidine (Hhpp). Fig. 1.3.9 shows the B^-–H···Na^+ interaction in the compound $Nb_2(hpp)_4\cdot2NaEt_3BH$.

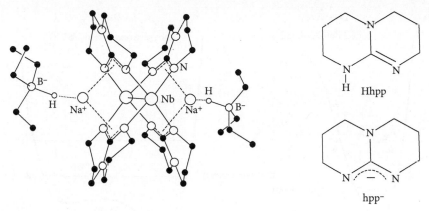

Fig. 1.3.9 Structure of the adduct $Nb_2(hpp)_4 \cdot 2NaEt_3BH$.

The concept of inverse hydrogen bond is a relatively recent development, and additional varieties of this novel type of interaction may be uncovered in the future.

1.3.6 **Nature of short and strong hydrogen bonds (SHBs)**

The $[F-H-F]^-$ ion is stabilized by a short and strong hydrogen bond (SHB), whose distinctive vibrational potential in water can be measured by femtosecond two-dimensional infrared spectroscopy. It shows superharmonic behavior of the proton motion, which is strongly coupled to the donor-acceptor stretching and disappears on hydrogen bending.

The vibrational transition frequencies depend on quantum numbers 0, 1, and 2 (Fig. 1.3.10(a)). Conventional H-bonds follow traditional rules for anharmonic vibrations, in which vibrational energy splitting decreases upon ascending the vibrational ladder (positive anharmonicity). The $|0\rangle$ to $|1\rangle$ transition frequency ω_{10} is larger than the $|1\rangle$ to $|2\rangle$ frequency ω_{21}, as characterized by their ratio $\gamma = \omega_{21}/\omega_{10} < 1$. In contrast, when proton confinement dictates the potential shape in strong hydrogen bonds, the spacing between states increases up the vibrational ladder (analogous to a particle-in-a-box) and $\gamma > 1$ (negative anharmonicity), manifesting an effect called superharmonicity.

The infrared spectrum of aqueous KHF_2 is shown in Fig. 1.3.11 with the assignment of vibrational modes that report on the principal geometric coordinates of any hydrogen bond (Fig. 1.3.10(b)). On the basis of gas phase, solid state, and theoretical studies, the bands at 1215 and 1521 cm^{-1} can be assigned to the F-H-F bending modes, δ, and the proton stretching (shuttling) mode, ν^H. The H-O-H bending vibration of water appears at 1652 cm^{-1}. The broad shoulder centered at 1835 cm^{-1} shows two components in the difference IR spectrum.

In summary, SHBs exhibit distinctive and counterintuitive characteristics: (a) delocalization of the proton in a flat-bottom confined potential that yields H-bond superharmonicity, and (b) a blue shift of the proton stretch frequency upon H-bond strengthening, and major electron-density redistribution leading to the emergence of hydrogen-mediated donor–acceptor bonding. The hydrogen bifluoride ion $[F-H-F]^-$ qualifies as the "strongest H-bonding system" that lies at the tipping point where hydrogen bonding ends and chemical bonding begins. Hence hydrogen bonding can be interpreted as a classical electrostatic interaction (in most cases), or as a covalent chemical bond if the interaction is strong enough.

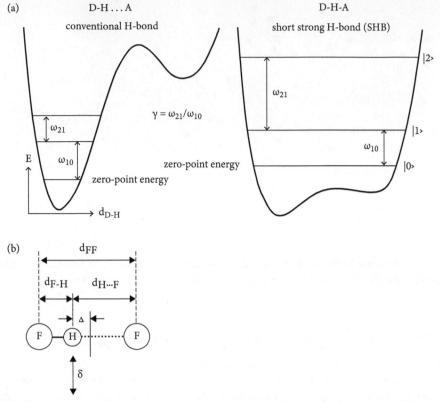

Fig. 1.3.10 Hydrogen bonds. (a) Types of hydrogen bonds depending on donor–acceptor distance. Potentials of proton motion are shown along with the first three quantum levels and associated lowest-energy transitions. (b) Principal coordinates of hydrogen bond in HF_2^-(aq): donor–acceptor distance (d_{FF}), proton asymmetry (Δ), and linearity (δ).

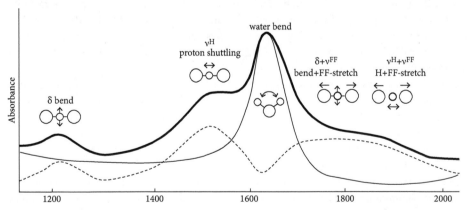

Fig. 1.3.11 IR spectroscopy of short and strong H-bonds. (A) Linear infrared spectrum of aqueous 3.6 M KHF_2 (solid line), water (thin line), and their difference (dotted). The blocks represent molecular motions of the respective vibrational modes.

Reference

B. Dereka, Q. Yu, N. H. C. Lewis, W. B. Carpenter, J. M. Bowman, and A. Tokmakoff, Crossover from hydrogen to chemical bonding, *Science*. **371**, 160–4 (2021).

1.4 Hydride Complexes

Hydrogen combines with many metals to form binary hydrides MH_x. The hydride ion H^- has two electrons with the noble gas configuration of He. Binary metal hydrides have the following characteristics:

(1) Most of these hydrides are non-stoichiometric, and their composition and properties depend on the purity of the metals used in the preparation.

(2) Many of the hydride phases exhibit metallic properties such as high electrical conductivity and metallic luster.

(3) They are usually produced by the reaction of metal with hydrogen. Besides the formation of true hydride phases, hydrogen also dissolves in the metal to give a solid solution.

Metal hydrides may be divided into two types: covalent and interstitial. They are discussed in the following two sections.

1.4.1 Covalent metal hydride complexes

In recent years, many crystal structures of transition metal hydride complexes have been determined. In compounds such as $CaMgNiH_4$, Mg_2NiH_4, Mg_2FeH_6, and K_2ReH_9, H^- behaves as an electron-pair donor covalently bonded to the transition metals. The transition metal in the NiH_4^{2-}, FeH_6^{4-}, and ReH_9^{2-} anions generally has the favored noble gas electronic configuration with 18 valence electrons. Fig. 1.4.1 shows the crystal structure of $CaMgNiH_4$.

The anion ReH_9^{2-} is a rare example of a central metal atom forming nine 2c-2e bonds, which are directed toward the vertices of a tricapped trigonal prism. Fig. 1.4.2 shows the structure of ReH_9^{2-}, and Table 1.4.1 lists the structures of a number of transition metal hydride complexes. In these complex anions, the distances of transition metal to hydrogen are in the range 150–160 pm for $3d$ metals and 170–180 pm for $4d$ and $5d$ metals, except for Pd (160–170 pm) and Pt (158–167 pm).

Low-temperature neutron diffraction analysis of $H_4Co_4(C_5Me_4Et)_4$ shows that the molecule consists of four face-bridging hydrides attached to a tetrahedral cobalt metal core, as shown in Fig. 1.4.3. The average distances (in pm) and angles in the core of the molecule are as follows:

Co–Co 257.1	Co–H 174.9	Co–C 215.8
H⋯H 236.6	Co–H–Co 94.6°	H–Co–H 85.1°

The hydride ligands are located off the Co–Co–Co planes by an average distance of 92.3 pm.

Polynuclear platinum and palladium carbonyl clusters containing the bulky tri-*tert*-butylphosphine ligand are inherently electron-deficient at the metal centers. The trigonal bipyramidal cluster $[Pt_3Re_2(CO)_6(P^tBu_3)_3]$, as shown in Fig. 1.4.4(a), is electronically unsaturated with a deficit of 10 valence electrons as it needs 72 valence electrons to satisfy an 18-electron configuration at each metal vertex. The cluster reacts with three equivalents of molecular hydrogen at room temperature to give the addition

Table 1.4.1 Structures of transition metal hydrides

Geometry		Example	Bond length*/pm
Tricapped trigonal-prismatic	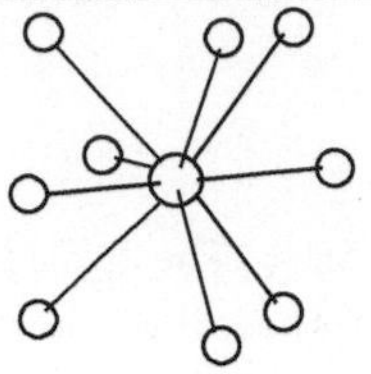	K_2ReH_9	ReH_9^{2-}: Re(1)–H (3×)172,(6×) 167 Re(2)–H (3×)161,(6×) 170
Octahedral	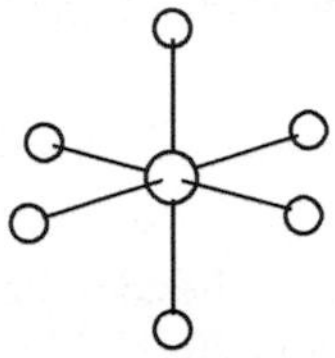	Na_3RhH_6 Mg_2FeH_6 Mg_3RuH_6	RhH_6^{3-}: Rh–H 163 –168 FeH_6^{4-}: Fe–H 156 RuH_6^{6-}: Ru–H (4×) 167, (2×) 173
Square-pyramidal	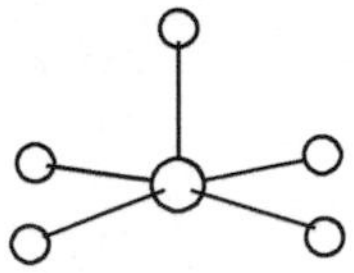	Mg_2CoH_5 Eu_2IrH_5	CoH_5^{4-}: Co–H (4×) 152, (1×) 159 IrH_5^{4-}: Ir–H (6×, disorder) 167
Square-planar	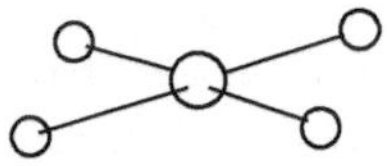	Na_2PtH_4 Li_2RhH_4	PtH_4^{2-}: Pt–H (4×) 164 RhH_4^{3-}: Rh–H (2×) 179, (2×) 175
Tetrahedral	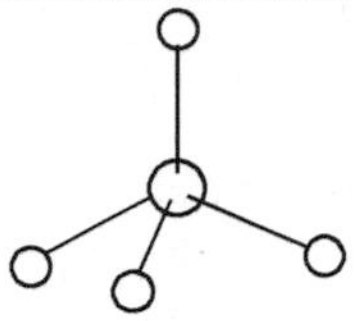	Mg_2NiH_4	NiH_4^{4-}: Ni–H 154–157
Saddle	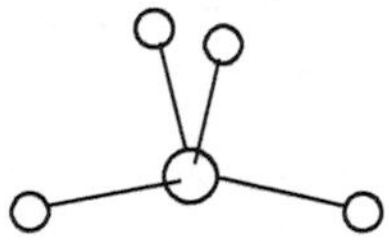	Mg_2RuH_4	RuH_4^{4-}: Ru–H (2×) 167, (2×) 168
T-shaped	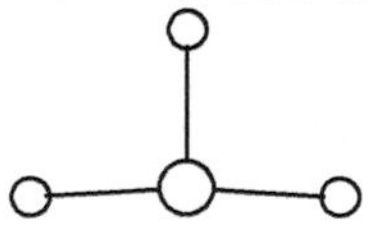	Mg_3RuH_3	RuH_3^{6-}: Ru–H 171
Linear	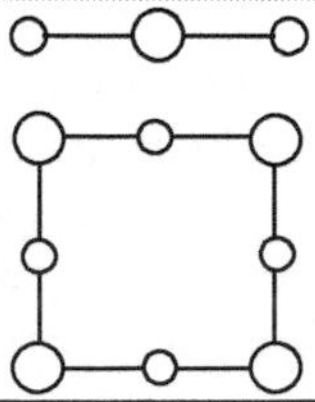	Na_2PdH_2 $MgRhH_{1-x}$	PdH_2^{2-}: Pd–H (2×) 168 $Rh_4H_4^{8-}$: Rh–H (2×) 171

* The bond lengths are determined by neutron diffraction for M–D.

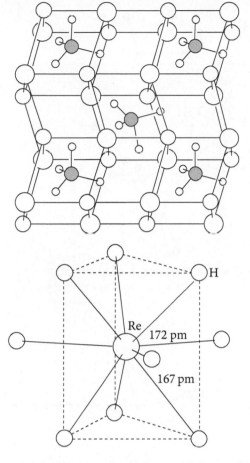

Fig. 1.4.1 Crystal structure of CaMgNiH$_4$ (large circles represent Ca, small circles represent Mg, and tetrahedral groups represent NiH$_4$).

Fig. 1.4.2 Structure of ReH$_9{}^{2-}$.

product [Pt$_3$Re$_2$(CO)$_6$(P^{t}Bu$_3$)$_3$(μ-H)$_6$] in 90% yield. Single-crystal X-ray analysis showed that the hexahydrido complex has a similar trigonal bipyramidal structure with one hydrido ligand bridging each of the six Pt–Re edges of the cluster core, as illustrated in Fig. 1.4.4(b). The Pt–Pt bond lengths are almost the same in both complexes (average values 272.25 pm versus 271.27 pm), but the Pt–Re bonds are much lengthened in the hexahydrido complex (average values 264.83 pm versus 290.92 pm). The Pt–H bonds are significantly shorter than the Re–H bonds (average values 160 pm versus 189 pm).

1.4.2 Interstitial and high-coordinate hydride complexes

Most interstitial metal hydrides have variable composition, for example, PdH$_x$ with $x < 1$. The hydrogen atoms are assumed to have lost their electrons to the d orbitals of the metal atoms and behave as mobile protons. This model accounts for the mobility of hydrogen in PdH$_x$, the fact that the magnetic susceptibility of palladium falls as hydrogen is added, and that if an electric potential is applied across a filament of PdH$_x$, hydrogen migrates toward the negative electrode.

A striking property of many interstitial metal hydrides is the high rate of hydrogen diffusion through the solid at slightly elevated temperatures. This mobility is utilized in the ultra-purification of H$_2$ by diffusion through a palladium-silver alloy tube.

Hydrogen has the potential to be an important fuel because it has an extremely high energy density per unit weight. It is also a non-polluting fuel, the main combustion product being water. The metallic

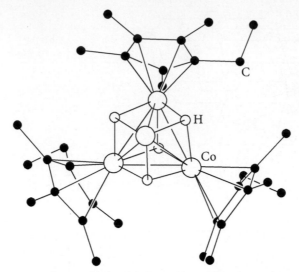

Fig. 1.4.3 Molecular structure of $H_4Co_4(C_5Me_4Et)_4$ by neutron diffraction analysis at 20 K.

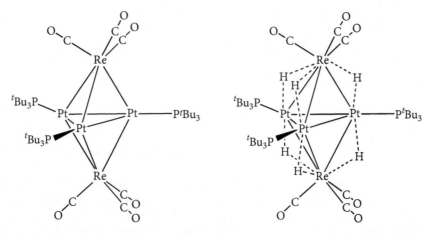

Fig. 1.4.4 Molecular structure of (a) $[Pt_3Re_2(CO)_6(P^tBu_3)_3]$ and its hydrogen adduct (b) $[Pt_3Re_2(CO)_6(P^tBu_3)_3(\mu\text{-}H)_6]$. (Ref. R. D. Adams and B. Captain, *Angew. Chem. Int. Ed.* **44**, 2531–3 (2005).)

hydrides, which decompose reversibly to give hydrogen gas and the metals, can be used for hydrogen storage. Table 1.4.2 lists the capacities of some hydrogen storage systems. It is possible to store more hydrogen in the form of these hydrides than in the same volume of liquid hydrogen.

The crystal structure of MgH_2 has been determined by neutron diffraction. It has the rutile structure, space group $P4_2/mnm$, with $a = 450.25$ pm, $c = 301.23$ pm, as shown in Fig. 1.4.5. The Mg^{2+} ion is surrounded octahedrally by six H^- anions at 194.8 pm. Taking the radius of Mg^{2+} as 72 pm (six-coordinate, Table 4.2.2), the radius of H^- (three-coordinate) is calculated to be 123 pm.

Fig. 1.4.6 shows the pressure (P) vs composition (x) isotherms for the hydrogen-iron-titanium system. This system is an example of the formation of a ternary hydride from an intermetallic compound.

In the majority of its metal complexes, the hydride ligand normally functions in the μ^1 (terminal), μ^2 (edge-bridging), and μ^3 (triangular face-capping) modes. Research efforts in recent years have led to the syntheses of an increasing number of high-coordinate (μ^4, μ^5, and μ^6) hydride complexes.

Table 1.4.2 Some hydrogen storage systems

Storage medium	Hydrogen percent by weight	Hydrogen density/kg dm^{-3}
MgH_2	7.6	0.101
Mg_2NiH_4	3.16	0.081
VH_2	2.07	0.095
$FeTiH_{1.95}$	1.75	0.096
$LaNi_5H_6$	1.37	0.089
liquid H_2	100	0.070
gaseous H_2 (10 MPa)	100	0.008

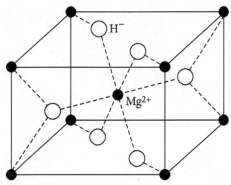

Fig. 1.4.5 Crystal structure of MgH_2.

The compounds $[Li_8(H)\{N(2\text{-Py})Ph\}_6]^+[Li(Me_2Al^tBu_2)_2]^-$ and $Li_7(H)[N(2\text{-Py})Ph]_6$ are examples of molecular species that contain a μ^6-hydride ligand surrounded by an assembly of main-group metal ions. The cation $[Li_8(H)\{N(2\text{-Py})Ph\}_6]^+$ encapsulates a H^- ion within an octahedron composed of Li atoms, the average Li–H distance being 201.5 pm, as shown in Fig. 1.4.7(a). In molecular $Li_7(H)[N(2\text{-Py})Ph]_6$, the H^- ion is enclosed in a distorted octahedral coordination shell, as shown in Fig. 1.4.7(b). The average Li–H distance is 206 pm. The seventh Li atom is located at a much longer Li$\cdots$H distance of 249 pm.

Neutron diffraction analysis of the hydrido cluster complex $[H_2Rh_{13}(CO)_{24}]^{3-}$ at low temperature has revealed a hexagonal close-packed metal skeleton in which each surface Rh atom is coordinated by one terminal and two bridging carbonyls. The two hydride ligands occupy two of the six square-pyramidal sites on the surface, each being slightly displaced from the plane of four basal Rh atoms toward the central Rh atom of the Rh_{13} cluster core (Fig. 1.4.8(a)). For either μ^5-hydride, the axial Rh(central)–H distance is shorter (average 184 pm) than the four Rh(surface)–H distances (average 197 pm).

DFT calculations, as wells as X-ray and neutron diffraction studies have established the first existence of a four-coordinate interstitial hydride ligand. In the isomorphous tetranuclear lanthanide polyhydride complexes $Ln_4H_8[C_5Me_4(SiMe_3)]_4$ (Ln = Lu, Y), the tetrahedral $[Ln_4H_8]^{4+}$ cluster core adopts a pseudo C_{3v} configuration with a body-centered μ^4, one face-capping μ^3, and six edge-bridging μ^2 hydride ligands. The molecular structure of the yttrium(III) complex is shown in Fig. 1.4.8(b).

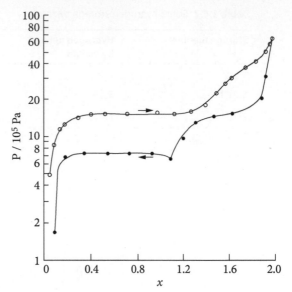

Fig. 1.4.6 P-x isotherms for the FeTiH$_x$ alloy. The upper curve corresponds to the equilibrium pressure as hydrogen is added stepwise to the alloy; the lower curve corresponds to the equilibrium pressure as hydrogen is removed stepwise from the hydride.

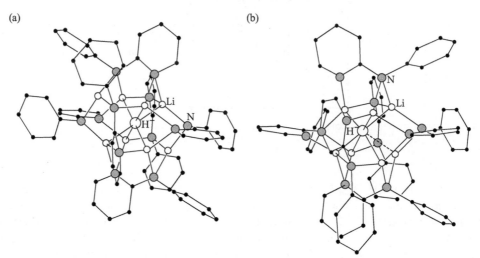

(a)   (b)

Fig. 1.4.7 Structure of main-group metal hydride complexes. (a) $[Li_8(H)\{N(2\text{-}Py)Ph\}_6]^+$. (b) $Li_7(H)[N(2\text{-}Py)Ph]_6$; the broken line indicates a weak interaction.

1.5 Molecular Hydrogen (H$_2$) Coordination Compounds and σ-Bond Complexes

1.5.1 Structure and bonding of H$_2$ coordination compounds

The activation of hydrogen by a metal center is one of the most important chemical reactions. The H–H bond is strong (436 kJ mol^{-1}), so that H$_2$ addition to unsaturated organic and other compounds must be mediated by metal centers whose roles constitute the basis of catalytic hydrogenation. In catalytic mechanisms, hydride complexes formed by the cleavage of H$_2$ are regarded as key intermediates.

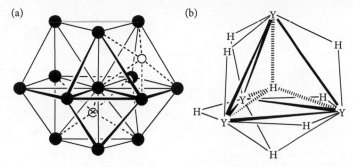

Fig. 1.4.8 (a) Molecular structure of [H$_2$Rh$_{13}$(CO)$_{24}$]$^{3-}$; the carbonyl groups are omitted for clarity. (b) Molecular structure of Y$_4$H$_8$[C$_5$Me$_4$(SiMe$_3$)]$_4$; the C$_5$Me$_4$(SiMe$_3$)$^-$ ligands each capping a metal center are omitted for clarity.

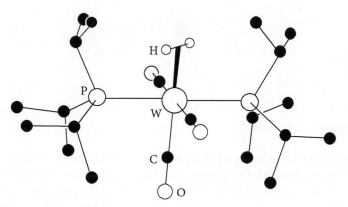

Fig. 1.5.1 Molecular structure of W(CO)$_3$[P(CHMe$_2$)$_3$]$_2$(H$_2$).

The first isolable transition metal complex containing a coordinated H$_2$ molecule is W(CO)$_3$[P(CHMe$_2$)$_3$]$_2$(H$_2$). X-ray and neutron diffraction studies and a variety of spectroscopic methods have confirmed that it possesses a η^2-H$_2$ ligand, as shown in Fig. 1.5.1.

The resulting geometry about the W atom is that of a regular octahedron. The H$_2$ molecule is symmetrically coordinated in an η^2 mode with an average W–H distance of 185 pm (X-ray) and 175 pm (neutron) at −100 °C. The H–H distance is 75 pm (X-ray) and 82 pm (neutron), slightly longer than that of the free H$_2$ molecule (74 pm).

The side-on bonding of H$_2$ to the metal involves the transfer of σ bonding electrons of H$_2$ to a vacant metal d orbital (or hybrid orbital), together with the transfer of electrons from a filled metal d orbital into the empty σ* orbital of H$_2$, as shown in Fig. 1.5.2. This synergistic (mutually assisting) bonding mode is similar to that of CO and ethylene with metal atoms. The π back donation from metal to σ* orbital on H$_2$ is consistent with the fact that H–H bond cleavage is facilitated by electron-rich metals.

The structure of *trans*-[Fe(η^2-H$_2$)(H)(PPh$_2$CH$_2$CH$_2$PPh$_2$)]BPh$_4$ has been determined by neutron diffraction at 20 K. The coordination environment about the Fe atom is shown in Fig. 1.5.3. This is the first conclusive demonstration of the expected difference between a hydride and an H$_2$ molecule in coordination to the same metal center. The H–H bond distance is 81.6 pm and the H–Fe bond length is 161.6 pm, which is longer than the terminal H–Fe distance of 153.5 pm.

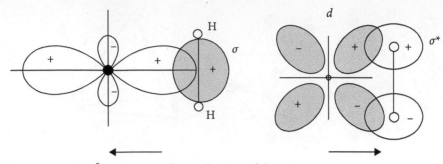

Fig. 1.5.2 Bonding in M-η^2-H_2. Arrow indicates direction of donation.

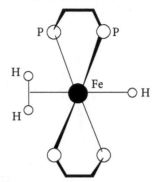

Fig. 1.5.3 Coordination of the Fe atom in *trans*-[Fe(η^2-H_2)(H)(PPh$_2$CH$_2$CH$_2$PPh$_2$)]BPh$_4$.

A large number of H_2 coordination compounds have been identified and characterized. Table 1.5.1 lists the H⋯H distances (d_{HH}) determined by neutron diffraction for some of these complexes. The H⋯H distances ranges from 82 to 160 pm, beyond which a complex is generally regarded to be a classical hydride. A "true" dihydrogen complex can be considered to have $d_{HH} < 100$ pm, and the complexes with $d_{HH} > 100$ pm are more hydride-like in their properties and have highly delocalized bonding.

The H_2 molecule can act as an electron donor in combination with a proton to yield a H_3^+ species:

$$\begin{array}{c}\mathrm{H} \\ | \\ \mathrm{H}\end{array} + \mathrm{H}^+ \implies \begin{array}{c}\mathrm{H} \\ | \longrightarrow \mathrm{H}^+ \\ \mathrm{H}\end{array} \implies \left[\begin{array}{c}\mathrm{H} \\ \diagup \diagdown \\ \mathrm{H} \text{------} \mathrm{H}\end{array}\right]^+$$

The existence of H_3^+ has been firmly established by its mass spectrum: H_3^+ has a relative molecular mass of 3.0235, which differs from those of HD (3.0219) and T (3.0160). The H_3^+ species has the shape of an equilateral triangle and is stabilized by a 3c-2e bond.

The higher homologs H_5^+, H_7^+, and H_9^+ have also been identified by mass spectroscopy. These species are 10^2 to 10^3 times more abundant than the even-member species H_{2n}^+ ($n = 2, 3, 4, ...$) in mass spectral measurements. The structures of H_5^+, H_7^+ and H_9^+ are shown below:

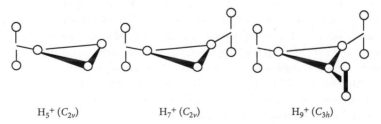

H_5^+ (C_{2v}) H_7^+ (C_{2v}) H_9^+ (C_{3h})

Table 1.5.1 H···H bond distances (d_{HH}) determined by neutron diffraction for H_2 coordination compounds

H₂ Coordination compound	$d_{HH}*$/pm
$Mo(Co)(dppe)_2(H_2)$	73.6 (84)
$[FeH(H_2)(dppe)_2]^+$	81.6
$W(CO)_3(P^iPr_3)_2(H_2)$	82
$[RuH(H_2)(dppe)_2]^+$	82 (94)
$FeH_2(H_2)(PEtPh_2)_3$	82.1
$[OsH(H_2)(dppe)_2]^+$	97
$[Cp*OsH_2(H_2)(PPh_3)]^+$	101
$[Cp*OsH_2(H_2)(AsPh_3)]^+$	108
$[Cp*Ru(dppm)(H_2)]^+$	108 (110)
cis-$IrCl_2H(H_2)(P^iPr_3)_2]$	111
$trans$-$[OsCl(H_2)(dppe)_2]^+$	122
$[Cp*OsH_2(H_2)(PCy_3)]^+$	131
$[Os(en)_2(H_2)(acetate)]^+$	134
$ReH_5(H_2)[P(p\text{-}tolyl)_3]_2$	135.7
$[OsH_3(H_2)(PPhMe_2)_3]^+$	149

* Uncorrected for effects of vibrational motion; the corrected values are given in parentheses.

1.5.2 **X–H σ-bond coordination metal complexes**

By analogy to the well-characterized η^2-H_2 complexes, methane and silane can coordinate to the metal in an μ^2-fashion via a C–H σ bond and a Si–H σ bond:

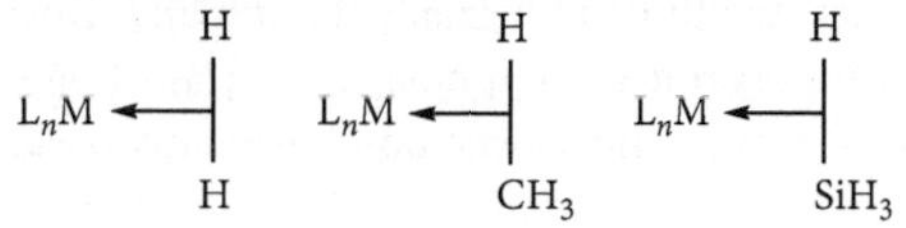

Transition metal complexes with a variety of σ-coordinated silane ligands have been extensively investigated. The molecular structure of $Mo(\eta^2$-H-$SiH_2Ph)(CO)(Et_2PCH_2CH_2PEt_2)_2$, as shown in Fig. 1.5.4(a), exhibits η^2-coordination of the Si–H bond at a coordination site cis to the CO ligand. The distances (in pm) and angles of the Si–H coordinate bond are: Si–H 177, Mo–H 170, Mo–Si 250.1, Mo–H–Si 92°, and Mo–Si–H 42.6°. The SiH_2Ph_2 analog, $Mo(\eta^2$-H-$SiHPh_2)(CO)(Et_2PCH_2CH_2PEt_2)_2$, as shown in Fig. 1.5.4(b), exhibits a similar structure with Si–H 166, Mo–H 204, and Mo–Si 256.4.

Two molecules can be combined to form an ion-pair through a σ coordination bond, in which one molecule provides its X–H (X = B, C, N, O, Si) σ bonding electrons to a transition metal atom (such as Zr) of another molecule. A good example is $[(C_5Me_5)_2Zr^+Me][B^-Me(C_6F_5)_3]$, whose structure is shown in Fig. 1.5.5. This bonding type is called an intermolecular pseudo-agostic (IPA) interaction.

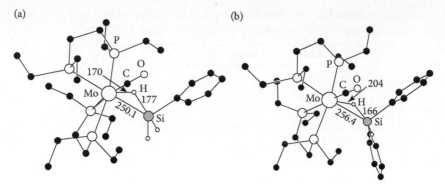

Fig. 1.5.4 Structure of (a) Mo(η^2–H–SiH$_2$Ph)(CO)(Et$_2$PCH$_2$CH$_2$PEt$_2$)$_2$ and (b) Mo(η^2–H–SiHPh$_2$)(CO) (Et$_2$PCH$_2$CH$_2$PEt$_2$)$_2$ (bond lengths in pm).

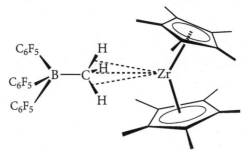

Fig. 1.5.5 Structure of [(C$_5$Me$_5$)$_2$Zr$^+$Me][B$^-$Me(C$_6$F$_5$)$_3$].

1.5.3 **Agostic bond**

The agostic bond is an intramolecular 3c-2e C–H ⟶ M bond, with the vacant metal orbital accepting an electron pair from the C–H σ bonding orbital. The term "agostic" denotes the intramolecular coordination of C–H bonds to transition metals and is derived from a Greek word meaning "to clasp, to draw toward, to hold to oneself." The term "agostic" should not be used to describe external ligand binding solely through a σ bond, which is best referred to as σ-bonded coordinate binding. Agostic bonds may be broadly understood in the following way. Many transition-metal compounds have less than an 18-electron count at the metal center, and thus are formally unsaturated. One way in which this deficiency may be alleviated is to increase the coordination number at the transition metal by clasping a H atom of a coordinated organic ligand. Therefore, the agostic bond generally occurs between carbon-hydrogen groups and transition metal centers in organometallic compounds, in which a H atom is covalently bonded simultaneously to both a C atom and to a transition metal atom. An agostic bond is generally written as C–H ⟶ M, where the "half arrow" indicates formal donation of two electrons from the C–H bond to the M vacant orbital. As in all 3c-2e bridging systems involving only three valence orbitals, the C–H ⟶ M fragment is bent and the agostic bond can be more accurately represented by $\overset{\displaystyle\frown}{\underset{H}{M}}\!\!\!\!\diagdown\,{}^{C}$. The agostic C–H distance is in the range 113–119 pm, about 5–10% longer than a non-bridging C–H bond; the M–H distance is also elongated by 10–20% relative to a normal terminal M–H bond. Usually NMR spectroscopy can be used for the diagnosis of static agostic systems which exhibit low J(C–H$_\alpha$) values owing to the reduced C–H$_\alpha$ bond order. Typical values of J(C–H$_\alpha$) are in the range of 60–90 Hz, which are significantly lower than those (120–130 Hz) expected for C(sp^3)–H bonds.

The C–H⟶M agostic bond is analogous to a X–H⋯Y hydrogen bond, in the sense that the strength of the M–H linkage, as measured by its internuclear distance, is variable, and correlates with the changes in C–H distance on the ligand. There are, however, two important differences. First, in a hydrogen bond the H atom is attracted by an electronegative acceptor atom, but the strength of an agostic bond is stronger for more electropositive metals. Secondly, the hydrogen bond is a 3c-4e system, but the agostic bond is a 3c-2e system.

The agostic interaction is considered to be important in such reactions as α-elimination, β-hydrogen elimination, and orthometalation. For example, in a β-hydrogen elimination, a H atom on the β-C atom of an alkyl group is transferred to the metal atom and an alkene is eliminated. This reaction proceeds through an agostic-bond intermediate, as shown below:

$$C_2H_5-\underset{\underset{H}{|}}{Pd}-PH_3 \longrightarrow \underset{\underset{H}{|}}{Pd}\overset{\overset{\displaystyle H_2}{\underset{H}{\diagup}\diagdown CH_2}}{-PH_3} \longrightarrow H-\underset{\underset{H}{|}}{Pd}-PH_3 \longrightarrow H-\underset{\underset{H}{|}}{Pd}-PH_3 + H_2C=CH_2$$

The discovery of agostic bonding has led to renewed interest in the ligand behavior of simple organic functional groups, such as the methyl group. The C–H group used to be regarded as an inert spectator, but now C–H containing hydrocarbon ligand systems are recognized as being capable of playing important roles in complex stereochemical reactions.

Agostic bonding has been extended to include general X–H⟶M systems, in which X may be B, N, Si, as well as C. The principal types of compounds with agostic bonds are shown in Fig. 1.5.6.

Fig. 1.5.7 shows the molecular structures of some transition metal complexes in which agostic bonding has been characterized by either X-ray or neutron diffraction. Structural data for these compounds are listed in Table 1.5.2.

Many complexes containing P-, N-, or O-containing ligands can also form agostic bonds:

$$\underset{\diagdown M \diagup}{E}\overset{\diagup (C)_n \diagdown}{H} \quad (E=P,N,O)$$

Agostic interactions involving phosphines are quite significant because they possess residual binding sites for other small molecules such as H₂, and phosphine groups play crucial roles in many homogeneous catalysts. Fig. 1.5.8(a) shows the structure of $W(CO)_3(PCy_3)_2$, in which the W⟶H distance is 224.0 pm

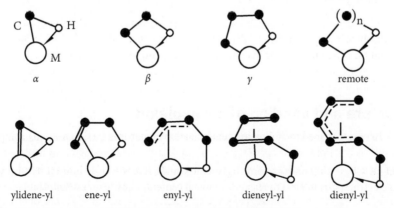

Fig. 1.5.6 Structural types of agostic alkyl and unsaturated hydrocarbonyl complexes.

Table 1.5.2 Structural data of some agostic-bonded X–H ⟶ M compounds

Compound	M–X/pm	X–H/pm	M–H/pm	X–H–M/°	Fig. 1.5.7
[Ta(CHCMe₃)(PMe₃)Cl₃]₂	Ta–C 189.8	C–H 113.1	Ta–H 211.9	C–H–Ta 84.8	(a)
[HFe₄(η^2-CH)(CO)₁₂]	Fe–C 182.7– 194.9	C–H 119.1	Fe–H 175.3	C–H–Fe 79.4	(b)
[Pd(H)(PH₃)(C₂H₅)]	Pd–C 208.5	C–H 113	Pd–H 213	C–H–Pd 88	(c)
{RuCl[S₂(CH₂CH₂)·C₂B₉H₁₀]-(PPh₃)₂}· Me₂CO		B–H 121	Ru–H 163		(d)
[Mn(HSiFPh₂)(η^5-C₅H₄Me)(CO)₂]	Mn–Si 235.2	Si–H 180.2	Mn–H 156.9		(e)

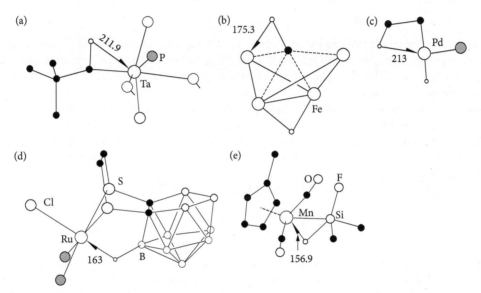

Fig. 1.5.7 Molecular structures of some complexes with agostic bonds: (a) [Ta(CHCMe₃)(PMe₃)Cl₃]₂, (b) [HFe₄(η^2-CH)(CO)₁₂], (c) [Pd(H)(PH₃)(C₂H₅)], (d) {RuCl[S₂(CH₂CH₂)·C₂B₉H₁₀](PPh₃)₂}·Me₂CO, (e) [Mn(HSiFPh₂)(η^5-C₅H₄Me)(CO)₂].

and the W⋯C distance is 288.4 pm. Fig. 1.5.8(b) shows the structure of [IrH(η^2-C₆H₄P^{t}Bu₂)(P^{t}Bu₂Ph)]$^+$ in its [BAr₄]$^-$ salt, in which the Ir⟵H distance is 203.2 pm and the Ir⋯C distance is 274.5 pm.

1.5.4 **Structure and bonding of σ complexes**

Recent studies have established the following generalizations regarding σ (sigma-bond) complexes:

(1) The X–H (X = B, C, Si) bonds in some ligand molecules, like the H–H bond in H₂, provide their σ bonding electron pairs to bind the ligands to metal centers. They form intermolecular coordination

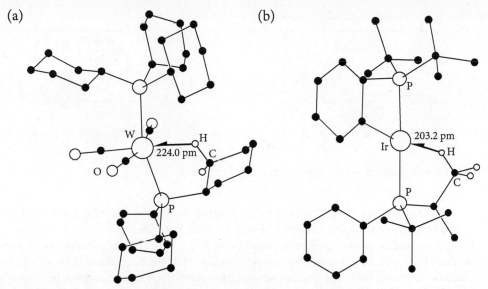

Fig. 1.5.8 Structure of (a) $W(CO)_3(PCy_3)_2$ and (b) $[IrH(\eta^2\text{-}C_6H_4P^tBu_2)(P^tBu_2Ph)]^+$.

compounds (i.e., sigma-bond complexes) or intramolecular coordination compounds (i.e., agostic-bond complexes). All such compounds involve non-classical 3c-2e bonding and are collectively termed σ complexes.

dihydrogen complex   (X = B, C, Si) σ complex   agostic complex

(2) An understanding of the bonding nature in the σ complexes has extended the coordination concept to complement the classical Werner-type donation of a lone pair and the π-electron donation of unsaturated ligands.

Werner-type complex   π complex   σ complex

(3) In σ complexes, the σ ligand is side-on (η^2 mode) bonded to the metal (M) to form a 3c-2e bond. The electron donation in σ complexes is analogous to that in π complexes. The transition metals can uniquely stabilize the σ ligands and π ligands due to back donation from their d orbitals, as shown in Fig. 1.5.9. The main-group metals, lacking electrons in their outer d orbitals, do not form stable σ complexes.

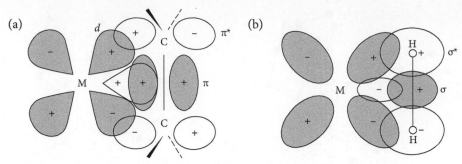

Fig. 1.5.9 Electron donation in (a) π complexes and (b) σ complexes.

(4) H–H and X–H are the strong covalent bonds. These bonds can be weakened and even broken by σ coordination to the transition metal and back donation from the metal to the σ* orbitals. The lengthening and eventual cleavage of a σ bond by binding it to M are dependent on the electronic character of M, which is influenced by the other ligands on M. In the σ complexes, the H···H and X···H distances can vary greatly. The gradual conversion of a metal dihydrogen σ complex to a dihydride is indicated by the following scheme:

$$
\begin{array}{c}
\text{H} \\
| \\
\text{H}
\end{array}
\xrightarrow{\text{M}}
\text{M}\!-\!\!\begin{array}{c}\text{H}\\|\\\text{H}\end{array}
\longrightarrow
\text{M}\!\cdots\!\begin{array}{c}\text{H}\\ \vdots \\\text{H}\end{array}
\longrightarrow
\text{M}\!\!<\!\!\begin{array}{c}\text{H}\\ \vdots \\\text{H}\end{array}
\longrightarrow
\text{M}\!\!<\!\!\begin{array}{c}\text{H}\\\text{H}\end{array}
$$

H_2 molecule	"true" H_2 σ complex	elongated H_2 complex		hydride
74 pm	80~90 pm	100~120 pm	130~150 pm	>160 pm

(5) There are two completely different pathways for the cleavage of H–H bonds: oxidative addition and heterolytic cleavage. Both pathways have been identified in catalytic hydrogenation and may also be applicable to other types of X–H σ bond activation such as C–H cleavage.

Back donation in the σ complexes is the crucial component in both aiding the binding of H_2 to M and activating the H–H bond toward cleavage. If the back donation becomes too strong, the σ bond breaks to form a dihydride due to overpopulation of the H_2 antibonding orbital. It is notable that back donation controls σ-bond activation toward cleavage and a σ bond cannot be broken solely by sharing its electron pair with a vacant metal d orbital. Although σ interaction is generally the predominant bonding component in a σ complex, it is unlikely to be stable at room temperature without at least a small amount of back donation.

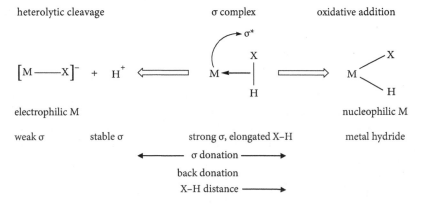

1.6 **Coinage Metal Hydrides**

Reference

A. J. Jordan, G. Lalic, and J. P. Sadighi, Coinage metal hydrides: synthesis, characterization, and reactivity, *Chem. Rev.* **116**, 8318–72 (2016).

1.6.1 **Copper hydrides**

The reaction of *meso*-bis[(diphenylphosphinomethyl)phenylphosphino]methane (dpmppm), with [(CH$_3$CN)$_4$Cu]X (X = BF$_4$ or PF$_6$) and sodium borohydride yielded two cationic copper hydride complexes: a dinuclear μ-hydrido complex, and a tetranuclear complex containing both μ_2- and μ_4-hydride bridges.

X-ray crystallography showed that the dinuclear μ-hydrido cation has C$_2$ symmetry, with each tetraphosphine chain chelates one copper center via P1 and P3, and bridges to the other copper center via P3 and P4, with P2 remaining unbound (Fig. 1.6.1(a)). The Cu$\cdots$Cu contact is rather long at 279.48(7) pm, and the hydride position determined from the difference Fourier map gives Cu−H = 162 pm, and Cu−H−Cu = 120°.

The tetranuclear complex is located at a crystallographic inversion center, with its Cu$_4$ rectangular framework bridged by two dpmppm ligands together with two μ_2- and one μ_4-hydride anions to exhibit a [Cu$_4$H$_3$]$^+$ core (Fig. 1.6.1(b)). Each pair of copper centers is bridged by a μ_2-hydride and the μ_4-hydride at a Cu(1)$\cdots$Cu(2) separation of 246.5(2) pm; the two pairs are connected via bridging dpmppm ligands and the μ_4-hydride at a Cu(1)$\cdots$Cu(2)* separation of 278.9(1) pm. The hydride positions located from the difference Fourier map gave Cu−H distances of 152 and 157 pm for the μ_2-hydride, and 182 and 190 pm for the μ_4-hydride.

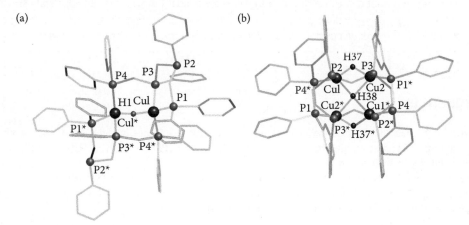

Fig. 1.6.1 Solid-state structures of phosphine-supported (a) dinuclear [Cu$_2$(μ-H)]$^+$ and (b) tetranuclear [Cu$_4$(μ_2-H)$_2$(μ_4-H)]$^+$ cations.

Reference

K. Nakamae, B. Kure, T. Nakajima, Y. Ura, and T. Tanase, Facile insertion of carbon dioxide into Cu$_2$(μ-H) dinuclear units supported by tetraphosphine ligands. *Chem. Asian J.* **9**, 3106–10 (2014).

1.6.2 **Silver hydrides**

The reduction of silver trifluoroacetate with sodium borohydride in the presence of 1,1-bis(diphenylphosphino)methane (L_2) led to condensed-phase synthesis of $[Ag_3H(Cl)(L_2)_3]^+BF_4^-$ from $AgBF_4$. X-ray and neutron diffraction established its distorted trigonal bipyramidal coordination core stabilized by μ_3-hydride and -chloride ligands, with the bisphosphine ligands bridging adjacent silver centers (Fig. 1.6.2(a)). The silver–hydrogen distances at 191.(2) pm are notably shorter than the silver–chlorine distances of 285.9(1) pm; the silver–silver distances within the Ag_3 triangle, 289.88(2) pm, are short enough to suggest argentophilic interactions between silver(I) centers.

Treatment of $AgBF_4$ with bisphosphine ligand and excess borohydride in acetonitrile solution, at or below −10 °C, led to the isolation of a hydride- and borohydride-bridged trisilver cation, $[Ag_3(\mu_3\text{-}H)(\mu_3\text{-}BH_4)(L_2)_3]^+BF_4^-$ (Fig. 1.6.2(b)). Selected bond distances: Ag(1)–H, 193.(3) pm; Ag(1)–H(1a), 217.(3) pm; B(1)–H(1a), 110.(3) pm; B(1)–H(1b), 107.(6) pm; argentophilic Ag(1)···Ag(1), 291.00(3) pm.

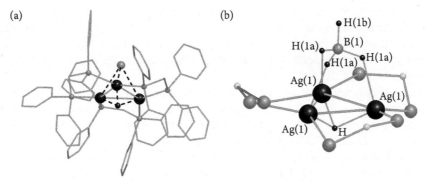

Fig. 1.6.2 (a) Solid-state structure of the cation $[Ag_3(\mu_3\text{-}H)(\mu_3\text{-}Cl)(L_2)_3]^+$ [L_2 = 1,1 bis(diphenylphosphino)methane] that contains a distorted trigonal bipyramidal $ClAg_3H$ core. (b) Solid-state structure of the cation $[Ag_3(\mu_3\text{-}H)(\mu_3\text{-}BH_4)(L_2)_3]^+$ [L_2 = 1,1 bis(diphenylphosphino)methane], with *P*-phenyl groups omitted for clarity. The hydride positions were established by neutron Laue diffraction analysis.

References

A. Zavras, G. N. Khairallah, T. U. Connell, J. M. White, A. J. Edwards, P. S. Donnelly, and R. A. J. O'Hair, Synthesis, structure and gas-phase reactivity of a silver hydride complex $[Ag_3\{(PPh_2)_2CH_2\}_3(\mu_3\text{-}H)(\mu_3\text{-}Cl)]BF_4$. *Angew. Chem. Int. Ed.* **52**, 8391–4 (2013).

A. Zavras, A. Ariafard, G. N. Khairallah, J. M. White, R. J. Mulder, A. J. Canty, and R. A. J. O'Hair, Synthesis, structure and gas-phase reactivity of the mixed silver hydride borohydride nanocluster $[Ag_3(\mu_3\text{-}H)(\mu_3\text{-}BH_4)LPh_3]BF_4$ (LPh = bis(diphenylphosphino)methane). *Nanoscale*. **7**, 18129–37 (2015).

1.6.3 **Gold hydrides**

The normally highly oxidizing Au(III) oxidation state has been employed to isolate a d^8 gold hydride. Treatment of (C∧N∧C)AuOH (where C∧N∧C is derived from 4′,4″-di-*tert*-butyl-2,6-diphenylpyridine by metalation at the 2′ and 2″ positions) with $Li^+[HBEt_3]^-$ in toluene solution affords the (C∧N∧C)gold(III) hydride, which can be isolated as yellow crystals of (C∧N∧C)Au^{III}H, the structure of which is shown in (Fig. 1.6.3(a)).

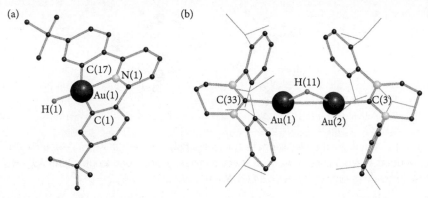

Fig. 1.6.3 (a) Structure of (C∧N∧C)Au^{III}H. Hydrogen atoms except hydride are omitted for clarity; the hydride position was located on the density map. Selected distances and angles: Au(1)–C(1), 207.4(4) pm; Au(1)–C(17), 207.3(4) pm; Au(1)–N(1), 203.5(3) pm; C(17)–Au(1)–C(1), 161.63(16)°. (b) Structure of the cation in {[(6Dipp)Au]$_2$(μH)}$^+$[BAr$^f_4$]$^-$. Selected bond lengths and angles: Au(1)–Au(2), 275.71(3) pm; Au(1)–C(33), 204.0(5) pm; Au(2)–C(3), 204.9(5) pm; Au(1)–Au(2)–C(3), 165.7(1)°; Au(2)–Au(1)–C(33), 164.6°. In both crystal structures the hydride ligand was located from difference Fourier maps.

Protonolysis of LAuH (L = 6Dipp) using Brookhart's acid {[(Et$_2$O)$_2$H]$^+$[BAr$^f_4$]$^-$, Ar^f = 3,5-bis-(trifluoromethylphenyl} at −80 °C in CD$_2$Cl$_2$ solution yielded the hydride-bridged dinuclear cation [(LAu)$_2$(μ-H)]$^+$, the structure of which is shown in (Fig. 1.6.3(b)).

References

D. A. Roşca, D. A. Smith, D. L. Hughes, and M. Bochmann, A thermally stable gold(III) hydride: synthesis, reactivity, and reductive condensation as a route to gold(II) complexes. *Angew. Chem. Int. Ed.* **51**, 10643–6 (2012).

N. Phillips, T. Dodson, R. Tirfoin, J. I. Bates, and S. Aldridge, Expanded-ring N-heterocyclic carbenes for the stabilization of highly electrophilic gold(I) cations. *Chem. Eur. J.* **20**, 16721–31 (2014).

1.6.4 **Molecular heterometallic polyhydride cluster**

Reaction of the tetranuclear yttrium octahydride complex [Cp′Y(μ-H)$_2$]$_4$(THF) (Cp′ = C$_5$Me$_5$SiMe$_3$, **1**) with 1 equivalent of M pentahydride complex [Cp*M(PMe$_3$)H$_5$] (M = molybdenum or tungsten, Cp* = C$_5$Me$_5$) afforded an almost quantitative yield of the corresponding heteropentametallic Y$_4$/M hendec-ahydride complexes [{(Cp′Y)$_4$(μ-H)$_7$}(μ-H)$_4$MCp*(PMe$_3$)] (**2a**, M = Mo; **2b**, M = W) with release of hydrogen. With ultraviolet irradiation in toluene at room temperature, dissociation of the PMe$_3$ ligand from the Mo (or W) atom in **2a** (or **2b**) took place to give the corresponding PMe$_3$-free Y$_4$MH$_{11}$-type complexes [{(Cp′Y)$_4$(μ-H)$_6$}(μ-H)$_5$MCp*] (**3a**: M = Mo, 64%; **3b**: M = W, 74%).

The X-ray structure of the hydride cluster Y$_4$Mo(PMe$_3$)H$_{11}$core of **2a** and neutron structure of the Y$_4$WH$_{11}$ core of **3b** are shown in Fig. 1.6.4(a) and (b), respectively.

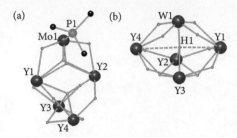

Fig. 1.6.4 Cluster structures of molecular heterobimetallic polyhydrides: (a) X-ray $Y_4Mo(PMe_3)H_{11}$ core of **2a** and (b) Neutron Y_4WH_{11} core of **3b**, with selected metal-hydride bond lengths (pm): W1–H1, 181.(3); Y2–H1, 229.(3); Y3–H1, 230.(3); Y1–H1, 254.(3); Y4–H1, 272.(3).

Reference

T. Shima, Y. Luo, T. Stewart, R. Bau, G. J. McIntyre, S. A. Mason, and Z. Hou, Molecular heterometallic hydride clusters composed of rare-earth and d-transition metals. *Nat. Chem.* **3**, 814–20 (2011).

General References

G. A. Jeffery, *An Introduction to Hydrogen Bonding*, Oxford University Press, Oxford, 1997.

F. D. Manchester (ed.), *Metal Hydrogen Systems: Fundamentals and Applications*, Vols I and II, Elsevier Sequoia, Lausanne, 1991.

G. R. Desiraju and T. Steiner, *The Weak Hydrogen Bond in Structural Chemistry and Biology*, Oxford University Press, Oxford, 1999.

G. A. Jeffrey and W. Saenger, *Hydrogen Bonding in Biological Structures*, Springer-Verlag, Berlin, 1991.

C. C. Wilson, *Single Crystal Neutron Diffraction from Molecular Materials*, World Scientific, Singapore, 2000.

M. Peruzzini and R. Poli (eds), *Recent Advances in Hydride Chemistry*, Elsevier, Amsterdam, 2001.

G. J. Kubas, *Metal Dihydrogen and σ-Bond Complexes*, Kluwer/Plenum, New York, 2001.

C.-K. Lam and T. C. W. Mak, Carbonate and oxalate dianions as prolific hydrogen-bond acceptors in supramolecular assembly. *Chem. Commun.* 2660–1 (2003).

G. R. Desiraju, Hydrogen bridges in crystal engineering: interactions without borders. *Acc. Chem. Res.* **35**, 565–73 (2002).

Chapter 2

Structural Chemistry of Alkali and Alkaline-Earth Metals

2.1 Survey of the Alkali Metals

The alkali metals: lithium, sodium, potassium, rubidium, cesium, and francium, are members of Group 1 of the periodic table, and each has a single ns^1 valence electron outside a rare gas core in its ground state. Some important properties of alkali metals are given in Table 2.1.1.

With increasing atomic number, the alkali metal atoms become larger and the strength of metallic bonding, enthalpy of atomization (ΔH_{at}), melting point (mp), standard enthalpy of fusion (ΔH_{fus}), and boiling point (bp) all progressively decrease. The elements show a regular gradation in physical properties down the group.

The radii of the metals increase with increasing atomic number and their atomic sizes are the largest in their respective periods. Such features lead to relatively small first ionization energy (I_1) for the atoms. Thus the alkali metals are highly reactive and form M^+ ions in the vast majority of their compounds. The very high second ionization energy (I_2) prohibits formation of the M^{2+} ions. Even though the electron affinities (Y) indicate only mild exothermicity, M^- ions can be produced for all the alkali metals (except Li) under carefully controlled conditions.

Table 2.1.1 Some properties of alkali metals[*]

Property	Li	Na	K	Rb	Cs
Atomic number, Z	3	11	19	37	55
Electronic configuration	[He]$2s^1$	[Ne]$3s^1$	[Ar]$4s^1$	[Kr]$5s^1$	[Xe]$6s^1$
ΔH_{at}^0/kJ mol^{-1}	159	107	89	81	76
mp/K	454	371	337	312	302
ΔH_{fus}^0/kJ mol^{-1}	3.0	2.6	2.3	2.2	2.1
bp/K	1620	1156	1047	961	952
r_M (CN = 8)/pm	152	186	227	248	265
r_{M+} (CN = 6)/pm	76	102	138	152	167
I_1/kJ mol^{-1}	520.3	495.8	418.9	403.0	375.7
I_2/kJ mol^{-1}	7298	4562	3051	2633	2230
Electron affinity (Y)/kJ mol^{-1}	59.8	52.7	48.3	46.9	45.5
Electronegativity (χ_s)	0.91	0.87	0.73	0.71	0.66

[*] Data are not available for francium, of which only artificial isotopes are known.

Structural Chemistry across the Periodic Table. Thomas Chung Wai Mak, Yu-San Cheung, Gong-Du Zhou, and Ying-Xia Wang, Oxford University Press. © Thomas Chung Wai Mak, Yu-San Cheung, Gong-Du Zhou, and Ying-Xia Wang (2023). DOI: 10.1093/oso/9780198872955.003.0002

The chemistry of the alkali metals has in the past attracted little attention as the metals have a fairly restricted coordination chemistry. However, interesting and systematic study has blossomed over the last 25 years, largely prompted by two major developments: the growing importance of lithium in organic synthesis and materials science, and the exploitation of macrocyclic ligands in the formation of complexed cations. Section 2.4 deals with the use of complexed cations in the generation of alkalides and electrides.

2.2 Structure and Bonding in Inorganic Alkali Metal Compounds

2.2.1 Alkali metal oxides

The following types of binary compounds of alkali metals (M) and oxygen are known:

(a) Oxides: M_2O.

(b) Peroxides: M_2O_2.

(c) Superoxides: MO_2 (LiO_2 is stable only in matrix at 15 K).

(d) Ozonides: MO_3 (except Li).

(e) Suboxides (low-valent oxides): Rb_6O, Rb_9O_2, Cs_3O, Cs_4O, Cs_7O, and $Cs_{11}O_3$.

(f) Sesquioxides: M_2O_3: These are probably mixed peroxide-superoxides in the form of $M_2O_2 \cdot 2MO_2$.

The structures and properties of the alkali metal oxides are summarized in Table 2.2.1.

Suboxides exist for the larger alkali metals. The crystal structures of all binary suboxides (except Cs_3O) as well as the structures of $Cs_{11}O_3Rb$, $Cs_{11}O_3Rb_2$, and $Cs_{11}O_3Rb_7$ have been determined by single crystal X-ray analysis. The structural data are listed in Table 2.2.2. All structures conform to the following rules:

(a) Each O atom occupies the center of an octahedron composed of Rb or Cs atoms.

(b) Face sharing of two such octahedra results in the cluster Rb_9O_2, and three equivalent octahedra form the cluster $Cs_{11}O_3$, as shown in Fig. 2.2.1.

(c) The O–M distances are near the values expected for M^+ and O^{2-} ions. The ionic character of the M atoms is reflected by the short intra-cluster M–M distances.

(d) The inter-cluster M–M distances are comparable to the distances in metallic Rb and Cs.

(e) The Rb_9O_2 or $Cs_{11}O_3$ clusters and additional alkali metals atoms form compounds of new stoichiometries. Some crystal structures have the inter-cluster space filled by metal atoms of the same kind, such as $(Cs_{11}O_3)Cs_{10}$ shown in Fig. 2.2.2(a), or by metal atoms of a different kind, such as $(Cs_{11}O_3)Rb$ and $(Cs_{11}O_3)Rb_7$ illustrated in Fig. 2.2.2(b) and (c), respectively.

The bonding within the alkali metal suboxides is illustrated by comparing the interatomic distances in the compounds Rb_9O_2 and $Cs_{11}O_3$ with those in the "normal" oxides and in metallic Rb and Cs. The M–O distances nearly match the sum of ionic radii. The large inter-cluster M–M distances correspond to the distances in elemental M (Rb and Cs). Therefore, the formulations $(Rb^+)_9(O^{2-})_2(e^-)_5$ and $(Cs^+)_{11}(O^{2-})_3(e^-)_5$, where e^- denotes an electron, represent a rather realistic description of the bonding

Table 2.2.1 Structures and properties of alkali metal oxides

	Li	Na	K	Rb	Cs
Oxides M_2O	Colorless mp 1843 K anti-CaF_2 structure	Colorless mp 1193 K anti-CaF_2 structure	Pale-yellow mp > 763 K anti-CaF_2 structure	Yellow mp > 840 K anti-CaF_2 structure	Orange mp 763 K anti-$CdCl_2$ structure
Peroxides M_2O_2	Colorless dec. > 473 K	Pale yellow dec. ≈ 948 K	Yellow dec. ≈ 763 K	Yellow dec. ≈ 843 K	Yellow dec. ≈ 863 K
Superoxides MO_2	–	Orange dec. ≈ 573 K NaCl-structure	Orange mp 653 K dec. ≈ 673 K CaC_2 structure	Orange mp 685 K CaC_2 structure	Orange mp 705 K CaC_2 structure
Ozonides MO_3	–	Red dec. < room temp.	Dark red dec. at room temp. CsCl structure	Dark red dec. ≈ room temp. CsCl structure	Dark red dec. > 323 K CsCl structure
Suboxides	–	–	–	Rb_6O Bronze color dec. 266 K	Cs_3O Blue green dec. 439 K
				Rb_9O_2 Copper color mp 313 K	Cs_4O Red-violet dec. 284 K
					Cs_7O Bronze mp 277 K
					$Cs_{11}O_3$ Violet mp 326 K

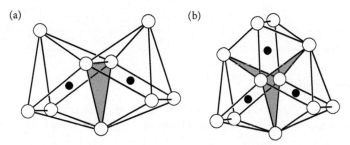

Fig. 2.2.1 Clusters in alkali suboxides: (a) Rb_9O_2 and (b) $Cs_{11}O_3$. The shared faces are shaded.

in these clusters. Their stability is attributable to consolidation by strong O–M and weak M–M bonds. All the alkali metal suboxides exhibit metallic luster and are good conductors. Thus the electrons in M–M bonding states do not stay localized within the clusters but are delocalized throughout the crystal due to the close contacts between the clusters.

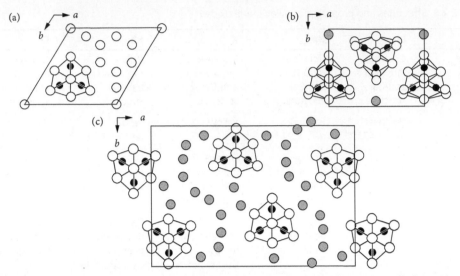

Fig. 2.2.2 Unit-cell projections showing the atomic arrangements in some low-valent alkali metal oxides: (a) $(Cs_{11}O_3)Cs_{10}$, (b) $(Cs_{11}O_3)Rb$, (c) $(Cs_{11}O_3)Rb_7$. The open, blackened, and shaded circles represent Cs, O, and Rb atoms, respectively.

Table 2.2.2 Structural data of alkali metal suboxides

Compounds	Space group	Structural feature
Rb_9O_2	$P2_1/m$	Rb_9O_2 cluster, Fig. 2.2.1(a)
Rb_6O	$P6_3/m$	$(Rb_9O_2)Rb_3$; three Rb atoms per Rb_9O_2 cluster
$Cs_{11}O_3$	$P2_1/c$	$Cs_{11}O_3$ cluster, Fig. 2.2.1(b)
Cs_4O	$Pna2_1$	$(Cs_{11}O_3)Cs$; one Cs atom per $Cs_{11}O_3$ cluster
Cs_7O	$P6m2$	$(Cs_{11}O_3)Cs_{10}$; ten Cs atoms per $Cs_{11}O_3$ cluster, Fig. 2.2.2(a)
$Cs_{11}O_3Rb$	$Pmn2_1$	One Rb atom per $Cs_{11}O_3$ cluster, Fig. 2.2.2(b)
$Cs_{11}O_3Rb_2$	$P2_1/c$	Two Rb atoms per $Cs_{11}O_3$ cluster
$Cs_{11}O_3Rb_7$	$P2_12_12_1$	Seven Rb atoms per $Cs_{11}O_3$ cluster, Fig. 2.2.2(c)

2.2.2 Lithium nitride

Among the alkali metals, only lithium reacts with N_2 at room temperature and normal atmospheric pressure to give red-brown, moisture-sensitive lithium nitride Li_3N (α-form), which has a high ionic conductivity.

The α form of Li_3N is made up of planar Li_2N layers, in which the Li atoms form a simple hexagonal arrangement, as in the case of the carbon layer in graphite, with a nitrogen atom at the center of each ring, as shown in Fig. 2.2.3(a). These layers are stacked and are linked by additional Li atoms midway between the N atoms of adjacent overlapping layers. Fig. 2.2.3(b) shows a hexagonal unit cell, with $a = 364.8$ pm, $c = 387.5$ pm. The Li–N distance is 213 pm within a layer, and 194 pm between layers.

α-Li_3N is an ionic compound, in which the N^{3-} ion has a much greater size than the Li^+ ion [$r_{N^{3-}}$ 146pm, r_{Li^+} 59pm (CN = 4)]. The crystal structure is very loosely packed, and conductivity arises from Li^+ vacancies within the Li_2N layers. The interaction between the carrier ions (Li^+) and the fixed ions (N^{3-}) is relatively weak, so that ion migration is quite effective.

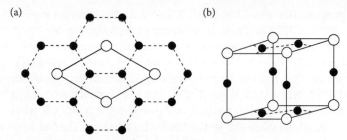

Fig. 2.2.3 Crystal structure of Li_3N: (a) Li_2N-layer, (b) the unit cell. The small black circle and large open circle represent Li and N atoms, respectively.

2.2.3 **Novel sodium compounds containing dinitrogen**

The first alkaline-earth diazenides SrN, SrN_2, and BaN_2 containing $[N_2]^{2-}$ anions were synthesized from the elements under high-pressure conditions in the early 2000s. A decade later, the diazenides Li_2N_2, CaN_2, SrN_2, and BaN_2 were obtained via controlled decomposition of alkali and alkaline-earth metal azides at 3–12 GPa.

There are at least three structurally characterized phases of NaN_3: α-NaN_3 (space group $C2/m$), β-NaN_3 ($R\bar{3}m$), and γ-NaN_3 ($I4/mcm$).

In a recent study, controlled decomposition of sodium azide NaN_3 at high-pressure conditions yielded two novel compounds containing dinitrogen anions: NaN_2 and Na_3N_8. The reaction scheme is shown below; note that Na_3N_8 is better formulated as $Na_3(N_2)_4$.

$$NaN_3 \xrightarrow{\text{28 GPa, T}} Na_3N_8 \xrightarrow{\text{4 GPa}} NaN_2 \xrightarrow{\text{0 GPa}} Na$$
$$\underset{\text{4.5 GPa, T}}{\underline{\qquad\qquad\qquad\qquad\qquad\uparrow}}$$

NaN_2 crystallizes in space group $P4/mmm$ (No. 123), and it is isostructural to α-$FeSi_2$. It exhibits a packed-layer structure, in which the layers are stacked along [001]. Each layer is composed of face-sharing, slightly distorted NaN_8 cubes that are interconnected through N–N bonds with d(N–N) = 116.1(9) pm (Fig. 2.2.4).

In the crystal structure of Na_3N_8 at 28 GPa, space group $I4_1/amd$ (No. 141), two symmetry-independent sodium atoms, Na1 and Na2, which occupy Wyckoff sites (4a) and (8e) respectively, form a

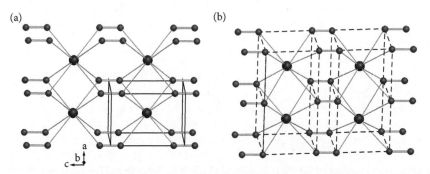

Fig. 2.2.4 (a) Crystal structure of NaN_2 at 4 GPa. Nitrogen and sodium atoms are shown as small and large spheres, respectively. (b) Crystal structure of NaN_2 at 4 GPa with the edges of NaN_8 coordination polyhedra shown in dashed lines. Bond distances: d(N–N) = 116.1(9) pm and d(Na–N) = 258.2(3) pm.

substructure isostructural to that of α-ThSi$_2$ (Fig. 2.2.5(a)). Each Na2 atom has three close Na2 neighbors with d(Na–Na) = 281 and 278 pm. These distances are close to those in bcc-Na at similar pressures (279 pm). With Na2–Na2–Na2 angles close to 120°, Na2 atoms form one of the basic 3-connected three-dimensional nets (ths) described by a vertex symbol 102104104. Nitrogen atoms N1 and N2 occupy Wyckoff sites 16(h) an 16(f) respectively, and they form dinitrogen dumbbells with d(N1–N1) = 114.7(3) pm and d(N2–N2) = 114.9(3) pm. The N1–N1 units are each surrounded by seven sodium atoms, forming a distorted pentagonal bipyramid, as shown in Fig. 2.2.5(b)–(d), while N2–N2 units are each surrounded by six sodium atoms that form a distorted octahedron. The Na1 atoms are each coordinated by eight N$_2$ units in a side-on manner, while the Na2 atoms are each coordinated by eight end-on N$_2$ units and one side-on N$_2$ unit.

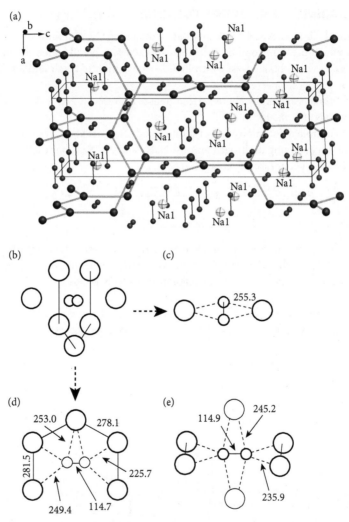

Fig. 2.2.5 (a) Crystal structure of Na$_3$N$_8$ [= Na$_3$(N$_2$)$_4$] at 28 GPa. The shortest Na–Na contacts are indicated by thick gray bonds. (b–e) Coordination environments of N$_2$ molecules. Big circle drawn in front ellipses: Na1, big circle drawn in thick line: Na2, small circle drawn in thin line: N1, small circle drawn in thin line: N2. Bond lengths are indicated in pm.

References

M. Bykov, K. R. Tasca, I. G. Batyrev, D. Smith, K. Glazyrin, S. Chariton, M. Mahmood, and A. F. Goncharov, Dinitrogen as a Universal electron acceptor in solid-state chemistry: an example of uncommon metallic compounds $Na_3(N_2)_4$ and NaN_2, *Inorg. Chem.* **59**, 14819–26 (2020).

S. B. Schneider, R. Frankovsky, and W. Schnick, High-pressure synthesis and characterization of the alkali diazenide Li_2N_2, *Angew. Chem. Int. Ed.* **51**, 1873–5 (2012).

S. B. Schneider, R. Frankovsky, and W. Schnick, Synthesis of alkaline earth diazenides $M_{AE}N_2$ (M_{AE} = Ca, Sr, Ba) by controlled thermal decomposition of azides under high pressure, *Inorg. Chem.* **51**, 2366–73 (2012).

M. O'Keeffe, M. Eddaoudi, H. Li, T. Reineke, and O. M. Yaghi, Frameworks for extended solids: geometrical design principles. *J. Solid State Chem.* **152**, 3–20 (2000).

2.2.4 Inorganic alkali metal complexes

(1) Structural features of alkali metal complexes

A growing number of alkali metal complexes $(MX·xL)_n$ (M = Li, Na, K; X = halogens, OR, NR_2, $\cdots$; L = neutral molecules) are known. It is now generally recognized that the M–X bonds in these complexes are essentially ionic or have a high degree of ionic character. This bonding character accounts for the following features of these complexes:

(a) Most alkali metal complexes have an inner ionic core, and at the same time frequently form molecular species. The ionic interaction leads to large energies of association that stabilize the formation of complexes. The outer molecular fragments prevent the continuous growth of MX to form an ionic lattice. This characteristic has been rationalized in terms of the relatively large size and low charge of the cations M^+. According to this view, the stability of alkali metal complexes should diminish in the sequence Li > Na > K > Rb > Cs, and this is frequently observed.

(b) The electrostatic association favors the formation of the M^+X^- ion pairs, and it is inevitable that such ion pairs will associate, so as to delocalize the electronic charges. The most efficient way of accomplishing this is by ring formation. If the X^- groups are relatively small and for the most part co-planar with the ring, then stacking of rings can occur. For example, two dimers are connected to give a cubane-like tetramer, and two trimers lead to a hexagonal prismatic hexamer.

(c) In the ring structures, the bond angles at X^- are acute. Hence the formation of larger rings would reduce repulsion between X^- ions and between M^+ ions.

(d) Since the alkali metal complexes are mostly discrete molecules with an ionic core surrounded by organic peripheral groups, they have relatively low melting points and good solubility in weakly polar organic solvents. These properties have led to some practical applications; for example, lithium complexes are used as low-energy electrolytic sources of metals, in the construction of portable batteries, and as halogenating agents and specific reagents in organic syntheses.

(2) Lithium alkoxides and aryloxides

Anionic oxygen donors display a strong attraction for Li^+ ions and usually adopt typical structures of their aggregation state, as shown in Fig. 2.2.6(a)–(c). Fig. 2.2.6(a) shows the dimeric structure of $[Li(THF)_2OC(NMe_2)(c\text{-}2,4,6\text{-}C_7H_6)]_2$, in which Li^+ is four-coordinated by the oxygen atoms from two THF ligands and two alkoxides $OC(NMe_2)(c\text{-}2,4,6\text{-}C_7H_6)$. The Li–O distances are 188 pm and 192 pm for the bridging oxygen of the enolate, and 199 pm for THF. Fig. 2.2.6(b) shows the tetrameric cubane-type structure of $[Li(THF)OC(CH^tBu)OMe]_4$. The Li–O distances are 196 pm (enolate) and 193 pm (THF). Fig. 2.2.6(c) shows the hexagonal-prismatic structure of $[LiOC(CH_2)^tBu]_6$. The Li–O distances

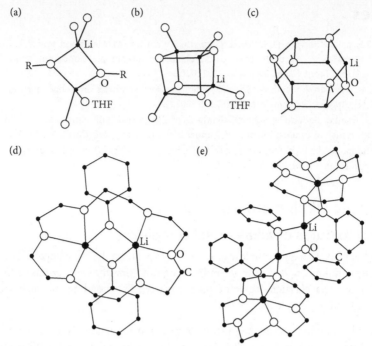

Fig. 2.2.6 Core structures of some lithium alkoxides and aryloxides: (a) [Li(THF)$_2$OC(NMe$_2$)(c-2,4,6-C$_7$H$_6$)]$_2$, (b) [Li(THF)OC(CH^{t}Bu)Ome]$_4$, (c) [LiOC(CH$_2$)^{t}Bu]$_6$, (d) (LiOPh)$_2$(18-crown-6), and (e) [(LiOPh)$_2$(15-crown-5)]$_2$.

are 190 pm (av.) in the six-membered ring of one stack, and 195 pm (av.) between the six-membered rings.

Figs. 2.2.4(d) and (e) show the structures of (LiOPh)$_2$(18-crown-6) and [(LiOPh)$_2$(15-crown-5)]$_2$. In the former, there is a central dimeric (Li–O)$_2$ unit. Owing to its larger size, 18-crown-6 can coordinate both Li$^+$ ions, each through three oxygen atoms (Li–O 215 pm, av.). These Li$^+$ ions are bridged by phenoxide groups (Li–O 188 pm). The structure of the latter has a very similar (LiOPh)$_2$ core (Li–O 187 pm and 190 pm). Each Li$^+$ ion is coordinated to another –OPh group as well as one ether oxygen from a 15-crown-5 donor. This ether oxygen atom also connected to a further Li$^+$ ion whose coordination is completed by the remaining four oxygen donors of the 15-crown-5 ligands.

(3) Lithium amides

The structures of lithium amides are characterized by a strong tendency for association. Monomeric structures are observed only in the presence of very bulky groups at nitrogen and/or donor molecules that coordinate strongly to lithium. In the dimeric lithium amides, the Li$^+$ ion is usually simultaneously coordinated by N atoms and other atoms. The trimer [LiN(SiMe$_3$)$_2$]$_3$ has a planar cyclic arrangement of its Li$_3$N$_3$ core. The Li–N bonds are 200 pm long with internal angles of 148° at Li and 92° at N (average values). Fig. 2.2.7 shows the structures of some higher aggregates of lithium amides.

(a) The tetramer (LiTMP)$_4$ (TMP = 2,2,6,6-tetramethylpiperidinide) has a planar Li$_4$N$_4$ core with a nearly linear angle (168.5°) at Li and an internal angle of 101.5° at N. The average Li–N distance is 200 pm, as shown in Fig. 2.2.7 (a).

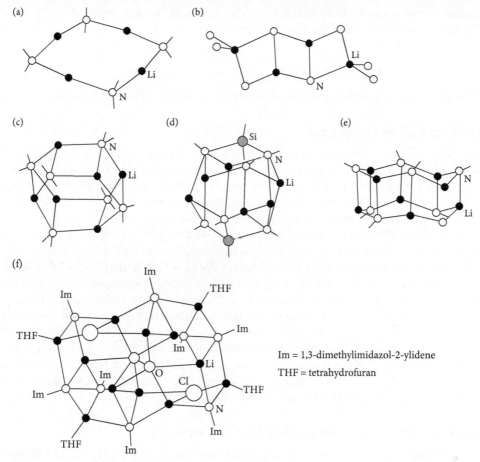

Fig. 2.2.7 Core structures of some lithium amides and imides: (a) $(LiTMP)_4$, (b) $Li_4(TMEDA)_2[c\text{-}N(CH_2)_4]$, (c) $[Li\{c\text{-}N(CH_2)_6\}]_6$, (d) $[\{LiN(SiMe_3)\}_3SiR]_2$, (e) $[LiNH^tBu]_8$, and (f) $[Li_{12}O_2Cl_2(ImN)_8(THF)_4]\cdot8(THF)$.

(b) The tetrameric molecule $Li_4(TMEDA)_2[c\text{-}N(CH_2)_4]$ ($TMEDA = N, N'$-tetramethylenediamine) has a ladder structure in which adjacent Li_2N_2 rings share edges with further aggregation blocked by TMEDA coordination, as shown in Fig. 2.2.7(b).

(c) The hexameric $[Li\{c\text{-}N(CH_2)_6\}]_6$ has a stacked structure, as shown in Fig. 2.2.7(c). This involves the association of two $[Li\text{-}c\text{-}N(CH_2)_6]_3$ units to form a hexagonal-prismatic Li_6N_6 unit, in which all N atoms have the relatively rare coordination number of five.

(d) The basic framework in $[\{LiN(SiMe_3)\}_3SiR]_2$ ($R = Me$, ^tBu and Ph) is derived from the dimerization of a trisamidosilane. It exhibits molecular D_{3d} symmetry, as shown in Fig. 2.2.7(d). Each of the three Li^+ ions in a monomeric unit is coordinated by two N atoms in a chelating fashion, and linkage of units occurs through single Li–N contacts. This causes all N atoms to be five-coordinate, which leads to weakening of the Li–N and Si–N bonds.

(e) Fig. 2.2.7(e) shows the core structure of $[LiNH^tBu]_8$, in which planar Li_2N_2 ring units are connected to form a discrete prismatic ladder molecule.

(f) Fig. 2.2.7(f) shows the centrosymmetric core structure of $[Li_{12}O_2Cl_2(ImN)_8(THF)_4]\cdot8(THF)$, which consists of a folded $Li_4N_2O_2$ ladder in which the central O_2^{2-} ion is connected to two Li centers. The two adjacent Li_4ClN_3 ladders are connected with the central $Li_4N_2O_2$ unit via Li–O, Li–Cl and Li–N interactions. The bond distances are:

Li–N	195.3–218.3 pm,	Li–O	191.7–259.3 pm
Li–Cl	238.5–239.6 pm,	O–O	154.4 pm

In this structure, Li atoms are four-coordinated, and N atoms are four- or five- coordinated.

(4) Lithium halide complexes

In view of the very high lattice energy of LiF, there is as yet no known complex that contains LiF and a Lewis base donor ligand. However, a range of crystalline fluorosilyl-amide and -phosphide complexes that feature significant Li⋯F contacts have been synthesized and characterized. For example, $\{[^tBu_2Si(F)]_2N\}Li \cdot 2THF$ has a heteroatomic ladder core, as shown in Fig.2.2.8(a), and dimeric $[^tBu_2SiP(Ph)(F)Li \cdot 2THF]_2$ contains an eight-membered heteroatomic ring.

A large number of complexes containing the halide salts LiX (X = Cl, Br and I) have been characterized in the solid state. The structures of some of these complexes are shown in Fig.2.2.8(b)-(f).

In the center of the cubane complex $[LiCl \cdot HMPA]_4$ (HMPA is $(Me_2N)_3P=O$), each Li^+ is bonded to three Cl^- and one oxygen of HMPA, as shown in Fig. 2.2.8(b). The complex $(LiCl)_6(TMEDA)_2$ has a complicated polymeric structure based on a $(LiCl)_6$ core, as shown in Fig. 2.2.8(c). Tetrameric $[LiBr]_4 \cdot 6[2,6-Me_2Py]$ has a staggered ladder structure, as shown in Fig. 2.2.8(d). In the complex $[Li_6Br_4(Et_2O)_{10}]^{2+} \cdot 2[Ag_3Li_2Ph_6]^-$, the hexametallic dication contains a Li_6Br_4 unit which can be regarded as two three-rung ladders sharing their terminal Br anions, as shown in Fig. 2.2.8(e). The structure of oligomeric $[LiBr \cdot THF]_n$ is that of a unique corrugated ladder, in which $[LiBr]_2$ units are linked together by μ_3-Br bridges, as shown in Fig. 2.2.8(f). Polymeric ladder structures have also been observed for a variety of other alkali metal organometallic derivatives.

(5) Lithium salts of heavier heteroatom compounds

Lithium can be coordinated by heavier main-group elements, such as S, Se, Te, Si, and P atoms, to form complexes. In the complex $[Li_2(THF)_2Cp^*TaS_3]_2$, the monomeric unit consists of a pair of four-membered Ta–S–Li–S rings sharing a common Ta–S edge. Dimerization occurs through further Li–S linkages, affording a hexagonal-prismatic skeleton, as shown in Fig. 2.2.9(a). The bond lengths are Li–S 248 pm(av.) and Ta–S 228 pm(av.). Fig. 2.2.9(b) shows the core structure of $Li(THF)_3SeMes^*$

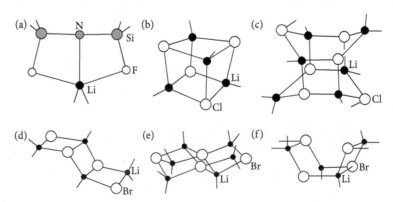

Fig. 2.2.8 Structures of some lithium halide complexes: (a) $\{[^tBu_2Si(F)]_2N\}Li \cdot 2THF$, (b) $[LiCl \cdot HMPA]_4$, (c) $(LiCl)_6(TMEDA)_2$, (d) $[LiBr]_4 \cdot 6[2,6-Me_2Py]$, (e) $[Li_6Br_4(Et_2O)_{10}]^{2+}$, and (f) $[LiBr \cdot THF]_n$.

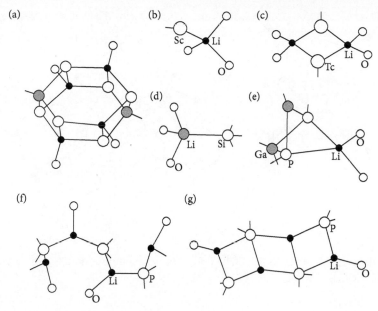

Fig. 2.2.9 Core structures of some lithium complexes bearing heavier heteroatom ligands: (a) [Li$_2$(THF)$_2$Cp*TaS$_3$]$_2$, (b) Li(THF)$_3$SeMes*, (c) [Li(THF)$_2$TeSi(SiMe$_3$)$_3$]$_2$, (d) Li(THF)$_3$SiPh$_3$, (e) Li(Et$_2$O)$_2$(P^{t}Bu)$_2$ (Ga$_2$^{t}Bu$_3$), (f) [Li(Et$_2$O)PPh$_2$]$_n$, (g) [LiP(SiMe$_3$)$_2$]$_4$(THF)$_2$.

(Mes* = 2,4,6-C$_6$H$_2$R$_3$, where R is a bulky group such as ^{t}Bu) which has Li–Se bond length 257 pm. Fig. 2.2.9(c) shows the core structure of the dimeric complex [Li(THF)$_2$TeSi(SiMe$_3$)$_3$]$_2$, in which the distances of bridging Li–Te bonds are 282 pm and 288 pm.

Some lithium silicide complexes have been characterized. The complex Li(THF)$_3$SiPh$_3$ displays a terminal Li–Si bond that has a length of 267 pm, which is the same as that in Li(THF)$_3$Si(SiMe$_3$). Fig. 2.2.9(d) shows the core structure of Li(THF)$_3$SiPh$_3$. There is no interaction between Li and the phenyl rings. The C–Si–C bond angles are smaller (101.3° av.) than the ideal tetrahedral value, while the Li–Si–C angles are correspondingly larger (116.8° av.). This indicates the existence of a lone pair on silicon.

Fig. 2.2.9(e)–(g) show the core structures of three lithium phosphide complexes. Li(Et$_2$O)$_2$ (P^{t}Bu)$_2$(Ga$_2$^{t}Bu$_3$) has a puckered Ga$_2$P$_2$ four-membered ring, in which one Ga atom is four- and the other three-coordinated, as shown in Fig. 2.2.9(e). The Li–P bond length is 266 pm (av.).

The polymeric complex [Li(Et$_2$O)PPh$_2$]$_n$ has a backbone of alternating Li and P chain, as shown in Fig.2.2.9(f). The bond length of Li–P is 248 pm (av.) and Li–O is 194 pm (av.). Fig. 2.2.9(g) shows the core structure of [LiP(SiMe$_3$)$_2$]$_4$(THF)$_2$, which has a ladder structure with Li–P bond length 253 pm (av.).

(6) Sodium coordination complexes

Recent development in the coordination chemistry of Group 1 elements has extended our knowledge of their chemical behavior and provided many interesting new structural types. Three examples among known sodium coordination complexes are presented below:

(a) A rare example of a bimetallic imido complex is the triple-stacked Li$_4$Na$_2$[N=C(Ph)(^{t}Bu)]$_6$. This molecule has six metal atoms in a triple-layered stack of four-membered M$_2$N$_2$ rings, with the outer

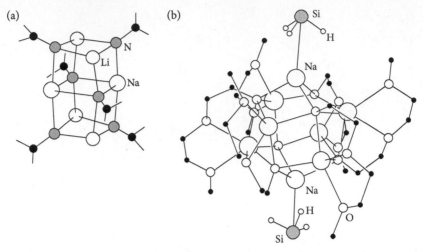

Fig. 2.2.10 Structures of (a) $Li_4Na_2[N=C(Ph)(^tBu)]_6$ (the Ph and ^tBu groups are not shown), and (b) $[Na_8(OCH_2CH_2OCH_2Ch_2OMe)_6(SiH_3)_2]$.

rings containing lithium and the central ring containing sodium. In the structure, lithium is three-coordinated and sodium is four-coordinated, as shown in Fig. 2.2.10(a).

(b)   The large aggregate $[Na_8(OCH_2CH_2OCH_2CH_2OMe)_6(SiH_3)_2]$ has been prepared and character-ized. The eight sodium atoms form a cube, the faces of which are capped by the alkoxo oxygen atoms of the six $(OCH_2CH_2OCH_2CH_2OMe)$ ligands, which are each bound to four sodium atoms with Na–O distances in the range 230–242 pm. The sodium and oxygen atoms constitute the vertices of an approximate rhombododecahedron. Six of eight sodium atoms are five-coordinated by oxygen atoms, and each of the other two Na atoms is bonded by a SiH_3^- group, which has inverted C_{3v} symmetry with Na–Si–H bond angles of 58° to 62°, as shown in Fig. 2.2.10(b).

Despite the strange appearance of the SiH_3^- configuration in this complex, *ab initio* calculations con-ducted on the simplified model compound $(NaOH)_3NaSiH_3$ indicate that the form with inverted hydrogens is 6 kJ mol^{-1} lower in energy than the uninverted form. The results suggest that electrostatic interaction, rather than agostic interaction between the SiH_3 group and its three adjacent Na neighbors, stabilizes the inverted form.

(c)   Mixed *tert*-butanolate/hydroxide aggregate $[Na_{11}(O^tBu)_{10}(OH)]$.

The complex $[Na_{11}(O^tBu)_{10}(OH)]$ is obtained from the reaction of sodium *tert*-butanolate with sodium hydroxide. Its structure features a 21-vertex cage constructed from 11 sodium cations and 10 *tert*-butanolate anions, with an encapsulated hydroxide ion in its interior, as shown in Fig. 2.2.11. In the lower part of the cage, eight sodium atoms constitute a square antiprism, and four of the lower triangular faces are each capped by a μ_3-O^tBu group. The Na_4 square face at the bottom is capped by a μ_4-O^tBu group, and the upper Na_4 square face is capped in an inverted fashion by the μ_4-OH^- group. Thus the OH^- group resides within a Na_8-core, and it also forms an O–H$\cdots$O hydrogen bond of length 297.5 pm. The Na–O distances in the 21-vertex cage are in the range of 219–243 pm.

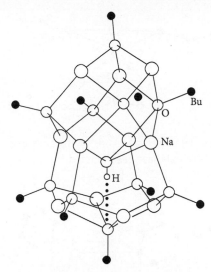

Fig. 2.2.11 Structure of $[Na_{11}(O^tBu)_{10}(OH)]$.

2.3 Structure and Bonding in Organic Alkali Metal Compounds

2.3.1 Methyllithium and related compounds

Lithium readily interacts with hydrocarbon π systems such as olefins, arenes, and acetylenes at various sites simultaneously. Lithium reacts with organic halides to give alkyllithium or aryllithium derivatives in high yield. Therefore, a wide variety of organolithium reagents can be made. These organolithium compounds are typically covalent species, which can be sublimed, distilled in a vacuum, and dissolved in many organic solvents. The Li–C bond is a strong polarized covalent bond, so that organolithium compounds serve as sources of anionic carbon, which can be replaced by Si, Ge, Sn, Pb, Sb, or Bi. When $Sn(C_6H_5)_3$ or $Sb(C_6H_{11})_2$ groups are used to replace CH_3, CMe_3 or $C(SiMe_3)_3$ in organolithium compounds, new species containing Li–Sn or Li–Sb bonds are formed.

Methyllithium, the simplest organolithium compound, is tetrameric. The $Li_4(CH_3)_4$ molecule with T_d symmetry may be described as a tetrahedral array of four Li atoms with a methyl C atom located above each face of the tetrahedron, as shown in Fig. 2.3.1(a). The bond lengths are Li–Li 268 pm, Li–C 231 pm, and the bond angle Li–C–Li is 68.3°.

Vinyllithium, $LiHC=CH_2$, has a similar tetrameric structure in the crystalline state with a THF ligand attached to each lithium atom [Fig. 2.3.1(b)]. In THF solution vinyllithium is tetrameric as well. The internuclear Li–H distances are in agreement with those calculated from the two-dimensional NMR [6]Li, ^{1}H HOSEY spectra.

When steric hindrance is minimal, organolithium compounds tend to form tetrameric aggregates with a tetrahedral Li_4 core. Figs 2.3.2(a)-(d) show the structures of some examples: (a) $(LiBr)_2 \cdot (CH_2CH_2CHLi)_2 \cdot 4Et_2O$, (b) $(PhLi \cdot Et_2O)_3 \cdot LiBr$, (c) $[C_6H_4CH_2N(CH_3)_2Li]_4$, and (d) $(^tBuC \equiv CLi)_4(THF)_4$. In these complexes, the Li–C bond lengths range from 219 pm to 253 pm, with an average value of 229 pm, while the Li–Li distances lie in the range 242–263 pm (average 256 pm). The Li–C bond lengths are slightly longer than those for the terminally bonded organolithium compounds. This difference may be attributed to multi-center bonding in the tetrameric structures.

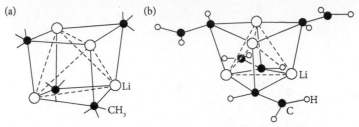

Fig. 2.3.1 Molecular structure of (a) methyllithium tetramer (Li_4Me_4), H atoms have been omitted; and (b) vinyllithium·THF tetramer ($LiHC=CH_2 \cdot THF)_4$, THF ligands have been omitted.

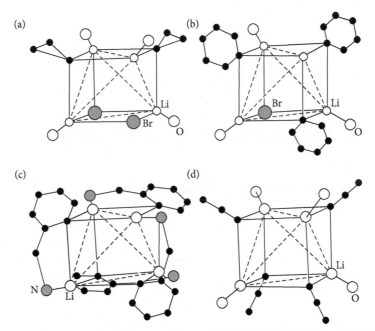

Fig. 2.3.2 Structures of some organolithium complexes: (a) $(LiBr)_2 \cdot (CH_2CH_2CHLi)_2 \cdot 4Et_2O$, (b) $(PhLi \cdot Et_2O)_3 \cdot LiBr$, (c) $[C_6H_4CH_2N(CH_3)_2Li]_4$, and (d) $(^tBuc \equiv CLi)_4(THF)_4$.

2.3.2 π-Complexes of lithium

The interactions between Li atoms and diverse π systems yield many types of lithium π complexes. Figs 2.3.3(a)-(c) show the schematic representation of the core structures of some π complexes of lithium, and Fig. 2.3.3(d) shows the molecular structure of $[C_2P_2(SiMe_3)_2]^{2-} \cdot 2[Li^+(DME)]$ (DME = dimethoxyethane).

(1) $Li[Me_2N(CH_2)_2NMe_2] \cdot [C_5H_2(SiMe_3)_3]$

In the cyclopentadienyllithium complex, the Li atom is coordinated by the planar Cp ring and by one chelating TMEDA ligand. The Li–C distances are 226 pm to 229 pm (average value 227 pm).

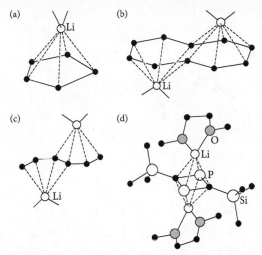

Fig. 2.3.3 Structures of some lithium π complexes: (a) Li[Me$_2$N(CH$_2$)$_2$NMe$_2$]·[C$_5$H$_2$(SiMe$_3$)$_3$], (b) Li$_2$[Me$_2$N(CH$_2$)$_2$NMe$_2$]·C$_{10}$H$_8$, (c) Li$_2$[Me$_2$N(CH$_2$)$_2$NMe$_2$][H$_2$C=CH–CH=CH–CH=CH$_2$], and (d) [C$_2$P$_2$(SiMe$_3$)$_2$]$^{2-}$·2[Li$^+$(DME)].

(2) Li$_2$[Me$_2$N(CH$_2$)$_2$NMe$_2$]·C$_{10}$H$_8$

In the naphthalene-lithium complex, each Li atom is coordinated by one η^6 six-membered ring and a chelating TEMDA ligand. The two Li atoms are not situated directly opposite to each other. The Li–C distances are in the range 226–266 pm; the average value is 242 pm.

(3) Li$_2$[Me$_2$N(CH$_2$)$_2$NMe$_2$][H$_2$C=CH–CH=CH–CH=CH$_2$]

In this complex, each Li atom is coordinated by the bridging bistetrahapto triene and by one chelating TMEDA ligand. The Li–C distances are 221 pm to 240 pm (average value 228 pm).

(4) [C$_2$P$_2$(SiMe$_3$)$_2$]$^{2-}$·2[Li$^+$(DME)]

In this complex, the C$_2$P$_2$ unit forms a planar four-membered ring, which exhibits aromatic properties with six π electrons. The two Li atoms are coordinated by one η^4-C$_2$P$_2$ ring from the two opposite directions. The distances of Li–C and Li–P are 239.1 pm and 245.8 pm, respectively. Each Li atom is also coordinated by one chelating DME ligand.

2.3.3 π-**Complexes of sodium and potassium**

Many π complexes of sodium and potassium have been characterized. The Na and K atoms of these complexes are polyhapto-bonded to the ring π systems, and frequently bonded to two or more rings to give infinite chain structures. Fig. 2.3.4 shows the structures of some π complexes of sodium and potassium.

In the structure of Na$_2$[Ph$_2$C=CPh$_2$]·2Et$_2$O, the two halves of the [Ph$_2$C=CPh$_2$]$^{2-}$ dianion twist through 56° relative to each other, and the central C=C bond is lengthened to 149 pm as compared to 136 pm in Ph$_2$C=CPh$_2$. The coordination of Na(1) involves the >C=C< π bond and two adjacent π bonds. The two Et$_2$O ligands are also coordinated to it. The Na(2) is sandwiched between two phenyl rings in a bent fashion, and further interacts with a π bond in the third ring, as shown in Fig. 2.3.4(a). The Na(1)–C and Na(2)–C distances lie in the ranges 270–282 pm and 276–309 pm, respectively.

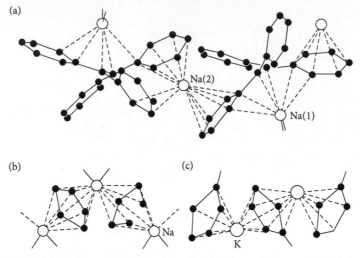

Fig. 2.3.4 Structures of some π complexes of sodium and potassium: (a) Na$_2$[Ph$_2$C=CPh$_2$]·2Et$_2$O, (b) Na(C$_5$H$_5$)(en), and (c) K[C$_5$H$_4$(SiMe$_3$)].

Figs. 2.3.4(b)-(c) show the structures of Na(C$_5$H$_5$)(en) and K[C$_5$H$_4$(SiMe$_3$)], respectively. In these complexes, each metal (Na or K) atom is centrally positioned between two bridging cyclopentadienyl rings, leading to a bent sandwich polymeric zigzag chain structure. The sodium atom is further coordinated by an ethylenediamine ligand, and the potassium atom interacts with an additional Cp ring of a neighboring chain.

2.4 **Alkalides and Electrides**

2.4.1 **Alkalides**

The alkali metals can be dissolved in liquid ammonia, and also in other solvents such as ethers and organic amines. Solutions of the alkali metals (except Li) contain solvated M$^-$ anions as well as solvated M$^+$ cations.

$$2M(s) \rightleftharpoons M^+(solv) + M^-(solv)$$

Successful isolation of stable solids containing these alkalide anions depends on driving the equilibrium to the right and then on protecting the anion from the polarizing effects of the cation. Both goals have been realized by using macrocyclic ethers (crown ethers and cryptands). The oxa-based cryptands such as 2.2.2-crypt (also written as C222) are particularly effective in encapsulating M$^+$ cations.

Alkalides are crystalline compounds that contain the alkali metal anions, M$^-$ (Na$^-$, K$^-$, Rb$^-$, or Cs$^-$). The first alkalide compound Na$^+$(C222)·Na$^-$ was synthesized and characterized by Dye in 1974.

Elemental sodium dissolves only very slightly ($\approx 10^{-6}$ mol L^{-1}) in ethylamine, but when C222 is added, the solubility increases dramatically to 0.2 mol L^{-1}, according to the equation:

$$2Na(s) + C(222) \longrightarrow Na^+(C222) + Na^-$$

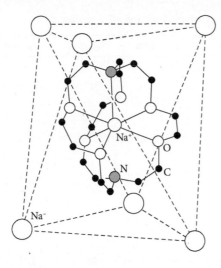

Fig. 2.4.1 The cryptated sodium cation surrounded by six natride anions in crystalline [Na⁺(C222)] Na⁻.

When cooled to −15 °C or below, the solution deposits shiny, gold-colored thin hexagonal crystals of [Na⁺(C222)]·Na⁻. Single-crystal X-ray analysis shows that two sodium atoms are in very different environments in the crystal. One is located inside the cryptand at distances from the nitrogen and oxygen atoms which are characteristic of a trapped Na⁺ cation. The other is a sodium anion (natride), Na⁻, which is located far away from all other atoms. Fig. 2.4.1 shows the cryptated sodium cation with its six nearest natrides in the crystals. The close analogy of the natride ion with an iodide ion is brought out clearly by comparing [Na⁺(C222)]·Na⁻ with [Na⁺(C222)]·I⁻.

More than 40 alkalide compounds that contain the anions Na⁻, K⁻, Rb⁻, or Cs⁻ have been synthesized, and their crystal structures have been determined. Table 2.4.1 lists the calculated radii of alkali metal anions from structures of alkalides and alkali metals. The values of $r_{M^-(av)}$ derived from alkalides and r_{M^-} from the alkali metals are in good agreement.

2.4.2 **Electrides**

The counterparts to alkalides are electrides, which are crystalline compounds with the same type of complexed M⁺ cations, but the M⁻ anions are replaced by entrapped electrons that usually occupy the same sites. The crystal structures of several electrides are known: Li⁺(C211)e⁻, K⁺(C222)e⁻, Rb⁺(C222)e⁻, Cs⁺(18C6)₂e⁻, Cs⁺(15C5)₂e⁻, and [Cs⁺(18C6)(15C5)e⁻]₆(18C6). Comparison of the structures of the complexed cation in the natride Li⁺(C211)Na⁻ and the electride Li⁺(C211)e⁻ shows that the geometric parameters are virtually identical, as shown in Fig. 2.4.2, supporting the assumption that the "excess" electron density in the electride does not penetrate substantially into the cryptand cage. In Cs⁺(18C6)₂Na⁻ and Cs⁺(18C6)₂e⁻, not only are the complexed cation geometries the same but the crystal structures are also very similar, the major difference being a slightly larger anionic site for Na⁻ than for e⁻.

Electrides made with crown ethers and oxa-based cryptands are generally unstable above −40 °C, since the ether linkages are vulnerable to electron capture and reductive cleavage. Using a specifically designed pentacyclic tripiperazine cryptand TripPip222 (molecular structure displayed in Fig. 2.4.3(a)), the pair of isomorphous compounds Na⁺(TripPip222)Na⁻ and Na⁺(TripPip222)e⁻ have been synthesized and fully characterized. This air-sensitive electride is stable up to about 40 °C before it begins to decompose into a mixture of the sodide and the free complexant.

While the trapped electron could be viewed as the simplest possible anion, there is a significant difference between alkalides and electrides. Whereas the large alkali metal anions are confined to the cavities,

Table 2.4.1 Radii of alkali metal anions from structure of alkalides and alkali metals*

Compound	$r_{M^-(min)}$/pm	$r_{M^-(av)}$/pm	d_{atom}/pm	r_{M^+}/pm	r_{M^-}/pm
Na metal	–	–	372	95	277
$K^+(C222) \cdot Na^-$	255	273 (14)			
$Cs^+(18C6) \cdot Na^-$	234	264 (16)			
$Rb^+(15C5)_2 \cdot Na^-$	260	289 (16)			
$K^+(HMHCY) \cdot Na^-$	248	277 (10)			
$Cs^+(HMHCY) \cdot Na^-$	235	279 (8)			
K metal	–	–	463	133	330
$K^+(C222) \cdot K^-$	294	312 (10)			
$Cs^+(15C5)_2 \cdot K^-$	277	314 (16)			
Rb metal	–	–	485	148	337
$Rb^+(C222) \cdot Rb^-$	300	321 (14)			
$Rb^+(18C6) \cdot Rb^-$	299	323 (9)			
$Rb^+(15C5)_2 \cdot Rb^-$	264	306 (16)			
Cs metal	–	–	527	167	360
$Cs^+(C222) \cdot Cs^-$	317	350 (15)			
$Cs^+(18C6)_2 \cdot Cs^-$	309	346 (15)			

* HMHCY stands for hexamethyl hexacyclen, which is the common name of 1,4,7,10.13,16-hexaaza-1,4,7,10,13, 16-hexamethyl cyclooctadecane. Here $r_{M^-(min)}$ is the distance between an anion and its nearest hydrogen atoms minus the van der Waals radius of hydrogen (120 pm), while $r_{M^-(av)}$ is the average radius over the nearest hydrogen atoms; the numbers in the brackets are the numbers of hydrogen atoms for averaging. Similarly, d_{atom} is the interatomic distance in the metal; r_{M^-} is equal to d_{atom} mimus r_{M^+}.

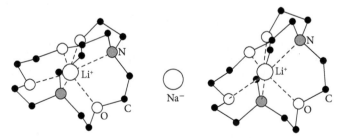

Fig. 2.4.2 Comparison of the structures of the complexed cation in $Li^+(C211)Na^-$ and $Li^+(C211)e^-$.

only the probability density of a trapped electron can be defined. The electronic wavefunction can extend into all regions of space, and electron density tends to seek out the void spaces provided by the cavities and by inter-cavity channels.

Fig. 2.4.3(b) displays the cavity-channel geometry in the crystal structure of $Rb^+(C222)e^-$, in which the dominant void space consists of parallel zigzag chain of cavities S and large channels A (diameter 260 pm) along the a-direction, which are connected at the "corners" by narrower channel B of diameter 115 pm to form "ladder like" one-dimensional chains along a. Adjacent chains are connected along b by

(a)

(b)

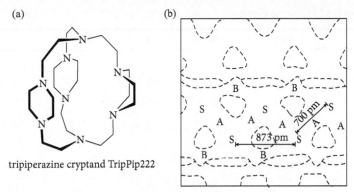

tripiperazine cryptand TripPip222

Fig. 2.4.3 (a) Structural formula of the complexant 1,4,7,10,13,16,21,24-octaazapentacyclo[8.8.8.24, 7.213,16.221,24]dotriacontane (TriPip222). (b) The "ladder-like" zigzag chain of cavities S and channels A and B, along the a-direction (horizontal), in the crystal structure of electride $Rb^+(C222)e^-$. Connection along b (near perpendicular to the plane of the figure) between chains involves channel C (not labeled). The c-axis is in the plane at 72° to the horizontal.

channels C of diameter 104 pm and along c by channels with diameter of 72 pm. Thus the void space in the crystal comprises a three-dimensional network of cavities and channels. The inter-electron coupling in adjacent cavities depends on the extent of overlap of the electronic wavefunction. The dimensionalities, diameters and lengths of the channels that connect the cavities play major roles. These structural features are consistent with the nature of wave-particle duality of the electron.

Removal of enclathrated oxygen ions from the cavities in a single crystal of $12CaO·7Al_2O_3$ leads to the formation of the thermally and chemically stable electride $[Ca_{24}Al_{28}O_{64}]^{4+}(e^-)_4$.

2.5 **Survey of the Alkaline-Earth Metals**

Beryllium, magnesium, calcium, strontium, barium, and radium constitute Group 2 in the Periodic Table. These elements (or simply the Ca, Sr, and Ba triad) are often called alkaline-earth metals. Some important properties of Group 2 elements are summarized in Table 2.5.1.

All Group 2 elements are metals, but an abrupt change in properties between Be and Mg occurs as Be shows anomalous behavior in forming mainly covalent compounds. Beryllium most frequently displays a coordination number of four, usually tetrahedral, in which the radius of Be^{2+} is 27 pm. The chemical behavior of magnesium is intermediate between that of Be and the heavier elements, and it also has some tendency for covalent bond formation.

With increasing atomic number, the Group 2 metals follow a general trend in decreasing values of I_1 and I_2 except for Ra, whose relatively high I_1 and I_2 are attributable to the $6s$ inert pair effect. Also, high I_3 values for all members of the group preclude the formation of the +3 oxidation state.

As the atomic number increases, the electronegativity decreases. In the organic compounds of Group 2 elements, the polarity of the M–C bond increases in the following order:

$$BeR_2 < MgR_2 < CaR_2 < SrR_2 < BaR_2 < RaR_2$$

This is due to the increasing difference between the χ_s of carbon (2.54) and those of the metals. The tendency of these compounds to aggregate also increases in the same order. The BeR_2 and MgR_2 compounds

Table 2.5.1 Properties of Group 2 elements

Property	Be	Mg	Ca	Sr	Ba	Ra
Atomic number, Z	4	12	20	38	56	88
Electronic configuration	[He]$2s^2$	[Ne]$3s^2$	[Ar]$4s^2$	[Kr]$5s^2$	[Xe]$6s^2$	[Rn]$7s^2$
ΔH_{at}^0/ kJ mol^{-1}	309	129	150	139	151	130
mp/K	1551	922	1112	1042	1002	973
ΔH_{fuse}(mp)/kJ mol^{-1}	7.9	8.5	8.5	7.4	7.1	–
bp/K	3243	1363	1757	1657	1910	1413
r_M (CN = 12)/pm	112	160	197	215	224	–
r_M2+ (CN = 6)/pm	45	72	100	118	135	148
I_1/kJ mol^{-1}	899.5	737.3	589.8	549.5	502.8	509.3
I_2/kJ mol^{-1}	1757	1451	1145	1064	965.2	979.0
I_3/kJ mol^{-1}	14,850	7733	4912	4138	3619	3300
χ_s	1.58	1.29	1.03	0.96	0.88	–
mp of MCl$_2$/K	703	981	1045	1146	1236	–

are linked in the solid phase via M–C–M 3c-2e covalent bonds to form polymeric chains, while CaR$_2$ to RaR$_2$ form three-dimensional network structures in which the M–C bonds are largely ionic.

All the M^{2+} ions are smaller and considerably less polarizable than the isoelectronic M$^+$ ions, as the higher effective nuclear charge binds the remaining electrons tightly. Thus the effect of polarization of cations on the properties of their salts are less important. Ca, Sr, Ba, and Ra form a closely allied series in which the properties of the elements and their compounds vary systematically with increasing size in much the same manner as in the alkali metals.

The melting points of Group 2 metal chlorides MCl$_2$ increase steadily, and this trend is in sharp contrast to the alkali metal chlorides: LiCl (883 K), NaCl (1074 K), KCl (1045 K), RbCl (990 K), and CsCl (918 K). This is due to several subtle factors: (a) the nature of bonding varies from covalent (Be) to ionic (Ba); (b) from Be to Ra the coordination number increases, so the Madelung constants, the lattice energies and the melting points also increase; (c) the radius of Cl$^-$ is large (181 pm) whereas the radii of M^{2+} are small, and in the case of metal coordination, an increase in the radius of M^{2+} reduces the Cl$^- \cdots$ Cl$^-$ repulsion.

2.6 Structure of Compounds of Alkaline-Earth Metals

2.6.1 Group 2 metal complexes

The coordination compounds of the alkaline-earth metals are becoming increasingly important to many branches of chemistry and biology. A considerable degree of structural diversity exists in these compounds, and monomers up to nonametallic clusters and polymeric species are known.

Beryllium, in view of its small size and simple set of valence orbitals, almost invariably exhibits tetrahedral four-coordination in its compounds. Fig. 2.6.1(a) shows the structure of Be$_4$O(NO$_3$)$_6$. The central oxygen atom is tetrahedrally surrounded by four Be atoms, and each Be atom is in turn tetrahedrally surrounded by four O atoms. The six nitrate groups are attached symmetrically to the six edges of the tetrahedron. This type of structure also appears in Be$_4$O(CH$_3$COO)$_6$.

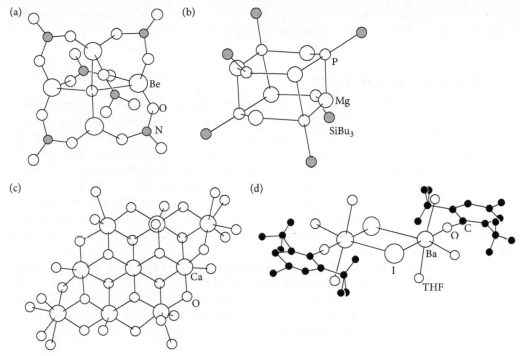

Fig. 2.6.1 Structures of some coordination compounds of alkaline-earth metals: (a) $Be_4O(NO_3)_6$, (b) $[MgPSi^tBu_3]_6$, (c) $Ca_9(OCH_2CH_2OMe)_{18}(HOCH_2CH_2OMe)_2$, and (d) $[BaI(BHT)(THF)_3]_2$.

Magnesium shows a great tendency to form complexes with ligands which have oxygen and nitrogen donor atoms, often displaying six-coordination, and its compounds are more polar than those of beryllium. The structure of hexameric $[MgPSi^tBu_3]_6$ is based on a Mg_6P_6 hexagonal drum, with Mg–P distances varying between 247 pm and 251 pm in the six-membered Mg_3P_3 ring, and 250-260 pm between the two rings, as shown in Fig. 2.6.1(b).

Calcium, strontium, and barium form compounds of increasing ionic character with higher coordination numbers, of which six to eight are particularly common. The complex $Ca_9(OCH_2CH_2OMe)_{18}(HOCH_2CH_2OMe)_2$ has an interesting structure: the central $Ca_9(\mu_3\text{-}O)(\mu_2\text{-}O)_8O_{20}$ skeleton is composed of three six-coordinate Ca atoms and six seven-coordinate Ca atoms that can be viewed as filling the octahedral holes in two close-packed oxygen layers, as shown in Fig. 2.6.1(c). The distances of Ca–O are: Ca–$(\mu_3\text{-}O)$ 239 pm, Ca–$(\mu_2\text{-}O)$ 229 pm, and Ca–O_{ether} 260 pm.

In the dimeric complex $[BaI(BHT)(THF)_3]_2$ [BHT = 2,6-di-t-butyl-4-methylphenol, $C_6H_2Me^tBu_2(OH)$], the coordination geometry around the Ba atom is distorted octahedral, with the two bridging iodides, the BHT, and a THF ligand in one plane, and two additional THF molecules lying above and below the plane, as shown in Fig. 2.6.1(d). The bond distances are Ba–I 344 pm, Ba–OAr 241 pm (av).

Many interesting coordination compounds of Group 2 elements have been synthesized in recent years. For example, the cation $[Ba(NH_3)_n]^{2+}$ is generated in the course of reducing the fullerenes C_{60} and C_{70} with barium in liquid ammonia. In the $[Ba(NH_3)_7]C_{60}\cdot NH_3$ crystal, the Ba^{2+} cation is surrounded by seven NH_3 ligands at the vertices of a monocapped trigonal antiprism, and the C_{60} dianion is well ordered. In the $[Ba(NH_3)_9]C_{70}\cdot 7NH_3$ crystal, the coordination geometry around Ba^{2+} is a distorted tricapped trigonal prism, with Ba–N distances in the range 289–297 pm; the fullerene C_{70} units are linked into slightly zigzag linear chains by single C–C bonds with length 153 pm.

2.6.2 **Group 2 metal nitrides**

The crystals of $M[Be_2N_2]$ (M = Mg, Ca, Sr) are composed of complex anion layers with covalent bonds between Be and N atoms. For example, $Mg[Be_2N_2]$ contains puckered six-membered rings in the chair conformation. These rings are condensed into single nets and are further connected to form double layers, in which the Be and N atoms are alternately linked. In this way, each beryllium atom is tetrahedrally coordinated by nitrogens, while each nitrogen occupies the apex of a trigonal pyramid whose base is made of three Be atoms, and a fourth Be atom is placed below the base. Thus the nitrogen atoms are each coordinated by an inverse tetrahedron of Be atoms and occupy positions at the outer boundaries of a double layer, as shown in Fig. 2.6.2(a). The Be–N bond lengths are 178.4 pm (3×) and 176.0 pm. The magnesium atoms are located between the anionic double-layers, and each metal center is octahedrally surrounded by nitrogens with Mg–N 220.9 pm.

The ternary nitrides $Ca[Be_2N_2]$ and $Sr[Be_2N_2]$ are isostructural. The crystal structure contains planar layers which consist of four- and eight-membered rings in the ratio of 1:2. The layers are stacked with successive slight rotation of each about the common normal, forming octagonal prismatic voids in the interlayer region. The eight-coordinated Ca^{2+} ions are accommodated in the void space, as shown in Fig. 2.6.2(b). The Be–N bond lengths within the layers are 163.2 pm (2×) and 165.7 pm. The Ca–N distance is 268.9 pm.

2.6.3 **Group 2 low-valent oxides and nitrides**

(1) **$(Ba_2O)Na$**

Similar to the alkali metals, the alkaline-earth metals can also form low-valent oxides. The first crystalline compound of this type is $(Ba_2O)Na$, which belongs to space group *Cmma* with a = 659.1, b = 1532.7, c = 693.9 pm, and Z = 4. In the crystal structure, the O atom is located inside a Ba_4 tetrahedron. Such $[Ba_4O]$ units share *trans* edges to form an infinite chain running parallel to the a axis. The $[Ba_{4/2}O]_∞$ chains are in parallel alignment and separated by the Na atoms, as illustrated in Fig. 2.6.3. Within each chain, the Ba–O bond length is 252 pm, being shorter than the corresponding distance of 277 pm in crystalline barium oxide. The interatomic distances between Ba and Na atoms lie in the range 421–433 pm, which are comparable with the values in the binary alloys BaNa and $BaNa_2$ (427 and 432 pm, respectively).

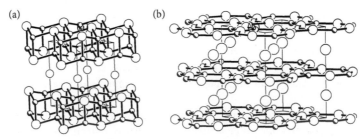

Fig. 2.6.2 Crystal structures of (a) $Mg[Be_2N_2]$ and (b) $Ca[Be_2N_2]$. Small circles represent Be, large circles N, while circles between layers are either Mg or Ca.

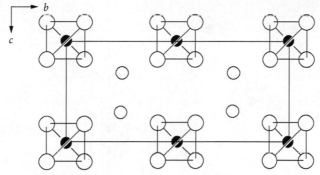

Fig. 2.6.3 Crystal structure of $(Ba_2O)Na$. The $[Ba_{4/2}O]_\infty$ chains are seen end-on.

(2) [Ba$_6$N] cluster and related compounds

When metallic barium is dissolved in liquid sodium or K/Na alloy in an inert atmosphere of nitrogen, an extensive class of mixed alkali metal-barium subnitrides can be prepared. They contain $[Ba_6N]$ octahedra that exist either as discrete entities or are condensed into finite clusters and infinite arrays.

Discrete $[Ba_6N]$ clusters are found in the crystal structure of $(Ba_6N)Na_{16}$, as shown in Fig. 2.6.4. $(Ba_6N)Na_{16}$ crystallizes in space group $Im3m$ with $a = 1252.7$ pm and $Z = 2$. The Na atoms are located between the cluster units, in an analogous manner to that in the low-valent alkali metal suboxides.

The series of barium subnitrides Ba_3N, $(Ba_3N)Na$, and $(Ba_3N)Na_5$ are characterized by parallel infinite $(Ba_{6/2}N)_\infty$ chains each composed of *trans* face-sharing octahedral $[Ba_6N]$ clusters. Fig. 2.6.5(a) shows the crystal structure of Ba_3N projected along a six-fold axis. In the crystal structure of $(Ba_3N)Na$, the sodium atoms are located in between the $(Ba_{6/2}N)_\infty$ chains, as shown in Fig. 2.6.5(b). In contrast, since $(Ba_3N)Na_5$ has a much higher sodium content, the $(Ba_{6/2}N)_\infty$ chains are widely separated from one another by the additional Na atoms, as illustrated in Fig. 2.6.5(c).

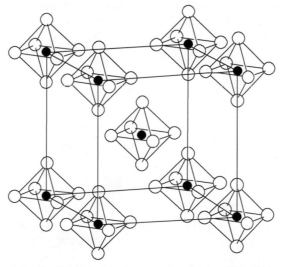

Fig. 2.6.4 Arrangement of the (Ba_6N) clusters in the crystal structure of $(Ba_6N)Na_{16}$.

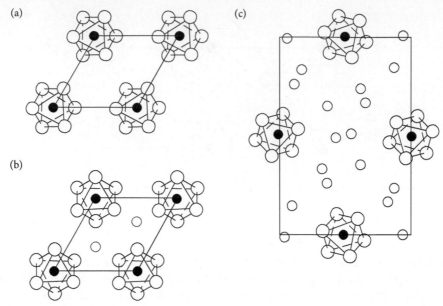

Fig. 2.6.5 Crystal structures of (a) Ba_3N, (b) $(Ba_3N)Na$, and (c) $(Ba_3N)Na_5$. The $(Ba_{6/2}N)_\infty$ chains are seen end-on.

(3) [Ba₁₄CaN₆] cluster and related compounds

The [$Ba_{14}CaN_6$] cluster can be considered as a Ca-centered Ba_8 cube with each face capped by a Ba atom, and the six N atoms are each located inside a Ba_5 tetragonal pyramid, as shown in Fig. 2.6.6. The structure can be viewed alternatively as a cluster composed of the fusion of six N-centered Ba_5Ca octahedra sharing a common Ca vertex. In the synthesis of this type of cluster compounds, variation of the atomic ratio of the Na/K binary alloy system leads to a series of compounds of variable stoichiometry: [$Ba_{14}CaN_6$]Na_x with $x = 7$, 8, 14, 17, 21, and 22. This class of compounds can be considered as composed of an ionically bonded [$Ba^{2+}{}_8Ca^{2+}N^{3-}{}_6$] nucleus surrounded by Ba_6Na_x units, which interact with the nucleus mainly via metallic bonding.

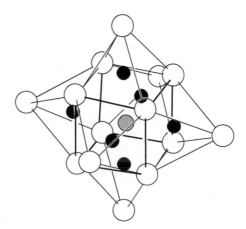

Fig. 2.6.6 Structure of the [$Ba_{14}CaN_6$] cluster.

2.7 **Organometallic Compounds of Group 2 Elements**

There are numerous organometallic compounds of Group 2 elements, and in this section only the following three typical species are described.

2.7.1 **Polymeric chains**

Dimethylberyllium, $Be(CH_3)_2$, is a polymeric white solid containing infinite chains, as shown in Fig. 2.7.1(a). Each Be center is tetrahedrally coordinated and can be considered to be sp^3 hybridized. As the CH_3 group only contributes one orbital and one electron to the bonding, there are insufficient electrons to form normal 2c-2e bonds between the Be and C atoms. In this electron-deficient system, the BeCBe bridges are 3c-2e bonds involving sp^3 hybrid atomic orbitals from two Be atoms and one C atom, as shown in Fig. 2.7.1(b).

The polymeric structures of BeH_2 and $BeCl_2$ are similar to that of $Be(CH_3)_2$. The 3c-2e bonding in BeH_2 is analogous to that in $Be(CH_3)_2$. But in $BeCl_2$ there are sufficient valence electrons to form normal 2c-2e Be–Cl bonds.

2.7.2 **Grignard reagents**

Grignard reagents are widely used in organic chemistry. They are prepared by the interaction of magnesium with an organic halide in ethers. Grignard reagents are normally assigned the simple formula RMgX, but this is an oversimplification as solvation is important. The isolation and structural determination of several crystalline RMgX compounds have demonstrated that the essential structure is RMgX·(solvent)$_n$. Fig. 2.7.2 shows the molecular structure of $(C_2H_5)MgBr·2(C_2H_5)_2O$, in which the central Mg atom is

(a)   (b)

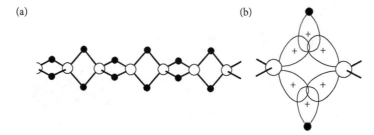

Fig. 2.7.1 (a) Linear structure of $Be(CH_3)_2$; (b) its BeCBe 3c-2e bonding in the polymer.

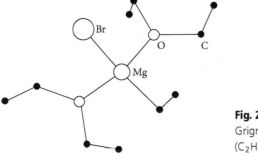

Fig. 2.7.2 Molecular structure of Grignard reagent $(C_2H_5)MgBr·2(C_2H_5)_2O$.

surrounded by one ethyl group, one bromine atom, and two diethyl ether molecules in a distorted tetrahedral configuration. The bond distances are Mg–C 215 pm, Mg–Br 248 pm, and Mg–O 204 pm. The latter is among the shortest Mg–O distances known.

In a Grignard compound the carbon atom bonded to the electropositive Mg atom carries a partial negative charge. Consequently a Grignard reagent is an extremely strong base, with its carbanion-like alkyl or aryl portion acting as a nucleophile.

2.7.3 **Alkaline-earth metallocenes**

The alkaline-earth metallocenes exhibit various structures, depending on the sizes of the metal atoms. In Cp_2Be, the Be^{2+} ion is η^5/η^1 coordinated between two Cp rings, as shown in Fig. 2.7.3(a). The mean Be–C distances are 193 pm and 183 pm for η^5 and η^1 Cp rings, respectively. In contrast, Cp_2Mg adopts a typical sandwich structure, as shown in Fig. 2.7.3(b). The mean Mg–C distance is 230 pm. In the crystalline state, the two parallel rings have a staggered conformation.

The compounds Cp_2Ca, Cp^*Sr, and Cp^*Ba exhibit a different coordination mode, as the two Cp rings are not aligned in a parallel fashion like Cp_2Mg, but are bent with respect to each other making a $Cp_{(centroid)}$–M–$Cp_{(centroid)}$ angle of 147°–154°. The positive charge of the M^{2+} ion is not only shared by two η^5-bound Cp rings, but also by other ligands. In Cp_2Ca, apart from the two η^5-Cp ligands, there are significant Ca–C contacts to η^3- and η^1-Cp rings, as shown in Fig. 2.7.3(c). In $Sr[C_5H_3(SiMe_3)_2]_2\cdot THF$, apart from the two η^5-Cp rings, a THF ligand is coordinated to the Sr^{2+} ion, as shown in Fig. 2.7.3(d). The Ba^{2+} ion has the largest size among the alkaline-earth metals, and accordingly it can be coordinated by a planar COT (C_8H_8) ring bearing two negative charges. Fig. 2.7.3(e) shows the structure of a triple-decker sandwich complex of barium, $Ba_2(COT)[C_5H(CHMe_2)_4]_2$, in which the mean Ba–C distances are 296 pm and 300 pm for the η^5-Cp ring and η^8-COT ring, respectively.

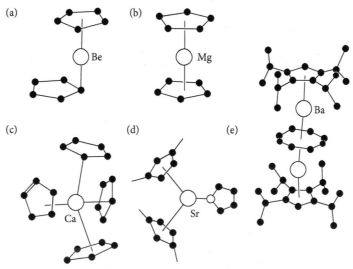

Fig. 2.7.3 Structures of some alkaline-earth metallocenes: (a) Cp_2Be, (b) Cp_2Mg, (c) Cp_2Ca, (d) $Sr[C_5H_3(SiMe_3)_2]_2\cdot THF$, (e) $Ba_2(COT)[C_5H(CHMe_2)_4]_2$.

2.8 Alkali and Alkaline-Earth Metal Complexes with Inverse Crown Structures

The term "inverse crown ether" refers to a metal complex in which the roles of the central metal core and the surrounding ether oxygen ligand sites in a conventional crown ether complex are reversed. An example of an inverse crown ether is the organosodium-zinc complex $Na_2Zn_2\{N(SiMe_3)_2\}_4(O)$ shown in Fig. 2.8.1(a). The general term "inverse crown" is used when the central core of this type of complex comprises non-oxygen atoms, or more than just a single O atom. Two related inverse crowns with cationic $(Na–N–Mg–N–)_2$ rings that encapsulate peroxide and alkoxide groups, respectively, are shown in Fig. 2.7.1(b) and (c). The octagonal macrocycle in $Na_2Mg_2\{N(SiMe_3)_2\}_4(OOct^n)_2$ takes the chair form with the $Na–OOct^n$ groups displaced on either side of the plane defined by the N and Mg atoms. Each alkoxide O atom interacts with both Mg atoms to produce a perfectly planar $(MgO)_2$ ring. The Mg–O and Na–O_{syn} bond lengths are 202.7(2)–203.0(2) pm and 256.6(1)–247.2(2) pm, respectively; the non-bonded Na$\cdots$$O_{anti}$ distances lie in the range 310.4(1)–320.4(2) pm.

Fig. 2.8.1(d) shows the structure of a hexameric "super" inverse crown ether $\{(THF)NaMg(Pr^i_2N)(O)\}_6$, which consists of a S_6-symmetric hexagonal prismatic Mg_6O_6 cluster with six external four membered rings constructed with Na–THF and Pr^i_2N appendages.

A remarkable series of inverse crown compounds featuring organic guest moieties encapsulated within host-like macrocyclic rings composed of sodium/potassium and magnesium ions together with anionic amide groups, such as TMP^- (TMPH = 2,2,6,6-tetramethylpiperidine) and $Pr^i_2N^-$ (Pr^i_2NH = diisopropylamine), have been synthesized and characterized. Table 2.8.1 lists some examples of this class of inverse crown molecules.

Fig. 2.8.2(a) shows the molecular structure of $Na_4Mg_2(TMP)_6(C_6H_4)$, in which the N atom of each tetramethylpiperidinide is bonded to two metal atoms to form a cationic 12-membered

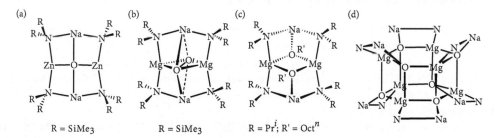

Fig. 2.8.1 Examples of some inverse crown compounds: (a) $Na_2Zn_2\{N(SiMe_3)_2\}_4(O)$, (b) $Na_2Mg_2\{N(SiMe_3)_2\}_4(O_2)$, (c) $Na_2Mg_2\{N(SiMe_3)_2\}_4(OOct^n)_2$, (d) cluster core of $\{(THF)NaMg(Pr^i_2N)(O)\}_6$; the THF ligand attached to each Na atom and the isopropyl groups are omitted for clarity.

Table 2.8.1 Compositions of some inverse crowns containing organic guest species

Group 1 metal	Group 2 metal	Amide ion	Core moiety	Host ring size
4Na	2Mg	6TMP	$C_6H_3Me^{2-}$	12
4Na	2Mg	6TMP	$C_6H_4^{2-}$	12 (Fig. 2.7.5(a))
6K	6Mg	12TMP	$6C_6H_5^-$	24
6K	6Mg	12TMP	$6C_6H_4Me^-$	24
4Na	4Mg	$8Pr^i_2N$	$Fe(C_5H_3)_2^{4-}$	12 (Fig. 2.7.5(b))

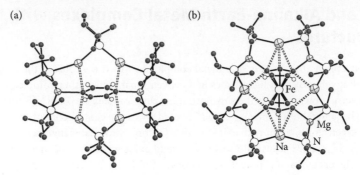

Fig. 2.8.2 Molecular structures of two centrosymmetric inverse crown molecules containing organic cores: (a) $Na_4Mg_2(TMP)_6(C_6H_4)$, (b) $Na_4Mg_4({}^iPr_2N)_8[Fe(C_5H_3)_2]$. The Na⋯C π interactions are indicated by broken lines.

$(Na–N–Na–N–Mg–N–)_2$ ring. The mixed metal macrocyclic amide acts as a host that completely encloses a 1,4-deprotonated benzenediide guest species whose naked carbon atoms are each stabilized by a covalent Mg–C bond and a pair of Na⋯C π interactions.

Fig. 2.8.2(b) shows the molecular structure of $Na_4Mg_4(Pr^i{}_2N)_8[Fe(C_5H_3)_2]$. Each amido N atom is bound to one Na and one Mg atom to form a 16-membered macrocyclic ring, which accommodates the ferrocene-1,1′,3,3′-tetrayl residue at its center. The deprotonated 1,3 positions of each cyclopentadienyl ring of the ferrocene moiety are each bound to a pair of Mg atoms by Mg–C covalent bonds, and the 1,2,3 positions further interact with three Na atoms via three different kinds of Na⋯C π bonds.

2.9 **Alkaline-Earth Metal Hydrides**

The lighter alkaline-earth metal hydrides BeH_2 is practically inert to hydrolysis, and MgH_2 reacts slowly with water. In contrast, CaH_2 serves as a universal drying agent for various solvents.

The reaction of MgN''_2 [N″ = N(SiMe₃)₂] with $PhSiH_3$ and the strongly coordinating NHC ligand gave $Mg_4H_6N''_2 \cdot (NHC)_2$ **1**. Using a bulky multidentate ligand such as the bidentate ß-diketiminate for kinetic stabilization, Ca hydride complex **2** has been synthesized. The hydride-bridged molecular structures of **1** and **2** are displayed in Fig. 2.9.1.

The reaction of BaN''_2 [N″ = N(SiMe₃)₂] with $PhSiH_3$ in toluene at –90 ºC, followed by slow warming to room temperature, gives $BaHN''$ as a white powder, which can be crystallized from benzene

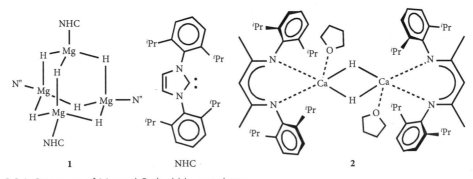

Fig. 2.9.1 Structures of Mg and Ca hydride complexes.

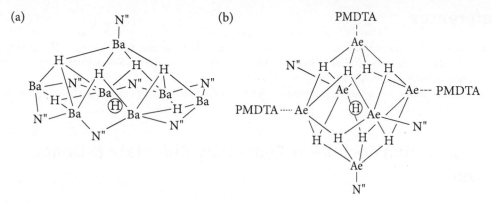

Fig. 2.9.2 Molecular structure of (a) hydride-bridged heptanuclear barium cluster $Ba_7H_7N''_7$ in its benzene solvate, and (b) $Ae_6H_9N''_3 \cdot (PMDTA)_3$ with Ae = Ca, Sr, N'' = $N(SiMe_3)_2$ and PMDTA = $MeN\{CH_2CH_2NMe_2)_2\}$.

as $Ba_7H_7N''_7 \cdot (C_6H_6)_2$ solvate. The structure of this hydride-bridged heptanuclear cluster is shown in Fig. 2.9.2(a).

References

M. Wiesinger, B. Maitland, C. Färber, G. Ballmann, C. Fischer, H. Elsen, and S. Harder, Simple access to the heaviest alkaline earth metal hydride: a strongly reducing hydrocarbon-soluble barium hydride cluster, *Angew. Chem. Int. Ed.* **56**, 16654–9 (2017).

D. Mukherjee, T. Höllerhage, V Leich, T P. Spaniol, U. Englert, L. Maron, and J. Okuda, The nature of the heavy alkaline earth metal–hydrogen bond: synthesis, structure, and reactivity of a cationic strontium hydride cluster, *J. Am. Chem. Soc.* **140**, 3403–11 (2018).

2.10 **Molecular Zero-Valent Beryllium Compounds**

The first homo- and heteroleptic zero-valent beryllium complexes bearing N-heterocyclic carbene ligands were synthesized and structurally characterized by X-ray crystallography in 2016. In the homoleptic complex $Be(^{Me}L)_2$ with ^{Me}L = 1-(2,6-diisopropylphenyl)-3,3,5,5-tetramethylpyrrolidine-2-ylidene shown in Fig. 2.10.1(a), the beryllium atom exhibits linear C–Be–C configuration with Be–C bond lengths of 166.4(2) and 165.9(5) pm, which are significantly shorter than those in similar Be^{2+} compounds (*ca.* 178 pm). This partial double-bond character arises from carbene-to-metal σ-donation abetted by metal-to-carbene π-back-bonding (Fig. 2.10.1(b)). The electronic configuration of the Be atom is therefore $1s^2 2s^0 2p^2$ with a delocalized three-center two-electron π-bond extended over the linear C–Be–C skeleton.

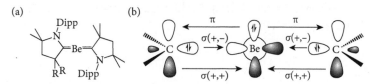

Fig. 2.10.1 (a) Molecular structure of $[Be(^{Me}L)_2]$; (b) Schematic representation of the orbitals involved in C-to-Be σ donation and Be-to-C π back donation.

References

M. Arrowsmith, H. Braunschweig, M. A. Celik, T. Dellermann, R. D. Dewhurst, W. C. Ewing, K. Hammond, T. Kramer, I. Krummenacher, J. Mies, K. Radacki, and J. K. Schuster, Neutral zero-valent s-block complexes with strong multiple bonding, *Nat. Chem.* **8**, 890–4 (2016).

K. M. Fromm, Chemistry of alkaline earth metals: it is not all ionic and definitely not boring!, *Coord. Chem. Rev.* **408** (2020) 213193.

2.11 Beryllium Complexes Containing Bidentate N-Donor Ligand

Complexes of this type are rare and sterically bulky ligands are required for their stability. Transmetalation reaction of beryllium chloride $BeCl_2$ with lithium amidinate $Li\{(NSiMe_3)_2CPh\}$ yielded monomeric beryllium bis(amidinate) $Be\{(NSiMe_3)_2CPh\}_2$, in which the Be center is coordinated by two bidentate amidinate ligands in a distorted tetrahedral fashion to form four-membered chelate rings (Fig. 2.11.1).

The crystal structure of $[Be\{(NSiMe_3)_2PPh_2\}_2]$ consists of two bidentate *N*, *N'*-bis(trimethylsilyl)aminoiminodiphenylphosphorane ligands (Fig. 2.11.2). In the core of the complex, the central beryllium(II) atom is coordinated by two chelating $Ph_2P(Me_3SiN)_2{}^-$ monoanions, resulting in an almost undistorted tetrahedral BeN_4 arrangement.

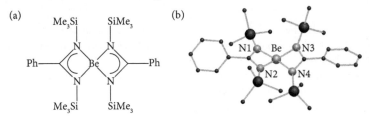

Fig. 2.11.1 (a) Chemical formula of beryllium bis(amidinate). (b) Its X-ray molecular structure with Be–N distances in the range 172.3(6)–173.0(6) pm. The chelating N–Be–N angles formed by the pair of bidentate ligands are N1–Be–N2 = 79.7(3)° and N3–Be–N4 = 79.6(3)°.

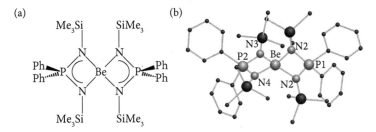

Fig. 2.11.2 (a) Chemical formula of $[Be\{(NSiMe_3)_2PPh_2\}_2]$. (b) Molecular structure in the crystalline state. Bond lengths and bond angles: Be–N1 177.1(4), Be–N2 175.3(4), Be–N3 177.7(3), Be–N4 176.0(3) pm; N1–P1–N2 99.1(1), N3–P2–N4 99.1(1)°.

References

M. Niemeyer and P. P. Power, Synthesis, ^{9}Be NMR spectroscopy, and structural characterization of sterically encumbered beryllium compounds, *Inorg. Chem.* **36**, 4688–96 (1997).

R. Fleischer and D. Stalke, Syntheses and structures of [(THF)$_n$M{(NSiMe$_3$)$_2$PPh$_2$}$_2$] complexes (M) Be, Mg, Ca, Sr, Ba; n = 0-2): deviation of alkaline earth metal cations from the plane of an anionic ligand, *Inorg. Chem.* **36**, 2413–9 (1997).

L.C. Perera, O. Raymond, W. Henderson, P. J. Brothers, and P. G. Plieger, Advances in beryllium coordination chemistry, *Coord. Chem. Rev.* **35**, 264–90 (2017).

2.12 **Stable Dimeric Magnesium(I) Compounds**

The first reported stable dimeric magnesium(I) compound is [{(Priso)Mg}$_2$] **1**, with Priso = [(DipN)$_2$CN^{i}Pr$_2$] and Dip = 2,6-diisopropylphenyl, which was synthesized by the following scheme.

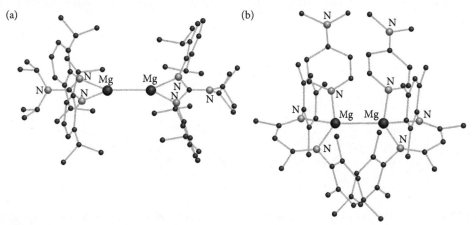

The X-ray molecular structure of **1** is shown is Fig. 2.12.1(a), which has a measured Mg–Mg bond length of 285.08(12) pm. Its structural analog [{(^{Mes}Nacnac)Mg(DMAP)}$_2$] **2** with ^{Mes}Nacnac =

Fig. 2.12.1 Molecular structure of (a) [{(^{Mes}Nacnac)Mg(DMAP)}$_2$] **1** and (b) [{(^{Mes}Nacnac)Mg(DMAP)}$_2$] [{(Priso)Mg}$_2$] **2**.

[(MesNCMe)$_2$CH], Mes = 2,4,6-trimethylphenyl), and DMAP = 4-(dimethylamino)pyridine is shown in Fig. 2.12.1(b), and the corresponding Mg–Mg bond length is 293.7(1) pm.

Reference

A. Stasch and C. Jones, Stable dimeric magnesium(I) compounds: from chemical landmarks to versatile reagents, *Dalton Trans.* **40**, 5659–72 (2011).

2.13 **Magnesium(II) Carbene Complex**

The monodeprotonation of [CH$_2$(PPh$_2$→BH$_3$)(PPh$_2$=E)] (E = S, O) with methyllithium afforded [CH(PPh$_2$→BH$_3$)(PPh$_2$=E)]$^-$ (E = S, O); the structures of their lithium salts were confirmed by X-ray crystallography.

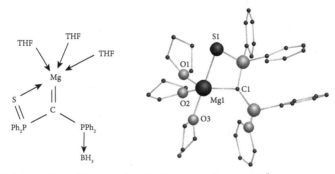

A rare Mg^{II} carbene complex [(THF)$_3$Mg{C(Ph$_2$P=S)(Ph$_2$P→BH$_3$)}] was obtained by the double deprotonation reaction of [CH$_2$(PPh$_2$→BH$_3$)(PPh$_2$=S)] with Mg(^{n}Bu)$_2$. It was isolated as a highly air- and moisture-sensitive yellow crystalline solid that is soluble in hydrocarbon solvents. X-ray crystal structure analysis showed that it is a monomeric molecule, which has a trigonal-bipyramidal coordination sphere composed of carbene C1, S1, and THF (O1, O2, O3) ligand atoms around the magnesium(II) center (Fig. 2.13.1).

Fig. 2.13.1 Chemical formula and X-ray molecular structure of a rare Mg^{II} carbene complex. Bond lengths in ppm: Mg(1)–C(1) = 211.3(4), Mg(1)–S(1) = 269.07(16), Mg(1)–O(1) = 205.3(3), Mg(1)–O(2) = 203.8(3), Mg(1)–O(3) = 210.3(3), C(1)–P(1) = 170.3(4), C(1)–P(2) = 166.6(4).

References

H. Heuclin, M. Fustier-Boutignon, S. Y.-F. Ho, X.-F. Le Goff, S. Carenco, C.-W. So, and N. Mézailles, Synthesis of phosphorus(V)-stabilized geminal dianions. the cases of mixed P=X/P→BH₃ (X = S, O) and P=S/SiMe₃ derivatives, *Organometallics* **32**, 498–508 (2013).

K. M. Fromm, Chemistry of alkaline earth metals: it is not all ionic and definitely not boring!, *Coord. Chem. Rev.* **408** (2020) 213193.

2.14 Low-Valent Monovalent Calcium Complexes Stabilized by Bulky β-diketiminate Ligand

Attempted preparation of monovalent calcium(I) complexes of the type (BDI)Ca–Ca(BDI) stabilized by bulky ligand BDI under dinitrogen atmosphere led to formation of (BDI)Ca(N₂)Ca(BDI), with BDI = β-diketiminate (HC{C(Me)N[2,6-(3-pentyl)-phenyl]}₂). Although this species could not be crystallized, addition of tetrahydrofuran (THF) or tetrahydropyran (THP) led to formation of red-brown solvate crystals which were structurally characterized by low-temperature X-ray crystallographic analysis. The structural formulas of the terminal BDI ligand and centrosymmetric {(THP)(BDI)}Ca(N₂)Ca{(BDI)(THP)} dimer bearing a central side-on bridging N₂ unit (Fig. 2.14.1).

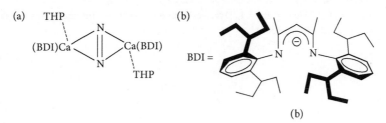

Fig. 2.14.1 (a) Chemical formula of dimeric {(THP)(BDI)}Ca(N₂)Ca{(BDI)(THP)}. (b) Structure of bulky β-diketiminate ligand BDI.

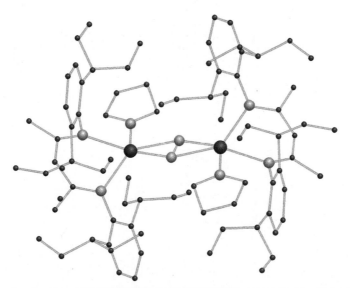

Fig. 2.14.2 Molecular structure of {(THP)(BDI)}Ca(N₂)Ca{(BDI)(THP)} determined by X-ray crystallography.

The N–N bond lengths in the iso-structural and symmetrical dimers [126.8(2) pm in THF complex, CCDC 2036242; 125.8(3) and 126.8(3) pm in THP complex, CCDC 2036243] are considerably elongated compared with the N–N bond in N_2 (109.8 pm) and consistent with N=N double-bond character in the N_2^{2-} anion, which is isoelectronic to O_2. This interpretation is supported by the observation that the highly charged N_2^{2-} anion is tightly sandwiched between Ca^{2+} cations with a very short mean Ca–N distance of 230.3 pm, in contrast with the mean Ca–N(BDI) bond distance of 239.4 pm (Fig. 2.14.2).

Reference

B. Rösch, T. X. Gentner, J. Langer, C. Färber, J. Eyselein, L. Zhao, C. Ding, G. Frenking, and S. Harder, Dinitrogen complexation and reduction at low-valent calcium, *Science* **371**, 1125–8 (2021).

General References

J. A. McCleverty and T. J. Meyer (Editors-in-chief), *Comprehensive Coordination Chemistry: From Biology to Nanotechnology*, Vol. 3, G. F. R. Parkin (volume ed.), *Coordination Chemistry of the s, p, and f Metals*, Elsevier-Pergamon, Amsterdam, 2004.

Q. Xie, R. H. Huang, A. S. Ichimura, R. C. Phillips, W. P. Pratt Jr, and J. L. Dye, Structure and properties of a new electride, Rb$^+$(cryptand[2.2.2])e$^-$. *J. Am. Chem. Soc.* **122**, 6971–8 (2000).

M. Y. Redko, J. E. Jackson, R. H. Huang, and J. L. Dye, Design and synthesis of a thermally stable organic electride. *J. Am. Chem. Soc.* **127**, 12416–22 (2005).

Chapter 3

Structural Chemistry of Group 13 Elements

3.1 Survey of the Group 13 Elements

Boron, aluminum, gallium, indium, and thallium are members of Group 13 of the Periodic Table. Some important properties of these elements are given in Table 3.1.1.

From Table 3.1.1, it is seen that the inner electronic configurations of the Group 13 elements are not identical. The ns^2np^1 electrons of B and Al lie outside a rare gas configuration, while the ns^2np^1 electrons of Ga and In are outside the d^{10} subshell, and those of Tl lie outside the $4f^{14}5d^{10}$ core. The increasing effective nuclear charge and size contraction, which occur during successive filling of d and f orbitals, combine to make the outer s and p electrons of Ga, In, and Tl more strongly held than expected by simple extrapolation from B to Al. Thus there is a small increase in ionization energy between Al and Ga, and between In and Tl. The drop in ionization energy between Ga and In mainly reflects the fact that both elements have a completely filled inner shell and the outer electrons of In are further from the nucleus.

The +3 oxidation state is characteristic of Group 13 elements. However, as in the later elements of this groups, the trend in I_2 and I_3 shows increases at Ga and Tl, leading to a marked increase in stability of their +1 oxidation state. In the case of Tl, this is termed the $6s$ inert-pair effect. Similar effects are observed for Pb (Group 14) and Bi (Group 15), for which the most stable oxidation states are +2 and +3, respectively, rather than +4 and +5.

Due to the $6s$ inert-pair effect, many Tl^+ compounds are found to be more stable than the corresponding Tl^{3+} compounds. The progressive weakening of M^{3+}–X bonds from B to Tl partly accounts for this. Furthermore, the relativistic effect (as described in Section 2.4.3) is another factor which contributes to the inert-pair effect.

3.2 Elemental Boron

The uniqueness of structure and properties of boron is a consequence of its electronic configuration. The small number of valence electrons (three) available for covalent bond formation leads to "electron deficiency," which has a dominant effect on boron chemistry.

Boron is notable for the complexity of its elemental crystalline forms. There are many reported allotropes, but most are actually boron-rich borides. Only two have been completely elucidated by X-ray diffraction, namely α-R12 and β-R105; here R indicates the rhombohedral system, and the numeral gives the number of atoms in the primitive unit cell. In addition, α-T50 boron (T denotes the tetragonal system) has been reformulated as a carbide or nitride, $B_{50}C_2$ or $B_{50}N_2$, and β-T192 boron is still not completely elucidated. These structures all contain icosahedral B_{12} units, which in most cases are accompanied by other boron atoms lying outside the icosahedral cages, and the linkage generates a three-dimensional framework.

The α-R12 allotrope consists of an approximately cubic-closest packing (ccp) arrangement of icosahedral B_{12} units bound to each other by covalent bonds. The parameters of the rhombohedral unit cell are $a = 505.7$ pm, $\alpha = 58.06°$ (60° for regular ccp). In space group $R\bar{3}m$, the unit cell contains one B_{12} icosahedron. A layer of interlinked icosahedra perpendicular to the three-fold symmetry axis is illustrated in

Structural Chemistry across the Periodic Table. Thomas Chung Wai Mak, Yu-San Cheung, Gong-Du Zhou, and Ying-Xia Wang, Oxford University Press. © Thomas Chung Wai Mak, Yu-San Cheung, Gong-Du Zhou, and Ying-Xia Wang (2023). DOI: 10.1093/oso/9780198872955.003.0003

Table 3.1.1 Some properties of Group 13 elements

Property	B	Al	Ga	In	Tl
Atomic number, Z	5	13	31	49	81
Electronic configuration	[He]$2s^2 2p^1$	[Ne]$3s^2 3p^1$	[Ar]$3d^{10} 4s^2 4p^1$	[Kr]$4d^{10} 5s^2 5p^1$	[Xe]$4f^{14} 5d^{10} 6s^2 6p^1$
ΔH^{θ}_{at}/kJ mol^{-1}	582	330	277	243	182
mp/K	2453*	933	303	430	577
ΔH^{θ}_{fuse}/kJ mol^{-1}	50.2	10.7	5.6	3.3	4.1
bp/K	4273	2792	2477	2355	1730
I_1/kJ mol^{-1}	800.6	577.5	578.8	558.3	589.4
I_2/kJ mol^{-1}	2427	1817	1979	1821	1971
I_3/kJ mol^{-1}	3660	2745	2963	2704	2878
$I_1 + I_2 + I_3$/kJ mol^{-1}	6888	5140	5521	5083	5438
χ_s	2.05	1.61	1.76	1.66	1.79
r_M/pm	–	143	153	167	171
r_{cov}/pm	88	130	122	150	155
$r_{ion, M^{3+}}$/pm	–	54	62	80	89
r_{ion, M^+}/pm	–	–	113	132	140
$D_{0, M-F (in MF_3)}$/kJ mol^{-1}	613	583	469	444	439 (in TlF)
$D_{0, M-Cl (in MCl_3)}$/kJ mol^{-1}	456	421	354	328	364 (in TlCl)

* For β-rhombohedral boron

Fig. 3.2.1. There are 12 neighboring icosahedra for each icosahedron: six in the same layer, and three others in each of the upper and lower layers, as shown in Fig. 9.6.21(b).

The B_{12} icosahedron is a regular polyhedron with 12 vertices, 30 edges, and 20 equilateral triangular faces, with B atoms located at the vertices, as shown in Fig. 3.2.2. The B_{12} icosahedron is a basic structural unit in all isomorphic forms of boron and in some polyhedral boranes such as $B_{12}H_{12}^{2-}$. It has 36 valence electrons; note that each line joining two B atoms (mean B–B distance 177 pm) in Fig. 3.2.2 does not represent a normal two-center two-electron (2c-2e) covalent bond.

When the total number of valence electrons in a molecular skeleton is less than the number of valence orbitals, the formation of normal 2c-2e covalent bonds is not feasible. In this type of electron-deficient compounds, three-center two-electron (3c-2e) bonds generally occur, in which three atoms share an electron pair. Thus one 3c-2e bond serves to compensate for the shortfall of four electrons and corresponds to a bond valence value of 2, as illustrated for a BBB bond in Fig. 3.2.3.

In a *closo*-B_n skeleton, there are only three 2c-2e covalent bonds and $(n-2)$ BBB 3c-2e bonds. (This will be derived in Section 3.4). For an icosahedral B_{12} unit, there are three B–B 2c-2e bonds and ten BBB 3c-2e bonds, as shown in Fig. 3.2.4.

The 36 electrons in the B_{12} unit may be partitioned as follow: 26 electrons are used to form bonds within an icosahedron, and the remaining 10 electrons to form the inter-icosahedral bonds. In the structure of α-R12 boron, each icosahedron is surrounded by six icosahedra in the same layer and forms six 3c-2e bonds [see Fig. 9.6.21(b)], in which every B atom contributes ⅔ electron, so each icosahedron uses $6 \times ⅔ = 4$

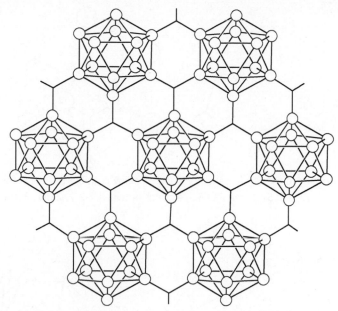

Fig. 3.2.1 A close-packed layer of interlinked icosahedra in the crystal structure of α-rhombohedral boron. The lines meeting at a node (not a B atom) represents a BBB 3c-2e bond linking three B_{12} icosahedra.

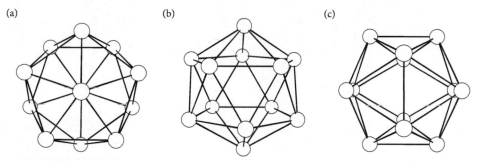

(a)   (b)   (c)

Fig. 3.2.2 From left to right, perspective view of the B_{12} icosahedron along one of its (a) six 5-fold, (b) 10 3-fold, and (c) 15 2-fold axes.

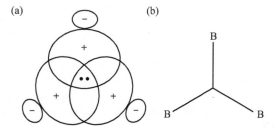

(a)   (b)

Fig. 3.2.3 The BBB 3c-2e bond: (a) three atoms share an electron pair, and (b) simplified representation of 3c-2e bond.

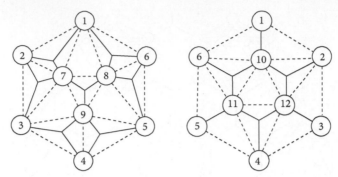

Fig. 3.2.4 Chemical bonds in the B_{12} icosahedron (one of many possible canonical forms): three 2c-2e bonds between 1-10, 3-12, and 5-11; ten BBB 3c-2e bonds between 1-2-7, 2-3-7, 3-4-9, ...

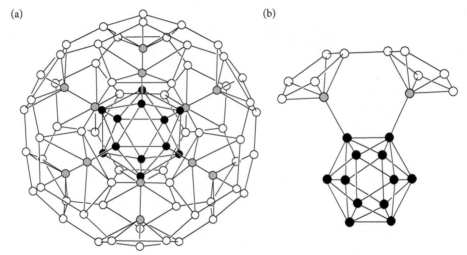

Fig. 3.2.5 Structure of β-R105 boron: (a) perspective view of the B_{84} unit [B_{12a}@B_{12}@B_{60}]. Black circles represent the central B_{12a} icosahedron, shaded circles represent the B atoms connecting the surface 60 B atoms to the central B_{12a} unit; (b) each vertex of inner B_{12a} is connected to a pentagonal pyramidal B_6 unit (only two adjoining B_6 units are shown).

electrons. Each icosahedron is bonded by normal 2c-2e B–B bonds to six icosahedra in the upper and lower layers, which need six electrons. The number of bonding electrons in an icosahedron of α-R12 boron is therefore 26 + 4 + 6 = 36.

The β-R105 boron allotrope has a much more complex structure with 105 B atoms in the unit cell (space group $R\bar{3}m$, a = 1014.5 pm, α = 65.28°). A basic building unit in the crystal structure is the B_{84} cluster illustrated in Fig. 3.2.5(a); it can be considered as a central B_{12a} icosahedron linked radially to 12 B_6 half-icosahedra (or pentagonal pyramids), each attached like an inverted umbrella to an icosahedral vertex, as shown in Fig. 3.2.5(b).

The basal B atoms of adjacent pentagonal pyramids are interconnected to form a B_{60} icosahedron that resembles fullerene-C_{60}. The large B_{60} icosahedron encloses the B_{12a} icosahedron, and the resulting B_{84} cluster can be formulated as B_{12a}@B_{12}@B_{60}. Additionally, a six-coordinate B atom lies at the center of

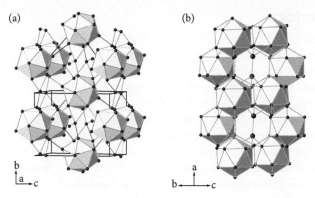

Fig. 3.2.6 Structure of γ-B$_{28}$: (a) Packing and Linkage of B$_{12}$ and B$_2$; (b) B$_2$ at the octahedral voids formed by B$_{12}$.

symmetry between two adjacent B$_{10}$ condensed units, and the crystal structure is an intricate coordination network resulting from the linkage of B$_{10}$–B–B$_{10}$ units and B$_{84}$ clusters, such that all 105 atoms in the unit cell except one are the vertices of fused icosahedra. Further details and the electronic structure of β-R105 boron are described in Section 3.4.6.

γ-Boron (also known as γ-B$_{28}$) is an orthorhombic phase formed under high temperature and pressure (2450 °C, 18–89 GPa), which can exist at room temperature and pressure after "quenching." The researchers obtained the basic information of the γ-B unit cell through X-ray diffraction and used simulation calculations to reproduce the pattern consistent with the diffraction experiment and hence determined the corresponding crystal structure. It crystallizes in the orthorhombic space group *Pnnm* with a = 505.44, b = 561.99, and c = 698.73 pm, as shown in Fig. 3.2.6(a). In the γ-B$_{28}$ structure established from simulation calculations that reproduced its powder diffraction pattern, the arrangement of B$_{12}$ icosahedra is similar to the cubic close packing of α-boron, but the difference is that all octahedral cavities produced by B$_{12}$ packing in the γ-B$_{28}$ structure are occupied by B$_2$ atomic pairs as shown in Fig. 3.2.6(b). γ-B$_{28}$ has 28 (= 12 × 2 + 2 × 2) B atoms in each unit cell, so its density is higher than that of α-B$_{12}$.

In the structure of γ-B$_{28}$, the average B–B bond length in the B$_{12}$ icosahedron is 180 pm, and it is 173 pm in the B$_2$ atomic pair. It is worthy to mention that there exists a partial charge transfer from B$_2$ to B$_{12}$, forming $(B_2)^{\delta+}(B_{12})^{\delta-}$ ($\delta \approx 0.34{\sim}0.48$), and then the interaction between the B$_{12}$ icosahedral cluster and the B$_2$ atomic pair shows part of ionic property.

Reference

A. R. Oganov, J. Chen, C. Gatti, Y. Ma, Y. Ma, C. W. Glass, Z. Liu, T. Yu, O. O. Kurakevych, and V. L. Solozhenko, Ionic high-pressure form of elemental boron, *Nature* **457**, 863–7 (2009).

3.3 **Borides**

Solid borides have high melting points, exceptionally high hardness, excellent wear resistance, and good immunity to chemical attack which make them industrially important with uses as refractory materials and in rocket cones and turbine blades. Some metal borides have been found to exhibit superconductivity.

3.3.1 **Metal borides**

A large number of metal borides have been prepared and characterized. Several hundred binary metal borides M_xB_y are known. With increasing boron content, the number of B–B bonds increases. In this manner, isolated B atoms, B–B pairs, fragments of boron chains, single chains, double chains, branched chains, and hexagonal networks are formed, as illustrated in Fig. 3.3.1. Table 3.3.1 summarizes the stoichiometric formulas and structures of metal borides.

In boron-rich compounds, the structure can often be described in terms of a three-dimensional network of boron clusters (octahedra, icosahedra, and cubo-octahedra), which are linked to one another directly or via non-cluster atoms. The metal atoms are accommodated in cages between the octahedra or icosahedra, thereby providing external bonding electrons to the electron-deficient boron network. The structures of some metal borides in Table 3.3.1 are discussed below:

(1) **MgB$_2$ and AlB$_2$**

Magnesium boride MgB$_2$ was discovered in 2001 to behave as a superconductor at T_c = 39 K, and its physical properties are similar to those of Nb$_3$Sn used in the construction of high-field superconducting magnets in NMR spectrometers. MgB$_2$ crystallizes in the hexagonal space group $P6/mmm$ with a = 308.6, c = 352.4 pm, and Z = 1. The Mg atoms are arranged in a close-packed layer, and the B atoms form a graphite-like layer with a B–B bond length of $a/\sqrt{3}$ = 178.2 pm These two kinds of layers are interleaved along the c axis, as shown in Fig. 3.3.1(f). In the resulting crystal structure, each B atom is located inside a Mg$_6$ trigonal prism, and each Mg atom is sandwiched between two planar hexagons of B atoms that constitute a B$_{12}$ hexagonal prism, as illustrated in Fig. 3.3.2. AlB$_2$ has the same crystal structure with a B–B bond length of 175 pm.

(2) **LaB$_4$**

Fig. 3.3.3(a) shows a projection of the tetragonal structure of LaB$_4$ along the c-axis. The chains of B$_6$ octahedra are directly linked along the c-axis and joined laterally by pairs of B$_2$ atoms in the ab plane to form a three-dimensional skeleton. In addition, tunnels accommodating the La atoms run along the c-axis.

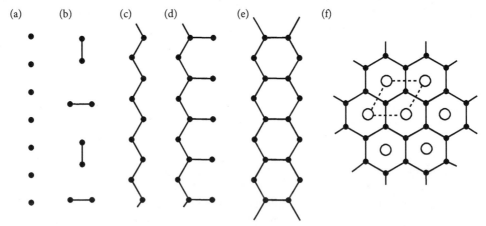

Fig. 3.3.1 Idealized patterns of boron catenation in metal-rich borides: (a) isolated boron atoms, (b) B–B pairs, (c) single chain, (d) branched chain, (e) double chain, (f) hexagonal net in MB$_2$.

Table 3.3.1 Stoichiometric formulas and structure of metal borides

Formula	Example	Catenation of boron and figure showing structure
M_4B	Mn_4B, Cr_4B	
M_3B	Ni_3B, Co_3B	
M_5B_2	Pd_5B_2	isolated B atom, Fig. 3.3.1(a)
M_7B_3	Ru_7B_3, Re_7B_3	
M_2B	Be_2B, Ta_2B	
M_3B_2	V_3B_2, Nb_3B_2	B_2 pairs, Fig. 3.3.1(b)
MB	FeB, CoB	single chains, Fig. 3.3.1(c)
$M_{11}B_8$	$Ru_{11}B_8$	branched chains, Fig. 3.3.1(d)
M_3B_4	Ta_3B_4, Cr_3B_4	double chains, Fig. 3.3.1(e)
MB_2	MgB_2, AlB_2	hexagonal net, Fig. 3.3.1(f) and Fig. 3.3.2
MB_4	LaB_4, ThB_4	B_6 octahedra and B_2 Fig. 3.3.3(a)
MB_6	CaB_6	B_6 octahedra, Fig. 3.3.3(b)
MB_{12}	YB_{12}, ZrB_{12}	B_{12} cubooctahedra, Fig. 3.3.4(a)
MB_{15}	NaB_{15}	B_{12} icosahedra, Fig. 3.3.4(b) and B_3 unit
MB_{66}	YB_{66}	$B_{12}C(B_{12})_{12}$ giant cluster

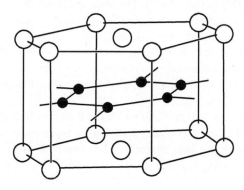

Fig. 3.3.2 Crystal structure of MgB_2.

(3) CaB_6

The cubic hexaboride CaB_6 consists of B_6 octahedra, which are linked directly in all six orthogonal directions to give a rigid but open framework, and the Ca atoms occupy cages each surrounded by 24 B atoms, as shown in Fig. 3.3.3(b).

The number of valence electrons for a B_6 unit can be counted as follows: six electrons are used to form six B–B bonds between the B_6 units, and the *closo*-B_6 skeleton is held by three B–B 2c-2e bonds and four BBB 3c-2e bonds, which require 14 electrons. The total number is 6 + 14 = 20 electrons per B_6 unit. Thus in the CaB_6 structure each B_6 unit requires the transfer of two electrons from metal atoms. However, the complete transfer of 2e per B_6 unit is not mandatory for a three-dimensional crystal structure, and theoretical calculations for MB_6 (M = Ca, Sr, Ba) indicate a net transfer of only 0.9e to 1.0e. This also explains why metal-deficient phases $M_{1-x}B_6$ remain stable and why the alkali metal K can also form a hexaboride.

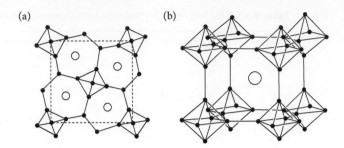

Fig. 3.3.3 Structures of (a) LaB_4 and (b) CaB_6.

(4) High boron-rich metal borides

In high boron-rich metal borides, there are two types of B_{12} units in their crystal structures: one is cubo-octahedral B_{12} found in YB_{12}, as shown in Fig. 3.3.4(a); the other is icosahedral B_{12} in NaB_{12}, as shown in Fig. 3.3.4(b). These two structural types can be inter-converted by small displacements of the B atoms. The arrows shown in Fig. 3.3.4(a) represent the directions of the displacements.

(5) Rare-earth metal borides

The ionic radii of the rare-earth metals have a dominant effect on the structure types of their crystalline borides, as shown in Table 3.3.2.

The "radius" of the 24-coordinate metal site in MB_6 is too large (M–B distances are in the range of 215–225 pm) to be comfortably occupied by the later (smaller) lanthanide elements Ho, Er, Tm, and Lu, and these elements form MB_4 compounds instead, where the M–B distances vary within the range of 185–200 pm.

YB_{66} crystallizes in space group $Fm3c$ with $a = 2344$ pm and $Z = 24$. The basic structural unit is a 13-icosahedra B_{156} giant cluster in which a central B_{12} icosahedron is surrounded by twelve B_{12} icosahedra; the B–B bond distances within the icosahedra are 171.9–185.5 pm, and the inter-icosahedra distances vary between 162.4 pm and 182.3 pm Packing of the eight $B_{12} \subset (B_{12})_{12}$ giant clusters in the unit cell generates channels and non-icosahedral bulges that accommodate the remaining 336 boron atoms in a statistical distribution. The yttrium atom is coordinated by 12 boron atoms belonging to four icosahedral faces and up to eight boron atoms located within a bulge.

$La_2Re_3B_7$ was prepared from a molten La/Ni eutectic at 1000 °C, in which the B atoms constitute an infinite zigzag chain with extensive B–B bonding (Fig. 3.3.5). The chain consists of two types of alternating "links": a B_3 acute isosceles triangle and a $trans$-B_4 zigzag chain fragment. These two types of links are connected through the corners of the bases of the B_3 triangles and the interior atoms of the B_4 fragments. The boron chains are connected by Re–B bonds, resulting in a 3D overall framework structure.

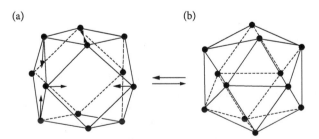

Fig. 3.3.4 Structures of (a) cubo-octahedral B_{12} unit and (b) icosahedral B_{12} unit. These two units can be inter-converted by the displacements of atoms. The arrows indicate shift directions of three pairs of atoms to form additional linkages (and triangular faces) during conversion from (a) to (b), and concurrent shifts of the remaining atoms are omitted for clarity.

Table 3.3.2 Variation of structural types with the ionic radii of rare-earth metals

Cation	radii/pm	Structural type				
		YB_{12}	AlB_2	YB_{66}	ThB_4	CaB_6
Eu^{2+}	130					+
La^{3+}	116.0				+	+
Ce^{3+}	114.3				+	+
Pr^{3+}	112.6				+	+
Nd^{3+}	110.9			+	+	+
Sm^{3+}	107.9			+	+	+
Gd^{3+}	105.3		+	+	+	+
Tb^{3+}	104.0	+	+	+	+	
Dy^{3+}	102.7	+	+	+	+	
Y^{3+}	101.9	+	+	+	+	
Ho^{3+}	101.5	+	+	+	+	
Er^{3+}	100.4	+	+	+	+	
Tm^{3+}	99.4	+	+	+	+	
Yb^{3+}	98.5	+	+	+	+	
Lu^{3+}	97.7	+	+	+	+	

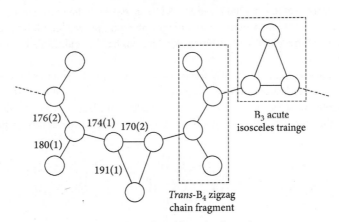

Fig. 3.3.5 The boron chain in $La_2Re_3B_7$ and B–B bond lengths (in pm).

Reference

D. E. Bugaris, C. D. Malliakas, D. Y. Chung, and M. G. Kanatzidis, Metallic borides, $La_2Re_3B_7$ and $La_3Re_2B_5$, featuring extensive boron-boron bonding. *Inorg. Chem.* **55**, 1664–73 (2016).

3.3.2 **Non-metal borides**

Boron forms a large number of non-metal borides with oxygen and other non-metallic elements. The principal oxide of boron is boric oxide, B_2O_3. Fused B_2O_3 readily dissolves many metal oxides to give borate glasses. Its major application is in the glass industry, where borosilicate glasses find extensive use because of their small coefficient of thermal expansion and easy workability. Borosilicate glasses include

Pyrex glass, which is used to manufacture most laboratory glassware. The chemistry of borates and related oxo complexes will be discussed in Section 3.5.

(1) **Boron carbides and related compounds**

The potential usefulness of boron carbides has prompted intensive studies of the system $B_{1-x}C_x$ with $0.1 \leq x \leq 0.2$. The carbides of composition $B_{13}C_2$ and B_4C, as well as the related compounds $B_{12}P_2$ and $B_{12}As_2$, all crystallize in the rhombohedral space group $R\bar{3}m$. The structure of $B_{13}C_2$ is derived from α-R12 boron with the addition of a linear C-B-C linking group along the [111] direction, as shown in Fig. 3.3.6. The structure of B_4C ($a = 520$ pm, $\alpha = 66°$) has been shown to be $B_{11}C(CBC)$ rather than $B_{12}(CCC)$, with statistical distribution of a C atom over the vertices of the $B_{11}C$ icosahedron. For boron phosphide and boron arsenide, a two-atom P-P or As-As link replaces the three-atom chain in the carbides.

(2) **Boron nitrides**

The common form of BN has an ordered layer structure containing hexagonal rings, as shown in Fig. 3.3.7(a). The layers are arranged so that a B atom in one layer lies directly over a N atom in the next, and vice versa. The B–N distance within a layer is 145 pm, which is shorter than the distance of 157 pm for a B–N single bond, implying the presence of π-bonding. The interlayer distance of 330 pm is consistent with van der Waals interactions. Boron nitride is a good lubricant that resembles graphite. However, unlike graphite, BN is white and an electrical insulator. This difference can be interpreted in terms of band theory, as the band gap in boron nitride is considerably greater than that in graphite because of the polarity of the B–N bond.

Heating the layered form of BN at ~2000 K and > 50 kbar pressure in the presence of a catalytic amount of Li_3N or Mg_3N_2 converts it into a more dense polymorph with the zinc blende structure, as shown in Fig. 3.3.7(b). This cubic form of BN is called borazon, which has hardness almost equal to that of diamond and is used as an abrasive.

As the BN unit is isoelectronic with a C_2 fragment, replacement of the latter by the former in various organic compounds leads to azaborane structural analogs. Some examples are shown below. Planar borazine (or borazole) $B_3N_3H_6$ is stabilized by π-delocalization, but it is much more reactive than benzene in view of the partial positive and negative charges on the N and B atoms, respectively.

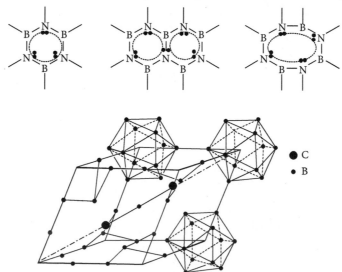

Fig. 3.3.6 Structure of $B_{13}C_2$. The small dark circles represent B atoms and large circles represent C atoms.

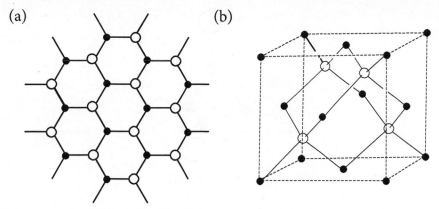

Fig. 3.3.7 Structures of BN: (a) graphite-like; (b) diamond-like (borazon).

Table 3.3.3 Some physical properties of BX_3

Molecule	B–X bond length/pm	$(r_B + r_X)^*$/pm	Bond energy/kJ mol^{-1}	mp/K	bp/K
BF_3	130	152	645	146	173
BCl_3	175	187	444	166	286
BBr_3	187	202	368	227	364
BI_3	210	221	267	323	483

* Quantities r_B and r_X represent the covalent radii of B and X atoms, respectively.

(3) Boron halides

Boron forms numerous binary halides, of which the monomeric trihalides BX_3 are volatile and highly reactive. These planar molecules differ from aluminum halides, which form Al_2X_6 dimers with a complete valence octet on Al. The difference between BX_3 and Al_2X_6 is due to the smallness of the B atom. The B atom uses its sp^2 hybrid orbitals to form σ bonds with the X atoms, leaving the remaining empty $2p_z$ orbital to overlap with three filled p_z orbitals of the X atoms. This generates delocalized π orbitals, three of which are filled with electrons. Thus there is some double bond character in the B–X bonding that stabilizes the planar monomer. Table 3.3.3 lists some physical properties of BX_3 molecules. The energy of the B–F bond in BF_3 is 645 kJ mol^{-1}, which is consistent with its multiple bond character.

All BX_3 compounds behave as Lewis acids, the strength of which is determined by the strength of the aforementioned delocalized π bonding; the stronger it is, the molecule more easily retains its planar configuration and the acid strength is weaker. Due to the steadily weakening of π bonding in going from BF_3 to BI_3, BF_3 is a weaker acid and BI_3 is stronger. This order is the reverse of that predicated when the electronegativity and size of the halogens are considered: the high electronegativity of fluorine could be expected to make BF_3 more receptive to receiving a lone pair from a Lewis base, and the small size of fluorine would least inhibit the donor's approach to the boron atom.

Salts of tetrafluoroborate, BF_4^-, are readily formed by adding a suitable metal fluoride to BF_3. There is a significant lengthening of the B–F bond from 130 pm in planar BF_3 to 145 pm in tetrahedral BF_4^-.

Among the boron halides, B_4Cl_4 is of interest as both B_4H_4 and $B_4H_4^{2-}$ do not exist. Fig. 3.3.8 shows the molecular structure of B_4Cl_4, which has a tetrahedral B_4 core consolidated by four terminal B–Cl

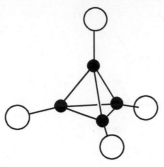

Fig. 3.3.8 Molecular structure of B_4Cl_4.

2c-2e bonds and four BBB 3c-2e bonds on four faces. This structure is further stabilized by σ–π interactions between the lone pairs of Cl atoms and BBB 3c-2e bonds. The bond length of B–Cl is 170 pm, corresponding to a single bond, and the bond length of B–B is also 170 pm.

3.3.3 Critical description of molecular compound containing a boron-boron multiple bond

Reduction of a bis(N-heterocyclic carbene)-stabilized tetrabromodiborane **1** with either two or four equivalents of sodium naphthalenide (a one-electron reducing agent) yields isolable diborene and diboryne compounds (Scheme 1).

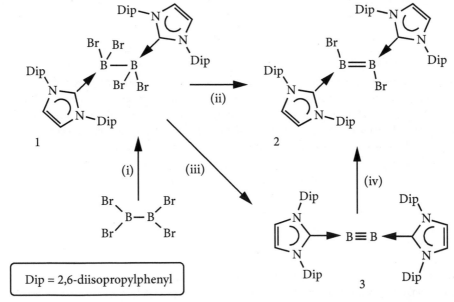

Scheme 1 Reaction conditions: (i) synthesis of diborane **1** from tetrabromodiborane: simple addition of 2 equiv. 1,3-bis-(2,6-diisopropylphenyl)imidazol-2-ylidene (IDip), pentane, −78 °C; (ii) synthesis of diborene **2** from diborane **1**: 2 equiv. sodium naphthalenide, THF, −78 °C; (iii) synthesis of diboryne **3** from diborane **1**: 4 equiv. sodium naphthalenide, THF, −78 °C; (iv) conversion of diboryne **3** to diborene **2**: comproportionation reaction involving mixing of equimolar amounts of **1** and **3** in C_6D_6 at room temperature.

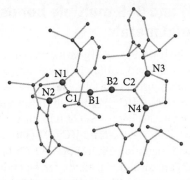

Fig. 3.3.9 Molecular structure of bis(NHC)-stabilized diboryne **3**, with (NHC = N-heterocyclic carbine $C_3N_2H_2(C_6H_3{}^iPr_2\text{-}2,6)_2$). Selected bond lengths (pm) and angles (°): B1≡B2, 144.9(3), B1–C1, 148.7(3), B2–C2, 149.5(3); B2–B1–C1, 173.0(2), B1–B2–C2, 173.3(2).

Bis(IDip)-stabilized diboryne **3** (with IDip = 1,3-bis-(2,6-diisopropylphenyl)imidazol-2-ylidene) crystallizes as a green solid, and its molecular structure elucidated by low-temperature X-ray crystallographic analysis is shown in Fig. 3.3.9. Spectroscopic characterization also supported that **3** is a halide-free linear system containing a boron-boron triple bond.

Notably, a more recent detailed study of both properties (ΔE and f) of bis(NHC)-stabilized diboryne **3** offers an alternative bonding interpretation based on detailed thermodynamic and force-field investigation of (1) fragmentation of the NHC⟶B≡B⟵NHC molecular skeleton into B_2 and two NHC units, and (2) extraction of vibrational spectrum force constants f_{BB} and f_{BC} for respective BB and BC bonds, which are classical procedures to ascertain single, double and triple C–C bonds in organic compounds.

In contrast to the previous assignment of a donor–acceptor triple bond in diboryne **3**, the bonding situation is now described by the resonance structures shown in Scheme 2 that involve 4 π electrons over the central BB bond and two peripheral BC bonds. In other words, the individual bond orders lie between one and two, and the molecular skeleton is stabilized by the delocalized 4-π bonding system shown in Scheme 3.

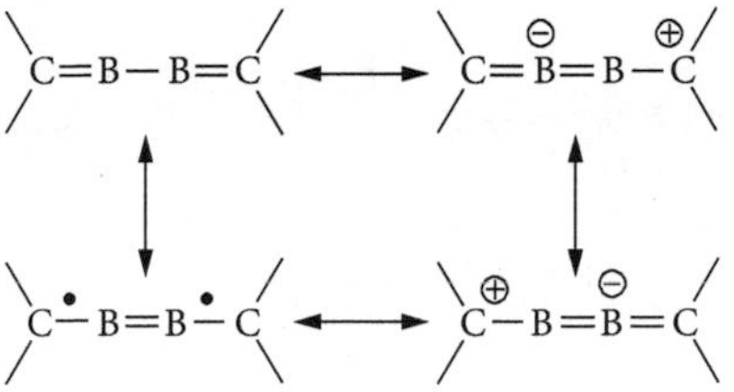

Scheme 2

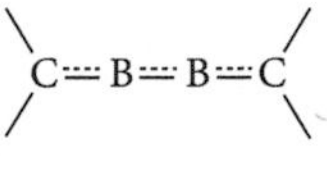

Scheme 3

References

H. Braunschweig, R. D. Dewhurst, K. Hammond, J. Mies, K. Radacki, and Alfredo Vargas, Ambient-temperature isolation of a compound with a boron-boron triple bond, *Science* **336**, 1420–2 (2012).

R. Köppe and H. Schnöckel, The boron–boron triple bond? A thermodynamic and force field based interpretation of the N-heterocyclic carbene (NHC) stabilization Procedure, *Chem. Sci.* **6**, 1199–205 (2015).

3.3.4 B≡E (E = N and O) and B=B multiple bonds in the coordination sphere of platinum group metals

Treatment of $[(Me_3Si)_2N=BBr_2]$ with $[(Me_3P)_3RhCl]$ gave the first Group 9 iminoboryl complex *cis, mer*-$[(Me_3P)_3Br_2Rh(B\equiv NSiMe_3)]$ shown in Fig. 3.3.10(a). Its boron–nitrogen bond distance of 125.5(7) pm determined from X-ray crystallography justifies the formulation of a B≡N triple bond.

In 2010, the first oxoboryl complex *trans*-$[(Cy_3P)_2BrPt(BO)]$ was isolated from the reaction between $[Br_2BOSiMe_3]$ and $[Pt(PCy_3)_2]$ in toluene, but it was not fully characterized because of its oily nature. Its subsequent reaction with $[Bu_4N]SPh$ yielded *trans*-$[(Cy_3P)_2(PhS)Pt(BO)]$ shown in Fig. 3.3.10(b), which exhibits Pt–B and B≡O bond lengths of 198.3(3) pm and 121.0(3) pm, respectively, from single-crystal X-ray analysis. These results are consistent with the corresponding Pt–B and B≡O bond lengths of 192.8(3) pm and 123.4(2) pm in *trans*-$[(Cy_3P)_2BrPt(B\equiv O\longrightarrow BAr^F_3)]$.

The reduction of a distorted σ-diboran(4)yl complex with $[LMg^I\text{–}Mg^IL]$ {L = (2,4,6-$Me_3C_6H_2NCMe)_2CH$} produced the first π-diborene stabilized by a $[Pt(P^iPr_3)_2]$ fragment (see Reaction Scheme below).

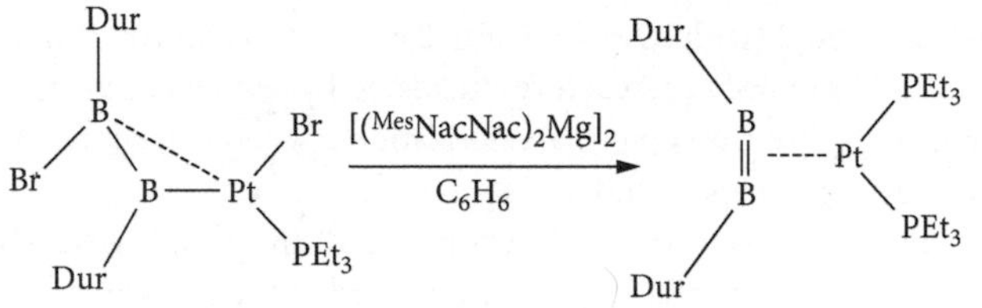

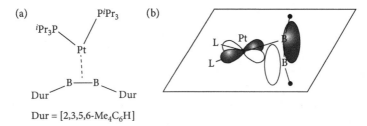

$$^{Mes}NacNac = [(2,4,6\text{-}Me_3\text{-}C_6H_2NCMe)_2CH]$$
$$Dur = [2,3,5,6\text{-}Me_4C_6H]$$

X-ray structure analysis of the first π-diborene stabilized by a $Pt(P^iPr_3)_2$ fragment (Fig. 3.3.11(a)) gave a B=B double bond distance of 151.0(14) pm. The B–B bond axis is orientated perpendicular to the P–Pt–P plane, and the two Pt–B bond distances of 206.9(9) and 207.2(9) pm indicate back-donation of the filled Pt $d_{x^2-y^2}$ orbital into the empty B–B π orbital (Fig. 3.3.11(b)).

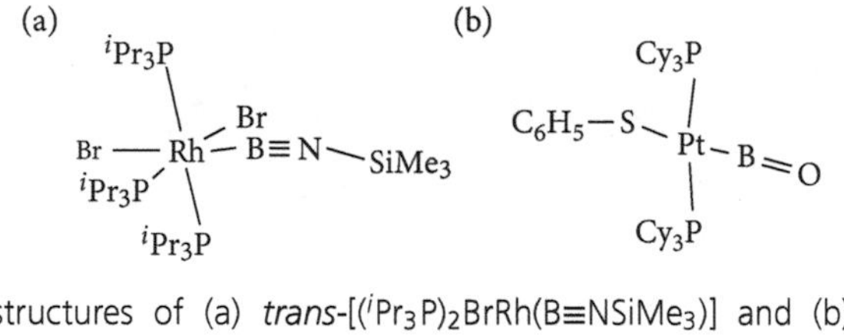

Fig. 3.3.10 Molecular structures of (a) *trans*-$[(^iPr_3P)_2BrRh(B\equiv NSiMe_3)]$ and (b) monocation in *trans*-$[(Cy_3P)_2(PhS)Pt(BO)]$ BAr^F_4.

Dur = $[2,3,5,6\text{-}Me_4C_6H]$

Fig. 3.3.11 (a) Molecular structure of Pt(II) diborene complex derived from X-ray crystallography. (b) The diborene ligand is orthogonal to the PtL_2 plane.

References

H. Braunschweig, K. Radacki, D. Rais, and K. Uttinger, Synthesis and characterization of palladium and platinum iminoboryl complexes, *Angew. Chem. Int. Ed.* **45**, 162–5 (2006).

H. Braunschweig, T. Kupfer, K. Radacki, A. Schneider, F. Seeler, K. Uttinger, and H. Wu, Synthesis and reactivity studies of iminoboryl complexes. *J. Am. Chem. Soc.* **130**, 7974–83 (2008).

H. Braunschweig, A. Damme, R. D. Dewhurst, and A.Vargas, Bond-strengthening π backdonation in a transition-metal π-diborene complex, *Nat. Chem.* **5**, 115–21 (2013).

General Reference

J. Brand, H. Braunschweig and S. S. Sen, B=B and B≡E (E = N and O) multiple bonds in the coordination sphere of late transition metals, *Acc. Chem. Res.* **47**, 180–91 (2014).

3.4 Boranes and Carboranes

3.4.1 Molecular structure and bonding

Boranes and carboranes are electron-deficient compounds with interesting molecular geometries and characteristic bonding features.

(1) Molecular structure of boranes and carboranes

Figs 3.4.1 and 3.4.2 show the structures of some boranes and carboranes, respectively, that have been established by X-ray and electron diffraction methods.

(2 Bonding in boranes

In simple covalent bonding theory, there are four bonding types in boranes: (a) normal 2c-2e B–B bond, (b) normal 2c-2e B–H bond, (c) 3c-2e BBB bond (described in Fig. 3.2.4), and (d) 3c-2e BHB bond.

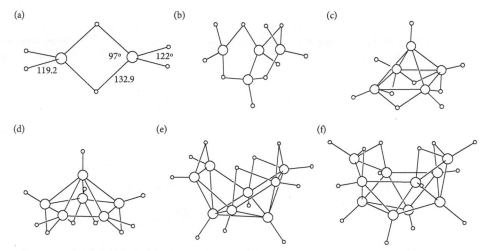

Fig. 3.4.1 Structures of boranes: (a) B_2H_6, (b) B_4H_{10}, (c) B_5H_9, (d) B_6H_{10}, (e) B_8H_{12}, (f) $B_{10}H_{14}$.

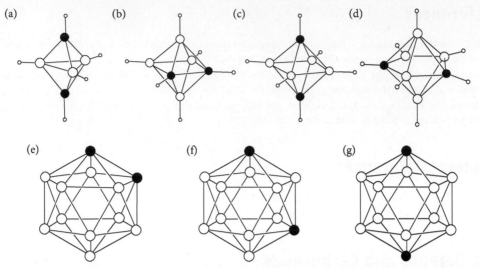

Fig. 3.4.2 Structures of carboranes: (a) 1,5-$C_2B_3H_5$, (b) 1,2-$C_2B_4H_6$, (c) 1,6-$C_2B_4H_6$, (d) 2,4-$C_2B_5H_7$, (e)–(g) three isomers of $C_2B_{10}H_{12}$ (H atoms are omitted for clarity).

The 3c-2e BHB bond is constructed from two orbitals of B_1 and B_2 atoms (ψ_{B1} and ψ_{B2}, which are sp^x hybrids) and ψ_H (1s orbital) of the H atom. The combinations of these three atomic orbitals result in three molecular orbitals:

$$\psi_1 = \frac{1}{2}\psi_{B1} + \frac{1}{2}\psi_{B2} + \frac{1}{\sqrt{2}}\psi_H$$

$$\psi_2 = \frac{1}{\sqrt{2}}(\psi_{B1} - \psi_{B2})$$

$$\psi_3 = \frac{1}{2}\psi_{B1} + \frac{1}{2}\psi_{B2} - \frac{1}{\sqrt{2}}\psi_H$$

The orbital overlaps and relative energies are illustrated in Fig. 3.4.3. Note that ψ_1, ψ_2, and ψ_3 are bonding, non-bonding, and anti-bonding molecular orbitals, respectively. For the 3c-2e BHB bond, only ψ_1 is filled with electrons.

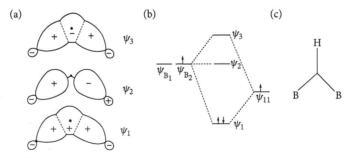

Fig. 3.4.3 The 3c-2e BHB bond: (a) orbitals overlap, (b) relative energies, and (c) simple representation.

(3) Topological description of boranes

The overall bonding in borane molecules or anions is sometimes represented by a four-digit code introduced by Lipscomb, the so-called *styx* number, where

   s is the number of 3c-2e BHB bonds,
   t is the number of 3c-2e BBB bonds,
   y is the number of normal B–B bonds, and
   x is the number of BH_2 groups.

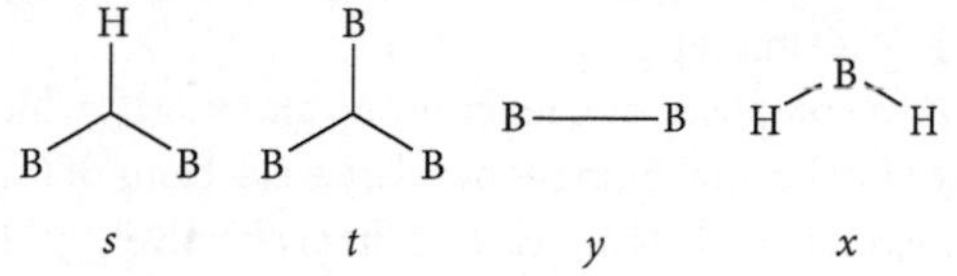

The molecular structural formulas of some boranes and their *styx* codes are shown in Fig. 3.4.4. The following simple rules must hold for the borane structures:

(a)   Each B atom has at least a terminal H atom attached to it.

(b)   Every pair of boron atoms which are geometric neighbors must be connected by a B–B, BHB, or BBB bond.

(c)   Every boron atom uses four valence orbitals in bonding to achieve an octet configuration.

(d)   No two boron atoms may be bonded together by both 2c-2e B–B and 3c-2e BBB bonds, or by both 2c-2e B–B and 3c-2e BHB bonds.

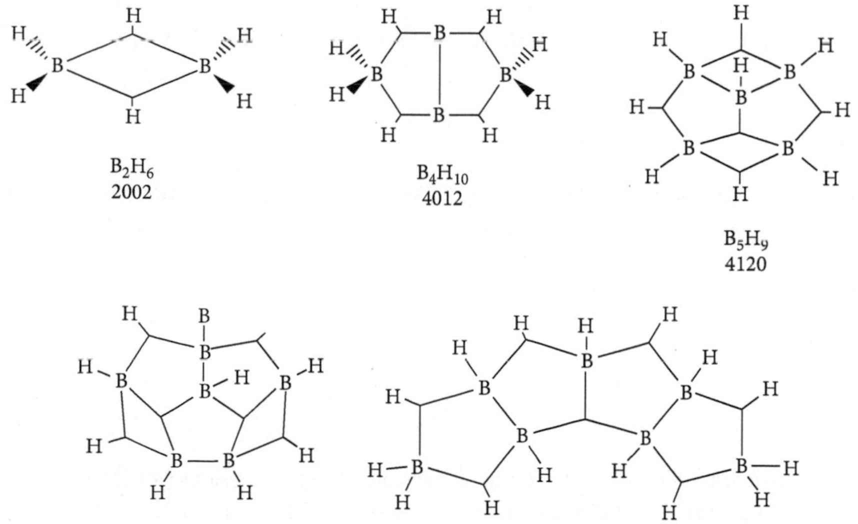

B_2H_6
2002

B_4H_{10}
4012

B_5H_9
4120

B_6H_{10}
4220

B_7H_{15}
6122

Fig. 3.4.4 Molecular structural formulas and *styx* codes of some boranes.

3.4.2 **Bond valence in molecular skeletons**

The geometry of a molecule is related to the number of valence electrons in it. According to the valence bond theory, organic and borane molecules composed of main-group elements owe their stability to the filling of all four valence orbitals (ns and np) of each atom, with eight valence electrons provided by the atom and those that are bonded to it. Likewise, transition-metal compounds achieve stability by filling the nine valence orbitals [$(n - 1)d$, ns and np] of each transition-metal atom with 18 electrons provided by the metal atom and the surrounding ligands. In molecular orbital language, the octet and 18-electron rules are rationalized in terms of the filling of all bonding and non-bonding molecular orbitals and the existence of a large HOMO-LUMO energy gap.

Consider a molecular skeleton composed of n main-group atoms, M_n, which takes the form of a chain, ring, cage, or framework. Let g be the total number of valence electrons of the molecular skeleton. When a covalent bond is formed between two M atoms, each of them effectively gains one electron in its valence shell. In order to satisfy the octet rule for the whole skeleton, $\frac{1}{2}(8n - g)$ electron pairs must be involved in bonding between the M atoms. The number of these bonding electron pairs is defined as the "bond valence" b of the molecular skeleton:

$$b = \tfrac{1}{2}(8n - g). \tag{3.4.1}$$

When the total number of valence electrons in a molecular skeleton is less than the number of valence orbitals, the formation of normal 2c-2e covalent bonds is insufficient to compensate for the lack of electrons. In this type of electron-deficient compounds there are usually found 3c-2e bonds, in which three atoms share an electron pair. Thus one 3c-2e bond serves to compensate for the lack of four electrons and corresponds to a bond valence value of 2, as discussed in Section 3.2.

There is an enormous number of compounds that possess metal-metal bonds. A metal cluster may be defined as a polynuclear compound in which there are substantial and directed bonds between the metal atoms. The metal atoms of a cluster are also referred to as skeletal atoms, and the remaining non-metal atoms and groups are considered as ligands.

According to the 18-electron rule, the bond valence of a transition metal cluster is

$$b = \tfrac{1}{2}(18n - g). \tag{3.4.2}$$

If the bond valence b calculated from (3.4.1) and (3.4.2) for a cluster M_n matches the number of connecting lines drawn between pairs of adjacent atoms in a conventional valence bond structural formula, the cluster is termed "electron-precise."

For a molecular skeleton consists of n_1 transition-metal atoms and n_2 main-group atoms, such as transition-metal carboranes, its bond valence is

$$b = \tfrac{1}{2}(18n_1 + 8n_2 - g). \tag{3.4.3}$$

The bond valence b of a molecular skeleton can be calculated from expressions (3.4.1) to (3.4.3), in which g represents the total number of valence electrons in the system. The g value is the sum of:

(1) the number of valence electrons of the n atoms that constitute the molecular skeleton M_n,

(2) the number of electrons donated by the ligands to M_n, and

(3) the number of net positive or negative charges carried by M_n, if any.

Table 3.4.1 Number of electrons supplied by ligands to a molecular skeleton (the skeletal atoms are considered to be uncharged)

Ligand	Coordinate mode*	No. of electrons	Ligand	Coordinate mode*	No. of electrons
H	μ_1, μ_2, μ_3	1	NR_3, PR_3	μ_1	2
B	int	3	NCR	μ_1	2
CO	μ_1, μ_2, μ_3	2	NO	μ_1	3
CR	μ_3, μ_4	3	OR, SR	μ_1	1
CR_2	μ_1	2	OR, SR	μ_2	3
CR_3, SiR_3	μ_1, μ_2	1	O, S, Se, Te	μ_2	2
η^2-C_2R_2	μ_1	2	O, S, Se, Te	μ_3	4
η^2-C_2R_4	μ_1	2	O, S	int	6
η^5-C_5R_5	μ_1	5	F, Cl, Br, I	μ_1	1
η^6-C_6R_6	μ_1	6	F, Cl, Br, I	μ_2	3
C, Si	int	4	Cl, Br, I	μ_3	5
N, P, As, Sb	int	5	PR	μ_3, μ_4	4

*μ_1 = terminal ligand, μ_2 = ligand bridging two atoms, μ_3 = ligand bridging three atoms, int = interstitial atom.

The simplest way to count the g value is to start from uncharged skeletal atoms and uncharged ligands. Ligands such as NH_3, PR_3, and CO each supplies two electrons. Non-bridging halogen atoms, H atom, CR_3, and SiR_3 groups are one-electron donors. A μ_2-bridging halogen atom contributes three electrons, and a μ_3-bridging halogen atom donates five electrons. Table 3.4.1 lists the number of electrons contributed by various ligands, depending on their coordination modes.

3.4.3 Structural types of borane and carborane and Wade's rules

In order to reasonably analyze the skeletal structure of borane, carborane and its derivatives, the number of bonding electron pairs can be related to the geometric configuration and bonding of the molecule. Based on the structural data available at the time, Wade used semi-empirical molecular orbital theory to process borane and carborane structures. He linked the number of skeleton electrons to the number of bonded molecular orbitals to develop a set of rules on the numbers of skeleton electron pairs for different structure types of borane and carborane, namely *closo*, *nido*, and *arachno*.

Taking $B_6H_6^{2-}$ as an example, the entire system has $3 \times 6 + 6 + 2 = 26$ electrons. Each B atom has four valence orbitals: one $2s$ and three $2p$. The skeleton B atoms adopt an octahedral arrangement to form a *closo* structure. The B atoms are sp hybridized. One of the sp-orbitals of a B atom and the s-orbital of a terminal H atom are overlapped to form a normal outward-oriented BH bond which is filled with one pair of electrons. Three orbitals are remained for each B atom: one sp-orbital pointing into the center of the octahedral skeleton and two p-orbitals which are perpendicular to each other and to the two sp-orbitals. The six sp-orbitals and 12 p-orbitals of the six B atoms are combined according to the principle of symmetry matching to form seven bonding orbitals and 11 anti-bonding orbitals, as shown in Fig. 3.4.5.

The number of remaining electrons in the whole system is $26 - 2 \times 6 = 14$. These seven pairs of electrons fill seven bonding orbitals to form a $B_6H_6^{2-}$ skeleton structure. There are six B atoms and seven pairs of electrons, and the number of bonding orbitals of the skeleton is $6 + 1$. Calculations show that the general formula of borane ions with a triangular polyhedron structure is $B_nH_n^{2-}$. A *closo* structure is formed and the number of bonding orbitals is $n + 1$.

Nido and *arachno* borane structures can be obtained from a *closo* structure such as $B_6H_6^{2-}$, as shown in Fig. 3.4.6. If one BH^{2+} group is removed from $B_6H_6^{2-}$, an additional pair of electrons is left and $B_5H_5^{4-}$ is

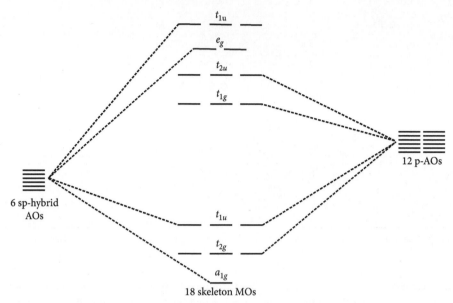

Fig. 3.4.5 Schematic diagram of molecular orbital formation in $B_6H_6^{2-}$.

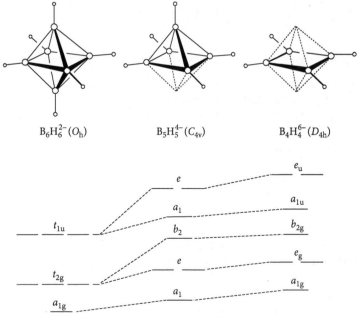

Fig. 3.4.6 Correlation diagram of molecular orbitals of *closo*-$B_6H_6^{2-}$, *nido*-$B_5H_5^{4-}$, and *arachno*-$B_4H_4^{6-}$.

resulted. The number of skeletal electrons is $n + 2$ now and the molecular orbital changes accordingly, but there are still seven bonding orbitals. Further removal of a BH^{2+} group leave another pair of electrons and $B_4H_4^{6-}$ is resulted. The number of skeleton electron pairs is $n + 3$ and the orbitals are adjusted accordingly.

On the basis of the above three types of skeleton structures, increasing the number of bonding electron pairs in the skeleton to $(n + 4)$ results in *hypho* structure for boranes, such as $B_5H_{11}^{2-}$. Therefore, the relationship between the number of skeleton electron pairs and the skeleton structure type can be summarized as follows:

(1)   A *closo* triangular cluster with n vertices is held together by $(n + 1)$ bonding electron pairs.

(2)   A *nido* triangular cluster with n vertices is held together by $(n + 2)$ bonding electron pairs.

(3)   A *arachno* triangular cluster with n vertices is held together by $(n + 3)$ bonding electron pairs.

(4)   A *hypho* triangular cluster with n vertices is held together by $(n + 4)$ bonding electron pairs.

According to the above classification, using the four numbers s, t, y, and x allows us to deeply understand the type of a borane skeleton and its bonding relationship. The four common structural types of borane and carborane are listed in Table 3.4.2. Fig. 3.4.7 shows the structures of the examples in the table: *closo-* $B_5H_5^{2-}$, *nido-* B_5H_9, *arachno-* B_5H_{11}, *hypho-* $B_5H_{11}^{2-}$, and their relationships.

It can be seen that *closo-*$B_5H_5^{2-}$ has trigonal bipyramidal geometry, the five B atoms in *nido-*B_5H_9 form a square pyramid, the B atoms in *aracho-*B_5H_{11} form a spread net, and the B atoms in *hypho-*$B_5H_{11}^{2-}$ constitute two open moieties connected together.

Boranes and carboranes of the *hypho* type are quite rare. The prototype *hypho* borane $B_5H_{11}^{2-}$ has been isolated, and its proposed structure is analogous to that of the isoelectronic analog $B_5H_9(PMe_3)_2$ (Fig. 3.4.8(a)), which was established by X-ray crystallography. The nine-vertex *hypho* tricarbaborane cluster anion $(NCCH_2)$-1,2,5-$C_3B_6H_{12}^-$ shown in Fig. 3.4.8(b) has 13 skeletal pairs, and accordingly it may be considered as derived from an icosahedron with three missing vertices. Similarly, in *endo-6-endo-7-*$[\mu^2-\{C(CN)_2\}_2]$-*arachno-6,8-*$C_2B_7H_{12}^-$ (Fig. 3.4.8(c)), the C_2B_7-fragment of the cluster anion can be viewed

Table 3.4.2 Structural types of borane and carborane and bonding

Structural type	Number of skeleton bonding electron pairs	Example	Bonding (*styx*)	Skeleton bonding valence $B = s + 2t + y$
Closo	6 ($n+1$)	$B_5H_5^{2-}$	(0330)	9
Nido	7 ($n+2$)	B_5H_9	(4120)	8
Aracho	8 ($n+3$)	B_5H_{11}	(3203)	7
Hypho	9 ($n+4$)	$B_5H_{11}^{2-}$	(2124)	6

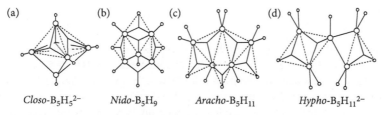

(a)   *Closo-*$B_5H_5^{2-}$ (b)   *Nido-*B_5H_9 (c)   *Aracho-*B_5H_{11} (d)   *Hypho-*$B_5H_{11}^{2-}$

Fig. 3.4.7 Structural evolution of pentaborane.

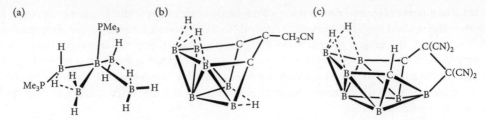

Fig. 3.4.8 Molecular structures of some *hypho* cluster systems: (a) $B_5H_9(PMe_3)_2$; (b) $(NCCH_2)C_3B_6H_{12}^-$; (c) $[\mu^2\text{-}\{C(CN)_2\}_2]C_2B_7H_{12}^-$. In carboranes (b) and (c), for clarity only bridging and extra H atoms are included, and all *exo* H atoms attached to the vertices are omitted.

as a 13-electron pair, nine-vertex *hypo* system with an *exo*-polyhedral TCNE substituent bridging a C atom and a B atom at the open side.

3.4.4 **Chemical bonding in *closo*-boranes**

Boranes of the general formula $B_nH_n^{2-}$ and isoelectronic carboranes such as $CB_{n-1}H_n^-$ and $CB_{n-2}H_n$ have *closo* structures, in which n skeletal B and C atoms are located at the vertices of a polyhedron bounded by trianglular faces (deltahedron). For the $B_nH_n^{2-}$ *closo*-boranes:

$$b = \tfrac{1}{2}[8n - (4n + 2)] = 2n - 1. \tag{3.4.4}$$

From the geometrical structures of *closo*-boranes, s and x of the *styx* code are both equal to zero. Each 3c-2e BBB bond corresponds to the value 2 of bond valence. Thus,

$$b = 2t + y. \tag{3.4.5}$$

Combining expressions (3.4.4) and (3.4.5) leads to

$$2t + y = 2n - 1. \tag{3.4.6}$$

When the number of electron pairs of the molecular skeleton is counted, one gets

$$t + y = n + 1. \tag{3.4.7}$$

Hence from (3.4.6) and (3.4.7),

$$t = n - 2, \tag{3.4.8}$$

$$y = 3, \tag{3.4.9}$$

so that each *closo*-borane $B_nH_n^{2-}$ has exactly three B–B bonds in its valence bond structural formula.

Table 3.4.3 lists the bond valence of $B_nH_n^{2-}$ and other parameters; n starts from 5, not from 4, as $B_4H_4^{2-}$ would need $t = 2$ and $y = 3$, which cannot satisfy rule (d) for a tetrahedron. Fig. 3.4.9 shows the localized bonding description (in one canonical form) for some *closo*-boranes. The charge of minus two in $B_nH_n^{2-}$ is required for consolidating the n BH units into a deltahedron. Except for $B_5H_5^{2-}$, the bond valence b is always smaller than the number of edges of the polyhedron.

Table 3.4.3 Bond valence and other parameters for $B_nH_n^{2-}$

n in $B_nH_n^{2-}$	Styx code	Bond valence b	No. of edges = 3t	No. of faces = 2t	$t + y$ (No. of skeletal electron pairs)
5	0330	9	9	6	6
6	0430	11	12	8	7
7	0530	13	15	10	8
8	0630	15	18	12	9
9	0730	17	21	14	10
10	0830	19	24	16	11
11	0930	21	27	18	12
12	0,10,3,0	23	30	20	13

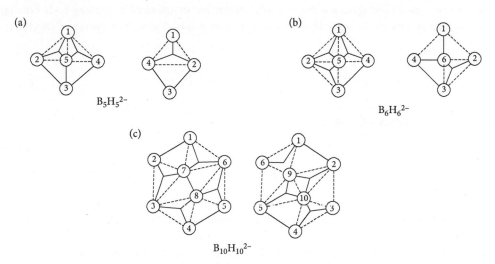

(a) $B_5H_5^{2-}$

(b) $B_6H_6^{2-}$

(c) $B_{10}H_{10}^{2-}$

Fig. 3.4.9 The B–B and BBB bonds in (a) $B_5H_5^{2-}$, (b) $B_6H_6^{2-}$, and (c) $B_{10}H_{10}^{2-}$; for each molecule both the front and back sides of a canonical structure are displayed. $B_{12}H_{12}^{2-}$ is shown in Fig. 3.2.5.

In the B_n skeleton of *closo*-boranes, there are $(n-2)$ 3c-2e BBB bonds and three normal B–B bonds, which amounts to $(n-2) + 3 = n + 1$ electron pairs.

Some isoelectronic species of $C_2B_{10}H_{12}$, such as $CB_{11}H_{12}^-$, $NB_{11}H_{12}$ and their derivatives are known. Since they have the same number of skeletal atoms and the same bond valence, these compounds are isostructural.

3.4.5 Chemical bonding in *nido*- and *arachno*-boranes

Using the *styx* code and bond valence to describe the structures of *nido* boranes or carboranes, the bond valence b is equal to:

$$b = \tfrac{1}{2}[8n - (4n + 4)] = 2n - 2, \tag{3.4.10}$$

$$b = s + 2t + y. \tag{3.4.11}$$

Combining expressions (3.4.10) and (3.4.11) lead to

$$s + 2t + y = 2n - 2. \tag{3.4.12}$$

The number of valence electron pairs for *nido* B_nH_{n+4} is equal to $\frac{1}{2}(4n + 4) = 2n + 2$, in which $n + 4$ pairs are used to form B–H bond and the remaining electron pairs used to form 2c-2e B–B bonds (y) and 3c-2e BBB bonds (t). Thus,

$$t + y = (2n + 2) - (n + 4) = n - 2. \tag{3.4.13}$$

The number of each bond type in *nido*-B_nH_{n+4} are obtained from expressions (3.4.12) and (3.4.13):

$$t = n - s, \tag{3.4.14}$$

$$y = s - 2, \tag{3.4.15}$$

$$x + s = 4. \tag{3.4.16}$$

The stability of a molecular species with open skeleton may be strengthened by forming 3c-2e bond(s). In a stable *nido*-borane, neighboring H–B–H and B–H groups always tend to be converted into the H–BHB–H system:

Thus, in *nido*-borane, $x = 0$, $s = 4$, and

$$t = n - 4 \tag{3.4.17}$$

$$y = 2 \tag{3.4.18}$$

Table 3.4.4 lists the *styx* code and bond valence of some *nido*-boranes, and Fig. 3.4.10 shows the chemical bonding in their skeletons.

For the *arachno*-boranes, B_nH_{n+6}, the relations between *styx* code and n are as follows:

$$t = n - s \tag{3.4.19}$$

$$y = s - 3 \tag{3.4.20}$$

$$x + s = 6 \tag{3.4.21}$$

Table 3.4.4 The *styx* codes of some *nido*-boranes

Borane	B_4H_8	B_5H_9	B_6H_{10}	B_7H_{11}	B_8H_{12}	B_9H_{13}	$B_{10}H_{14}$
s	4	4	4	4	4	4	4
t	0	1	2	3	4	5	6
y	2	2	2	2	2	2	2
x	0	0	0	0	0	0	0
$b = s + 2t + y$	6	8	10	12	14	16	18

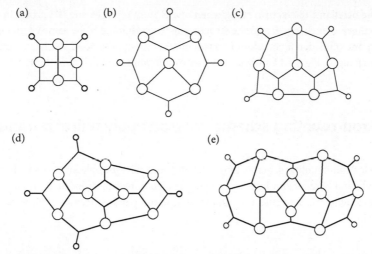

Fig. 3.4.10 Chemical bonding in the skeletons of some *nido*-boranes (large circles represent BH group, small circles represent H atoms in BHB bonds): (a) B_4H_8 (4020), (b) B_5H_9 (4120), (c) B_6H_{10} (4220), (d) B_8H_{12} (4420), (e) $B_{10}H_{14}$(4620).

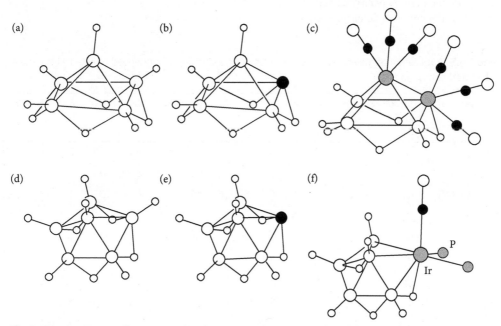

Fig. 3.4.11 Structures of corresponding boranes, carboranes, and metalloboranes having the same bond valence: (a) B_5H_9, (b) CB_4H_8, and (c) $B_3H_7[Fe(CO)_3]_2$, $b = 8$; (d) B_6H_{10}, (e) CB_5H_9, and (f) $B_5H_8Ir(CO)(PPh_3)_2$, $b = 10$.

According to the octet and 18-electron rules, when one C atom replaces one BH group in a borane, the g value and b value are unchanged, and the structure of the carborane is the same as that of the borane. Likewise, when one transition metal atom replaces one BH group, the bond valence counting of (3.4.3) also remains unchanged. Fig. 3.4.11 shows two series of compounds, which have the same b value in each series.

3.4.6 Electron-counting scheme for macropolyhedral boranes: *mno* rule

A generalized electron-counting scheme, known as the *mno* rule, is applicable to a wide range of poly-condensed polyhedral boranes and heteroboranes, metallaboranes, metallocenes, and any of their combinations. According to this *mno* rule, the number of electron pairs N necessary for a macropolyhedral system to be stable is

$$N = m + n + o + p - q$$

where m = the number of polyhedra,
   n = the number of vertices,
   o = the number of single-vertex-sharing connections,
   p = the number of missing vertices,
and q = the number of capping vertices.

Table 3.4.5 Application of the *mno* rule for electron counting in some condensed polyhedral boranes and related compounds*

Formula	Structure in Fig. 3.4.12	m	n	o	p	q	N	BH	B	CH	H_b	α	β	x
$B_{20}H_{16}$	(a)	2	20	0	0	0	22	16	6	0	0	0	0	0
$(C_2B_{10}H_{11})_2$	(b)	2	24	0	0	0	26	18	2	6	0	0	0	0
$[(C_2B_9H_{11})_2Al]^-$	(c)	2	23	1	0	0	26	18	0	6	0	1.5	0	0.5
$[B_{21}H_{18}]^-$	(d)	2	21	0	0	0	23	18	4.5	0	0	0	0	0.5
Cp_2Fe	(e)	2	11	1	2	0	16	0	0	15	0	1	0	0
$Cp^*IrB_{18}H_{20}$	(f)	3	24	1	3	0	31	15	4.5	7.5	2.5	1.5	0	0
$[Cp^*IrB_{18}H_{19}S]^-$	(g)	3	25	1	3	0	32	16	3	7.5	1.5	1.5	2	0.5
$Cp^*_2Rh_2S_2B_{15}H_{14}(OH)$	(h)	4	29	2	5	0	40	13	3	15	2	3	4	0
$(CpCo)_3B_4H_4$	(i)	4	22	3	3	1	31	4	0	22.5	0	4.5	0	0

* BH = number of electron pairs from BH groups, B = number of electron pairs from B atoms, CH = number of electron pairs from CH groups, α = number of electron pairs from metal center(s), β = number of electron pairs from main-group hetero atom, H_b = number of electron pairs from bridging hydrogen atoms, x = number of electron pairs from the charge. Also, BH (or CH) can be replaced by BR (or CR).

For a *closo* macropolyhedral borane cluster, the rule giving the required number of electron pairs is $N = m + n$.

Some examples illustrating the application of the *mno* rule is given in Table 3.4.5, which makes reference to the structures of compounds shown in Fig. 3.4.12.

$B_{20}H_{16}$ is composed of two polyhedra sharing four atoms. The $(m + n)$ electron pair count is $2 + 20 = 22$, which is appropriate for a *closo* structure stabilized by 22 skeletal electron pairs (16 from 16 BH groups and six from four B atoms). In $(C_2B_{10}H_{11})_2$, which consists of two icosahedral units connected by a B–B single bond, the electron count is simply twice that of a single polyhedron. The 26 skeletal electron pairs are provided by 18 BH groups (18 pairs), two B atoms (two pairs, B–B bond not involved in cluster bonding), and four CH groups (six pairs, three electrons from each CH group). The structure of $[(C_2B_9H_{11})_2Al]^-$ is constructed from condensation of two icosahedra through a common vertex. The required skeletal electron pairs are contributed by 18 BH groups (18 pairs), four CH groups (six pairs), the Al atom (1.5 pairs), and the anionic charge (0.5 pair). The polyhedral borane anion $B_{21}H_{18}^-$ has a

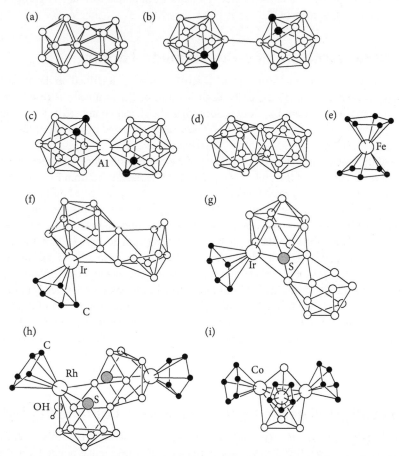

Fig. 3.4.12 Structure of condensed polyhedral boranes, metallaboranes, and ferrocene listed in Table 3.4.5: (a) $B_{20}H_{16}$, (b) $(C_2B_{10}H_{11})_2$, (c) $[(C_2B_9H_{11})_2Al]^-$, (d) $[B_{21}H_{18}]^-$, (e) Cp^*_2Fe, (f) $Cp^*IrB_{18}H_{20}$, (g) $[Cp^*IrB_{18}H_{19}S]^-$, (h) $Cp^*_2Rh_2S_2B_{15}H_{14}(OH)$, and (i) $(CpCo)_3B_4H_4$. (The Me group of Cp* and H atoms are not shown.)

shared triangular face. Its 18 BH groups, three B atoms and the negative charge provide $18 + 4.5 + 0.5 = 23$ skeletal pairs, which fit the *mno* rule with $m + n = 2 + 21$. For ferrocene, which has 16 electron pairs (15 from ten CH groups and one from Fe), the *mno* rule suggests a molecular skeleton with two open (*nido*) faces ($m + n + o + p = 2 + 11 + 1 + 2 = 16$). The metalloborane $[Cp^*IrB_{18}H_{20}]$ is condensed from three *nido* units. The *mno* rule gives 31 electron pairs, and the molecular skeleton is stabilized by electron pairs from 15 BH groups (15) five CH groups (7.5), three shared B atoms (4.5), five bridging H atoms (2.5), and the Ir atom $[½(9 – 6) = 1.5$; six electrons occupying three non-bonding metal orbitals]. Addition of a sulfur atom as a vertex to the above cluster leads to $[Cp^*IrB_{18}H_{19}S]^-$, which requires 32 electron pairs. The electrons are supplied by the S atom (four electrons, as the other two valence electrons occupy an *exo*-cluster orbital), 16 BH groups, two shared B atoms, three bridging H atoms, the Ir atom, and the negative charge. The cluster compound $[Cp^*_2Rh_2S_2B_{15}H_{14}(OH)]$ contains a *nido* $\{RhSB_8H_7\}$ unit and an *arachno* $\{RhSB_9H_8(OH)\}$ unit conjoined by a common B–B edge. The *mno* rule leads to 40 electron pairs, which are supplied by 13 BH groups (13), two B atoms (3), ten CH groups (15), four bridging H atoms (2), two Rh atoms (3), and two S atoms (4). In the structure of $(CpCo)_3B_4H_4$, one B atom caps the Co_3 face of a Co_3B_3 octahedron, and the 31 electron pairs are supplied by four BH groups (4), 15 CH groups (22.5), and three Co atoms (4.5).

3.4.7 Electronic structure of β-rhombohedral boron

In Section 3.2 and Fig. 3.2.6, the structure of β-R105 boron is described in terms of large B_{84} ($B_{12a}@B_{12}@B_{60}$) clusters and B_{10}–B–B_{10} units. Each B_{10} unit of C_{3v} symmetry is fused with three B_6 half-icosahedra from three adjacent B_{84} clusters, forming a B_{28} cluster composed of four fused icosahedra, as shown in Fig. 3.4.13(a). A pair of such B_{28} clusters are connected by a six-coordinate B atom to form a B_{57} (B_{28}–B–B_{28}) unit.

To understand the subtlety and electronic structure of the complex covalent network of β-R105 boron, it is useful to view the contents of the unit cell (Fig. 3.4.13(b)) as assembled from B_{12a} (at the center of a B_{84} cluster) and B_{57} (lying on a C_3-axis) fragments that are connected by 2c-2e bonds.

The electronic requirement of the B_{57} unit can be assessed by saturating the dangling valencies with hydrogen atoms, yielding the molecule $B_{57}H_{36}$. According to the *mno* rule, $8 + 57 + 1 = 66$ electron pairs are required for stability, but the number available is 67.5 (36 (from 36 BH) + (21 x 3)/2 (from 21B)). Thus the $B_{57}H_{36}$ polyhedral skeleton needs to get rid of three electrons to achieve stability, whereas the $B_{12a}H_{12}$ skeleton requires two additional electrons for sufficiency. As the unit cell contains four B_{12a} and one B_{57} fragments, the idealized structure of β-rhombohedral boron has a net deficiency of five electrons.

X-ray studies have established that β-R105 boron has a very porous (only 36% of space is filled in the idealized model) and defective structure with the presence of interstitial atoms and partial occupancies. The B_{57} fragment can dispose of excess electrons by removal of some vertices to form *nido* or *arachno* structures, and individual B_{12a} units can gain electrons by incorporating capping vertices that are accommodated in interstitial holes (see Fig. 3.4.13(b)).

3.4.8 Persubstituted derivatives of icosahedral borane $B_{12}H_{12}{}^{2-}$

The compound $Cs_8[B_{12}(BSe_3)_6]$ is prepared from the solid-state reaction between Cs_2Se, Se, and B at high temperature. In the anion $[B_{12}(BSe_3)_6]^{8-}$, each trigonal planar Bse_3 unit is bonded to a pair of neighboring B atom in the B_{12} skeleton, as shown in Fig. 3.4.14. It is the first example of bonding between a chalcogen and an icosahedral B_{12} core. The point group of $[B_{12}(BSe_3)_6]^{8-}$ is D_{3d}.

Icosahedral $B_{12}H_{12}{}^{2-}$ can be converted to $[B_{12}(OH)_{12}]^{2-}$, in which the 12 hydroxy groups can be substituted to form carboxylic acid esters $[B_{12}(O_2CMe)_{12}]^{2-}$, $[B_{12}(O_2CPh)_{12}]^{2-}$ and $[B_{12}(OCH_2Ph)_{12}]^{2-}$.

(a)

(b)

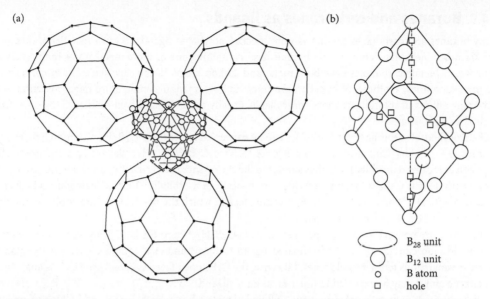

⬭	B_{28} unit
◯	B_{12} unit
∘	B atom
▢	hole

Fig. 3.4.13 Structure of β-R105 boron. (a) Three half-icosahedra of adjacent B_{84} clusters fuse with a B_{10} unit to form a B_{28} unit. (b) Unit cell depicting the B_{12} units at the vertices and edge centers, leaving the B_{28}–B–B_{28} unit along the main body diagonal. Holes of three different types are also indicated.

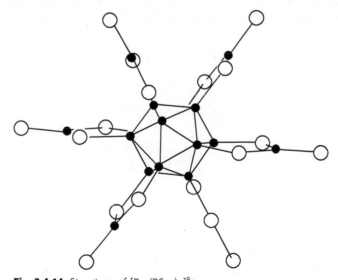

Fig. 3.4.14 Structure of $[B_{12}(BSe_3)_6]^{8-}$.

The sequential two-electron oxidation of $[B_{12}(OCH_2Ph)_{12}]^{2-}$ with Fe^{3+} in ethanol affords *hypercloso*-$B_{12}(OCH_2Ph)_{12}$ as a dark-orange crystal. An X-ray diffraction study of *hypercloso*-$B_{12}(OCH_2Ph)_{12}$ revealed that the molecule has distorted icosahedral geometry (D_{3d}). The term *hypercloso* refers to an n-vertex polyhedral structure having fewer skeletal electron pairs than ($n + 1$) as required by Wade's rule.

3.4.9 **Boranes and carboranes as ligands**

Many boranes and carboranes can act as very effective polyhapto ligands to form metallaboranes and metallacarboranes. Metallaboranes are borane cages containing one or more metal atoms in the skeletal framework. Metallacarboranes have both metal and carbon atoms in the cage skeleton. In contrast to the metallaboranes, syntheses of metallacarboranes via low- or room-temperature metal insertion into carborane anions in solution are more controllable, usually occurring at a well-defined C_2B_n open face to yield a single isomer.

The icosahedral carborane cage $C_2B_{10}H_{12}$ can be converted to the *nido*- $C_2B_9H_{11}^{2-}$, *nido*-$C_2B_{10}H_{12}^{2-}$, or *arachno*-$C_2B_{10}H_{12}^{4-}$ anions, as shown in Fig. 3.4.15. *Nido*-$C_2B_9H_{11}^{2-}$ has a planar five-membered C_2B_3 ring with delocalized π-orbitals, which is similar to the cyclopentadienyl ring. Likewise, *nido*-$C_2B_{10}H_{12}^{2-}$ has a nearly planar C_2B_4 six-membered ring and delocalized π orbitals, which is analogous to benzene. *Arachno*-$C_2B_{10}H_{12}^{4-}$ has a boat-like C_2B_5 bonding face in which the five B atoms are coplanar and the two C atoms lie *ca.* 60 pm above this plane.

The similarity between the C_2B_3 open face in *nido*-$C_2B_9H_{11}^{2-}$ and the $C_5H_5^-$ anion implies that the dicarbollide ion can act as an η^5-coordinated ligand to form sandwich complexes with metal atoms, which are analogous to the metallocenes. Likewise, the C_2B_4 open face in *nido*-$C_2B_{10}H_{12}^{2-}$ is analogous to benzene and it may be expected to function as an η^6-ligand.

A large number of metallacarboranes and polyhedral metallaboranes of *s*-, *p*-, *d*-, and *f*-block elements are known. In these compounds, the carboranes and boranes act as polyhapto ligands. The domain of metallaboranes and metallacarboranes has grown enormously and engenders a rich structural chemistry. Some new advances in the chemistry of metallacarboranes of *f*-block elements are described below.

The full-sandwich lanthanacarborane $[Na(THF)_2][(\eta^5\text{-}C_2B_9H_{11})_2La(THF)_2]$ has been prepared by direct salt metathesis between $Na_2[C_2B_9H_{11}]$ and $LaCl_3$ in THF. The structure of the anion is shown in Fig. 3.4.16(a). The average La-cage atom distances is 280.4 pm, and the ring centroid-La-ring centroid angle is 132.7°.

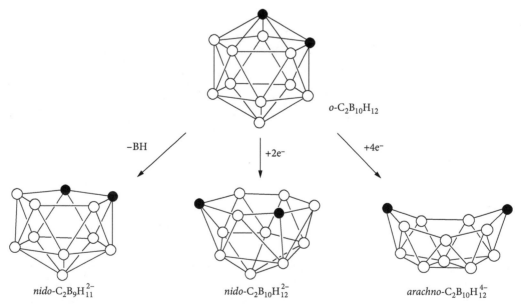

Fig. 3.4.15 Conversion of icosahedral carborane *o*-$C_2B_{10}H_{12}$ to *nido*-$C_2B_9H_{11}{}^{2-}$, *nido*-$C_2B_{10}H_{12}{}^{2-}$, and *arachno*-$C_2B_{10}H_{12}{}^{4-}$.

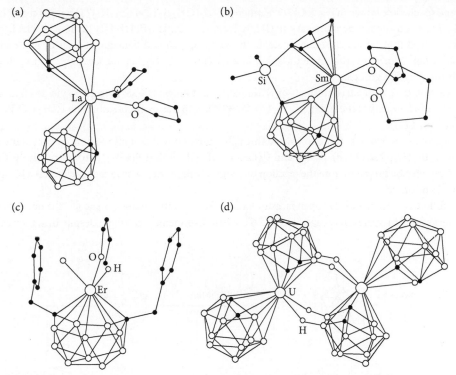

Fig. 3.4.16 Structure of some metallacarboranes of *f*-block elements: (a) $[\eta^5\text{-}C_2B_9H_{11}]_2La(THF)_2]^-$, (b) $[\eta^5{:}\eta^6\text{-}Me_2Si(C_5H_4)(C_2B_{10}H_{11})]Sm(THF)_2$, (c) $[\eta^7\text{-}C_2B_{10}H_{10}(CH_2C_6H_5)_2]Er(THF)^-$, (d) $[(\eta^7\text{-}C_2B_{10}H_{12})(\eta^6\text{-}C_2B_{10}H_{12})U]_2^{4-}$.

The structure of samaracarborane $[\eta^5{:}\eta^6\text{-}Me_2Si(C_5H_4)(C_2B_{10}H_{11})]Sm(THF)_2$ is shown in Fig. 3.4.16(b), in which the Sm^{3+} ion is surrounded by an η^5-cyclopentadienyl ring, η^6-hexagonal C_2B_4 face of the $C_2B_{10}H_{11}$ cage, and two THF molecules in a distorted tetrahedral manner. The ring-centroid-Sm-ring-centroid angle is 125.1° and the average Sm-cage atom and Sm-C(C_5 ring) distances are 280.3 pm and 270.6 pm, respectively.

Fig. 3.4.16(c) shows the coordination environment of the Er atom in the $\{\eta^7\text{-}[(C_6H_5CH_2)_2(C_2B_{10}H_{10})]Er(THF)\}_2\cdot\{Na(THF)_3\}_2\cdot2THF$. The Er atom is coordinated by $\eta^7\text{-}arachno\text{-}(C_6H_5CH_2)_2C_2B_{10}H_{10}]^{4-}$ and σ-bonded to two B–H groups from a neighboring $[arachno\text{-}(C_6H_5CH_2)_2C_2B_{10}H_{10}]^{4-}$ ligand and one THF molecule. The average Er–B(cage) distance is 266.5 pm, and that for the Er–C(cage) is 236.6 pm.

The compound $[\{(\eta^7\text{-}C_2B_{10}H_{12})(\eta^6\text{-}C_2B_{10}H_{12})U\}_2\{K_2(THF)_5\}]_2$ has been prepared from the reaction of $o\text{-}C_2B_{10}H_{12}$ with excess K metal in THF, followed by treatment with a THF suspension of UCl_4. In this structure each U^{4+} ion is bonded to $\eta^6\text{-}nido\text{-}C_2B_{10}H_{12}^{2-}$, $\eta^7\text{-}arachno\text{-}C_2B_{10}H_{12}^{4-}$, and coordinated by two B–H groups from the C_2B_5 bonding face of a neighboring $arachno\text{-}C_2B_{10}H_{12}^{4-}$ ligand, as shown in Fig. 3.4.16(d). This is the first example of an actinacarborane bearing a $\eta^6\text{-}C_2B_{10}H_{12}^{2-}$ ligand.

3.4.10 Carborane skeletons beyond the icosahedron

Recent synthetic studies have shown that carborane skeletons larger than the icosahedron can be constructed by the insertion of additional BII units. The reaction scheme and structure motifs of these carborane species are shown in Fig. 3.4.17. Treatment of $1,2\text{-}(CH_2)_3\text{-}1,2\text{-}C_2B_{10}H_{10}$ (**1**) with excess lithium metal in THF at room temperature gave $\{[(CH_2)_3\text{-}1,2\text{-}C_2B_{10}H_{10}][Li_4(THF)_5]\}_2$ (**2**) in 85% yield. X-ray analysis showed that, in the crystalline state, both the hexagonal and pentagonal faces of the "carbon-atoms-adjacent" *arachno*-carborane tetra-anion in **2** are capped by lithium ions, which may in principle be substituted by BH groups.

The reaction of **2** with 5.0 equivalents of $HBBr_2 \cdot SMe_2$ in toluene at $-78\,°C$ to $25\,°C$ yielded a mixture of a 13-vertex *closo*-carborane $(CH_2)_3C_2B_{11}H_{11}$ (**3**, 32%), a 14-vertex *closo*-carborane $(CH_2)_3C_2B_{12}H_{12}$ (**4**, 7%), and **1** (2%).

Closo-carboranes **3** and **4** can be converted through reduction reaction into their *nido*-carborane salts $\{[(CH_2)_3C_2B_{11}H_{11}][Na_2(THF)_4]\}_n$ (**5**) and $\{[(CH_2)_3C_2B_{12}H_{12}][Na_2(THF)_4]\}_n$ (**6**), respectively. Compound **4** can also be prepared from the reaction of **5** with $HBBr_2 \cdot SMe_2$, whose cage skeleton is a bicapped hexagonal antiprism.

Quite different from the planar open faces in *nido*-$C_2B_9H_{11}{}^{2-}$ and *nido*-$C_2B_{10}H_{12}{}^{2-}$, the open faces of the 13-vertex and 14-vertex *nido* cages in **5** and **6** are bent five-membered rings. Despite this, they can be

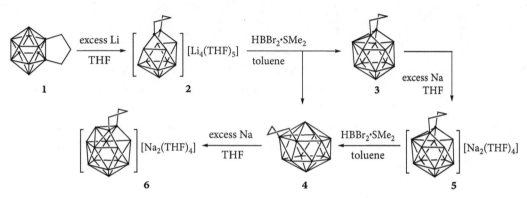

Fig. 3.4.17 Synthesis of "carbon-atoms-adjacent" 13-vertex and 14-vertex *closo*-carboranes and *nido*-carborane anions. (L. Deng, H.-S. Chan, and Z. Xie, *Angew. Chem. Int. Ed.* **44**, 2128–31 (2005).)

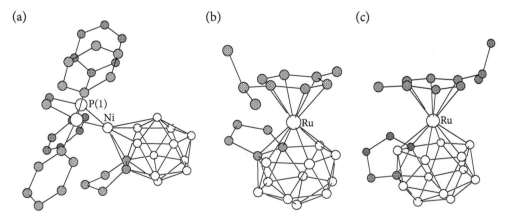

Fig. 3.4.18 Structure of some metallacarboranes with 14 and 15 vertices: (a) $[\eta^5\text{-}(CH_2)_3C_2B_{11}H_{11}]Ni(dppe)$; (b) $[\eta^6\text{-}(CH_2)_3C_2B_{11}H_{11}]Ru(p\text{-cymene})$; (c) $[\eta^6\text{-}(CH_2)_3C_2B_{12}H_{12}]Ru(p\text{-cymene})$. (L. Deng, H.-S. Chan, and Z. Xie, *J. Am. Chem. Soc.* **128**, 5219–30 (2006); L. Deng, J. Zhang, H.-S. Chan, and Z. Xie, *Angew. Chem. Int. Ed.* **45**, 4309–13 (2006).)

capped by metal atoms to give metallacarboranes with 14 and 15 vertices, respectively. Fig. 3.4.18 shows some examples of these novel metallacarboranes.

The complexes $[\eta^5\text{-}(CH_2)_3C_2B_{11}H_{11}]Ni(dppe)$ and $[\eta^6\text{-}(CH_2)_3C_2B_{11}H_{11}]Ru(p\text{-cymene})$ were prepared by the reactions of **5** with (dppe)NiCl$_2$ and $[(p\text{-cymene})RuCl_2]_2$, respectively. Although both of them are 14-vertex metallacarboranes having a similar bicapped hexagonal anti-prismatic cage geometry, the coordination modes of their carborane anions are quite different: the carborane anion $[(CH_2)_3C_2B_{11}H_{11}]^{2-}$ is η^5-bound to the Ni atom in the nickelacarborane (Fig. 3.4.18(a)), whereas an η^6-bonding fashion is observed in the ruthenacarborane with an average Ru-cage atom distance of 226.6 pm (Fig. 3.4.18(b)).

Fig. 3.4.18(c) shows the structure of a 15-vertex ruthenacarborane, $[(CH_2)_3C_2B_{12}H_{12}]Ru(p\text{-cymene})$, which was synthesized by the reaction of **6** with $[(p\text{-cymene})RuCl_2]_2$. Similar to that in the above-mentioned 14-vertex ruthenacarborane, the carborane moiety $[(CH_2)_3C_2B_{12}H_{12}]^{2-}$ in this 15-vertex ruthenacarborane is also η^6-bound to the Ru atom, forming a hexacosahedral structure. The distances of Ru-arene(cent) and Ru-CB$_5$(cent) are 178 pm and 141 pm, respectively. This 15-vertex ruthenacarborane has the largest vertex number in the metallacarborane family presently known.

3.5 **Boric Acid and Borates**

3.5.1 **Boric acid**

Boric acid, B(OH)$_3$, is the archetype and primary source of oxo-boron compounds. It is also the normal end product of hydrolysis of most boron compounds. It forms flaky white and transparent crystals, in which the BO$_3$ units are joined to form planar layers by O–H···O hydrogen bonds, as shown in Fig. 3.5.1.

Boric acid is a very weak monobasic acid (pK_a = 9.25). It generally behaves not as a Brønsted acid with the formation of the conjugate-base $[BO(OH)_2]^-$, but rather as a Lewis acid by accepting an electron pair from an OH$^-$ anion to form the tetrahedral species $[B(OH)_4]^-$.

(HO)$_3$B	+	:OH$^-$	$\longrightarrow$	B(OH)$_4^-$
$[sp^2+p(\text{unoccupied})]$				sp^3

Thus this reaction is analogous to the acceptor–donor interaction between BF$_3$ and NH$_3$.

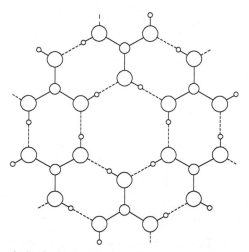

Fig. 3.5.1 Structure of layer of B(OH)$_3$ (B–O 136.1 pm, O–H···O 272 pm).

F$_3$B	+	:NH$_3$	$\longrightarrow$	F$_3$B	$\longleftarrow$ NH$_3$
[sp^2+p(unoccupied)]				sp^3	

In the above two reactions, the hybridization of the B atom changes from sp^2(reactant) to sp^3(product).

Partial dehydration of B(OH)$_3$ above 373 K yields metaboric acid, HBO$_2$, which consists of trimeric units B$_3$O$_3$(OH)$_3$. Orthorhombic metaboric acid is built of discrete molecules B$_3$O$_3$(OH)$_3$, which are linked into layers by O–H···O bonds, as shown in Fig. 3.5.2.

In dilute aqueous solution, there is an equilibrium between B(OH)$_3$ and B(OH)$_4^-$:

$$B(OH)_3 + 2H_2O \rightleftharpoons H_3O^+ + B(OH)_4^-$$

At concentrations above 0.1 M, secondary equilibria involving condensation reactions of the two dominant monomeric species give rise to oligomers such as the triborate monoanion $[B_3O_3(OH)_4]^-$, the triborate dianion $[B_3O_3(OH)_5]^{2-}$, the tetraborate $[B_4O_5(OH)_4]^{2-}$, and the pentaborate $[B_5O_6(OH)_4]^-$.

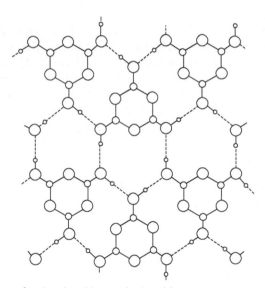

Fig. 3.5.2 Structure of layer of ortho-rhombic metaboric acid.

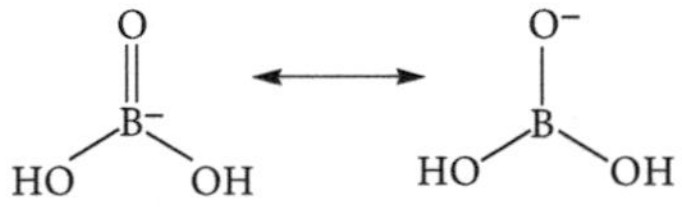

	I		II	
B–O	132.3(5) pm	O–B–OH	124.3(2)°
B–OH	139.5(3) pm	HO–B–OH	111.3(3)°

Fig. 3.5.3 Valence-bond structural formulas for the dihydrogen borate ion, [BO(OH)$_2$]$^-$, and its measured dimensions in [(CH$_3$)$_4$N]$^+$[BO(OH)$_2$]$^-$·2(NH$_2$)$_2$CO·H$_2$O.

In $(Et_4N)_2[BO(OH)_2]_2 \cdot B(OH)_3 \cdot 5H_2O$, the conjugate acid-base pair $B(OH)_3$ and dihydrogen borate $[BO(OH)_2]^-$ coexist in the crystalline state. In the crystal structure of the inclusion compound $Me_4N^+[BO(OH)_2]^- \cdot 2(NH_2)_2CO \cdot H_2O$, the $BO(OH)_2^-$, $(NH_2)_2CO$ and H_2O molecules are linked by O–H···O bonds to form a host lattice featuring a system of parallel channels, which accommodate the guest Me_4N^+ cations (see Section 10.4.4 for further details). The measured dimensions of the $[BO(OH)_2]^-$ anion, which has crystallographically imposed symmetry m, show that the valence tautomeric form **I** with a formal B=O double bond makes a makes a more important contribution than **II** to its electronic structure (Fig. 3.5.3).

3.5.2 **Structure of borates**

(1) **Structural units in borates**

The many structural components that occur in borates consist of planar triangular BO_3 units and/or tetrahedral BO_4 units with vertex-sharing linkage, and may be divided into three kinds. Some examples are given below (Fig. 3.5.4):

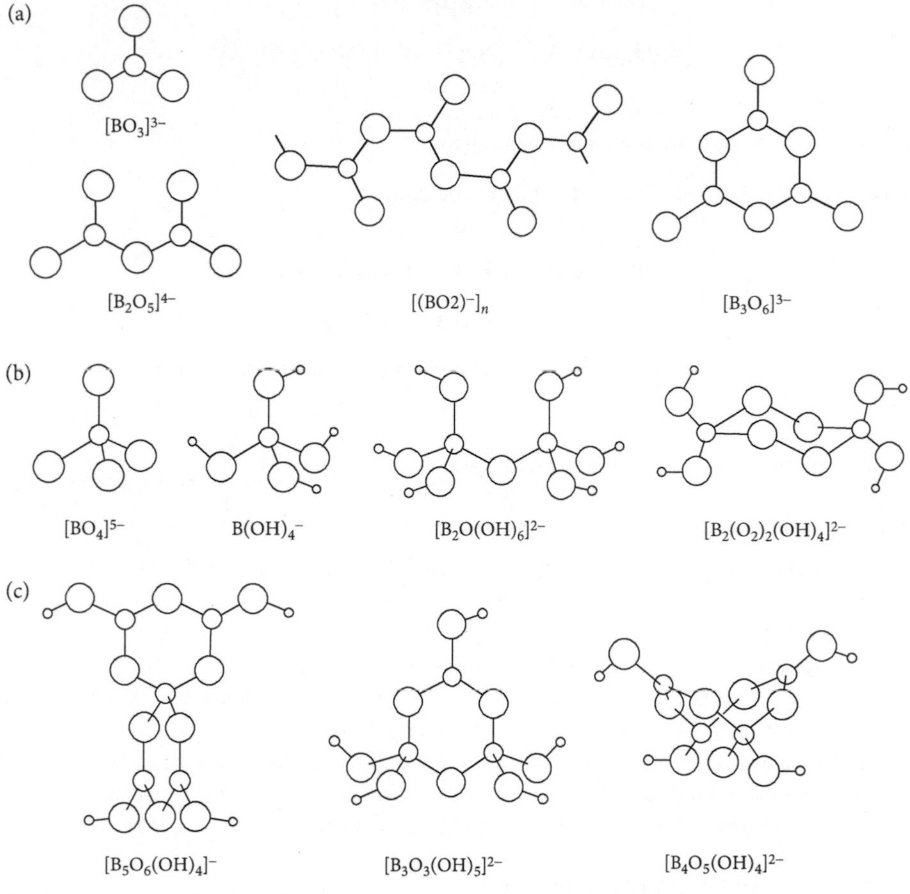

Fig. 3.5.4 Structural units in borates.

(a)   Units containing B in planar BO_3 coordination only:

$$BO_3^{3-} : Mg_3(BO_3)_2, LaBO_3$$

$$[B_2O_5]^{4-} : Mg_2B_2O_5, Fe_2B_2O_5$$

$$[B_3O_6]^{3-} : K_3B_3O_6, Ba_3(B_3O_6)_2 \equiv BaB_2O_4 \text{ (BBO)}$$

$$[(BO_2)^-]_n : Ca(BO_2)_2$$

The structures of these units are shown in Fig. 3.5.4(a).

(b)   Units containing B in tetrahedral BO_4 coordination only:

$$BO_4^{5-} : TaBO_4$$

$$B(OH)_4^- : Na_2[B(OH)_4Cl]$$

$$[B_2O(OH)_6]^{2-} : Mg[B_2O(OH)_6]$$

$$[B_2(O_2)_2(OH)_4]^{2-} : Na_2[B_2(O_2)_2(OH)_4]\cdot 6H_2O$$

The structures of these units are shown in Fig. 3.5.4(b).

(c)   Units containing B in both BO_3 and BO_4 coordination:

$$[B_5O_6(OH)_4]^- : K[B_5O_6(OH)_4]\cdot 2H_2O$$

$$[B_3O_3(OH)_5]^{2-} : Ca[B_3O_3(OH)_5]\cdot H_2O$$

$$[B_4O_5(OH)_4]^{2-} : Na_2[B_4O_5(OH)_4]\cdot 8H_2O$$

The structures of these units are shown in Fig. 3.5.4(c).

(2) Structure of borax

The main source of boron comes from borax, which is used in the manufacture of borosilicate glass, borate fertilizers, borate-based detergents and flame-retardants. The crystal structure of borax was determined by X-ray diffraction in 1956, and by neutron diffraction in 1978. It consists of $B_4O_5(OH)_4^{2-}$, Na^+ ions and water molecules, so that its chemical formula is $Na_2[B_4O_5(OH)_4]\cdot 8H_2O$, not $Na_2B_4O_7\cdot 10H_2O$ as given in many older books. The structure of $[B_4O_5(OH)_4]^{2-}$ is shown in Fig. 3.5.5(a) and (b). In the crystal, the anions are linked by O–H$\cdots$O hydrogen bonds to form an infinite chain, as shown in Fig. 3.5.5(c). The sodium ions are octahedrally coordinated by water molecules, and the octahedra share edges to form another chain. These two kinds of chains are linked by weaker hydrogen bonds between the OH groups of $[B_4O_5(OH)_4]^{2-}$ anions and the aqua ligands, and this accounts for the softness of borax.

(3) Structural principles for borates

Several general principles have been formulated in connection with the structural chemistry of borates:

(a) In borates, boron combines with oxygen either in trigonal planar or tetrahedral coordination. In addition to the mononuclear anions BO_3^{3-} and BO_4^{5-} and their partially or fully protonated forms, there

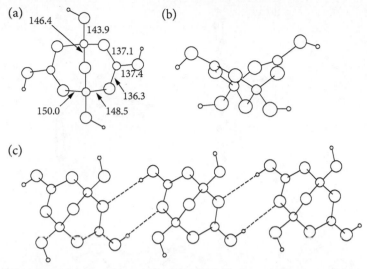

Fig. 3.5.5 Structure of borax: (a) bond lengths in $[B_4O_5(OH)_4]^{2-}$ (in pm), (b) perspective view of the anion, and (c) $[B_4O_5(OH)_4^{2-}]_n$ chain.

exists an extensive series of polynuclear anions formed by corner-sharing of BO_3 and BO_4 units. The polyanions may form separate boron-oxygen complexes, rings, chains, layers and frameworks.

(b) In boron-oxygen polynuclear anions, most of the oxygen atoms are linked to one or two boron atoms, and a few can bridge three boron atoms.

(c) In borates, the hydrogen atoms do not attach to boron atoms directly, but always to oxygens which are not shared by two borons, thereby generating OH groups.

(d) Most boron-oxygen rings are six-membered comprising three B atoms and three oxygen atoms.

(e) In a borate crystal there may be two or more different boron-oxygen units.

3.6 **Organometallic Compounds of Group 13 Elements**

3.6.1 **Compounds with bridged structure**

The compounds $Al_2(CH_3)_6$, $(CH_3)_2Al(\mu\text{-}C_6H_5)_2Al(CH_3)_2$ and $(C_6H_5)_2Al(\mu\text{-}C_6H_5)_2Al(C_6H_5)_2$ all have bridged structures, as shown in Fig. 3.6.1(a), (b), and (c) respectively. The bond distances and angles are listed in Table 3.6.1.

The bonding in the Al–C–Al bridges involves a 3c-2e molecular orbital formed from essentially sp^3 orbitals of the C and Al atoms. This type of 3c-2e interaction in the bridge necessarily results in an acute Al–C–Al angle (75° to 78°), which is consistent with the experimental data. This orientation is sterically favored and places each *ipso*-carbon atom in an approximately tetrahedral configuration.

In the structure of $Me_2Ga(\mu\text{-}C{\equiv}C\text{-}Ph)_2GaMe_2$ (Fig. 3.3.1(d)), each alkynyl bridge leans over toward one of the Ga centers. The organic ligand forms an Ga–C σ-bond and interacts with the second Ga center by using its C≡C π bond. Thus each alkynyl group is able to provide three electrons for bridge bonding, in contrast to one electron as normally supplied by an alkyl or aryl group.

Compounds $In(Me)_3$ and $Tl(Me)_3$ are monomeric in solution and the gas phase. In the solid state, they also exist as monomers essentially, but close intermolecular contacts become important. In crystalline

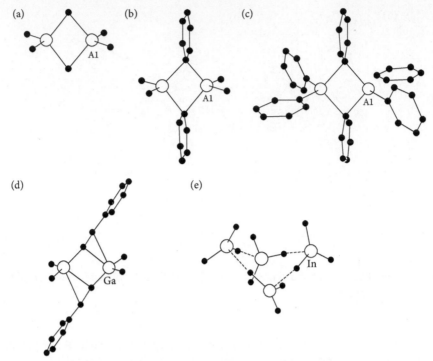

Fig. 3.6.1 Structure of some Group 13 organometallic compounds: (a) $Al_2(Me)_6$, (b) $Al_2(Me)_4(Ph)_2$, (c) $Al_2(Ph)_6$, (d) $Ga_2Me_4(C_2Ph)_2$, and (e) $In(Me)_3$ tetramer.

Table 3.6.1 Structural parameters found in some carbon-bridged organoaluminum compounds

Compound	Distance/pm				Angle/degrees	
	$Al–C_b$	$Al–C_t$	$Al\cdots Al$	$Al–C–Al$	$C–Al–C$ (internal)	$C–Al–C$ (external)
Al_2Me_6	214	197	260	74.7	105.3	123.1
$Al_2Me_4Ph_2$	214	197	269	77.5	101.3	115.4
Al_2Ph_6	218	196	270	76.5	103.5	121.5

$In(Me)_3$, significant $In\cdots C$ intermolecular interactions are observed, suggesting that the structure can be described in terms of cyclic tetramers, as shown in Fig. 3.6.1(e), with interatomic distances of In–C 218 pm and $In\cdots C$ 308 pm The corresponding bond distances in isostructural $Tl(Me)_3$ are Tl–C = 230 pm and $Tl\cdots C$ = 316 pm

3.6.2 Compounds with π bonding

Many organometallic compounds of Group 13 are formed by π bonds. The π ligands commonly used are olefins, cyclopentadiene, or their derivatives.

The structure of the dimer $[ClAl(Me–C=C–Me)AlCl]_2$ is of interest. There are 3c-2e π bonds between the Al atoms and the C=C bonds, as shown in Fig. 3.6.2(a). The Al–C distances in the Al-olefin units are

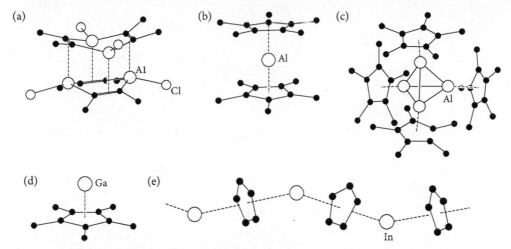

Fig. 3.6.2 Structure of some organometallic compounds formed by π bonds:
(a) [ClAl(Me–C≡C–Me)AlCl]$_2$, (b) [Al(Cp*)$_2$]$^+$, (c) Al$_4$Cp*$_4$, (d) GaCp*, and (e) InCp (or TlCp).

relatively long, *ca.* 235 pm, but the sum of the four interactions lead to remarkable stability of the dimer. The energy gained through each Al-olefin interaction is calculated to be about 25 to 40 kJ mol^{-1}.

The [Al(η^5-Cp*)$_2$]$^+$ (Cp* = C$_5$Me$_5$) ion is at present the smallest sandwich metallocene of any main-group element. The Al–C bond length is 215.5 pm Fig. 3.6.2(b) shows the structure of this ion.

The molecular structure of Al$_4$Cp*$_4$ in the crystal is shown in Fig. 3.6.2(c). The central Al$_4$ tetrahedron has Al–Al bond length 276.9 pm, which is shorter than those in Al metal (286 pm). Each Cp* ring is η^5-coordinated to an Al atom, whereby the planes of the Cp* rings are parallel to the opposite base of the tetrahedron. The average Al–C distances is 233.4 pm.

The half-sandwich structure of GaCp* in the gas phase, as determined by electron diffraction, is shown in Fig. 3.6.2(d). This compound has a pentagonal pyramidal structure with an η^5-C$_5$ ring. The Ga–C distance is 240.5 pm.

Both CpIn and CpTl are monomeric in the gas phase, but in the solid they possess a polymeric zigzag chain structure, in which the In (or Tl) atoms and Cp rings alternate, as shown in Fig. 3.6.2(e).

3.6.3 Compounds containing M–M bonds

The compounds R$_2$Al–AlR$_2$, R$_2$Ga–GaR$_2$, and R$_2$In–InR$_2$ [R=CH(SiMe$_3$)$_2$] have been prepared and characterized. The bond lengths in these compounds are: Al–Al 266.0 pm, Ga–Ga 254.1 pm, and In–In 282.8 pm; the M$_2$C$_4$ frameworks are planar as shown in Fig. 3.6.3(a). In Al$_2$(C$_6$H$_2$^{i}Pr$_3$)$_4$, Ga$_2$(C$_6$H$_2$^{i}Pr$_3$)$_4$, and In$_2$[C$_6$H$_2$(CF$_3$)$_3$]$_4$, the bond lengths are: Al–Al 265 pm, Ga–Ga 251.5 pm, and In–In 274.4 pm, and here the M$_2$C$_4$ framework are non-planar, as shown in Fig. 3.6.3(b). In these structures the Ga–Ga bond lengths are shorter than the Al–Al bond lengths. The anomalous behavior of gallium among Group 13 elements is due to the insertion of electrons into the *d* orbitals of the preceding 3*d* elements in the Periodic Table and the associated contraction of the atomic radius.

The structures of compounds Na$_2$[Ga$_3$(C$_6$H$_3$Mes$_2$)$_3$] and K$_2$[Ga$_3$(C$_6$H$_3$Mes$_2$)$_3$] are worthy of note. In these compounds, the Na$_2$Ga$_3$ and K$_2$Ga$_3$ cores have trigonal bipyramidal geometry, as shown in Fig. 3.6.3(c). The bond lengths are: Ga–Ga 244.1 pm (Na salt), 242.6 pm (K salt), Ga–Na 322.9 pm, and Ga–K 355.4 pm The planarity of the Ga$_3$ ring and the very short Ga–Ga bonds implies electron delocalization in the three-membered ring, the requisite two π electrons being provided by the two K (or Na) atoms (one electron each) to the empty p_z orbitals of the three sp^2-hybridized Ga atoms.

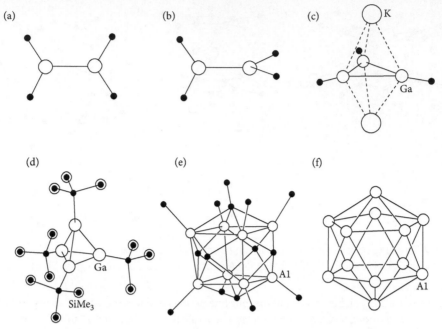

Fig. 3.6.3 Structures of some compounds containing M–M bonds: (a) planar M_2R_4 (M = Al, Ga, In), (b) non-planar M_2R_4 (M = Al, Ga, In), (c) K_2Ga_3 core of $K_2[Ga_3(C_6H_3Mes_2)_3]$, (d) $[GaC(SiMe_3)_3]_4$, (e) $(AlMe)_8(CCH_2Ph)_5(C{\equiv}C-Ph)$, and (f) icosahedral Al_{12} core in $K_2[Al_{12}{}^iBu_{12}]$.

Fig. 3.6.3(d)–(f) show the structures of three compounds which contain metallic polyhedral cores. In $[GaC(SiMe_3)_3]_4$, the Ga–Ga bond length is 268.8 pm, in $(AlMe)_8(CCH_2Ph)_5(C{\equiv}C-Ph)$ the Al–Al bond lengths are observed to be close to two average values: 260.9 and 282.9 pm, and in $K_2[Al_{12}{}^iBu_{12}]$ the Al–Al bond length is 268.5 pm.

$Al_{50}Cp^*{}_{12}$ (or $Al_{50}C_{120}H_{180}$) is a giant molecule, whose crystal structure shows a distorted square-antiprismatic Al_8 moiety at its center, as shown in Fig. 3.6.4. This Al_8 core is surrounded by 30 Al atoms that form an icosidodecahedron with 12 pentagonal faces and 20 trigonal faces. Each pentagonal face is capped by an $AlCp^*$ unit, and the set of 12 Al atoms form a very regular icosahedron, in which the $Al{\cdots}Al$ average distance is 570.2 pm. Each of the peripheral 12 Al atoms is coordinated by 10 atoms (five Al and five C) in the form of a staggered "mixed sandwich." In the molecule, the average Al–Al bond length is 277.0 pm (257.8–287.7 pm). The 60 CH_3 groups of the 12 Cp^* ligands at the surface exhibits a topology resembling that of the carbon atoms in fullerene-C_{60}. The average distance between a pair of nearest methyl groups of neighboring Cp^* ligands (386 pm) is approximately twice the van der Waals radius of a methyl group (195 pm). The entire $Al_{50}Cp^*{}_{12}$ molecule has a volume which is about five times large than that of a C_{60} molecule.

3.6.4 **Linear catenation in heavier Group 13 elements**

For the heavier congeners of boron, the occurrence of well-characterized compounds possessing a linear extended skeleton containing two or more unsupported two-electron E–E bonds are quite rare. Some discrete molecules exhibiting this feature include the open-chain trigallium subhalide complex $I_2(PEt_3)Ga-GaI(PEt_3)-GaI_2(PEt_3)$ (Fig. 3.6.5(a)) and the trigonal tetranuclear indium complex $\{In[In(2,4,6-{}^iPr_3C_6H_2)_2]_3\}$ (Fig. 3.6.5(b)).

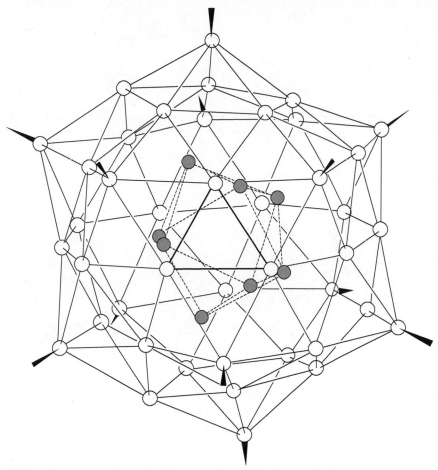

Fig. 3.6.4 Molecular structure of $Al_{50}Cp^*_{12}$ (Cp^* = pentamethylcyclopentadienide). For clarity, the Cp^* units are omitted, and only the bonds from the outer 12 Al atoms toward the ring centers are shown. The shaded atoms represent the central distorted square-antiprismatic Al_8 moiety.

Using a chelating ligand of the β-diketiminate class, a novel linear homocatenated hexanuclear indium compound has been synthesized (Fig. 3.6.5(c)). The four internal indium atoms are in the +1 oxidation state, and the terminal indium atoms, each carrying an iodo ligand, are both divalent. The coordination geometry at each indium center is distorted tetrahedral. As shown in Fig. 3.6.6, the mixed-valent molecule has a pseudo C_2 axis with a β-diketiminate ligand chelated to each metal center, and its zigzag backbone is held together by five unsupported In–In single bonds constructed from sp^3 hybrid orbitals.

3.7 **Structure of Naked Anionic Metalloid Clusters**

The term metalloid cluster is used to describe a multi-nuclear molecular species in which the metal atoms exhibit closest packing (and hence delocalized intermetallic interactions) like that in bulk metal, and the metal-metal contacts outnumber the peripheral meta-ligand contacts. Most examples are found in the field of precious-metal cluster chemistry. In recent years, an increasing number of cluster species of Group

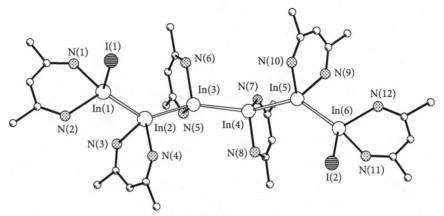

Fig. 3.6.5 Structural formulas of some open-chain homocatenated compounds of heavier Group 13 elements.

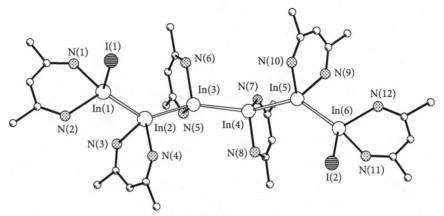

Fig. 3.6.6 Molecular structure of a linear homocatenated compound containing six indium centers. The 3,5-dimethylphenyl groups of the β-diketiminate ligands have been omitted for clarity. Bond lengths (pm): In(1)–In(2) 281.2, In(2)–In(3) 283.5, In(3)–In(4) 285.4, In(4)–In(5) 284.1, In(5)–In(6) 282.2; In(1)–I(1) 279.8, In(6)–I(2) 278.0; standard deviation 0.1 pm Average In–In–In bond angle at In(2) to In(5) is 139.4°.

13 elements have been synthesized with cores consisting of Al_n (n = 7, 12, 14, 50, 69) and Ga_m (m = 9, 10, 19, 22, 24, 26, 84) atoms, for example $[Al_{77}\{N(SiMe_3)_2\}_{20}]^{2-}$ and $[Ga_{84}\{N(SiMe_3)_2\}_{20}]^{4-}$.

3.7.1 **Structure of $Ga_{84}[N(SiMe_3)_2]_{20}^{4-}$**

The largest metalloid cluster characterized to date is $Ga_{84}[N(SiMe_3)_2]_{20}^{4-}$, the structure of which is illustrated in Fig. 3.7.1. It comprises four parts: a Ga_2 unit, a Ga_{32} shell, a Ga_{30} "belt," and 20 $Ga[N(SiMe_3)_2]$ groups. The Ga_2 unit (as shown in (a)) is located at the center of the 64 naked Ga atoms; the Ga–Ga bond distance is 235 pm, which is almost as short as the "normal" triple Ga–Ga bond (232 pm). The Ga_2 unit is encapsulated by a Ga_{32} shell in the form of a football with icosahedral caps, as shown in (b). The $Ga_2@Ga_{32}$ aggregate is surrounded by a belt of 30 Ga atoms that are also naked, as shown in (c). Finally the entire Ga_{64} framework is protected by 20 $Ga[N(SiMe_3)_2]$ groups to form $Ga_{84}[N(SiMe_3)_2]_{20}^{4-}$, as shown in (d). This large anionic cluster has a diameter of nearly 2 nm.

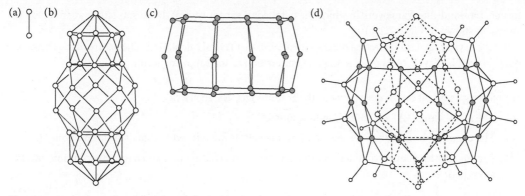

Fig. 3.7.1 Structure of $Ga_{84}[N(SiMe_3)_2]_{20}{}^{4-}$ (the larger circles (shaded and unshaded) represent Ga atoms, and the small circles represent N atoms): (a) central Ga_2 unit, (b) Ga_{32} shell, (c) a "belt" of 30 Ga atoms, and (d) the front of $Ga_{84}[N(SiMe_3)_2]_{20}{}^{4-}$. The Ga_2 and N atoms at the back are not shown, the Ga_{32} shell is emphasized by the broken lines, the "belt" of 30 Ga atoms are shaded, and for the ligands $N(SiMe_3)_2$ only the N atoms directly bonded to the Ga atoms are shown.

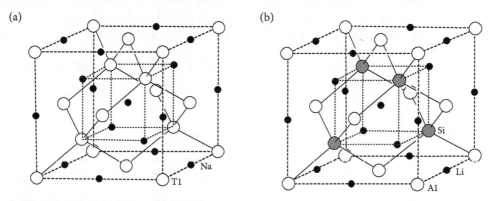

Fig. 3.7.2 Structure of (a) NaTl and (b) LiAlSi.

3.7.2 **Structure of NaTl**

The structure of sodium thallide NaTl can be understood as a diamond-like framework of Tl atoms, whose vacant sites are completely filled with Na atoms. Fig. 3.7.2(a) shows the structure of NaTl, in which the Tl–Tl covalent bonds are represented by solid lines. The Tl atom has three valence electrons, which are insufficient for the construction of a stable diamond framework. The deficit can be partially compensated by the introduction of Na atoms. The effective radius of the Na atom is considerably smaller than that in pure metallic sodium. Therefore, the chemical bonding in NaTl is expected to be a mixture of covalent, ionic, and metallic interactions.

The NaTl type structure is the prototype for Zintl phases, which are intermetallic compounds which crystallize in typical "non-metal" crystal structures. Binary AB compounds LiAl, LiGa, LiIn and NaIn are both isoelectronic (iso-valent) and isostructural with NaTl. In the LiAlSi ternary compound, Al and Si form a diamond-like framework in which the octahedral vacant sites of the Al sub-lattice are filled by Li atoms, as shown in Fig. 3.7.2(b).

From the crystal structure, physical measurements and theoretical calculations, the nature of the chemical bond in the NaTl-type compound AB can be understood in the following terms:

(a) Strong covalent bonds exist between the B atoms (Al, Ga, In, Tl, Si).

(b) The alkali atoms (A) are not in bonding contact.

(c) The chemical bond between the A and B framework is metallic with a small ionic component.

(d) For the upper valence/conduction electron states a partial metal-like charge distribution can be identified.

3.7.3 Naked $Tl_n{}^{m-}$ anion clusters

The heavier elements of Group 13, in particular thallium, are able to form discrete naked clusters with alkali metals. Table 3.7.1 lists some examples of $Tl_n{}^{m-}$ naked anion clusters, and Fig. 3.7.3 shows their structures.

The bonding in these $Tl_n{}^{m-}$ anion clusters is similar to that in boranes. For example, the bond valences (b) of $Tl_5{}^{7-}$ and centered $Tl_{13}{}^{11-}$ clusters are equal to those of $B_5H_5{}^{2-}$ and $B_{12}H_{12}{}^{2-}$, respectively. Note that, in $Tl_{13}{}^{11-}$, the central Tl atom contributes all three valence electrons to cluster bonding, so that the total number of bonding electrons is $(3 + 12 \times 1 + 11) = 26$. This cluster is thus consolidated by ten Tl–Tl–Tl 3c-2e and three Tl–Tl 2c-2e bonds. The $Tl_9{}^{9-}$ cluster has 36 valence electrons, and its bond valence

Table 3.7.1 Some examples of $Tl_n{}^{m-}$ anion clusters

Composition	Anion	Structure in Fig. 3.7.3	Cluster symmetry	Bond valence (b)	Bonding
Na_2Tl	$Tl_4{}^{8-}$	(a)	T_d	6	6 Tl–Tl 2c-2e bonds
$Na_2K_{21}Tl_{19}$	$2Tl_5{}^{7-}$	(b)	D_{3h}	9	3 Tl–Tl–Tl 3c-2e bonds 3 Tl–Tl 2c-2e bonds
	$Tl_9{}^{9-}$	(e)	defective I_h	18	3 Tl–Tl–Tl 3c-2e bonds 12 Tl–Tl 2c-2e bonds
KTl	$Tl_6{}^{6-}$	(c)	D_{4h}	12	12 Tl–Tl 2c-2e bonds
$K_{10}Tl_7$	$Tl_7{}^{7-}$, $3e^-$	(d)	$\sim D_{5h}$	14	5 Tl–Tl–Tl 3c-2e bonds 4 Tl–Tl 2c-2e bonds
K_8Tl_{11}	$Tl_{11}{}^{7-}$, e^-	(f)	$\sim D_{3h}$	24	3 Tl–Tl–Tl 3c-2e bonds 18 Tl–Tl 2c-2e bonds
$Na_3K_8Tl_{13}$	$Tl_{13}{}^{11-}$	(g)	Centered $\sim I_h$	23	10 Tl–Tl–Tl 3c-2e bonds 3 Tl–Tl 2c-2e bonds

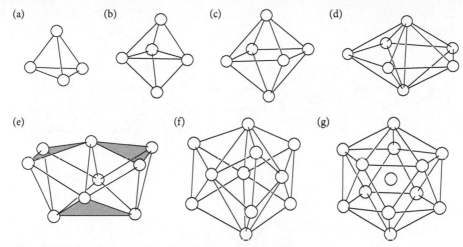

Fig. 3.7.3 Structures of some Tl_n^{m-} anion clusters: (a) Tl_4^{8-}, (b) Tl_5^{7-}, (c) Tl_6^{6-}, (d) Tl_7^{7-}, (e) Tl_9^{9-} (shaded faces represent three Tl–Tl–Tl 3c-2e bonds), (f) Tl_{11}^{7-}, (g) Tl_{13}^{11-}.

b is 18. The cluster is stabilized by three Tl–Tl–Tl 3c-2e bonds, as labeled by the three shaded faces in Fig. 3.7.3(e), and the remaining 12 edges represent 12 Tl–Tl 2c-2e bonds.

The compound $K_{10}Tl_7$ is composed of 10 K^+, a Tl_7^{7-} and three delocalized electrons per formula unit and exhibits metallic properties. The Tl_7^{7-} cluster is an axially compressed pentagonal bipyramid conforming closely to D_{5h} symmetry (Fig. 3.7.3(d)). The apex–apex bond distance of 346.2 pm is slightly longer than the bonds in the pentagonal waist (318.3–324.7 pm). Comparison between the structures of Tl_7^{7-} and $B_7H_7^{2-}$ (Fig. 3.7.6(a)) shows that both are pentagonal bipyramidal but Tl_7^{7-} is compressed along the C_5 axis for the formation of a coaxial 2c-2e Tl–Tl bond, so the bond valences of Tl_7^{7-} and $B_7H_7^{2-}$ are 14 and 13, respectively.

3.8 Ligand "Brackets" for Ga–Ga Bond

The reaction of digallane (dpp-Bian)Ga–Ga(dpp-Bian) (**1**) (dpp-Bian = 1,2-bis[(2,6-diisopropylphenyl)imino]-acenaphthene) with acenaphthenequinone (AcQ) in 1:1 molar ratio proceeds via two-electron reduction of AcQ to give (dpp-Bian)Ga(μ_2-AcQ)Ga(dpp-Bian) (**2**), in which diolate $[AcQ]^{2-}$ acts as a "bracket" for the Ga–Ga bond. The interaction of **1** with AcQ in 1:2 molar ratio involves oxidation of both dpp-Bian ligands, as well as of the Ga–Ga bond, to form (dpp-Bian)Ga(μ_2-AcQ)$_2$Ga(dpp-Bian) (**3**). At 330 K in toluene complex **2** decomposes to give **3** and **1**. The reaction of **2** with atmospheric oxygen results in oxidation of the Ga–Ga bond to afford (dpp-Bian)Ga(μ_2-AcQ)(μ_2-O)Ga(dpp-Bian) (**4**).

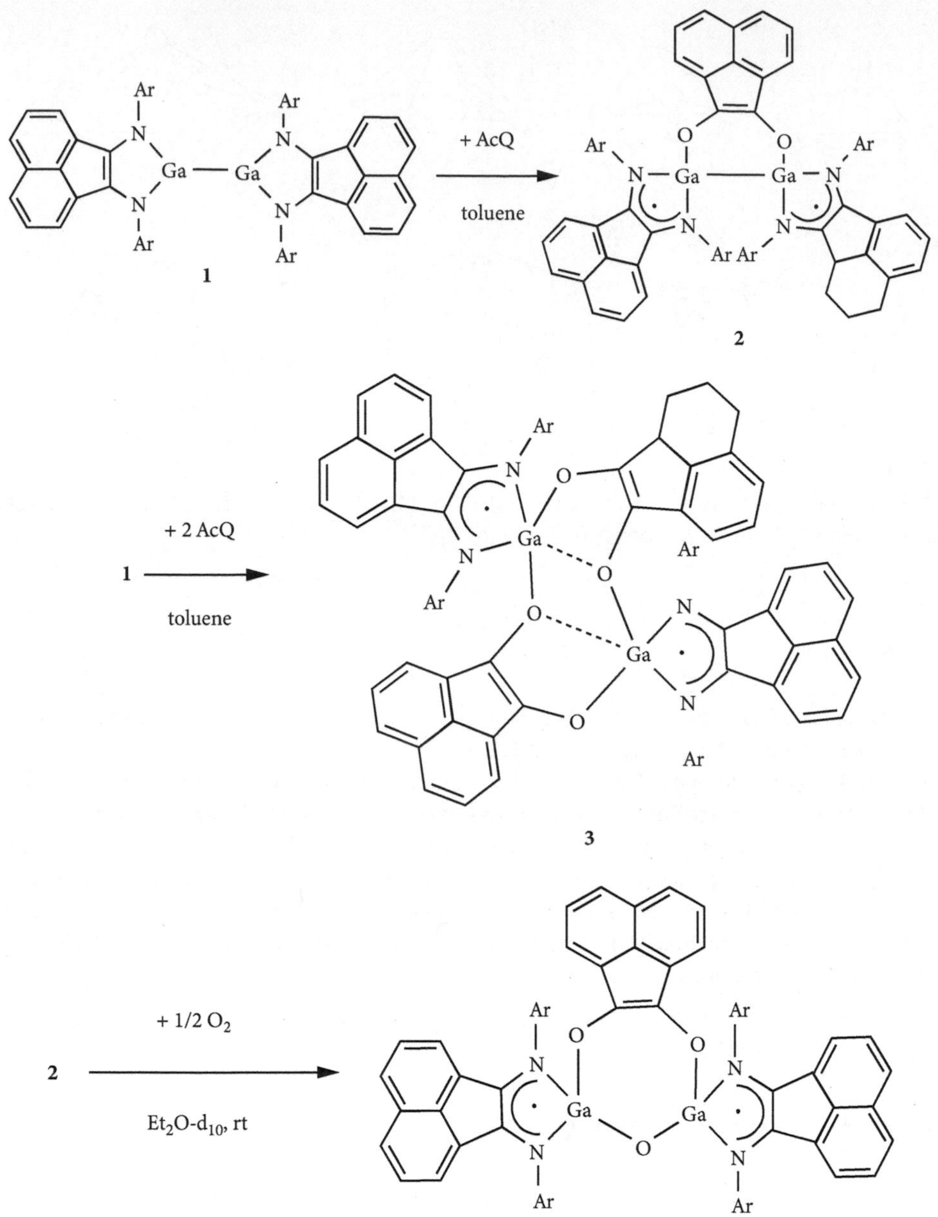

The reaction of digallane **1** with SO_2 produces, depending on the ratio (1:2 or 1:4), dithionites (dpp-Bian)Ga(μ_2-O$_2$S–SO$_2$)Ga(dpp-Bian) (**5**) and (dpp-Bian)Ga(μ_2-O$_2$S–SO$_2$)$_2$Ga(dpp-Bian) (**6**). In compound **5** the Ga–Ga bond is preserved and supported by a dithionite dianionic bracket. In compound **6** the gallium centers are bridged by two dithionite ligands. Both **5** and **6** contain dpp-Bian radical anionic ligands. Four-electron reduction of azobenzene with 1 mol equiv of digallane **1** gives complex (dpp-Bian)Ga(μ_2-NPh)$_2$Ga(dpp-Bian) (**7**).

The molecular structures of compounds **2–7** determined by X-ray crystallography are shown in Fig. 3.8.1. In compound **2** the diolate [AcQ]$^{2-}$ ligand acts as a "bracket" for the Ga–Ga bond. In com-

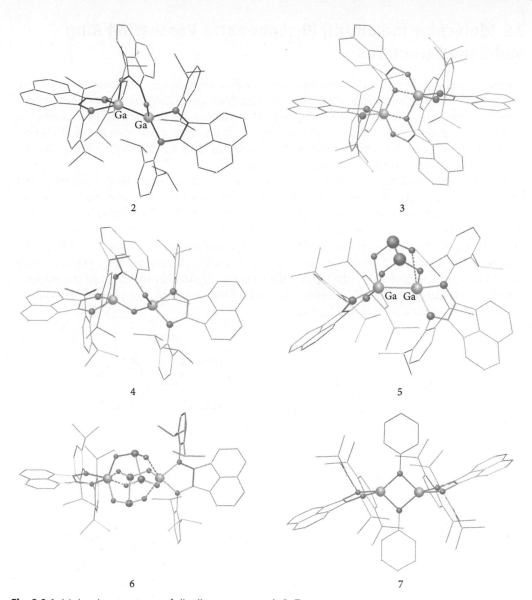

Fig. 3.8.1 Molecular structures of digallane compounds **2**–**7**.

pound **5** the Ga–Ga bond is preserved and supported by a dithionite dianionic bracket. The Ga–Ga bond lengths in **2** and **5** are notably different: 251.57(7) pm versus 246.59(5) pm, respectively.

Reference

I. Fedushkin, A. A. Skatova, V. A. Dodonov, X.-J. Yang, V. A. Chudakova, A. V. Piskunov, S. Demeshko, and E. V. Baranov, Ligand "brackets" for Ga–Ga bond. *Inorg. Chem.* **55**, 9047–56 (2016).

3.9 **Molecular Indium(III) Phosphonates Possessing Ring and Cage Structures**

The reactions of organotin oxides/hydroxides/oxide-hydroxides with phosphonic/phosphinic acids have yielded a rich variety of novel compounds that exhibit much structural diversity. In the case of indium, some phosphates, phosphites, and phosphate-phosphites are known, but structural characterization was mostly done by powder X-ray diffraction. The first single crystal X-ray structure of an indium(III) phosphonate that possesses an extended structure was reported in 2012, and the first report on discrete molecular indium(III) phosphonates appeared four years later.

Indium(III) phosphonates, $[In_2(^tBuPO_3H)_4(phen)_2Cl_2]$ (**1**) with phen = 1,10-phenanthroline and $[In_3(C_5H_9PO_3)_2(C_5H_9PO_3H)_4(phen)_3]\cdot NO_3\cdot 3.5H_2O$ (**2**), were synthesized by solvothermal reactions involving indium(III) salts and organophosphonic acids (Fig. 3.9.1). The X-ray crystal structures of **1** and **2** are shown in Fig. 3.9.2. Compound **1** is a dinuclear compound with the indium centers bridged by a pair of isobidentate phosphonate ligands, $[^tBuP(O)_2OH]^-$, resulting in an eight-membered $(In_2P_2O_4)$ puckered ring. Compound **2** is trinuclear, in which the In_3 platform is held together by two bicapping tripodal phosphonate ligands from the top and bottom of the In_3 plane. In addition, two monoanionic phosphonate bridging ligands each serves to bind a pair of indium centers. Both **1** and **2** also contain monoanionic unidentate phosphonate ligands.

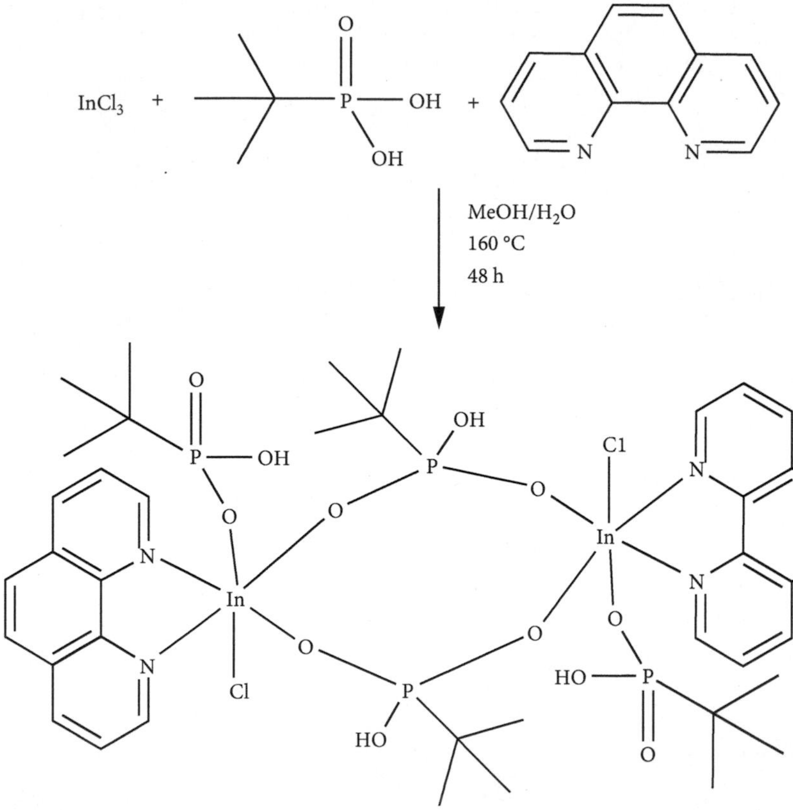

Fig. 3.9.1 Synthetic scheme for molecular indium(III) phosphonates **1** and **2**.

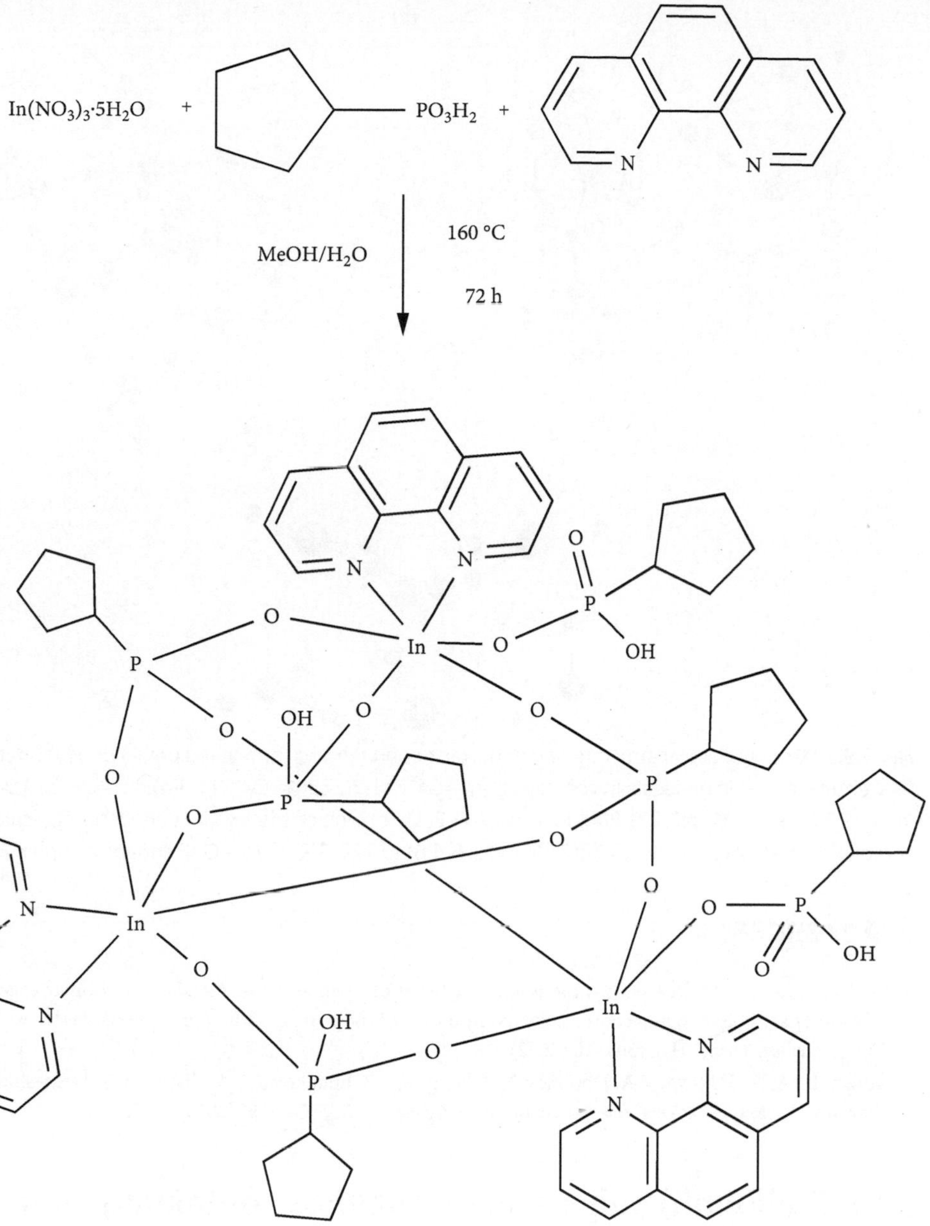

Fig. 3.9.1 (Continued)

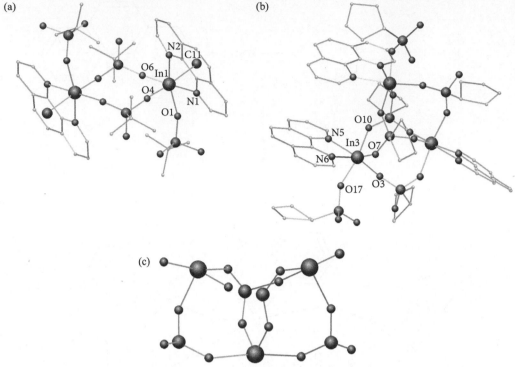

Fig. 3.9.2 X-ray molecular structure and important bond lengths (pm) in indium(III) phosphonates. (a) Centrosymmetric molecule in complex **1**: In1–O1 211.7(3), In1–O4 211.3(4), In1–O6 209.9(3), In1–N1 229.8(4), In1–N2 231.2(4), In1–Cl1 243.2(1). (b) One of the two independent molecules in complex **2**: In–O 213.4(7) to 207.6(7), In–N 226.4(8) to 227.9(9). (c) In–O–P core of complex **2**.

References

A. F. Richards and C. M. Beavers, Synthesis and structures of a pentanuclear Al(III) phosphonate cage, an In(III) phosphonate polymer, and coordination compounds of the corresponding phosphonate ester with GaI₃ and InCl₃. *Dalton Trans.* **41**, 11305–10 (2012).

I. Fedushkin, A. A. Skatova, V. A. Dodonov, X.-J. Yang, V. A. Chudakova, A. V. Piskunov, S. Demeshko, and E. V. Baranov, Ligand "brackets" for Ga–Ga bond. *Inorg. Chem.* **55**, 9047–56 (2016).

3.10 Thallium(I) polymeric compounds containing linear Tl(I)–M arrays [M = Pt(II), Au(I)]

The Pt(II), Au(I) and Tl(I) ions, which have closed shell and pseudo-closed shell d^8, d^{10}, and s^2 electronic structures, respectively, can interact to form supramolecular compounds stabilized by weak Tl–M (M = Pt, Au) metal-metal bonds.

3.10.1 Supramolecular compounds with Tl–Pt polymeric chains

In supramolecular s^2-d^8 (Tl^I–Pt^{II}) molecular compounds composed of linear Pt–Tl–Pt–Tl polymeric chains, the bonding interactions between the metal centers result from a combination of metallophillic

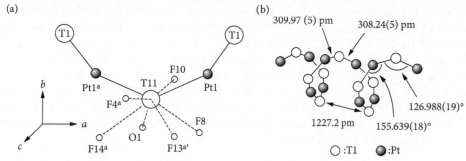

Fig. 3.10.1 (a) Tl(1) environment. (b) Core of helical chain composed of Pt–Tl(1) units.

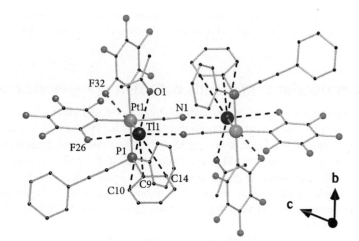

Fig. 3.10.2 Molecular structure of the centrosymmetric Tl^I–Pt^{II} dimer $[Tl\{cis\text{-}Pt(C_6F_5)_2(CN)(PPh_2 C\equiv CPh\}\cdot CH_3COCH_3]_2$.

and coulombic factors. The metallophillic interaction in these complexes results from σ-bonding that involves the filled $6s$ and empty $6p_z$ orbitals on thallium and the filled $5d_{z^2}$ and empty $6p_z$ orbitals on platinum. In complexes that involve Tl^I···Pt^{II} interactions, the Tl–Pt distances fall in the 277–289 pm range; in complexes with Tl^I···Pt^0 interactions, the Tl–Pt distances fall in the 279–344 pm range. Thus the range of distances in complexes with the s^2–d^8 (Tl^I···Pt^{II}) and s^2–d^{10} (Tl^I···Pt^0) electronic structures are similar. This might be anticipated, since platinum uses filled d_{z^2} and empty p_z orbitals to interact with the filled s^2 and empty p_z orbitals on thallium in both cases. While Pt(0) is generally larger than Pt(II) in size, Coulombic repulsion may also contribute to lengthening of the Tl(I)–Pt(II) bond.

In the complex $\{Tl[Tl\{cis\text{-}Pt(C_6F_5)_2(CN)_2\}]\cdot(H_2O)\}_n$, the asymmetric unit contains two independent Tl centers and one water molecule. As shown in Fig. 3.10.1(a), Tl(1) interacts directly with two adjacent platinum(II) centers to generate an extended $[-Pt-Tl(1)-Pt-Tl(1)-]_n^{n-}$ helical chain by connection of anionic $[Pt-Tl(1)(R_F)_2(CN)_2]^-$ motifs (Fig. 3.10.1(a)). In each chain, the Pt atom is coordinated to two thallium centers, with Pt–Tl(1) bond distances of 308.24(5) pm and 309.97(5) pm and angles at Tl–Pt–Tl and Pt–Tl–Pt metal centers of 155.64(2)° and 126.99(2)°, respectively. In contrast, Tl(2) contacts through electrostatic interactions with the N(1) atom of one cyanide group. In this unit, the Pt–Tl(1) vector is perceptibly bent by 13.6(2)° toward the cyanide ligands.

The Tl(2) center inside the $[-Pt-Tl(1)-Pt-Tl(1)-]_n^{n-}$ helical chain is mainly stabilized by electrostatic interaction with four cyanide groups [Tl···N 269.9(9)–287.9(9) pm], and o-fluorine atoms of two adjacent C_6F_5 rings and a p-fluorine of an adjacent helical chain [Tl···F 319.8(8)–344.1(7) pm].

In the tetranuclear [Tl{cis-Pt$(C_6F_5)_2$(CN)(PPh$_2$C≡CPh)}·CH$_3$COCH$_3$]$_2$ dimeric Tl^I–Pt^{II} complex shown in Fig. 3.10.2, the thallium(I) center is stabilized by weak bonding interactions with o-fluoro substituents on two phenyl groups [Tl(1)–F(26) at 329.4(3) and F(32) at 315.9(4) pm], an acetone solvate molecule [Tl(1)–O1 282.5(6) pm], and the π electronic density of a phenyl ring. The distances between phenyl carbon atoms C(10), C(9), C(14), and Tl(1) [344.3(8)–370.6(8) pm] are within the range observed for other compounds with π–Tl interactions.

Reference

J. Forniés, A. García, E. Lalinde, and M. T. Moreno, Luminescent one- and two-dimensional extended structures and a loosely associated dimer based on Platinum(II)–Thallium(I) backbones, *Inorg. Chem.* **47**, 3651–60 (2008).

3.10.2 Supramolecular compounds with Tl–Au polymeric chains

The Tl^I–Au^I heteropolynuclear complexes [AuTl$(C_6X_5)_2]_n$ (X = F, Cl) react with dimethylsulfoxide (DMSO) in different molar ratios, yielding stoichiometric products [Tl$_2${Au$(C_6F_5)_2$}$_2${μ-DMSO}$_3]_n$ (**1**) and [Tl$_2${Au$(C_6Cl_5)_2$}$_2${μ-DMSO}$_2]_n$ (**2**). X-ray crystallographic analysis showed that they can be viewed as extended polymeric chains built with Tl–Au–Tl units in which the thallium atoms are bridged by the oxygen atoms of DMSO ligands (Fig. 3.10.3 and 3.10.4). Notably, there is a Tl···Tl interaction of 375.62(6) pm in **2** that is not observed in complex **1**. Additional [Au$(C_6X_5)_2]^-$ fragments interact with one or two thallium centers, respectively, giving rise to two different types of metal–metal interactions in each molecule (Fig. 3.10.3 and 3.10.4).

The Au···Tl distances range from 322.25(6) to 351.82(8) pm for complex **1** (the longest one corresponding to the terminal [Au$(C_6F_5)_2]^-$ unit) and from 312.20(5) to 328.39(6) pm for complex **2**, and they are slightly longer than most of the Au···Tl contacts observed in other polymeric species with unsupported Au···Tl interactions [290.78(3)–348.99(6) pm].

Each metallic center displays several metal–halogen contacts within the ranges 322.3–324.7 pm for Au···F and 286.5–338.2 pm for Tl···F in complex **1**, and 328.2–342.9 pm for Au···Cl and 322.8–358.8 pm for Tl···Cl in complex **2**, that contribute to stabilization of their zigzag polymeric structures.

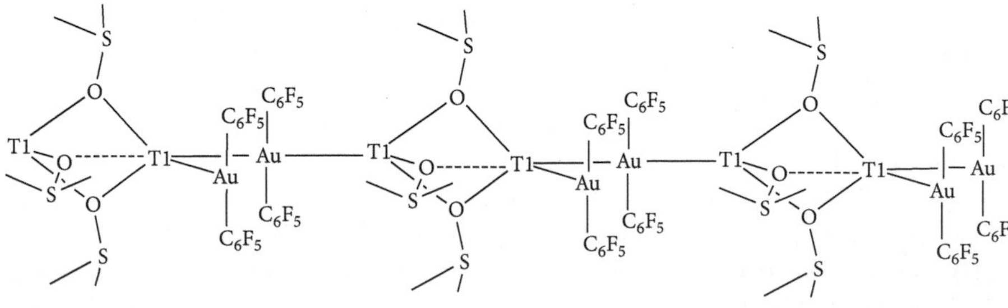

Fig. 3.10.3 Polymeric-chain crystal structure of [Tl$_2${Au$(C_6F_5)_2$}$_2${μ-DMSO}$_3]_n$ (complex **1**).

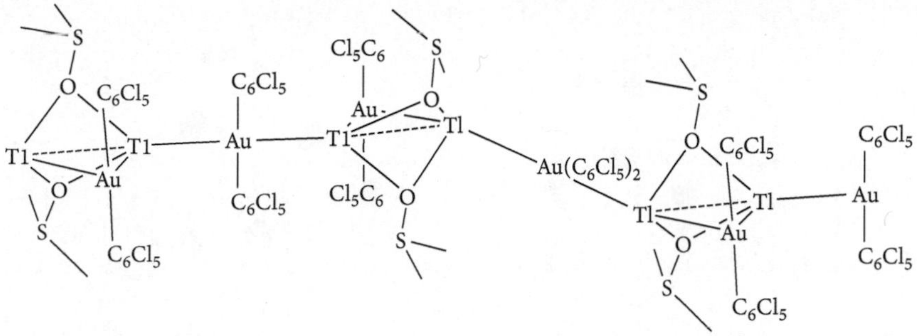

Fig. 3.10.4 Polymeric-chain crystal structure of $[Tl_2\{Au(C_6Cl_5)_2\}_2\{\mu\text{-DMSO}\}_2]_n$ (complex **2**).

References

E. J. Fernández, A. Laguna, J. M. López-de-Luzuriaga, M. Montiel, M. E. Olmos, and J. Pérez, Dimethylsulfoxide gold–thallium complexes. Effects of the metal–metal interactions in the luminescence, *Inorg. Chim. Acta* **358**, 4293–300 (2005).

K. Akhbari and A. Morsali, Thallium(I) supramolecular compounds: structural and properties consideration, *Coord. Chem. Rev.* **254**, 1977–2006 (2010).

3.11 **Platinum–Thallium Clusters**

Metallophilic interactions between closed or pseudo closed-shell transition metals (d^{10}, d^8, d^{10}, s^2) have been used as a tool in the field of crystal or molecular engineering for linking different sub-assemblies.

Thallium (I) having a $d^{10}s^2$ electron configuration is a fairly representative acceptor in the assembly of heteropolynuclear clusters. It is a typical "inert pair" species in forming donor–Tl(I) couples such as $M(d^8) \longrightarrow Tl(I)$ with M = Ru(0), Rh(I), Pd(II), Ir(I), Pt(II), and $M(d^{10}) \longrightarrow Tl(I)$ with M = Pt(0) and Au(I). In practice, Pt(II) and Au(I) are the two most used donor centers.

The use of Pt precursors containing the tridentate cyclometallated 2,6-diphenylpyridinate (CNC) ligand has led to the preparation of two cluster complexes containing unsupported Pt(II)$\longrightarrow$Tl(I) bonds that exhibit square-planar molecular geometries, namely [Pt(CNC)L] (in which CNC = C,N,C-2,6-$NC_5H_3(C_6H_4\text{-}2)_2$, with L = tetrahydrothiophene (tht = SC_4H_8) for **1** and L = CN^tBu for **2**). These two compounds react with $TlPF_6$ in different Pt:Tl molar ratios (3:1 in the case of **1**, and 1:1 in the case of **2**), yielding the complexes [{Pt(CNC)(tht)}$_3$Tl](PF$_6$) (**3**) and [Pt(CNC)(CN^{t}Bu)Tl](PF$_6$) (**4**), respectively.

X-Ray crystal structure determination of **1** showed the expected square-planar "Pt(CNC)S" coordination mode with expected structural parameters. The molecular structures of **3** and **4** are shown in Fig. 3.11.1. Tetranuclear complex **3** has a Pt$_3$Tl cluster core with its Tl(I) center lying on a six-fold symmetry axis, so that the three Pt atoms form a perfect equilateral triangle consolidated by unsupported Pt$\longrightarrow$Tl dative bonds of 290.88(5) pm. Complex **4** is built of three "Pt(CNC)(CN^{t}Bu)Tl" subunits arranged in a triangular cycle and connected by η^6-Tl–arene interactions, together with intermetallic Pt–Tl bonds of *ca.* 304 pm.

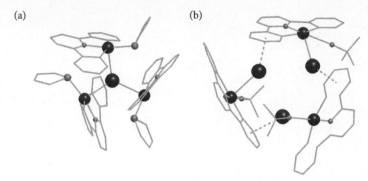

Fig. 3.11.1 Molecular structures of (a) complex **3** and (b) complex **4**.

Reference

Ú. Belío, S. Fuertes, and A. Martín, Preparation of Pt–Tl clusters showing new geometries. X-ray, NMR and luminescence studies. *Dalton Trans.* **43**, 10828–43 (2014).

3.12 Triel Bonding and Coordination of Triel Centers

The triel bond is defined as the interaction of a Group 13 atom (B, Al, Ga, In, or Tl) in its compound that acts as a Lewis acid center, with the electron-rich site of its bonded species playing the role of a Lewis base. This interaction is often classified as a π-hole bond. Fig. 3.12.1 compares such bonding interactions in three pairs of 1:1 electron acceptor–donor molecules: HO–H:NH$_3$ hydrogen bond in the water-ammonia complex, F–Cl:NH$_3$ halogen bond in the chlorine monofluoride-ammonia donor–acceptor complex, and F$_3$B:N≡CH triel bond in the boron trifluoride-acetonitrile complex.

The triel centre, T, may interact with electron-rich sites so that T···B, T···π, and T···σ triel bonds may be identified. Tetrahedral molecular structures that obey the octet rule occur most frequently, as they are characterized by energetic stability. Trivalent and hypovalent triel structures that feature trigonal coordination, as well as hypervalent triel structures exhibiting trigonal-bipyramidal and octahedral configurations, also occur in the crystal structures of Group 13 compounds.

Cambridge Structural Database (CSD) searches (May 2019 release) performed for Group 13 triel centres in their crystalline compounds confirmed that coordination number 4 occurs most often. For the lighter elements boron, aluminium and gallium, coordination number 4 occurs around 70% of their compounds. For the heavier elements indium and thallium, such occurrence drops to 39.0% and 25.5%, respectively, but these are still the highest compared with 26.9% for octahedral coordination in indium compounds and 22.5% for trigonal pyramidal coordination in thallium(I) compounds.

Examples of compounds that feature heavy triel (indium and thallium) centers are shown in Fig. 3.12.2 and Fig. 3.12.3, respectively.

Fig. 3.12.2(a) shows an In(III) triel center coordinated by two (L^{CF3})$^{2-}$ complex anions from a (HNEt$_3$)[In(L^{CF3})$_2$] salt, and Fig. 3.2(b) indicates that H$_2$L^{CF3} is a partially fluorinated *S,N,S*-tridentate thiosemicarbazone.

Fig. 3.12.3 shows the coordination mode of the Tl(III) triel center in the (1,4,7-trithiacyclononane)-thallium(I) cation of its hexafluorophosphate salt.

Type of interaction (up)
Orbital-orbital overlap (below)

Type of interaction (up)
Orbital-orbital overlap (below)

Hydrogen bond

A-H⋯B
$n_B \rightarrow \sigma_{AH}^*$

O-H⋯N
$n_N \rightarrow \sigma_{OH}^*$

Halogen bond

A-X⋯B
$n_B \rightarrow \sigma_{AX}^*$

F-Cl⋯N
$n_N \rightarrow \sigma_{FCl}^*$

Triel bond

A-T⋯B
$n_B \rightarrow \sigma_{AT}^*$
$n_B \rightarrow n_T^*$

F-B⋯N
$n_N \rightarrow n_B^*$

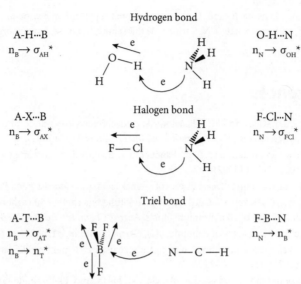

Fig. 3.12.1 Complexes that exemplify three kinds of electron acceptor–donor interactions: A–H⋯B hydrogen bond, A–X⋯B halogen bond, and A–T⋯B triel bond.

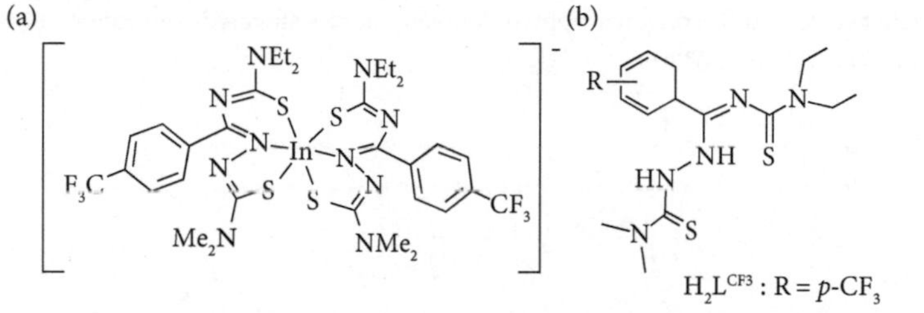

(a)

(b)

H_2L^{CF3} : R = p-CF$_3$

Fig. 3.12.2 (a) Coordination of In(III) triel center in the $[In(L^{CF3})_2]^-$ complex anion determined from crystal structure analysis of the $(HNEt_3)[In(L^{CF3})_2]$ salt. Bond lengths (pm): In–S 253.4(1)–261.1(1), In–N 226.9(4)–229.5(4). (b) Potentially tridentate S,N,S-thiosemicarbazone ligand H_2L^{CF3} derived from N,N-dialkyl-N'-benzoylthiourea.

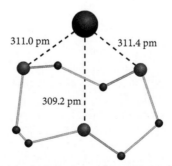

311.0 pm   311.4 pm

309.2 pm

Fig. 3.12.3 Coordination of Tl(III) triel center in (1,4,7-trithiacyclononane)-thallium(I) hexafluorophosphate.

References

S. J. Grabowski, Triel bond and coordination of triel centres—comparison with hydrogen bond interaction, *Coord. Chem. Rev.* **407**, 213171 (2020).

F. Salsi, M. R. Jungfer, A. Hagenbach, and U. Abram, Trigonal-bipyramidal vs. octahedral coordination in indium(iii) complexes with potentially *S, N, S*-tridentate thiosemicarbazones, *Eur. J. Inorg. Chem.* **2020**, 1222–9 (2020).

General References

T. Peymann, C. B. Knobler, S. I. Khan, and M. F. Hawthorne, Dodeca(benzyloxy)-dodecaborane $B_{12}(OCH_2Ph)_{12}$: a stable derivative of hypercloso-$B_{12}H_{12}$. *Angew. Chem. Int. Ed.* **40**, 1664–7 (2001).

E. D. Jemmis, M. M. Balakrishnarajan, and P. D. Pancharatna, Electronic requirements for macropolyhedral boranes. *Chem. Rev.* **102**, 93–144 (2002).

Z. Xie, Advances in the chemistry of metallacarboranes of *f*-block elements. *Coord. Chem. Rev.* **231**, 23–46 (2002).

J. Vollet, J. R. Hartig, and H. Schnöckel, $Al_{50}C_{120}H_{180}$: A pseudofullerene shell of 60 carbon atoms and 60 methyl groups protecting a cluster core of 50 aluminium atoms. *Angew. Chem. Int. Ed.* **43**, 3186–9 (2004).

H. W. Roesky and S. S. Kumar, Chemistry of aluminium(I). *Chem. Commun.* **41**, 4027–38 (2005).

A. Schnepf and H. Schnöckel, Metalloid aluminum and gallium clusters: element modifications on the molecular scale? *Angew. Chem. Int. Ed.* **41**, 3532–52 (2002).

W. Uhl, Organoelement compounds possessing Al–Al, Ga–Ga, In–In, and Tl–Tl single bonds. *Adv. Organomet. Chem.* **51**, 53–108 (2004).

S. Kaskel and J. D. Corbett, Synthesis and structure of $K_{10}Tl_7$: the first binary trielide containing naked pentagonal bipyramidal Tl_7 clusters. *Inorg. Chem.* **39**, 778–82 (2000).

M. S. Hill, P. B. Hitchcock, and R. Pongtavornpinyo, A Llinear homocatneated compound containing six indium centers. *Science* **311**, 1904–7 (2006).

Chapter 4

Structural Chemistry of Group 14 Elements

4.1 Allotropic Modifications of Carbon

With the exception of the gaseous low-carbon molecules: $C_1, C_2, C_3, C_4, C_5, \ldots$, the carbon element exists in the diamond, graphite, fullerene, and amorphous allotropic forms.

4.1.1 Diamond

Diamond forms beautiful, transparent, and highly refractive crystals, and has been used as a noble gem since antiquity. It consists of a three-dimensional network of carbon atoms, each of which is bonded tetrahedrally by covalent C–C single bonds to four others, so that the whole diamond crystal is essentially a "giant molecule." Nearly all naturally occurring diamonds exist in the cubic form, space group $Fd\bar{3}m$ (No. 227), with $a = 356.688$ pm and $Z = 8$. The C–C bond length is 154.45 pm, and the C–C–C bond angle is 109.47°. In the crystal structure, the carbon atoms form six-membered rings that take the all-chair conformation (Fig. 4.1.1(a)). The mid-point of every C–C bond is located at an inversion center, so that the six nearest carbon atoms about it are in a staggered arrangement, which is the most stable conformation.

In addition to cubic diamond, there is a metastable hexagonal form, which has been found in aerorite and can be prepared from graphite at 13 GPa above 1300 K. Hexagonal diamond crystallizes in space group $P6_3/mmc$ (No. 194) with $a = 251$ pm, $c = 412$ pm, and $Z = 4$, as shown in Fig. 4.1.1(b). The bond type and the C–C bond length are the same as in cubic diamond. The difference between the two forms is the orientation of two sets of tetrahedral bonds about neighboring carbon atoms: cubic diamond has the staggered arrangement about each C–C bond at an inversion center, but hexagonal diamond takes the eclipsed arrangement related by mirror symmetry. As repulsion between non-bonded atoms in the eclipsed arrangement is greater than that in the staggered arrangement, hexagonal diamond is much rarer than cubic diamond.

Diamond is the hardest natural solid known and has the highest melting point, 4400 ± 100 K (12.4 GPa). It is an insulator in its pure form. Since the density of diamond (3.51 g cm^{-3}) far exceeds that of graphite (2.27 g cm^{-3}), high pressures can be used to convert graphite into diamond even though graphite is thermodynamically more stable by 2.9 kJ mol^{-1}. To attain commercially viable rates for the pressure-induced conversion of graphite to diamond, a transition metal catalyst such as iron, nickel, or chromium is generally used. Recently, the method has been employed to deposit thin films of diamond onto a metallic or other material surface. The extreme hardness and high thermal conductivity of diamond find applications in numerous areas, notably the hardening of surfaces of electronic devices and as cutting and/or grinding materials.

Elemental silicon, germanium, and tin have the cubic diamond structure with unit cell edge $a = 543.072$ pm, 565.754 pm, and 649.12 pm (α-Sn), respectively.

Structural Chemistry across the Periodic Table. Thomas Chung Wai Mak, Yu-San Cheung, Gong-Du Zhou, and Ying-Xia Wang. Oxford University Press. © Thomas Chung Wai Mak, Yu-San Cheung, Gong-Du Zhou, and Ying-Xia Wang (2023). DOI: 10.1093/oso/9780198872955.003.0004

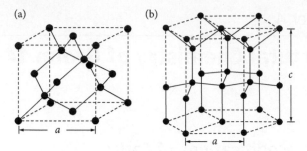

Fig. 4.1.1 Crystal structure of diamond: (a) cubic form and (b) hexagonal form.

4.1.2 **Graphite**

Graphite is the common modification of carbon that is stable under normal conditions. Its crystal structure consists of planar layers of hexagonal carbon rings. Within the layer each carbon atom is bond covalently to three neighboring carbon atoms at 141.8 pm. The σ bonds between neighbors within a layer are formed from the overlap of sp^2 hybrids, and overlap involving the remaining electron and perpendicular p_z orbital on every atom generates a network of π bonds that are delocalized over the entire layer. The relatively free movement of π electrons in the layers leads to abnormally high electrical conductivity for a non-metallic substance.

The layers are stacked in a staggered manner with half of the atoms of one layer situated exactly above atoms of the layer below, and the other half over the rings centers. There are two crystalline forms that differ in the sequence of layer stacking.

(1) Hexagonal graphite or α-graphite. The layers are arranged in the sequence ...ABAB..., as shown in Fig. 4.1.2(a). The space group is $P6_3/mmc$ (No. 194), and the unit cell has dimension $a = 245.6$ pm. and $c = 669.4$ pm, so that the interlayer distance is $c/2 = 334.7$ pm.

(2) Rhombohedral graphite or β-graphite. The layers are arranged in the sequence ...ABCABC..., as shown in Fig. 4.1.2(b). The space group is $R\bar{3}m$ (No. 166), and the unit cell has dimension $a = 246.1$ pm and $c = 1006.4$ pm The interlayer separation $c/3 = 335.5$ pm is similar to that of α-graphite.

The enthalpy difference between hexagonal and rhombohedral graphite is only 0.59 ± 0.17 kJ mol^{-1}. The two forms are inter-convertible by grinding (hexagonal → rhombohedral) or heating above 1025 °C (rhombohedral → hexagonal). Partial conversion leads to an increase in the average spacing between layers; this reaches a maximum of 344 pm for turbostratic graphite in which the stacking sequence of the parallel layers is completely random.

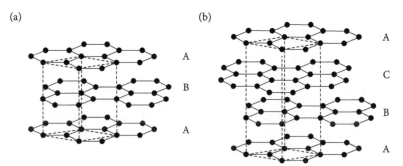

Fig. 4.1.2 Structure of graphite: (a) α-graphite (b) β-graphite.

In graphite the layers are held together by van der Waals forces. The relative weak binding between layers is consistent with its softness and lubricity, as adjacent layers are able to easily slide by each other. Graphite mixed with clay constitute pencil "lead," which should not be confused with metallic lead or dark-gray lead sulfide.

4.1.3 **Fullerenes**

The third allotropic modification of carbon, the fullerenes, consists of a series of discrete molecules of closed cage structure that are bounded by planar faces, whose vertices are made up of carbon atoms. If the molecular cage consists of pentagons and hexagons only, the number of pentagon must always be equal to 12, while the number of hexagons may vary. Fullerenes are discrete globular molecules that are soluble in organic solvents, and their structure and properties are different from diamond and graphite.

Fullerenes can be obtained by passing an electric arc between two graphite electrodes in a controlled atmosphere of helium, or by controlling the helium:oxygen ratio in the incomplete combustion of benzene, followed by evaporation of the carbon vapor and recrystallization from benzene. The main product of the preparation is fullerene-C_{60}, and the next abundant product is fullerene-C_{70}.

The structure of C_{60} has been determined by single-crystal neutron diffraction and electron diffraction in the gas phase. It has icosahedral (I_h) symmetry with 60 vertices, 90 edges, 12 pentagons, and 20 hexagons. The C–C bond lengths are 139 pm for 6/6 bonds (fusion of two six-membered rings) and 144 pm for 6/5 bonds (fusion between five- and six-membered rings). The 60 carbon atoms all lie within a shell of mean diameter 700 pm. Fig. 4.1.3(a) shows the soccer-ball shape of the C_{60} molecule. Each carbon atom forms three σ bonds with three neighbors, and the remaining orbitals and electrons of the 60 C atoms form delocalized π bonds. This structure may be formulated by the valence-bond formula shown in Fig. 4.1.3(b). The C_{60} molecule can be chemically functionalized, and an unambiguous atom-numbering scheme is required for systematic nomenclature of its derivatives. Fig. 4.1.3(c) shows the planar formula with carbon atom-numbering scheme of the C_{60} molecule.

In the C_{60} molecule, the sum of the σ bond angles at each C atom is 348° (= 120° + 120° + 108°), and the mean C–C–C angle is 116°. The π atomic orbital lies normal to the convex surface, the angle between the σ and π orbitals being 101.64°. It may be approximately calculated that each σ orbital has *s* component 30.5% and *p* 69.5%, and each π orbital has *s* 8.5% and *p* 91.5%.

Fullerene-C_{60} is a brown-black crystal in which the nearly spherical molecules rotate continuously at room temperature. The structure of the crystal can be considered as a stacking of spheres of diameter

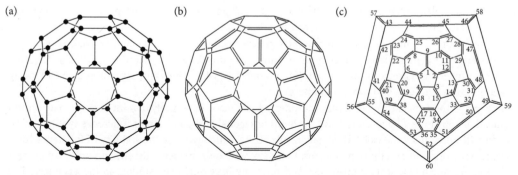

Fig. 4.1.3 Structure of fullerene-C_{60}: (a) molecular shape, (b) valence-bond formula, and (c) planar Schlegel diagram with carbon atoms numbered.

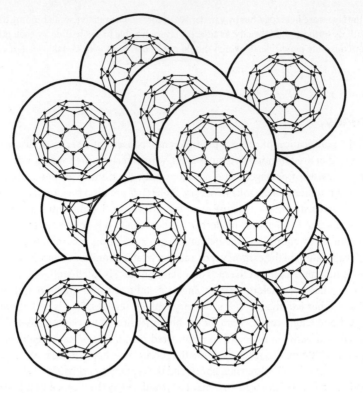

Fig. 4.1.4 The cubic face-centered structure of fullerene-C_{60}.

1000 pm in cubic closest packing (ccp, a = 1420 pm) or hexagonal closest packing (hcp, a = 1002 pm, c = 1639 pm). Fig. 4.1.4 shows the crystal structure of fullerene-C_{60}.

Below 249 K, the molecules are orientated in an ordered fashion, and the symmetry of the crystal is reduced from a face-center cubic lattice to a primitive cubic lattice. At 5 K, the crystal structure determined by neutron diffraction yielded the following data: space group $Pa\bar{3}$ (No. 205), a = 1404.08(1) pm; C–C bond lengths: (6/6) 139.1 pm, (6/5) 144.4 pm, and 146.6 pm (mean 145.5 pm).

In the C_{60} molecule, the mean distance from the center to every C vertex is 350 pm, so the molecule has a spherical skeleton with diameter 700 pm. Allowing for the van der Waals radius of the C atom (170 pm), the C_{60} molecule has a central cavity of diameter 360 pm, which can accommodate a foreign atom. Some physical properties of C_{60} are summarized in Table 4.1.1.

In addition to C_{60}, many other higher homologs have been prepared and characterized. Several synthetic routes to fullerenes have yielded gram quantities of pure C_{60} and C_{70}, and C_{76}, C_{78}, C_{80}, C_{82}, C_{84}, and other fullerenes have been isolated as minor products. Fig. 4.1.5 shows the structures of the following fullerenes: C_{20}, C_{50}, C_{70}, and two C_{78}-isomers.

In general, the polyhedral closed cages of fullerenes are made up entirely of n three-coordinate carbon atoms that constitute 12 pentagonal and ($n/2-10$) hexagonal faces. The larger fullerenes synthesized so far do faithfully satisfy the isolated pentagon rule (IPR), which governs the stability of fullerenes comprising hexagons and exactly 12 pentagons. On the other hand, smaller fullerenes do not obey the IPR rule and are so labile that their properties and reactivity have only been studied in the gas phase. The smallest fullerene that can exist theoretically is C_{20}. It has been synthesized from dodecahedrane $C_{20}H_{20}$ by replacing the hydrogen atoms with relatively weakly bound bromine atoms to form a triene precursor

Table 4.1.1 Some physical properties of fullerene-C_{60}

Density/g cm^{-3}	1.65
Bulk modulus/GPa	18
Refractive index (630 nm)	2.2
Heat of combustion (crystalline C_{60})/kJ mol^{-1}	2280
Electron affinity/eV	2.6-2.8
First ionization energy/eV	7.6
Band gap/eV	1.9
Solubility (303K)/g dm^{-3}	
CS_2	5.16
toluene	2.15
benzene	1.44
CCl_4	0.45
hexane	0.04

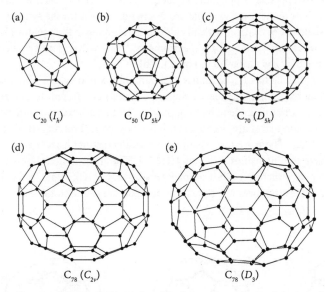

(a) (b) (c)

C_{20} (I_h) C_{50} (D_{5h}) C_{70} (D_{5h})

(d) (e)

C_{78} (C_{2v}) C_{78} (D_3)

Fig. 4.1.5 Structures of some fullerenes. Note that (b) shows the carbon skeleton of $C_{50}Cl_{10}$, whose 10 equatorial chloro substituents have been omitted.

of average composition [$C_{20}HBr_{13}$], which was then subjected to debromination in the gas phase. The bowl isomer of C_{20} is likewise generated by gas-phase debromination of a [$C_{20}HBr_9$] precursor prepared from bromination of corannulene $C_{20}H_{10}$. Identification of the two C_{20} isomers was achieved by mass-selective anion-photoelectron spectroscopy. Fig. 4.1.5(a) shows the molecular structure of fullerene C_{20}.

Dodecahedrane $C_{20}H_{20}$ $\longrightarrow$ $[C_{20}HBr_{13}]$ $\longrightarrow$ Fullerence C_{20}

Corannulene $C_{20}H_{10}$ $\longrightarrow$ $[C_{20}HBr_9]$ $\longrightarrow$ Bowl isomer of C_{20}

[H. Prinzbach, A. Weiler, P. Landenberger, F. Wahl, J. Wrth, L. T. Scott, M. Gelmont, D. Olevano, and B. v. Issendorff, *Nature* **407**, 6–3 (2000).]

Fullerene-C_{50} has been trapped as its perchloro adduct $C_{50}Cl_{10}$, which was obtained in milligram quantity from the addition of CCl_4 to the usual graphite arc-discharge process for the synthesis of C_{60} and larger fullerenes. The D_{5h} structure of $C_{50}Cl_{10}$, with all chlorine atoms lying in the equatorial plane and attached to sp^3 carbon atoms, was established by mass spectrometry, ^{13}C NMR, and other spectroscopic methods. The idealized structure of C_{50} is displayed in Fig. 4.1.5(b).

The fullerene-C_{70} molecule has D_{5h} symmetry and an approximately ellipsoidal shape, as shown in Fig. 4.1.5(c). As in C_{60}, the 12 five-membered rings in C_{70} are not adjacent to one another. In contrast to C_{60}, C_{70} has an equatorial phenylene belt comprising 15 fused hexagons and two polar caps each assembled from a pentagon that shares its edges with five hexagons. The curvature at the polar region is very similar to that of C_{60}. There are five sets of geometrically distinct carbon atoms, and the observed ^{13}C NMR signals have the intensities ratios of 1:1:2:2:1. The measured carbon-carbon bond lengths (Fig. 4.1.6) in the crystal structure of the complex $(\eta^2\text{-}C_{70})Ir(CO)Cl(PPh_3)_2$ suggest that two equivalent lowest-energy Kekulé structures per equatorial hexagon are required to describe the structure and reactivity properties of C_{70}.

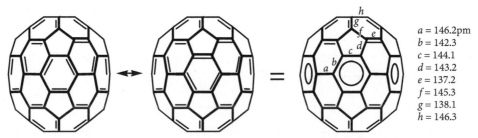

$a = 146.2\,\text{pm}$
$b = 142.3$
$c = 144.1$
$d = 143.2$
$e = 137.2$
$f = 145.3$
$g = 138.1$
$h = 146.3$

Fig. 4.1.6 Two equivalent valence bond structures of C_{70} and measured bond lengths of the polyhedral cage in $(\eta^2\text{-}C_{70})Ir(CO)Cl(PPh_3)_2$.

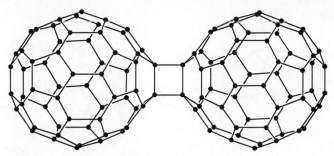

Fig. 4.1.7 Molecular structure of the dimer $[C_{60}]_2$.

The structures of two geometric isomers of fullerene-C_{78} have been elucidated by ^{13}C NMR spectroscopy: one has C_{2v} symmetry, as shown in Fig. 4.1.5(d); the other has D_3 symmetry, as shown in Fig. 4.1.5(e). Reliable to certain structural assignments have also been made for $C_{74}(D_{3h})$, $C_{76}(D_2)$, C_{78}' [a new isomer of C_{2v} symmetry, which has a more spherical shape compared to the $C_{78}(C_{2v})$ isomer shown in Fig. 4.1.5(d)], $C_{80}(D_2)$, $C_{82}(C_2)$, $C_{84}(D_2)$, and $C_{84}(D_{2d})$.

In addition to the single globular species, fullerenes can be formed by joining two or more carbon cages, as found for the dimer C_{120} shown in Fig. 4.1.7. X-ray diffraction showed that the dimer is connected by a pair of C–C bonds linking the edges of hexagonal faces (6/6 bonds) in two C_{60} units to form a central four-membered ring, in which the bond lengths are 157.5 pm (connecting the two cages) and 158.1 pm (6/6 bonds).

Fullerene-C_{60} was selected as "Molecule of the Year 1991" by the journal *Science (Washington)*. Discovery of this modification of carbon in the mid-1980s created great excitement in the scientific community and popular press. Some of this interest undoubtedly stemmed from the fact that carbon is a common element that had been studied since ancient times, but still an entirely new field of fullerene chemistry suddenly emerged with great potential for exciting research and practical applications.

4.1.4 **Fullerenes with fused pentagonal faces**

The geometries of most fullerenes are governed by the *isolated pentagon rule* (IPR), which states that stable fullerenes have each of their 12 pentagons surrounded by five hexagons. Among the first reported non-*IPR* fullerenes are the endohedral $Sc_2@C_{66}$ (Fig. 4.1.8) and exohedral $C_{66}Cl_{10}$ (Fig. 4.1.9) molecules

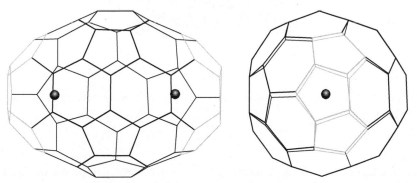

Fig. 4.1.8 Two orthogonal views of an endohedral $Sc_2@C_{66}$ molecule. Thick lines indicate bonds close to the observer and thin lines are for bonds behind. Fused pentagons are indicated by gray lines.

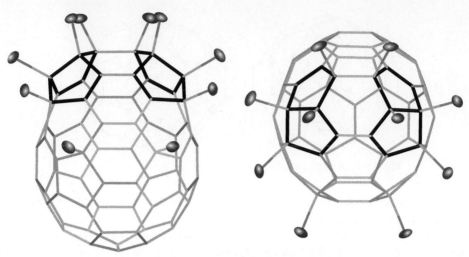

Fig. 4.1.9 Side (left) and top (right) views of fused-pentagon fullerene $C_{66}Cl_{10}$ stabilized by exohedral chlorination. Fused pentagons are indicated by thick black lines.

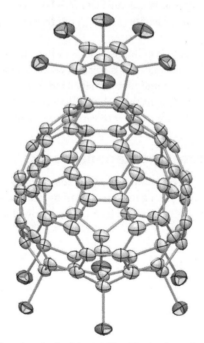

Fig. 4.1.10 $C_{2v}(2)$-$C_{78}Cl_6(C_5Cl_6)$ molecule doubly stabilized by both exohedral chlorine atoms and perchlorinated cyclopentadiene.

that feature fused pentagonal faces. A recent example is a $C_{2v}(2)$-C_{78} cage doubly functionalized by exohedral attachment of six chlorine atoms and a perchlorinated cyclopentadiene group (Fig. 4.1.10).

References

M. Yamada, H. Kurihara, M. Suzuki, J. D. Guo, M. Waelchli, M. M. Olmstead, A. L. Balch, S. Nagase, Y. Maeda, T. Hasegawa, X. Lu, and T. Akasaka, $Sc_2@C_{66}$ revisited: an endohedral fullerene with scandium ions nestled within two unsaturated linear triquinanes. *J. Am. Chem. Soc.* **136**, 7611–4 (2014).

C.-L. Gao, X. Li, Y.-Z. Tan, X.-Z. Wu, Q. Zhang, S.-Y. Xie, and R.-B. Huang, Synthesis of long-sought C_{66} with exohedral stabilization. *Angew. Chem. Int. Ed.* **53**, 7853–5 (2014).

C.-L. Gao, L. Abella, H.-R. Tian, X. Zhang, Y.-Y. Zhong, Y.-Z. Tan, X.-Z. Wu, A. Rodríguez-Fortea, J. M. Poblet, S.-Y. Xie, R.-B. Huang, and L.-S. Zheng, Double functionalization of a fullerene in drastic arc-discharge conditions: synthesis and formation mechanism of $C_{2v}(2)$-$C_{78}Cl_6(C_5Cl_6)$. *Carbon* **129**, 286–92 (2018).

4.1.5 **Amorphous carbon**

Amorphous carbon is a general term that covers non-crystalline forms of carbon such as coal, coke, charcoal, carbon black (soot), activated carbon, vitreous carbon, glassy carbon, carbon fiber, carbon nanotubes, and carbon onions, which are important materials and widely used in industry. The arrangements of the carbon atoms in amorphous carbon are different from those in diamond, graphite, and fullerenes, but the bond types of carbon atoms are the same as in these three crystalline allotropes. Most forms of amorphous carbon consist of graphite scraps in irregularly packing.

Coal is by far the world's most abundant fossil fuel, with a total recoverable resource of about 1000 billion (10^{12}) tons. It is a complex mixture of many compounds that contain a high percentage of carbon and hydrogen, but many other elements are also present as impurities. The composition of coal varies considerably depending on its age and location. A typical bituminous coal has the approximate composition 80% C, 5% H, 8% O, 3% S, and 2% N. The structures of coal are very complex and not clearly defined.

The high-temperature carbonization of coal yields coke, which is a soft, poorly graphitized form of carbon, most of which is used in steel manufacture.

Activated carbon is a finely divided form of amorphous carbon manufactured from the carbonization of an organic precursor, which possesses a microporous structure with a large internal surface area. The ability of the hydrophobic surface to adsorb small molecules accounts for the widespread applications of activated carbon as gas filters, de-coloring agents in the sugar industry, water purification agents, and heterogeneous catalysts.

Carbon black (soot) is made by the incomplete combustion of liquid hydrocarbons or natural gas. The particle size of carbon black is exceedingly small, only 0.02–0.3 µm, and its principal application is in the rubber industry where it is used to strengthen and reinforce natural rubber.

Carbon fibers are filaments consisting of non-graphitic carbon obtained by carbonization of natural or synthetic organic fibers, or fibers drawn from organic precursors such as resins or pitches, and subsequently heat treated up to temperatures of about 300 ºC. Carbon fibers are light and exceeding strong, so that they are now important industrial materials that have gained increasing applications ranging from sport equipment to aerospace strategic uses.

From observations in transmission electron micrography (TEM), carbon soot particles are found to have an idealized onion-like shelled structure, as shown in Fig. 4.1.11. Carbon onions varying from 3 nm to 1000 nm in diameter have been observed experimentally. In an idealized model of a carbon onion, the first shell is a C_{60} core of I_h symmetry, the second shell is C_{240} (comprising $2^2 \times 60$ atoms), and in general the number of carbon atom in the nth shell is $n^2 \times 60$. The molecular formula of a carbon onion is $C_{60}@C_{240}@C_{540}@C_{960}@...$, and the inter-shell distance is always ~350 pm.

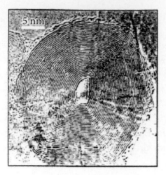

Fig. 4.1.11 Section of a carbon onion.

Recent work based on HRTEM (high-resolution transmission electron microscopy) and simulations has established that carbon onions are spherical rather than polyhedral, and the inter-shell spacing increases gradually from a value well below that of graphite at the center to the expected value at the outermost pair. The individual shells are not aligned in any regular fashion and do not rotate relative to each other. The innermost core can be a smaller fullerene (e.g., C_{28}) or a diamond fragment varying in size from 2 nm to 4.5 nm, and the internal cavity can be as large as 2 nm in diameter.

4.1.6 **Graphene**

Graphene is the general name for single-layered graphite molecules, multi-layered graphite molecules, and their derivatives. The structure of single-layer graphene is shown in Fig. 4.1.12. In this structure, each carbon atom uses sp^2 hybrid orbitals to bond to other three surrounding C atoms through σ bonds in a trigonal-planar manner. One p_z orbital, which is perpendicular to the layer, and one electron are available for additional bonding. The wave functions of all p_z orbitals are superposed to form a delocalized π bond. The electrons fill up the π orbitals and can move freely in the layer. The bonding force between atoms is increased and the chemical stability of graphene is enhanced. Graphene has relatively high chemical inertness and oxidation resistance. The single layer of this hexagonal network structure is only 0.34 nm thick. It is flexible, foldable, and highly stable, and does not allow penetration by He atoms. The mass of a single layer of graphene of 1 m² is only 0.77 mg, which contains about 3.8×10^{19} C atoms. Due to the existence of delocalized π-bonds, graphene has a metallic character and low electrical resistance. When it conducts electricity, almost no heat is generated. Capacitors made with it have large capacitance and high charge and discharge speeds. Single-layer graphene is almost transparent, with a white light transmission of 97.7% and an absorption rate of only 2.3%. Graphene has unique thermal properties and is an excellent thermal conductor; at 27 °C, its thermal conductivity is 5×10^6 W/K.

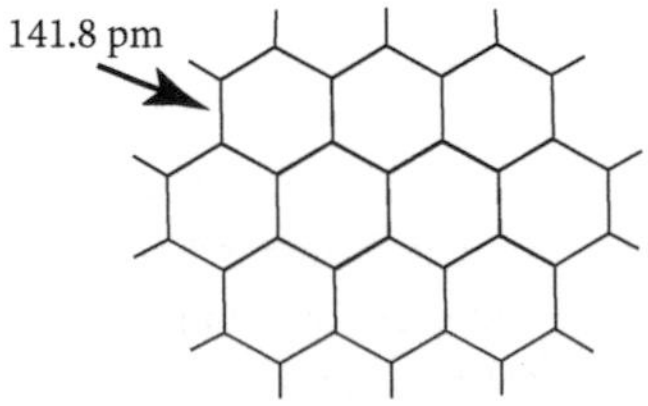

141.8 pm

Fig. 4.1.12 Structure of single-layer graphene

Double-layer and multi-layer graphene have different characteristics from single-layer graphene and graphite. Both sides of single-layer graphene are exposed, and the specific surface area can reach 2610 m^2/g. The surface area of multi-layer and single-layer graphene is the same, but the surface layer and the internal layer are in different environments and behave differently during chemical reaction treatment, so different compositions and properties can be achieved.

First, the bonding force between graphene layers is weaker van der Waals action, so ions or molecules can be inserted between the layers by chemical or physical processes to form a graphene intercalation compound with a sandwich structure. The properties of the intercalation compound can be adjusted according to the size, nature, and amount of intercalation ions or molecules between the layers. For more than a decade, one of the research focuses of graphene-derived materials is to obtain graphene with different layers, introduce various species with different properties between layers, and develop intercalation materials with desirable properties.

Second, on the edges and surfaces of multilayer graphene, unsaturated carbon atoms can chemically react with a variety of reagents, undergo various chemical modifications, and form materials with different properties.

Although it is difficult to obtain completely pure structural graphene, it is still an important area for scientific exploration in view of its predicted good performance. Related to this, other layered compounds, such as BN and MoS_2, have also attracted much attention. New advance in the synthesis, structural analysis, properties and applications of graphene can be found in the following literature: Y. Zhong, Z. Zhen, and H. Zhu, Graphene: Fundamental research and potential applications, *FlatChem* **4**, 20–32 (2017).

4.1.7 **New planar *sp²*-hybridized carbon allotrope**

The quest for planar sp^2-hybridized carbon allotropes apart from graphene (structure **A**) has stimulated substantial research efforts because of their predicted mechanical, electronic, and transport properties for applications. In May 2021 the bottom-up growth of an ultra-flat biphenylene (4-6-8) network (structure **B**) composed of periodically arranged four-, six-, and eight-membered sp^2 carbon rings (structure **C**, bottom reaction, see Fig. 4.1.13) was achieved by a two-step sequence consisting of linear polymerization of DHTP monomer (Ullmann coupling) followed by an on-surface inter-chain dehydrofluorination (HF-zipping) reaction.

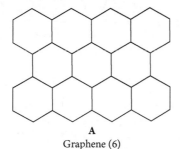
A
Graphene (6)

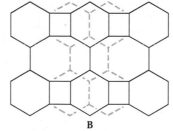
B
Biphenylene (4–6–8) network
(broken lines indicate size
matching with Graphene)

Characterization of this biphenylene network by scanning probe methods revealed that it exhibits metallic properties, making it a candidate for conducting wires in future carbon-based circuitry.

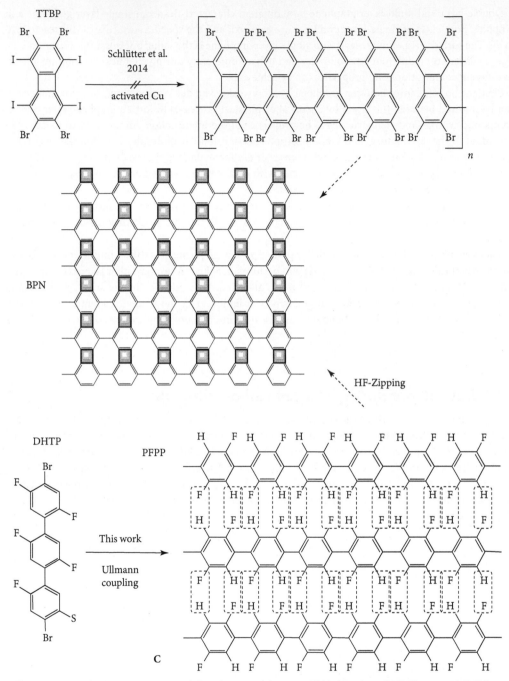

Fig. 4.1.13 Synthetic strategy toward the planar non-benzenoid biphenylene (4-6-8) network (BPN).

Reference

Q. Fan, L. Yan, M. W. Tripp, O. Krejčí, S. Dimosthenous, S. R. Kachel, M. Chen, A. S. Foster, U. Koert, P Liljeroth, and J. M. Gottfried, Biphenylene network: a nonbenzenoid carbon allotrope, *Science* **372**, 852–6 (2021).

4.1.8 **Carbon nanotubes**

Carbon nanotubes were discovered in 1991 and consist of elongated cages bounded by cylindrical walls constructed from rolled graphene (graphite-like) sheets. In contrast to the fullerenes, nanotubes possess a network of fused six-membered rings, and each terminal of the long tube is closed by a half-fullerene cap. Single-walled carbon nanotubes (SWNTs) with diameters in the range 0.4–3.0 nm have been observed experimentally; most of them lie within the range 0.6–2.0 nm, and those with diameter 0.7, 0.5, and 0.4 nm correspond to the fullerenes C_{60}, C_{36}, and C_{20}, respectively. Electron micrographs of the smallest 0.4 nm nanotubes prepared by the arc-discharge method showed that each is capped by half of a C_{20} dodecahedron and has an antichiral structure.

A carbon SWNT can be visualized as a hollow cylinder formed by rolling a planar sheet of hexagonal graphite (unit-cell parameters $a = 0.246$, $c = 0.669$ nm). It can be uniquely described by a vector $\mathbf{C} = n\mathbf{a_1} + m\mathbf{a_2}$, where $\mathbf{a_1}$ and $\mathbf{a_2}$ are reference unit vectors as defined in Fig. 4.1.14. The SWNT is generated by rolling up the sheet such that the two end-points of the vector $\mathbf{C}$ are superimposed. The tube is denoted as (n, m) with $n \geq m$, and its diameter D is given by:

$$D = |C|/\pi = a\left(n^2 + nm + m^2\right)^{1/2}/\pi.$$

The tubes with $m = n$ are called "armchair" and those with $m = 0$ are referred to as "zigzag." All others are chiral with the chiral angle θ defined as that between the vectors $\mathbf{C}$ and $\mathbf{a_1}$; θ can be calculated from the equation:

$$\theta = \tan^{-1}\left[3^{1/2}m/(m + 2n)\right]$$

The values of θ lies between 0° (for a zigzag tube) and 30° (for an armchair tube). Note that the mirror image of a chiral (n, m) nanotube is specified by $(n + m, -m)$. The three types of SWNTs are illustrated in Fig. 4.1.15.

There are multi-walled carbon nanotubes (MWNTs) each consisting of 10 inner tubes or more. In a carbon MWNT, the spacing between two adjacent co-axial zigzag tubes $(n_1, 0)$ and $(n_2, 0)$ is $\Delta d/2 = (0.123/\pi)(n_2 - n_1)$. However, this cannot be made to be close to $c/2 = 0.335$ nm (the interlayer separation in graphite) by any reasonable combination of n_2 and n_1, and hence no zigzag nanotube can exist as a component of a MWNT. On the other hand, a MWNT can be constructed for all armchair tubes $(5m,$

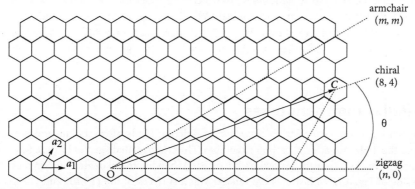

Fig. 4.1.14 Generation of the chiral (8,4) carbon nanotube by rolling a graphite sheet along the vector $\mathbf{C}$ = $n\mathbf{a_1} + m\mathbf{a_2}$, and definition of the chiral angle θ. The reference unit vectors $\mathbf{a_1}$ and $\mathbf{a_2}$ are shown, and the broken lines indicate the directions for generating achiral zigzag and armchair nanotubes.

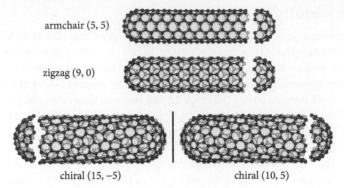

armchair (5, 5)

zigzag (9, 0)

chiral (15, −5)   chiral (10, 5)

Fig. 4.1.15 Lateral view of three kinds of carbon nanotubes with end caps: (a) armchair (5,5) capped by one-half of C_{60}, (b) zigzag (9,0) capped by one-half of C_{60}, and (c) an enantiomorphic pair of chiral SWNTs each capped by a hemisphere of icosahedral fullerene C_{140}.

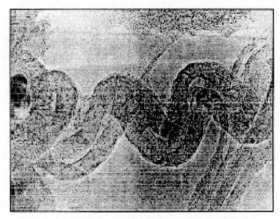

Fig. 4.1.16 A helical multi-walled carbon nanotube.

$5m$) with $m = 1, 2, 3$, etc., for which the inter-tube spacing is $(0.123/\pi)3^{1/2}(5) = 0.339$ nm, which satisfies the requirement.

In practice, defect-free co-axial nanotubes rarely occur in experimental preparations. The observed structures include the capped, bent, and toroidal SWNTs, as well as the capped and bent, branched, and helical MWNTs. Fig. 4.1.16 shows the HRTEM micrograph of a MWNT which incorporates a small number of five- and seven-membered rings into the graphene sheets of the nanotube surfaces.

4.1.9 **Filled carbon nanotubes**

Much research has been devoted to the insertion of different kinds of crystalline and non-crystalline material into the hollow interior of carbon nanotubes. The encapsulated species include fullerenes, clusters, one-dimensional (1D) metal nanowires, binary metal halides, metal oxides, and organic molecules.

The left side of Fig. 4.1.17 illustrates the van der Waals surfaces of the (10,10) and (12,12) armchair SWNTs with diameters of 1.36 and 1.63 nm and corresponding internal diameters of *ca.* 1.0 and 1.3 nm, respectively, as specified by the van der Waals radii of the sp^2 carbon atoms forming the walls. The (10,10) tube has the right size to accommodate a linear array of C_{60} molecules, as shown by the HRTEM images

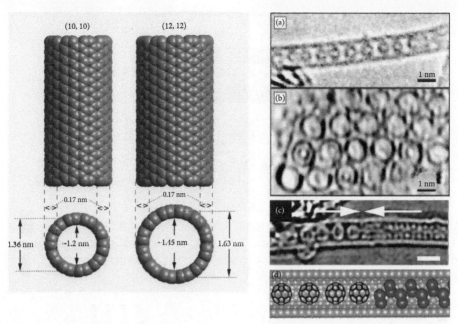

Fig. 4.1.17 Left: Schematic representations of the van der Waals surfaces of (10,10) and (12,12) armchair SWNTs. Right: (a) HRTEM image showing a (10,10) SWNT filled with C_{60} molecules. (b) Second image from the same specimen as in (a), showing a cross-sectional view of an ordered bundle of SWNTs, some of which are filled with fullerene molecules. (c) HRTEM image (scale bar = 1.5 nm) of a filled SWNT showing an interface between four C_{60} molecules (a possible fifth molecule is obscured at the left) and a 1D FeI_2 crystal. (d) Schematic structural representation of (c) (Fe atoms = small spheres; I atoms = large spheres). Courtesy of Prof. M. L. H. Green.

marked (a) and (b) on the right side of Fig. 4.1.17. Part (c) and (d) show the HRTEM image and modeling of an interface between four fullerene molecules and a 1D FeI_2 crystal.

The formation of an ordered 2×2 KI crystal column within a (10,10) SWNT with $D \sim 1.4$ nm is shown in Fig. 4.1.18. The structural model is illustrated in (a), and a cross-sectional view is shown in (b). The coordination numbers of the cation and anion are changed from 6:6 in bulk KI to 4:4 in the 2×2 column. Each dark spot in the HRTEM image (c) represents an overlapping I—K or K—I arrangement viewed in projection, which matches the simulated image (d). The spacing between spots along the SWNT is ~ 0.35 nm, corresponding to the {200} spacing in bulk KI, whereas across the SWNT capillary the spacing increases to ~ 0.4 nm, representing a $\sim 17\%$ tetragonal expansion.

The incorporation of a 3×3 KI crystal in a wider SWNT ($D \sim 1.6$ nm) is depicted in Fig. 4.1.19. In this case, there are three different coordination types (6:6, 5:5, and 4:4) are exhibited by atoms forming the central, face, and corner $\cdots$I—K—I—K$\cdots$ rows of the 3×3 column, respectively, along the tube axis. The <110> projection shows the K^+ and I^- sub-lattices as pure element columns, which are distinguishable by their scattering powers. The iodine atoms located along <110> all show a slight inward displacement relative to their positions in bulk KI, whereas the K atoms located along the same cell diagonal exhibit a small expansion.

It was found that KI can also be used to fill a DWNT, as shown in Fig. 4.1.20. The HRTEM image (a) and reconstructed image (b) of the KI@DWNT composite was analysed using a model composed of an inner (11,22) SWNT ($D \sim 2.23$ nm) and an outer (18,26) SWNT ($D \sim 3.04$ nm). The measured averaged spacings between the atomic columns gave values of 0.37 nm across the DWNT axis and 0.36 nm along

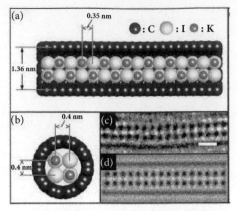

Fig. 4.1.18 A 2 × 2 KI crystal column filling a (10,10) SWNT. (a) Cutaway structural representation of composite model used in the simulation calculations. (b) End-on view of the model, showing an increased lattice spacing of 0.4 nm across the capillary in two directions (assuming a symmetrical distortion). (c) HRTEM image. (d) Simulated Image. Courtesy of Prof. M. L. H. Green.

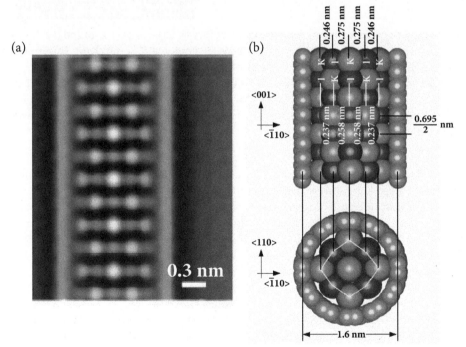

Fig. 4.1.19 (a) Reconstructed HRTEM image (averaged along the tube axis) of the <110> projection of a 3 × 3 KI crystal in a D ~1.6 nm diameter SWNT. Note that the contrast in this image is reversed so that regions of high electron density appear bright and low electron density appear dark. (b) Structural model derived from (a). Courtesy of Prof. M. L. H. Green.

it, which agree well with the {200} d-spacing of rock-salt KI (0.352 nm). In the simulation calculation, the encapsulated fragment was divided into three sections in order to model the lattice defects such as plane shear and plane rotation.

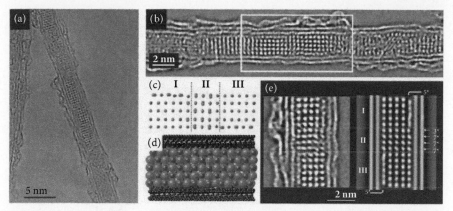

Fig. 4.1.20 (a) An isolated DWNT filled with KI. (b) Reconstructed image of the KI@DWNT composite; the marked area was structurally analyzed in detail. (c) Crystal model showing the distortions imposed and the three sub-sections considered for its construction. (d) Structural model of the marked section in (b). (e) Reconstructed versus simulated image of the three sections. Courtesy of Prof. M. L. H. Green.

Reference

P. M. F. L. Costa, S. Friedrichs J. Sloan, and M. L. H. Green, Imaging lattice defects and distortions in alkali-metal iodides encapsulated within double-walled carbon nanotubes, *Chem. Mater.* **17**, 3122–9 (2005).

4.2 **Compounds of Carbon**

More than 40 million compounds containing carbon atoms are now known, the majority of which are organic compounds that contain carbon-carbon bonds.

From the perspective of structural chemistry, the modes of bonding, coordination, and the bond parameters of a particular element in its allotropic modifications may be further extended to its compounds. Thus organic compounds can be conveniently divided into three families that originate from their prototypes: aliphatic compounds from diamond, aromatic compounds from graphite, and fullerenic compounds from fullerenes.

4.2.1 **Aliphatic compounds**

Aliphatic compounds comprise hydrocarbons and their derivatives in which the molecular skeletons consist of tetrahedral carbon atoms connected by C–C single bonds. These tetrahedral carbon atoms can be arranged as chains, rings, or finite frameworks, and often with an array of functional groups as substituents on various sites. The alkanes C_nH_{2n+2} and their derivatives are typical examples of aliphatic compounds.

Some frameworks of alicyclic compounds are derived from fragments of diamond, as shown in Fig. 4.2.1. In these molecules, all six-membered carbon rings have the chair conformation. Diamantane $C_{10}H_{20}$ is also named congressane as it was chosen as the logo of the XIXth Conference of IUPAC in London in 1963 as a challenging target for the participants; the successful synthesis was accomplished two years later. There are three structural isomers of tetramantane $C_{22}H_{28}$. X-ray analysis of *anti*-tetramantane has revealed an interesting bond-length progression: CH–CH$_2$ = 152.4 pm,

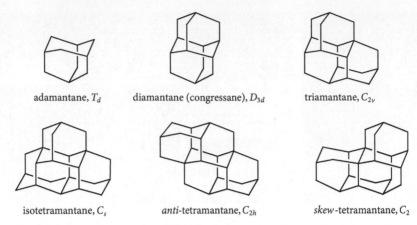

adamantane, T_d diamantane (congressane), D_{3d} triamantane, C_{2v}

isotetramantane, C_s *anti*-tetramantane, C_{2h} *skew*-tetramantane, C_2

Fig. 4.2.1 Some frameworks of alicyclic compounds as fragments of diamond.

C–CH$_2$ = 152.8 pm, CH–CH = 153.7 pm, and C–CH = 154.2 pm, approaching toward the limit of 154.45 pm in diamond as the number of bonded H atoms decreases.

4.2.2 **Aromatic compounds**

Graphite typifies the basic structural unit present in aromatic compounds, in which the planar carbon skeletons of these molecules and their derivatives can be considered as fragments of graphite, each consisting of carbon atoms which use their sp^2 hybrids to form σ bonds to one another, and overlap between the remaining parallel p_z orbitals gives rise to delocalized π bonding. The aromatic compounds may be divided into the following four classes.

(1) **Benzene and benzene derivatives**

Up to six hydrogen atoms in benzene can be mono- or poly-substituted by other atoms or groups to give a wide variety of derivatives. Up to six sterically bulky groups such as SiMe$_3$ and ferrocenyl ($C_5H_5FeC_5H_4$,Fc) groups can be substituted into a benzene ring. Hexaferrocenylbenzene, C_6Fc_6, is of structural interest as a supercrowded arene, a metalated hexakis(cyclopentadienylidene)radialene, and the core for the construction of "Ferris wheel" supermolecules. In the crystalline solvate $C_6Fc_6 \cdot C_6H_6$, the C_6Fc_6 molecule adopts a propeller-like configuration with alternating up and down Fc groups around the central benzene ring. Perferrocenylation causes the benzene ring to take a chair conformation with alternating C–C–C–C dihedral angles of ± 14º, and the elongated C–C bonds exhibit noticeable bond alternation averaging 142.7/141.1 pm The C_{ar}–C_{Fc} bonds average 146.9(5) pm.

(2) **Polycyclic benzenoid aromatic compounds**

These compounds consist of two or more benzene rings that are fused together, and the number of delocalized π electrons conforms to the Hückel ($4n + 2$) rule for aromaticity. Fig. 4.2.2(a) shows the carbon skeletons of some typical examples.

(3) **Non-benzenoid aromatic compounds**

Many aromatic compounds have considerable resonance stabilization but do not possess a benzene nucleus, or in the case of a fused polycyclic system, the molecular skeleton contains at least one ring that is not a benzene ring. The cyclopentadienyl anion $C_5H_5^-$, the cycloheptatrienyl cation $C_7H_7^+$, the aromatic annulenes (except for [6]annulene, which is benzene), azulene, biphenylene, and acenaphthylene (see Fig. 4.2.2(b)) are common examples of non-benzenoid aromatic hydrocarbons. The cyclic oxocarbon dianions $C_nO_n^{2-}$ (n = 3, 4, 5, 6) constitute a class of non-benzenoid aromatic compounds that are stabilized by two delocalized π electrons. Further details are given in Section 10.4.4.

(4) Heterocyclic aromatic compounds

In many cyclic aromatic compounds an element other than carbon (commonly N, O, and S) is also present in the ring. These compounds are called heterocycles. Fig. 4.2.3 shows some nitrogen heterocycles that are commonly used as ligands and planar, sunflower-like octathio[8]circulene, $C_{16}S_8$, that can be regarded as a novel form of carbon sulfide.

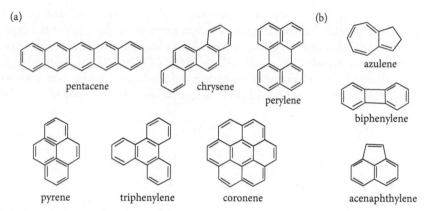

Fig. 4.2.2 Carbon skeletons of some polycyclic aromatic compounds of the (a) benzenoid type and (b) non-benzenoid type.

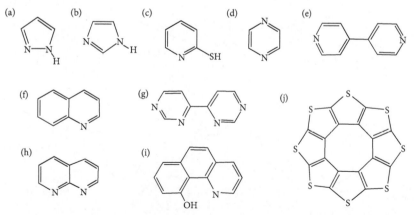

Fig. 4.2.3 Some commonly used heterocyclic nitrogen ligands: (a) pyrazole, (b) imidazole, (c) pyridine-2-thiol, (d) pyrazine, (e) 4,4′-bipyridine, (f) quinoline, (g) 4,4′-bipyrimidine, (h) 1,8-naphthyridine, (i) 1,10-phenanthroline. Compounds (a), (b), and (c) occur in metal complexes in the anionic (deprontonated) form with delocalization of the negative charge. Octathio[8]circulene (j) has a highly symmetric planar structure.

4.2.3 **Fullerenic compounds**

The derivatives of fullerenes are called fullerenic compounds, which are now mainly prepared from C_{60} and, to a lesser extent, from C_{70} and C_{84}.

Since efficient methods for the synthesis and purification of gram quantities of C_{60} and C_{70} become available in the early 1990s, fullerene chemistry has developed at a phenomenal pace. There are many reactions which can generate fullerenic compounds, as shown in Fig. 4.2.4. Unlike the aromatics, fullerenes have no hydrogen atoms or other groups attached, and so are unable to undergo substitution reactions. However, the globular fullerene carbon skeleton gives rise to an unprecedented diversity of derivatives. This unique feature leads to a vast number of products that may arise from addition of just one reagent. Substitution reactions can take place on derivatives, once these have been formed by addition. Some fullerenic compounds are briefly described below.

(1) **Fullerenes bonded to non-metallic elements**

This class of compounds consists of fullerene adducts with covalent bonds formed between the fullerene carbon atoms and non-metallic elements. Since all carbon atoms lie on the globular fullerene surface, the number of sites, as well as their positions, where additions take place vary from case to case. Examples of these compounds include $C_{60}O$, $C_{60}(CH_2)$, $C_{60}(CMe_3)$, $C_{60}Br_6$, $C_{60}Br_8$, $C_{60}Br_{24}$, $C_{50}Cl_{10}$, and $C_{60}[OsO_4(PyBu)_2]$. In the first two examples, the oxygen atom and the methylene carbon atom are each bonded to two carbon atoms on a (6/6) edge in the fullerene skeleton. The structure of $C_{60}O$ is shown in

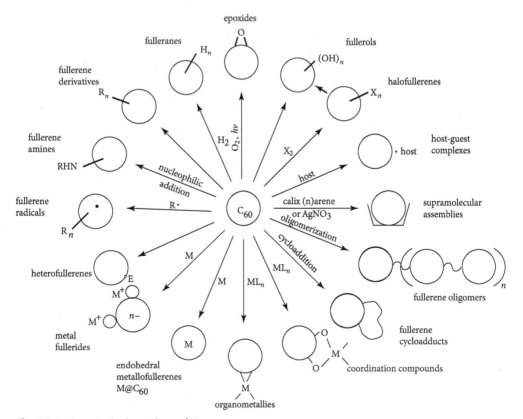

Fig. 4.2.4 The principal reactions of C_{60}.

Fig. 4.2.5(a). In the next five examples, the C atom of CMe_3, as well as Br and Cl, are each bonded to only one carbon atom of fullerene. The structure of $C_{60}Br_6$ is displayed in Fig. 4.2.5(b). In the final example, the six-coordinate osmium(VI) moiety $OsO_4(PyBu)_2$ is linked to C_{60} through the formation of a pair of O–C single bonds with two neighboring carbon atoms in the fullerene skeleton, as shown in Fig. 4.2.5(c).

The crystal structures of three spherical carbon chlorides $C_{50}Cl_{10}$, $C_{64}Cl_4$, and $C_{66}Cl_{10}$ have been determined by X-ray diffraction, as shown in Fig. 4.2.6. It can be seen that the three types of spherical carbon groups C_{50}, C_{64}, and C_{66} do not follow the rule of five-membered ring separation and hence they are unstable. They only exist as parts of compounds and it is difficult to form the corresponding elementary spherical carbon molecules.

(2) Coordination compounds of fullerene

This class of coordination compounds features direct covalent bonds between complexed metal groups and the carbon atoms of fullerene systems. Monoadducts such as $C_{60}Pt(PPh_3)_2$ each has only one group bonded to a fullerene, as shown in Fig. 4.2.7(a). Multiple adducts are formed when several groups are attached to the same fullerene nucleus. Typical examples are $C_{60}[Pt(PPh_3)_2]_6$ and $C_{70}[Pt(PPh_3)_2]_4$, whose structures are displayed in Fig. 4.2.7(b) and 4.2.7(c), respectively.

(3) Fullerenes as π-ligands

In this class of metal complexes, there is delocalized π bonding between fullerene and the metal atom. The structures of $(\eta^5\text{-}C_5H_5)Fe(\eta^5\text{-}C_{60}Me_5)$ and $(\eta^5\text{-}C_5H_5)Fe(\eta^5\text{-}C_{70}Me_3)$, each containing a fused ferrocene moiety, are shown in Fig. 4.2.8. The shared pentagonal carbon ring of the C_{60} (or C_{70}) skeleton acts as a 6π-electron donor ligand to the Fe(II) atom of the $Fe(C_5H_5)$ fragment. In $(\eta^5\text{-}C_5H_5)Fe(\eta^5\text{-}C_{60}Me_5)$, the five methyl groups attached to five sp^3 carbon atoms protrude outward at an angle of 42° relative to the symmetry axis of the molecule. The C_5H_5 group and cyclopentadienide in $Fe(C_{60}Me_5)$ are arranged in a

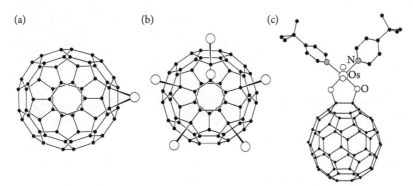

(a)   (b)   (c)

Fig. 4.2.5 Fullerenes bonded to non-metallic elements: (a) $C_{60}O$, (b) $C_{60}Br_6$, and (c) $C_{60}[OsO_4(PyBu)_2]$.

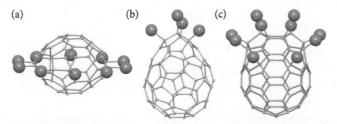

(a)   (b)   (c)

Fig. 4.2.6 Structures of three spherical carbon chlorides: (a) $C_{50}Cl_{10}$, (b) $C_{64}Cl_4$, and (c) $C_{66}Cl_{10}$.

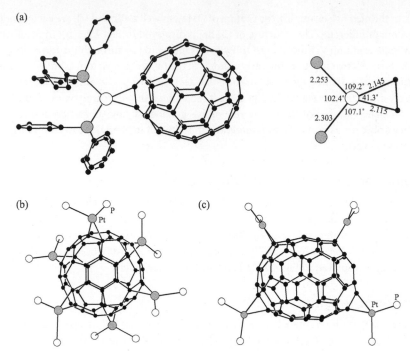

Fig. 4.2.7 Molecular structure of (a) $C_{60}Pt(PPh_3)_2$, (b) $C_{60}[Pt(PPh_3)_2]_6$, and (c) $C_{70}[Pt(PPh_3)_2]_4$.

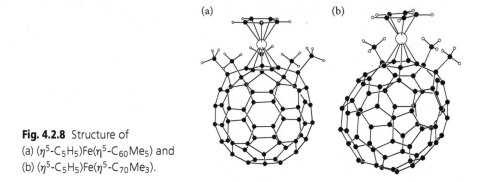

Fig. 4.2.8 Structure of
(a) $(\eta^5\text{-}C_5H_5)Fe(\eta^5\text{-}C_{60}Me_5)$ and
(b) $(\eta^5\text{-}C_5H_5)Fe(\eta^5\text{-}C_{70}Me_3)$.

staggered manner; the C–C bond lengths are 141.1 pm (averaged for C_5H_5) and 142.5 pm (averaged for $C_{60}Me_5$). The Fe–C distances are 203.3 pm (for C_5H_5) and 208.9 pm (for $C_{60}Me_5$), which are comparable to those in known ferrocene derivatives. The structural features of $(\eta^5\text{-}C_5H_5)Fe(\eta^5\text{-}C_{70}Me_3)$ are similar to those of $(\eta^5\text{-}C_5H_5)Fe(\eta^5\text{-}C_{60}Me_5)$. The C–C bond lengths in the shared pentagon are 141–143 pm. The Fe–C bond distances are 205.4 pm (averaged for C_5H_5) and 208.3 pm (averaged for $C_{70}Me_3$).

(4) Metal fullerides

Fullerenes exhibit an electron-accepting nature and react with strong reducing agents, such as the alkali metals, to yield metal fulleride salts. The compounds $Li_{12}C_{60}$, $Na_{11}C_{60}$, M_6C_{60} (M = K, Rb, Cs), K_4C_{60} and M_3C_{60} (M_3 = K_3, Rb_3, $RbCs_2$) have been prepared.

The M_3C_{60} compounds are particularly interesting as they become superconducting materials at low temperature, with transition temperatures (T_c) of 19 K for K_3C_{60}, 28 K for Rb_3C_{60}, and 33 K for $RbCs_2C_{60}$.

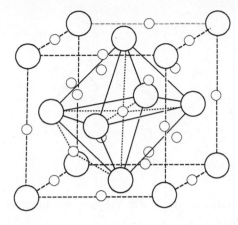

Fig. 4.2.9 Crystal structure of K_3C_{60} (large circles represent C_{60} and small circles represent K).

Fulleride K_3C_{60} is a face-center cubic crystal, space group $Fm\bar{3}m$, with $a = 1424(1)$ pm and $Z = 4$. The C_{60}^{3-} ions form a ccp structure with K^+ ions filling all the tetrahedral interstices (radius 112 pm) and octahedral interstices (radius 206 pm), as shown in Fig. 4.2.9.

Both K_4C_{60} and Rb_4C_{60} are tetragonal, and the structure of M_6C_{60} at room temperature is body-centered cubic. These metal fullerides are all insulators.

(5) Fullerenic supramolecular adducts

Fullerenes C_{60} and C_{70} form supramolecular adducts with a variety of molecules, such as crown ethers, ferrocene, calixarene, and hydroquinone. In the solid state, the intermolecular interactions may involve ionic interaction, hydrogen bonding, and van der Waals forces. Fig. 4.2.10 shows a part of the structure of $[K(18C6)]_3 \cdot C_{60} \cdot (C_6H_5CH_3)_3$, in which C_{60}^{3-} is surrounded by a pair of $[K^+(18C6)]$ complexed cations.

(6) Fullerene oligomers and polymers

This class of compounds contain two or more fullerenes and they may be further divided into the following three categories.

(a) Dimers and polymers containing two or more fullerenes linked together through C–C covalent bonds formed by the carbon atoms on the globular surface. Fig. 4.2.11(a) shows the structure of

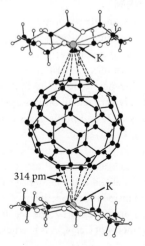

Fig. 4.2.10 A part of the structure of $[K(18C6)]_3 \cdot C_{60} \cdot (C_6H_5CH_3)_3$.

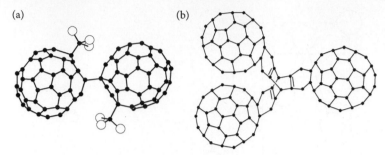

(a)   (b)

Fig. 4.2.11 Structures of (a) $[C_{60}(^tBu)]_2$ and (b) $[C_{60}]_3(C_{14}H_{14})$.

dimeric $[C_{60}(^tBu)]_2$. In addition, chain-like polymeric fullerenes with the following structure have been proposed, although none has yet been prepared.

$$[C_{60}H_2 - C_{60}H_2 - C_{60}H_2]_n$$

(b) Two or more fullerenes are linked together through the formation of C–C covalent bonds with an organic functional group. Fig. 4.2.11(b) shows the structure of $[C_{60}]_3(C_{14}H_{14})$, in which three fullerenes form an adduct through C–C linkage with the central $C_{14}H_{14}$ unit.

(c) Fullerenes are linked as pendants at regular intervals to a skeleton of a polymeric chain, such as:

$$\sim\!\!\sim\!\!\sim\ \underset{\underset{C_{60}H}{|}}{CH\!-\!CR\!=\!CH\!-\!CH_2}\ \sim\!\!\sim\!\!\sim\ \underset{\underset{C_{60}H}{|}}{CH\!-\!CR\!=\!CH\!-\!CH_2}\ \sim\!\!\sim\!\!\sim$$

(7) Heterofullerenes

Heterofullerenes are fullerenes in which one or more carbon atoms in the cage are replaced by other main-group atoms. Compounds containing boron, such as $C_{59}B$, $C_{58}B_2$, $C_{69}B$, $C_{68}B_2$ have been detected in the mass spectra of the products obtained using boron/graphite rods in arc discharge synthesis. Since nitrogen has one more electron than carbon, the azafullerenes are radicals ($C_{59}N\cdot$, $C_{69}N\cdot$), which can either dimerise to give $(C_{59}N)_2$ and $(C_{69}N)_2$, or take up hydrogen to give $C_{59}NH$.

(8) Endohedral fullerenes (incar-fullerenes)

Fullerenes can encapsulate various atoms within the cages, and these compounds have been referred to as endohedral fullerenes. For example, the symbolic representations La@C_{60} and La$_2$@C_{80} indicate that the fullerene cage encapsulates one and two lanthanum atom(s), respectively. The IUPAC description refers to these fullerenes species as incar-fullerenes, and the formulas are written as iLaC_{60} and iLa$_2C_{80}$, (i is derived from *incarcerane*). Some metal endohedral fullerenes are listed in Table 4.2.1. The endohedral fullerenes are expected to have interesting and potentially very useful bulk properties as well as a fascinating chemistry. Some non-metallic elements, such as N, P, and noble gases can be incarcerated into fullerenes to form N@C_{60}, P@C_{60}, N@C_{70}, Sc$_3$N@C_{80}, Ar@C_{60}, etc.

Theoretical and experimental studies of the structures and electronic properties of endohedral fullerenes have yielded many interesting results. The enclosed N and P atoms of N@C_{60} and P@C_{60} retain their atomic ground state configuration and are localized at the center of the fullerenes, as shown in Fig. 4.2.12(a). The atoms are almost freely suspended inside the respective molecular cages and exhibit properties resembling those of ions in electromagnetic traps. In Ca@C_{60}, the Ca atom lies at an off-center position by 70 pm, as shown in Fig. 4.2.12(b). The symmetry of Ca@C_{60} is predicted to be C_{5v}, implying

Table 4.2.1 Endohedral fullerenes

Fullerene	Metallic atom	Fullerene	Metallic atom
C_{36}	U	C_{56}	U_2
C_{44}	K, La, U	C_{60}	Y_2, La_2, U_2
C_{48}	Cs	C_{66}	Sc_2
C_{50}	U, La	C_{74}	Sc_2
C_{60}	Li, Na, K, Rb, Cs, Ca, Ba, Co,	C_{76}	La_2
	Y, La, Ce, Pr, Nd, Sm, Eu, Gd,	$C_{79}N$	La_2
	Tb, Dy, Ho, Lu, U	C_{80}	La_2, Ce_2, Pr_2
C_{70}	Li, Ca, Y, Ba, La, Ce, Gd, Lu, U	C_{82}	Er_2, Sc_2, Y_2, La_2, Lu_2,
C_{72}	U		Sc_2C_2
C_{74}	Ca, Sc, La, Gd, Lu	C_{84}	Sc_2, La_2, Sc_2C_2
C_{76}	La	C_{68}	Sc_3N
C_{80}	Ca, Sr, Ba	C_{78}	Sc_3N
$C_{81}N$	La	C_{80}	Sc_3N, $ErSc_2N$, Sc_2LaN,
C_{82}	Ca, Sr, Ba, Sc, Y, La, Ce, Pr,		$ScLa_2N$, La_3N
	Nd, Sm, Eu, Gd, Tb, Dy, Ho,	C_{82}	Sc_3
	Er, Tm, Yb, Lu	C_{84}	Sc_3
C_{84}	Ca, Sr, Ba, Sc, La	C_{82}	Sc_4

that the Ca atom lies on a five-fold rotation axis of the C_{60} cage. The predicted distances between the Ca atom and the first and second set of nearest C atoms are 279 pm and 293 pm, respectively.

A single crystal X-ray diffraction study of $[Sc_3N@C_{78}]\cdot[Co(OEP)]\cdot 1.5(C_6H_6)\cdot 0.3(CHCl_3)$ (OEP is the dianion of octaethylporphyrin) showed that the fullerene is embraced by the eight ethyl groups of the porphyrin macrocycle. The structure of $[Sc_3N@C_{78}]$ is shown in Fig. 4.2.12(c). The N–Sc distances range from 198 pm to 212 pm, and the shortest C–Sc distances fall within a narrow range of 202–211 pm. The flat Sc_3N unit is oriented so that it lies near the equatorial mirror plane of the C_{78} cage.

A synchrotron X-ray powder diffraction study of $(Sc_2C_2)@C_{84}$ showed that the lozenge-shaped Sc_2C_2 unit is encapsulated by the D_{2d}-C_{84} fullerene, as shown in Fig. 4.2.12(d). The Sc–Sc, Sc–C, and C–C distances in the Sc_2C_2 unit are 429 pm, 226 pm, and 142 pm, respectively.

(9) Open-cage fullerenes

An open-cage fullerene compound is formed by breaking part of the moleculear polyhedron in a chemical reaction, such that the C atoms at the opening are connected to other atoms like N and O of molecular groups. Small molecules such as H_2O, H_2, etc. can be encapsulated in the opened cage. Fig. 4.2.13 shows the structure of the two open-cage fullerene compounds entrapping the H_2O molecule. Successful preparation of various types of the above-mentioned spherical carbon-cage compounds has inspired the pursuit of synthetic chemists. Based on the six-membered backbone and other aromatic species, fruitful synthetic research has been carried out to yield a variety of new products. With spherical C_{60} as the primary skeleton, plus other fullerene molecules of different sizes and shapes, a variety of componds of fullerene-type can be synthesized to broaden a new field in the forseeable future.

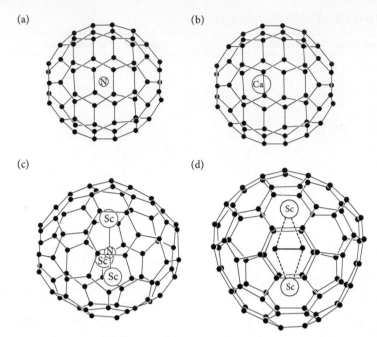

Fig. 4.2.12 Structures of some endofullerenes: (a) N@C$_{60}$, (b) Ca@C$_{60}$, (c) Sc$_3$N@C$_{78}$, and (d) Sc$_2$C$_2$@C$_{84}$.

Fig. 4.2.13 Structures of open-cage fullerence. (a) H$_2$O@C$_{59}$O$_6$(NC$_6$H$_5$) and (b) H$_2$O@C$_{59}$O$_5$(OH)(OO^{t}Bu)(NC$_6$H$_4$Br). For clarity H atoms are not shown. "●*" represents H$_2$O molecule inside the cages in (a) and (b), whereas "●**" represents OH group in (b).

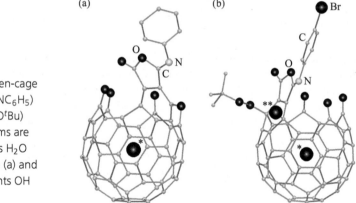

References

X. Han, S.-J. Zhou, Y.-Z. Tan, X. Wu, F. Gao, Z.-J. Liao, R.-B. Huang, Y.-Q. Feng, X. Lu, S.-Y. Xia, and L-S. Zhen, Crystal structures of saturn-like C$_{50}$Cl$_{10}$ and pineapple-shaped C$_{64}$Cl$_4$: geometric implications of double- and triple-pentagon-fused chlorofullerenes. *Angew. Chem. Int. Ed.* **47**, 5340–3 (2008).

C.-L. Gao, X. Li, Y.-Z. Tan, X.-Z. Xu, Q.-Y. Zhang, S.-Y. Xie, and R.-B. Huang, Synthesis of long-sought C$_{66}$ with exohedral stabilization, *Angew. Chem. Int. Ed.* **53**, 7853–5 (2014).

Z. Xiao, J. Yao, D. Yang, F. Wang, S. Huang, L. Gan, Z. Jia, Z. Jiang, X. Yang, B. Zheng, G. Yuan, S. Zhang, and Z. Wang, Synthesis of [59]fullerenones through peroxide-mediated stepwise cleavage of fullerene skeleton bonds and X-ray structures of their water-encapsulated open-cage complexes. *J. Am. Chem. Soc.* **129**, 16149–62 (2007).

4.3 **Bonding in Carbon Compounds**

4.3.1 **Types of bonds formed by the carbon atom**

The capacity of the carbon atom to form various types of covalent bonds is attributable to its unique characteristics as an element in the Periodic Table. The electronegativity of the carbon atom is 2.5, which means that the carbon atom cannot easily gain or lose electrons to form an anion or cation. As the number of valence orbitals is exactly equal to the number of valence electrons, the carbon atom cannot easily form a lone pair or electron deficient bonds. Carbon has a small atomic radius, so its orbitals can overlap effectively with the orbitals of neighbor atoms in a molecule.

For simplicity, we use the conventional concept of hybridization to describe the bond types, but bonding is frequently more subtle and more extended than implied by this localized description. The parameters of typical hybridization schemes of the carbon atom are listed in Table 4.3.1.

The hybrid orbitals of carbon always overlap with orbitals of other atoms in a molecule to form σ bonds. The remaining p orbitals can then be used to form π bonds, which can be classified into two categories: localized and delocalized. The localized π bonds of carbon form double and triple bonds are illustrated as follows:

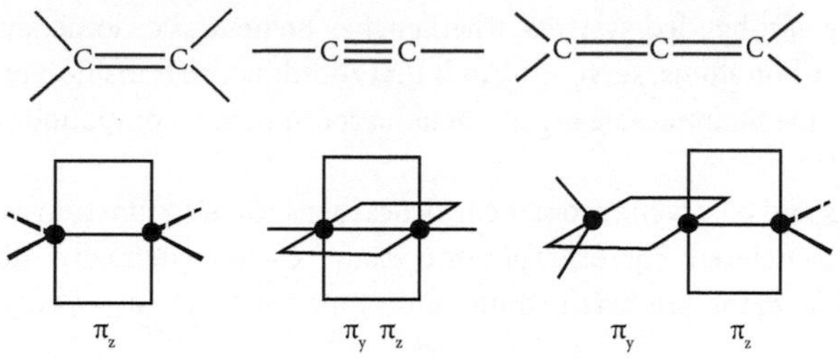

The carbon atom can also form σ and π bonds to metal atoms in various fashions. For example:

(a) 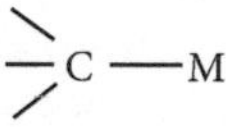 $\overset{\diagdown}{\underset{\diagup}{C}}$—M   single bond

(b) 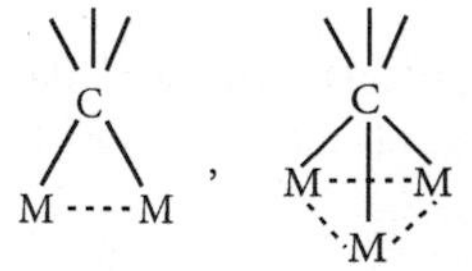 , ... polycenter metal-carbon bonds

(c) 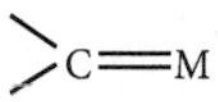 $\overset{\diagdown}{C}$=M   double bond

(d) 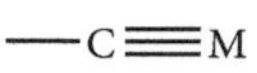 —C≡M   triple bond

Table 4.3.1 Hybridization schemes of carbon atom

	sp	*sp²*	*sp³*
Number of orbitals	2	3	4
Inter-orbital angle	180°	120°	109.47°
Geometry	linear	trigonal	tetrahedral
% *s* character	50	33	25
% *p* character	50	67	75
Electronegativity of carbon	3.29	2.75	2.48
Remaining *p* orbitals	2	1	0

A particularly interesting example is the tungsten complex

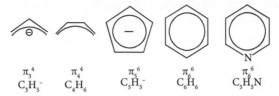

which contains C≡W, C=W, and C−W bonds with lengths 179 pm, 194 pm, and 226 pm, respectively. The delocalized π bonds involve three or more carbon atoms or heteroatoms. For instance:

π_3^4 π_4^4 π_5^6 π_6^6 π_6^6

$C_3H_5^-$ C_4H_6 $C_5H_5^-$ C_6H_6 C_5H_5N

An enormous variety of π-bonded systems, whether they be neutral or ionic, cyclic or linear, odd or even in the number of carbon atoms, serve as ligands that coordinate to transition metals. Fig. 4.3.1 shows some representatives of the innumerable organometallic coordination compounds stabilized by metal-π bonding.

Larger aromatic rings and polycyclic aromatic hydrocarbons can also function as π ligands in forming sandwich-type metal complexes. For example, the planar cyclooctatetraenyl dianion $C_8H_8^{2-}$ functions as a η^8-ligand to form the sandwich compound uranocene, $U(C_8H_8)_2$, which takes the eclipsed configuration.

Recently the "sandwich" motif has been extended to the case of two cyclic aromatic ligands flanking a small planar aggregate of metal atoms. In $[Pd_3(\eta^7-C_7H_7)_2Cl_3](PPh_4)$, a triangular unit of palladium(0) atoms each coordinated by a terminal chloride ligand is sandwiched between two planar cycloheptatrienyl $C_7H_7^+$ rings, as shown in Fig. 4.3.2(a, b). The measured Pd–Pd (274.5 to 278.9 pm) and Pd–Cl (244.2 to 247.1 pm) bonds are within the normal ranges.

In $[Pd_5(naphthacene)_2(toluene)][B(Ar_f)_4]_2 \cdot 3toluene$, where $Ar_f = 3,5$-$(CF_3)_2C_6H_3$, the sheet-like pentapalladium(0) core is composed of a triangle sharing an edge with a trapezoid; the innermost Pd–Pd

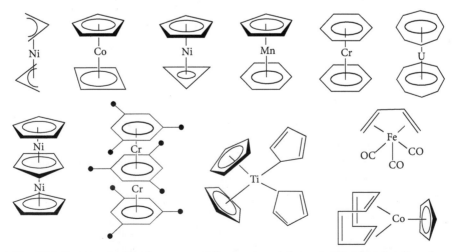

Fig. 4.3.1 Some examples of organometallic compounds formed with π bonding ligands.

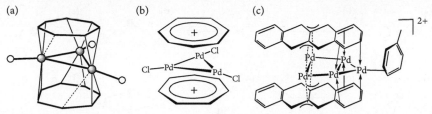

Fig. 4.3.2 (a) Molecular geometry and (b) structural formula of the $[Pd_3(\eta^7\text{-}C_7H_7)_2Cl_3]^-$ ion; the Pd–C and Pd$\cdots$C bond lengths are in the ranges 215–247 and 253–261 pm, respectively. (c) Structure of the $[Pd_5(C_{18}H_{12})_2(C_7H_8)]^{2+}$ ion; coordination of an electronically delocalized C_3 fragment to a Pd center is represented by a broken line.

distance (291.6 pm) is relatively long. This metal monolayer is sandwiched between two naphthacene radical cations, each of which coordinates to the Pd$_5$ sheet through 12 carbons via the $\mu_5\text{-}\eta^2\text{:}\eta^2\text{:}\eta^2\text{:}\eta^3\text{:}\eta^3$ mode, as illustrated in Fig. 4.3.2(c). One of the four independent toluene molecules in the unit cell is located near the apex Pd atom at a closest contact of 252 pm, but its coordination mode (either η^2 or η^1) cannot be definitively assigned owing to disorder.

4.3.2 Coordination numbers of carbon

Carbon is known with all coordination numbers from 0 to 8. Some typical examples are given in Table 4.3.2, and their structures are shown in Fig. 4.3.3. In these examples, the compounds with the

Table 4.3.2 Coordination numbers of carbon

Coordination number	Examples	Structure in Fig. 4.3.2
0	C atoms, gas phase	
1	$\dot{C}O$, stable gas	(a)
2	CO_2, stable gas	(b) linear
2	HCN, stable gas	(c) linear
2	$:CX_2$ (carbene), X = H, F, OH	(d) bent
3	C=OXY (oxohalides, ketones)	(e) planar
3	CH_3^-, CPh_3^-	(f) pyramidal
3	$Ta(=CHCMe_3)_2(Me_3C_6H_2)(PMe_3)_2$	(g) T-shaped*
4	CX_4 (X = H, F, Cl)	(h) tetrahedral
4	$Fe_4C(CO)_{13}$	(i) C capping Fe_4
5	Al_2Me_6	(j) bridged dimer
5	$(Ph_3PAu)_5C^+ \cdot BF_4^-$	(k) trigonal bipyramidal
6	$[Ph_3PAu]_6C^{2+}$	(l) octahedral
6	$C_2B_{10}H_{12}$	(m) pentagonal pyramidal
7	$[LiMe]_4$ crystal	(n) †
8	$[(Co_8C(CO)_{18}]^{2-}$	(o) cubic

* The unique H is equatorial and angle Ta=C–CMe$_3$ is 169°.

† The distance of intramolecular C–Li is 231 pm, (a C atom caps 3 Li atoms in each face of the Li$_4$ tetrahedron), C–H is 96 pm, and intermolecular C–Li is 236 pm.

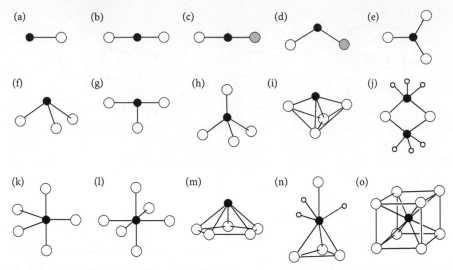

Fig. 4.3.3 Coordination numbers of carbon in its compounds.

high coordination numbers, such as ≥ 5, do not belong to the class of hypervalent compounds but rather to electron-deficient systems. Hypervalent molecules usually have a central atom which requires the presence of more than an octet of electrons to form more than four 2c-2e bonds, such as the S atom in SF_6.

4.3.3 Bond lengths of C–C and C–X bonds

The bond lengths of carbon-carbon bonds are listed in Table 4.3.3. The bond lengths of some important bond types of carbon-heteroatom bonds are given in Table 4.3.4. The values given here are average values from experimental determinations and do not necessarily exactly apply to a particular compound.

4.3.4 Factors influencing bond lengths

The measured bond lengths in a molecule provide valuable information on its structure and properties. Some factors that influence the bond lengths are discussed below:

(1) Bonds between atoms of different electronegativities

It is generally found that the greater the difference in electronegativity between the bonding atoms, the greater the deviation from the predicted bond distance based on covalent radii or the mean values given in Tables 4.3.3 and 4.3.4. Empirical methods have been proposed for the adjustment of covalent bond lengths by a factor that is dependent on the electronegativity differences between the atoms.

(2) Steric strain

Steric strain exists in a molecule when bonds are forced to make abnormal angles. There are in general two kinds of structural features that result in sterically caused abnormal bond angles. One of these is common to small ring compounds, where the bond angles must be less than those resulting from normal orbital

Table 4.3.3 Bond lengths (in pm) of carbon-carbon bonds

Bond	Bond length	Example
C–C		
sp^3–sp^3	153	ethane (H_3C–CH_3)
sp^3–sp^2	151	propene (H_3C–CH=CH_2)
sp^3–sp	147	propyne (H_3C–C≡CH)
sp^2–sp^2	148	butadiene (H_2C=CH–CH=CH_2)
sp^2–sp	143	vinylacetylene (H_2C=CH–C≡CH)
sp–sp	138	butadiyne (HC≡C–C≡CH)
C=C		
sp^2–sp^2	132	ethylene (H_2C=CH_2)
sp^2–sp	131	allene (H_2C=C=CH_2)
sp–sp	128	butatriene (H_2C=C=C=CH_2)
C≡C		
sp–sp	118	acetylene (HC≡CH)

Table 4.3.4 Bond lengths (in pm) of C–X bonds

C–H			C–N			C–S		
				sp^3–N	147		sp^3–S	182
	sp^3–H	109		sp^2–N	138		sp^2–S	175
	sp^2–H	108	C=N				sp–S	168
	sp–H	108		sp^2–N	128	C=S		
C–O			C≡N				sp^2–S	167
	sp^3–O	143		sp–N	114	C–Si		
	sp^2–O	134	C–P				sp^3–Si	189
C=O				sp^3–P	185	C=Si		
	sp^3–O	121	C=P				sp^2–Si	170
	sp^2–O	116		sp^2–P	166			
			C≡P					
				sp–P	154			
C–X	X =		F	Cl		Br	I	
	sp^3–X		140	179		197	216	
	sp^2–X		134	173		188	210	
	sp–X		127	163		179	199	

overlap. Such strain is called small-angle strain. The other arises when non-bonded atoms are forced into close proximity by the geometry of the molecule. This effect is known as steric overcrowding.

Steric strain generally leads to elongated bond lengths as compared to the expected values given in Tables 4.3.3 and 4.3.4.

(3) Conjugation

A hybrid atomic orbital of higher s content has a smaller size and lies closer to the nucleus. Accordingly, carbon-carbon bonds are shortened by increasing s character of the overlapping hybrid orbitals. The $C(sp^3)$–$C(sp^3)$ single bond is generally longer than other single bonds involving sp^2 and sp carbon atoms. This general rule arises mainly from conjugation between the bonded atoms. When mean values are used to estimate bond lengths, the conjugation factor must be taken into account. For example, the C–C bond length in benzene is 139.8 pm, which is virtually equal to the mean value of the $C(sp^2)$–$C(sp^2)$ and C=C bond lengths: ½(148 + 132) = 140 pm. Hence the conjugation of alternating single and double bonds in benzene can be expressed as resonance between two limiting valence bond (canonical) structures:

(4) Hyperconjugation

The overlap of a C–H σ orbital with the π (or p) orbital on a directly-bonded carbon atom is termed hyperconjugation. This interaction has a shortening effect on the C–C bond length, a good example being the structure of the *tert*-butyl cation $[C(CH_3)_3]^+$.

The structure of $[C(CH_3)_3]^+$ in the crystalline salt $[C(CH_3)_3]Sb_2F_{11}$ is shown in Fig. 4.3.4(a). The carbon skeleton is planar with an average C–C bond length of 144.2 pm and approximate D_{3h} molecular symmetry. The experimental C–C bond length is shorter by 6.8 pm than the normal $C(sp^3)$–$C(sp^3)$ bond length of 151 pm This is due to hyperconjugative interaction between three filled C–H σ bond orbitals and the empty p orbital on the central carbon atom that leads to partial C–C π bonding, as shown in Fig. 4.3.4 (b).

(5) Surroundings of bonding atoms

The geometry and connected groups of bonding atoms usually influence the bond lengths. For example, an analysis of C–OR bond distances in more than 2000 ethers and carboxylic esters (all with sp^3 carbon) showed that this distance increases with increasing electron-withdrawing power of the R group, and also

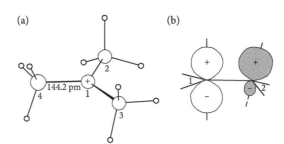

Fig. 4.3.4 Hyperconjugation in $[C(CH_3)_3]^+$: (a) structure of $[C(CH_3)_3]^+$ and (b) hyperconjugation between central carbon atom C(1) and one of the C–H σ bonds.

when the C atom changes from primary through secondary to tertiary. For such compounds mean C–O bond lengths range from 141.8 pm to 147.5 pm.

As an illustrative example taken from the current literature, consider the variation of C–C and C–O bond lengths in the deltate species $C_3O_3{}^{2-}$ held within a dinuclear organometallic uranium(IV) complex. In a remarkable synthesis, this cyclic aromatic oxocarbon dianion is generated by the metal-mediated reductive cyclotrimerization of carbon monoxide, as indicated in the following reaction scheme:

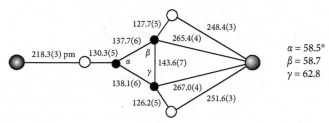

A strongly reducing organouranium(III) complex stabilized by η^5-cyclopentadienide and η^8-cyclooctatetraenediide ligands, together with tetrahydrofuran, is used to crack the robust C≡O triple bond at room temperature and ambient pressure. Low-temperature X-ray analysis revealed the presence of a reductively homologated CO trimer, $C_3O_3{}^{2-}$, as a $\eta^1{:}\eta^2$ bridging ligand between two organouranium(IV) centers. The measured dimensions of the central molecular skeleton are shown below:

127.7(5) 248.4(3)
137.7(6) 265.4(4)
218.3(3) pm 130.3(5) α β 143.6(7)
138.1(6) 267.0(4)
126.2(5) 251.6(3)

$\alpha = 58.5°$
$\beta = 58.7$
$\gamma = 62.8$

The noticeable distortion of the C_3 ring and measured C–C, C–O, and U–O bond lengths reflect the conjugation effect, the chemical environment of individual atoms, steric congestion around the respective uranium centers, as well as variation caused by experimental errors.

4.3.5 **Abnormal carbon-carbon single bonds**

(1) **Unusually long C–C bonds**

Examples of organic molecules containing very long carbon-carbon single bonds have been reported in recent years. The extended $C(sp^3)$–$C(sp^3)$ bond in the hexaarylethane shown in Fig. 4.3.5(a) is caused by severe steric repulsion between the bulky substituted phenyl groups. In the bi(anthracene-9,10-dimethylene) photodimer shown in Fig. 4.3.5(b), the bridge bond of the cyclobutane ring has a longer length than the rest.

(a)

$CAr_3 \underset{167(3)}{\overline{\quad\quad\quad}} CAr_3$

$Ar =$ 'Bu 'Bu

(b)

... 164.8(3)

Fig. 4.3.5 Molecules containing very long C–C single bonds. The bond lengths are shown in pm.

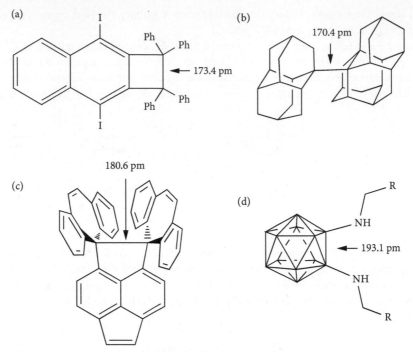

Fig. 4.3.6 Some compounds with ultra-long C–C bonds.

Recent studies have revealed the existence of a number of exceptionally long C–C single bonds in compounds having extraordinary ring strain, extreme steric congestion, or electronic perturbations. Some examples reported in the past decade are shown in Fig. 4.3.6. The current record is held by a diamine-substituted *ortho*-carborane, which exhibits an inner-cluster C–C single-bond length of 193.1 pm (see Fig. 4.3.6(d)).

(2) Unusually short bonds between tetracoordinate carbon atoms

Fig. 4.3.7 shows some organic molecules containing abnormally short single bonds between two four-coordinate carbon atoms.

The bond between two bridgehead carbon atoms in bicyclo[1.1.0]butane exhibits the properties of a carbon-carbon multiple bond, although it is formally a single bond. The dihedral angle δ between the three-membered rings in 1,5-dimethyltricyclo[2.1.0.0]pentan-3-one is made small by the short span of the carbonyl linkage, as shown in Fig. 4.3.7(a). The bridgehead bond has a pronounced π character with a length of 140.8(2) pm. In the 1,5-diphenyl analog, the two aromatic rings are oriented almost perpendicular (at 93.6° and 93.6°) to the plane bisecting the angle δ. There is optimal conjugation between the phenyl groups via the π-population of the bridge bond, and the conjugation effect leads to its lengthening of the latter to 144.4(3) pm. The central exocyclic $C(sp^3)$–$C(sp^3)$ bond connecting two bicyclobutane moieties is quite short (Fig. 4.3.7(b)), as are the related linkages in bicubyl (Fig. 4.3.7(c)) and hexakis(trimethylsilyl)bitetrahedryl (Fig. 4.3.7(d)). The calculated s character of the linking C–C bond in the bitetrahedryl molecule is $sp^{1.53}$, which is consistent with its significant shortening.

In the crystal structure of the *in*-isomer of the methylcyclophane shown in Fig. 4.3.7(e), there are two independent molecules with measured C–Me bond distances of 147.5(6) pm and 149.5(6) pm.

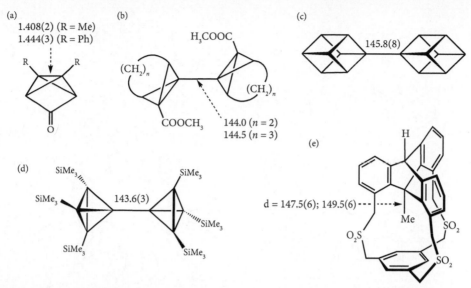

Fig. 4.3.7 Molecules containing very short C–C single bonds. The bond lengths are shown in pm. Values are given for two independent molecules of compound (e).

The inward-pointing methyl group is forced into close contact with the basal aromatic ring, and the steric congestion accounts for compression of the C–Me bond.

References

J. Li, R. Pang, Z. Li, G. Lai, X.-Q. Xiao, and T. Müller, Exceptionally long C–C single bonds in diamino-*o*-carborane as induced by negative hyperconjugation. *Angew. Chem. Int. Ed.* **58**, 1397–401 (2019).

Y. Ishigaki, T. Shimajiri, T. Takeda, R. Katoono, and T. Suzuki, Longest C–C single bond among neutral hydrocarbons with a bond length beyond 1.8 Å. *Chem* **4**, 795–806 (2018).

4.3.6 **Highly twisted polyhexacycles**

Acenes (tetracene, pentacene, hexacene, etc.), which are composed of (4, 5, 6, respectively) linearly fused benzene rings, possess small band gaps that quickly decrease along the length of their planar conjugated π system to make them promising high-conductive materials. Among them, pentacene (reported by E. Clar in 1929) has been employed as a semiconductor in organic electronics, but the synthesis of hexacene and higher homologs remains a challenging target for chemists as they are susceptible to two problems, namely poor solubility and high reactivity, with linear extension of the π system.

Highly twisted polyhexacycles can be prepared by benzannulation of the ends of acenes. The current "world record-holder" is the longitudinally twisted octaphenyltetrabenzo[*a,c,n,p*]hexacene **1**, which displays a remarkable end-to-end twist angle of 184°.

Dodecaphenyltetracene **2** is the largest perphenylacene known up to 2019. X-ray structure analysis established that it exhibits D_2 molecular symmetry with an end-to-end twist of 97°. Its central acene moiety is encapsulated by the peripheral phenyl substituents, and consequently it is relatively unreactive and even displays reversible electrochemical oxidation and reduction.

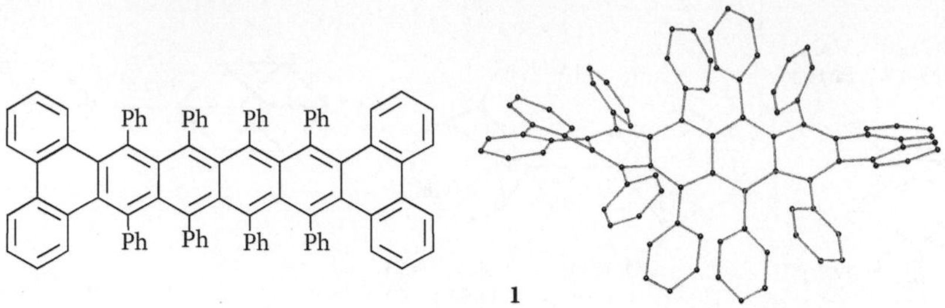

1

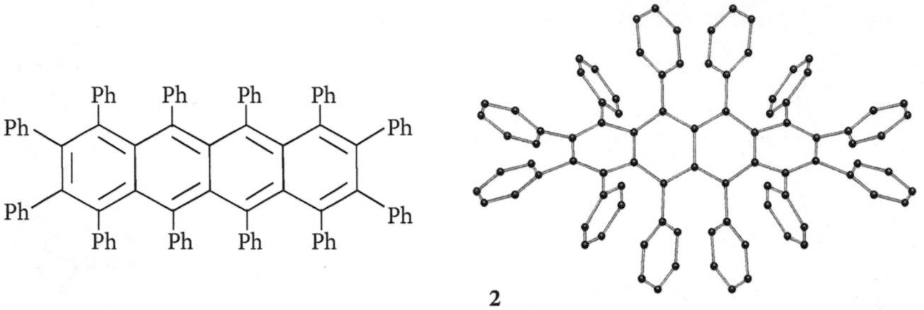

2

References

M. Rickhaus, M. Mayor, and M. Juriček, Strain-induced helical chirality in polyaromatic systems, *Chem. Soc. Rev.* **45**, 1542–56 (2016).

R. G. Clevenger, B. Kumar, E. M. Menuey, and K. V. Kilway, Synthesis and structure of a longitudinally twisted hexacene, *Chem. Eur. J.* **24**, 3113–6 (2018).

Y. Xiao, J. T. Mague, R. H. Schmehl, F. M. Haque, and R. A. Pascal, Jr, Dodecaphenyltetracene. *Angew. Chem. Int. Ed.* **58**, 2831–3 (2019).

K. Ota and R. Kinjo, Inorganic benzene valence isomers, *Chem. Asian J.* **15**, 2558 –74 (2020).

4.3.7 **Topological molecular nanocarbons: all-benzene catenanes and trefoil knot**

The synthesis of all-benzene catenanes **3A**, **3B**, and molecular trefoil knot **4** that consist solely of *para*-connected benzene rings have been achieved in 2019.

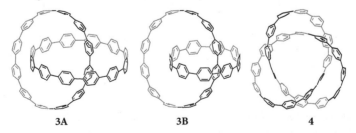

3A   **3B**   **4**

Reference

Y. Segawa, M. Kuwayama, Y. Hijikata, M. Fushimi, T. Nishihara, J. Pirillo, J. Shirasaki, N. Kubota, and K. Itami, Topological molecular nanocarbons: all-benzene catenane and trefoil knot, *Science* **3365**, 272–6 (2019).

4.3.8 **Complexes containing a naked carbon atom**

In the realm of all-carbon ligands in the formation of transition metal complexes, the naked carbon atom holds a special position. Based on the geometry of metal-carbon interaction, these compounds can be divided into four classes: terminal carbide (I), 1,3-dimetallallene (II), C-metalated carbyne (III), and carbide cluster (IV):

$$:C{\equiv}M \qquad M{=}C{=}M \qquad M{\equiv}C{-}M \qquad C@M_n$$

$$\text{(I)} \qquad\qquad \text{(II)} \qquad\qquad \text{(III)} \qquad\qquad \text{(IV)}$$

There are two well-characterized examples of a naked carbon atom bound by a triple bond to a metal center (Fig. 4.3.8). The molybdenum carbide anion $[CMo\{N(R)Ar\}_3]^-$ (R = $C(CD_3)_2(CH_3)$, Ar = $C_6H_3Me_2$-3,5), an isoelectronic analog of $NMo\{N(R)Ar\}_3$, can be prepared in a multi-step procedure via deprotonation of the d^0 methylidyne complex $HCMo\{N(R)Ar\}_3$. The Mo≡C distance of 171.3(9) pm is at the low end of the known range for molybdenum-carbon multiple bonds. In the diamagnetic, air-stable terminal ruthenium carbide complex $Ru(\equiv C:)Cl_2(LL')$ (L = L' = PCy_3, or L = PCy_3 and L' = 1,3-dimesityl-4,5-dihydroimidazol-2-ylidene), the measured Ru–C distance of 165.0(2) pm is consistent with the existence of a very short Ru≡C triple bond.

Many complexes containing a M–C–M' bridge have been reported. The earliest know example of a 1,3-dimetallallene complex, $\{Fe(tpp)\}_2C$ (tpp = tetraphenylporphyrin), was synthesized by the reaction of Fe(tpp) with CI_4, CCl_3SiMe_3, CH_2Cl_2, and BuLi. The single carbon atom bridges the two Fe(tpp) moieties with Fe–C 167.5(1) pm in the linear Fe–C–Fe unit. Thermal decomposition of the olefin metathesis catalyst $(IMesH_2)(PCy_3)(Cl)_2Ru=CH_2$ ($IMesH_2$ = 1,3-dimesityl-4,5-dihydroimidazol-2-ylidene) results in the formation of a C-bridged dinuclear ruthenium complex, as shown in the following scheme. The Ru≡C–Ru bond angle is 160.3(2)°. The measured Ru≡C bond distance of 169.8(4) pm is slightly longer than those in reported μ-carbide ruthenium complexes such as $(PCy_3)_2(Cl)_2Ru{\equiv}C{-}Pd(Cl)_2(SMe_2)$ (166.2(2) pm), and the Ru–C distance of 187.5(4) pm is much shorter than the usual R–C single bonds in ruthenium complexes such as $(Me_3CO)_3W{\equiv}C{-}Ru(CO)_2(Cp)$ (209(2) pm).

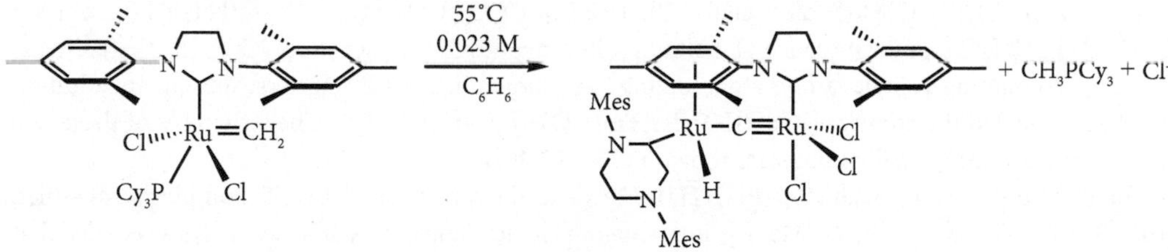

The carbide-centered polynuclear transition metal carbonyl clusters exhibit a rich variety of structures. A common feature to this class of carbide complexes is that the naked carbon is wholly or partially enclosed in a metal cage composed of homo/hetero metal atoms, and there is also a sub-class that can be considered as tetra-metal-substituted methanes. The earliest known compound of this kind is $Fe_5C(CO)_{15}$, in which the carbon atom is located at the center of the base of a square pyramid with Fe(CO)$_3$ groups occupying its five vertices. Carbido carbonyl clusters of Ru and Os are well documented.

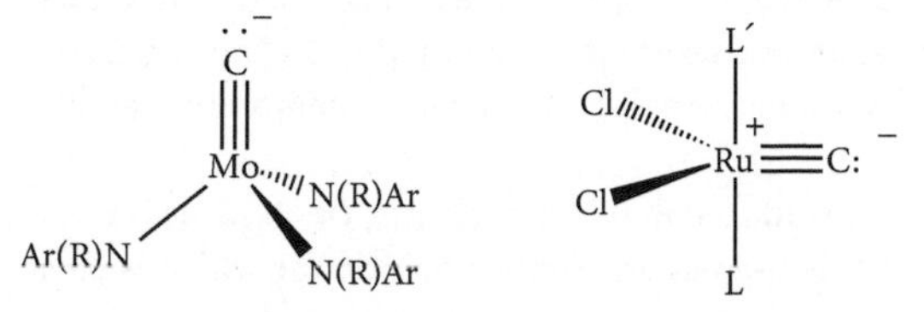

Fig. 4.3.8 Complexes containing a metal-terminal carbon triple bond.

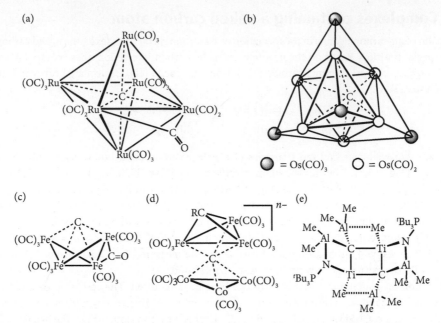

Fig. 4.3.9 Molecular structures of some transition metal clusters containing a naked carbon atom.

Octahedral $Ru_6C(CO)_{17}$ is composed of four $Ru(CO)_3$ and two cis-$Ru(CO)_2$ fragments, the latter being bridged by a carbonyl group (Fig. 4.3.9(a)). $Os_{10}C(CO)_{24}$ is built of an octahedral arrangement of $Os(CO)_2$ groups with four of its eight faces each capped by an $Os(CO)_3$ group, as shown in Fig. 4.3.9(b). The exposed carbon atom in $Fe_4C(CO)_{13}$ shows the greatest chemical activity, and the Fe_4C system serves as a plausible model for a surface carbon atom in heterogeneous catalytic processes (Fig. 4.3.9(c)). Addition of $Co_3(\mu_3\text{-}CCl)(CO)_9$ to a solution of $(PPh_4)_2[Fe_3(\mu\text{-}CCO)(CO)_9]$ (CCO is the ketenylidene group C=C=O) in CH_2Cl_2, in the presence of thallium salt, generates the species $[\{Co_3(CO)_9\}C\{Fe_3(CO)_9(\mu\text{-}CCO)\}]^-$ containing a single carbon atom linking two different trimetallic clusters; subsequent addition of ethanol yielded the complex $[\{Co_3(CO)_9\}C\{Fe_3(CO)_9(\mu\text{-}C\text{-}CO_2Et)\}]^{2-}$. The structures of these two hexanuclear hetereometallic anions are shown in Fig. 4.3.9(d).

The reaction of $AlMe_3$ with $(^tBu_3PN)_2TiMe_2$ leads to the formation of two Ti complexes, of which $[(\mu^2\text{-}^tBu_3PN)Ti(\mu\text{-}Me)(\mu^4\text{-}C)(AlMe_2)_2]_2$ is the major product. Single-crystal X-ray analysis revealed that it has a saddle-like structure, with two phosphinimide ligands lying on one side and four $AlMe_2$ groups on the other (Fig. 4.3.9(e)). The Ti and carbide C atoms in the central Ti_2C_2 ring both exhibit distorted tetrahedral coordination geometry.

4.3.9 Carbon atom functioning as an electron-pair donor

When a carbon atom forms a bond with an electron-rich group L such as sulfur, selenium, or phosphorus, L will supply electrons to the C atom, causing the C atom to appear as a π orbital containing a lone pair of electrons, becoming a four-electron donor, as shown in Fig. 4.3.10(a). Further studies showed that when a ligand containing a C atom donor and a metal atom cluster form a bond, the C atom provides a lone pair of electrons to the metal atom cluster to form a covalently bonded cluster compound, as shown in Fig. 4.3.10(b) and (c).

In the silver-carbon coordination compounds introduced in this section, Ag_n (n = 1, 2, 4) is used as the core to form the cluster skeleton, and the C atom is used as an electron pair donor which is covalently bonded to the Ag_n cluster. As shown in Fig. 4.3.11, the arrows indicate the donation of electron pairs. In Fig. 4.3.11(a) and (b), the coordinating C atom only provides a pair of lone pair electrons to one Ag atom and the remaining electrons form a polycentric delocalized bond. In Fig. 4.3.11(c), the C atom provides a lone pair of electrons to each of the two Ag atoms. The experimentally measured C–Ag bond lengths in

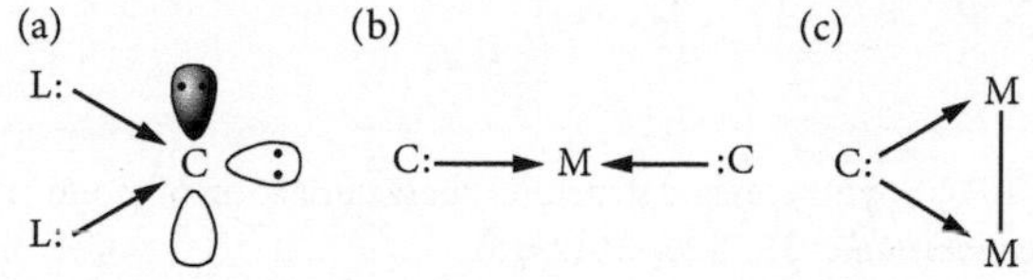

Fig. 4.3.10 Schematic diagram of carbon atom donor and its coordination modes with electron-rich group and metal.

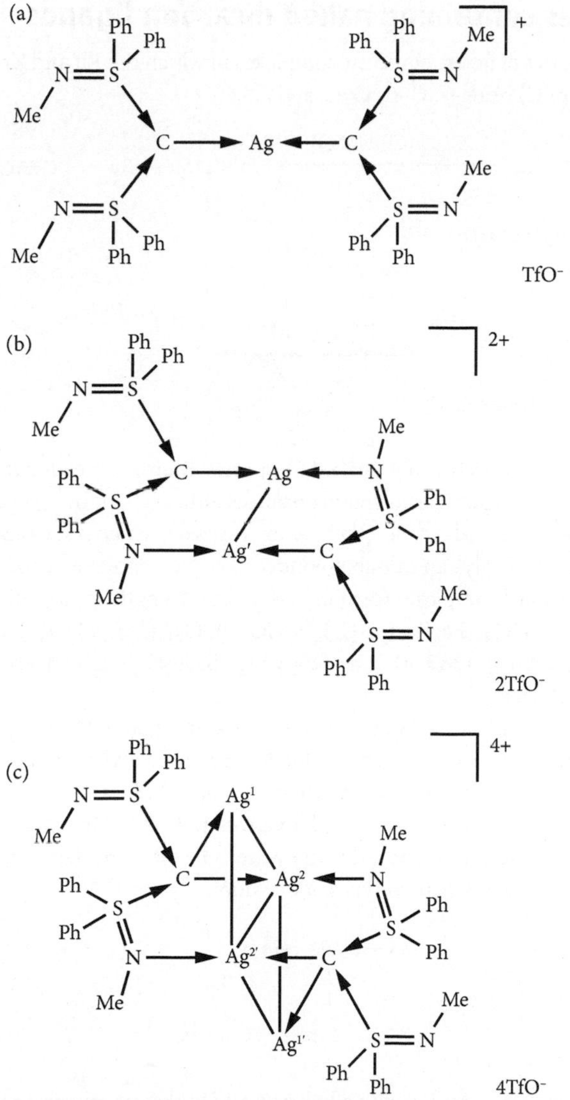

Fig. 4.3.11 Structures of silver-carbon coordination compounds.

Fig. 4.3.11(a) and (b) are 212 pm and 215 pm, respectively, which is slightly shorter than the C–Ag bond length in Fig. 4.3.11(c), which is 221 pm, indicating that the C–Ag bonds in Fig. 4.3.11(a) and (b) are strengthened by the formation of delocalized bonds. Au_n clusters are similar to the Ag_n clusters described above and can also form complexes with this kind of structure.

Reference

T. Morosaki, T. Suzuki, and T Fujii, Syntheses and structural characterization of mono-, di-, and tetranuclear silver carbone complexes. *Organometallics* **35**, 2715–21 (2016).

4.3.10 **Complexes containing naked dicarbon ligands**

There is an interesting series of heterobinuclear complexes in which the Ru and Zr centers are connected by three different types of C_2 bridges: C–C, C=C, and C≡C.

In polynuclear metal complexes bearing a naked C_2 species, multiple metal-carbon interactions generally occur, and the measured carbon-carbon bond distances indicate that the C_2 ligand may be considered to originate from fully deprotonated ethane, ethylene, or acetylene, which is stabilized in a "permetallated" coordination environment. Singly- and doubly-bonded dicarbon moieties are found in some polynuclear transition metal carbonyl complexes (carbon-carbon bond length in pm): $Rh_{12}(C_2)(CO)_{25}$, 148(2); $[Co_6Ni_2(C_2)_2(CO)_{16}]^{2-}$, 149(1); $Fe_2Ru_6(\mu_6\text{-}C_2)_2(\mu\text{-}CO)_3(CO)_{14}Cp_2$, 133.4(8) and 135.4(7); $Ru_6(\mu_6\text{-}C_2)(\mu\text{-}SMe_2)_2(\mu\text{-}PPh_2)_2(CO)_{14}$, 138.1(8). The following discussion is concerned only with anionic species derived from acetylene.

Acetylene is a Brønsted acid ($pK_a \sim 25$). Its chemistry is associated with its triple-bond character and the labile hydrogen atoms. It can easily lose one proton to form the acetylide monoanion HC≡C⁻ (IUPAC name acetylenide) or release two to give the acetylide dianion ⁻C≡C⁻ (C_2^{2-}, IUPAC name acetylenediide). The acetylenide H–C≡C⁻ and substituted derivatives R–C≡C⁻ form organometallic compounds with the alkali metals. In these compounds, the interactions of the π orbitals of the C≡C fragment with metal orbitals may lead to many structural types, for example,

$$R–C≡C–Li$$
$$| \qquad |$$
$$Li–C≡C–R$$

The acetylenediide C_2^{2-} can combine with alkali and alkaline-earth metals to form ionic salts, which are readily decomposed by water. Four modifications of CaC_2 (commonly known as calcium carbide) are known: room-temperature tetragonal CaC$_2$ **I**, high-temperature cubic CaC$_2$ **IV**, low-temperature CaC$_2$ **II**, and a fourth modification CaC$_2$ **III** (Fig. 4.3.12). MgC$_2$, SrC$_2$, and BaC$_2$ adopt the tetragonal CaC$_2$ **I** structure, which consists of a packing of Ca^{2+} and discrete C$_2^{2-}$ ions in a distorted NaCl lattice, with the C$_2^{2-}$ dumbbell (bond length 119.1 pm from neutron powder diffraction) aligned parallel to the c axis.

Ternary metal acetylenediides of composition AM^{I}C$_2$ (A = Li to Cs and M^I = Ag(I), Au(I); or A = Na to Cs and M^I = Cu(I)) have been prepared; NaAgC$_2$, KAgC$_2$, and RbAgC$_2$ are isomorphous (Fig. 4.3.13(a)), but LiAgC$_2$ (Fig. 4.3.13(b)) and CsAgC$_2$ (Fig. 4.3.13(c)) belong to different structural types. The ternary acetylenediides A$_2$MC$_2$ (A = Na to Cs, M = Pd, Pt) crystallize in the same structure type, which is characterized by [M(C$_2$)$_{2/2}^{2-}$]$_\infty$ chains separated by alkali metal ions (Fig. 4.3.13(d)). The ternary alkaline-earth acetylenediide Ba$_3$Ge$_4$C$_2$ can be synthesized from the elements or by the reaction of BaC$_2$ with BaGe$_2$ at 1530 K; it consists of slightly compressed tetrahedral [Ge$_4$]$^{4-}$ anions inserted into a twisted octahedral Ba$_{6/2}$ three-dimensional framework. The Ba$_6$ octahedra are centered by C$_2^{2-}$ dumbbells (C–C bond length 120(6) pm), which are statistically oriented in two directions.

The Group 11 (Cu$_2$C$_2$, Ag$_2$C$_2$, and Au$_2$C$_2$) and Group 12 (ZnC$_2$, CdC$_2$, and Hg$_2$C$_2\cdot$H$_2$O and Hg$_2$C$_2$) acetylenediides exhibit properties that are characteristic of covalent polymeric solids, but their tendency to detonate upon mechanical shock and insolubility in common solvents present serious difficulties in structural characterization. The earliest known and most studied non-ionic acetylenediide is Ag$_2$C$_2$ (commonly known as silver acetylide or silver carbide), which forms a series of double salts of the general formula Ag$_2$C$_2\cdot m$AgX, where X$^-$ = Cl$^-$, I$^-$, NO$_3^-$, H$_2$AsO$_4^-$, or ½EO$_4^{2-}$ (E = S, Se, Cr, or W), and m is

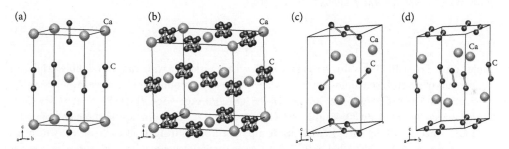

Fig. 4.3.12 Crystal structure of (a) tetragonal CaC$_2$ **I** (*I 4/mmm*, Z = 2); (b) cubic CaC$_2$ **IV** (*Fm3m*, Z = 4), the C$_2^{2-}$ dumbbell exhibits orientational disorder; (c) low-temperature CaC$_2$ **II** (*C2/c*, Z = 4); (d) meta-stable CaC$_2$ **III** (*C2/m*, Z = 4).

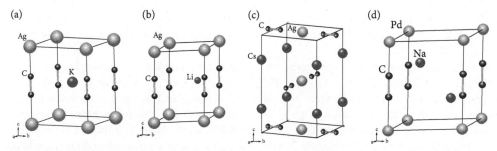

Fig. 4.3.13 Crystal structure of (a) KAgC$_2$ (*P4/mmm*, Z = 1); (b) LiAgC$_2$ (*P6m2*, Z = 1); (c) CsAgC$_2$ (*P4$_2$/mmc*, Z = 2); (d) Na$_2$PdC$_2$ (*P3m1*, Z = 1).

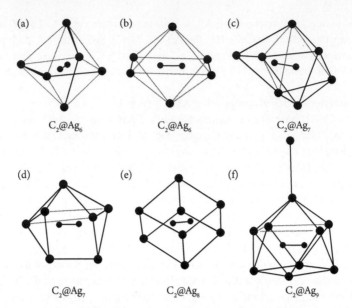

Fig. 4.3.14 Polyhedral Ag_n (n = 6–9) cages with encapsulated C_2^{2-} species found in various silver(I) double salts: (a) $Ag_2C_2 \cdot 2AgClO_4 \cdot 2H_2O$; (b) $Ag_2C_2 \cdot AgNO_3$; (c) $Ag_2C_2 \cdot 5.5AgNO_3 \cdot 0.5H_2O$; (d) $Ag_2C_2 \cdot 5AgNO_3$; (e) $Ag_2C_2 \cdot 6AgNO_3$ (the C_2^{2-} group is shown in one of its three possible orientations); (f) $Ag_2C_2 \cdot 8AgF$. Other ligands bonded to the silver vertices are not shown. The dotted lines represent polyhedral edges that exceed 340 pm (twice the van der Waals radius of the Ag atom).

the molar ratio. From 1998 onward, systematic studies have yielded a wide range of double, triple, and quadruple salts of silver(I) containing silver acetylenediide as a component, for example, $Ag_2C_2 \cdot mAgNO_3$ (m = 1, 5, 5.5, and 6), $Ag_2C_2 \cdot 8AgF$, $Ag_2C_2 \cdot 2AgClO_4 \cdot 2H_2O$, $Ag_2C_2 \cdot AgF \cdot 4AgCF_3SO_3 \cdot RCN$ (R = CH_3, C_2H_5) and $2Ag_2C_2 \cdot 3AgCN \cdot 15AgCF_3CO_2 \cdot 2AgBF_4 \cdot 9H_2O$. A structural feature common to these compounds is that the C_2^{2-} species is fully encapsulated inside a polyhedron with Ag(I) at each vertex, which may be represented as $C^2@Ag_n$. Note that each $C^2@Ag_n$ cage carries a charge of $(n-2)^+$, and such cages are linked by anionic (and coexisting neutral) ligands to from a two- or three-dimensional coordination network. Some polyhedral Ag_n cages with encapsulated C_2^{2-} species found in various silver(I) double salts are shown in Fig. 4.3.14. In $Ag_2C_2 \cdot 6AgNO_3$, the dumbbell-like C_2^{2-} moiety is located inside a rhombohedral silver cage whose edges lie in the range of 295–305 pm, but it exhibits orientational disorder about a crystallographic three-fold axis (Fig. 4.3.14(e)). The utilization of $C_2@Ag_n$ polyhedra as building blocks for the supramolecular assembly of new coordination frameworks has resulted in a series of discrete, 1-D, 2-D, and 3-D complexes bearing interesting structural motifs; further details are presented in section 10.4.5.

To date, there are only two well-characterized copper(I) acetylenediide complexes. In $[Cu_4(\mu-\eta^1:\eta^2-C\equiv C)(\mu-dppm)_4](BF_4)_2$ (dppm = $Ph_2PCH_2PPh_2$), the cation contains a saddle-like $Cu_4(\mu-dppm)_4$ system, with the C_2 unit surrounded by a distorted rectangular Cu_4 array and interacting with the Cu atoms in η^1 and η^2 modes (Fig. 4.3.15(a)). In comparison, the tetranuclear, C_2-symmetric cation of $[Cu_4(\mu-Ph_2Ppypz)_4(\mu-\eta^1:\eta^2-C\equiv C)](ClO_4)_2 \cdot 3CH_2Cl_2$ [Ph_2Ppypz = 2-(diphenylphosphino-6-pyrazolyl)pyridine] consists of a butterfly-shaped Cu_4C_2 core in which the acetylenediide anion bridges a pair of Cu_2 sub-units in both η^1 and η^2 bonding modes; the $C\equiv C$ bond length is 126(1) pm (Fig. 4.3.15(b)).

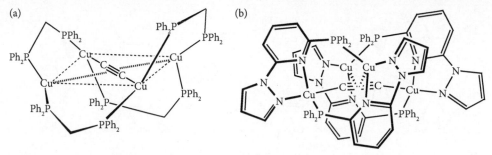

Fig. 4.3.15 Structure of the tetranuclear molecular cation in (a) $[Cu_4(\mu\text{-}\eta^1:\eta^2\text{-}C\equiv C)(\mu\text{-dppm})_4](BF_4)_2$ and (b) $[Cu_4(\mu\text{-Ph}_2Ppypz)_4(\mu\text{-}\eta^1:\eta^2\text{-}C\equiv C)](ClO_4)_2\cdot 3CH_2Cl_2$.

4.4 **Gold and Silver Carbone Complexes**

A carbone is a divalent carbon(0) species CL_2 that exhibits unique bonding and donating characteristics at the central carbon atom, which possesses all four valence electrons as two lone pairs and is bonded to the two σ-donor ligands L through donor–acceptor interactions (L→C←L). An important difference between carbone and carbene is that the former is a four-electron donor while the latter is usually a two-electron donor. The most prominent example of the four-electron-donating ability of carbones is the first X-ray structurally characterized C-dimetalated solvated complex $[(AuCl)_2\{i\text{-}C\{PPh_3\}_2\}]\cdot 3CH_2Cl_2$.

In the molecular structure of $[(AuCl)_2\{i\text{-}C\{PPh_3\}_2\}]$ shown in Fig. 4.4.1, the central carbon atom binds two Au^I atoms, clearly demonstrating that it donates an electron pair to each gold(I) center. Hence dimetalation of a central carbon atom can be used as a criterion for the definition of divalent carbon(0) character. Further theoretical and experimental work has revealed that the σ- and π-lone pairs of carbones can be utilized to bind double-Lewis-acid-bearing vacant σ- and π-orbitals.

Carbodiphosphorane as a ligand is also termed bis(phosphane)carbon(0) and abbreviated as (BPC:**A**) in Fig. 4.4.2. Also shown are the structural formulas of other types of bischalcogenane-stabilized carbon(0) species (BChCs), including bis(iminosulfane)carbon(0) (BiSC: **B**), iminosulfane-(sulfane)carbon(0) (iSSC: **C**), iminosulfane(selenane)-carbon(0) (iSSeC: **D**), and phosphorus- and sulfur-stabilized carbones (iminosulfane(phosphane)carbon(0), iSPC: **E–G**).

The σ- and π-donating abilities of BChCs and imine nitrogen on the iminosulfane substituent are expected to potentially stabilize silver clusters. The molecular structures of a group of synthesized mono-, di-, and tetranuclear silver complexes stabilized by BChCs are shown in Fig. 4.4.3.

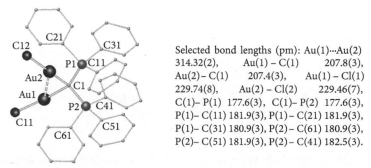

Selected bond lengths (pm): Au(1)···Au(2) 314.32(2), Au(1) – C(1) 207.8(3), Au(2) – C(1) 207.4(3), Au(1) – Cl(1) 229.74(8), Au(2) – Cl(2) 229.46(7), C(1)– P(1) 177.6(3), C(1)– P(2) 177.6(3), P(1)– C(11) 181.9(3), P(1)– C(21) 181.9(3), P(1)– C(31) 180.9(3), P(2)– C(61) 180.9(3), P(2)– C(51) 181.9(3), P(2)– C(41) 182.5(3).

Fig. 4.4.1 Molecular structure of dinuclear gold(I) carbone complex with TfO anions.

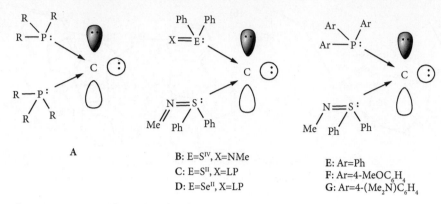

A

B: E=S^{IV}, X=NMe
C: E=S^{II}, X=LP
D: E=Se^{II}, X=LP

E: Ar=Ph
F: Ar=4-MeOC$_6$H$_4$
G: Ar=4-(Me$_2$N)C$_6$H$_4$

Fig. 4.4.2 Structural formulas of carbone ligands **A** to **G**.

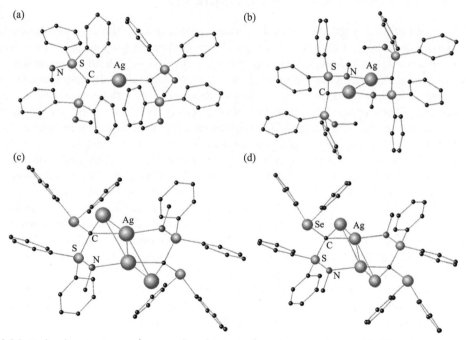

Fig. 4.4.3 Molecular structures of mono, di and tetranuclear silver complexes with TfO anions.

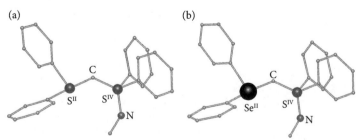

Fig. 4.4.4 Comparison of molecular structures of (a) Ph$_2$S^{II}⟶C⟵S^{IV}Ph$_2$(NMe) and (b) Ph$_2$Se^{II}⟶ C⟵S^{IV}Ph$_2$(NMe).

The carbone $Ph_2Se^{II} \rightarrow C \leftarrow S^{IV}Ph_2(NMe)$ is first carbone containing selenium(II) as the coordinated atom. In its molecular structure determined by X-ray crystallography, the $S^{IV} \rightarrow C$ and $Se^{II} \rightarrow C$ bond lengths are 165.4(2) pm and 187.6(3) pm, respectively. Its molecular structure is compared with that of $Ph_2S^{II} \rightarrow C \leftarrow S^{IV}Ph_2(NMe)$ in Fig. 4.4.4. Their respective $E^{II} \rightarrow C \leftarrow S^{IV}$ fragments ((a) 106.7°, (b) 105.5°) are significantly bent.

References

T. Fujii, T. Ikeda, T. Mikami, T. Suzuki, and T. Yoshimura, Synthesis and structure of (MeN)Ph2S=C=SPh2(NMe). *Angew. Chem. Int. Ed.* **41**, 2576–8 (2002).

T. Morosaki, T. Suzuki, W. W. Wang, S. Nagase, and T.Fujii, Syntheses, structures, and reactivities of two chalcogen-stabilized carbones. *Angew. Chem. Int. Ed.* **53**, 9569–71 (2014).

T. Morosaki, T. Suzuki, and T. Fujii, Syntheses and structural characterization of mono-, di-, and tetranuclear silver carbone complexes. *Organometallics* **35**, 2715–21 (2016).

J. Vicente, A. R. Singhal, and P. G. Jones, New ylide–, alkynyl–, and mixed alkynyl/ylide–gold(i) complexes. *Organometallics* **21**, 5887–900 (2002).

L. Zhao, C. Chai, W. Petz, and G. Frenking, Carbones and carbon atom as ligands in transition metal complexes. *Molecules* **25**, 4943 (2020).

4.5 Phosphine-Sulfoxide Carbon(0) Complexes

A new type of carbon(0) ligand with a phosphine and a sulfoxide coordination sites was synthesized and fully characterized (**Scheme 1**). It exhibits excellent coordination ability, in contrast to the related phosphine/sulfide-supported carbon(0) complexes. The synthetic scheme is shown below:

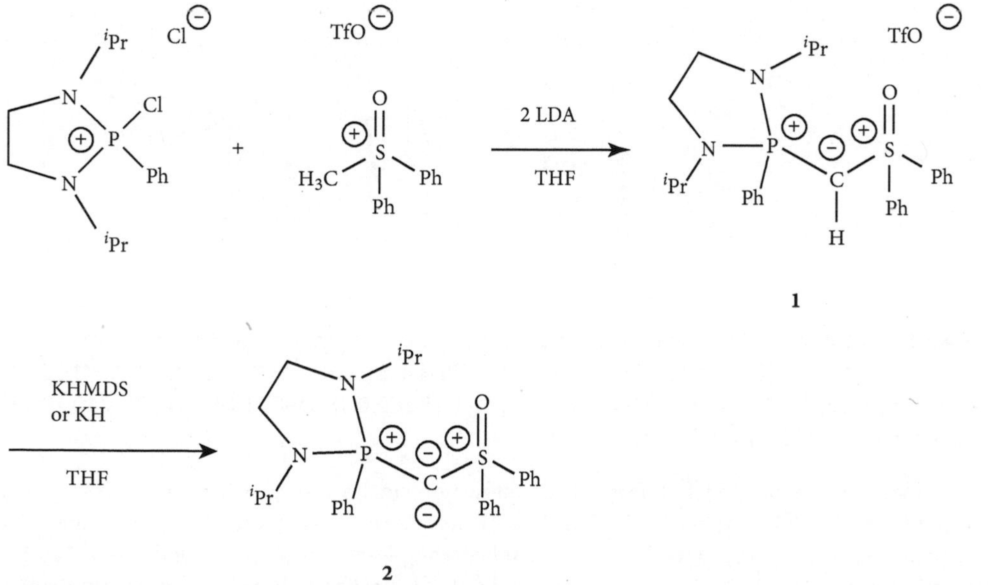

Scheme 1. Synthesis of ylide **1** and phosphine/sulfoxide carbon(0) complex **2**.

The nucleophilic character of **2** was experimentally confirmed by an immediate reaction with methyl iodide, giving raise to the corresponding C-methylated salt **3** (**Scheme 2**).

The molecular structures of compounds **1** and **2** (see Fig. 4.5.1) were confirmed by X-ray diffraction analysis.

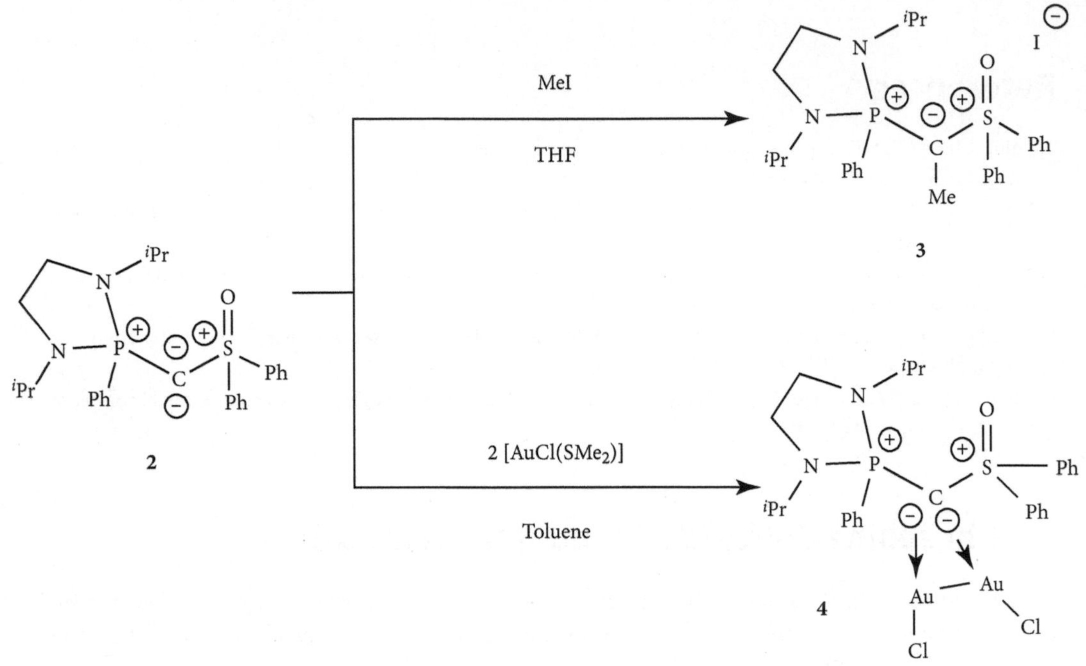

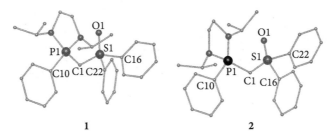

Scheme 2. Methylation and complexation of **2**.

Fig. 4.5.1 X-ray molecular structures of **1** and **2**. Selected bond lengths (pm) and angles (°): for **1**, S1–C1 165.4(2), P1–C1 171.9(2), S1–C16 177.6 (2), S1–C22 177.7(2), P1–C10 178.9(2), S1–O1 145.3(1), S1–C1–P1 121.0(1); for **2**, S1–C1 159.3(1), P1–C1 165.6(1), S1–C16 180.0(1), S1–C22 180.0(1), P1–C10 180.8(1), S1–O1 146.8(1), S1–C1–P1 120.7(1).

The molecular structure of **4** from X-ray diffraction analysis is shown in Fig. 4.5.2. The C–Au bond lengths (205.6(4) pm and 207.1(4) pm) are shorter than those observed in the related carbodisphosphorane- and carbodicarbene-diaurated complexes (207.4, 207.8, and 208.0, 210.3 pm respectively). The Au1-Au2 aurophilic interaction at 301.8(1) pm lies in the range of other *gem*-diaurated carbones (295.2–314.3 pm).

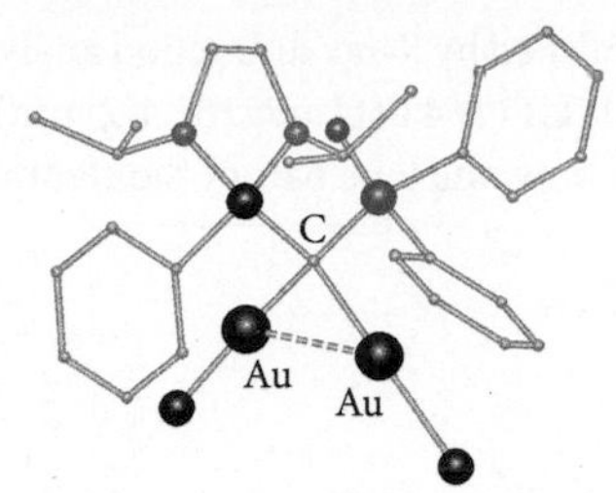

Fig. 4.5.2 Molecular structure of **4**.

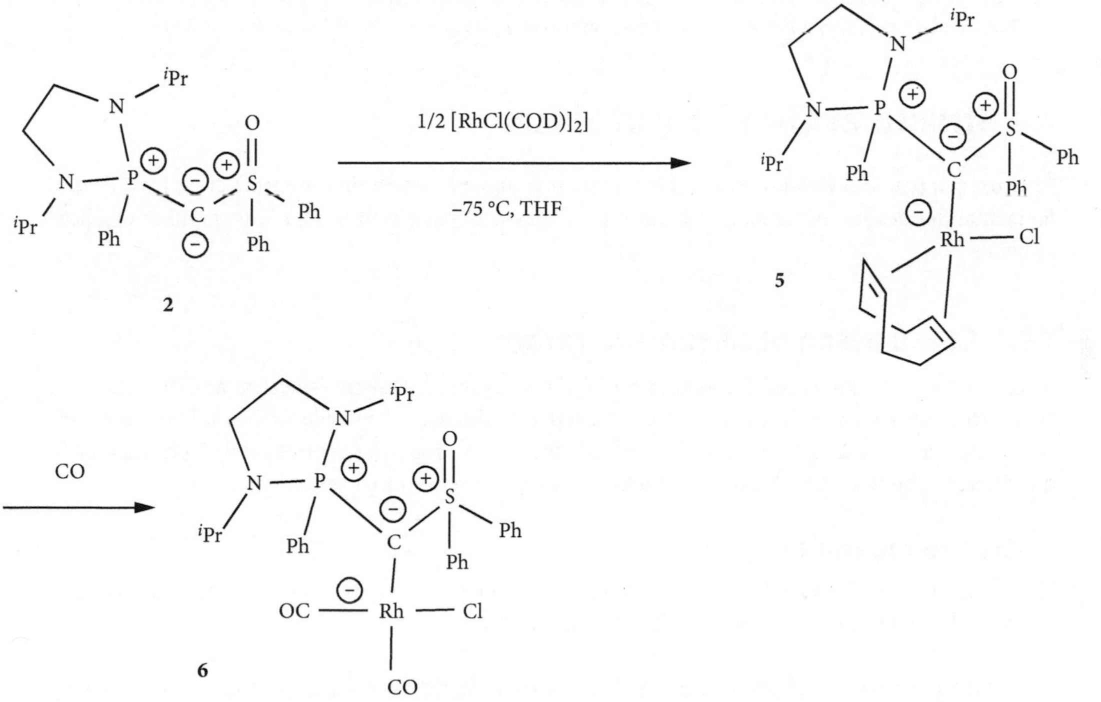

Scheme 3. Formation of rhodium(I) complexes **5** and **6**.

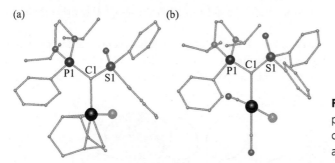

(a)

(b)

Fig. 4.5.3 Molecular structure of phosphine/sulfide-supported carbon(0) rhodium(I) complexes (a) **5** and (b) **6**.

The reaction of phosphine/sulfoxide-carbone **2** with 0.5 equivalents of [RhCl(COD)]$_2$ cleanly forms rhodium(I) *cis, cis*-1,5-cyclooctadiene complex **5**, which remains stable in solution at low temperature but decomposes slowly at RT. The corresponding Rh(I) dicarbonyl complex **6** was prepared by bubbling carbon monoxide gas through a THF solution of **5** at –78 °C (**Scheme 3**).

The molecular structures of both complexes **5** and **6** established by X-ray diffraction analysis (Fig. 4.5.3) showed that their P1–C1 (170.7(1) and 170.9(2) pm) and S1–C1 (164.0(1) and 165.4(2) pm) bond lengths are very similar to those observed in **1** and **3**, meaning that only one lone pair of the central carbon atom interacts with the Rh center.

Reference

M. L. González, L Bousquet, S. Hameury, C. A. Toledano, N. Saffon-Merceron, V. Branchadell, E. Maerten, and A. Baceiredo, Phosphine/sulfoxide-supported carbon(0) complex. *Chem. Eur. J.* **24**, 2570–4 (2018).

4.6 **Structural Chemistry of Silicon**

After oxygen (*ca.* 45.5 wt%), silicon is the next most abundant element in the earth's crust (*ca.* 27 wt%). Elemental Si does not occur naturally, but it combines with oxygen to form a large number of silicate minerals.

4.6.1 **Comparison of silicon and carbon**

Silicon and carbon command dominant positions in inorganic chemistry (silicates) and organic chemistry (hydrocarbons and their derivatives), respectively. Although they have similar valence electronic configurations: $[He]2s^2 2p^2$ for C and $[Ne]3s^2 3p^2$ for Si, their properties are not similar. The reasons for the difference between the chemistry of the two elements are elaborated below.

(1) **Electronegativity**

The electronegativity of C is 2.54, as compared with 1.92 for Si. Carbon is strictly nonmetallic whereas Si is essentially a non-metallic element with some metalloid properties.

(2) **Configuration of valence shell and multiplicity of bonding**

Unlike carbon, the valence shell of the silicon atom has available *d* orbitals. In many silicon compounds, the *d* orbitals of Si contribute to the hybrid orbitals and Si forms more than four 2c-2e covalent bonds. For example, SiF_5^- uses $sp^3 d$ hybrid orbitals to form five Si–F bonds, and SiF_6^{2-} uses $sp^3 d^2$ hybrid orbitals to form six Si–F bonds.

Furthermore, silicon can use its *d* orbitals to form *d*π–*p*π multiple bonds, whereas carbon only uses its *p* orbitals to form *p*π–*p*π multiple bonds. Trisilylamines such as $N(SiH_3)_3$ is a planar molecule, as shown in Fig. 4.6.1(a), which differs from the pyramidal $N(CH_3)_3$ molecule as shown in Fig. 4.6.1(b).

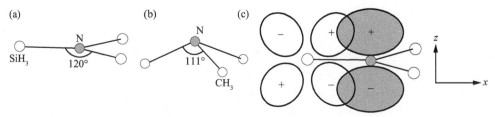

Fig. 4.6.1 (a) Structure of planar $N(SiH_3)_3$, (b) pyramidal $N(CH_3)_3$, and (c) the *d*π–*p*π bonding in $N(SiH_3)_3$. The occupied $2p_z$ orbital of the N atom is shaded, and only one Si $3d_{xz}$ orbital is shown for clarity.

In the planar configuration of $N(SiH_3)_3$, the central N atom uses trigonal planar sp_xp_y hybrid orbitals to form N–Si bonds, with the non-bonding electron pair of N residing in the $2p_z$ orbital. Silicon has empty, relatively low-lying $3d$ orbitals that are able to interact appreciably with the $2p_z$ orbital of the N atom. This additional delocalized $p\pi$–$d\pi$ interaction shown in Fig. 4.6.1(c) enhances the strength of the bonding that causes the NSi_3 skeleton to adopt a planar configuration.

(3) Catenation and silanes

The term catenation is used to describe the tendency for covalent bond formation between atoms of a given element to form chains, cycles, layers, or 3D frameworks. Catenation is common in carbon compounds, but it only occurs to a limited extent in silicon chemistry. The reason can be deduced from the data listed in Table 4.6.1.

Inspection of Table 4.6.1 shows that $E(C–C) > E(Si–Si)$, $E(C–H) > E(Si–H)$ and $E(C–C) > E(C–O)$, but $E(Si–Si) \ll E(Si–O)$. Thus alkanes are much more stable than silanes, and silanes react readily with oxygen to convert the Si–Si bonds to stronger Si–O bonds. Although silanes do not exist in nature, some compounds with Si–Si and Si=Si bonds have been synthesized in the absence of air and in non-aqueous solvents. The silanes Si_nH_{2n+2} ($n = 1$–8), cyclic silanes Si_nH_{2n} ($n = 5, 6$) and some polyhedral silanes are known. The structures of tetrahedral $Si_4(Si^tBu_3)_4$, trigonal-prismatic $Si_6(2,6\text{-}^iPrC_6H_3)_6$, and cubane-like $Si_8(2,6\text{-}Et_2C_6H_3)_8$ are shown below:

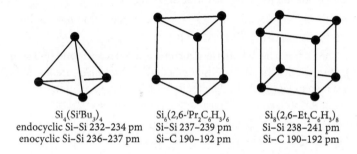

$Si_4(Si^tBu_3)_4$
endocyclic Si–Si 232–234 pm
enocyclic Si–Si 236–237 pm

$Si_6(2,6\text{-}^iPr_2C_6H_3)_6$
Si–Si 237–239 pm
Si–C 190–192 pm

$Si_8(2,6\text{-}Et_2C_6H_3)_8$
Si–Si 238–241 pm
Si–C 190–192 pm

The structures of other examples of oligocyclosilanes are shown in Fig. 4.6.2. The adamantane-like cluster $(SiMe_2)_6(Si–SiMe_3)_4$ with pendant methyl and trimethylsilyl groups was prepared in 2005.

Table 4.6.1 Comparison of chemical bonds formed by C and Si

Bond	Bond length/pm	Bond energy/kJ mol^{-1}
C–C	154	356
C–H	109	413
C–O	143	336
Si–Si	235	226
Si–H	148	318
Si–O	166	452

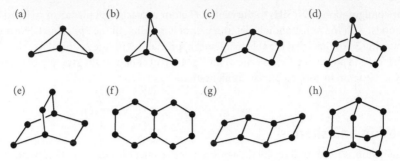

Fig. 4.6.2 The structures of the molecular skeletons of some oligocyclosilanes: (a) $Si_4{}^tBu_2(2,6\text{-}Et_2C_6H_3)_4$, (b) $Si_5H_2{}^tBu_4(2,4,6\text{-}{}^tBu^iPr_2C_6H_2)_2$, (c) $Si_6{}^iPr_{10}$, (d) Si_7Me_{12}, (e) Si_8Me_{14}, (f) $Si_{10}Me_{18}$, (g) $Si_8{}^iPr_{12}$, (h) $Si_{10}Me_{16}$.

4.6.2 Metal silicides

There exist a number of metal silicides, in which silicon anions formally occur. The structural types of the silicides in the solid state are diverse:

(1) Isolated units

    (a) Si^{4-}: Examples are found in Mg_2Si, Ca_2Si, Sr_2Si, Ba_2Si.

    (b) Si_2 units: The $Si_2{}^{6-}$ ion occurs in U_3Si_2 (Si–Si distance is 230 pm) and in the series Ca_5Si_3, Sr_5Si_3, and Ba_5Si_3 [as $(M^{2+})_5(Si_2)^{6-}(Si)^{4-}$].

    (c) Si_4 units: The $Si_4{}^{6-}$ ion has a butterfly structure; examples are found in Ba_3Si_4, as shown in Fig. 4.6.3 (a). On the other hand, $Si_4{}^{4-}$ is tetrahedral and examples are found in $NaSi$, KSi, $CsSi$, $BaSi_2$, as shown in Fig. 4.6.3 (b). Finally, in K_7LiSi_8, pairs of Si_4-units are connected by Li^+, as shown in Fig. 4.6.3(c), with additional interactions involving K^+ ions.

(2) Si_n-chains: Polymeric unit $[Si^{2-}]_n$ has a planar zigzag shape. Examples are found in USi and $CaSi$, as shown in Fig. 4.6.3(d). The Si–Si distances are 236 (USi) and 247 pm ($CaSi$).

(3) Si_n-layers: Polymeric unit $[Si^-]_n$ in $CaSi_2$ consists of corrugated layers of six-membered rings, as shown in Fig. 4.6.3(e). In β-$ThSi_2$, the $[Si^-]_n$ units form planar hexagonal nets, which are similar to the B_n layers in AlB_2 (Fig. 13.3.1(f)).

(4) Si_n-three dimensional network, for example, $SrSi_2$, α-USi_2, in which $[Si^-]_n$ constitutes a network with metal cations occupying the interstitial sites.

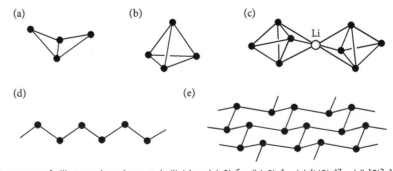

Fig. 4.6.3 Structure of silicon anions in metal silicides: (a) $Si_4{}^{6-}$, (b) $Si_4{}^{4-}$, (c) $[LiSi_8]^{7-}$, (d) $[Si^{2-}]_n$, (e) $[Si^-]_n$.

4.6.3 **Stereochemistry of silicon**

(1) **Silicon molecular compounds with coordination numbers from three to six**

The compounds formed by one Si atom and one to two other non-metallic atoms, such as SiX or SiX_2 (X = H, F, O, and Cl), are not stable species.

A number of silenes ($R_2Si=CR_2$) and disilenes ($R_2Si=SiR_2$) containing double bonds, where each Si atom is three-coordinated, are known.

The structure of the free silylium ion $(mes)_3Si^+$ has been characterized by X-ray analysis of $[(mes)_3Si][HCB_{11}Me_5Br_6] \cdot C_6H_6$. The $(mes)_3Si^+$ cation is well separated from the carborane anion and benzene solvate molecule. As shown in Fig. 4.6.4, the Si atom has trigonal planar coordination geometry expected of an sp^2 silylium center, with Si–C bond length 181.7 pm (av.), and the mesityl groups have a propeller-like arrangement around the Si center with an averaged twist angle $\tau = 49.2°$.

In the vast majority of its compounds, Si is tetrahedrally coordinated, which include SiH_4, SiX_4, SiR_4, SiO_2, and silicates. Five-coordinate Si can be either trigonal bipyramidal or square pyramidal, the former being more stable than the latter. In $[Et_4N][SiF_5]$ the anion SiF_5^- is trigonal bipyramidal with $Si–F_{(ax)}$ 165 pm and $Si–F_{(eq)}$ 159 pm (av). Among the stable compounds of penta-coordinate silicon, the most studied are the silatranes, an example of which is shown below:

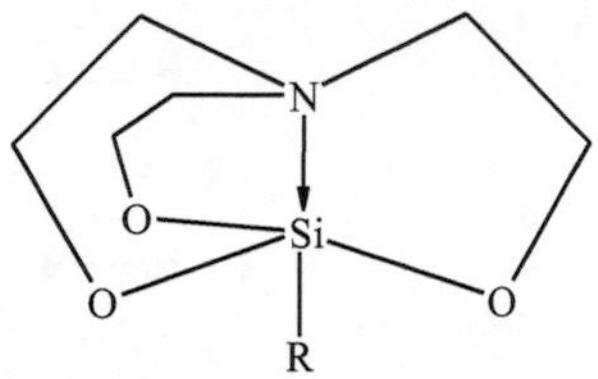

In this molecule, the axial nitrogen atom is linked through three $(CH_2)_2$ tethers to oxygen atoms at the triangular base. The tetradentate tripodal ligand occupies four coordination positions around the central Si atom, forming an intra-molecular transannular dative bond of the type N→Si. The observed N→Si bond distances in about 50 silatranes containing three condensed rings vary within the range 200–240 pm The longer the N→Si bond, the more planar the configuration of the NC_3 moiety, and the structure of the $RSiO_3$ fragment approaches tetrahedral.

Numerous compounds of octahedral hexa-coordinate Si are known. Some examples are shown below.

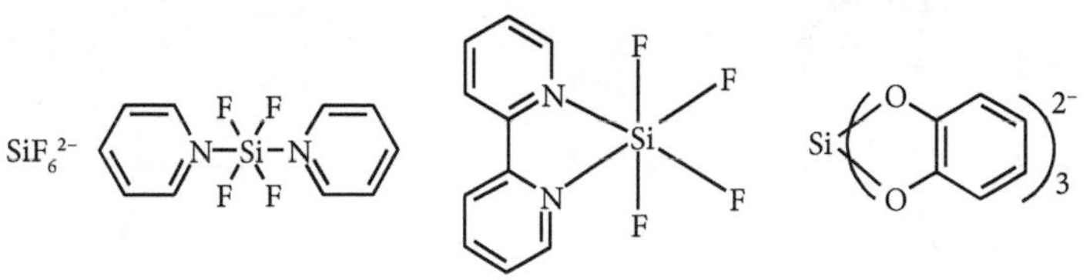

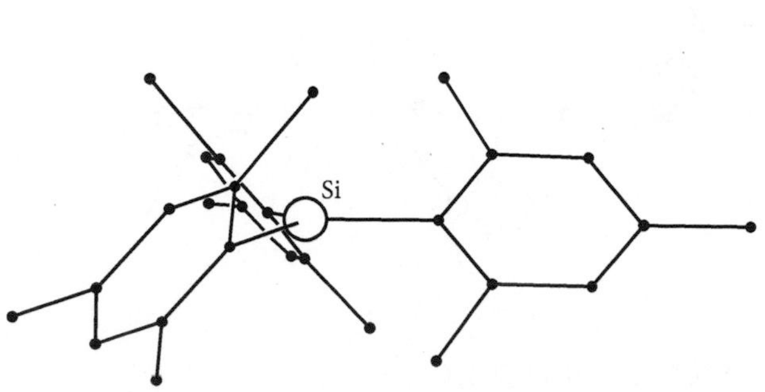

Fig. 4.6.4 Structure of silylium ion $(mes)_3Si^+$.

In the salt $[C(NH_2)_3]_2[SiF_6]$, the Si–F bond length in octahedral SiF_6^{2-} is 168 pm, which is longer than the Si–F bonds in SiF_5^-.

(2) Structure of Si(IV) compounds with SiO_5 and SiO_6 skeletons

The chemistry of silicon oxygen compounds with SiO_5 and SiO_6 skeletons in aqueous solution is of special interest. It has been speculated that such Si(IV) complexes with ligands derived from organic hydroxy compounds (such as pyrocatechol derivatives, hydroxycarboxylic acids, and carbohydrates) may play a significant role in silicon biochemistry by controlling the transport of silicon.

Several zwitterionic neutral compounds and anionic species with the SiO_5 skeleton have been synthesized and characterized by X-ray diffraction. The Si coordination polyhedra of these compounds are typically distorted trigonal bipyramids, with the carboxylate oxygen atoms of the two bidentate ligands in the axial positions. Fig. 4.6.5 shows the structures of (a) $Si[C_2O_3(Me)_2]_2[O(CH_2)_2NHMe_2]$ with Si–$O_{(ax)}$ 177.3 pm and 179.8 pm, and Si–$O_{(eq)}$ 164.3 pm to 165.9 pm, and (b) the anion in crystalline $[Si(C_2O_3Ph_2)_2(OH)]^-[H_3NPh]^+$, where Si–$O_{(ax)}$ 179.8 pm and Si–$O_{(eq)}$ 165.0 pm to 166.0 pm

Some neutral compounds, cationic species, and anionic species with distorted octahedral SiO_6 skeletons have been established. Fig. 4.6.6 shows the structures of (a) $Si[C_2O_3Ph_2][C_3HO_2Ph_2]_2$ in which

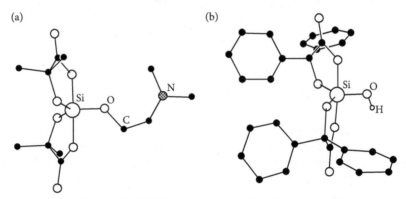

Fig. 4.6.5 Structures of (a) $Si[C_2O_3Me_2]_2[O(CH_2)_2NHMe_2]$ and (b) $[Si(C_2O_3Ph_2)_2(OH)]^-$.

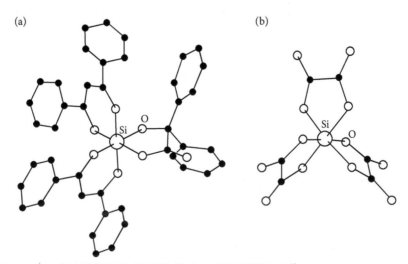

Fig. 4.6.6 Structures of (a) $Si[C_2O_3Ph_2][C_3HO_2Ph_2]_2$ and (b) $[Si(C_2O_4)_3]^{2-}$.

the Si–O bond lengths are 169.3 pm to 182.1 pm, and (b) the dianion in $[Si(C_2O_4)_3]^{2-}[HO(CH_2)_2$ $NH(CH_2)_4^+]_2$, in which the Si–O bond lengths vary from 176.7 pm to 178.8 pm.

(3) Structural features of Si=Si double bond in disilenes

The first reported disilene is $(mes)_2Si=Si(mes)_2$, which was isolated and characterized in 1981. The characteristic structural features of disilenes are the length of the Si=Si double bond d, the twist angle τ, and the *trans*-bent angle θ, as shown below.

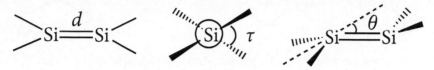

In contrast to the C=C double bond of sterically crowded alkenes, in which variations of the bond lengths are small, the Si=Si bond lengths of disilenes vary between 214 pm and 229 pm The twist angle τ of the two SiR_2 planes varies between $0°$ and $25°$. A further peculiarity of the disilenes, not observed in alkenes, is the possibility of *trans*-bending of the substituents, which is described by the *trans*-bent angle θ between the SiR_2 planes and the Si=Si vector. The θ values can reach up to $34°$. These differences can be rationalized as follows: carbenes have either a triplet ground state (T), as shown in Fig. 4.6.7(a), or a singlet ground state (S), as shown in Fig. 4.6.7(b), with relatively low S→T transition energies. The familiar picture of a C=C double bond results from the approach of two triplet carbenes (Fig. 4.6.7(c)). In contrast, silylenes have singlet ground states with a relatively large S→T excitation energy. Approach of two singlet silylenes should usually result in repulsion rather than bond formation (Fig. 4.6.7(d)). However, if the two silylene molecules are rotated with respect to each other, interaction between the doubly occupied sp^2-type orbitals and the vacant p orbitals can form a double donor-acceptor bond. This is accompanied by a *trans*-bending of the substituents about the Si=Si vector, as illustrated in Fig. 4.6.7(e).

(4) Stable silyl radicals

Silyl radicals stabilized with bulky trialkylsilyl groups can be isolated in crystalline form. X-ray analysis showed that the $(^tBu_2MeSi)_3Si^{\cdot}$ radical has trigonal planar geometry about the central sp^2 Si atom, with an averaged Si–Si bond length of 242(1) pm. Interestingly, all the methyl substituents at the α-Si atoms lie in the plane of the polysilane skeleton so that steric hindrance is minimized. Reaction of this radical with lithium metal in hexane at room temperature afforded $[(^tBu_2MeSi)_3Si]Li$, the crystal structure of which showed a planar configuration (sum of Si–Si–Si angles = $119.7°$ for the central anionic Si atom) and an averaged Si–Si bond length of 236(1) pm.

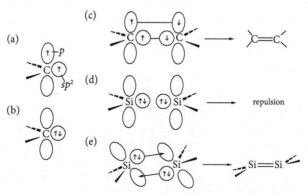

Fig. 4.6.7 Double bond formation from two triplet carbenes and singlet silylenes.

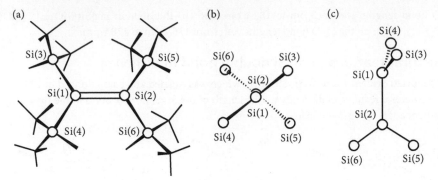

Fig. 4.6.8 (a) Molecular structure of disilene ($^tBu_2MeSi)_2Si=Si(SiMe^tBu_2)_2$. Si($m$)–Si($n$) bond lengths (in pm): 1–2, 226.0(1); 1–3, 241.6(2); 1–4, 241.9(2); 2–5, 241.3(2); 2–6, 242.3(2). (b) Twisted configuration of the $Si_2Si=SiSi_2$ skeleton. (c) Nearly orthogonal configuration of the silyl anion radical $(^tBu_2MeSi)_2Si^-$– $Si^{\cdot}(SiMe^tBu_2)_2$. Bond lengths Si(m)–Si(n) (in pm): 1–2, 234.1(5); 1–3, 239.0(5); 1–4, 239.2(5); 2–5, 241.2(1); 2–6, 240.1(1).

Crystal structure analysis of the disilene $(^tBu_2MeSi)_2Si=Si(SiMe^tBu_2)_2$ has revealed a highly twisted configuration about the central Si=Si double bond, as shown in Fig. 4.6.8(a). The Si(1) and Si(2) atoms forming the double bond are both sp^2 hybridized, and the twist angle τ has an exceptionally large value of 54.5° (see Fig. 4.6.8(b)). The disilene reacts with ^tBuLi to produce the $[Li(THF)_4]^+$ salt of the silyl anion radical $(^tBu_2MeSi)_2Si^-$-$Si^{\cdot}(SiMe^tBu_2)_2$, in which the anionic Si(1)$^-$ atom has flattened pyramidal geometry (sum of bond angles = 352.7°) whereas the radical Si(2)$^{\cdot}$ atom retains its sp^2 configuration. The mean planes of the Si(3)-Si(1)-Si(4) and Si(5)-Si(2)-Si(6) fragments are nearly orthogonal, with a τ angle of 88° about the central Si–Si single bond, as illustrated in Fig. 4.6.8(c).

4.6.4 **Silicates**

(1) **Classification of silicates**

Silicates constitute the largest part of the earth's crust and mantle. They play a dominant role as raw materials as well as products in technological processes, such as building materials, cements, glasses, ceramics, and refractory materials.

In the silicates, the tetrahedral SiO_4 units are either isolated or share corners with other tetrahedra, giving rise to an enormous variety of structures. In many silicates, silicon may be replaced to a certain extent by other elements, such as aluminium, so the structures of silicates are further extended to cover cases where such partial substitution occurs.

The kind and the degree of linkage of SiO_4 tetrahedra constitute a basis for the classification of natural silicates, as shown in Table 4.6.2.

(a) **Silicates containing discrete SiO_4^{4-} or $Si_2O_7^{6-}$**

The discrete SiO_4^{4-} anion [as shown in Fig. 4.6.9(a)] occurs in the orthosilicates, such as phenacite Be_2SiO_4, olivine $(Mg,Fe,Mn)_2SiO_4$, and zircon $ZrSiO_4$. Another important group of orthosilicates is the garnets, $M_3^{II}M_2^{III}(SiO_4)_3$, in which M^{II} are eight-coordinate (e.g., Ca, Mg, Fe) and M^{III} are six-coordinate (e.g., Al, Cr, Fe). Orthosilicates are also vital components of Portland cements: β-Ca_2SiO_4 has a discrete $[SiO_4]$ group with Ca in six- or eight-coordination.

Table 4.6.2 Classification of natural silicates

Number of shared O atoms in SiO$_4$ unit	Structure	Name
0 (no O atom shared)	Discrete SiO$_4$ unit	neso-silicates
1	Discrete Si$_2$O$_7$ unit	soro-silicates
2	Cyclic (SiO$_3$)$_n$ structures	cyclo-silicates
2	Infinite chains or ribbons	ino-silicates
3	Infinite layers	phyllo-silicates
4	Infinite 3D frameworks	tecto-silicates

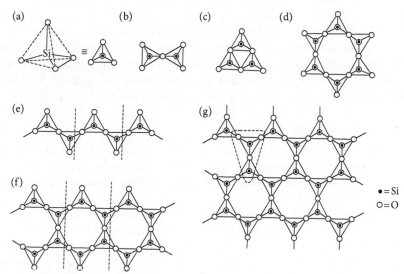

Fig. 4.6.9 Structures of some silicates: (a) SiO$_4^{4-}$, (b) Si$_2$O$_7^{6-}$, (c) cyclic [Si$_3$O$_9$]$^{6-}$, (d) cyclic [Si$_6$O$_{18}$]$^{12-}$, (e) single chain [SiO$_3$]$_n^{2n-}$, (f) double chain [Si$_4$O$_{11}$]$_n^{6n-}$, (g) layer [Si$_2$O$_5$]$_n^{2n-}$.

An example of a disilicate containing the discrete Si$_2$O$_7^{6-}$ ion [as shown in Fig. 4.6.9(b)] is the mineral thortveitite Sc$_2$Si$_2$O$_7$, in which Sc^{III} is octahedral and Si–O–Si is linear with staggered conformation. There is also a series of lanthanoid disilicates Ln$_2$Si$_2$O$_7$, in which the Si–O–Si angle decreases progressively from 180° to 133°, and the coordination number of Ln increases from 6 through 7 to 8 as the size of Ln increases from six-coordinated Lu^{III} to eight-coordinated Nd^{III}. In the Zn mineral hemimorphite the angle is 150°, but the conformation of the two tetrahedra is eclipsed rather than staggered.

Discrete chains of three or four linked SiO$_4$ tetrahedra are extremely rare, but they exist in aminoffite Ca$_3$(BeOH)$_2$[Si$_3$O$_{10}$], kinoite Cu$_2$Ca$_2$[Si$_3$O$_{10}$]·2H$_2$O, and vermilion Ag$_{10}$[Si$_4$O$_{13}$].

(b) Silicates containing cyclic [SiO$_3$]$_n^{2n-}$ ion

Cyclic [SiO$_3$]$_n^{2n-}$, having 3, 4, 6, or 8 linked tetrahedra are known, though 3 and 6 (as shown in Fig. 4.6.9(c)–(d)) are the most common. These anions are exemplified by the minerals benitoite BaTi[Si$_3$O$_9$], catapleite Na$_2$Zr[Si$_3$O$_9$]·2H$_2$O, beryl Be$_3$Al$_2$[Si$_6$O$_{18}$], tourmaline (Na,Ca)(Li,Al)$_3$Al$_6$(OH)$_4$(BO$_3$)$_3$[Si$_6$O$_{18}$], and murite Ba$_{10}$(Ca,Mn,Ti)$_4$[Si$_8$O$_{24}$](Cl,OH,O)$_{12}$·4H$_2$O.

(c) Silicates with infinite chain or ribbon structure

Single chains formed by tetrahedral SiO_4 groups sharing two vertices adopt various configurations in the crystal. There are different numbers of tetrahedra in the repeat unit for different minerals; for example, both diopside $CaMg[SiO_3]_2$ and enstatite $MgSiO_3$ have two repeat units, as shown in Fig. 4.6.9(e).

In the double chains or ribbons, there are different kinds of tetrahedra sharing two and three vertices. The most numerous amphiboles and asbestos minerals, such as tremolite, $Ca_2Mg_5(Si_4O_{11})_2(OH)_2$, adopt the $[Si_4O_{11}]_n^{6n-}$ double chain structure, as shown in Fig. 4.6.9(f).

(d) Silicates with layer structures

Silicates with layer structures include clay minerals, micas, asbestos, and talc. The simplest single layer has composition $[Si_2O_5]_n^{2n-}$, in which each SiO_4 tetrahedron shares three vertices with three others, as shown in Fig. 4.6.9(g). Muscovite $KAl_2[AlSi_3O_{10}](OH)_2$ is a layer silicate of the mica group. There are many and complex variations of structure based on one or more of the layer types. The vertices of the tetrahedra can form part of a hydroxide-like layer. Their bases can either be directly opposed or can be fitted to hydroxyl or water layers. When layers of different types succeed each other with more or less regularity, the changes in stacking these layers together occur in an endless variety of ways. This irregularity of structure is no doubt related to their capacity to take up or lose water and mediate ion exchange. In many cases the structures swell or contract in different stages of hydration, and can take up replaceable ions.

(e) Silicates with framework structures

In silicates, the most important and widespread substitution is that of Al for Si in tetrahedral coordination to form $(Si,Al)O_4$ tetrahedra, so that most of the "silicates" that occur in nature are in fact aluminosilicates. This substitution must be accompanied by the incorporation of cations to balance the charge on the Si–O framework.

Sharing of all vertices of each $(Si,Al)O_4$ tetrahedron leads to infinite 3D framework aluminosilicates, such as felspars, ultramarines, and zeolites. The felspars are the most abundant of all minerals and comprise about 60% of the Earth's crust, and are subdivided into two groups according to the symmetry of their structures. Typical members of the groups are: (a) orthoclase $KAlSi_3O_8$ and celsian $BaAl_2Si_2O_8$; and (b) the plagioclase felspar: albite $NaAlSi_3O_8$ and anorthite $CaAl_2Si_2O_8$. Examples of ultramarine are sodalite $Na_8Cl_2[Al_6Si_6O_{24}]$ and ultramarine $Na_8(S_2)[Al_6Si_6O_{24}]$.

(2) Zeolites

Zeolites are crystalline, hydrated aluminosilicates that possess framework structures containing regular channels and/or polyhedral cavities. Zeolites generally contain loosely held water, which can be removed by heating the crystals and subsequently regained on exposure to a moist atmosphere. When occluded water is removed, it can be replaced by other small molecules. There is a close relationship between the size of molecules that can diffuse through the framework and the dimensions of the aperture or "bottleneck" connecting one cavity to the next. This capacity of zeolites to admit molecules below a certain limiting size, while blocking passage to larger molecules, has led to their being termed "molecular sieves."

There are more than 60 natural zeolites and more than 1000 synthesized zeolites. However, there are 255 kinds of framework types (to July 2022) among the zeolites. Table 4.6.3 lists the ideal composition and the member of rings of some zeolites. Fig. 4.6.10(a) shows the construction of a truncated octahedron (β-cage) formed from 24 linked $(Si,Al)O_4$ tetrahedra, and Fig. 4.6.10(b) shows a simplified representation of this cavity formed by joining the Si(Al) atom positions. Several other types of polyhedra have also been observed.

Zeolite A is a synthetic zeolite that has not been found in nature. In the framework structure of dehydrated zeolite 4A, $Na_{12}[Al_{12}Si_{12}O_{48}]$, truncated octahedra lie at the corners of the unit cell, generating

Table 4.6.3 Ideal composition and member of rings in some zeolites

Zeolite	Ideal composition	Member of rings
Zeolite A	$Na_{12}[Al_{12}Si_{12}O_{48}]\cdot27H_2O$	4,6,8
Faujasite (Zeolite X and Y)	$Na_{58}[Al_{58}Si_{134}O_{384}]\cdot240H_2O$	4,6,12
Zeolite ZSM-5	$Na_3[Al_3Si_{93}O_{192}]\cdot16H_2O$	4,5,6,7,8,10
Chabazite	$Na_4Ca_8[Al_{20}Si_{52}O_{144}]\cdot56H_2O$	4,5,6,8,10
Mordenite	$Na_8[Al_8Si_{40}O_{96}]\cdot24H_2O$	4,5,6,8,12

(a)　　　　　　　　　　　　(b)

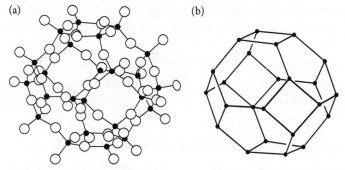

Fig. 4.6.10 Structure of truncated octahedron cage (β-cage): (a) arrangement of $(Si,Al)O_4$ (black circles represent Si and Al), (b) simplified representation of the polyhedral cavity.

(a)　　　　　　　　　　　　(b)

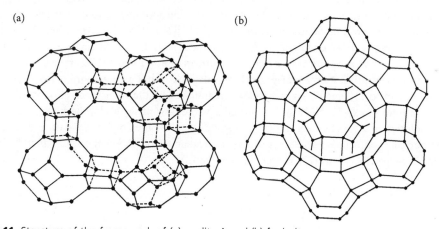

Fig. 4.6.11 Structure of the framework of (a) zeolite A and (b) faujasite.

small cubes at the centers of the cell edges and eight-membered rings at the face centers. Access through these eight-membered rings to a very large cavity (truncated cubo-octahedron, α-cage) at the center of the unit cell is facilitated. Fig. 4.6.11(a) shows the structure of zeolite A framework.

Faujasite or zeolite X and Y exists in nature. Its framework can be formed as follows. The truncated octahedra are placed at the positions of the carbon atoms in the diamond structure, and they are joined by hexagonal prisms through four of the eight hexagonal faces of each truncated octahedron, as shown in Fig. 4.6.11(b).

(3) **General rules governing the structures of natural silicates**

The structures of natural silicates obey the following general rules:

(a) With some exceptions, such as stishovite, the Si atoms form SiO_4 tetrahedra with little deviation of bond lengths and bond angles from the mean values: Si–O = 162 pm, O–Si–O = 109.5°.

(b) The most important and widespread substitution is that of Al for Si in tetrahedral coordination, which is accompanied by the incorporation of cations to balance the charge. The mean Al–O bond length in AlO_4 is 176 pm. In almost all silicates that occur as minerals, the anions are tetrahedral $(Si,Al)O_4$ groups. The Al atoms also occupy positions of octahedral coordination in aluminosilicates.

(c) The $(Si,Al)O_4$ tetrahedra are linked to one another by sharing common corners rather than edges or faces. Since two Si–O–Al groups have a lower energy content than one Al–O–Al plus one Si–O–Si group, the AlO_4 tetrahedra do not share corners in tetrahedral framework structures if this can be avoided.

(d) One oxygen atom can belong to no more than two SiO_4 tetrahedra.

(e) If s is the number of oxygen atoms of a SiO_4 tetrahedron shared with other SiO_4 tetrahedra, then for a given silicate anion the difference between the s values of all SiO_4 tetrahedra tends to be small.

4.6.5 **Square-planar silicon(IV) complex**

Synthesis and structural characterization of the first square-planar tetra-coordinated silicon(IV) crystalline complex was achieved in 2021.

The synthetic route is shown in Fig. 4.6.12. Reaction of the tetralithium salt of *meso*-octaethylcalix[4]pyrrole with $SiCl_4$ in dimethoxyethane, followed by salt metathesis with PPh_4Cl, provided tetraphenylphosphonium *meso* octaethyl-calix[4]pyrrolato chlorido silicate, abbreviated as $[PPh_4][\mathbf{1}]$, in an overall isolated yield of 36%. Subsequent reaction of $[PPh_4][\mathbf{1}]$ with $Na[B(C_6H_5)_4]$ in CH_2Cl_2 afforded a macrocyclic, tetrameric sodium salt of the chlorido silicate $[Na]_4[\mathbf{1}]_4$ by salt metathesis. Stirring a solution of $[Na]_4[\mathbf{1}]_4$ in *n*-hexane induced its clean transformation into silicon(IV) complex **2** in 78% yield. The ^{29}Si-NMR chemical shift of **2** (–55.6 ppm) indicated that it has a tetra-coordinated square-planar silicon center, and its D_{2d} molecular symmetry is manifested by the occurrence of only one peak for pyrrole-ring protons in the ^{1}H NMR spectrum, and a two-fold set of signals for the ethyl groups. Storing a CH_2Cl_2 solution of **2** for one week at –40°C deposited block-shaped orange crystals, and XRD analysis of compound **2** confirmed the presence of a square-planar silicon(IV) center with typical Si–N bond lengths in the range 178.7(1) to 179.7(1) pm (see Fig. 4.6.13).

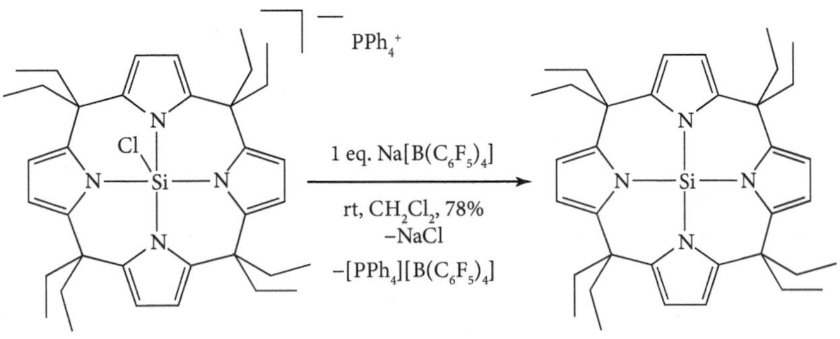

Fig. 4.6.12 Synthesis of crystalline square-planar silicon(IV) complex **2**.

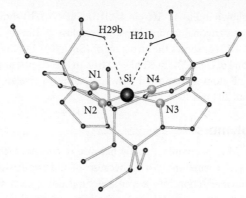

Fig. 4.6.13 X-Ray molecular structure of complex **2**, which features novel square-planar silicon(IV) coordination. Selected bond distances (pm) and angles (°): Si–N1 178.84(5), Si–N2 179.26(5), Si–N3 179.90(4), Si–N4 178.81(4); N1–Si–N2 90.15(2), N2–Si–N3 89.28(2), N3–Si–N4 90.43(2), N4–Si–N1 90.26(2); Si–H21b 273.3, Si–H29b 240.3.

Reference

F. Ebner and Lutz Greb, An isolable, crystalline complex of square-planar silicon(IV), *Chem* **7**, 1–9 (2021). <https://doi.org/10.1016/j.chempr.2021.05.002>

4.7 Structures of Halides and Oxides of Heavier Group 14 Elements

4.7.1 Subvalent halides

Subvalent halides of heavier Group 14 elements MX_2 (M = Ge, Sn and Pb; X = F, Cl, Br, and I) and their complexes exhibit the following structural features and properties.

(1) Stereochemically-active lone pair

In the dihalides of Ge, Sn, and Pb and their complexes, the metal atom always has one lone pair. The discrete, bent MX_2 molecules are only present in the gas phase with bond angles less than 120°. Fig. 4.7.1(a) shows the structure of $SnCl_2$, which has bond angle 95° and bond distance 242 pm.

In the crystalline state, the coordination numbers of metal atoms are usually increased to three or four. Due to the existence of the lone pair, the MX_3^- or MX_4^{2-} units all adopt the trigonal pyramidal or square

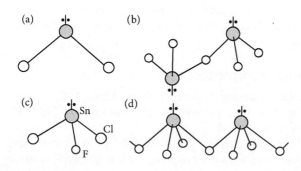

Fig. 4.7.1 Structure of subhalides of tin: (a) $SnCl_2$, (b) $(Sn_2F_5)^-$ in $NaSn_2F_5$, (c) $(SnCl_2F)^-$ in $[Co(en)_3][SnCl_2F][Sn_2F_5]Cl$, and (d) $(SnF_3^-)_\infty$ in $KSnF_3 \cdot \frac{1}{2}H_2O$.

pyramidal configuration, as shown in Fig. 4.7.1(b) to 4.7.1(d). Fig. 4.7.1(b) shows the structure of $Sn_2F_5^-$ in $NaSn_2F_5$, in which each Sn is trigonal pyramidal with two close F_t ($Sn-F_t$ 207 pm and 208 pm) and one F_b ($Sn-F_b$ 222 pm). Fig. 4.7.1(c) shows the structure of $(SnCl_2F)^-$ in $[Co(en)_3](SnCl_2F)[Sn_2F_5]Cl$. Fig. 4.7.1(d) shows the structure of $(SnF_3^-)_\infty$ in $KSnF_3 \cdot \frac{1}{2}H_2O$. In this compound, the Sn atoms are square pyramidal and each bridging F atom is connected to two Sn atoms to give an infinite chain with $Sn-F_b$ 227 pm and $Sn-F_t$ 201 pm and 204 pm.

(2) Oligomers and polymers

The structural chemistry of MX_2 is complex, partly because of the stereochemical activity (or non-activity) of the lone pair of electrons and partly because of the propensity of M^{II} to increase its coordination numbers by polymerization into larger structural units, such as rings, chains, or layers. The M^{II} center rarely adopts structures typical of spherically symmetrical ions because the lone pair of electrons, which is ns^2 in the free gaseous ion, is readily distorted in the condensed phase. This can be described in terms of ligand-field distortions or the adoption of some p-orbital character. The lone pair can act as a donor to vacant orbitals, and the vacant np orbitals and nd orbitals can act as acceptors in forming extra covalent bonds.

In the solid state, GeF_2 has a unique structure in which trigonal pyramidal GeF_3 units share two F atoms to form an infinite spiral chain. Reaction between GeF_2 and F^- gives GeF_3^- which is a trigonal pyramidal ion.

Crystalline SnF_2 is composed of Sn_4F_8 tetramers, which are puckered eight-membered rings of alternating Sn and F atoms, with $Sn-F_b$ 218 pm and $Sn-F_t$ 205 pm The tetramers are interlinked by weaker Sn–F interactions. Bridge formation is observed in $Na(Sn_2F_5)$, $Na_4(Sn_3F_{10})$, and other salts such as $[Co(en)_3][SnCl_2F][Sn_2F_5]Cl$.

The structure of $SnCl_2$ is shown in Fig. 4.7.1(a). In the crystalline state it has a layer structure with chains of corner-shared, trigonal-pyramidal $SnCl_3$ units. The commercially available solid hydrate $SnCl_2 \cdot 2H_2O$ also has a puckered-layer structure.

The dihalides PbX_2 is much more stable thermally and chemically than PbX_4. Structurally, α-PbF_2, $PbCl_2$, and $PbBr_2$ all form colorless orthorhombic crystals in which Pb^{II} is surrounded by nine X ligands (seven closer and two farther away) at the corners of a tricapped trigonal prism.

(3) Stability

The stability of the dihalides of Ge, Sn and Pb steady increases in the sequence $GeX_2 < SnX_2 < PbX_2$. Thus PbX_2 is more stable than PbX_4, whereas GeX_4 is more stable than GeX_2. At 978 K, SnF_4 sublimes to give a vapor containing SnF_4 molecules, which is thermally stable, but PbF_4 (prepared by the action of F_2 on Pb compounds) decomposes into PbF_2 and F_2 when heated.

The preference for the +2 over +4 oxidation state increases down the group, the change being due to relativistic effects that make an important contribution to the inert pair. The "inert pair" concept holds only for the lead ion Pb^{2+}(aq), which could have a $6s^2$ configuration. In more covalent Pb^{II} compounds and most Sn^{II} compounds there are stereochemically active lone pairs. In some MX_2 (M = Ge or Sn) compounds Ge and Sn can act as donor ligands.

(4) Mixed halides and mixed-valence complexes

Many mixed halides are known, such as all ten trihalogenostannate(II) anions $[SnCl_xBr_yI_z]^-$, (xyz = 300, 210, 201, 120, 102, 111, 021, 012, 030, 003) have been observed and characterized by ^{119}Sn NMR spectroscopy. An example $(SnCl_2F)^-$ is shown in Fig. 4.7.1(c).

Many mixed halides of Pb^{II} have also been characterized, including PbFCl, PbFBr, PbFI, and $PbX_2 \cdot 4PbF_2$. Of these, PbFCl has an important tetragonal layer structure, which is frequently adopted by large cations in the presence of two anions of differing size; its sparing solubility in water forms the basis of a gravimetric method for the determination of F.

In halides, some mixed-valence complexes are known, such as $Ge_5F_{12} = (Ge^{II}F_2)_4(Ge^{IV}F_4)$, α-$Sn_2F_6 = Sn^{II}Sn^{IV}F_6$, and $Sn_3F_8 = Sn_2^{II}Sn^{IV}F_8$. In these compounds, the M^{II}–X bond distances are longer than the corresponding M^{IV}–X bond distances.

The coordination geometries of M^{IV} in compounds or mixed-valence complexes, in contrast to M^{II}, tend to exhibit high symmetries; the MX_4 molecules are tetrahedral and $[MX_6]^{2-}$ units adopt the octahedral (or slightly distorted) configuration.

4.7.2 Oxides of Ge, Sn, and Pb

(1) Subvalent oxides

Both SnO and PbO exist in several modifications. The most common blue-black modification of SnO and red PbO (litharge) have a tetragonal layer structure, in which the M^{II} (Sn^{II} and Pb^{II}) atom is bonded to four oxygen atoms arranged in a square to one side of it, with the lone pair of electrons presumably occupying the apex of the tetragonal pyramid (Sn–O 221 pm and Pb–O 230 pm). Each oxygen atom is surrounded tetrahedrally by four M^{II} atoms. Fig. 4.7.2 shows the crystal structure of SnO (and PbO).

Dark-brown crystalline GeO is obtained when germanium powder and GeO_2 are heated or $Ge(OH)_2$ is dehydrated, but this compound is not well characterized.

(2) Dioxides

Germanium dioxide, GeO_2, closely resembles SiO_2 and exists in both α-quartz and tetragonal rutile (Fig. 4.2.6) forms. In the latter, Ge^{IV} is octahederally coordinated with Ge–O 188 pm (mean). SnO_2 occurs naturally as cassiterite, adopting the rutile structure with Sn–O 209 pm (mean). There are two modifications of PbO_2: the tetragonal maroon form has the rutile structure with Pb–O 218 pm (mean), whereas α-PbO_2 is an orthorhombic black form whose structure is derived from HCP layers with half of the octahedral sites filled.

(3) Mixed-valence oxides

Pb_3O_4 ($Pb_2^{II}Pb^{IV}O_4$) and Pb_2O_3 ($Pb^{II}Pb^{IV}O_3$) are the two well-known mixed-valence oxides of lead. Red lead Pb_3O_4 is important commercially as a pigment and primer. Its tetragonal crystal structure consists of chains of $Pb^{IV}O_6$ octahedra sharing opposite edges (mean Pb^{IV}–O 214 pm). These chains are aligned parallel to the c-axis and linked by the Pb^{II} atoms, each being pyramidally coordinated by three oxygen atoms ($Pb^{II}O_3$ unit, Pb^{II}–O two at 218 pm and one at 213 pm). Fig. 4.7.3 shows a portion of the tetragonal unit cell of Pb_3O_4.

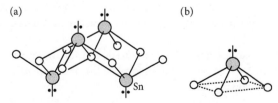

(a)   (b)

Fig. 4.7.2 Crystal structure of SnO (and PbO): (a) viewed from the side of a part of one layer, (b) the arrangement of bonds and lone pair at a metal atom.

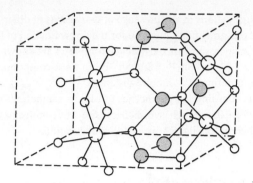

Fig. 4.7.3 Crystal structure of Pb_3O_4: (large circles represent Pb atom, white Pb^{IV} and shaded Pb^{II}; small circles represent O atom).

The sesquioxide Pb_2O_3 is a black monoclinic crystal, which consists of $Pb^{IV}O_6$ octahedra (Pb^{IV}–O 218 pm, mean) and very irregular six-coordinate $Pb^{II}O_6$ units (Pb^{II}–O 231, 243, 244, 264, 291, and 300 pm). The Pb^{II} atoms are situated between layers of distorted $Pb^{IV}O_6$ octahedra.

Among the various mixed-valence oxides of tin that have been reported, the best characterized is Sn_3O_4, that is, $Sn_2^{II}Sn^{IV}O_4$, the structure of which is similar to that of Pb_3O_4. Mixed oxides of Pb^{IV} (or Pb^{II}) with other metals find numerous applications in technology and industry. Specifically, $M^{II}Pb^{IV}O_3$ and $M^{II}Pb^{IV}O_4$ (M^{II} = Ca, Sr, Ba) are important materials; for example, $CaPbO_3$ is a priming pigment to protect steel corrosion by salt water. Mixed oxides of Pb^{II} are also important. $PbTiO_3$, $PbZrO_3$, $PbHfO_3$, $PbNb_2O_6$, and $PbTi_2O_6$ are ferroelectric materials. The high Curie temperature of many Pb^{II} ferroelectrics makes them particularly useful for high-temperature applications.

4.8 **Polyatomic Anions of Ge, Sn, and Pb**

Germanium, tin, and lead dissolve in liquid ammonia in the presence of alkali metals to give highly colored anions, which are identified as polyatomic anions (e.g., Sn_5^{2-}, Pb_5^{2-}, Sn_9^{4-}, Pb_9^{4-} and Ge_{10}^{2-}) in salts containing cryptated cations, for example, $[Na(C222)]_2Pb_5$, $[Na(C222)]_4Sn_9$, and $[K(C222)]_3Ge_9$.

In an unusual synthesis, the compound $[K(C222)]_2Pb_{10}$ was obtained by oxidation of a solution of Pb_9^{4-} ions in ethylenediamine with Au^I in the form of $[P(C_6H_5)_3AuCl]$ as the oxidizing agent:

$$2Pb_9^{4-} + 6Au^+ \longrightarrow Pb_{10}^{2-} + 8Pb^0 \downarrow + 6Au^0$$

A single-crystal X-ray structure analysis revealed that the intensely brown-colored Pb_{10}^{2-} anion has a bicapped square-antiprismatic structure of nearly perfect D_{4d} symmetry.

Since the homopolyatomic anions and cations (Zintl ions) are devoid of ligand attachment, they are sometimes referred to as "naked" clusters. The structures of some naked anionic clusters are shown in Fig. 4.8.1: (a) tetrahedral Ge_4^{4-}, Sn_4^{4-}, Pb_4^{4-}, $Pb_2Sb_2^{2-}$; (b) trigonal bipyramidal Sn_5^{2-}, Pb_5^{2-}; (c) octahedral Sn_6^{2-}; (d) tricapped-trigonal prismatic Ge_9^{2-} and paramagnetic Ge_9^{3-}, Sn_9^{3-}; (e) monocapped square-antiprismatic Ge_9^{4-}, Sn_9^{4-}, Pb_9^{4-}; (f) bicapped square-antiprismatic Ge_{10}^{2-}, Pb_{10}^{2-}, Sn_9Tl^{3-}.

The bonding in these polyatomic anions is delocalized, and for the diamagnetic species the bond valence method (see section 3.4) can be used to rationalize the observed structures. This method is to be applied according to the following rules:

(a) Each pair of neighbor M atoms is connected by a M–M 2c-2e bond or MMM 3e-2e bond to form the cluster.

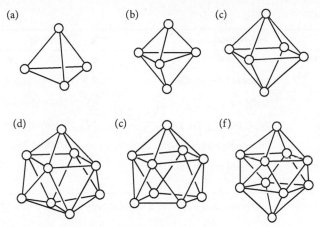

Fig. 4.8.1 Structure of polyatomic anions of Ge, Sn and Pb: (a) Ge_4^{4-}, Sn_4^{4-}, Pb_4^{4-} and $Pb_2Sb_2^{2-}$, (b)Sn_9^{2-}, Pb_9^{2-}, (c) Sn_6^{2-} (d) Ge_9^{2-}, (e) Ge_9^{4-}, Sn_9^{4-}, Pb_9^{4-}, (f) Ge_{10}^{2-}, Pb_{10}^{2-}, Sn_9Tl^{3-}.

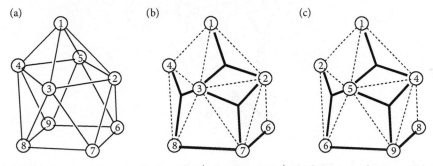

Fig. 4.8.2 (a) Structure and atom numbering of Ge_9^{4-}. Bonding in Ge_9^{4-}: (b) front view of a canonical structure, in which a M–M 2c-2e bond is represented by a solid line joining two atoms, and a MMM 3c-2e bond by three solid lines connecting atoms to the bond center; (c) rear view of the canonical structure.

(b) Each M atom use its four valence orbitals to form chemical bonds to attain the octet configuration.

(c) Two M atoms cannot participate in forming both a M–M 2c-2e bond and a MMM 3c-2e bond.

For example, the bond valence (b) of Ge_9^{4-} is $b = \frac{1}{2}(8n - g) = \frac{1}{2}[9 \times 8 - (9 \times 4 + 4)] = 16$, and the monocapped square-antiprismatic anion is stabilized by six MMM 3c-2e bonds and four M–M 2c-2e bonds. Bonding within the polyhedron involves a resonance hybrid of several canonical forms of the type shown in Fig. 4.8.2. The structures of the polyatomic anions are summarized in Table 4.8.1.

The polystannide cluster Sn_9^{4-} (or the related Pb_9^{4-}) reacts with $Cr(CO)_3(mes)$ to form a bicapped square antiprismatic cluster $[Sn_9Cr(CO)_3]^{4-}$.

$$Cr(CO)_3\,(mes) + K_4Sn_9 + 4C222 \xrightarrow{\text{en / toluene}} [K\,(C222)]_4\,[Sn_9Cr(CO)_3]$$

The nonastannide cluster is slightly distorted to accommodate the chromium carbonyl fragment at the open face. Fig. 4.8.3 shows the structure of $[M_9Cr(CO)_3]^{4-}$ (M = Sn, Pb).

Table 4.8.1 Structure of polyatomic anions

Anion	Polyhedron	b	Bonding
Ge_4^{4-}, Sn_4^{4-}, Pb_4^{4-}, $Pb_2Sb_2^{2-}$	tetrahedral	6	6[M–M 2c-2e]
Ge_5^{2-}, Pb_5^{2-}, Sn_5^{2-}	trigonal bipyramidal	9	9[M–M 2c-2e]
Sn_6^{2-}	octahedral	11	4[MMM 3c-2e] 3[M–M 2c-2e]
Ge_9^{2-}, (Sn_9^{3-}, Ge_9^{3-})	tricapped-trigonal prismatic	17	7[MMM 3c-2e] 3[M–M 2c-2e]
Ge_9^{4-}, Pb_9^{4-}, Sn_9^{4-}*	monocapped square-antiprismatic	16	6[MMM 3c-2e] 4[M–M 2c-2e]
Ge_{10}^{2-}, Pb_{10}^{2-}, Sn_9Tl^{3-}	bicapped square-antiprismatic	19	8[MMM 3c-2e] 3[M–M 2c-2e]

* The cation Bi_9^{5+} is isoelectronic and isostructural with these anions, and has the same kind of bonding.

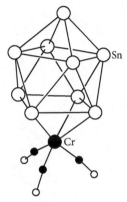

Fig. 4.8.3 Structure of $[M_9Cr(CO)_3]^{4-}$ (M = Sn, Pb).

The bond valence (b) in the ten-atom cluster $[Sn_9Cr(CO)_3]^{4-}$ is:

$$b = \frac{1}{2}[8n_1 + 18n_2 - g]$$
$$= \frac{1}{2}[8 \times 9 + 18 - (9 \times 4 + 6 + 6 + 4)]$$
$$= 19$$

This anion has eight MMM 3c-2e bonds and three M–M 2c-2e bonds, just as in $B_{10}H_{10}^{2-}$, and the scheme is illustrated in Fig. 3.4.9(c).

4.9 Organometallic Compounds of Heavier Group 14 Elements

Elements Ge, Sn, and Pb form many interesting varieties of organometallic compounds, some of which are important commercial products. For example, Et_4Pb is an anti-knock agent, and organotin compounds are employed as polyvinylchloride (PVC) stabilizers against degradation by light and heat. Some examples and advances are discussed in the present section.

4.9.1 Cyclopentadienyl complexes

Cyclopentadienyl complexes of Ge, Sn, and Pb exist in a wide variety of composition and structure, such as half sandwiches CpM and CpMX, bent and parallel sandwiches Cp_2M and polymeric $(Cp_2M)_x$.

(1) $(\eta^5\text{-}C_5Me_5)Ge^+$ and $(\eta^5\text{-}C_5Me_5)GeCl$

In crystalline $[\eta^5\text{-}(C_5Me_5)Ge]^+[BF_4]^-$ the cation has a half-sandwich structure, as shown in Fig. 4.9.1(a). In the chloride $(\eta^5\text{-}C_5Me_5)GeCl$, the Ge atom is bonded to the Cl atom in a bent configuration, as shown in Fig. 4.9.1(b).

(2) $(\eta^5\text{-}C_5R_5)_2M$ (R = H, Me, Ph)

The compounds $(\eta^5\text{-}C_5H_5)_2M$, (M = Ge, Sn and Pb) are angular molecules in the gas phase, as shown in Fig. 4.9.1(c). The ring centroid-M-ring centroid angles of $(\eta^5\text{-}C_5H_5)_2M$ are: –Ge– 130°, –Sn– 134°, and –Pb– 135°.

However, as the H atoms in Cp is substituted by other groups, the angles will be changed. For example, the angle in $(\eta^5\text{-}C_5Me_5)_2$ Pb is 151°. For $(\eta^5\text{-}C_5Ph_5)_2Sn$, the two planar C_5 rings are exactly parallel and staggered, as shown in Fig. 4.9.1(d), and the opposite canting of the phenyl rings with respect to the C_5 rings results in overall S_{10} molecular symmetry.

(3) Polymeric $[(\eta^5\text{-}C_5H_5)_2Pb]_x$

There are three known crystalline modifications of $(\eta^5\text{-}C_5H_5)_2Pb$. When crystals are grown by sublimation, the structure features a zigzag chain with alternating bridging and non-bridging C_5H_5 groups; however, crystallization from toluene results in modification of the conformation of the polymeric chain (as shown in Fig. 4.9.1(e)), and also cyclization into a hexamer $[(\eta^5\text{-}C_5H_5)_2Pb]_6$.

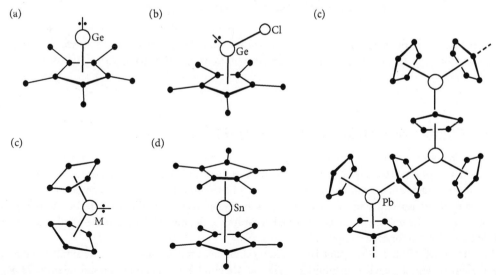

Fig. 4.9.1 Structures of some cyclopentadienyl complexes: (a) $(\eta^5\text{-}C_5Me_5)Ge^+$, (b) $(\eta^5\text{-} C_5Me_5)GeCl$, (c) $(\eta^5\text{-}C_5H_5)_2M$, (M = Ge, Sn and Pb), (d) $(\eta^5\text{-}C_5Ph_5)_2Sn$ (only one C atom of each phenyl group is shown) and (e) $[(\eta^5\text{-}C_5H_5)_2Pb]_x$ chain.

4.9.2 Sila- and germa-aromatic compounds

Sila- and germa-aromatic compounds are Si- and Ge-containing $(4n + 2)$ π electron ring systems that represent the heavier congeners of carbocyclic aromatic compounds. Several kinds of sila- and germa-aromatic compounds bearing a bulky hydrophobic 2,4,6-tris[bis{trimethylsilyl)methyl]phenyl group [Tbt = 2,4,6,-{CH(SiMe$_3$)$_2$}$_3$C$_6$H$_2$] on the sp^2 Si or Ge atom have been prepared and isolated as stable crystalline solids, as shown in Fig. 4.9.2(a)–(e).

X-ray diffraction analyses showed that all these compounds have an almost planar aromatic ring, and the geometry around the central Si or Ge atom is completely trigonal planar. The measured bond lengths of Si–C or Ge–C are almost equal, the average values being Si–C 175.4 pm and Ge–C 182.9 pm, which lie between those of typical double and single bonds.

The germabenzene shown in Fig. 4.9.2(d) serves as an η^6-arene ligand in reacting with [M(CH$_3$CN)$_3$(CO)$_3$] (M = Cr, Mo, W) to form sandwich complexes [(η^6-C$_5$H$_5$GeTbt)M(CO)$_3$]. Fig. 4.9.2(f) shows a part of the structure of [(η^6-C$_5$H$_5$GeTbt)Cr(CO)$_3$]. This complexation of germa-aromatic compounds indicates that sila- and germa-benzenes have a π electron delocalized structure with considerable aromaticity.

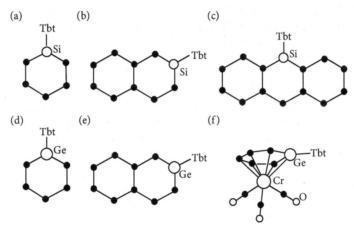

Fig. 4.9.2 Structure of some sila- and germa-aromatic compounds: (a) silabenzene, (b) 2-silanaphthalene, (c) 9-silaanthracene, (d) germabenzene, (e) 2-germanaphthalene, and (f) a part of (η^6-C$_5$H$_5$GeTbt)Cr(CO)$_3$.

4.9.3 Cluster complexes of Ge, Sn, and Pb

Extensive research over the past decade has greatly increased the number of cluster complexes of Ge, Sn, and Pb. The properties of M–M bonds and structures of these compounds are similar to the C–C bonds and carbon skeletons in organic compounds. Table 4.9.1 lists some of these cluster compounds that have been synthesized, isolated, and characterized by X-ray crystallography. In the table, the bond lengths are average values. The b value of M$_n$ is calculated according to formula (3.4.1), and the bond valence per M–M bond is listed for each compound.

In [GeSi^{t}Bu$_3$]$_3^+$ [BPh$_4$]$^-$, the cyclotrigermanium cation has a π delocalized bond π_3^2 similar to that of the carbon analog, the cyclopropenium ion. The Ge–Ge bond in this cation has bond valence 5/3. Fig. 4.9.3(a) shows the structure of the cluster.

The mixed-metal compound [Ge(SiMe$_3$)$_3$]$_2$SnCl$_2$ is a bent molecule (as shown in Fig. 4.9.3(b)), which forms two Ge–Sn bonds with bond lengths 263.6 pm and 262.6 pm.

Table 4.9.1 Cluster complexes of Ge and Sn

Compound	Bond length/pm	Bond valence	Fig. 4.9.3
$[Ge(Si^tBu_3)]_3^+ \ [BPh_4]^-$	Ge Ge, 232.6	$1\tfrac{2}{3}$	(a)
$[Ge(SiMe_3)_3]_2SnCl_2$	Ge—Sn, 263.1	1	(b)
$[Ge(Cl)Si(SiMe_3)_3]_4$	Ge—Ge, 253.4	1	(c)
$Ge_4(Si^tBu_3)_4$	Ge—Ge, 244	1	(d)
$Ge_6[CH(SiMe_3)_2]_6$	Ge—Ge, 256	1	(e)
$[Sn(Ph_2)]_6$	Sn—Sn, 277.5	1	(f)
$Sn_5(2,6\text{-}Et_2C_6H_3)_6$	Sn—Sn, 285.8	1	(g)
	Sn- - -Sn, 336.7		
$Sn_7(2,6\text{-}Et_2C_6H_3)_8$	Sn—Sn, 284.5	1	(h)
	Sn- - -Sn, 334.8		
$Ge_8(CMeEt_2)_8$	Ge—Ge, 249.0	1	(i)
$Ge_8{}^tBu_8Br_2$	Ge—Ge, 248.4	1	(j)
$Sn_{10}(2,6\text{-}Et_2C_6H_3)_{10}$	Sn—Sn, 285.6	1	(k)

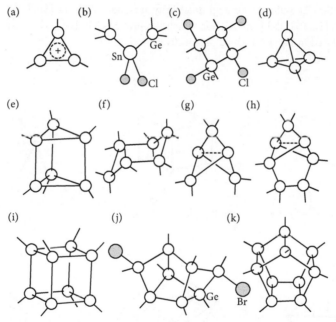

Fig. 4.9.3 Structures of metallic cluster complexs of Ge and Sn (for clarity only the metal and halogen atoms are shown, and the organic groups have been omitted): (a) $[Ge(Si^tBu_3)]_3^+$, (b) $[Ge(SiMe_3)_3]_2SnCl_2$, (c) $[Ge(Cl)Si(SiMe_3)_3]_4$, (d) $Ge_4(Si^tBu_3)_4$, (e) $Ge_6[CH(SiMe_3)_2]_6$, (f) $[SnPh_2]_6$, (g) $Sn_5(2,6\text{-}Et_2C_6H_3)_6$, (h) $Sn_7(2,6\text{-}Et_2C_6H_3)_8$, (i) $Ge_8(CMeEt_2)_8$, (j) $Ge_8{}^tBu_8Br_2$, (k) $Sn_{10}(2,6\text{-}Et_2C_6H_3)_{10}$.

The tetramer $[Ge(Cl)Si(SiMe_3)_3]_4$ forms a four-membered ring with Ge–Ge bond lengths 255.8 pm and 250.9 pm (as shown in Fig. 4.9.3(c)), while $Ge_4(Si^tBu_3)_4$ is the first molecular germanium compound with a Ge_4 tetrahedron, as shown in Fig. 4.9.3(d).

The hexamer $Ge_6[CH(SiMe_3)_2]_6$ has a prismane structure as shown in Fig. 4.9.3(e). The Ge–Ge distances within the two triangular faces are 258 pm, longer than those in the quadrilateral edges, which are 252 pm. On the other hand, the hexamer $(SnPh_2)_6$ exists in the chair conformation, (as shown in Fig. 4.9.3(f)) with Sn–Sn distances 277 pm to 278 pm

Both $Sn_5(2,6-Et_2C_6H_3)_6$ and $Sn_7(2,6-Et_2C_6H_3)_8$ have a pentastanna[1.1.1]propellane structure, which consists of three fused three-membered rings, as shown in Fig. 4.9.3(g) and (h). In these two compounds, the bridgehead Sn atoms each has all four bonds directed to one side of the relevant atom, thus forming "inverted configuration" Sn⋯Sn bonds, the distances of which are 336.7 pm and 334.8 pm, respectively. These distances far exceed the longest known Sn–Sn single bond length of 280 pm. These structures are also similar to that of 1-cyanotetracyclo-decane (Fig. 4.3.4(d)), in which the two bridgehead C atoms form a very long (164.3 pm) C–C bond.

The octagermacubanes $Ge_8(CMeEt_2)_8$ and $Ge_8(2,6-Et_2C_6H_3)_8$ have a cubic structure, as shown in Fig. 4.9.3(i), with Ge–Ge bond lengths 247.8 pm to 250.3 pm, with a mean value of 249.0 pm. The polycyclic octagermane $Ge_8{}^tBu_8Br_2$ forms a chiral C_2 symmetric skeleton, as shown in Fig. 4.9.3(j). The average Ge–Ge bond length is 248.4 pm.

The pentaprismane skeleton of $Sn_{10}(2,6-Et_2C_6H_3)_{10}$ is composed of five four-membered rings and two five-membered rings, as shown in Fig. 4.9.3(k). The average Sn–Sn bond length is 285.6 pm.

4.9.4 Metalloid clusters of Sn

Literature information is available for high-nuclearity metalloid clusters $[Sn_xR_y]$ ($x > y$), including $[Sn_8R_4]$, $[Sn_8R_6]^{2-}$, $[Sn_9R_3]$, and $[Sn_{10}R_3]^-$, in which R represents a bulky aryl or silyl ligand, and their structures are displayed in Fig. 4.9.4(a)–(d), respectively.

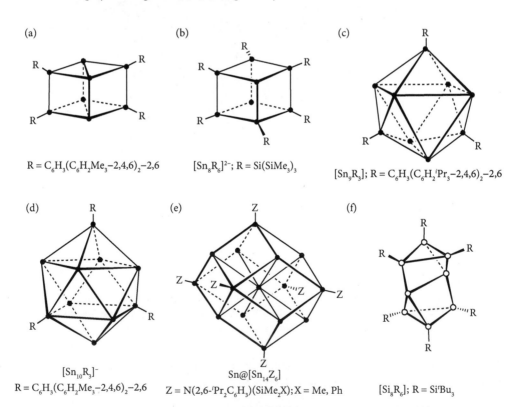

Fig. 4.9.4 Structures of some metalloid clusters of tin and the isoleptic compound $Si_8(Si^tBu_3)_6$.

The largest tin clusters known to date are the pair of isoleptic amido complexes $[Sn_{15}Z_6]$ (Z = N(2,6-$^iPr_2C_6H_3$)(SiMe$_2$X); X = Me, Ph). Both possess a common polyhedral core consisting of 14 tin atoms that fully encapsulates a central tin atom. The eight ligand-free peripheral tin atoms constitute the corners of a distorted cube, with each of its six faces capped by a Sn{N(2,6-$^iPr_2C_6H_3$)(SiMe$_2$X)} moiety, as shown in Fig. 4.9.4(e). The edges of the Sn_{14} polyhedron have an average length of 302 pm. The Sn(central)–Sn bonds in this high-nuclearity $Sn@Sn_{14}$ metalloid cluster have an average length of 315 pm, which is close to the corresponding value of 310 pm in gray tin (α-Sn, diamond lattice) but considerably longer than that in white tin (β-Sn).

Interestingly, the isoleptic compound Si_8R_6 (R = Si^{t}Bu$_3$; "supersilyl") does not exhibit a cubanoid cluster skeleton but instead contains an unsubstituted Si_2 dumbbell with a short Si–Si single bond of 229(1) pm, which is sandwiched between two almost parallel Si_3R_3 rings, as illustrated in Fig. 4.9.4(f). Each terminal of the Si_2 dumbbell acts as a bridgehead with an "inverted" tetrahedral bond configuration analogous to that found in 1-cyano-tetracyclododecane (see Section 4.3.5), forming two normal Si–Si bonds of 233 pm to one Si_3R_3 ring and a longer one of 257 pm to the other. In the crystal structure, the atoms of the Si_2 dumbbell are disordered over six sites.

An intermetalloid cluster is a metalloid cluster composed of more than one metal, for example $[Ni_2@Sn_{17}]^{4-}$. Known intermetalloid clusters involving other Group 14 elements and various transition metals include $[Pd_2@Ge_{18}]^{4-}$, $[Ni(Ni@Ge_9)_2]^{4-}$, $[Ni@Pb_{10}]^{2-}$, and $[Pt@Pb_{12}]^{2-}$; in the latter complex, the Pt atom is encapsulated by a Pb_{12} icosahedron (see Fig. 9.6.18).

4.9.5 Donor–acceptor complexes of Ge, Sn, and Pb

The system $R_2Sn{\rightarrow}SnCl_2$, with R = CH(SiMe$_3$)C$_9$H$_6$N, provides the first example of a stable donor–acceptor complex between two tin centers, as shown in Fig. 4.9.5. The alkyl ligand R is bonded in a *C,N*-chelating fashion to a Sn atom which adopts a pentacoordinate square-pyramidal geometry. This Sn atom is bonded directly to the other Sn atom of the SnCl$_2$ fragment with a Sn–Sn distance of 296.1 pm, which is significantly longer than the similar distance of 276.8 pm in $R_2Sn{=}SnR_2$ [R = CH(SiMe$_3$)$_2$], and much shorter than the similar distance of 363.9 pm in $Ar_2Sn–SnAr_2$ [Ar = 2,4,6-(CF$_3$)$_3$C$_6$H$_2$]. The fold angle defined as the angle between the Sn–Sn vector and the SnCl$_2$ plane is 83.3°.

Some donor–acceptor complexes containing Ge–Ge and Sn–Sn bonds are listed in Table 4.9.2.

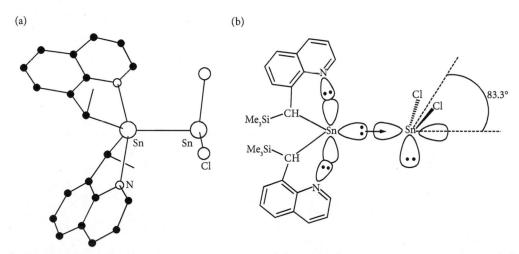

Fig. 4.9.5 Structure of the $R_2Sn{\rightarrow}SnCl_2$ (R = CH(SiMe$_3$)C$_9$H$_6$N). (a) Molecular structure, (the SiMe$_3$ groups have been omitted for clarity), (b) the lone pair forming a donor–acceptor bond.

The compounds listed in Table 4.9.2 have the formal formula:

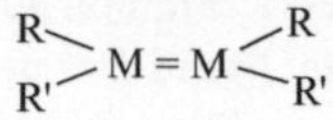

$$\begin{array}{c} R \\ R' \end{array} M = M \begin{array}{c} R \\ R' \end{array}$$

which indicates the presence of a M=M double bond, but the measured M–M bond lengths vary over a wide range. The M_2C_4 (or M_2Si_4) skeleton is not planar, in contrast with that of an olefin. Thus the properties of the formal M=M bonds are diverse and interesting. Two examples listed in this table are discussed below.

As reference data, the "normal" values M–M single bond lengths are calculated from the covalent radii (Table 3.4.3) and the "normal" bond lengths of M=M double bonds are estimated as 0.9 × (single bond lengths). Thus the "normal" bond lengths are:

$$\text{Ge–Ge 244 pm, Ge=Ge 220 pm}$$
$$\text{Sn–Sn 280 pm, Sn=Sn 252 pm}$$

Compound A in Table 4.9.2, $(Me_3SiN=PPh_2)_2C=Ge \rightarrow Ge=C(Ph_2P=NSiMe_3)_2$, comprises two germavinylidene units $Ge=C(Ph_2P=NSiMe_3)_2$ that are bonded together in a head-to-head manner. The molecule is asymmetrical with two different Ge environments: four-coordinate Ge and two-coordinate Ge, and shows *trans* linkage of the C=Ge–Ge=C skeleton, the torsion angle about the Ge–Ge axis being 43.9°. The Ge–Ge bond distance of 248.3 pm is consistent with a single bond, and both observed Ge–C bond lengths 190.5 pm and 190.8 pm are between those of standard Ge–C and Ge=C bonds. Therefore, the Ge–Ge bond in compound A is more approximately described as a donor-acceptor interaction similar

Table 4.9.2 Some donor–acceptor complexes with Ge–Ge and Sn–Sn bonds

Compound	M–M bond length/pm
$[Ge(2,6-Et_2C_6H_3)_2]_2$	Ge–Ge, 221.3
$[Ge(2,6-i-C_3H_7)_2C_6H_3]\{2,4,6-Me_3C_6H_2\}_2$	Ge–Ge, 230.1
$[Ge\{CH(SiMe_3)_2\}_2]_2$	Ge–Ge, 234.7
A ¶	Ge–Ge, 248.3
$[Sn\{CH(SiMe_3)_2\}_2]_2$	Sn–Sn, 276.8
$[Sn\{Si(SiMe_3)_3\}_2]_2$	Sn–Sn, 282.5
$[Sn(4,5,6-Me_3-2-{}^tBuC_6H)_2]_2$	Sn–Sn, 291.0
B ¶	Sn–Sn, 300.9
C ¶	Sn–Sn, 308.7
$[Sn\{2,4,6-(CF_3)_3C_6H_2\}_2]_2$	Sn–Sn, 363.9

A

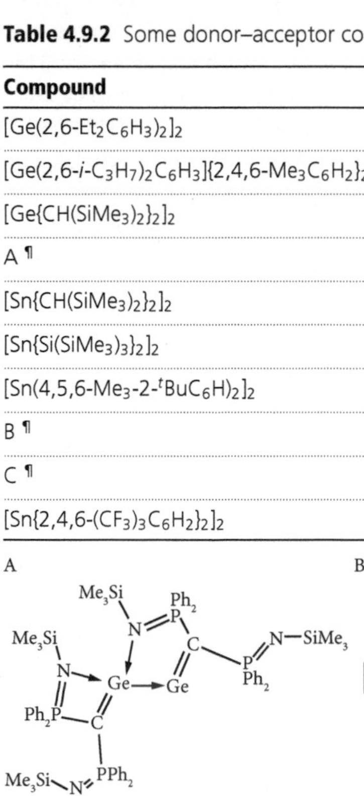

B

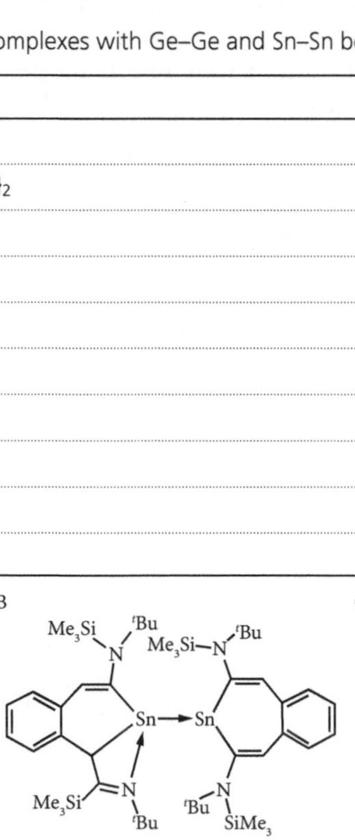

C

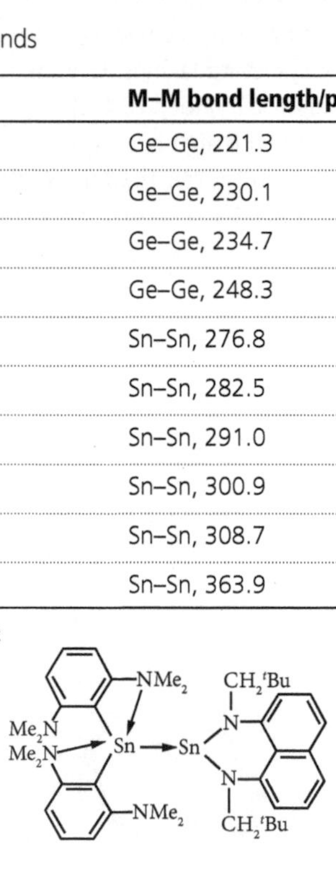

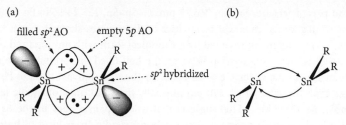

Fig. 4.9.6 Sn–Sn bonding model in [Sn{CH(SiMe$_3$)$_2$}$_2$]$_2$: (a) overlap of atomic orbitals, (b) representation of the donor–acceptor bonding.

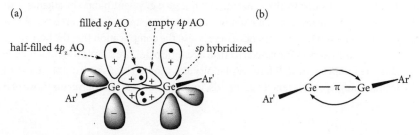

Fig. 4.9.7 Ge–Ge bonding model in Ar'Ge≡GeAr' (Ar' = 2,6-(C$_6$H$_3$-2,6-^{i}Pr$_2$)$_2$C$_6$H$_3$)): (a) overlap of atomic orbitals, (b) representation of the multiple bonding.

to that of R$_2$Sn→SnCl$_2$ (as shown in Fig. 4.9.5), and the four-coordinate Ge behaves as the donor and the two-coordinate Ge as a Lewis acid center.

The compound [Sn{CH(SiMe$_3$)$_2$}$_2$]$_2$ does not possess a planar Sn$_2$C$_4$ framework, and the Sn–Sn bond length (276.8 pm) is too long to be consistent with a "normal" double bond. A bonding model involving overlap of filled sp^2 hybrids and vacant $5p$ atomic orbitals has been suggested, as shown in Fig. 4.9.6. The structure is similar to that in Fig. 4.4.9(e), but the *trans*-bent angle is larger.

The digermanium alkyne analog 2,6 Dipp$_2$H$_3$C$_6$Ge≡GeC$_6$H$_3$-2,6-Dipp$_2$ (Dipp = C$_6$H$_3$-2,6-^{i}Pr$_2$) was synthesized by the reaction of Ge(Cl)C$_6$H$_3$-2,6-Dipp$_2$ with potassium in THF or benzene. It was isolated as orange-red crystals and fully characterized by spectroscopic methods and X-ray crystallography. The digermyne molecule is centrosymmetric with a planar *trans*-bent C(*ipso*)–Ge–Ge–C(*ipso*) skeleton, and the central aryl ring of the terphenyl ligand is virtually co-planar with the molecular skeleton, each flanking aryl ring being oriented at ~82° with respect to it. The structural parameters are: C–Ge, 199.6 pm; Ge–Ge, 228.5 pm (considerably shorter than the Ge–Ge single bond distance of ca. 244 pm); C–Ge–Ge, 128.67°. The measured Ge–Ge distance lies on the short side of the known range (221–246 pm) for digermenes, the digermanium analogs of alkenes.

By analogy to the model used to describe a Sn=Sn double bond, the observed geometry of the digermyne molecule and its multiple bonding character can be rationalized as shown in Fig. 4.9.7. Each germanium atom is considered to be sp hybridized in the C–Ge–Ge plane; a singly filled sp hybrid orbital is used to form a covalent bond with the terphenyl ligand, and the other sp hybrid is completely filled and forms a donor bond with an empty in-plane $4p$ orbital on the other germanium atom. Two half-filled $4p_z$ orbitals, one on each germanium atom and lying perpendicular to the molecular skeleton, then overlap to form a much weaker π bond. This simplified bonding description is consistent with the *trans*-bent molecular geometry and the fact that the formal Ge≡Ge bond length is not much shorter than that of the Ge=Ge bond.

The distannyne Ar'SnSnAr' and diplumbyne Ar*PbPbAr* (Ar* = 2,4,6-(C$_6$H$_3$-2,6-^{i}Pr$_2$)$_3$C$_6$H$_2$) have also been synthesized as crystalline solids and fully characterized. They are isostructural with digermyne, and their structural parameters are: Sn–Sn, 266.75 pm; C–Sn–Sn, 125.24°; Pb–Pb, 318.81 pm; C–Pb–Pb, 94.26°. Recent studies of the chemical reactivities of digermynes showed that it has considerable diradical character, which can be represented by a canonical form with an unpaired electron on each germanium atom, that is, negligible overlap between the half-filled $4p_z$ orbitals. The reactivity of the alkyne analogs decreases in the order Ge > Sn > Pb. In the diplumbyne, the Pb–Pb bond length is significantly longer than the value of *ca.* 290 pm normally found in organometallic lead-lead bonded species, for example, $Me_3PbPbMe_3$. This suggests that a lone pair resides in the $6s$ orbital of each lead atom in the dilead compound, and a single σ bond results from head-to-head overlap of $6p$ orbitals.

In summary, it is noted that multiple bonding between the heavier Group 14 elements E (Ge, Sn, Pb) differs in nature in comparison with the conventional σ and π covalent bonds in alkenes and alkynes. In an E=E bond, both components are of the donor-acceptor type, and a formal E≡E bond involves two donor–acceptor components plus a *p-p* π bond. There is also the complication that the bond order may be lowered when each E atom bears an unpaired electron or a lone pair. The simple bonding models provide a reasonable rationale for the marked difference in molecular geometries, as well as the gradation of bond properties in formally single, double, and triple bonds, in compounds of carbon versus those of its heavier congeners.

4.9.6 **Structure of germylene and stannylene compounds**

When $^tBu–N=S=N–^tBu$ reacts with PhLi in ether at low temperature (–78 °C) and then with $GeCl_2$ and $SnCl_2$ separately, structurally analogous germylene and stannylene compounds are obtained, as shown in Fig. 4.9.8 (a) and (b). The central four-membered GeN_2S/SnN_2S ring has an η^2-chelating diimido-sulfinate ligand exhibiting non-planar geometry. The Ge/Sn atoms are three-coordinated and exhibit a distorted trigonal-pyramidal environment. The Cl atom on the Ge/Sn and the phenyl group on the S atom take a *trans*-configuration with respect to the GeN_2S/SnN_2S ring.

When $^tBu–N=S=N–^tBu$ in the above reaction is replaced by its silicon analog, $(CH_3)_3Si–N=S=N–Si(CH_3)_3$, the products formed are significantly different in structure. The products obtained from $GeCl_2$ and $SnCl_2$ are shown in Fig. 4.9.8(c) and (d), respectively. Both are similar in structure and they contain a six-membered ring of distorted boat conformation. The : Ge–Cl and :Sn–Cl moieties at the 1,3-positions adopt an *anti*-configuration. The two Ge/Sn atoms exhibit distorted trigonal-pyramidal geometry.

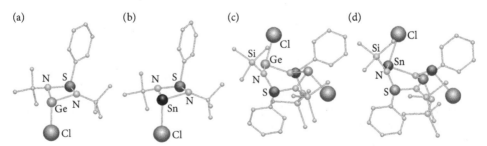

Fig. 4.9.8 Structures of germylenes and stannylenes: (a) $PhS(N^tBu)_2GeCl$, (b) $PhS(N^tBu)_2SnCl$, (c) $PhS(NSiMe_3)_2(GeCl)_2$, and (d) $PhS(NSiMe_3)_2(SnCl)_2$.

Reference

N. Nakata, N. Hosada, S. Takashi, and A. Ishii, Chlorogermylenes and-stannylenes stabilized by diimidosulfinate ligands: synthesis, structures, and reactivity. *Dalton Trans.* **47**, 481–90 (2018).

4.10 **Tetrel Bonding Interactions in Group 14 Compounds**

A tetrel bond is defined as an interaction between a Group 14 element acting as Lewis acid and an electron donating entity. Tin and lead form the strongest tetrel bonds in their compounds.

4.10.1 σ-Hole interactions and Si–Ge tetrel bonds (TrB)

The σ-hole interaction refers to the change in electron distribution when atoms of Groups 14 to 17 elements form a covalent bond. The probability density becomes anisotropic and an electron-deficient region is generated. This region and the electron-attracting group of another atom can form covalent interaction as secondary bonding to increase the binding between atoms. In the literature, such secondary bonding types are given different names according to their families in the Periodic Table: tetrel bond (TrB) for silicon, germanium, tin, and lead (Group 14); pnictogen bond (PnB) for arsenic, antimony, bismuth (Group 15); chalcogen bond (ChB) for sulfur, selenium, and tellurium (Group 16); halogen bond (XB) for chlorine, bromine, iodine (Group 17). These secondary bond types will be discussed separately in sections in the corresponding chapters. This section illustrates the basic concepts with two examples of tetrel bonding involving the light Group 14 elements silicon and germanium.

Example 1. $[Si_6X_{12}I_2]^{2-}$ (X = Cl or Br), which is produced in the reaction of Si_6X_{12} and I^-, has a planar hexagonal Si_6 ring structure with one I^- ion positioned on each side, as illustrated in Fig. 4.10.1. The I^- ions and Si_6X_{12} unit are bound by the TrB interaction between them.

Example 2. In the cavity structure of $[Ge_8O_{12}(OH)_8F]$ shown in Fig. 4.10.2, the encapsulated F^- ion is stabilized by TrB interaction between it and eight Ge(OH) units.

Tetrel bonding in compounds of the heavy Group 14 elements tin and lead is presented in Section 4.10.

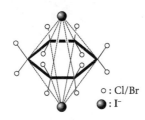

○ : Cl/Br
● : I^-

Fig. 4.10.1 TrB in $[Si_6X_{12}I_2]^{2-}$ (indicated by light dashes).

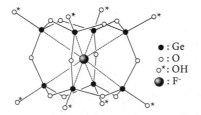

● : Ge
○ : O
○* : OH
● : F^-

Fig. 4.10.2 Solid-state structure showing encapsulation of F^- within a Ge_8O_{12} cage and its stabilization by eight TrB interactions (indicated by light dashes).

References

J. Y. C. Lim and P. D. Beer, Sigma hole interactions in anion recognition. *Chem.* **1**, 731–83 (2018).

I. Alkorta, J. Elguero, and A. Frontera, Not only hydrogen bonds: other noncovalent interactions, *Crystals.* **10**, 180 (2020). doi:10.3390/cryst10030180.

4.10.2 **Interactions involving tin**

In the crystal structure of tris(trimethylstannyl)acetonitrile (**1**), the nitrile group acts as lone-pair donor in the tetrel bond that controls the formation of infinite polymeric chains.

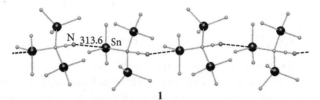

1

Dibromo-dimethyl-tin and 1,4-dithiane form a 2:1 co-crystal (**2**), which is a linear supramolecular polymer assembled by two different tetrel bonding interactions: one between Sn and an axial S lone pair, and the other between Sn and Br.

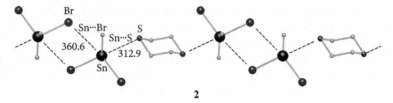

2

The sheet structure of 1,3-bis(trimethyltin)-1,3,5,7-tetraaza-2,4,6,8-tetrathiocin-2,2-dioxide (**3**) is con-solidated by two different tetrel bonding interactions: one with a sulfoxide O atom as the lone-pair donor (309.1 pm), and the other with a N atom of the tetraaza-tetrathiocin ring (338.0 pm).

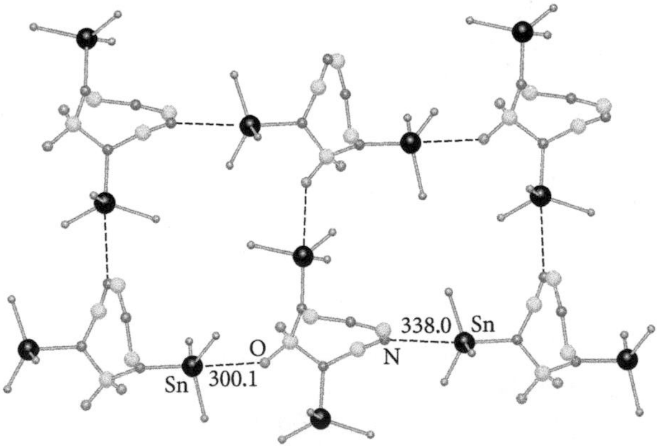

3

In the crystal structure of (dicycnoethene-1,2-dithiolate)-dimethyltin (**4**), supramolecular sheets are formed by means of cooperative tetrel Sn···N and chalcogenide S···N bonding interactions. Tetrel bonds

(306.8 pm) exist between the Sn atom and two N atoms from two adjacent molecules, and chalcogen bonds (317.3 pm) exist between two S atoms of the same molecule and two N atoms from two adjacent molecules.

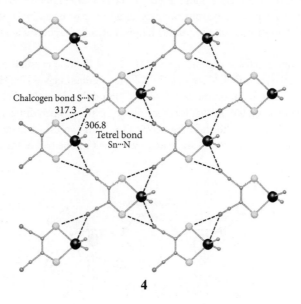

4

4.10.3 Interactions involving lead

Recent studies have shown that tetrel bonding plays a significant role in the supramolecular chemistry of lead.

5   **6**

The crystal structure of diphenyl-(2,2′-sulfanediyldibenzenethiolato-*S*, *S*′)-lead (**5**) features a centrosymmetric dimeric ring in which the Pb⋯S tetrel bond is opposite to the most polarized Pb–S bond, as expected for σ-hole interactions. The linear polymeric crystal structure of triphenyl-(2-bromophenyl-thiolato)-lead (**6**) is stabilized by the Pb⋯Br tetrel bonding interaction, such that the tetracoordinated Pb(IV) atom acts as the Lewis acid.

Crystallization of Pb(SCN)$_2$ with organic ligand HL from methanol yielded solvated [Pb(HL)(SCN)$_2$]·CH$_3$OH (**7**), and changing the solvent to ethanol gave [Pb(HL)(SCN)$_2$] (**8**). In both crystal structures the hemi-directional coordination mode of the Pb^{2+} ion facilitates its tetrel bonding with a thiocyanate coordinated to an adjacent Pb^{2+} ion.

4.10.4 **Supramolecular assembly of Pb(II) compounds with tetrel bonding**

The most common valence of lead is +2, and the Pb^{2+} ion has a large ionic radius and variable coordination number. In particular, its $6s^2$ electrons can exist as a lone pair or an inert pair. Fig. 4.10.3 shows the corresponding Pb(II) coordination modes: (a) hemispherical tetracoordinate and (b) symmetrical octahedral modes.

$Pb(SCN)_2$ and the unsymmetrical organic ligand HL (Fig. 4.10.4(a)) crystallize in methanol to give the solvate $[Pb(HL)(SCN)_2]\cdot CH_3OH$. Changing the solvent to ethanol gives $[Pb(HL)(SCN)_2]$. In these two crystal structures, the coordination modes of the Pb^{2+} ion differ. In molecule **7** (Fig. 4.10.4(b)), the coordination sphere is not saturated, so that further linkage occurs to form dimers (Fig. 4.10.4(c)). The Pb^{2+} cation is hexa-coordinated by the chelating *N,N,O* sites, with two thiocyanate N and the S atom of the third thiocynate to form a dimer. As all six bonds are located in one half of the coordination sphere,

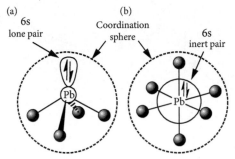

Fig. 4.10.3 Pb(II) coordination modes: (a) hemispherical tetracoordinate and (b) symmetrical octahedral modes.

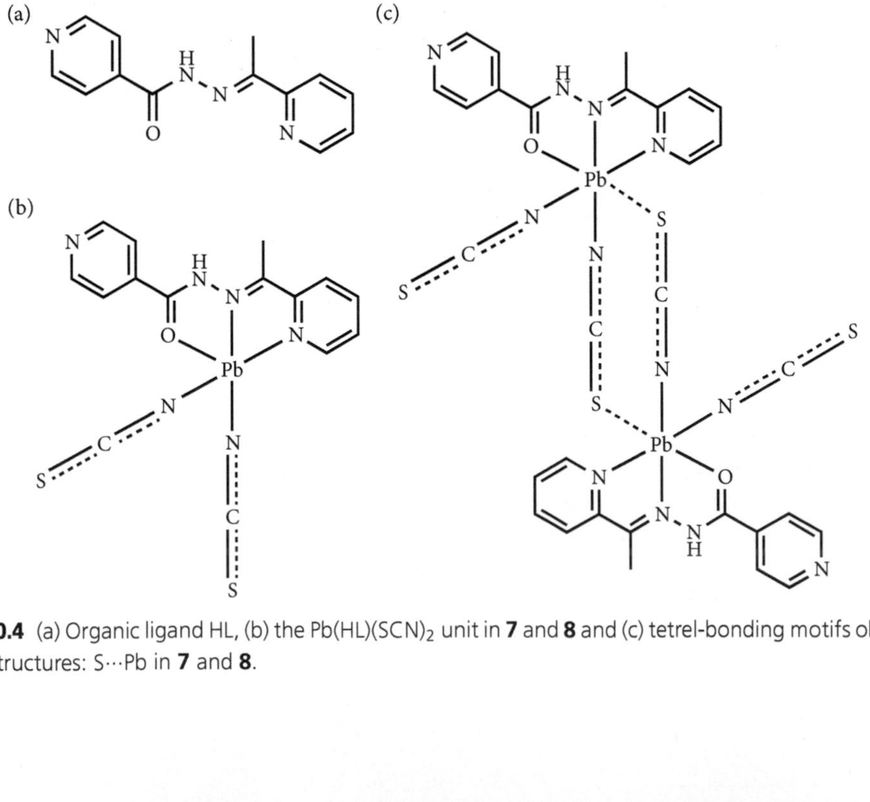

Fig. 4.10.4 (a) Organic ligand HL, (b) the $Pb(HL)(SCN)_2$ unit in **7** and **8** and (c) tetrel-bonding motifs observed in the structures: S···Pb in **7** and **8**.

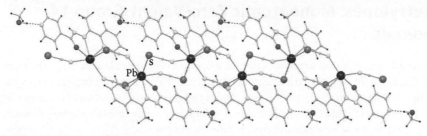

Fig. 4.10.5 Crystal structure of **7**.

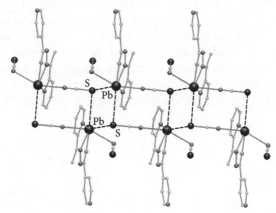

Fig. 4.10.6 Crystal structure of **8**.

the dimer can be connected with the S atom of a thiocyanate, thereby generating a polymeric chain via Pb···S tetrel bonding (Fig. 4.10.5). The methanol solvate molecule is connected with the pryidyl atom via hydrogen bonding.

In the crystal structure of **7** (Fig. 4.10.5), the Pb^{2+} ion is six-coordinated by the chelating N,N,O set of ligand HL, two thiocyanate N atoms, and a S atom of a third thiocyanate ligand. All six bonds are concentrated on one hemisphere of the coordination sphere, enabling attachment of a second thiocyanate S atom. Such Pb···S binding further interconnect the dimers into a polymeric chain. The solvated methanol is hydrogen-bonded to the pyridyl N atom.

In the crystal structure of **8** (Fig. 4.10.6), the Pb^{2+} ion is penta-coordinated by the chelating N,N,O set of the organic ligand and two thiocyanate N atoms in a pronounced hemispherical fashion, thus allowing it to form two Pb···S tetrel bonds, leading to interconnection of the monomeric coordination units to form a zigzag double chain.

References

A. Bauzá, S. K. Seth, and A. Frontera, Tetrel bonding interactions at work: impact on tin and lead coordination compounds. *Coord. Chem. Rev.* **384**, 107–25 (2019).

M. S. Gargari, V. Stilinović, A. Bauzá, A. Frontera, P. MaArdle, D. V. Derveer, S. W. Ng, and G. Mahmoudi, Design of lead(II) metal-organic frameworks based on covalent and tetrel bonding. *Chem. Eur. J.* **21**, 17951–8 (2015).

4.11 Tetrylones: Monoatomic Zero-Valent Group 14 Compounds

Tetrylones (ylidones) represent a class of zero-valent Group 14 compounds with the general formula EL_2 (E = C, Si, Ge, Sn, or Pb; L = neutral σ-donating ligand), wherein the tetrel atom E(0) has its four valence electrons functioning as two electron lone pairs, and is moreover coordinated by two L donor ligands via donor–acceptor interactions (L→E←L). Tetrylones differ significantly from tetrelenes (ylidenes, >E:), in which E contains only one lone pair and two electron-sharing bonds.

Tristannaallene **1** synthesized by Wiberg and colleagues in 1999 is the first example of a heavier tetrylone (stannylone). X-ray structural analysis revealed a bent Sn–Sn–Sn moiety (156.01(3)°) with short Sn–Sn bonds of average value 268.3 pm. In 2003 Kira and co-workers reported the synthesis of the thermally stable, air-sensitive, crystalline trisilaallene **2** as a green solid. Its molecular structure at –150 °C revealed a significantly bent Si=Si=Si skeleton [bond angle 136.49(6)°] with disilene Si=Si bond lengths of 217.7(1) and 218.8(1) pm. The molecular structures of **1** and **2** are shown in Fig. 4.11.1.

Further work by Kira and co-workers reported the first examples of 1,3-disilagermaallene **3** with a Si–Ge–Si bond angle of 132.38(2)°, trigermaallene **4** with a Ge–Ge–Ge bond angle of 122.61(6)°, and 1,3-digermasilaallene **5**, which exhibits a Ge–Si–Ge bond angle of 125.71(7)° (Fig. 4.11.2).

(a)　　　　　　　　　　　　　　　(b)

Fig. 4.11.1 Molecular structure of (a) tristannaallene **1** and (b) trisilaallene **2**.

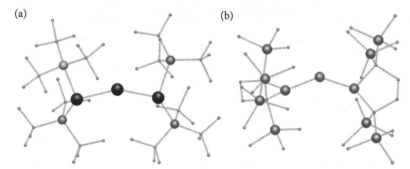

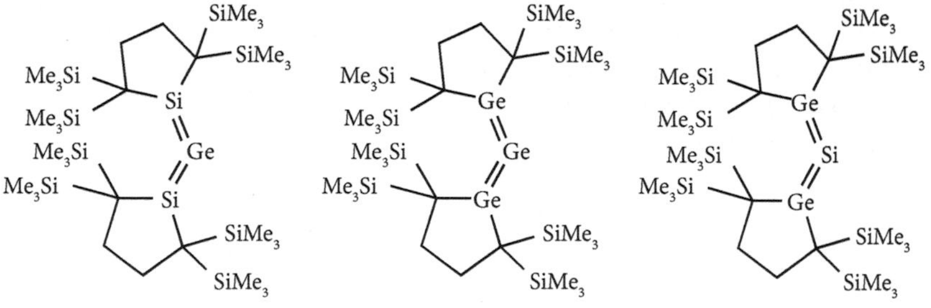

Fig. 4.11.2 Structural formulas of 1,3-disilagermaallene **3**, trigermaallene **4**, and 1,3-digermasilaallene **5**.

Reference

P. K. Majhi and T. Sasamori, Tetrylones: an intriguing class of monoatomic zero-valent group 14 compounds. *Chem. Eur. J.* **24**, 9441–55 (2018).

General References

J. Baggott, *Perfect Symmetry, The Accidental Discovery of Buckminster Fullerene*, Oxford University Press, Oxford, 1994.

H. W. Kroto, J. E. Fischer, and D. E. Cox (eds.), *Fullerenes*, Pergamon Press, Oxford, 1993.

A. Hirsch and M. Brettreich, *Fullerenes: Chemistry and Reactions*, Wiley-VCH, Weinheim, 2005.

T. Akasaka and S. Nagase (eds.), *Endofullerenes: A New Family of Carbon Clusters*, Kluwer, Dordrecht, 2002.

M. Meyyappan (ed.), *Carbon Nanotubes: Science and Applications*, CRC Press, Boca Raton, 2005.

M. A. Brook, *Silicon in Organic, Organometallic, and Polymer Chemistry*, Wiley, New York, 2000.

P. Jutzi and U. Schubert (eds.), *Silicon Chemistry: From the Atom to Extended Systems*, Wiley-VCH, Weinheim, 2003.

W. M. Meier and D. H. Olson, *Atlas of Zeolite Structure Types*, Butterworths, London, 1987.

M. Sawamura, Y. Kuninobu, M. Toganoh, Y. Matsuo, M. Yamanaka, and E. Nakamura, Hybrid of ferrocene and fullerene. *J. Am. Chem. Soc.* **124**, 9354–5 (2002).

S. Y. Xie, F. Gao, X. Lu, R.-B. Huang, C.-R. Wang, X. Zhang, M.-L. Liu, S.-L. Deng, and L.-S. Zheng, Capturing the labile fullerene[50] as $C_{50}Cl_{10}$. *Science* **304**, 699 (2004).

O. T. Summerscales, F. G. N. Cloke, P. B. Hitchcock, J. C. Green, and N. Hazari, Reductive cyclotrimerization of carbon monoxide to the deltate dianion by an organometallic uranium complex. *Science* **311**, 829–31 (2006).

D. R. Huntley, G. Markopoulos, P. M. Donovan, L. T. Scott, and R. Hoffmann, Squeezing C–C bonds. *Angew. Chem. Int. Ed.* **44**, 7549–53 (2005).

U. Ruschewitz, Binary and ternary carbides of alkali and alkaline-earth metals. *Coord. Chem. Rev.* **244**, 115–36 (2003).

G.-C. Guo, G.-D. Zhou, and T. C. W. Mak, Structural variation in novel double salts of silver acetylide with silver nitrate: fully encapsulated acetylide dianion in different polyhedral silver cages. *J. Am. Chem. Soc.* **121**, 3136–41 (1999).

H.-B. Song, Q.-M. Wang, Z.-Z. Zhang, and T. C. W. Mak, A novel luminescent copper(I) complex containing an acetylenediide-bridged, butterfly-shaped tetranuclear core. *Chem. Commun.* **37**, 1658–9 (2001).

A. Sekiguchi, S. Inoue, M. Ichinoche, and Y. Arai, Isolable anion radical of blue disilene $({}^{t}Bu_2MeSi)_2Si=Si(SiMe^{t}Bu_2)_2$ formed upon one-electron reduction: synthesis and characterization. *J. Am. Chem. Soc.* **126**, 9626–9 (2004).

S. Nagase, Polyhedral compounds of the heavier group 14 elements: silicon, germanium, tin and lead. *Acc. Chem. Res.* **28**, 469–76 (1995).

M. Brynda, R. Herber, P. B. Hitchcock, M. F. Lappert, I. Nowik, P. P. Power, A. V. Protchenko, A. Ruzicka, and J. Steiner, Higher-nuclearity group 14 metalloid clusters: $[Sn_9\{Sn(NRR')\}_6]$. *Angew. Chem. Int. Ed.* **45**, 4333–7 (2006).

G. Fischer, V. Huch, P. Maeyer, S. K. Vasisht, M. Veith, and N. Wiberg, $Si_8(SitBu_3)_6$: a hitherto unknown cluster structure in silicon chemistry. *Angew. Chem. Int. Ed.* **44**, 7884–7 (2005).

W. P. Leung, Z. X. Wang, H. W. Li, and T. C. W. Mak, Bis(germavinylidene)-$[(Me_3SiN=PPh_2)_2 C=Ge\rightarrow Ge=C(Ph_2P=NSiMe_3)]$ and 1,3-dimetallacyclobutanes $[M\{\mu^2-C(Ph_2PSiMe_3)_2\}]_2$ (M = Sn, Pb). *Angew. Chem. Int. Ed.* **40**, 2501–3 (2001).

W. P. Leung, W. H. Kwok, F. Xue, and T. C. W. Mak, Synthesis and crystal structure of an unprecedented tin(II)-tin(II) donor-acceptor complex, $R_2Sn\rightarrow SnCl_2$, $[R = CH(SiMe_3)C_9H_6N-8]$. *J. Am. Chem. Soc.* **119**, 1145–6 (1997).

N. Tokitoh, New progress in the chemistry of stable metalla-aromatic compounds of heavier group 14 elements. *Acc. Chem. Res.* **37**, 86–94 (2004).

M. Stender, A. D. Phillips, R. J. Wright, and P. P. Power, Synthesis and characterization of a digermanium analogue of an alkyne. *Angew. Chem Int. Ed.* **41**, 1785–7 (2002).

P. P. Power, Synthesis and some reactivity studies of germanium, tin and lead analogues of alkyne. *Appl. Organometal. Chem.* **19**, 488–93 (2005).

Chapter 5

Structural Chemistry of Group 15 Elements

5.1 The N_2 Molecule, All-Nitrogen Ions, and Dinitrogen Complexes

5.1.1 The N_2 molecule

Nitrogen is the most abundant uncombined element in the earth's surface. It is one of the four essential elements (C, H, O, N) that support all forms of life. It constitutes, on average, about 15% by weight in proteins. The industrial fixation of nitrogen in the production of agricultural fertilizers and other chemical products is now carried out on a vast scale.

The formation of a triple bond, $N \equiv N$, comprising one σ and two π components with bond length 109.7 pm, accounts for the extraordinary stability of the dinitrogen molecule N_2. The energy level diagram for N_2 is shown in Fig. 3.3.3(a). Gaseous N_2 is rather inert at room temperature mainly because of the great strength of the $N \equiv N$ bond and the large energy gap between the HOMO and LUMO (≈ 8.6 eV), as well as the absence of bond polarity. The high bond dissociation energy of the N_2 molecule, 945 kJ mol^{-1}, accounts for the following phenomena:

(a) Molecular nitrogen constitutes 78.1% by volume (about 75.5% by weight) of the Earth's atmosphere.

(b) It is difficult to "fix" nitrogen, that is, to convert molecular nitrogen into other nitrogen compounds by means of chemical reactions.

(c) Chemical reactions that release N_2 as a product are highly exothermic and often explosive.

5.1.2 Nitrogen ions and catenation of nitrogen

There are three molecular ions which consist of nitrogen atoms only:

(1) Nitride ion N^{3-}

The N^{3-} ion exists in salt-like nitrides of the types M_3N and M_3N_2. In M_3N compounds, M is an element of Group 1 (Li) or Group 11 (Cu, Ag). In M_3N_2 compounds, M is an element of Group 2 (Be, Mg, Ca, Sr, Ba) or Group 12 (Zn, Cd, Hg). In these compounds, the bonding interactions between N and M atoms are essentially ionic, and the radius of N^{3-} is 146 pm. Nitride Li_3N exhibits high ionic conductivity with Li^+ as the current carrier. The crystal structure of Li_3N has been discussed in Section 2.2.

Structural Chemistry across the Periodic Table. Thomas Chung Wai Mak, Yu-San Cheung, Gong-Du Zhou, and Ying-Xia Wang. Oxford University Press. © Thomas Chung Wai Mak, Yu-San Cheung, Gong-Du Zhou, and Ying-Xia Wang (2023). DOI: 10.1093/oso/9780198872955.003.0005

(2) **Azide N_3^-**

The azide ion N_3^- is a symmetrical linear group that can be formed by neutralization of hydrogen azide HN_3 with alkalis. The bent configuration of HN_3 is shown below:

H
114°
98 pm
N————N————N
124 pm   113 pm

The Group 1 and 2 azides NaN_3, KN_3, $Sr(N_3)_2$, and $Ba(N_3)_2$ are well-characterized colorless crystalline salts which can be melted with little decomposition. The corresponding Group 11 and 12 metal azides such as AgN_3, $Cu(N_3)_2$ and $Pb(N_3)_2$ are shock-sensitive and detonate readily, and they are far less ionic with more complex structures. The salt $(PPh_4)^+(N_3HN_3)^-$ has been synthesized from the reaction of Me_3SiN_3 with $(PPh_4)(N_3)$ in ethanol; the $(N_3HN_3)^-$ anion has a non-planar bent structure consisting of distinct N_3^- and HN_3 units connected by a hydrogen bond, as shown in Fig. 5.1.1(a). In $Cu_2(N_3)_2(PPh_3)_4$ and $[Pd_2(N_3)_6]^{2-}$, N_3^- acts either as a terminal or a bridging ligand, as shown in Fig. 5.1.1(b) and (c), respectively. The azide ion N_3^- can also be combined with a metalloid ion such as As^{5+} to form $[As(N_3)_6]^-$, as shown in Fig. 5.1.1(d). Table 5.1.1 lists the coordination modes of N_3^- in its metal complexes; the highest ligation μ-1,1,1,3,3,3 mode (h) was found to exist in the double salt $AgN_3 \cdot 2AgNO_3$.

The uranium(IV) heptaazide anion $U(N_3)_7^{3-}$ has been synthesized as the n-tetrabutylammonium salt. This is the first homoleptic azide of an actinide as well as the first structurally characterized heptaazide. Two polymorphs of $(Bu_4N)_3[U(N_3)_7]$ were obtained from crystallization in $CH_3CN/CFCl_3$ (form A) or CH_3CH_2CN (form B). Form A belongs to space group $Pa3$ with $Z = 8$, and hence the U atom and one of the three independent azide groups are located on a crystallographic 3 axis. This results in a 1:3:3 monocapped octahedral arrangement of the azide ligands around the central uranium atom, as illustrated in Fig. 5.1.2(a). Form B crystallizes in space group $P2_1/c$ with $Z = 4$, and the azides ligands exhibit a

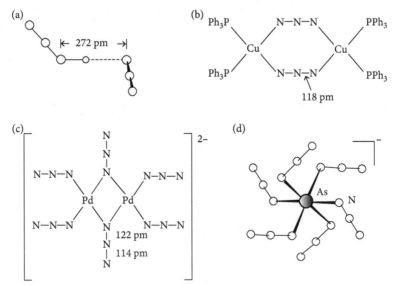

Fig. 5.1.1 Structures of some azide compounds: (a) $(N_3HN_3)^-$, (b) $Cu(N_3)_2(PPh_3)_4$, (c) $[Pd_2(N_3)_6]^{2-}$, (d) $[As(N_3)_6]^-$.

Table 5.1.1 Coordination modes of N_3^- in metal complexes

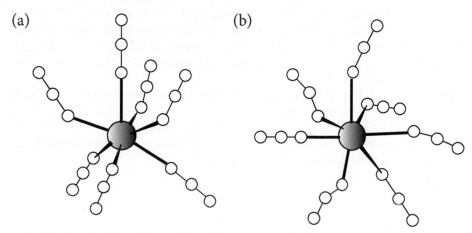

Fig. 5.1.2 Molecular structure of $U(N_3)_7^{3-}$ in (a) form A and (b) form B of the Bu_4N^+ salt. (Ref. M.-J. Crawford, A. Ellern, and P. Meyer, *Angew. Chem. Int. Ed.* **44**, 7874–8 (2005).)

distorted 1:5:1 pentagonal-bipyramidal coordination mode, as shown in Fig. 5.1.2(b). The $U-N_\alpha$ bond lengths range from 232 pm to 243 pm.

Two isomeric polymorphs of $[(C_5Me_5)_2U(\mu\text{-}N)U(\mu\text{-}N_3)(C_5Me_5)_2]_4$ have been synthesized and structurally characterized by X-ray crystallography. In the tetrameric macrocycle, eight $(\eta^5\text{-}C_5Me_5)_2U^{2+}$ units are charged balanced by four N_3^- (azide) ligands and four formally N^{3-} (nitride) ligands. Isomer A has a $(UNUN_3)_4$ ring in a pseudo-crown (or chair) conformation, whereas that of isomer B exhibits a pseudo-saddle (or boat) geometry, as shown in Fig. 5.1.3. The observed U–N(azide) bond distances in the range 246.7–252.5 pm are longer than typical U(IV)–N single bonds. The U–N(nitride) bond lengths in the range 204.7–209.0 pm are consistent with a bond order of two in a symmetrical bonding scheme for each UNU segment, rather than an alternating single/triple bond pattern, as shown below:

$$U \!=\!\!=\! N \!\rightleftharpoons\! U \quad \longleftrightarrow \quad U \!\rightleftharpoons\! N \!=\!\!=\! U \qquad\qquad U \!-\! N \!\equiv\!\!\equiv\! U$$

symmetrical nitride bridging   asymmetrical nitride bridging

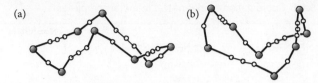

Fig. 5.1.3 Molecular skeleton of the $(UNUN_3)_4$ ring in two crystalline polymorphs of $[(C_5Me_5)_2U(\mu\text{-}N)U(\mu\text{-}N_3)(C_5Me_5)_2]_4$: pseudo-crown conformation observed for isomeric form A and pseudo-saddle observed for form B. The pair of η^5-C_5Me_5 groups bonded to each uranium(IV) atom is not shown. (Ref. W. J. Evans, S. A. Kozimor, and J. W. Ziller, *Science* **309**, 1835–8 (2005).)

(3) **Pentanitrogen cation N_5^+**

The pentanitrogen cation N_5^+ was first synthesized in 1999 as the white solid $N_5^+AsF_6^-$, which is not very soluble in anhydrous HF but stable at –78 °C. The N_5^+ cation has a closed-shell singlet ground state, and its C_{2v} symmetry was characterized by Raman spectroscopy and NMR of ^{14}N and ^{15}N nuclei. A series of 1:1 salts of N_5^+ have also been prepared with the monoanions HF_2^- (as the adduct $N_5^+HF_2^-\cdot n$HF), BF_4^-, PF_6^-, SO_3F^-, $(Sb_2F_{11})^-$, $B(N_3)_4^-$, and $P(N_3)_6^-$. A determination of the crystal structure of $N_5^+(Sb_2F_{11})^-$, which is stable at room temperature, yielded the structural parameters shown in Fig. 5.1.4. Note that each terminal N–N–N segment in N_5^+ deviates from exact linearity. Section Section 5.8.3 of *Advanced Structural Inorganic Chemistry* (2008).

The catenation of nitrogen refers to the tendency of N atoms to be connected to each other, and is far lower than those of C and P. This is because the repulsion of lone pairs on adjacent N atoms weakens the N–N single bond, and the lone pairs can easily react with electrophilic species. The structures and examples of known compounds that contain chains and rings of N atoms are listed in Table 5.1.2. Note that none of these has a linear configuration of nitrogen atoms.

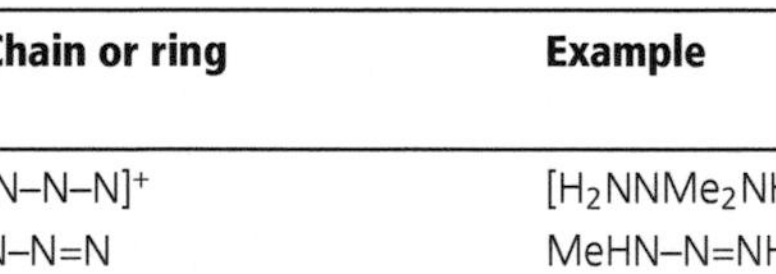

$$168° \qquad 130 \text{ pm} \qquad 111° \qquad 111 \text{ pm}$$

Fig. 5.1.4 Molecular dimensions of the pentanitrogen cation N_5^+ from crystal structure analysis.

Table 5.1.2 Species containing catenated N atoms

Number of N atoms	Chain or ring	Example
3	$[N–N–N]^+$	$[H_2NNMe_2NH_2]Cl$
3	N–N=N	MeHN–N=NH
4	N–N=N–N	$H_2N–N=N–NH_2$, $Me_2N–N=N–NMe_2$
4	N–N–N–N	$(CF_3)_2N–N(CF_3)–N(CF_3)–N(CF_3)_2$
5	N=N–N–N=N	PhN=N–N(Me)–N=NPh
6	N=N–N–N–N=N	PhN=N–N(Ph)–N(Ph)–N=NPh
8	N=N–N–N=N–N–N=N	PhN=N–N(Ph)–N=N–N(Ph)–N=NPh
5	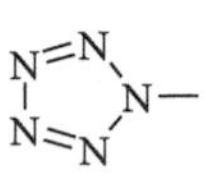 N=N / N=N ring with N–	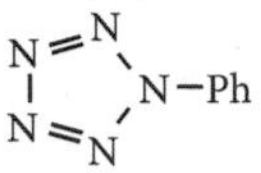 N=N / N=N ring with N–Ph

5.1.3 **Dinitrogen complexes**

Molecular nitrogen can react directly with some transition metal compounds to form dinitrogen complexes, the structure and properties of which are of considerable interest because they may serve as models for biological nitrogen fixation and as intermediates in synthetic applications. The known coordination modes of dinitrogen are discussed below and summarized in Table 5.1.3. The skeletal views of dinitrogen coordination modes in some metal complexes are shown in Fig. 5.1.5.

(1) η^1–N$_2$

The first complex containing molecular nitrogen as a ligand, [Ru(NH$_3$)$_5$(N$_2$)]Cl$_2$, was synthesized and identified in 1965 (Fig. 5.1.5(a)). The complex (Et$_2$PCH$_2$CH$_2$PEt$_2$)$_2$Fe(N$_2$) exhibits trigonal bipyramidal coordination geometry, with N$_2$ lying in the equatorial plane, as shown in Fig. 5.1.5(a'). Most examples of stable dinitrogen complexes have been found to belong to the η^1–N$_2$ category, in which the dinitrogen ligand binds in a linear, end-on mode with only a slightly elongated N–N bond length (112–124 pm) as compared to that in gaseous dinitrogen (109.7 pm). The stable monomeric titanocene complexes {(PhMe$_2$Si)C$_5$H$_4$}$_2$TiX (X = N$_2$, CO) are isomorphous, with a crystallographic C_2 axis passing through the Ti atom and the η^1–X ligand. The measured bond distances (in pm) are Ti–N = 201.6(1), N–N = 111.9(2) for the dinitrogen complex and Ti–C = 197.9(2), C–O = 115.1(2) for the carbonyl complex.

(2) μ-(bis-η^1)-N$_2$

As a bridging ligand in dinuclear systems, dinitrogen may formally be classified into three types.

(a) M–N≡N–M: This dinitrogen ligand corresponds to a neutral N$_2$ molecule, which uses its lone pairs to coordinate to two M atoms. Complexes of this type show a relatively short N–N distance of 112–120 pm. In [(η^5-C$_5$Me$_5$)$_2$Ti]$_2$(N$_2$), the binuclear molecular skeleton consists of two (η^5-C$_5$Me$_5$)$_2$Ti moieties bridged by the N$_2$ ligand in an essentially linear Ti–N≡N–Ti arrangement, as shown in Fig. 5.1.5(b). The N–N distance is 116 pm (av.).

(b) M=N=N=M: This dinitrogen ligand corresponds to diazenido(–2). The (N$_2$)$^{2-}$ anion coordinates to two M atoms. In (Mes)$_3$Mo(N$_2$)Mo(Mes)$_3$ (where Mes = 2,4,6-Me$_3$C$_6$H$_2$), Mo=N=N=Mo forms a linear chain. The length of N–N distance is 124.3 pm.

(c) M≡N–N≡M: This dinitrogen ligand corresponds to hydrazido(–4). The (N$_2$)$^{4-}$ anion coordinates to two M atoms. In [PhP(CH$_2$SiMe$_2$NPh$_2$)$_2$NbCl]$_2$(N$_2$), the Nb≡N–N≡Nb moiety is linear, and the distance of N–N is 123.7 pm.

(3) μ_3-η^1:η^1:η^2-N$_2$

In the [(C$_{10}$H$_8$)(C$_5$H$_5$)$_2$Ti$_2$](μ_3–N$_2$)[(C$_5$H$_4$)(C$_5$H$_5$)$_3$Ti$_2$] complex, the dinitrogen ligand is coordinated simultaneously to three Ti atoms, as shown in Fig. 5.1.5(c). The N–N distance is 130.1 pm.

(4) μ_3-η^1:η^1: η^1-N$_2$

In the mixed-metal complex [WCl(py)(PMePh)$_3$(μ_3-N$_2$)]$_2$(AlCl$_2$)$_2$, both WNN linkages are essentially linear, and the four metal atoms and two μ_3–N$_2$ ligands almost lie in the same plane, as shown in Fig. 5.1.5(d).

(5) μ-(η^1: η^2)-N$_2$

In [PhP(CH$_2$SiMe$_2$NPh)$_2$]$_2$Ta$_2$(μ-H)$_2$(N$_2$), the dinitrogen moiety is end-on bound to one Ta atom and side-on bound to the other, as shown in Fig. 5.1.5(e). The N–N distance of 131.9 pm is consistent with a

Table 5.1.3 Coordination modes of dinitrogen

Coordination mode	Example	d_{N-N}/pm	Structure in Fig. 5.1.5
(1) η^1-N_2 M—N≡N	$[Ru(NH_3)_5(N_2)]^{2+}(depe)_2Fe(N_2)$	112 113.9	(a) (a')
(2) μ-(bis-η^1)-N_2 (a) M—N≡N—M (b) M≡N—N≡M (c) M≡N—N≡M	$[(C_5Me_5)_2Ti]_2(N_2)$ $(Mes)_3Mo(N_2)Mo(Mes)_3$ $[PhP(CH_2SiMe_2NPh_2)_2NbCl]_2(N_2)$	116(av.) 124.3 123.7	(b)
(3) μ_3- η^1:η^1:η^2-N_2 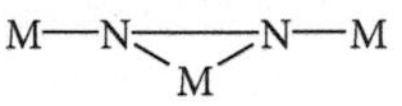 M—N⟍⟋N—M M	$[(C_{10}H_8)(C_5H_5)_2Ti_2]^-[(C_5H_4)(C_5H_5)_3Ti_2](N_2)$	130.1	(c)

(4) μ_3- η^1:η^1:η^2-N_2 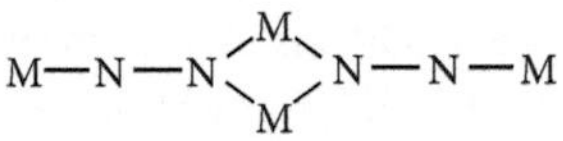
      M
 M—N—N⟍⟋N—N—M
      M
	$[WCl(py)(PMePh)_3(N_2)]^{2-}(AlCl_2)_2$	125	(d)

(5) μ-(η^1:η^2)-N_2 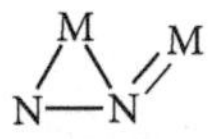 M M N—N	$[PhP(CH_2SiMe_2NPh)_2]_2Ta_2(\mu\text{-}H)_2(N_2)$	131.9	(e)

(6) μ-(bis-η^2)-N_2 
    N
 M⟍|⟋M
    N
 (planar)
	$[Cp''_2Zr]_2(N_2)$	147	(f)

(7) μ-(bis-η^2)-N_2  N 	 M—N—M (non-planar)	$\{[(SiMe_3)_2N]_2Ti(N_2)\}_2^-$	135	(g)

(8) μ_4- η^1:η^1:η^2:η^2-N_2 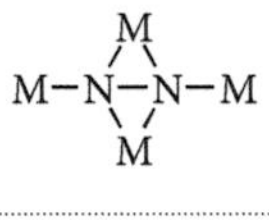
    M
 M—N—N—M
    M
	$[Ph_2C(C_4H_3N)_2Sm]_4(N_2)$	141.2	(h)

(9) μ_5- η^1:η^1:η^2:η^2:η^2-N_2 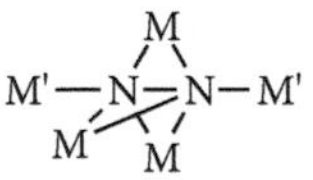
    M
 M'—N⟍⟋N—M'
   M M
	$\{[(-CH_2-)_5]_4\text{calix-tetrapyrrole}\}_2Sm_3Li_2(N_2)^-$	150.2	(i)

(10) μ_6- η^1:η^1:η^2:η^2:η^2:η^2-N_2 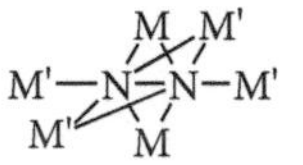
   M M'
 M'—N≡N—M'
  M' M
	$[(THF)_2Li(OEPG)Sm]_2Li_4(N_2)$	152.5	(j)

formal assignment of the bridging dinitrogen moiety as $(N_2)^{4-}$. The shortest distance of the end-on Ta–N bond is 188.7 pm, which is consistent with its considerable double-bond character.

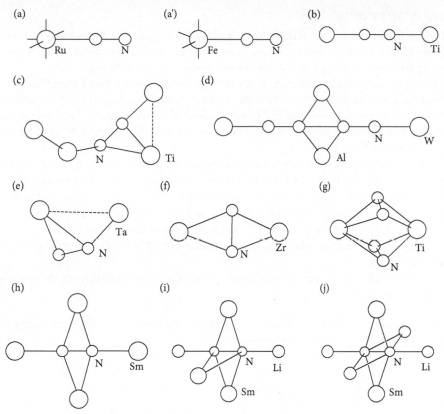

Fig. 5.1.5 Dinitrogen coordination modes in metal complexes.

(6) μ-(bis-η^2)-N$_2$ (planar)

Fig. 5.1.5(f) shows a planar, side-on bonded N$_2$ ligand between two Zr atoms in the compound [Cp"$_2$Zr]$_2$(N$_2$) [Cp" = 1,3-(SiMe$_2$)$_2$C$_5$H$_3$]. The N–N bond length is 147 pm. Side-on coordination of the dinitrogen ligand appears to be important for its reduction.

(7) μ-(bis-η^2)-N$_2$ (non-planar)

In the compound Li{[(SiMe$_3$)$_2$N]$_2$Ti(N$_2$)}$_2$, each Ti atom is side-on bound to two N$_2$ molecules, as shown in Fig. 5.1.5(g). The N–N distance is 137.9 pm.

(8) μ_4-η^1:η^1:η^2:η^2-N$_2$

In [Ph$_2$C(C$_4$H$_3$N)$_2$Sm]$_4$(N$_2$), the N$_2$ moiety is end-on bound to two Sm atoms and side-on bound to the other two, as shown in Fig. 5.1.5(h). The bond lengths are N–N 141.2 pm, Sm(terminal)–N 217.7 pm, and Sm(bridging)–N 232.7 pm.

(9) μ_5-η^1:η^1:η^2:η^2:η^2-N$_2$

In {[(–CH$_2$–)$_5$]$_4$calix-tetrapyrrole}$_2$Sm$_3$Li$_2$(N$_2$)[Li(THF)$_2$]·(THF), the N$_2$ moiety is end-on bound to two Li atoms and side-on bound to three Sm atoms, as shown in Fig. 5.1.5(i). The bond lengths are N–N 150.2 pm, Sm–N 233.3 pm(av.), and Li–N 191.0 pm(av.).

(10) $\mu_6\text{-}\eta^1\text{:}\eta^1\text{:}\eta^2\text{:}\eta^2\text{:}\eta^2\text{:}\eta^2\text{-}N_2$

In $[(THF)_2Li(OEPG)Sm]_2Li_4(N_2)$ (OEPG = octaethylporphyrinogen), the N_2 moiety is end-on bound to two Li atoms and side-on bound to two Sm atoms and two Li atoms, as shown in Fig. 5.1.5(j). The bond lengths are N–N 152.5 pm, Sm–N 235.0 pm (av.), and Li–N 195.5 pm(av.).

The bonding of dinitrogen to transition metals may be divided into the "end-on" and "side-on" categories. In the end-on arrangement, the coordination of the N_2 ligand is accomplished by a σ bond between the $2\sigma_g$ orbital of the nitrogen molecule and a hybrid orbital of the metal, and by π back-bonding from a doubly degenerate metal $d\pi$ orbitals (d_{xz}, d_{yz}) to the vacant $1\pi^*_g$ orbitals (π^*_{xz}, π^*_{yz}) of the nitrogen molecule, as shown in Fig. 5.1.6.

In the side-on arrangement, the bonding is considered to arise from two interdependent components. In the first part, σ overlap between the filled π orbital of N_2 and a suitably directed vacant hybrid metal orbital forms a donor bond. In the second part, the M atom and N_2 molecule are involved in two back-bonding interactions, one having π-symmetry as shown in Fig. 5.1.7(a), and the other with δ-symmetry as shown in Fig. 5.1.7(b). These π- and δ-back bonds synergically reinforce the σ bond.

Since the overlap of a δ-bond should be less effective than that of a π-bond, the end-on mode is generally preferred over the side-on form.

The transition metal dinitrogen complexes have been investigated theoretically, and the results lead to the following generalizations:

(a) Both σ donation and π or δ back donation are related to the formation of the metal-nitrogen bond, the former interaction being more important.

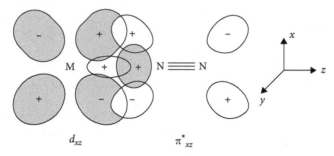

Fig. 5.1.6 Orbital interactions of N_2 and transition metal M in end-on coordinated dinitrogen complexes. The $d\pi$-$1\pi^*_g$ overlap shown here also occurs in the yz plane.

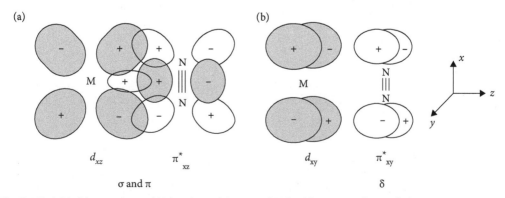

Fig. 5.1.7 Orbital interactions of N_2 and transition metal M in side-on coordinate dinitrogen complexes.

(b) The N–N bond of the side-on complex is appreciably weakened by electron donation from the bonding π and σ orbitals of the N_2 ligand to the vacant orbitals of the metal.

(c) The end-on coordination mode takes precedence over the side-on one. The weak N–N bond in the side-on complex indicates that the N_2 ligand in this type of compound is fairly reactive. The reduction of the coordinated nitrogen molecule may proceed through this activated form.

5.2 **Compounds of Nitrogen**

5.2.1 **Molecular nitrogen oxides**

Nitrogen displays nine oxidation states, ranging from –3 to +5. Since it is less electronegative than oxygen, nitrogen forms oxides and oxidized compounds with an oxidation number between +1 and +5. Eight oxides of nitrogen, N_2O, NO, N_2O_2, N_2O_3, NO_2, N_2O_4, N_2O_5, and N_4O, are known, and the ninth, NO_3, exists as an unstable intermediate in various reactions involving nitrogen oxides. Their structures and some properties are presented in Table 5.2.1 and Fig. 5.2.1.

Table 5.2.1 Structure and properties of the nitrogen oxides

Formula	Name	Oxidation number	Structure (in Fig. 5.2.1)	ΔH_f^0/kJ mol^{-1}	Properties
N_2O	dinitrogen monoxide (dinitrogen oxide, nitrous oxide, laughing gas)	+1 or (0,+2)	(a) linear $C_{\infty v}$	82.0	mp 182.4 K, bp 184.7 K colorless gas, fairly unreactive with pleasuring odor and sweet taste
NO	nitrogen oxide (nitric oxide, nitrogen monoxide)	+2	(b) linear $C_{\infty v}$	90.2	mp 109 K, bp 121.4 K colorless, paramagnetic gas
$(NO)_2$	dimer of nitrogen oxide	+2	(c), (d)	–	–
N_2O_3	dinitrogen trioxide	+3 or (+2,+4)	(d) planar C_s	80.2	mp 172.6 K, dec. 276.7 K, dark blue liquid, pale blue solid, reversibly dissociates to NO and NO_2
NO_2	nitrogen dioxide	+4	(f) C_{2v}	33.2	orange brown, paramagnetic gas, reactive
N_2O_4	dinitrogen tetroxide	+4	(g) planar D_{2h}	9.16	mp 262.0 K, bp 294.3 K, colorless liquid, reversibly dissociates to NO_2
N_2O_5	dinitrogen pentoxide	+5	(h) planar C_{2v}	11.3 (gas) −43.1 (cryst.)	sublimes at 305.4 K, colorless, volatile solid consisting of NO_2^+ and NO_3^-; exists as N_2O_5 in gaseous state
N_4O	nitrosyl azide	*	(i) planar C_s	−297.3	pale yellow solid (188 K)

The assignment of an oxidation number is not appropriate as only one of the four nitrogen atoms is bonded to the oxygen atom in the N_4O molecule.

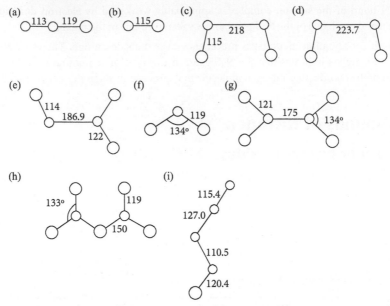

Fig. 5.2.1 Structure of nitrogen oxides (bond length in pm): (a) N_2O, (b) NO, (c) N_2O_2 (in crystal), (d) N_2O_2 (in gas), (e) N_2O_3, (f) NO_2, (g) N_2O_4, (h) N_2O_5, (i) N_4O.

All nitrogen oxides have planar structures. Nitrogen displays all its positive oxidation states in these compounds, and in N_2O, N_2O_3, and N_4O the N atoms in molecule have different oxidation states. In the gaseous state, six stable nitrogen oxides exist, each with a positive heat of formation primarily because the N≡N bond is so strong. The structure and properties of nitrogen oxides are presented as follows:

(1) N_2O

Dinitrogen oxide (nitrous oxide), N_2O, is a linear unsymmetrical molecule with a structure similar to its isoelectronic analog CO_2:

$$\bar{N} \leftarrow \overset{+}{N} = O \longleftrightarrow N \equiv \overset{+}{N} - \bar{O}$$

Dinitrogen oxide is unstable and undergoes dissociation when heated to about 870 K:

$$N_2O \longrightarrow N_2 + \tfrac{1}{2}O_2$$

The activation energy for this process is high (≈ 520 kJ mol^{-1}), so N_2O is relatively unreactive at room temperature. Dinitrogen oxide has a pleasant odor and sweet taste, and its past use as an anaesthetic with undesirable side effect accounts for its common name as "laughing gas."

(2) NO and $(NO)_2$

The simplest thermally stable odd-electron molecule known is nitrogen monoxide, NO, which is discussed in the next section. High purity nitrogen monoxide partially dimerizes when it liquefies to give a colorless liquid. The heat of dissociation of the dimer is 15.5 kJ mol^{-1}. The structure of the $(NO)_2$ dimer in the crystalline and vapor states are shown in Figs. 5.2.1(c) and 5.2.1(d), respectively. The $(NO)_2$ dimer adopts the *cis* arrangement with C_{2v} symmetry. In the crystalline state, the N–N distance is 218 pm (223.7 pm in the gas phase), and the O···O distance is 262 pm. The very long N–N distance has not been accounted for satisfactorily in bonding models.

(3) N_2O_3

Dinitrogen trioxide is formed by the reaction of stoichiometric quantities of NO and O_2:

$$2NO + \tfrac{1}{2}O_2 \longrightarrow N_2O_3$$

At temperatures below 172.6 K, N_2O_3 crystallizes as a pale blue solid. On melting it forms an intensely blue liquid which, as the temperature is raised, is increasingly dissociated into NO and an equilibrium mixture of NO_2 and N_2O_4. This dissociation occurs significantly above 243 K and the liquid assumes a greenish hue resulting from the brown color of NO_2 mixed with the blue. The N–N distance of the N_2O_3 molecule, 186.9 pm, is considerably longer than the typical N–N single bond (145 pm) in hydrazine, $H_2N\ NH_2$.

(4) NO_2 and N_2O_4

Nitrogen dioxide and dinitrogen tetroxide are in rapid equilibrium that is highly dependent on temperature. Below the melting point (262.0 K) the oxide consists entirely of colorless, diamagnetic N_2O_4 molecules. As the temperature is raised to the boiling point (294.3 K), the liquid changes to an intense red-brown colored, highly paramagnetic phase containing 0.1% NO_2. At 373 K, the proportion of NO_2 increases to 90%. The configuration, bond distances and bond angles of NO_2 and N_2O_4 are given in Figs. 5.2.1(f) and 5.2.1(g). In view of the large bond angle of NO_2 and its tendency for dimerization, bonding in the molecule can be described in terms of resonance between the following canonical structures:

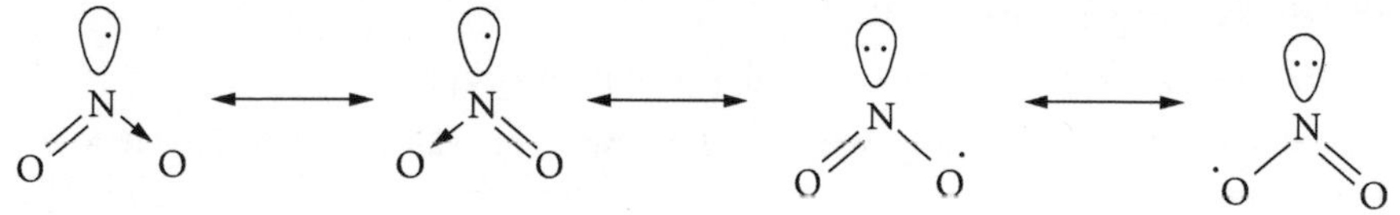

The N–N distance in planar N_2O_4 is 175 pm, with a rotation barrier of about 9.6 kJ mol^{-1}. Although the N–N bond is of the σ type, it is lengthened because the bonding electron pair is delocalized over the entire N_2O_4 molecule with a large repulsion between the doubly occupied MOs on the two N atoms.

(5) N_2O_5 and NO_3

Nitrogen pentoxide, N_2O_5, is a colorless, light- and heat-sensitive crystalline compound that consists of linear NO_2^+ cations (N–O 115.4 pm) and planar NO_3^- anions (N–O 124 pm). In the gas phase, N_2O_5 is molecular but its configuration and dimensions have not been reliably measured. Also, N_2O_5 is the anhydride of nitric acid and can be obtained by carefully dehydrating the concentrated acid with P_4O_{10} at low temperatures:

$$4HNO_3 + P_4O_{10} \overset{-10°C}{\longrightarrow} 2N_2O_5 + 4HPO_3$$

The existence of the fugitive, paramagnetic trioxide NO_3 is also implicated in the $N_2O_5^-$ catalyzed decomposition of ozone, and its concentration is sufficiently high for its absorption spectrum to be recorded. It has not been isolated as a pure compound, but probably has a symmetrical planar structure like that of NO_3^-.

(6) **N₄O**

Nitrosyl azide, N_4O, is a pale yellow solid formed by the reaction of activated, anhydrous NaN_3 and NOCl, followed by low-temperature vacuum sublimation. The Raman spectrum and *ab initio* calculation characterized the structure of N_4O as that shown in Fig. 5.2.1(i). The bonding in the N_4O molecule can be represented by the following resonance structures:

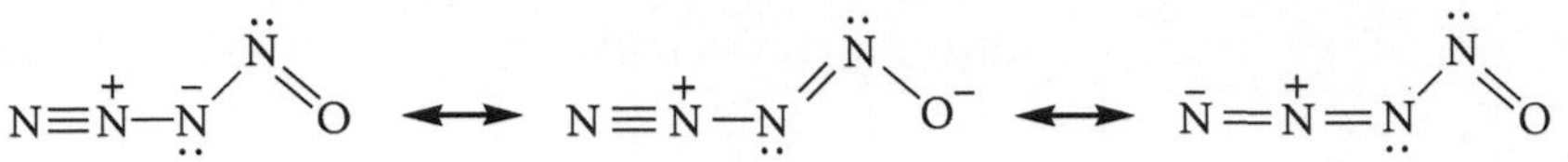

5.2.2 Molecule of the year for 1992: nitric oxide

Nitric oxide, NO, was named molecule of the year for 1992 by the journal *Science*. It is one of the most extensively investigated molecules in inorganic and bioinorganic chemistry.

Nitric oxide is biosynthesized in animal species, and its polarity and small molecular dimensions allow it to diffuse readily through cell walls, acting as a messenger molecule in biological systems. It plays an important role in the normal maintenance of many important physiological functions, including neurotransmission, blood clotting, regulation of blood pressure, muscle relaxation, and annihilation of cancer cells.

The valence shell electron configuration of NO in its ground state is: $(1\sigma)^2(1\sigma^*)^2(1\pi)^4(2\sigma)^2(1\pi^*)^1$, which accounts for the following properties:

(a) The NO molecule has a net bond order of 2.5, bond energy 627.5 kJ mol^{-1}, bond length 115 pm, and an infrared stretching frequency of 1840 cm^{-1}.

(b) The molecule is paramagnetic in view of its unpaired π^* electron.

(c) NO has a much lower ionization energy (891 kJ mol^{-1} or 9.23 eV) than N_2 (15.6 eV) or O_2 (12.1 eV).

(d) NO has a dipole moment of 0.554×10^{-30} C m, or 0.166 Debye.

(e) Employing its lone-pair electrons, NO serves as a terminal or bridging ligand in forming numerous coordination compounds.

(f) NO is thermodynamically unstable ($\Delta G^\circ = 86.57$ kJ mol^{-1}, $\Delta S^\circ = 217.32$ kJ mol^{-1} K^{-1}) and decomposes to N_2 and O_2 at high temperature. It is capable of undergoing a variety of redox reactions. The electrochemical oxidation of NO around 1.0 V has been used to devise NO-selective amperometric micro-probe electrodes to detect its release in biological tissues.

As a result of its unique structure and properties, the NO molecule exhibits a wide variety of reactions with various chemical species. In particular, it readily releases an electron in the antibonding π^* orbital to form the stable nitrosyl cation NO$^+$, increasing the bond order from 2.5 to 3.0, so that the bond distance decreases by 9 pm. The NO species in an aqueous solution of nitrous acid, HONO, is NO$^+$:

$$HONO + H^+ \longrightarrow NO^+ + H_2O$$

There are many nitrosyl salts, including $(NO)HSO_4$, $(NO)ClO_4$, $(NO)BF_4$, $(NO)FeCl_4$, $(NO)AsF_6$, $(NO)PtF_6$, $(NO)PtCl_6$, and $(NO)N_3$.

Nitrosyl halides are formed when NO reacts with F_2, Cl_2, and Br_2, and they all have a bent structure:

Table 5.2.2 Two types of MNO coordination geometry

	M–N–O	M–N⟍O
Bond angle	165°–180°	120°–140°
M-N distance	≈ 160 pm	> 180 pm
NO frequency	1650–1985 cm^{-1}	1525–1590 cm^{-1}
Chemical properties	electrophilic;	nucleophilic;
	NO$^+$ similar to CO:	NO$^-$ similar to O$_2$:
	$\overset{-}{M} = \overset{+}{N} = O$	M—N⟍O
Ligand behavior	three-electron donor	one-electron donor

$$X{-}N{=}O$$

X = F, X–N = 152 pm, N = O = 114 pm, X–N=O = 110°
X = Cl, X–N = 197 pm, N = O = 114 pm, X–N=O = 113°
X = Br, X–N = 213 pm, N = O = 114 pm, X–N=O = 117°

Due to its close resemblance to O_2 and its paramagnetism, NO has been used extensively as an O_2 surrogate to probe the metal environment of various metalloproteins. In particular, NO has been shown to bind to Fe^{2+} in heme-containing oxygen transporting proteins, in a fashion nearly identical to O_2, with the important consequence of rendering the complex paramagnetic and thus detectable by EPR spectroscopy. Since the unpaired electron assumes appreciable iron d-orbital character, analysis of the EPR characteristics of hemo-protein nitrosyl complexes yields valuable information on both the ligand-binding environment around the heme and on the conformational state of the protein.

Nitric oxide can bind to metals in both terminal and bridging modes to give metal nitrosyl complexes. Depending upon the stereochemistry of the complexes, NO may exhibit within one given complex either a NO$^+$ or a NO$^-$ character, as illustrated in Table 5.2.2.

Nitric oxide contains one more electron than CO and generally behaves as a three-electron donor in metal nitrosyl complexes. Formally this may be regarded as the transfer of one electron to the metal atom, thereby reducing its oxidation state by one, followed by coordination of the resulting NO$^+$ to the metal atom as a two-electron donor. This is in accordance with the general rule that three terminal CO groups in a metal carbonyl compound may be replaced by two NO groups. In this type of bonding the M–N–O bond angle would formally be 180°. However, in many instances this bond angle is somewhat less than 180°, and slight bent M–N–O groups with angles in the range 165° to 180° are frequently found.

In a second type of bonding in nitrosyl coordination compounds, the M–N–O bond angle lies in the range of 120° to 140° and the NO molecule acts as a one-electron donor. Here the situation is analogous to XNO compounds and the M–N bond order is one.

It should be emphasized that the NO$^+$ and NO$^-$ "character" of NO and the linear or bent angle in a nitrosyl complex do not necessarily imply that NO would be released from the complex in the free form NO or as NO$^+$ or NO$^-$.

5.2.3 Oxoacids and oxo-ions of nitrogen

The oxoacids of nitrogen known either as the free acids or in the form of their salts are listed in Tables 5.2.3 and 5.2.4. The structures of these species are shown in Figs 5.2.2 and 5.2.3.

Table 5.2.3 Oxoacids of nitrogen

Formula	Name	Properties	Structure
HON=NOH	hyponitrous acid	weak acid; salts are known	*trans* form, Fig. 5.2.3 (a)
H_2NNO_2	nitramide	isomeric with hyponitrous acid	Fig. 5.2.2 (a)
HNO	nitroxyl	reactive intermediate; salt known	Fig. 5.2.2 (b)
$H_2N_2O_3$	hyponitric acid	known only in solution and as salts	
HNO_2	nitrous acid	unstable, weak acid	Fig. 5.2.2 (c)
HNO_3	nitric acid	stable, strong acid	Fig. 5.2.2 (d)
H_3NO_4	orthonitric acid	acid unknown; Na_3NO_4 and K_3NO_4 have been prepared	Tetrahedral $NO_4{}^{3-}$

Table 5.2.4 The nitrogen oxo-ions

Oxiation number	Formula	Name	Structure (Fig. 5.2.3)	Properties
+1	$N_2O_2{}^{2-}$	hyponitrite	*trans*, C_{2h}, (a) *cis*, C_{2v} (not shown)	reducing agent
+2	$N_2O_3{}^{2-}$	hyponitrate	*trans*, C_s, (b) *cis*, C_s, (c)	reducing agent
+3	$NO_2{}^-$	nitrite	bent, C_{2v}, (d), bond angle 115°	oxidizing or reducing agent
+5	$NO_3{}^-$	nitrate	planar, D_{3h}, (e)	oxidizing agent
+5	$NO_4{}^{3-}$	orthonitrate	tetrahedral, T_d, (f)	oxidizing agent
+3	NO^+	nitrosonium	$C_{\infty v}$, (g)	oxidizing agent
+5	$NO_2{}^+$	nitronium	$D_{\infty v}$, (h)	oxidizing agent

(1) Hyponitrous acid

Spectroscopic data indicate that hyponitrous acid, HON=NOH, has the planar *trans* configuration. Its structural isomer nitramide, $H_2N–NO_2$, is a weak acid. The structure of nitramide is shown in Fig. 5.2.2(a); in this molecule, the angle between the NNO_2 plane and the H_2N plane is 52°. Both *trans* and *cis* forms of the hyponitrite ion, $(ONNO)^{2-}$, are known; the *trans* isomer, as illustrated in Fig. 5.2.3(a), is the stable form with considerable π-bonding over the molecular skeleton.

(2) Nitroxyl, HNO

This is a transient species whose structure is shown in Fig. 5.3.2(b). The existence of the NO^- anion has been established by single crystal X-ray analysis of $(Et_4N)_5[(NO)(V_{12}O_{32})]$, in which the NO^- group lies inside the cage of $(V_{12}O_{32})^{4-}$. The bond length of NO^- is 119.8 pm, which is longer than that of the neutral molecule NO, 115.0 pm.

(3) Hyponitric acid, $H_2N_2O_3$

Hyponitric acid has not been isolated in pure form, but its salts are known. In the anion $N_2O_3{}^{2-}$, both

$$\left(O = \overset{..}{N} - \overset{..}{N} \diagup_{O^-}^{O^-} \right)$$

N atoms each bears a lone pair. The structure of the anion adopts a non-planar configuration, and has *cis*- and *trans*-forms, as shown in Figs 5.2.3(b) and 5.2.3(c).

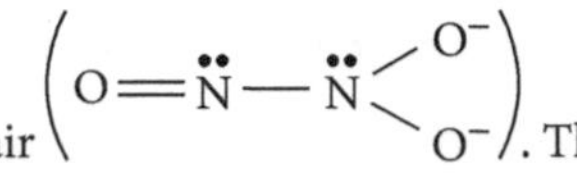

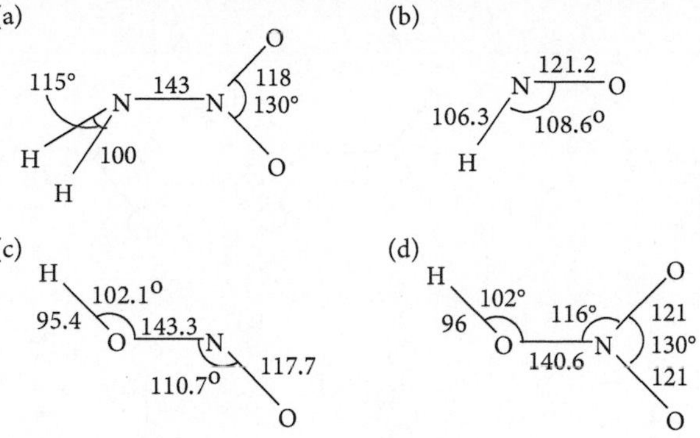

Fig. 5.2.2 Structures of two amides and two oxoacids of nitrogen: (a) $H_2N_2O_2$, (b) HNO, (c) HNO_2, (d) HNO_3.

(4) **Nitrous acid, HNO_2**

Although nitrous acid has never been isolated as a pure compound, its aqueous solution is a widely used reagent. Nitrous acid is a moderately weak acid with $pK_a = 3.35$ at 291 K. In the gaseous state, it adopts the *trans*-planar structure, as shown in Fig. 5.2.2(c).

The nitrite ion, NO_2^- is bent with C_{2v} symmetry, and its structure can be represented by two simple resonance formulas:

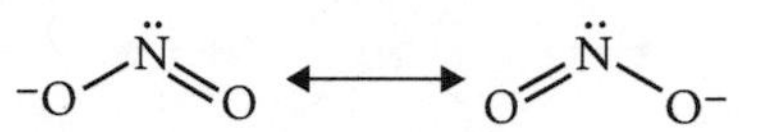

Many stable metal nitrites (Li^+, Na^+, K^+, Rb^+, Cs^+, Ag^+, Tl^+, Ba^{2+}, NH_4^+) contain the bent $(O-N-O)^-$ anion with N–O bond length in the range of 113–123 pm, and the angle 116°–132°, as shown in Fig. 5.2.3(d).

(5) **Nitric acid, HNO_3**

Nitric acid is one of the three major inorganic acids in the chemical industry. Its structural parameters in the gaseous state are shown in Fig. 5.2.2(d). The crystals of nitric acid monohydrate consist of H_3O^+ and NO_3^-, which are connected by strong hydrogen bonds. In the acid salts, HNO_3 molecules are bound to nitrate ions by strong hydrogen bonds. For example, the structures of $[H(NO_3)_2]^-$ in $K[H(NO_3)_2]$ and $[H_2(NO_3)_3]^-$ in $(NH_4)H_2(NO_3)_3$ are as follows:

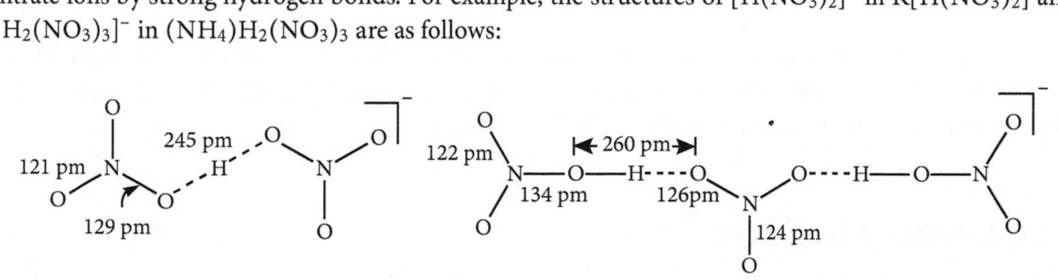

The nitrate ion NO_3^- has D_{3h} symmetry, as shown in Fig. 5.2.3(e), and its bonding can be represented by resonance structures:

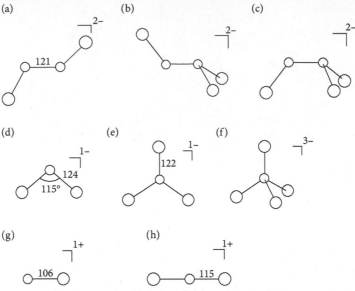

Fig. 5.2.3 Structure of nitrogen oxo-ions (bond lengths in pm): (a) $(ON=NO)^{2-}$, (b) cis-$N_2O_3^{2-}$, (c) $trans$-$N_2O_3^{2-}$, (d) NO_2^-, (e) NO_3^-, (f) NO_4^{3-}, (g) NO^+, (h) NO_2^+.

(6) **Orthonitric acid, H$_3$NO$_4$**

Orthonitric acid is still unknown, but its salts Na_3NO_4 and K_3NO_4 have been characterized by X-ray crystallography. The NO_4^{3-} ion has regular T_d symmetry, and its bonding can be represented by resonance structures:

The nitrosonium NO^+ and nitronium NO_2^+ ions, which are mentioned in previous sections, have $C_{\infty v}$ and $D_{\infty h}$ symmetry, respectively. Their structures are shown in Figs 5.2.3(g) and 5.2.3(h), and can be represented by the familiar Lewis structures: $N\equiv O^+$ and $O=N^+=O$.

5.2.4 **Nitrogen hydrides**

(1) **Ammonia, NH$_3$**

The most important nitrogen hydride is ammonia, which is a colorless, alkaline gas with a unique odor. Its melting point is 195 K. Its boiling point 240 K is far higher than that of PH_3 (185.4 K). This is due to the strong hydrogen bonds between molecules in liquid ammonia. Liquid ammonia is an excellent solvent and a valuable medium for chemical reactions, as its high heat of vaporization ($23.35 \text{ kJ mol}^{-1}$) makes it relatively easy to handle. As its dielectric constant ($\varepsilon = 22$ at 239 K) and self-ionization are both lower than those of water, liquid ammonia is a poorer ionizing solvent but a better one for organic compounds. The self-ionization of ammonia is represented as:

$$2NH_3 \rightleftharpoons NH_4^+ + NH_2^-, \ K = 10^{-33} \ (223 \text{ K}),$$

and ammonium compounds behave as Lewis acids while amides are bases.

Ammonia is an important industrial chemical used principally (over 80%) as fertilizers in various forms, and is employed in the production of many other compounds such as urea, nitric acid, and explosives.

(2) Hydrazine, H_2NNH_2

Hydrazine is an oily, colorless liquid in which nitrogen has an oxidation number of –2. The length of the N–N bond is 145 pm, and there is a lone pair on each N atom. Its most stable conformer is the *gauche* form, rather than the *trans* or *cis* form. The rotational barrier through the *trans* or staggered position is 15.5 kJ mol^{-1}, and through the *cis* or eclipsed position is 49.7 kJ mol^{-1}. These numbers reflect the modest repulsion of the non-bonding electrons for a neighboring bond pair and the significantly greater repulsion of the non-bonding pairs for each other in the *cis* arrangement. The melting point of hydrazine is 275 K, and the boiling point is 387 K. The very high exothermicity of its combustion makes it a valuable rocket fuel.

(3) Diazene, HN=NH

Diazene (or diimide) is a yellow crystalline compound that is unstable above 93 K. In the molecule, each N atom uses two sp^2 hybrids for σ bonding with the neighboring N and H atoms, and the lone pair occupies the remaining sp^2 orbital. The molecule adopts the *trans* configuration:

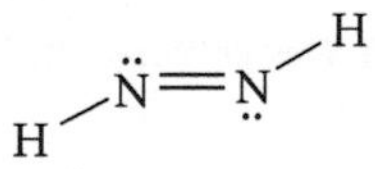

(4) Hydroxylamine, NH_2OH

Anhydrous NH_2OH is a colorless, thermally unstable hygroscopic compound which is usually handled as an aqueous solution or in the form of its salts. Pure hydroxylamine melts at 305 K and has a very high dielectric constant (77.6–77.9). Aqueous solutions are less basic than either ammonia or hydrazine:

$$NH_2OH \, (aq) + H_2O \rightleftharpoons NH_3OH^+ \, (aq) + OH^- \, (aq), K = 6.6 \times 10^{-9} \, (298K)$$

Hydroxylamine can exist as two configurational isomers (*cis* and *trans*) and in numerous intermediate *gauche* conformations. In the crystalline form, hydrogen bonding tends to favor packing in the *trans* conformation. The N–O bond length is 147 pm, consistent with its formulation as a single bond. Above room temperature the compound decomposes by internal oxidation-reduction reactions into a mixture of N_2, NH_3, N_2O, and H_2O. Aqueous solutions are much more stable, particularly acid solutions in which the protonated species $[NH_3(OH)]^+$ is generally used as a reducing agent.

5.3 Polymeric Nitrogen with the Cubic Gauche (cg-N) Structure

At temperatures above 2000 K and pressures over 110 GPa, molecular dinitrogen N_2 transforms to a transparent allotropic form that adopts the theoretically predicted cubic gauche (cg-N) structure. It crystalizes in the space group $I2_13$ (No. 199) with $a = 345.42(9)$ pm. The N atoms locate at Wyckoff position $8(a)$ with $x = 0.067(1)$, giving rise to Z = 4 in the unit cell. In the crystal structure, each nitrogen atom is covalently bound to three nearest neighbors, and the measured bond length is 134.6(4) pm (Fig. 5.3.1)

(a)   (b)

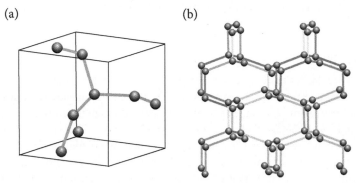

Fig. 5.3.1 Cubic gauge (cg-N) crystal structure of dinitrogen: (a) cubic unit cell, in which each nitrogen atom is connected to three nearest neighbors by covalent N–N single bonds; (b) extended three-dimensional polymeric network.

Reference

M. I. Eremets, A. G. Gavriliuk, I. A. Trojan, D. A. Dzivenko, and R. Boehler, Single-bonded cubic form of nitrogen, *Nat. Mater.* **3**, 558–63 (2004).

5.4 Reductive Coupling of Dinitrogen

Recent work has achieved organoboron-mediated catenation of two N_2 molecules under near-ambient conditions to generate a complex in which a $[N_4]^{2-}$ chain bridges two boron centers.

Reduction of the stable [(CAAC)BClTip] radical (**1**), where CAAC = 1-(2,6-di*iso*propylphenyl)-3,3,5,5-tetramethylpyrrolidin-2-ylidene and Tip = 2,4,6-tri*iso*propylphenyl, using KC_8 (10 eq.) in toluene under 4 atm of N_2 yielded a turquoise-colored solution from which dark blue crystals of **2** were deposited. X-ray crystallographic study revealed that it contains an entirely planar $(NCB)N_4(BCN)$ core. The measured B–N bond lengths of 144.2(2) pm suggest a degree of multiple bonding between boron and nitrogen, and all N–N bond lengths in the central N_4K_2 moiety [134.9(2) pm for N1–N2 and N3–N4 and 133.5(2) pm for N2–N3] lie between the ranges of N–N double and single bonds.

Upon addition of an excess of degassed water to the turquoise solution of **2**, the mixture immediately turned dark blue, from which slow evaporation deposited dark blue crystals of {[(CAAC)TipB]$_2$(μ^2-N_4H_2)} **3** in 75% yield. Single-crystal X-ray diffraction analysis of **3** showed that while the tetrazene core is preserved, the B–N bonds [143.2(2) pm] are close to those of covalent single bonds. In addition, the terminal N–N bonds [137.1(1) pm] lie within the range of single bonds, and a conventional double bond [127.2(2) pm] exists between the central two N atoms.

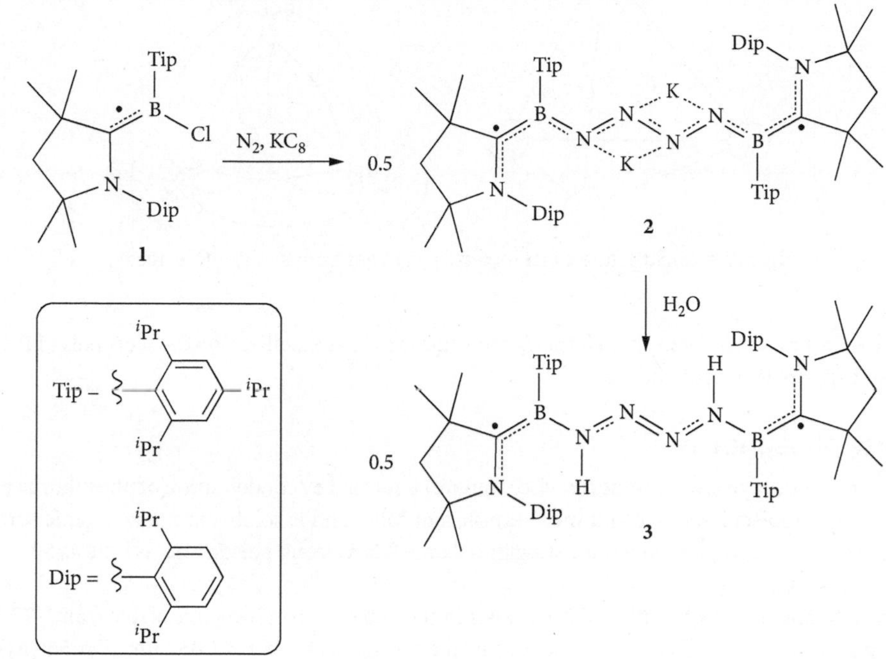

Reference

M.-A. Légaré, M. Rang, G. Bélanger-Chabot, J. I. Schweizer, I. Kummenacher, R. Bertermann, M. Arrowsmith, M. C. Holthausen, and H. Braunschweig, The reductive coupling of dinitrogen. *Science* **363**, 1329–32 (2019).

5.5 Structure and Bonding of Elemental Phosphorus and P$_n$ Groups

Homonuclear aggregates of phosphorus atoms exist in many forms: discrete molecules, covalent networks in crystals, polyphosphide anions, and phosphorus fragments in molecular compounds.

5.5.1 Elemental phosphorus

(1) P$_4$ and P$_2$ molecules

Elemental phosphorus is known in several allotropic forms. All forms melt to give the same liquid which consists of tetrahedral P$_4$ molecules, as shown in Fig. 5.5.1(a). The same molecular entity exists in the gas phase, the P–P bond length being 221 pm. At high temperature (> 800 °C) and low pressure P$_4$ is in equilibrium with P$_2$ molecules, in which the P≡P bond length is 189.5 pm.

The bonding between phosphorous atoms in the P$_4$ molecule can be described by a simple bent bond model, which is formed by the overlap of sp^3 hybrids of the P atoms. Maximum overlap of each pair of sp^3 orbitals does not occur along an edge of the tetrahedron. Instead, the P–P bonds are bent, as shown in

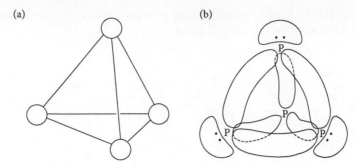

Fig. 5.5.1 (a) Structure of P_4 molecule, (b) bent bonds in P_4 molecule.

Fig. 5.5.1(b). In a more elaborate model, the P_4 molecule is further stabilized by the d orbitals of P atoms which also participate in the bonding.

(2) White phosphorus

White phosphorus (or yellow phosphorus when impure) is formed by condensation of phosphorus vapor. Composed of P_4 molecules, it is a soft, waxy, translucent solid, and is soluble in many organic solvents. It oxidizes spontaneously in air, often bursting into flame. It is a strong poison and as little as 50 mg can be fatal to humans.

At normal temperature, white phosphorus exists in the cubic α-form, which is stable from –77 °C to its melting point (44.1 °C). The crystal data of α-white phosphorus are $a = 1.851$ nm, $Z = 56$ (P_4) and $D = 1.83$ g cm^{-3}, but its crystal structure is still unknown. At –77 °C the cubic α-form transforms to a hexagonal β-form with a density of 1.88 g cm^{-3}.

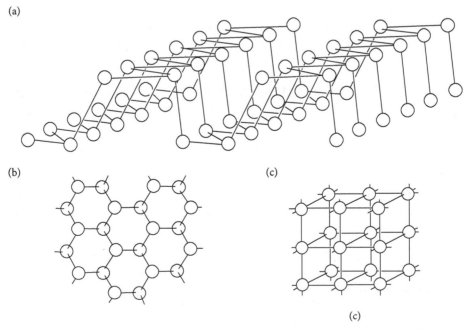

Fig. 5.5.2 Structure of black phosphorus: (a) orthorhombic, (b) rhombohedral, and (c) cubic black phosphorus.

(3) **Black phosphorus**

Black phosphorus is thermodynamically the most stable form of the element and exists in three known crystalline modifications: orthorhombic, rhombohedral, and cubic, as well as in an amorphous form. Unlike white phosphorus, the black forms are all highly polymeric, insoluble, practically non-flammable, and have comparatively low vapor pressures. The black phosphorus varieties represent the densest and chemically the least reactive of all known forms of the element.

Under high pressure, orthorhombic black phosphorus undergoes reversible transitions to produce denser rhombohedral and cubic forms. In the rhombohedral form the simple hexagonal layers are not as folded as in the orthorhombic form, and in the cubic form each atom has an octahedral environment, as shown in Figs 5.5.2(a)–(c).

(4) **Violet phosphorus**

Violet phosphorus (Hittorf's phosphorus) is a complex three-dimensional polymer in which each P atom has a pyramidal arrangement of three bonds linking it to neighboring P atoms to form a series of inter-connected tubes, as shown in Fig. 5.5.3. These tubes lie parallel to each other, forming double layers, and in the crystal structure one layer has its tubes packed at right angles to those in adjacent layers.

(5) **Red phosphorus**

Red phosphorus is a term used to describe a variety of different forms, some of which are crystalline and all of which are more or less red in color. They show a range of densities from 2.0 g cm^{-3} to 2.4 g cm^{-3}, and melting points in the range of 585–610 °C. Red phosphorus is a very insoluble species. It behaves as a high polymer that is inflammable and almost non-toxic.

5.5.2 **Polyphosphide anions**

Almost all metals form phosphides and over 200 different binary compounds are now known. In addition, there are many ternary mixed-metal phosphides. These phosphides consist of metal cations and phosphide anions. In addition to some simple anions (P^{3-}, P_2^{4-}, P_3^{5-}), there are many polyphosphide anions that exist in the form of rings, cages, chains, and sheets, as shown in Fig. 5.5.4.

In some metal phosphides, the polyphosphide anions constitute infinite chains and sheets, as shown in Fig. 5.5.5.

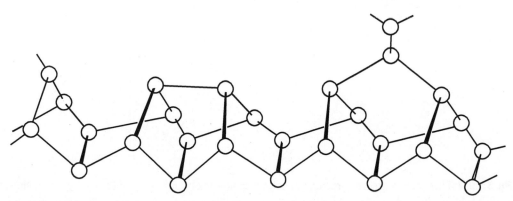

Fig. 5.5.3 Structure of the tube in violet phosphorus.

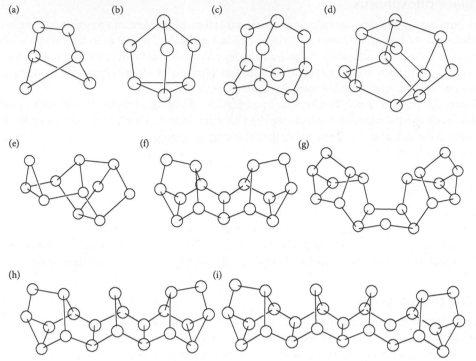

Fig. 5.5.4 The structures of some polyphosphide anions: (a) P_6^{4-} in [Cp''Th(P_6)ThCp''] (Cp'' =1,3-$Bu^+_2C_5H_3$), (b) P_7^{3-} in Li_3P_7, (c) P_{10}^{6-} in Cu_4SnP_{10}, (d) P_{11}^{3-} in Na_3P_{11}, (e) P_{11}^{3-} in [$Cp_3(CO)_4Fe_3$]P_{11} (f) P_{16}^{2-} in $(Ph_4P)_2P_{16}$, (g) P_{19}^{3-} in Li_3P_{19}, (h) P_{21}^{3-} in $K_4P_{21}I$, (i) P_{26}^{4-}.

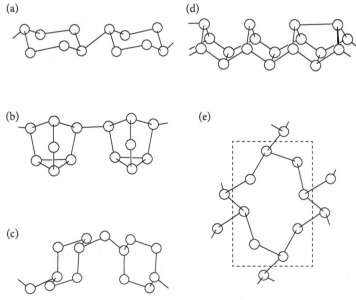

Fig. 5.5.5 Infinite chain and sheet structures of polyphosphide anions: (a) [P_6^{4-}]$_n$ in BaP_3, (b) [P_7^-]$_n$ in RbP_7, (c) [P_7^{5-}]$_n$ in Ag_3SnP_7, (d) [P_{15}^-]$_n$ in KP_{15}, (e) [P_8^{4-}]$_n$ in CuP_2.

5.5.3 **Structure of P_n groups in transition metal complexes**

Transition metal complexes with phosphorus bonded to metal atoms have been investigated extensively. They include single P atoms encapsulated in cages of metal atoms, and various P_n groups where $n = 2$ to at least 12. These P_n groups can be chains, rings, or fragments which are structurally related to their valence electron numbers. Fig. 5.5.6 shows some skeletal structures of the transition metal complexes with P_n groups.

In Figs. 5.5.6(b), (f), (j), and (n), the P_n groups P_3, P_4, P_5, and P_6 form planar three-, four-, five-, and six-membered rings, respectively. They can be considered as isoelectronic species of planar $(CH)_3$, $(CH)_4$, $(CH)_5$, and $(CH)_6$ molecules. The $[(P_5)_2Ti]^{2-}$ anion (Fig. 5.5.6(j)) is the first entirely inorganic metallocene, the structure of which has a pair of parallel and planar P_5 rings symmetrically positioned about the central Ti atom. The average P–P bond distance is 215.4 pm, being intermediate between those of P–P single (221 pm) and P=P double (202 pm) bonds. The average Ti–P distance is 256 pm The Ti–P_5 (center) distance is 179.7 pm.

The P_n groups in the complexes shown in Figs 5.5.6(e), (k), and (o) are four-, five-, and six-membered rings, respectively. Open chains P_4 and P_5 as multidentate ligands are exemplified by the structures shown in Figs 5.5.6(g), (h), (l), and (m). Metal complexes containing bi- and tri-cyclic P_n rings are shown in Figs 5.5.6(d) and (p), respectively. The skeletons of metal complexes of polyphosphorus P_n ligands with $n > 6$ are shown in Figs 5.5.6(q)–(v).

The reaction of Cp^*FeP_5 (see Fig. 5.5.6(e)) with CuCl in CH_2Cl_2/CH_3CN solvent leads to the formation of $[Cp^*FeP_5]_{12}(CuCl)_{10}(Cu_2Cl_3)_5\{Cu(CH_3CN)_2\}_5$. In this large molecule, the *cyclo*-P_5 rings of Cp^*FeP_5 are surrounded by six-membered P_4Cu_2 rings that result from the coordination of each of the P atomic lone pairs to CuCl metal centers, which are further coordinated by P atoms of other *cyclo*-P_5 rings. Thus five- and six-membered rings are fused in a manner reminiscent of the formation of the fullerene-C_{60} molecule. Fig. 5.5.7 shows the structure of a hemisphere of this globular molecule. The two hemispheres are joined by $[Cu_2Cl_3]^-$ as well as by $[Cu(CH_3CN)_2]^+$ units, and this inorganic fullerene-like molecule has an inner diameter of 1.25 nm and an outer diameter of 2.13 nm, making it about three times as large as C_{60}.

5.5.4 **Bond valence in P_n species**

The bonding in P_n species can be expressed by their bond valence b, which corresponds to the number of P–P bonds. Let g be the total number of valence electrons in P_n. When a covalent bond is formed between two P atoms, each of them gains one electron in its valence shell. In order to satisfy the octet rule for P_n, $\frac{1}{2}(8n - g)$ electron pairs must be involved in bonding between the P atoms. The number of these bonding electron pairs is defined as the bond valence b of the P_n species:

$$b = \tfrac{1}{2}(8n - g)$$

Applying this simple formula:

P_4: $b = \frac{1}{2}(4 \times 8 - 4 \times 5) = 6$, 6 P–P bonds;

P_7^{3-}: $b = \frac{1}{2}[7 \times 8 - (7 \times 5 + 3)] = 9$, 9 P–P bonds;

P_{11}^{3-}: $b = \frac{1}{2}[11 \times 8 - (11 \times 5 + 3)] = 15$, 15 P–P bonds;

P_{16}^{2-}: $b = \frac{1}{2}[16 \times 8 - (16 \times 5 + 2)] = 23$, 23 P–P bonds.

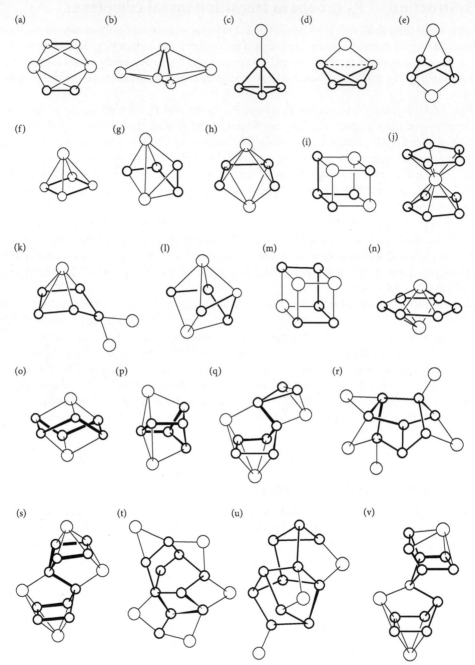

Fig. 5.5.6 Structure of P_n groups bonded to metal atoms in transition metal complexes: (a) $(Cp''Co)_2(P_2)_2$, (b) $(Cp_2Th)_2P_3$, (c) $W(CO)_3(PCy_3)_2P_4$, (d) $RhCl(PPh_3)_2P_4$, (e) $[Cp*Co(CO)]_2P_4$, (f) $Cp*Nb(CO)_2P_4$, (g) $[Ni(CO)_2Cp]_2P_4$, (h) $(CpFe)_2P_4$, (i) $[(Cp*Ni)_3P]P_4$, (j) $[Ti(P_5)_2][K(18-C-6)]$, (k) $[Cp*Fe]P_5[Cp*Ir(CO)]_2$, (l) $[Cp*Fe]P_5[TaCp'']$, (m) $[Cp*Fe]P_5[TaCp'']_2$, (n) $Cp*Mo_2P_6$, (o) $(Cp*Ti)_2P_6$, (p) $(Cp'_2Th)_2P_6$, (q) $[Cp'''Co(CO)_2]_3P_8$, (r) $[Cp*Ir(CO)]_2P_8[Cr(CO)_5]_3$, (s) $(Cp'Rh)_4P_{10}$, (t) $[CpCr(CO)_2]_5P_{10}$, (u) $[Cp^{Pr}Fe(CO)_2]P_{11}[Cp^{Pr}Fe(CO)_2]$, (v) $[CpCo(CO)_2]_3P_{12}$.

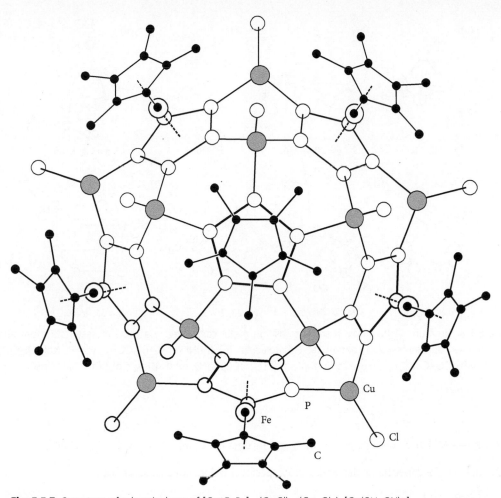

Fig. 5.5.7 Structure of a hemisphere of $[Cp^*FeP_5]_{12}(CuCl)_{10}(Cu_2Cl_3)_5[Cu(CH_3CN)_2]_5$.

In these P$_n$ species, the b value is exactly equal to the bond number in the structural formula, as shown in Fig. 5.5.1, Fig. 5.5.4, and Fig. 5.5.5. But for the planar ring P$_6$ (Fig. 5.5.6(n)),

$$b = \tfrac{1}{2}[6 \times 8 - (6 \times 5)] = 9$$

which implies the existence of three P–P bonds and three P=P bonds, and hence aromatic behavior as in the benzene molecule.

The bond valence of a P$_n$ group changes with its number of valence electrons. The P$_4$ species is a good example, as shown in Fig. 5.5.8. In these structures, each transition metal atom also conforms to the 18-electron rule.

The bonding structure of transition metal complexes with P$_n$ group can be classified into four types:

(1) Covalent P–M σ bond: each atom donates one electron to bonding and the g value of P$_n$ increases by one, as shown in Fig. 5.5.8(e).

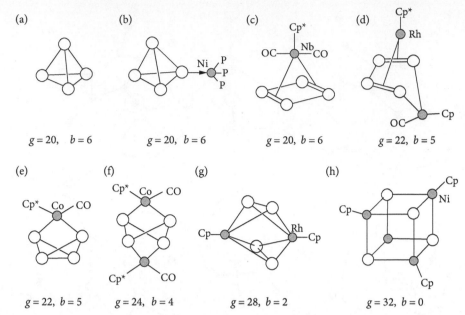

Fig. 5.5.8 Structure of P_4 group in some transition metal complexes (big circle represents P atom and arrow $\longrightarrow$ represents dative bond): (a) P_4 molecule, (b) $(P_4)Ni(PPh_2CH_2)_3CCH_3$, (c) $(P_4)Nb(CO)_2Cp^*$, (d) $(P_4)Rh_2(CO)(Cp)(Cp^*)$, (e) $(P_4)Co(CO)Cp^*$, (f) $(P_4)[Co(CO)Cp^*]_2$, (g) $(P_2)_2Rh_2(Cp)_2$, (h) $(P)_4(NiCp)_4$.

(2) $P \longrightarrow M$ dative bond: the g value of P_n species does not change, as shown in Fig. 5.5.8(b).

(3) $\overset{P}{\underset{P}{||}} \longrightarrow M \pi$ dative bond: the g value of P_n species also does not change, as shown in Fig. 5.5.8(c).

(4) $(\eta^n-P_n) \longrightarrow M$ dative bond: the P_n ring ($n = 3–6$) donates its delocalized π electrons to the M atom, and the g value of P_n group does not change.

Some molecular transition metal complexes of P_7^{3-}, As_7^{3-}, and Sb_7^{3-} have been isolated as salts of cryptated alkali metal ions. The structures of the complex anions are shown in Fig. 5.5.9. The bond valence b and bond number of these complex anions are as follows:

(a) $[P_7Cr(CO)_3]^{3-}$ in $[Rb\cdot crypt]_3[P_7Cr(CO)_3]$:
     $b = 12$, eight P–P and four P–Cr bonds.

(b) $[As_7Mo(CO)_3]^{3-}$ in $[Rb\cdot crypt]_3[As_7Mo(CO)_3]$:
     $b = 12$, eight As–As and four As–Mo bonds.

(c) $[P_7Ni(CO)]^{3-}$ in $[Rb\cdot crypt]_3[P_7Ni(CO)]$:
     $b = 12$, eight P–P and four P–Ni bonds.

(d) $[Sb_7Ni_3(CO)_3]^{3-}$ in $[K\cdot crypt]_3[Sb_7Ni_3(CO)_3]$:
     $b = 18$, four Sb–Sb, five SbSbNi 3c-2e, and two SbNiNi 3c-2e bonds.

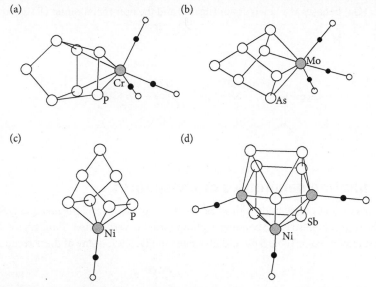

Fig. 5.5.9 Structure of transition metal complexes of P_7^{3-}, As_7^{3-}, and Sb_7^{3-}: (a) $[P_7Cr(CO)_3]^{3-}$, (b) $[As_7Mo(CO)_3]^{3-}$, (c) $[P_7Ni(CO)]^{3-}$, (d) $[Sb_7Ni_3(CO)_3]^{3-}$.

5.6 Bonding Type and Coordination Geometry of Phosphorus

5.6.1 Potential bonding types of phosphorus

To illustrate the potential diversity of structure and bonding of phosphorus, the classic Lewis representations for each possible coordination number 1 to 6 are shown in Fig. 5.6.1. Many of these bonding types have been observed in stable compounds, which are discussed in the following sections.

Phosphines are classical Lewis bases or ligands in transition metal complexes, but the cationic species shown in Fig. 5.6.1(i) are likely to exhibit Lewis acidity by virtue of the positive charge. In spite of their electron-rich nature, an extensive coordination chemistry has been developed for Lewis acidic phosphorus. For example, the compound shown below has a coordinatively unsaturated Ga(I) ligand bonded to a phosphenium cation; it can be considered as a counter-example of the traditional coordinate bond since

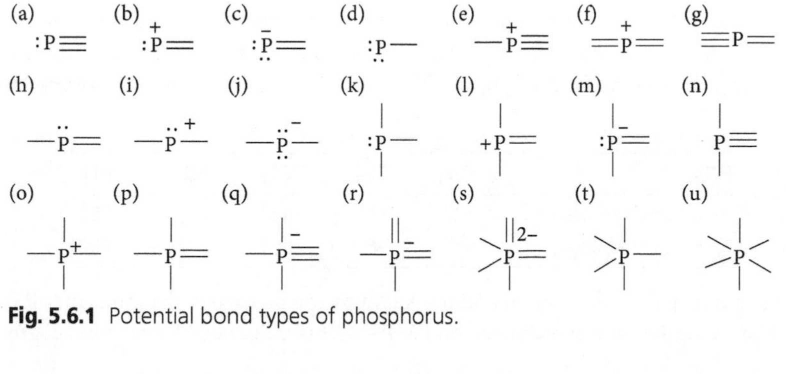

Fig. 5.6.1 Potential bond types of phosphorus.

the metal center (Ga) behaves as a Lewis donor (ligand) and the non-metal center (P) behaves as a Lewis acceptor.

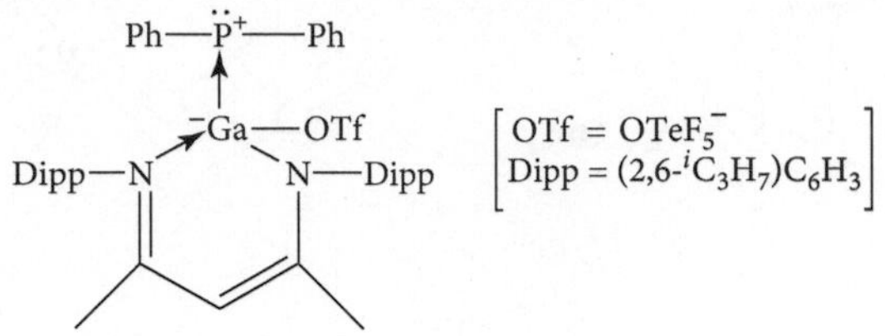

5.6.2 Coordination geometries of phosphorus

Phosphorus forms various compounds with all elements except Sb, Bi, and the inert gases for the binary compounds. The stereochemistry and bonding of phosphorus are very varied. Some typical coordination geometries are summarized in Table 5.6.1 and illustrated in Fig. 5.6.2. Many of these compounds will be discussed below.

(1) Coordination number 1

Coordination number 1 is represented by the compounds $P \equiv N$, $P \equiv C-H$, $P \equiv C-X$ (X = F, Cl) and $P \equiv C-Ar$ (Ar = $^tBu_3C_6H_2$). In these compounds, the P atom forms a triple bond with the N or C atom. The bond length are $P \equiv N$ 149 pm, $P \equiv C$ 154 pm.

(2) Coordination number 2

Compounds of this coordination number have three bond types:

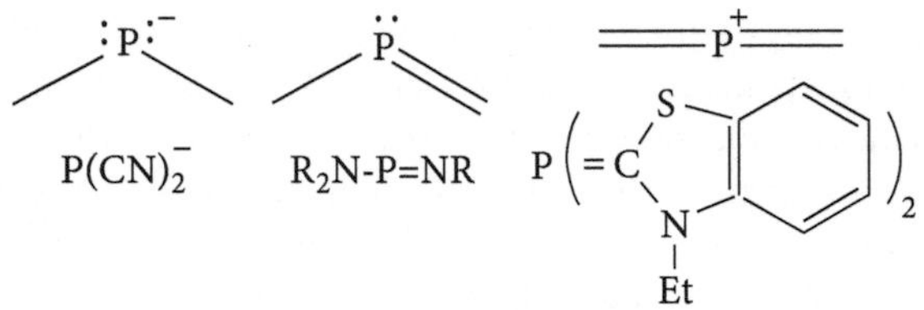

In $P(CN)_2^-$, the bond length P–C is 173 pm, $C \equiv N$ is 116 pm, and bond angle C–P–C is 95°.

(3) Coordination number 3

In compounds of this coordination number, the pyramidal configuration is the most common type, and the planar one is much less favored. In pyramidal PX_3, the three X groups are either the same or different, X = F, Cl, Br, I, H, OR, OPh, Ph, ^tBu, etc. The observed data of some PX_3 compounds are listed follow:

PX_3	PH_3	PF_3	PCl_3	PBr_3	PI_3
P–X bond length/pm	144	157	204	222	252
X–P–X bond angle	94°	96°	100°	101°	102°

Since PX_3 has a lone pair at the P atom and the X groups can be varied, the molecules PX_3 are important ligands. The strength of the coordinate bond $X_3P \longrightarrow M$ is influenced by different X groups in three aspects:

(a) σ P⟶M bonding

The stability of the σ P⟶M interaction, which uses the lone pair of electrons on P atom and a vacant orbital on M atom, is influenced by different X groups in the sequence:

$$P^tBu_3 > P(OR)_3 > PR_3 \approx PPh_3 > PH_3 > PF_3 > P(OPh)_3$$

(4) π back donation

The possibility of synergic π back donation from a non-bonding d_π pair of electrons on M into a vacant $3d_\pi$ orbital on P varies in the sequence:

$$PF_3 > P(OPh)_3 > PH_3 > P(OR)_3 > PPh_3 \approx PR_3 > P^tBu_3$$

(5) Steric interference

The stability of the P–M bonds are influenced by steric interference of the X groups, in accordance with the sequence:

$$P^tBu_3 > PPh_3 > P(OPh)_3 > PMe_3 > P(OR)_3 > PF_3 > PH_3$$

In the $PhP\{Mn(C_5H_5)(CO)_2\}_2$ molecule, the P atom is bonded to two Mn atoms and one phenyl C atom by single bonds in a planar configuration, as shown in Fig. 5.6.2(d). The bond angle Mn–P–Mn is 138°.

Table 5.6.1 Coordination geometries of phosphorus atoms in compounds

CN	Coordination geometry	Example	Structure (in Fig. 5.6.1)
1	linear	$P\equiv N$, $F-C\equiv P$	(a)
2	bent	$[P(CN)_2]^-$	(b)
3	pyramidal	PX_3 (X = H, F, Cl, Br, I)	(c)
	planar	$PhP\{Mn(C_5H_5)(CO)_2\}_2$	(d)
4	tetrahedral	P_4O_{10}	(e)
	square	$[P\{Zr(H)Cp_2\}_4]^+$	(f)
5	square pyramidal	$Os_5(CO)_{15}(\mu_4\text{-POMe})$	(g)
	trigonal bipyramidal	PF_5	(h)
6	octahedral	PCl_6^-	(i)
	trigonal prismatic	$(\mu_6\text{-P})[Os(CO)_3]_6^-$	(j)
7	monocapped trigonal prismatic	Ta_2P	(k)
8	cubic	Ir_2P	(l)
	bicapped trigonal prismatic	Hf_2P	(m)
9	monocapped square antiprismatic	$[Rh_9(CO)_{21}P]^{2-}$	(n)
	tricapped trigonal prismatic	Cr_3P	(o)
10	bicapped square antiprismatic	$[Rh_{10}(CO)_{22}P]^{3-}$	(p)

(6) **Coordination number 4**

This very common coordination number usually leads to a tetrahedral configuration for phosphoric acid, phosphates, and many phosphorus(V) compounds. In these compounds, the P(V) atom forms one double bond and three single bonds with other atoms. Fig. 5.6.2(e) shows the structure of P_4O_{10}, whose symmetry is T_d. The length of the terminal P=O bond is 143 pm, and the bridging P–O bond is 160 pm. The bond angle O–P–O is 102°, and P–O–P is 123°.

The compound P_4O_{10}, known as "phosphorus pentoxide," is the most common and most important oxide of phosphorus. This compound exists in three modifications. When phosphorus burns in air and condenses rapidly from the vapor, the common hexagonal (H) form of P_4O_{10} is obtained. The H form is metastable and can be transformed into a metastable orthorhombic (O) form by heating for 2h at 400 °C, and into a stable orthorhombic (O') form by heating for 24h at 450 °C. All three modifications undergo hydrolysis in cold water to give phosphoric acid, H_3PO_4.

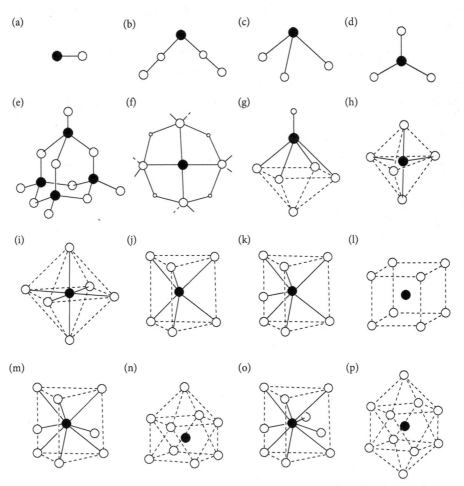

Fig. 5.6.2 Coordination geometries of the phosphorus atom in some compounds (black circles represent P, open circles represent other atoms): (a) P≡N, (b) [P(CN)$_2$]$^-$, (c) PX$_3$, (d) PhP{Mn(C$_5$H$_5$)(CO)$_2$}$_2$, (e) P$_4$O$_{10}$, (f) [P{Zr(H)Cp$_2$}$_4$]$^+$, (g) Os$_5$(CO)$_{15}$(μ_4-POMe), (h) PF$_5$, (i) [PCl$_6$]$^-$, (j) [Os(CO)$_3$]$_6$P, (k) Ta$_2$P, (l) Ir$_2$P, (m) Hf$_2$P, (n) [Rh$_9$(CO)$_{21}$P]$^{2-}$, (o) Cr$_3$P, (p) [Rh$_{10}$(CO)$_{22}$P]$^{3-}$.

Square-planar geometry of the P atom occurs in the compound $[P\{Zr(H)Cp_2\}_4][BPh_4]$. Fig. 5.6.2(f) shows the structure of $P\{Zr(H)Cp_2\}_4^+$, in which the Zr–P–Zr angles are very close to 90°, and the bridging hydrogens and the Zr atoms form a nearly planar eight-membered ring that encircles the central P atom.

(7) Coordination number 5

Fig. 5.6.2 (g) shows the structure of $Os_5(CO)_{15}(\mu_4\text{-POMe})$, in which five Os atoms constitute a square-pyramidal cluster. Each Os atom is coordinated by three CO ligands, and only the four basal Os atoms are bonded to the apical P atom.

Various structures have been found for phosphorus pentahalides:

(a) The molecular structure of PF_5 is trigonal bipyramidal, as shown in Fig. 5.6.2(h). The axial P–F bond length is 158 pm, which is longer than the equatorial P–F bond length of 153 pm.

(b) In the gaseous phase, PCl_5 is trigonal bipyramidal with axial $P-Cl_{ax}$ bond length 214 pm and equatorial $P-Cl_{eq}$ bond length 202 pm. In the crystalline phase, PCl_5 consists of a packing of tetrahedral $[PCl_4]^+$ and octahedral $[PCl_6]^-$ ions; in the latter anion shown in Fig. 5.6.2(i), the P–Cl bond length is 208 pm.

(c) In the crystalline phase, PBr_5 consist of a packing of $[PBr_4]^+$ and Br^-. Owing to steric overcrowding, $[PBr_6]^-$ cannot be formed by grouping six bulky Br atoms surround a relatively small P atom.

(d) There is as yet no evidence for the existence of PI_5.

(8) Coordination number ≥ 6

In each of the structures shown in Figs 5.6.2(j)–5.6.2(p), the transition metal atoms form a polyhedron around the P atom, which donates five valence electrons to stabilize the metal cluster.

5.7 Structure and Bonding in Phosphorus-Nitrogen and Phosphorus-Carbon Compounds

5.7.1 Types of P–N bonds

When phosphorus and nitrogen atoms are directly bonded, they form one of the most intriguing and chemically diverse linkages in inorganic chemistry. A convenient classification of phosphorus-nitrogen compounds can be made on the basis of formal bonding. Azaphosphorus compounds containing the P–N group are known as phosphazanes, those containing the P=N group are phosphazenes, and those with the P≡N group are phosphazynes. The common types of phosphorus-nitrogen bonds of these three classes of compounds are displayed in Table 5.7.1.

Phosphazynes are rare. The bond length of the diatomic molecule P≡N is 149 pm. The first stable compound containing the P≡N group is $[P\equiv N-R]^+[AlCl_4]^-$, $[R = C_6H_2(2,4,6\text{-}^tBu_3)]$, in which the length of the P≡N bond is 147.5 pm.

5.7.2 Phosphazanes

Some examples of phosphazanes are described below according to the types of P–N bonds.

(1)

Table 5.7.1 The common types of phosphorus-nitrogen bonds.*

Phosphazanes	Phosphazanes	Phosphazynes
pᴵᴵᴵ $\ddot{P}_{+}$—N $(sp^2)\sigma^2, \lambda^2$ $\ddot{P}$—N $(sp^3)\sigma^3, \lambda^3$	$\ddot{P}$=N $(sp^2)\,\sigma^2, \lambda^3$	$:P\equiv N$ σ^1, λ^3
pⱽ P—N $(sp^3)\,\sigma^4, \lambda^4$ P—N $(sp^3)\,\sigma^4, \lambda^5$ P—N $(sp^3d)\,\sigma^5, \lambda^5$ P←N $(sp^3d)\,\sigma^5, \lambda^5$ P←N $(sp^3d^2)\,\sigma^6, \lambda^6$	P=N $(sp^3)\,\sigma^4, \lambda^5$ N P—N $(sp^3)\,\sigma^4, \lambda^5$ N P=N $(sp^3)\,\sigma^4, \lambda^5$ N P=N $(sp^2)\,\sigma^3, \lambda^5$	P≡N $(sp^2)\,\sigma^3, \lambda^5$

* The hybridization of the P atom is enclosed in parentheses. The symbols σ and λ represent the coordination number and bonding number of P atoms.

In this bonding type, the P atom uses its sp^2 hybrid orbitals, one of which accommodates a lone pair. A bent molecular structure of this type is found for $(^tPr_2N)_2P^+$:

$$R_2\ddot{N}\diagdown{}^{\overset{+}{\ddot{P}}}\diagup\ddot{N}R_2 \longleftrightarrow R_2\overset{+}{N}\diagup\diagup{}^{\ddot{P}}\diagdown\ddot{N}R_2 \longleftrightarrow R_2\ddot{N}\diagup{}^{\ddot{P}}\diagdown\diagdown\overset{+}{N}R_2$$

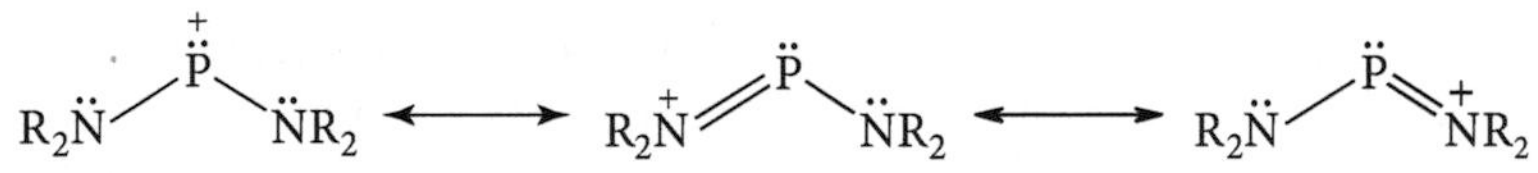

The bond length of P–N is 161.2 pm, shorter than the standard P–N single bond distance of 177 pm observed in H_3N–PO_3, which does not have a lone pair on the N atom. The bond angle N–P–N is 115°, smaller than idealized 120°, due to the repulsion of the lone pairs.

(2) 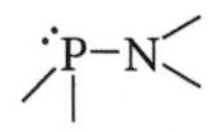 $\quad\ddot{P}$–N

The configuration of most R_2P–NR_2 compounds is represented by F_2P–NMe_2, which features a short P–N distance (162.8 pm) and trigonal planar arrangement (sp^2 hybridization) at the N atom. The plane defined by the C_2N unit bisects the F–P–F angle. This configuration minimizes steric repulsion between the substituents on P and N and also orients the lone pairs on the P and N atoms at a dihedral angle of about 90°, as shown in Fig. 5.7.1(a).

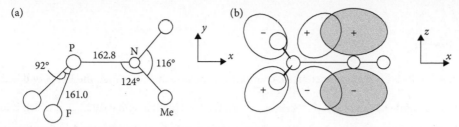

Fig. 5.7.1 (a) Molecular structure of $F_2P–NMe_2$, (b) overlap between the d_{xz} orbital of the P atom and the p_z orbital of the N atom.

The geometry of the molecule and the short P–N distance indicate that there is π bonding between the P and N atoms. In the molecule, the z axis is perpendicular to the PNC_2 plane and the x axis lies parallel to the P–N bond. The $p_z(N)$ orbital with a lone pair and the empty $d_{xz}(P)$ orbital overlap to form a π bond, as shown in Fig. 5.7.1(b).

The structure of F_2PNH_2 resembles that of F_2PNMe_2 with a P–NH_2 bond distance of 166 pm.

(3) $\overset{\diagdown}{\diagup}\overset{+}{P}-N\overset{\diagup}{\diagdown}$

Many compounds contain formally a single-bonded P–N group, such as the cation $[PCl_2(NMe_2)_2]^+$ in crystalline $[PCl_2(NMe_2)_2](SbCl_6)$, the cation $[P(NH_2)_4]^+$, and the anion $[P(NR)_4]^{3-}$. The bond length of P–N in $[P(NH_2)_4]^+$ is 160 pm, and that in $[P(NR)_4]^{3-}$ is 164.5 pm.

(4) $\overset{\diagdown\!\!\diagdown}{\diagup}P-N\overset{\diagup}{\diagdown}$

An important group of compounds such as $(Me_2N)_2POCl$, $(Me_2N)POCl_2$, and $(R_2N)_3PO$ possess this bonding type. For example, $(Me_2N)_3PO$ is a colorless mobile liquid which is miscible with water in all proportions. It forms an adduct with $HCCl_3$, dissolves ionic compounds, and can dissolve alkali metals to give blue paramagnetic solutions which are strong reducing agents.

(5) $\overset{|}{\underset{|}{\diagup}}\overset{\diagdown}{P}-N\overset{\diagup}{\diagdown}$

The compound $F_4P–NEt_2$ has this bonding type, in which the P atom uses sp^3d hybrid orbitals.

(6) $\overset{|}{\underset{|}{\diagup}}\overset{\diagdown}{P}\!\leftarrow\!N\overset{\diagup}{\diagdown}$

The phosphatranes that have this bonding type are analogs of silatranes. The cage molecule shown on the right is a trigonal bipyramidal five-coordinated phosphazane with a rather long P–N bond.

(7)

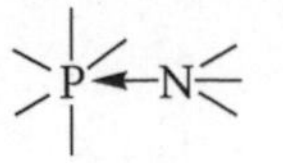

The adduct $F_5P{\leftarrow}NH_3$ is an octahedral six-coordinated phosphanane, in which the P–N bond length is 184.2 pm. In another example, $Cl_5P \longleftarrow N \bigcirc N$, the P–N bond length is 202.1 pm, which is longer than the former. This difference is due to the high electronegativity of F, which renders the F_5P group a better acceptor as compared to Cl_5P.

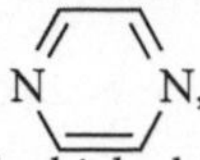

5.7.3 Phosphazenes

Phosphazenes, formerly known as phosphonitrilic compounds, are characterized by the presence of the group P=N. Known compounds, particularly those containing the $\diagdown P{=}N{-}$ group, are very numerous and they have important potential applications.

(1) Bonding types of phosphazenes

(a) Containing –P=N– bonding type

In this bonding type, the P atom uses sp^2 hybrid orbitals, one of which contains the lone pair electrons. This type of phosphazene compounds exists in a bent configuration. For example, the structure of $(SiMe_3)_2NPN(SiMe_3)$ is shown below:

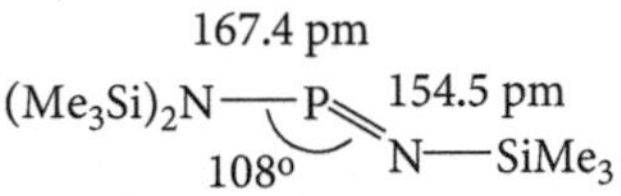

$$(Me_3Si)_2N \overset{167.4\ pm}{\underset{108°}{-\!\!\!-}} P \overset{154.5\ pm}{\underset{N-SiMe_3}{=\!\!=}}$$

(b) Containing $\diagdown P{=}N{-}$ bonding type

The compound $(Me_3Si)_2N\text{-}P(NSiMe_3)_2$ belongs to this type, as shown below:

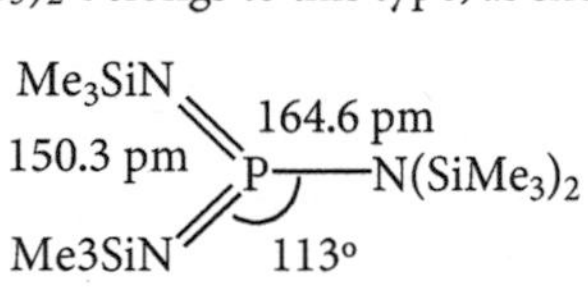

$$\begin{array}{c} Me_3SiN \diagdown \\ {}^{150.3\ pm} \diagdown \\ Me3SiN \diagup \end{array} P \overset{164.6\ pm}{\underset{113°}{-\!\!\!-}} N(SiMe_3)_2$$

(C) Containing $\diagdown P{=}N{-}$ bonding type

The simplest compound of this type is iminophosphorane, $H_3P{=}NH$, whose derivatives are very numerous, including $R_3P{=}NR'$, $Cl_3P{=}NR$, $(RO)_3P{=}NR'$, and $Ph_3P{=}NR$. In these compounds the P atom uses its sp^3 hybrid orbitals to form four σ bonds, and is also strengthened by $d\pi\text{-}p\pi$ overlap with N and other atoms.

(2) Structure and bonding of cyclic phosphazenes

The main products of refluxing a mixture of PCl_5 and NH_4Cl using tetrachloroethane as solvent are the cyclic trimer $(PNCl_2)_3$ and tetramer $(PNCl_2)_4$, which are stable white crystalline compounds that can be isolated and purified by recrystallization from non-polar solvents.

(a)   (b)

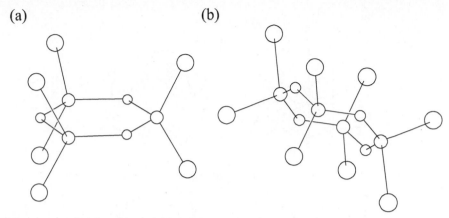

Fig. 5.7.2 Structure of (a) $(PNCl_2)_3$ and (b) $(PNCl_2)_4$ (circles of decreasing sizes represent Cl, P, and N atoms, respectively).

The trimer $(PNCl_2)_3$ has a planar six-membered ring structure of D_{3h} symmetry with the six Cl atoms disposed symmetrically above and below the plane of the ring. The P–N bonds within the ring are of the same length, 158 pm, and the interior angles are all close to 120°. The Cl–P–Cl planes are perpendicular to the plane of the central ring, and the Cl–P–Cl angle is 120°. Fig. 5.7.2 shows the structures of $(PNCl_2)_3$ and chair-like $(PNCl_2)_4$.

The shortness and equality of the P–N bond lengths in $(PNCl_2)_3$ arise from electron delocalization involving the d orbitals of the P atoms and the p orbitals of the N atoms. In such a system, the σ bonds formed from phosphorus sp^3 orbitals overlapping with the nitrogen sp^2 orbitals are enhanced by π bonding between the nitrogen p_z orbitals and the phosphorus d orbitals.

In $(PNCl_2)_3$, the π bonding occurs over the entire ring. First, $d\pi$-$p\pi$ overlap occurs between the nitrogen p_z orbital (z axis perpendicular to the ring plane) and the d_{xz} and d_z orbitals of phosphorus. Fig. 5.7.3(a) shows the orientation of one d_{xz} orbital of the P atom and two p_z orbitals of neighbor N atoms. Fig. 5.7.3(b) shows a projection of the overlap of d_{xz} with the p_z orbitals. Secondly, "in-plane" electron delocalization probably arises from overlap of the lone-pair orbitals on nitrogen with the d_{xy} orbitals on phosphorus, forming additional π bonds in the plane of the ring. Fig. 5.7.3(c) shows the $d_{x^2-y^2}$ of P overlapping with the lone-pair (sp^2) orbitals of two adjacent N atoms. Thirdly, the d_{z^2} orbitals of P atoms overlap with the p orbitals of the exocyclic Cl atoms.

The geometric disposition of the d orbitals in $d\pi$-$p\pi$ systems allows puckering and accounts for the variety of ring conformations which are found among larger cyclic phosphazene compounds, such as $(PNCl_2)_4$.

Of the five binary phosphorus-nitrogen molecules described in the literature, namely P_4N_4, $P(N_3)_3$, $P(N_3)_5$, the anion in the ionic compound $(N_5)^+[P(N_3)_6]^-$ and the phosphazene derivative $[PN(N_3)_2]_3$, only the last has been fully structurally characterized. These compounds are difficult to isolate and handle owing to their highly endothermic character and extremely low energy barriers, which often lead to uncontrollable explosive decomposition. Single-crystal X-ray analysis of $[PN(N_3)_2]_3$ conducted in 2006 showed that it is a structural analog of $[PNCl_2]_3$, with three azide groups oriented nearly parallel to the phosphazene ring and the other three nearly perpendicular to the ring.

A hybrid borazine-phosphazine ring system has been found in $[\overline{ClBNMePCl_2NPCl_2NMe}](GaCl_4)$; an X-ray study revealed that the 6π aromatic cation (structural formula shown below) is virtually planar with B–N bond lengths of 143.6(9) pm and 142.2(10) pm, which are close to those found in borazines (143 pm).

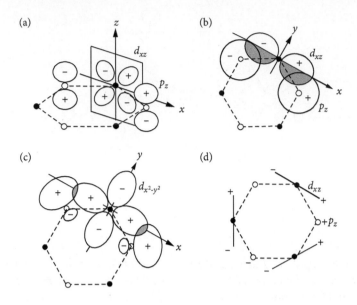

Fig. 5.7.3 $d\pi$-$p\pi$ Bonding in the ring of $(PNCl_2)_3$: (a) orientation of d_{xz} of P atom and p_z of N atom; (b) overlap of d_{xz} and p_z; (c) overlap of $d_{x^2-y^2}$ of P atom and lone pair (sp^2) of N atom; (d) possible mismatch of the d_{xz} orbital of P atom and p_z orbital of the N atom.

All phosphazenes, whether cyclic or chain-like, contain the formally unsaturated group P=N with four-coordinate P and two-coordinate N atoms. Based on available experimental data, the following generalizations may be made in regard to their structure and properties:

(a) The rings and chains are very stable.

(b) The skeletal interatomic distances are equal within the ring or along the chain, unless there is different substitution at various P atoms.

(c) The P–N distances are shorter than expected for a covalent single bond (≈ 177 pm) and are usually in the range 158 ± 2 pm.

(d) The N–P–N angles are usually in the range $120\pm2°$; but the P–N–P angles in various compounds span the range 120–$148°$.

Table 5.7.2 Types of phosphinidene complexes

Types	Structure	Structure and properties
Two-electron complexes	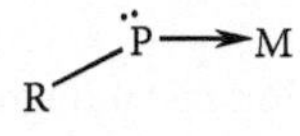	η^1-bent, electrophilic
	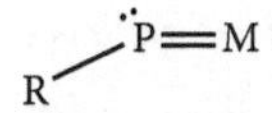	η^1-bent, nucleophilic*
	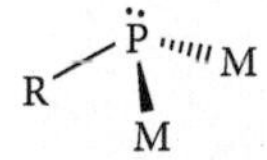	μ_2-pyramidal
Four-electron complexes	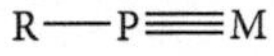	η^1-linear**
	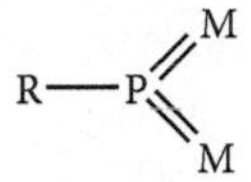	μ_2-planar
	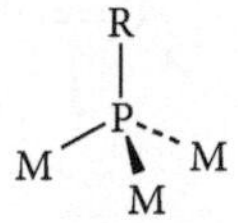	μ_3-tetrahedral
	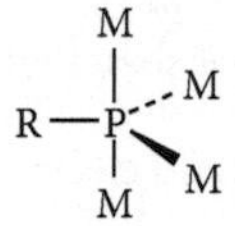	μ_4-bipyramidal

* In the complex Mes–P=Mo(Cp)$_2$, P=Mo 237.0 pm, Mo –P –C 115.8°.

** In the complex Mes–P≡WCl$_2$(CO)(PMePh$_2$), P≡W 216.9 pm, C–P–W 168.2°.

(e) Skeletal N atoms are weakly basic and can coordinate to metals or be protonated, especially when the P atoms carry electron-releasing groups.

(f) Unlike many aromatic systems, the phosphazene skeleton is difficult to reduce electrochemically.

(g) Spectral effects associated with organic π systems are not exhibited.

5.7.4 **Bonding types in phosphorus-carbon compounds**

Phosphorus and carbon are diagonal relatives in the periodic table. The diagonal analogy stresses the electronegativity of the element (C 2.5 vs P 2.2) which governs its ability to release or accept electrons. This property controls the reactivity of any species containing the element. This section covers the types of phosphorus-carbon bonds and the structures of representative species.

(1) **Phosphinidenes (C—P̈) and phosphinidene complexes**

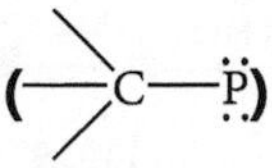

Phosphinidenes (recommended IUPAC name: phosphanylidenes) are unstable species and analogous to the carbenes. The parent compound H–P is a six-electron species that is still unknown, but it can give rise to seven different types of complexes, as listed in Table 5.7.2.

(2) **Phosphaalkenes, $R^1R^2C=PR^3$**

Phosphaalkenes are tervalent phosphorus derivatives with a double bond between carbon and phosphorus. The observed P=C bond lengths range from 161 pm to 171 pm (average 167 pm), appreciably shorter than the single P–C bond length of 185 pm.

Phosphaalkenes may coordinate to transition metal fragments in various ways:

(a) η^1 mode via the lone pair on P atom. An example is:

Ph, Ph C=P, Me, Cr(CO)$_5$ P=C 167.9 pm

(b) η^2 mode via the π bond electrons. An example is:

Tms, Cp* C=P, Cp*, Ni(PEt$_3$)$_2$

In η^2 complexes the P–C bond length is longer than that in η^1 complexes or the free ligands.

(c) η^1, η^2 mode via both the lone pair and π bond electrons. An example is:

H$_2$C=P, Mes*, Fe(CO)$_4$, Fe(CO)$_4$ P=C 173.7 pm

(3) **Phosphaalkynes, RC≡P**

Phosphaalkynes are compounds of tervalent phosphorus which contain a P≡C triple bond. Fig. 5.7.4 shows the structure of ^tBu–C≡P, whose P≡C bond is very short (154.8 pm), and the electronic ionization energies [I_1(π MO) = 9.61 eV, I_2(P lone pair) = 11.44 eV] are low, suggesting that it may have a chemistry closely related to that of the alkynes.

Phosphaalkynes have a rich coordination chemistry in which both the triple bond and the lone pair of the P atom can participate. Table 5.7.3 lists the coordination modes of phosphaalkynes, and two structures are shown in Fig. 5.7.5.

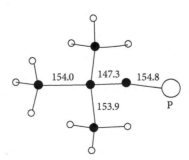

Fig. 5.7.4 Structure of ^tBu–C≡P
(bond lengths in pm).

Table 5.7.3 Coordination modes of phosphaalkynes

Coordination mode		Example and structure
η^1	R—C≡P⟶M	[^{t}Bu–C≡P–Fe(H)(dppe)$_2$][BPh$_4$] C≡P 151.2 pm
η^2	R—C≡P ↓ M	^{t}Bu—C≡P ↓ Pt(PR$_3$)
η^1, η^2	R—C≡P⟶M ↓ M	^{t}Bu—C≡P⟶Cr(CO)$_5$ ↓ Pt(Ph$_2$PCH$_2$CH$_2$PPh$_2$)
4e	R—C–P (with M above and below)	^{t}BuCP[Fe$_2$(CO)$_5$(PPh$_2$CH$_2$PPh$_2$)] (Fig. 5.7.5(a))
6e	R—C–P⟶M (with M above and below)	^{t}BuCP[W(CO)$_5$][Co$_2$(CO)$_6$] (Fig. 5.7.5(b))

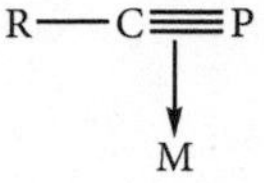

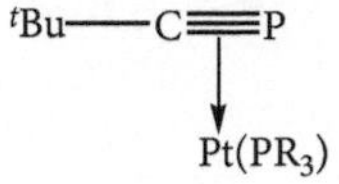

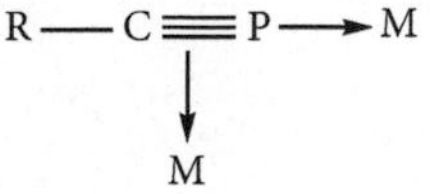

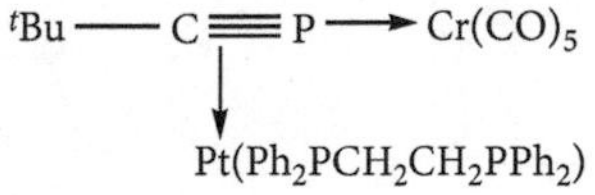

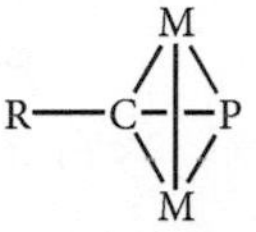

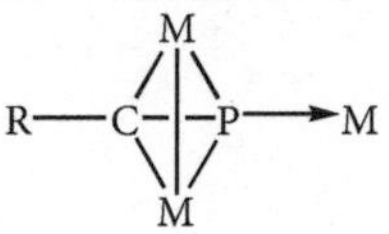

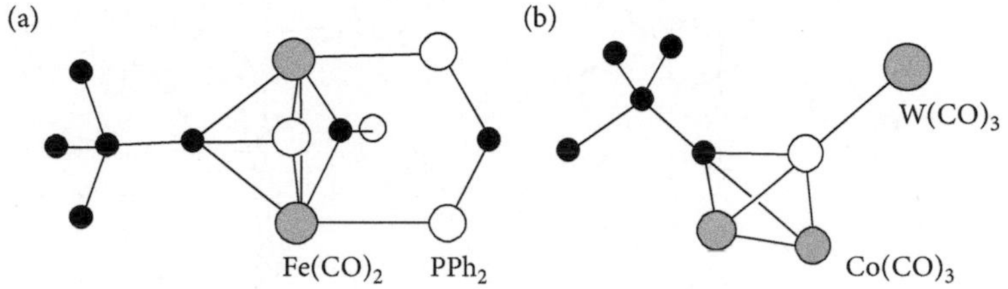

Fig. 5.7.5 Structure of (a) ^{t}Bu-CP[Fe$_2$(CO)$_5$(PPh$_2$CH$_2$PPh$_2$)] and (b) ^{t}Bu-CP[W(CO)$_5$][Co$_2$(CO)$_6$].

Some oligomers of the phosphaalkynes ^{t}BuCP have been characterized. The phosphaalkyne cyclotetramer exists in several isomeric forms, whose structures are shown in Fig. 5.7.6(a) to Fig. 5.7.6(e). In cubane-like P$_4$C$_4$^{t}Bu$_4$, the P–C bond lengths are all identical (188 pm), and are typical for single bonds. The angle at P is reduced from the idealized 90° to 85.6°, while that at C is widened to 94.4°.

The phosphaalkyne pentamer P$_5$C$_5$^{t}Bu$_5$ has the cage structure shown in Fig. 5.7.6(f), which can be derived from the tetramer by replacing one corner C atom of the "cube" by a C$_2$P triangular fragment.

The phosphaalkyne hexamer P$_6$C$_6$^{t}Bu$_6$ consists of a lantern-like cage constructed from the linkage of a chair-like P$_4$C$_2$ ring with a pair of C$_2$P rings above and below it, as shown in Fig. 5.7.6(g).

(4) Cyaphide P≡C⁻

The quest for cyaphide, the phosphorus homolog of cyanide, as a ligand in a stable metal complex came to a satisfactory conclusion in 2006. The pair of related complexes [RuH(dppe)$_2$(Ph$_3$SiC≡P)]OTf and [RuH(dppe)$_2$(C≡P)] were obtained via a new synthetic route and structurally characterized by X-ray crystallography. Their molecular geometries are displayed in Fig. 5.7.7. As expected, the Si–C≡P and P≡C⁻ ligands are *P*-coordinated and *C*-coordinated to the Ru(II) center, respectively. The long C≡P bond in the cyaphide complex likely arises from back donation from Ru to the π* orbitals of the ligand.

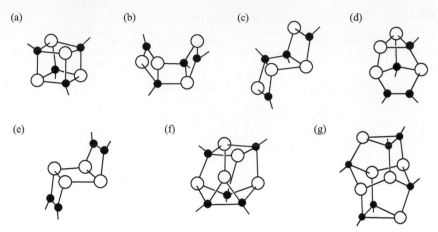

Fig. 5.7.6 Structure of oligomers of phosphaalkyne: (a)–(e) cyclotetramer, (f) pentamer, (g) hexamer.

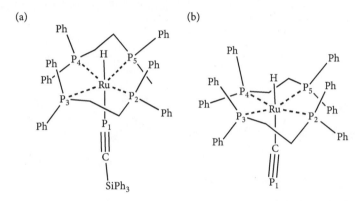

Fig. 5.7.7 Molecular structures and bond lengths (pm) of (a) [RuH(dppe)$_2$(Ph$_3$SiC≡P)] $^+$, C≡P$_1$ 153.0(3), Ru-P$_1$ 224.85(8), Ru-P$_2$ 238.11(7), Ru-P$_3$ 237.21(7), Ru-P$_4$ 235.59(7), Ru-P$_5$ 236.94(7); (b) [RuH(dppe)$_2$(C≡P)], C≡P$_1$ 157.3(2), Ru-C 205.7(2), Ru-P$_2$ 233.42(5), Ru-P$_3$ 233.15(5), Ru-P$_4$ 232.22(5), Ru-P$_5$ 233.96(4).

5.7.5 π-Coordination complexes of phosphorus-carbon compounds

(1) Diphosphenes (R–P=P–R)

Diphosphenes contain the –P=P– group, the majority of which adopt a *trans*-configuration. For example, the stable compound

$$\underset{Ar}{\overset{}{\diagdown}}\ddot{P} = \ddot{P}\overset{Ar}{\diagup} \qquad (Ar = 2,4,6\text{-}^tBu_3C_6H_2)$$

exhibits the *trans* form with a P=P bond length of 203.4 pm.

Various coordination modes of diphosphenes toward transition metals involve σ- and π-interaction with the metal center, as listed in Table 5.7.4.

Table 5.7.4 Coordination modes of diphosphenes

Coordination mode	Example and structure	
η^1	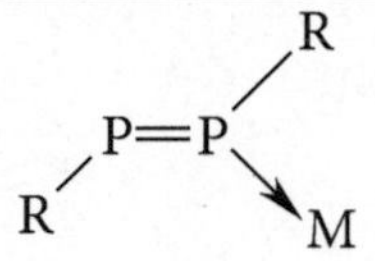	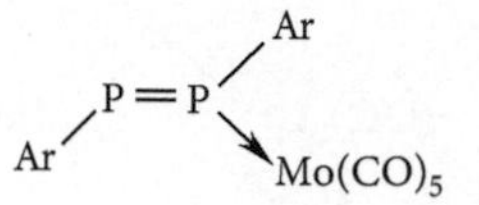 (Ar = 2,4,6-^{t}Bu$_3$C$_6$H$_2$)
η^2 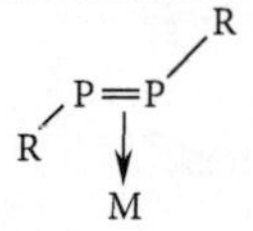		(η^2-^{t}BuP=P^{t}Bu)Zr(Cp*)$_2$
$\mu(\eta^1:\eta^1)$	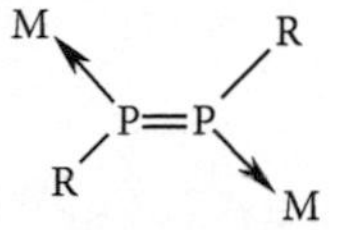	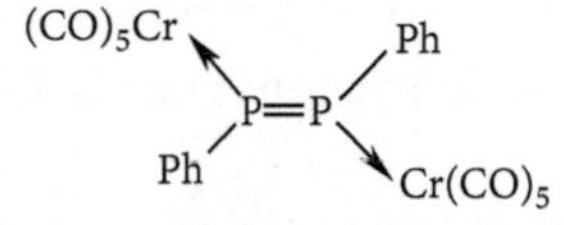
$\mu(\eta^1:\eta^2)$ 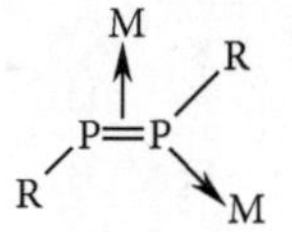		—
$\mu_3(\eta^1:\eta^1:\eta^2)$	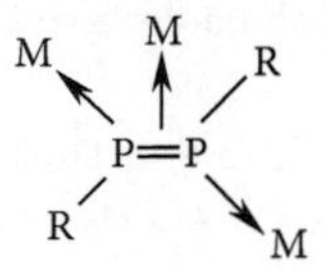	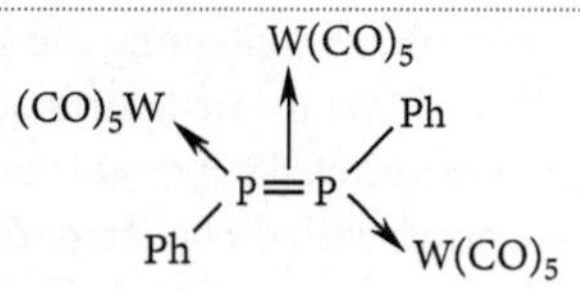
$\mu(\eta^2:\eta^2)$		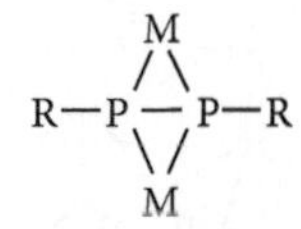 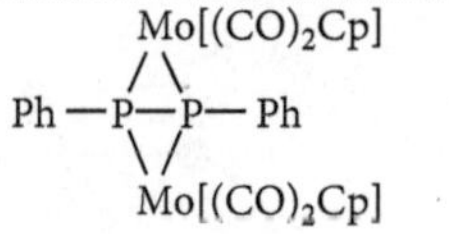 Mo$_2$P$_2$ as a butterfly
$\mu(\eta^2:\eta^2)$		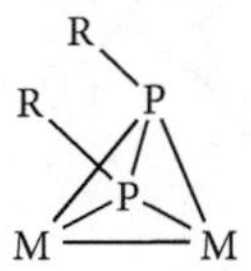 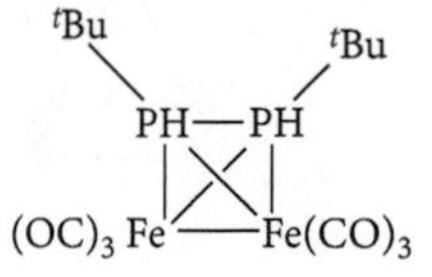 Fe$_2$P$_2$ as a tetrahedral

(2) η^3-Phosphaallyl and η^3-phosphirenes

The phosphaallyl anions $\left(\text{—} \overset{\overset{\displaystyle |}{\underset{\ominus}{C}}}{P} \text{—} \right)$ and the phosphirenes which contain the group

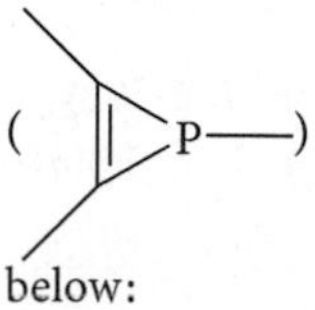

$\left(\triangleright\!\!\!\text{P} \text{—} \right)$ are η^3-ligands that can form complexes with transition metals. Some examples are shown

below:

(3) η^4-Phosphadienes and diphosphacyclobutadienes

Phosphadienes, $-P{=}C{-}C{=}C\big\langle$ or $\big\rangle C{=}P{-}C{=}C\big\langle$, are η^4-ligands that can form complexes with transition metals, for example:

Many metal complexes containing the 1,2 or 1,3-diphosphacyclobutadiene ring have been characterized, and Fig. 5.7.8 shows the structures of some examples.

Interestingly, it has proved impossible to displace the η^4-ligated $(P_2C_2{}^tBu_2)$ rings from any of the above complexes, in contrast with the behavior of the analogous η^4-ligated cyclobutadiene ring complexes. This

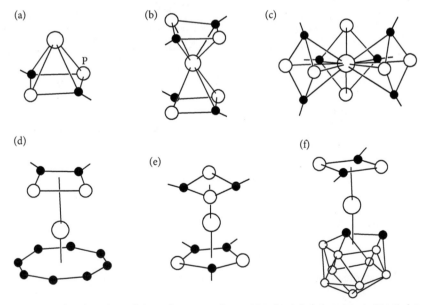

Fig. 5.7.8 Structure of diphosphacyclobutadiene complexes: (a) Fe(CO)$_3$[η^4-P$_2$C$_2{}^t$Bu$_2$], (b) Ni[η^4-P$_2$C$_2{}^t$Bu$_2$]$_2$, (c) Mo[η^4-P$_2$C$_2{}^t$Bu$_2$]$_3$, (d) Ti(η^4-P$_2$C$_2{}^t$Bu$_2$)(η^8-COT), (e) Co(η^4-P$_2$C$_2{}^t$Bu$_2$)(η^5-P$_2$C$_3{}^t$Bu$_3$), and (f) Rh(η^4-P$_2$C$_2{}^t$Bu$_2$)(η^5-C$_2$B$_9$H$_{11}$).

may be attributed to the significantly stronger π-interaction between the metal and the phosphorus-containing ring system.

(4) η^5-Phospholyl complexes

The phospholyl unit, which contains one to five P atoms, in an analog of the cyclopentadienyl ligand. Figs 5.7.9(a)–(f) show some phosphametallocenes, including sandwich, half-sandwich, and tilted structures. Figs 5.7.9(g)–(j) show some complex phosphametallocenes, in which only essential parts of the structure are displayed. Fig. 5.7.9(k) shows the supramolecular structure of $[Sm(\eta^5\text{-}PC_4Me_4)_2(\eta^1\text{-}PC_4Me_4)K(\eta^6\text{-}C_6H_5Me)]Cl$.

(5) η^6-Phosphinine complexes

The phosphinine rings (PC_5R_5, $P_2C_4R_4$,···P_6) are analogs of benzene and form sandwich structures with the transition metals. In η^6-phosphinine complexes, the phosphinine rings are all planar.

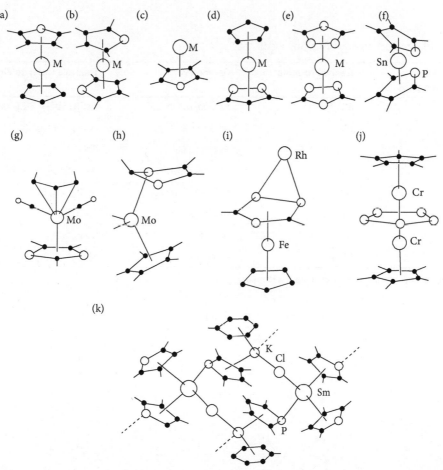

Fig. 5.7.9 Structure of phospholyl π-complexes: (a)–(e) sandwich-type phosphametallocenes, (f) $Sn[\eta^5\text{-}PC_4(TMS)_2Cp_2]_2$, (g) $(\eta^3\text{-}C_9H_7)Mo(CO)_2(\eta^5\text{-}P_2C_3{}^tBu_3)$, (h) $(\eta^3\text{-}P_2C_3{}^tBu_3)Mo(CO)_2(\eta^5\text{-}Cp^*)$, (i) $[(\eta^5\text{-}Cp^*)(CO)]Rh[\eta^5\text{-}P_3C_2{}^tBu_2]Fe(\eta^5\text{-}Cp)$, (j) $(\eta^5\text{-}Cp^*)Cr(\eta^5\text{-}P_5)Cr(\eta^5\text{-}Cp^*)$, (k) $[Sm(\eta^5\text{-}PC_4Me_4)_2(\eta^1\text{-}PC_4Me_4)K(\eta^6\text{-}C_6H_5Me)]Cl$.

5.8 **Structural Chemistry of As, Sb, and Bi**

5.8.1 **Stereochemistry of As, Sb, and Bi**

The series As, Sb, and Bi show a gradation of properties from non-metallic to metallic, but the discrete molecules and ions of these elements exhibit similar stereochemistry, as listed in Table 5.8.1 and shown in Fig. 5.8.1. The presence of a lone pair (denoted by E in the table) in these atoms implies M^{III}, otherwise it is M^V.

Many compounds of the MX_3E type have been prepared, and all 12 trihalides of As, Sb, and Bi are well known and available commercially. In either the gaseous or solid state, the lone pair causes the bond angles to be less than the ideal tetrahedral angle in every case. For example, $SbCl_3$ in the gas phase has bond length 233 pm and bond angle 97.1°, and in the crystal it has three short Sb–Cl 236 pm and three long Sb···Cl ≥ 350 pm, and the bond angle Cl–Sb–Cl is 95°. The cations of MX_4^+ are all tetrahedral. The anion SbF_4^- is known in the monomeric form and has the MX_4E type disphenoidal geometry. In the dimer $Sb_2F_7^-$ both Sb atoms have a disphenoidal geometry with the bridging fluorine in one axial position.

Table 5.8.1 Stereochemistry of As, Sb, and Bi

Total number of electron pairs	General formula*	Geometry	Example (refer to Fig. 5.8.1)
4	MX_3E	*trigonal pyramidal*	$AsCl_3$, $SbCl_3$, $BiCl_3$ (a)
4	MX_4	tetrahedral	$AsCl_4^+$, $SbCl_4^+$ (b)
5	MX_4E	disphenoidal	SbF_4^- (c)
5	MX_5	trigonal bipyramidal	AsF_5, $SbCl_5$, BiF_5 (d)
5	MX_5	square pyramidal	$Sb(C_6H_5)_5$, $Bi(C_6H_5)_5$ (e)
6	MX_5E	square pyramidal	$SbCl_5^{2-}$ (f)
6	MX_6	octahedral	$SbBr_6^-$ (g)
7	MX_6E	octahedral	$SbBr_6^{3-}$, $BiBr_6^{3-}$ (g)

* M = As, Sb, or Bi, X = ligand atom or group, E = lone pair.

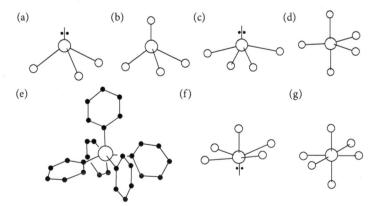

Fig. 5.8.1 Stereochemistry of As, Sb and Bi: (a) $AsCl_3$, (b) $AsCl_4^+$, (c) SbF_4^-, (d) $SbCl_5$, (e) $Bi(C_6H_5)_5$, (f) $SbCl_5^{2-}$, (g) $SbBr_6^{3-}$ and $BiBr_6^{3-}$.

Compounds of the MX_5 type exhibit two geometries, trigonal bipyramidal (more common) and square pyramidal. In the trigonal bipyramidal molecules, the axial bonds are longer than the equatorial bonds. If there are different ligands, the more electronegative ones usually occupy the axial positions. The compounds $Bi(C_6H_5)_5$ and $Sb(C_6H_5)_5$ have a square pyramidal shape, as shown in Fig. 5.8.1(e), for which the bond lengths and bond angles are:

$Bi(C_6H_5)_5$: $Bi–C_{ax}$ 222.1 pm $Bi–C_{ba}$ 233.6 pm $C_{ax}–Bi–C_{ba}$ 101.6°
$Sb(C_6H_5)_5$: $Sb–C_{ax}$ 211.5 pm $Sb–C_{ba}$ 221.6 pm $C_{ax}–Sb–C_{ba}$ 105.4°

The molecules of MX_5E type, such as SbF_5^{2-}, $BiCl_5^{2-}$, and some oligolymeric anions $(SbF_4)_4^{4-}$ and $(BiCl_4)_2^{2-}$ have square pyramidal geometry at each M atom. In these cases, the four ligands in the base of the square pyramid lie in a plane slightly above the central M atom, and the bond angles are all less than 90° caused by the greater lone-pair repulsions.

The anions of the type MX_6, such as SbF_6^-, $SbBr_6^-$, and $Sb(OH)_6^-$, have the expected octahedral geometry. The anions of the type MX_6E, such as $SbBr_6^{3-}$ and $BiBr_6^{3-}$, frequently also have a regular octahedral structure. The undistorted nature of the $SbBr_6^{3-}$ octahedral suggests that the lone pair is predominantly $5s^2$, but in a sense it is still stereochemically active since the Sb–Br distance in $Sb^{III}Br_6^{3-}$ is 279.5 pm, which is longer than the distance 256.4 pm in $Sb^VBr_6^-$.

The bismuthonium ylide 4,4-dimethyl-2,6-dioxo-1-triphenylbismuthoniocyclohexane exhibits a distorted tetrahedral geometry with a $Bi–C_{ylide}$ bond length of 215.6 pm, $Bi–C_{Ph}$ bond lengths in the range 221–222 pm, and a weak Bi···O interaction of 301.9 pm with one of the carbonyl oxygen atoms (the other Bi···O separation is 335.2 pm). The X-ray data are consistent with the expectation that the negative charge resides mainly on a deprotonated enolic oxygen atom rather than on the ylidic carbon atom, whose $2p$ orbital does not overlap effectively with the $6d$ orbital of bismuth. Accordingly, formula I is a faithful representation of the structure in preference to II, III, and IV, as displayed below.

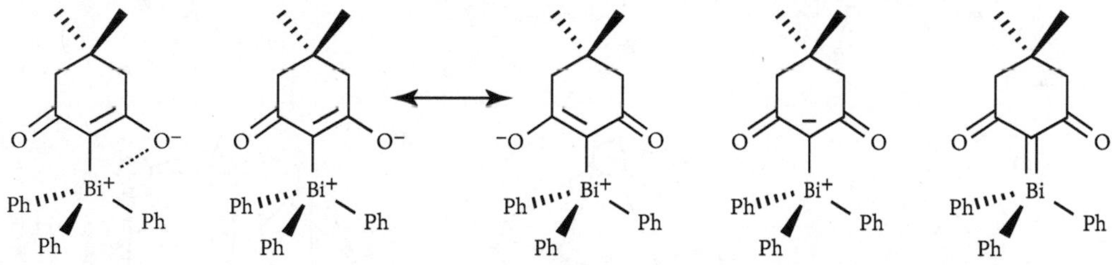

(I)   (II)   (III)   (IV)

From 1995 onward, studies have led to the synthesis and structural characterization of stable tungsten complexes with a heavier Group 15 element functioning as a triple-bonded terminal ligand. In the series of complexes $[(CH_2CH_2NSiMe_3)_3N]W\equiv E$ (E = P, As, Sb), the tungsten atom exhibits a distorted trigonal bipyramidal coordination geometry with three equatorial N atoms and one N atom and the E atom occupying the axial positions. The molecular structure and $W\equiv E$ bond distances are shown below.

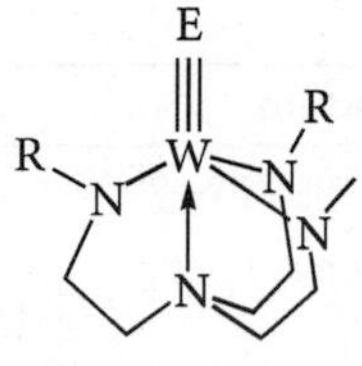

R = SiMe$_3$
$W\equiv P$ 216.2(4) pm
$W\equiv As$ 229.0(1)
$W\equiv Sb$ 252.6(2) pm

5.8.2 **Metal–metal bonds and clusters**

Many compounds containing M–M bonds or stable rings and clusters of Group 15 elements are known. Figs 5.8.2(a)–(c) show the structures of some organometallic compounds of As, Sb and Bi, which contain M–M bonds, and Figs. 5.8.2(d)–(e) show the structures of naked cluster cations Bi_n^{m+}, which are the components of some complex salts of bismuth. The structure of $As_6(C_6H_5)_6$ illustrates the typical trigonal pyramidal environment of the As atom; the As–As bond length is 246 pm, in which the As_6 ring adopts a chair conformation. In $Sb_4(\eta^1\text{-}C_5Me_4)_4$, Sb_4 forms a twisted ring, with Sb–Sb 284 pm, and all Sb–Sb–Sb bond angles are acute. Tetrameric bis(trimethylsilyl)methylbismuthine $[(Me_3Si)_2CHBi]_4$ contains a folded four-membered metallacycle with fold angles of 112.6° and 112.9°, Bi–Bi bond lengths in the range 297.0–304.4 pm, and Bi–Bi–Bi angles in the range 79.0–79.9°. The discrete molecules As_2Ph_4, Sb_2Ph_4, and Bi_2Ph_4 all adopt the staggered conformation, and the bond lengths are As–As 246 pm, Sb–Sb 286 pm, and Bi–Bi 299 pm

Dibismuthenes of the general formula LBi=BiL can be synthesized when L is a bulky aryl ligand. The Lbi=BiL molecule is centrosymmetric and therefore exists in the *trans* configuration. For L = 2,4,6-tris[bis(trimethylsilyl)methyl]phenyl, X-ray analysis of the dibismuthene yielded a Bi=Bi double bond length of 282.1 pm and a Bi=Bi–C angle of 100.5°.

The structures of cationic bismuth clusters Bi_5^{3+}, Bi_8^{2+}, and Bi_9^{5+} are listed in Table 5.8.2.

Metalloid and intermetalloid clusters of Group 13 and 14 elements have been described in the two preceding chapters. For Group 15 elements, the ligand-free intermetalloid clusters $[As@Ni_{12}@As_{20}]^{3-}$

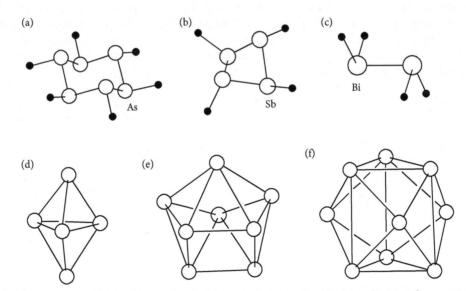

Fig. 5.8.2 Structures of (a) $As_6(C_6H_5)_6$, (b) $Sb_4(\eta^1\text{-}C_5Me_4)_4$, (c) $Bi_2(C_6H_5)_4$, (d) Bi_5^{3+}, (e) Bi_8^{2+}, and (f) Bi_9^{5+}. In (a)–(c) the phenyl ligands are represented by C atoms that are bonded to the metal skeleton.

Table 5.8.2 Cationic bismuth clusters

Cation	Crystal	Structure	Symmetry
Bi_5^{3+}	$Bi_5(AlCl_4)_3$	trigonal bipyramidal	D_{3h}
Bi_8^{2+}	$Bi_8(AlCl_4)_2$	square antiprism	D_{4h}
Bi_9^{5+}	$Bi_{24}Cl_{28}$, or $(Bi_9^{5+})_2(BiCl_5^{2-})_4(Bi_2Cl_8^{2-})$	tricapped trigonal prism	C_{3h} (~D_{3h})

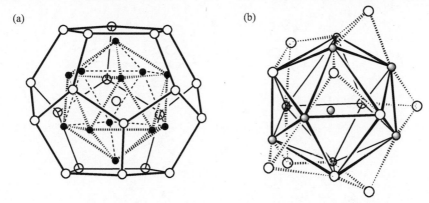

Fig. 5.8.3 (a) Structure of the $[As@Ni_{12}@As_{20}]^{3-}$ cluster; the solid and open circles represent Ni and As atoms, respectively. (b) Structure of the $[Zn@Zn_8Bi_4@Bi_7]^{5-}$ cluster; the shaded and open circles represent Zn and Bi atoms, respectively. Interatomic distances: Zn_{cent}–Zn_{ico} 283.2(2)–359.4(3), Zn_{cent}–Bi_{ico} 282.2(2)–292.8(2), Zn_{ico}–Bi_{ico} 287.6(2)–375.5(2), Bi_{ico}–Bi_{ico} 320.5(1), Zn_{ico}–Zn_{ico} 289.4(2)–376.2(2), Bi_{cap}–Zn_{ico} 259.8(1)–276.8(1), Bi_{cap}–Bi_{ico} 310.7(1)–331.2(1) pm; the subscripts cent, ico, and cap denote atoms occupying the central, icosahedral and capping sites, respectively.

and $[Zn@Zn_8Bi_4@Bi_7]^{5-}$ are known. In $[As@Ni_{12}@As_{20}]^{3-}$, a central As atom is located inside a Ni_{12} icosahedron, which is in turn enclosed by a As_{20} pentagonal dodecahedron, as shown in Fig. 5.8.3(a).

The first example of an intermetalloid cluster that almagamates eleven bismuth and nine zinc atoms has been found in the complex $[K(2.2.2\text{-crypt})]_5[Zn_9Bi_{11}]\cdot 2en\cdot toluene$ (2.2.2-crypt = 4,7,13,16,21,24-hexaoxa-1,10-diazabicyclo[8.8.8]hexacosane). The naked $[Zn@Zn_8Bi_4@Bi_7]^{5-}$ cluster consists of a central Zn atom trapped inside a distorted icosahedron whose vertices are 8 Zn atoms and 4 Bi atoms, with seven of the 20 triangular faces each capped by a Bi atom, as shown in Fig. 5.8.3(b).

A simplified model may be used to rationalize the bonding in the heteroatomic species $[Zn@Zn_8Bi_4@Bi_7]^{5-}$. According to the electron counting theory proposed by Wade, the formation of a *closo* deltahedra of 12 vertices is stabilized by 13 skeletal electron pairs. The total of 26 electrons required for skeletal bonding may be considered to be provided as follows: two from the interstitial Zn atom, $(8 \times 0 + 4 \times 3 = 12)$ from the Zn_8Bi_4 icosahedral unit (each vertex atom carries an *exo* lone pair or bond pair), seven from the capping Bi atoms, and five from the negative charges. Note that each vertex Bi atom is considered to retain one lone pair, whereas each capping Bi atom withholds two lone pairs.

5.8.3 Intermolecular interactions in organoantimony and organobismuth compounds

Intermolecular interactions have long been known to exist in inorganic compounds of antimony and bismuth. For example, $SbCl_3$ has three Sb–Cl bonds of length 234 pm and five Sb$\cdots$Cl distances in the range 346–374 pm, thus enlarging the coordination sphere from trigonal pyramidal to (3 + 5) bicapped trigonal prismatic geometry.

The organoantimony and organobismuth compounds in oxidation states I–III also exhibit strong intermolecular interactions, which lead to secondary bonds between molecules in forming chain-like, layer-type, and three-dimensional supramolecular structures. Fig. 5.8.4 shows the structures of three organometallic compounds of Sb and Bi.

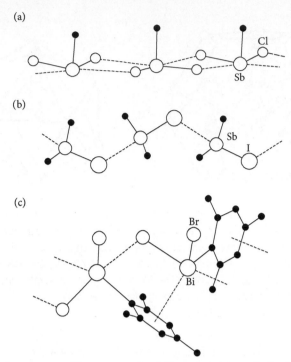

Fig. 5.8.4 Structures of organometallic compounds of Sb and Bi: (a) MeSbCl$_2$, (b) Me$_2$SbI, (c) MesBiBr$_2$ (Mes = mesityl).

(1) MeSbCl$_2$

Crystalline MeSbCl$_2$ contains alternating layers each consisting of double chains of MeSbCl$_2$ molecules. Along a MeSbCl$_2$ chain, the bond lengths and angles are: Sb–Cl, 236.8 and 243.0 pm; Sb⋯Cl, 333.7 and 386.5 pm; Sb–Cl⋯Sb (within chain), 102.9°.

(2) Me$_2$SbI

In the Me$_2$SbI chain, the bond lengths are Sb–I 279.9 pm and Sb⋯I 366.6 pm, so that the short and long Sb–I bonds are clearly different. The chain atoms lie almost in a plane, while the methyl groups are directed to one side of this plane. The reverse side is exposed to neighboring chains with very weak inter-chain Sb⋯I contacts (402.4–416.7 pm) that are close to the van der Waals separation. This type of chain packing leads to the formation of double layers.

(3) MesBiBr$_2$

In the chain structure of MesBiBr$_2$, the Bi and bridging Br atoms constitute a zigzag chain, with the non-bridging Br atoms lying on one side and the mesityl groups on the other side of the plane as defined by the positions of the chain atoms. The structure is stabilized by π-interaction between the Bi atom and the non-coordinated mesityl group. The bond lengths are: Bi–Br, 261.9 pm and 281.8 pm; Bi⋯Bi, 301.7 pm and 302.2 pm; Bi–mesityl ring centroid, 319.5 pm and 330.1 pm

5.9 σ-Hole Interaction in Pnictogen Compounds

Pnictogen bond (abbreviated as PnB) is mentioned when σ-hole interaction is discussed in Section 4.10.1. This section provides real examples to illustrate such intermolecular interactions in organophosphorus compounds.

TMEDA (N, N, N', N'-tetramethylethylenediamine) and PBr_3 molecules undergo donor-receptor complexation to form a 1:1 complex, as shown in Fig. 5.9.1(a). Shortening of the P–Br^- distance is related to the action of PnB. It can be seen that one P–Br bond maintains its original covalent character, but there is P⋯Br^- contact involving the two bromide ions. The distances are only 77% and 84% of the sum of van der Waals radii of normal P⋯Br^-, indicating that the three atoms N^+–P⋯Br^- adopt a closely linear arrangement caused by the PnB interaction.

The heavier pnictogen element Sb(III) also tends to form supramolecular interactions rather than donor covalent bonds. In the structure shown in Fig. 5.9.1(b), the host molecule containing the bicentered Sb(III) cations interacts with the oxygen atoms of the anionic triflate to form two C–Sb⋯O PnB interactions. The distances of Sb(III)⋯O^- are only 81% and 85% of the sum of their van der Waals radii, and the three atoms (C–Sb⋯O) are close to a linear arrangement.

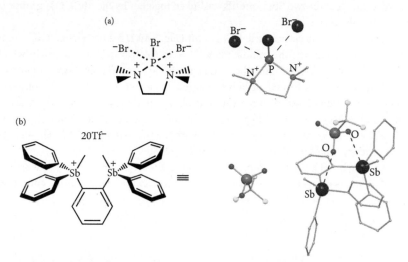

Fig. 5.9.1 (a) The donor–acceptor complex composed of TMEDA and PBr_3 molecules. (b) Pnictogen interaction between bicenterd Sb(III) cation complex and a pair of triflate oxygen atoms.

5.10 Bismuth and Antimony Polyanions

In 2000, the first naked double bond between two heavy atoms was found in Bi_2^{2-}, which does not need stablilization by bulky supporting ligands.

A precursor compound $K_5In_2Bi_4$ made by heating a stoichiometric mixture of the elements, together with crypt [crypt = {2,4,6-$Me_3C_6H_2$}$_2H_3C_6$], were dissolved into ethylenediamine to form a green-blue solution, which was then carefully layered with toluene. In time crystals of (K-crypt)$_2Bi_2$ (**1**) grew on the walls as toluene slowly diffused into ethylenediamine, and the product usually contained two other phases: (K-crypt)$_2$(InBi$_3$) (**2**) with [InBi$_3$]$^{2-}$ tetrahedra (~80%) and (K-crypt)$_6$(In$_4Bi_5$)$_2$ (**3**) with [In$_4Bi_5$]$^{3-}$ monocapped square antiprisms (~10%). The synthesis is reproducible and can be successfully carried out repeatedly.

X-Ray crystallography established that **1** features the first naked diatomic dianion of bismuth Bi_2^{2-} with a short Bi=Bi double bond of 283.77(7) pm. This discovery violates the commonly recognized dictum that double bonds between heavy elements require bulky ligands for kinetic stabilization. Of note is that the kinetic stabilization of $[Bi=Bi]^{2-}$ is provided by the bulky counterion $(K\text{-crypt})^+$ in the salt $(K\text{-crypt})_2Bi_2$, rather than bulky ligands that bind dibismuthene.

Pnictogens (group 15 elements Pn) excluding nitrogen are notable for their capability to form (poly)cyclic polyanions such as Pn_4^{2-}, Sb_5^{5-}, Pn_7^{3-}, Sb_8^{8-}, Pn_{11}^{3-}, Pn_{14}^{4-}, and Pn_{21}^{3-}. Square aromatic Pn_4^{2-} has a Pn–Pn bond order of 1.25. The polyantimonide anions Sb_{10}^{2-} and Sb_2^{2-} coexist in $[K(18\text{-crown-6})]_6[Sb_{10}][Sb_2Mo(CO)_3]_2$ (**2**). The doubly bonded Sb_2^{2-} anion is kinetically stabilized by metal carbonyl in $[Sb_4\{Mo(CO)_3\}_2]^{4-}$, rather than by bulky counterion as in the case of $(K\text{-crypt})_2Bi_2$.

Compound **2** was obtained from the reaction of KMnSb with $(C_7H_8)Mo(CO)_3$ in the presence of 18-crown-6, and a little amount of 2,2,2-cryptand was added to promote dissolution in ethylenediamine (en) at room temperature. X-ray structural analysis of **2** revealed a Sb_2Sb_2 rectangle, which is bicapped by two $Mo(CO)_3$ fragments from above and below it to form centrosymmetric $[Sb_2\{Mo(CO)_3\}]_2^{4-}$ (Fig. 5.10.1(a)). Unlike square-planar Sb_4^{2-}, the pair of Sb_2 units in **2** has a rectangular configuration $[2 \times 90.26(2)°, 2 \times 89.74(2)°]$ but remarkably different side lengths [Sb1–Sb2, 271.1(2) pm, Sb1–Sb2′, 364.5(2) pm], revealing that the two Sb_2^{2-} motifs are linked together by the $Mo(CO)_3$ groups positioned above and below the Sb_4 plane.

The crystallographically centrosymmetric Sb_{10}^{2-} unit (Fig. 5.10.1(b)) has virtual C_{2h} symmetry with the Sb3–Sb4–Sb5 triangle lying in the middle plane. It is an electron-precise cluster in which eight 3-bonded vertices (Sb1, Sb2, Sb4, Sb5 and their equivalents) are neutral, and two 2-connected atoms (Sb3 and its equivalent) carry formal negative charges. It can be considered as composed of two norbornadiene-like Sb_7 units fused through the four Sb atoms through which they characteristically would datively bond to $Mo(CO)_3$. The Sb–Sb bonds fall into three categories: three short bonds (Sb1–Sb2, Sb1–Sb5′, Sb2–Sb5); two middle ones associated with the two-connected Sb3 (Sb3–Sb4, Sb3–Sb5); and two long secondary bonds involving Sb4 (Sb1–Sb4, Sb2–Sb4′). Such categories may be understood on the basis that a long bond is usually coupled with a short one for a single atom. It is worth noting that the formation of Sb_{10}^{2-} is energetically not favorable compared to the known Sb_{11}^{3-} cluster anion, as in the case of As_{10}^{2-}.

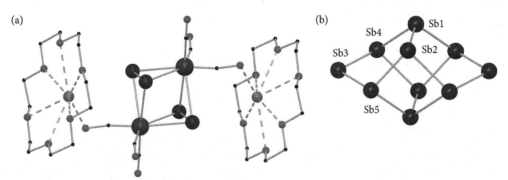

Fig. 5.10.1 (a) Centrosymmetric $[Sb_4\{Mo(CO)_3\}_2]^{4-}$ anion in $[K(18\text{-crown-6})]_6[Sb_{10}]$ $[Sb_4\{Mo(CO)_3\}_2]$ (**2**); it is sandwiched by two $[K(18\text{-crown-6})]^+$ units. Selected interatomic distances (pm) and angles (°): Sb1–Sb2 2.711(2), Mo1–Sb1 306.0(2), Mo1–Sb2 296.6(2), Mo–C 194.(2), 179.(2), 183.(3). (b) Centrosymmetric Sb_{10}^{2-} anion in **2**.

References

L. Xu, S. Bobev, J. El-Bahraoui, and S. C. Sevov, A naked diatomic molecule of bismuth, $[Bi_2]^{2-}$, with a short Bi-Bi bond: synthesis and structure. *J. Am. Chem. Soc.* **122**, 1838–9 (2000).

H. Ruan, L. Wang, Z. Li, and L. Xu, Sb_{10}^{2-} and Sb_2^{2-} found in $[K(18\text{-}crown\text{-}6)]_6[Sb_{10}][Sb_4\{Mo(CO)_3\}_2]\cdot 2en$: two missing family members. *Dalton Trans.* **46**, 7219–22 (2017).

5.11 **Triarylhalostibonium Cations**

$Mes_3Sb(OTf)_2$ (**1**) was conveniently obtained by the reaction of Mes_3SbBr_2 with AgOTf in CH_2Cl_2. It reacts with Mes_3SbF_2 in CH_2Cl_2 to afford $Mes_3SbF(OTf)$ (**2**). The antimony center in **1** (Fig. 5.11.1a) adopts trigonal-bipyramidal geometry with the triflate anions occupying the apical sites. The trigonal-pyramidal structure of **2** has the fluoride and triflate ligands occupying the apical sites (Fig. 5.11.1b). Notably, the Sb–O bond distance in **2** [217.8(2) pm] is almost equal to those in **1** [217.1(1) pm] despite the larger steric demand of two mesityl substituents in the latter complex.

The reaction of Mes_3SbCl_2 with $SbCl_5$ in CH_2Cl_2 gave $[Mes_3SbCl]SbCl_6$ (**3**) in 75% yield. X-ray crystal structure analysis revealed that **3** exists as a salt with no short contact between the trimesitylchlorostibonium cation and the hexachloroantimonate anion.

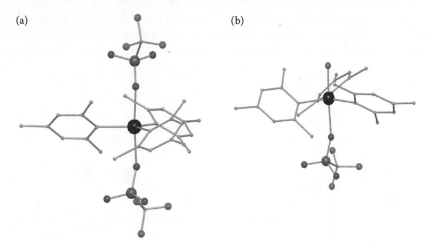

(a)   (b)

Fig. 5.11.1 Molecular structure of (a) $Mes_3Sb(OTf)_2$ (**1**) and (b) $Mes_3SbF(Otf)$ (**2**).

Reference

M. Yang and F. P. Gabbaï, Synthesis and properties of triarylhalostibonium cations, *Inorg. Chem.* **56**, 8644–50 (2017).

5.12 **Paddlewheel 1,2,4-diazaphospholide Distibines**

Room-temperature reaction between $SbCl_3$ and potassium 1,2,4-diazaphospholide $K[3,5\text{-}R_2dp]$ (R = ^tBu(**1**), ^iPr(**2**), or Cy(**3**)) at a ratio of 1:3 in THF resulted in a wine-red solution, from which the respective paddlewheel distibine $\{(\eta^1,\eta^1\text{-}3,5\text{-}R_2dp)_2(Sb\text{–}Sb)(\eta^1,\eta^1\text{-}3,5\text{-}R_2dp)_2\}$ ($R = {}^tBu(\mathbf{4})$, ${}^iPr(\mathbf{5})$, or $Cy(\mathbf{6})$) was readily isolated as colorless crystals in moderate yield (46% for **4**, 42% for **5**, and 38% for **6**) (**Scheme 1**).

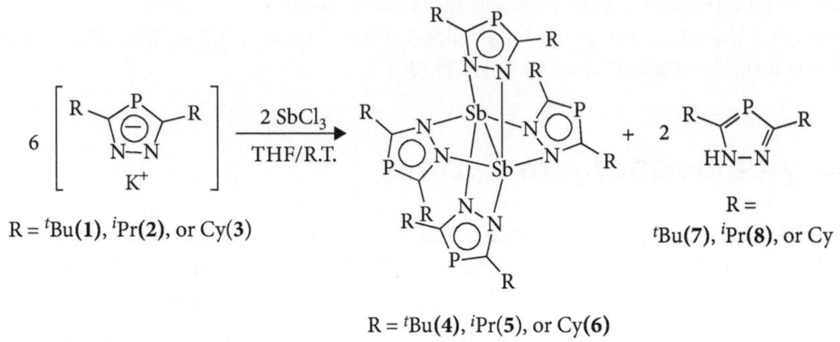

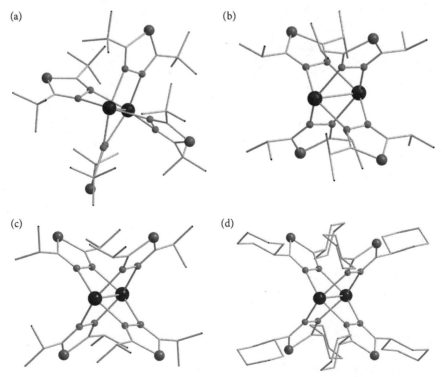

Scheme 1 Preparation of the paddlewheel distibines **4**, **5**, and **6**.

The distibine compounds crystallize in the centrosymmetric monoclinic space group $P2_1/c$ for **4** and triclinic space group $P\text{–}1$ for **5** and **6**. Complex **5** is composed of a packing of two similar but distinct Sb_2 molecules ($\mathbf{5\alpha}$, $\mathbf{5\beta}$). The molecular structures of **4**, $\mathbf{5\alpha}$, $\mathbf{5\beta}$, and **6** determined by X-ray crystallography are shown in Fig. 5.12.1, with an inversion center located at the midpoint of the Sb–Sb bond in each distibine.

The overall ligand arrangement around the Sb_2 core is an almost perfect paddlewheel for $\mathbf{5\alpha}$, $\mathbf{5\beta}$, and **6** but a twisted structure (owing to steric substituent repulsion) for **4**. Each antimony atom is coordinated

Fig. 5.12.1 Molecular structures of distibines (a) **4**, (b) $\mathbf{5\alpha}$, (c) $\mathbf{5\beta}$, and (d) **6**.

by the other Sb atom and four N atoms in slightly distorted square-pyramidal geometry. The average Sb–N bond distance was *ca.* 239.3(6) pm for **4**, 239.9(5) pm for **5α**, 237.1(6) pm for **5β**, and 234.8(8) pm for **6**. These Sb–N bond lengths are close to that found for the distibenium dication (Sb–N 237.2(2) pm). The Sb–Sb distances (266.91(8) pm in **4**, 274.51(8) pm in **5α**, 274.07(8) pm in **5β**, and 273.99(5) pm in **6**) are much shorter than that found in [Sb_2Me_4] (286.2(2) pm) and other distibines (*ca.* 285 pm). Strikingly, the very short Sb–Sb bond (266.91(8) pm) in **4** is even close to the Sb=Sb double bond distances of *ca.* 265 pm in distibenes TbtSb=SbTbt (264.2(1) pm), 2,6-Mes_2–H_3C_6Sb=SbC_6H_3–2,6-Mes_2 (265.58(5) pm), and $L^\dagger$SbSb$L^\dagger$ [$L^\dagger$ = –N($Ar^\dagger$)-(Si^{i}Pr$_3$), $Ar^\dagger$ = C_6H_2(C(H)Ph$_2$)$_2$^{i}Pr-2,6,4) (271.04(5) pm].

Reference

M. Zhao L. Wang, X. Zhanga, and W. Zheng, Paddlewheel 1,2,4-diazaphospholide distibines with the shortest antimony–antimony single bonds. *Dalton Trans.* **45**, 10505–9 (2016).

General References

D. E. C. Corbridge, *Phosphorus: An Outline of its Chemistry, Biochemistry and Technology*, 5th ed., Elsevier, Amsterdam, 1995.

M. Regitz, and O. J. Scherer (eds.), *Multiple Bonds and Low Coordination in Phosphorus Chemistry*, Verlag, Stuttgart, 1990.

A. Durif, *Crystal Chemistry of Condensed Phosphate*, Plenum Press, New York, 1995.

K. B. Dillon, F. Mathey, and J. F. Nixon, *Phosphorus: The Carbon Copy*, Wiley, Chichester, 1998.

J.–P. Majoral (ed.), *New Aspects in Phosphorus Chemistry I*, Springer, Berlin, 2002.

G. Meyer, D. Neumann, and L. Wesemann (eds.), *Inorganic Chemistry Highlights*, Wiley-VCH, Weimheim, 2002.

S. M. Kauzlarich (ed.), *Chemistry, Structure and Bonding of Zintl Phases and Ions*, VCH, New York, 1996.

M. Gielen, R. Willem, and B. Wrackmeyer (eds.), *Unusual Structures and Physical Properties in Organometallic Chemistry*, Wiley, West Sussex, 2002.

H. Suzuki and Y. Matano (eds.), *Organobismuth Chemistry*, Elsevier, Amsterdam, 2001.

K. O. Christe, W. W. Wilson, J. A. Sheehy, and J. A. Boatz, N_5^+: a novel homoleptic polynitogen ion. *Angew. Chem. Int. Ed.* **38**, 2004–9 (1999).

B. A. Mackay and M. D. Fryzuk, Dinitrogen coordination chemistry: on the biomimetic borderlands. *Chem. Rev.* **104**, 385–401 (2004).

E. Urnezius, W. W. Brennessel, C. J. Cramer, J. E. Ellis, and P. von R. Schleyer, A carbon-free sandwich complex [(P_5)$_2$Ti]$^{2-}$. *Science* **295**, 832–4 (2002).

M. Peruzzini, L. Gonsalvia, and A. Romerosa, Coordination chemistry and functionalization of white phosphorus via transition metal complexes. *Chem. Soc. Rev.* **34**, 1038–47 (2005).

J. G. Cordaro, D. Stein, H. Rüegger, and H. Grützmacher, Making the true "CP" ligand. *Angew. Chem. Int. Ed.* **45**, 6159–62 (2006).

J. Bai, A. V. Virovets, and M. Scheer, Synthesis of inorganic fullerene-like molecules. *Science* **300**, 781–3 (2003).

M. J. Moses, J. C. Fettinger, and B. W. Eichhorn, Interpenetrating As_{20} fullerene and Ni_{12} icosahedra in the onion-skin [As@Ni_{12}@As_{20}]$^{3-}$ ion. *Science* **300**, 778–80 (2003).

J. M. Goicoechea and S. C. Sevov, [Zn_9Bi_{11}]$^{5-}$: a ligand-free intermetalloid cluster. *Angew. Chem. Int. Ed.* **45**, 5147–50 (2006).

Chapter 6

Structural Chemistry of Group 16 Elements

6.1 Dioxygen and Ozone

Oxygen is the most abundant element on the earth's surface. It occurs both in the free state and as a component in innumerable compounds. The common allotrope of oxygen is dioxygen (O_2) or oxygen gas; the other allotrope is ozone (O_3).

6.1.1 Structure and properties of dioxygen

Molecular oxygen (or dioxygen) O_2 and related species are involved in many chemical reactions. The valence molecular orbitals and electronic configurations of the homonuclear diatomic species O_2 and O_2^- are shown in Fig. 6.1.1.

In the ground state of O_2, the outermost two electrons occupy a doubly degenerate set of antibonding π^* orbitals with parallel spins. Dioxygen is thus a paramagnetic molecule with a triplet ground state ($^3\Sigma$), and its formal double bond has a length of 120.752 pm.

The first electronic excited state ($^1\Delta$) is 94 kJ mol^{-1} above the ground state and is a singlet with no unpaired electrons. This very reactive species, commonly called singlet oxygen, has a half-life of 2 μs in water. Its bond length is 121.563 pm. Another singlet ($^1\Sigma$) species, with paired electrons in the π_g orbitals, is higher in energy than the ground state by 158 kJ mol^{-1}, and its bond length is 122.688 pm. Of the two

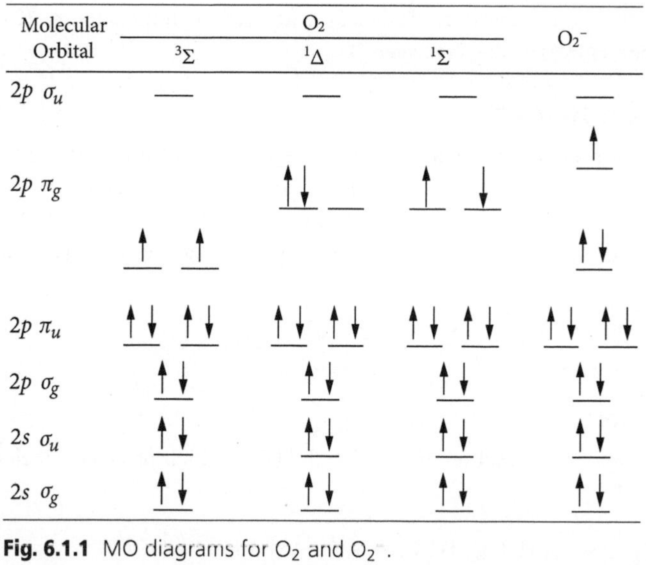

Molecular Orbital	O_2			O_2^-
	$^3\Sigma$	$^1\Delta$	$^1\Sigma$	
$2p\ \sigma_u$	—	—	—	—
$2p\ \pi_g$	↑ ↑	↑↓ __	↑ ↓	↑ / ↑↓
$2p\ \pi_u$	↑↓ ↑↓	↑↓ ↑↓	↑↓ ↑↓	↑↓ ↑↓
$2p\ \sigma_g$	↑↓	↑↓	↑↓	↑↓
$2s\ \sigma_u$	↑↓	↑↓	↑↓	↑↓
$2s\ \sigma_g$	↑↓	↑↓	↑↓	↑↓

Fig. 6.1.1 MO diagrams for O_2 and O_2^-.

Structural Chemistry across the Periodic Table. Thomas Chung Wai Mak, Yu-San Cheung, Gong-Du Zhou, and Ying-Xia Wang. Oxford University Press. © Thomas Chung Wai Mak, Yu-San Cheung, Gong-Du Zhou, and Ying-Xia Wang (2023). DOI: 10.1093/oso/9780198872955.003.0006

Table 6.1.1 Bond properties of dioxygen species

Species	Compound	Bond order	O–O/pm	Bond energy/kJ mol^{-1}	ν(O–O)/cm^{-1}
Oxygenyl O_2^+	$O_2[AsF_6]$	2.5	112.3	625.1	1858
Triplet $O_2(^3\Sigma)$	$O_2(g)$	2	120.752	490.4	1554.7
Singlet $O_2(^1\Delta)$	$O_2(g)$	2	121.563	396.2	1483.5
Singlet $O_2(^1\Sigma)$	$O_2(g)$	2	122.688	–	–
Superoxide O_2^-	KO_2	1.5	128	–	1145
Peroxide O_2^{2-}	Na_2O_2	1	149	204.2	842
—O—O—	H_2O_2 (cryst.)	1	145.8	213	882

singlet states only the lower $^1\Delta$ survives long enough to participate in chemical reactions. The higher $^1\Sigma$ state is deactivated to the O_2 ($^1\Delta$) state so rapidly that it has no significant chemistry.

Simple reduction of dioxygen to the superoxide ion, O_2^-, is thermodynamically unfavorable as it involves adding an electron to a half-filled antibonding orbital. Reduction to the peroxide ion, O_2^{2-}, however, is favorable from either superoxide or dioxygen. Owing to the increased population of antibonding orbitals in the reduced O_2 species (O_2^- and O_2^{2-}), both the bond order and O–O IR stretching frequency decrease, while the O–O distance increases. Table 6.1.1 lists relevant data on the bond properties of dioxygen species.

Owing to its non-polar nature, O_2 is more soluble in organic solvents than in H_2O. Since the $^3\Sigma$ ground state of dioxygen is a spin triplet, concerted oxygenation reaction is subject to spin restriction. As a result, many types of dioxygen reactions proceed slowly, even in cases where such reactions are strongly favored thermodynamically, such as a mixture of O_2 and H_2. The reactivity of dioxygen with singlet molecules can be increased by exciting it to its $^1\Delta$ singlet state, thereby removing the spin restriction. This energy corresponds to an infrared wavelength of 1270 nm. As singlet oxygen in solution deactivates by transferring its electronic energy to vibration of solvent molecules, its lifetime depends strongly on the medium. Solvents with high vibrational frequencies provide the most efficient relaxation. For this reason, the lifetime (about 2–4 μs) is shortest in water, which has a strong OH vibration near 3600 cm^{-1}. Solvents with CH groups ($\approx$ 3000 cm^{-1}) are the next most efficient, with lifetime in the range of 30–100 μs.

There are two major sources of singlet oxygen O_2 ($^1\Delta$):

(1) Photochemical sources

One of the most common sources of singlet oxygen is energy transfer from an excited "sensitizer" (Sens), such as a dye or natural pigment, to the ground state of the oxygen molecule. The role of the sensitizer is to absorb irradiation and be converted to an electronically excited state (Sens*). This state, normally a triplet, then transfers its energy to the oxygen molecule, producing singlet oxygen and regenerating the sensitizer, as indicated below:

$$\text{Sens} \xrightarrow{h\nu} \text{Sens}^* \xrightarrow{O_2} \text{Sens} + O_2(^1\Delta)$$

(2) Chemical sources

An early chemical source is the reaction of HOCl with H_2O_2, which produces singlet $O_2(^1\Delta)$ in nearly quantitative yield:

$$H_2O_2 + HOCl \longrightarrow O_2\left(^1\Delta\right) + H_2O + HCl$$

Another source is derived from phosphite ozonides as in the following reaction:

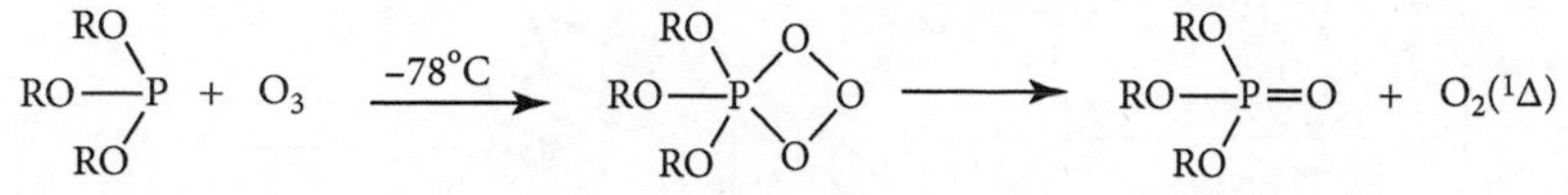

In addition to the above sources, a number of other chemical reactions have been suggested for the production of singlet oxygen.

Singlet oxygen $O_2(^1\Delta)$ is a useful synthetic reagent, which allows stereospecific and regiospecific introduction of O_2 into organic substrates. Since singlet oxygen $O_2(^1\Delta)$ has a vacant π_g orbital, similar to the frontier MO of ethylene, it behaves as an electrophilic reagent, with a reactivity pattern reminiscent of that exhibited by ethylene with electronegative substituents. For example, it undergoes [4+2] cycloadditions with 1,3-dienes, and [2+2] cycloadditions and ene reactions with isolated double bonds. Some examples are given below.

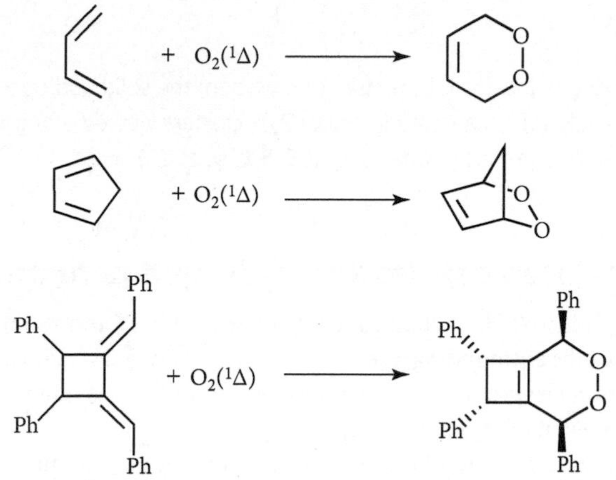

6.1.2 Crystalline phases of solid oxygen

X-ray studies of solid oxygen began in the 1920s, and presently six distinct crystallographic forms have been unequivocally identified. The α, β, and γ phases exist at ambient pressure and low temperature. At T = 295 K, oxygen solidifies to form the β phase at 5.4 GPa, which transforms into the orange δ phase at 9.6 GPa, which in turn is converted to the red ε phase at 10 GPa. Above 96 GPa, the ε phase undergoes transformation to the metallic ς phase.

The monoclinic α, rhombohedral β, and orthorhombic γ phases all have layered structures composed of O_2 molecules. Recently, single-crystal X-ray analysis has established that in the crystal structure of the ε phase at 13.2–17.6 GPa (space group $C2/m$ with $Z = 8$), four O_2 molecules associate into a rhombohedral $(O_2)_4$ cluster by serving as the short, parallel edges of a rhombic prism located at a site of symmetry $2/m$, which is held together by weak chemical bonds. The layer-type crystal structure is shown in Fig. 6.1.2. The O···O···O angles at the rhombic face O2-O3-O1A-O3B are 84.5° and 95.8°. The two independently measured O–O bond lengths are 120 pm and 121 pm, which are in agreement with that in the gas phase. The shortest intra-cluster, inter-cluster, and inter-layer O···O contacts are 218 pm, 259 pm, and 250 pm, respectively.

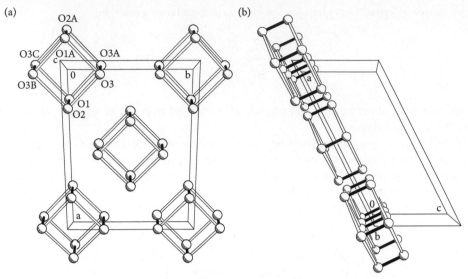

Fig. 6.1.2 Molecular packing in the crystal structure of ε-oxygen; the solid and open lines indicate covalent and weak bonds, respectively. (a) *C*-centered layer of $(O_2)_4$ clusters viewed along the *c*-axis; (b) the same layer viewed along the *b*-axis. Symmetry codes: A ξ, *y*, ζ; B *x*, ψ, *z*; C ξ, ψ, ζ.

6.1.3 **Dioxygen-related species and hydrogen peroxide**

The oxygenyl cation, O_2^+, is well characterized in both the gas phase and in salts with non-oxidizable anions. Removal of one of the antibonding π_g electrons from O_2 gives O_2^+ with a bond order of 2.5 and, consequently, a shorter bond length of 112.3 pm. The chemistry of O_2^+ is quite limited in scope because of the high ionization potential of O_2 (1163 kJ mol^{-1}).

The superoxide ion O_2^- is formed by adding one electron into the antibonding orbital π_g of O_2, thereby reducing the bond order to 1.5 and increasing the bond length to 128 pm, as shown in Table 6.1.1.

The peroxide ion, O_2^{2-}, has two more electrons than neutral dioxygen, and these additional electrons completely fill the π_g MO, resulting in a diamagnetic molecule with an O–O bond order of 1. The single bond character is evident from the longer bond length (about 149 pm) and from the lower rotational barrier of the peroxide bond. The peroxide anion is found in ionic salts with the alkali metals and heavier alkaline-earth metals such as calcium, strontium, and barium. These ionic peroxides are powerful oxidants, and hydrogen peroxide, H_2O_2, is formed when they are dissolved in water. Fig. 6.1.3 shows the molecular structure of H_2O_2.

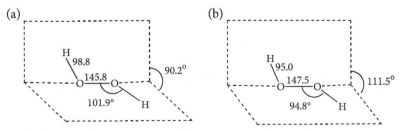

Fig. 6.1.3 Molecular structure of H_2O_2: (a) in the crystal and (b) in the gas phase (bond lengths in pm).

Table 6.1.2 Dihedral angle of H_2O_2 in some crystalline phases

Compound	Dihedral angle	Compound	Dihedral angle
H_2O_2 (s)	90.2°	$Li_2C_2O_4 \cdot H_2O_2$	180°
$K_2C_2O_4 \cdot H_2O_2$	101.6°	$Na_2C_2O_4 \cdot H_2O_2$	180°
$Rb_2C_2O_4 \cdot H_2O_2$	103.4°	$NH_4F \cdot H_2O_2$	180°
$H_2O_2 \cdot 2H_2O$	129°	Theoretical	90°–120°

Table 6.1.2 lists the values of the dihedral angle of H_2O_2 in some crystalline phases (a value of 180° corresponds to a planar *trans* configuration). This large range of values indicates that the rotational barrier is low and the molecular configuration of H_2O_2 is very sensitive to its surroundings.

Hydrogen peroxide has a rich and varied chemistry which arises from its versatility: (a) It can act either as an oxidizing or a reducing agent in both acid and alkaline media. (b) It undergoes proton acid/base reactions to form peroxonium salts $(H_2OOH)^+$, hydroperoxides $(OOH)^-$ and peroxides (O_2^{2-}). (c) It gives rise to peroxometal complexes and peroxoacid anions. (d) It can form crystalline addition compounds with other molecules through hydrogen bonding, such as $Na_2C_2O_4 \cdot H_2O_2$, $NH_4F \cdot H_2O_2$, and $H_2O_2 \cdot 2H_2O$.

The O–O bond length of 134 pm for the superoxide ion, O_2^-, was determined accurately for the first time in its $[1,3-(Me_3N)_2C_6H_4][O_2]_2 \cdot 3NH_3$ salt. Orientational disorder of the O_2^- ion that often occurs in crystals is avoided by using the bulky, non-spherical organic counter cation, which imposes order on the anionic sites.

6.1.4 Ozone

Ozone is a diamagnetic gas with a distinctive strong odor. The O_3 molecule (Fig. 6.1.4(a)) is angular, with an O–O bond length of 127.8 pm, which is longer than that of O_2 (120.8 pm.); the bond energy of 297 kJ mol^{-1} is lesser than that of O_2 (490 kJ mol^{-1}). The central O atom uses its sp^2 hybrid orbitals to form σ bonds with the terminal O atoms and to accommodate a lone pair, leaving the remaining p orbital for π bonding, as shown in Fig. 6.1.4(b). Each O–O bond has a bond order of 1.5. Ozone has a measured dipole moment of 1.73×10^{-30} C m (0.52 Debye) due to the uneven electron distribution over the central and terminal O atoms.

The deep red ozonide ion O_3^- has a bent structure, and sodium ozonide is isostructural with sodium nitrite ($NaNO_2$). The X-ray structural data of the alkali metal ozonides show that an increase in cationic size corresponds to a decrease in O–O bond length and an increase in O–O–O angle, as summarized in the following table.

MO_3	O–O/pm	O–O–O/°
NaO_3	135.3(3)	113.0(2)
KO_3	134.6(2)	113.5(1)
RbO_3	134.3(7)	113.7(5)
CsO_3	133.3(9)	114.6(6)

Since the additional electron enters an antibonding π∗ orbital, the bond order of O–O in O_3^- is 1.25 and the bond length increases. Ionic ozonides such as KO_3 are thermodynamically metastable and extremely sensitive to moisture and carbon dioxide.

Ozone is a major atmospheric pollutant in urban areas. In addition to its damaging effect on lung tissue and even on exposed skin surfaces, ozone attacks the rubber of tires, causing them to become brittle and

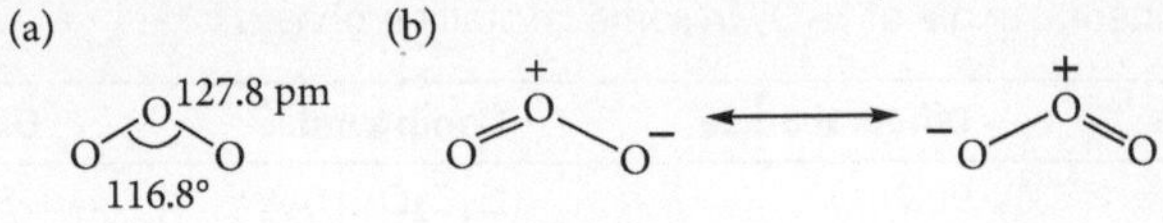

(a)   (b)

Fig. 6.1.4 The O_3 molecule: (a) geometry and (b) bonding.

crack. But in the stratosphere, where ozone absorbs much of the short wavelength UV radiation from the sun, it provides a vital protective shield for life forms on Earth.

Ozone has proven to be useful as a reagent to oxygenate organic compounds, a good example being the conversion of olefins to cleaved carbonyl products (Fig. 6.1.5). When O_3 reacts with olefins, the first product formed is a molozonide which in turn undergoes an O–O bond cleavage to give an aldehyde or a ketone plus a carbonyl oxide. The carbonyl oxide then decomposes to yield the products. But under appropriate conditions, the aldehyde or ketone will recombine with the carbonyl oxide to generate an ozonide, which then decomposes in the presence of H_2O to give two carbonyl compounds and hydrogen peroxide.

There exist only a few compounds which contain a linear chain of three oxygen atoms. The unstable compound bis(fluoroformyl)trioxide, F–C(O)–O–O–O–C(O)–F exists as the *trans-syn-syn* rotamer with C_2 symmetry in the crystalline state (with respect to the central O_3 fragment, the C–O bonds are *trans* and both C=O bonds are *cis*). The C–O–O–O torsion angle and O–O–O bond angle are 99.0(1) and 104.0(1)°, respectively, as shown in Fig. 6.1.6. In the stable molecule bis(trifluoromethyl)trioxide F_3C–O–O–O–CF_3, the corresponding values are 95.9(8) and 106.4(1)°, respectively.

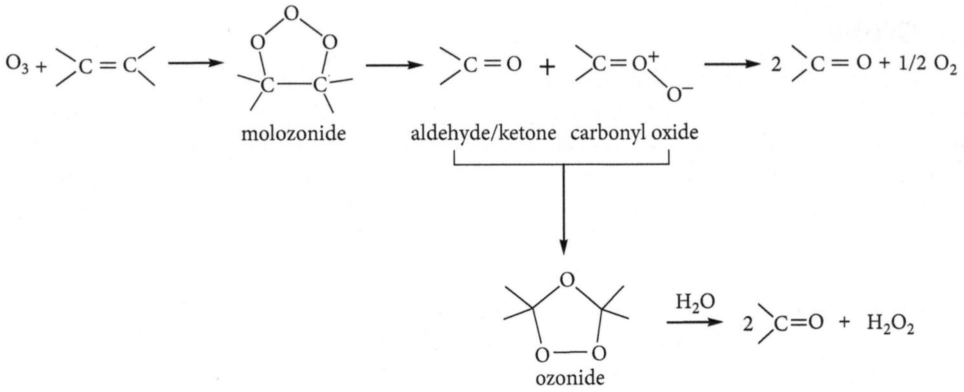

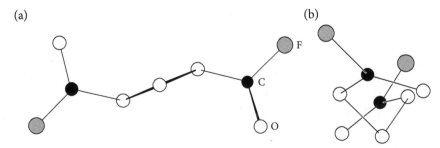

Fig. 6.1.5 Mechanism of reaction between O_3 and an olefin.

(a)   (b)

Fig. 6.1.6 Structure of the molecule FC(O)OOOC(O)F showing (a) its C_2 symmetry axis and (b) left-handed chiral conformation.

6.2 Oxygen and Dioxygen Metal Complexes

6.2.1 Coordination modes of oxygen in metal-oxo complexes

The structures of inorganic crystals are usually described by the packing of anions and metal coordination geometries. In this section, we place particular emphasis on the coordination environment of the oxygen atom. The oxygen atom (oxo ligand) exhibits various ligation modes in binding to metal centers, as shown in Table 6.2.1.

6.2.2 Ligation modes of dioxygen in metal complexes

In general, dioxygen metal complexes can be classified in terms of their mode of binding. With non-bridging O_2 ligands, both η^1 (superoxo) and η^2 (peroxo) complexes are known. Bridging dioxygen groups

Table 6.2.1 Coordination geometry of oxo ligand

Coordination geometry	Examples and notes
$M=O$	$[VO]^{2+}$, V=O bond length (155 pm to 168 pm) has partial triple bond character.
$M-O-M$	linear, ReO_3: Re–O–Re forms the edges of cubic unit cell.
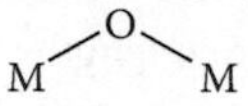 $\begin{array}{c}\quad O \\ M\diagup\;\diagdown M\end{array}$	bent, $Cr_2O_7^{2-}$: Cr–O bond length 177 pm, Cr–O–Cr bond angle 123°.
$O=M-O-M=O$	linear (in the presence of strong π bonding) and bent: these types exist in the compounds of Re and Tc; bond length M=O 165 pm to 170 pm, and M–O 190 pm to 192 pm.
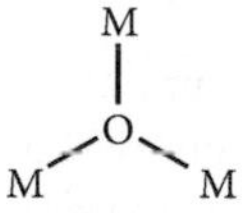 $\begin{array}{c}M \\ \mid \\ O \\ M\diagup\;\diagdown M\end{array}$	pyramidal, as in H_3O^+ and $[O(HgCl)_3]^+$. trigonal planar, as in $[Fe_3(\mu_3\text{-}O)(O_2CCMe_3)_6(MeOH)_3]^+$.
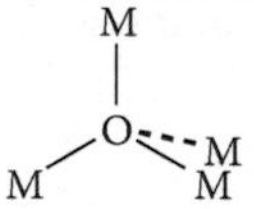 $\begin{array}{c}M \\ \mid \\ O\cdots M \\ M\diagup\;\diagdown M\end{array}$	tetrahedral, as in Cu_2O and $Be_4(\mu_4\text{-}O)(O_2CMe)_6$.
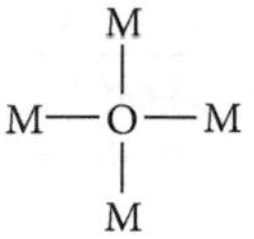 $\begin{array}{c}M \\ \mid \\ M-O-M \\ \mid \\ M\end{array}$	square planar, as in $[Fe_8(\mu_4\text{-}O)(\mu_3\text{-}O)_4(OAc)_8(tren)_4]^{6+}$.
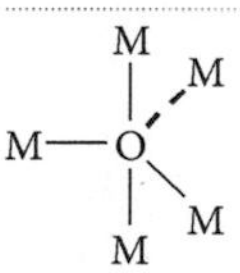 $\begin{array}{c}M\quad\;\;M \\ \;\diagdown\mid\diagup \\ M-O \\ \mid \\ M\end{array}$	trigonal bipyramidal, as in $[Fe_5(\mu_5\text{-}O)(O_2CMe)_{12}]^+$.
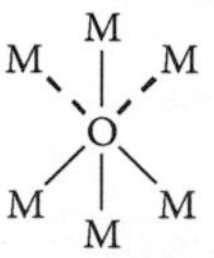 $\begin{array}{c}M\quad M\quad M \\ \;\diagdown\mid\diagup \\ O \\ \diagup\mid\diagdown \\ M\quad M\quad M\end{array}$	octahedral, as in $[Fe_6(\mu_6\text{-}O)(\mu_2\text{-}OMe)_{12}(OMe)_6]^{2-}$.

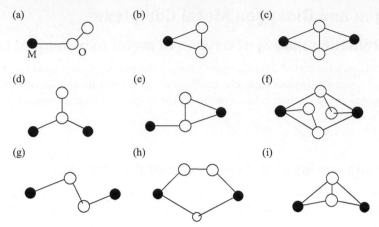

Fig. 6.2.1 Coordination modes of dioxygen metal complexes: (a) η^1-superoxo, (b) η^2-superoxo, (c) μ-η^2: η^2 peroxo, symmetric, (d) μ-η^1-superoxo, symmetrical, (e) μ-η^1: η^2 peroxo, (f) μ-η^1: η^2-peroxo, centrosymmetric, (g) *trans* μ-η^1: η^1, superoxo or peroxo, (h) *cis* μ-η^1: η^1, superoxo or peroxo, (i) μ-η^2: η^2 peroxo, unsymmetric.

can adopt η^1:η^1, η^1:η^2, and η^2:η^2 configurations. Fig. 6.2.1 shows the observed coordination modes in dioxygen metal complexes.

Another commonly adopted classification of dioxygen metal complexes is based on O–O bond lengths (referring to Table 6.2.2): superoxo complexes of types Ia and Ib, in which the O–O distance is roughly constant and close to the value reported for the superoxide anion (~130 pm); and types IIa and IIb, in which the O–O distance is close to the values reported for H_2O_2 and O_2^{2-} (~148 pm). The *a* or *b* classification distinguishes complexes in which the dioxygen is bound to one metal atom (type *a*) or bridging two metal atoms (type *b*).

The stretching frequencies attributed to the O–O vibration are closely related to the structural type. Type I complexes show O–O stretching vibrations around 1125 cm^{-1} and type II around 860 cm^{-1}. This sharp difference enables the O–O stretching frequency as measured by infrared or Raman spectroscopy to be used for structure type classification. Table 6.2.2 lists the properties of dioxygen metal complexes.

Table 6.2.2 Properties of dioxygen metal complexes

Complex	Type	O_2:M ratio	Structure	O–O/pm (normal range)	ν (O–O)/cm^{-1} (normal range)
Superoxo	I*a*	1:1	$M\!-\!O\overset{\cdots}{=}\!O$	125–135	1130–1195
Superoxo	I*b*	1:2	$M\diagdown O\overset{\cdots}{=}O\diagup M$	126–136	1075–1122
Peroxo	II*a*	1:1	$M\langle{}^{O}_{O}$	130–155	800–932
Peroxo	II*b*	1:2	$M\diagdown O\!-\!O\diagup M$	144–149	790–884

The first example of an end-on (η^1) (superoxo)copper(II) complex **1** that has its molecular structure determined by X-ray diffraction was reported in 2006. Bond lengths [pm] and angles [°] involving the monodentate superoxo ligand: Cu–O 192.7(2), O–O 128.0(3), Cu···O (non-bonded) 284.2(7), Cu–O–O 123.53(2).

1

2   AgOTf or Cu^{II} salts   **3**

In 2021, synthesis and structural characterization of the superoxo dicopper(II) species **3** together with its one-electron–reduced peroxo congener **2** (a rare cis-μ-1,2-peroxo dicopper(II) complex) was accomplished. Interconversion of **2** and **3** occurs at low potential (−0.58 V versus Fc/Fc$^+$) and is both chemically and electrochemically reversible. Comparison of X-ray crystallographic bond lengths [d(O–O) = 144.1(2) pm for **2** and 132.9(7) pm for **3**], and Raman spectra of crystalline **2**(BPh$_4$) [$\tilde{\nu}$ (^{16}O–^{16}O) = 793 cm^{-1}] versus 1073 cm^{-1} for a solution of **3** in MeCN, showed that the redox process occurs at the bridging O$_2$-derived unit. The superoxo Cu^{II}–O$_2$$^{·-}$–Cu^{II} complex **3** has a S = 1/2 spin ground state according to magnetic and EPR data, being in agreement with density-functional theory calculations. Further computation showed that the potential associated with changes of the Cu–O–O–Cu dihedral angle φ is shallow for both **2** (φ = 55.3°) and **3** (φ = 75.4°).

References

C. Würtele, E. Gaoutchenova, K. Harms, M. C. Holthausen, J. Sundermeyer, and S. Schindler, Crystallographic characterization of a synthetic 1:1 end-on copper dioxygen adduct complex, *Angew. Chem. Int. Ed.* **45**, 3867–9 (2006).

A. Brinkmeier, R. A. Schulz, M. Buchhorn, C.-J. Spyra, S. Dechert, S. Demeshko, V. Krewald, and F. Meyer, Structurally characterized μ-1,2-peroxo/superoxo dicopper(II) pair, *J. Am. Chem. Soc.* **143**, 10361–6 (2021).

6.2.3 **Biological dioxygen carriers**

The proteins hemoglobin (Hb) and myoglobin (Mb) serve to transport and to store oxygen, respectively, in all vertebrates. The active site in both proteins is a planar heme group (Fig. 6.6.2). Myoglobin is composed of one globin (globular protein molecule) and one heme group. Hemoglobin consists of four myoglobin-like subunits, two α and two β. In each subunit, the heme group is partially embedded in the globin, and an oxygen molecule can bond to the iron atom on the side of the porphyrin opposite the proximal histidine, thus forming a hexa-coordinate iron complex, as shown in Fig. 6.2.3.

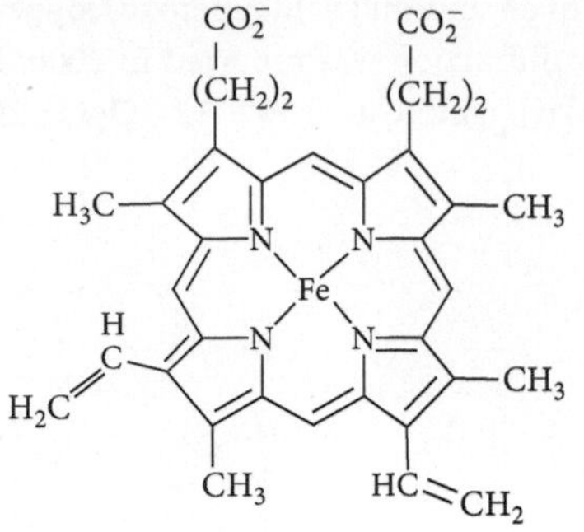

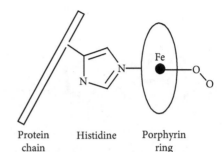

Fig. 6.2.2 The heme group.

Fig. 6.2.3 The binding of dioxygen by heme.

Protein chain   Histidine   Porphyrin ring

Myoglobin contains Fe^{2+} (d^6 configuration) in a high-spin state, which has a radius of about 78 pm in a pseudo-octahedral environment. The Fe^{2+} ion is too large to fit into the central cavity of the porphyrin ring, and lies some 42 pm above the plane of the N donor atoms. When a dioxygen molecule binds to Fe^{2+}, the Fe^{2+} becomes low-spin d^6 with a shrunken radius of only 61 pm, which enables it to slip into the porphyrin cavity.

In hemoglobin, O_2 binding to Fe^{2+} does not oxidize it to Fe^{3+}, as it is protected by protein units around the heme group. A non-aqueous environment is required for reversible O_2 binding.

The $O_2(^3\Sigma)$ and the high spin Fe^{2+} atom of the heme combine to form a spin-paired diamagnetic system. The resulting Fe–O–O is bent, with a bond angle varying from 115° to 153°. The stronger σ interaction is between the d_{z^2}(Fe) and $\pi_g(O_2)$ orbitals, and the weak π interaction is between d_{xz}(Fe) and $\pi_g{}^*(O_2)$. The increased ligand effect results in pairing of electrons and a weakened O–O bond.

Oxygen enters the blood in the lungs, where the partial pressure of O_2 is relative high (2.1×10^4 Pa) under ideal conditions; in the lungs with mixing of inhaled and non-exhaled gases, the O_2 partial pressure is ~1.3×10^4 Pa. It is then carried by red blood cells to the tissues where the O_2 partial pressure is lowered to $\approx 4 \times 10^3$ Pa. The reactions that occur are represented as follows:

$$\text{Lungs:} \quad \text{Hb} + 4O_2 \longrightarrow \text{Hb}(O_2)_4$$
$$\text{Tissues:} \quad \text{Hb}(O_2)_4 + 4\text{Mb} \longrightarrow 4\text{Mb}(O_2) + \text{Hb}$$

The four subunits ($2\alpha + 2\beta$) of hemoglobin function cooperatively. When one O_2 is bound to one heme group, the conformation of Hb subtly changes to make binding of additional oxygen molecules easier. Thus the four irons can each carry one O_2 with steadily increasing equilibrium constants. As a result, as soon as some O_2 has been bound to the molecule, all four irons readily become oxygenated (oxyHb). In a similar fashion, initial removal of oxygen triggers the release of the remainder, and the entire load of O_2 is delivered at the required site. This effect is also facilitated by pH changes caused by increased CO_2 concentration in the capillaries. As the CO_2 concentration increases, the pH decreases due to formation of $HCO_3{}^-$, and the increased acidity favors release of O_2 from oxyHb. This is known as the Bohr effect.

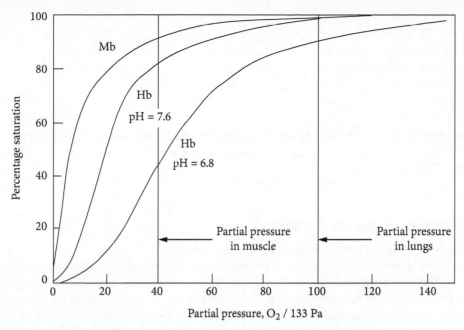

Fig. 6.2.4 The oxygen binding curves for Mb and Hb, showing the pH dependence for the latter.

Hemoglobin carries oxygen *via* the circulatory system from the lungs to body tissues. There the oxygen can be transferred to myoglobin, where it is stored for oxidation of foodstuffs. The properties of Hb and Mb can be represented by the oxygen-binding curves of Fig. 6.2.4. At high oxygen concentrations, Hb and Mb bind O_2 with approximately equal affinity. However, at low oxygen concentrations, such as in muscle tissue during or immediately after muscular activity, Hb is a poor oxygen binder and therefore can deliver its oxygen to Mb. The difference in O_2-binding ability between Mb and Hb is accentuated at low pH values; hence the transfer from Hb to Mb has a great driving force where it is most needed, viz., in tissues where oxygen has been converted to CO_2.

6.3 **Structure of Water and Ices**

Water is the most common substance on the earth's surface, and it constitutes about 70% of the human body and the food it consumes. It is also the smallest molecule with the greatest potential for hydrogen bonding. Water is the most important species in chemistry because a majority of chemical reactions can be carried out in its presence.

6.3.1 **Water in the gas phase**

The H_2O molecule has a bent structure, with O–H bond length 95.72 pm and bond angle H–O–H 104.52°; the O–H bonds and the lone pairs form a tetrahedral configuration, as shown in Fig. 6.3.1(a). They are capable of forming donor O–H···X and acceptor O···H–X hydrogen bonds (X is a highly electronegative atom).

In the gas phase, the adducts of H_2O with HF, HCl, HBr, HCN, HC≡CH, and NH_3 have been studied by microwave spectroscopy to determine the relationship of hydrogen-bond donor and acceptor. The structures of the adducts are H_2O:···H–F, H_2O:···H–Cl, and H_2O:···H–CN. Only with NH_3 is water a H donor: H_3N:···H–OH.

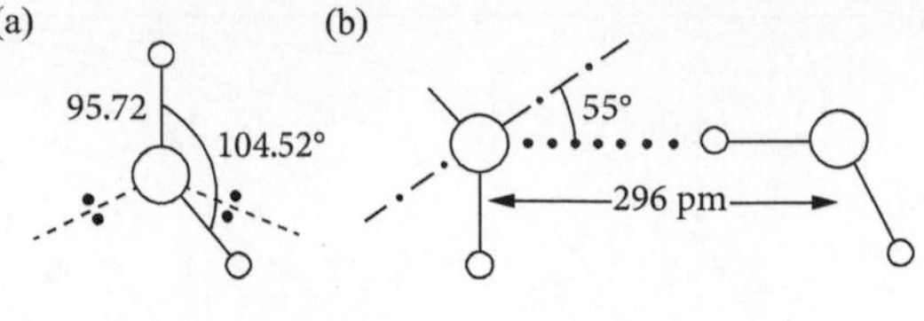

(a)   (b)

Fig. 6.3.1 Structures of (a) gaseous H_2O molecule and (b) water dimer (bond length in pm).

95.72   104.52°   55°

296 pm

Water forms a hydrogen-bonded dimer, the structure of which has been determined by microwave spectroscopy, as shown in Fig. 6.3.1(b). The experimental value of the binding energy of this dimeric system is 22.6 kJ mol^{-1}.

6.3.2 **Water in the solid phase: ices**

Solid H_2O has 11 known crystalline forms, as listed in Table 6.3.1.

Ordinary hexagonal ice I_h and cubic ice I_c are formed at normal pressures, the latter being stable below −120 °C. Their hydrogen bonding patterns are very similar, and only the arrangements of the oxygen atoms differ. In I_c, the oxygens are arranged like the carbon atoms in cubic diamond, whereas in I_h the atomic arrangement corresponds to that in hexagonal diamond. Fig. 6.3.2 shows the structure of I_h.

In both I_h and I_c, the hydrogen-bonded systems contain cyclohexane-like buckled hexagonal ring motifs: chair form in I_c and boat form in I_h. An important distinction is that in I_c all four hydrogen bonds are equivalent by symmetry. In I_h, the hydrogen bond in the direction of the hexagonal axis can be differentiated from the other three.

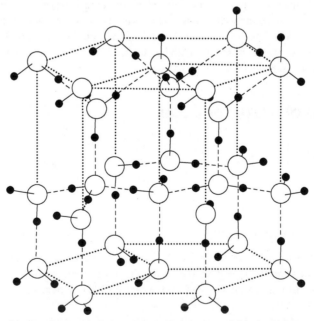

Fig. 6.3.2 Structure of ice I_h. (Large circle represents O atom, small black circle represents H atom, and broken line represents a disordered hydrogen bond, for which only one of the two arrangements, O–H⋯O and O⋯H–O, is shown).

A neutron diffraction analysis of H_2O and D_2O ice I_h at different temperatures has shown that, in addition to the disordered hydrogens, the oxygen atoms are also disordered about sites 6 pm off the hexagonal axis.

In the high-pressure ices the hydrogen-bonding patterns become progressively more complex with increasing pressure until the ultra-high-pressure ice polymorphs VII and VIII are formed. Ice II contains nearly planar hexagonal rings of hydrogen-bond O atoms arranged in columns. The O···O···O angles are greatly distorted from tetrahedral, with some close non-bonded O···O separation of 324 pm

Ice III is composed mainly from hydrogen-bonded pentagons with some heptagons, but with no hexagonal rings. Ice IX has the same structure as ice III, except that the hydrogen atoms are ordered in IX and disordered in III. The transformation III → IX occurs between 208 and 165 K.

Ice IV, which is metastable, is even more distorted with O···O···O angles between 88° and 128° and a large number of non-hydrogen-bonded O···O contacts between 314 pm and 329 pm. Some oxygen atoms

Table 6.3.1 Crystal structures of ice polymorphs

Ice	Space group	Parameters of unit cell/pm	Z	Hydrogen positions	Density /g cm^{-3}	Nearest non-bonding distance/ pm	Hydrogen bond distance/pm	Nearest non-bonding distance/ pm
I_h (273K)	$P6_3/mmc$ (No. 194)	$a = 451.35(14)$ $c = 735.21(12)$	4	disordered	0.92	4	275.2–276.5	450
I (5K)	$Cmc2_1$ (No. 36)	$a = 450.19(5)$ $b = 779.78(8)$ $c = 732.80(2)$	12	ordered	0.93	4	273.7	450
I_c (143K)	$Fd3m$ (No. 227)	$a = 635.0(8)$	8	disordered	0.93	4	275.0	450
II	$R\bar{3}$ (No. 148)	$a = 779(1)$ $\alpha = 113.1(2)°$	12	ordered	1.18	4	275–284	324
III	$P4_12_12$ (No. 92)	$a = 673(1)$ $c = 683(1)$	12	disordered	1.16	4	276–280	343
IV	$R\bar{3}c$ (No. 167)	$a = 760(1)$ $\alpha = 70.1(2)°$	16	disordered	1.27	4	279–292	314
V	$A2/a$ (No. 15)	$a = 922(2)$ $b = 754(1)$ $c = 1035(2)$ $\beta = 109.2(2)°$	28	disordered	1.23	4	276–287	328
VI	$P4_2/nmc$ (No. 137)	$a = 627$ $c = 579$	10	disordered	1.31	4	280–282	351
VII	$Pn3m$ (No. 224)	$a = 343$	2	disordered	1.49	8	295	295
VIII	$I4_1/amd$ (No. 141)	$a = 467.79(5)$ $c = 680.29(10)$	8	ordered	1.49	6	280–296	280
IX	$P4_12_12$ (No. 92)	$a = 673(1)$ $c = 683(1)$	12	ordered	1.16	4	276–280	351

lie at the centers of hexagonal rings. Ice V contains quadrilaterals, pentagons, and hexagons, in which respect it resembles some of the clathrate hydrate host frameworks.

It is of interest to note that ice VI is the first of the ice structures composed of two independent interpenetrating hydrogen-bonded frameworks (e.g., it is a self-clathrate).

Ice VII and VIII are constructed from much more regular interpenetrating frameworks, each with the I_c structure with little distortion from tetrahedral bonding configuration. In ice VII, each oxygen has eight nearest neighbors, four hydrogen-bonded at O···O = 288 pm, and four non-bonded at a shorter distance of 274 pm, demonstrating that it is easier to compress a van der Waals O···O distance than a O–H···O distance. Until the formation of interpenetrating frameworks, the principal effect of increasing pressure is to produce more complex hydrogen-bond patterns with large departures from tetrahedral coordination around the oxygen atoms; with the interpenetrating hydrogen bond frameworks, the patterns become much simpler.

A very high pressure form, ice X, which was predicated to have a H atom located midway between every pair of hydrogen-bonded O atoms, was identified by infrared spectroscopy with a transition at 44 GPa.

In addition to the various forms of ice, there are a large number of ice-like structures with four-coordinated water molecules that constitute a host framework, which are only stable when voids in the host framework are occupied by other guest molecules. Such compounds are known as clathrate hydrates.

6.3.3 **Structural model of liquid water**

Water is a unique substance in view of its unusual physical and chemical properties:

(a) The density of liquid water reaches a maximum at 4 °C.

(b) Water has a high dielectric constant associated with the distortion or breaking of hydrogen bonds.

(c) Water has relatively high electrical conductivity associated with the transfer of H_3O^+ and OH^- ions through the hydrogen-bonded structure.

(d) Water can be super-cooled and its fluidity increases under pressure.

(e) Water is an essential chemical for all life processes.

Many models for the structure of liquid water have been proposed. One of the most useful is the polyhedral model.

In liquid water, the thermal motions of molecules are perpetual, and the relative positions of the molecules are changing all the time. Although the structure of liquid water has no definite pattern, the hydrogen bonds between molecules still exist in large numbers. Thus liquid water is a dynamic system in which the H_2O molecules self-assemble in perfect, imperfect, isolated, linked, and fused polyhedra (Fig. 6.3.3), among which the pentagonal dodecahedron takes precedence.

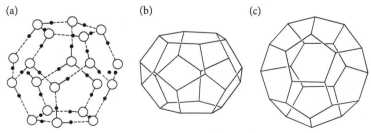

Fig. 6.3.3 Structures of some polyhedra for liquid water: (a) pentagonal dodecahedron (5^{12}), (b) 14-hedron ($5^{12}6^2$), (c) 15-hedron ($5^{12}6^3$).

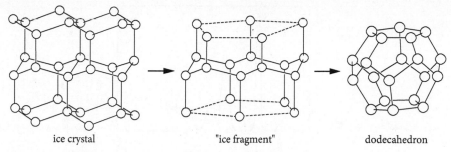

ice crystal   "ice fragment"   dodecahedron

Fig. 6.3.4 Schematic diagram for melting of ice to form water.

The ideal process of the melting of ice may be visualized as a partial (~15%) breakage of hydrogen bonds to form "ice fragments," which are then converted into a polyhedral system, as indicated in Fig. 6.3.4.

The polyhedral model is useful for understanding the properties of liquid water:

(1) Hydrogen bonds in liquid water

The geometry of a pentagonal dodecahedron is well suited to the formation of hydrogen bonds between molecules. The interior angle of a five-membered ring (108°) is close to the H–O–H bond angle (104.5°) of the water molecule and its tetrahedral charge distribution.

(2) Density of water

When "ice fragments" convert to the dodecahedral form, the packing of molecules becomes compact, so liquid water is denser than solid ice. In other circumstances, the density of water decreases when the temperature is raised. Water has its highest density at 4 °C.

(3) X-ray diffraction of liquid water

The "$g(r)$ vs r" diagrams from X-ray diffraction of ice and liquid water are shown in Fig. 6.3.5, in which $g(r)$ represents the oxygen atom-pair correlation function and r represents the distance between an oxygen atom pair. The diagram of liquid water shows that the strongest peak occurs at $r = 280$ pm, which corresponds to the distance of the hydrogen bond in the dodecahedral system. The next peak is at 450 pm, and the others lie in the range of 640–780 pm. There is no peak in the ranges of 280–450 pm and 450–640 pm. In the pentagonal dodecahedron the distances between a vertex and its neighbors are 1 (length of the edge, which is taken as the unit length), τ, $\sqrt{2}\tau$, τ^2, $\sqrt{3}\tau$, with $\tau = 1.618$. The calculated distances of oxygen atom pairs are 280, 450, 640, 730, and 780 pm, which fit the X-ray diffraction data of liquid water.

The diagram of ice-I_h has a distinct peak at $r = 520$ pm, which does not occur in liquid water. At 77 K, the "$g(r)$ vs r" diagrams of amorphous water solid and liquid water are very similar to each other.

(4) Thermal properties

The relations between the thermal properties of ice, liquid water and steam are indicated below:

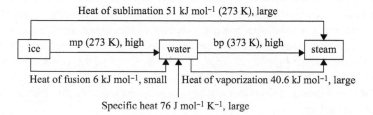

Heat of sublimation 51 kJ mol^{-1} (273 K), large

| ice | mp (273 K), high | water | bp (373 K), high | steam |

Heat of fusion 6 kJ mol^{-1}, small   Heat of vaporization 40.6 kJ mol^{-1}, large

Specific heat 76 J mol^{-1} K^{-1}, large

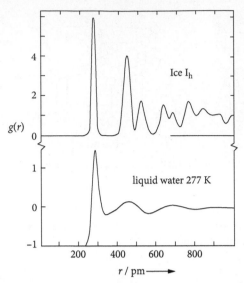

Fig. 6.3.5 The "*g(r)~r*" diagrams of ice-I$_h$ (top) and liquid water at 277 K (bottom). [Here *g(r)* represents the oxygen atom-pair correlation function and *r* represents the distance of an oxygen atom-pair. (Taken from A. H. Narten, C. A. Venkatesh, and S. A. Rice, *J. Chem Phys.* **64**, 1106 (1976).)

These properties show that in liquid water there still exist a large number of intermolecular hydrogen bonds, thus lending support to the polyhedral model.

(5) Formation of clathrate hydrates

Many gases, such as Ar, Kr, Xe, N$_2$, O$_2$, Cl$_2$, CH$_4$, and CO, can be crystallized with water to form ice-like clathrate hydrates. The basic structural components of these hydrates are the (H$_2$O)$_{20}$ pentagonal dodecahedron and other larger polyhedra bounded by five- and six-membered hydrogen-bonded rings, which can accommodate the small neutral molecules. The inclusion properties of water imply that such polyhedra are likely to be present in liquid water as its structural components.

Recently there have been extensive studies on the structure and property of water. In December, 2004, upon assessing the major achievements of the past 12 months, *Science* declared that the study of water as one of the Breakthroughs of the Year. (Specifically, it was listed as No. 8.)

In the structural study of gaseous protonated water clusters, H$^+$(H$_2$O)$_n$, with n = 6 – 27, it was found that the mass peak with n = 21 is particularly intense, signifying a "magic number" in the spectrum. Furthermore, infrared spectroscopic studies revealed that this cluster resembles the pentagonal dodecahedral cage that encapsulates a methane molecule in a Type I clathrate hydrate. In the H$^+$(H$_2$O)$_{21}$ cluster, the hydronium ion H$_3$O$^+$ can either occupy the (methane) position inside the clathrate cage or take up a site on the cage's surface, thereby pushing a neutral water molecule to the center of the cage.

Another ongoing debate is concerned with the structure of liquid water. In the conventional century-old picture, liquid water is simply an extended network where each water molecule is linked (or hydrogen bonded) to four others in a tetrahedral pattern. Contrary to this, there are recent synchrotron X-ray results which suggest that many water molecules are linked to only two neighbors. Yet there are also more-recent X-ray data which support the "original" structure, and so this controversy is by no means resolved. Water clearly plays indispensable roles in essentially all fields of science, most notably in chemistry and the life sciences. These roles will be better understood once the structure of liquid water is unequivocally determined.

6.3.4 Protonated water species, H_3O^+ and $H_5O_2^+$

The oxonium (or hydronium) ion H_3O^+ is a stable entity with a pyramidal structure, in which the O–H bond has the same length as that in ice, and the apical angle lies between 110° and 115°. Fig. 6.3.6(a) shows the structure of H_3O^+.

The $H_5O_2^+$ ion is a well-documented chemical entity in many crystal structures. Some relevant data are listed in Table 6.3.2. The common characteristic of these compounds is a very short O–H⋯O hydrogen bond (240–245 pm.). Fig. 6.3.6(b) shows the structure of $H_5O_2^+$ in the crystalline dihydrate of hydrochloric acid $H_5O_2^+ \cdot Cl^-$.

Besides H_3O^+ and $H_5O_2^+$, several other hydronium ion complexes have been proposed, for example $H_7O_3^+$, $H_9O_4^+$, $H_{13}O_6^+$, and $H_{14}O_6^{2+}$. If 245 pm is taken as the upper limit for the O–H⋯O distance within a hydronium ion, longer O–H⋯O distances represent hydrogen bonds between a hydronium ion and neighboring water molecules. According to this criterion, reasonable and unambiguous structural formulas can be assigned to many acid hydrates, as listed in Table 6.3.3.

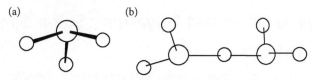

Fig. 6.3.6 Structure of (a) H_3O^+ and (b) $H_5O_2^+$.

Table 6.3.2 Some compounds containing the $H_5O_2^+$ ion

Compound	Diffraction method	O⋯O/pm	Geometry of hydrogen bond
$(H_5O_2)Cl \cdot H_2O$	X-ray	243.4	
$(H_5O_2)ClO_4$	X-ray	242.4	
$(H_5O_2)Br$	neutron	240	117 122 O——H——O 174.7°
$(H_5O_2)_2SO_4 \cdot 2H_2O$	X-ray	243	
$(H_5O_2)[trans\text{-}Co(en)_2Cl_2]_2$	neutron	243.1	Centered
$(H_5O_2)[C_6H_2(NO_2)_3SO_3] \cdot 2H_2O$	neutron	243.6	112.8 131.0 O——H——O 175°
$(H_5O_2)[C_6H_4(COOH)SO_3] \cdot H_2O$	X-ray, neutron	241.4	120.1 121.9 O——H——O
$(H_5O_2)[PW_{12}O_{40}]$	X-ray, neutron	241.4	Centered
$(H_5O_2)[Mn(H_2O)_2(SO_4)_2]$	X-ray	242.6	Centered

Table 6.3.3 Structural formulas for some acid hydrates

Acid hydrate	Empirical formula	Structural formula
$HNO_3 \cdot 3H_2O$	$(H_7O_3)NO_3$	$(H_3O)NO_3 \cdot 2H_2O$
$HClO_4 \cdot 3H_2O$	$(H_7O_3)ClO_4$	$(H_3O)ClO_4 \cdot 2H_2O$
$HCl \cdot 3H_2O$	$(H_7O_3)Cl$	$(H_5O_2)Cl \cdot H_2O$
$HSbCl_6 \cdot 3H_2O$	$(H_7O_3)SbCl_6$	$(H_5O_2)SbCl_6 \cdot H_2O$
$HCl \cdot 6H_2O$	$(H_9O_4)Cl \cdot 2H_2O$	$(H_3O)Cl \cdot 5H_2O$
$CF_3SO_3H \cdot 4H_2O$	$(H_9O_4)CF_3SO_3$	$(H_3O)(CF_3SO_3) \cdot 3H_2O$
$2[HBr \cdot 4H_2O]$	$(H_9O_4)(H_7O_3)Br_2 \cdot H_2O$	$2[(H_3O)Br \cdot 3H_2O]$
$[(C_9H_{18})_3(NH)_2Cl] \cdot Cl \cdot HCl \cdot 6H_2O$	$[(C_9H_{18})_3(NH)_2Cl] \cdot H_{13}O_6 \cdot Cl_2$	$[(C_9H_{18})_3(NH)_2Cl \cdot (H_5O_2) \cdot Cl_2 \cdot 4H_2O]$
$HSbCl_6 \cdot 3H_2O$	$(H_{14}O_6)_{0.5}(SbCl_6)$	$(H_5O_2)(SbCl_6) \cdot H_2O$

6.4 Allotropes of Sulfur and Polyatomic Sulfur Species

6.4.1 Allotropes of sulfur

The allotropy of sulfur is far more extensive and complex than that of any other element. This arises from the following factors.

(1) A great variety of molecules can be generated by –S–S– catenation. This becomes evident by comparing the experimental bond enthalpies of O_2 and S_2, which are 498 and 431 kJ mol^{-1}, respectively. On the other hand, the S–S single bond enthalpy, 263 kJ mol^{-1}, is much greater than the value of 142 kJ mol^{-1} for the O–O single bond. In view of the difference in atomic size, the repulsion energies of lone pairs on two O atoms are larger than that on two S atoms. Thus elemental sulfur readily forms rings or chains with S–S single bonds, whereas oxygen exists primarily as a diatomic molecule.

(2) The S–S bonds are very variable and flexible, generating large S_n molecules or extended structures, and hence the relevant compounds are solids at room temperature. The interatomic distances cover the range of 198 pm to 218 pm, depending mainly on the extent of multiple bonding. Bond angles S–S–S are in the range 101° to 111° and the dihedral angles S–S–S–S vary from 74° to 100°.

(3) The S_n molecules can be packed in numerous ways in the crystal lattice.

All allotropic forms of sulfur that occur at room temperature consist of S_n rings, with n = 6 to 20 (Table 6.4.1). Many homocyclic sulfur allotropes have been characterized by X-ray crystallography: S_6, S_7 (two modifications), S_8 (three modifications), S_9, S_{10}, $S_6 \cdot S_{10}$, S_{11}, S_{12}, S_{13}, S_{14}, S_{18} (two forms), S_{20} and polymer chain S_∞. Their structural data are listed in Table 6.4.2, and the molecular structures are shown in Fig. 6.4.1.

The most common allotrope of sulfur is orthorhombic α-S_8. At about 95.3 °C, α-S_8 transforms to monoclinic β-S_8, such that the packing of S_8 molecules is altered and their orientation becomes partly disordered. This leads to a lower density of 1.94 g cm^{-3} to 2.01 g cm^{-3}, but the dimensions of S_8 rings in the two allotropes are very similar. Monoclinic γ-S_8 also comprises cyclo-S_8 molecules, but the packing is more efficient and leads to a higher density of 2.19 g cm^{-3}. It reverts slowly to α-S_8 at room temperature, but rapid heating gives a melting point of 106.8 °C.

Allotrope S_7 is known in four crystalline modifications, one of which (δ form) is obtained by crystallization from CS_2 at −78 °C. It features one long bond (bond S^6–S^7 in Fig. 6.4.1) of 218.1 pm, which probably arises from the virtually coplanar set of atoms S^4-S^6-S^7-S^5, which leads to maximum repulsion

Table 6.4.1 Some properties of sulfur allotropes

Allotrope	Color	Density/g cm^{-3}	mp or dp/ °C
S_2(g)	blue-violet	–	very stable at high temperature
S_3(g)	cherry-red	–	stable at high temperature
S_6	orange-red	2.209	dp > 50
S_7	yellow	2.182 (–110°C)	dp > 39
S_8 (α)	yellow	2.069	112.8
S_8 (β)	yellow	1.94-2.01	119.6
S_8 (γ)	light-yellow	2.19	106.8
S_9	intense yellow	–	stable below room temp.
S_{10}	pale yellow green	2.103 (–110°C)	dp > 0
S_{11}	–	–	–
S_{12}	pale yellow	2.036	148
S_{14}	yellow	–	113
S_{18}	lemon yellow	2.090	mp 128 (dec)
S_{20}	pale yellow	2.016	mp 124 (dec)
S_x	yellow	2.01	104 (dec)

mp = melting point; dp = decomposition temperature; dec = with decomposition

Table 6.4.2 Structural data of S_n molecules in the sulfur allotropes

Molecule	Bond length/pm	Bond angle/°	Torsion angle/°
S_2 (matrix at 20K)	188.9	–	–
S_6	206.8	102.6	73.8
γ-S_7	199.8–217.5	101.9–107.4	0.4 108.8
δ-S_7	199.5–218.2	101.5–107.5	0.3–108.0
α-S_8	204.6–205.2	107.3–109.0	98.5
β-S_8	204.7–205.7	105.8–108.3	96.4–101.3
γ-S_8	202.3–206.0	106.8–108.5	97.9–100.1
α-S_9	203.2–206.9	103.7–109.7	59.7–115.6
S_{10}	203.3–207.8	103.3–110.2	75.4–123.7
S_{11}	203.2–211.0	103.3–108.6	69.3–140.5
S_{12}	204.8–205.7	105.4–107.4	86.0–89.4
S_{13}	197.8–211.3	102.8–111.1	29.5–116.3
S_{14}	204.7–206.1	104.0–109.3	72.5–101.7
α-S_{18}	204.4–206.7	103.8–108.3	79.5–89.0
β-S_{18}	205.3–210.3	104.2–109.3	66.5–87.8
S_{20}	202.3–210.4	104.6–107.7	66.3–89.9
catena-S_x	206.6	106.0	85.3

between non-bonding lone pairs on adjacent S atoms. As a result of this weakening of the S^6–S^7 bond, the adjacent S^4–S^6 and S^5–S^7 bonds are strengthened (199.5 pm), and there are further alternations of bond lengths (210.2 pm and 205.2 pm) throughout the molecule.

Cyclo-S_9 crystallizes in two allotropic forms: α and β. α-S_9 belongs to space group $P2_1/n$ with two independent molecules of similar structure in the unit cell, but the molecular symmetry is approximately C_2, as shown in Fig. 6.4.1. The bond lengths lie between 203.2 pm and 206.9 pm The bond angles vary in the range of 103.7° to 109.7°, and torsion angles in the range of 59.7° to 115.6°. The conformation of a sulfur homocycle is best described by its "motif," that is, the order of the signs of the torsion angles around the ring. The motif of α-S_9 is + + − − + + − + −.

A new allotrope of sulfur, *cyclo*-S_{14}, has been isolated as yellow rod-like crystals (mp 113 °C) from the reaction of [(tmeda)ZnS$_6$] (tmeda = N, N, N', N'-tetra-methylethylenediamine) with S_8Cl_2. The bond distances vary from 204.7 pm to 206.1 pm, the bond angles from 104.0° to 109.3°, and the torsion angles are in the range of 72.5° to 101.7°. The motif of *cyclo*-S_{14} is + + − − + + − − + + − − + −, in which the first 12 signs are the same as those of *cyclo*-S_{12}, as shown in Fig. 6.4.1. Hence the structure of S_{14} can be obtained by opening a bond in S_{12} and inserting an S_2 fragment (i.e., atoms S^9 and S^{10} in the figure).

Solid polycatena sulfur comes in many forms as described by their names: rubbery S, plastic S, lamina S, fibrous S, polymeric S, and insoluble S which is also termed S_μ or S_ω. Fibrous S consists of parallel helical chains of sulfur atoms whose axes are arranged on a hexagonal close packed net 463 pm apart. The structure contains both left- and right-handed helices of radius 95 pm (Fig. 6.4.1, S_∞) with a repeat

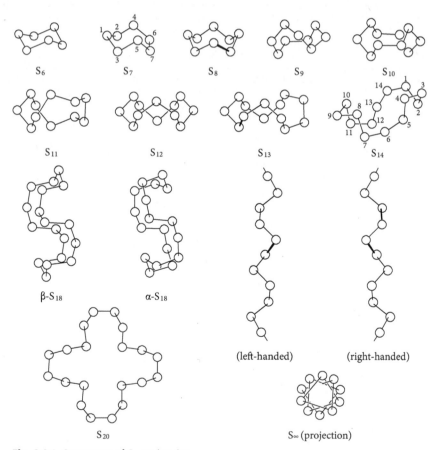

Fig. 6.4.1 Structures of S_n molecules.

distance of 1380 pm comprising 10 S atoms in three turns. Within each helix the metric parameters are: bond distance S–S 206.6 pm, bond angle S–S–S 106.0°, and dihedral angle S–S–S–S 85.3°

6.4.2 **Polyatomic sulfur ions**

(1) **Cations**

Sulfur may be oxidized by different oxidizing agents to yield a variety of polyatomic cations. The currently known species that have been characterized by X-ray diffraction are S_4^{2+}, S_8^{2+}, and S_{19}^{2+}, the structures of which are shown in Fig. 6.4.2.

In crystalline $(S_4^{2+})(S_7I^+)_4(AsF_6^-)_6$, the cation S_4^{2+} takes the form of a square-planar ring of edge 198 pm. As this is significantly shorter than a typical S–S single bond (206 pm), the S–S bonds in S_4^{2+} have some double bond character.

In the salt $(S_8^{2+})(AsF_6^-)_2$, the eight-membered ring of S_8^{2+} has an *exo-endo* conformation and a rather long transannular bond of 283 pm. This structure may be regarded as being half-way between those of the cage molecule S_4N_4 and the crown-shaped ring of S_8. Since S_4N_4 has two electrons less than S_8^{2+} and is isoelectronic with the unknown S_8^{4+}, it is conceivable that, as two electrons are removed from S_8, one end folds up to generate a transannular bond. Then, with the removal of two more electrons, the other end also folds up and another bond is formed to give the S_4N_4 structure, as shown in Fig. 6.4.3.

In crystalline $(S_{19}^{2+})(AsF_6^-)_2$, the S_{19}^{2+} cation consists of two seven-membered rings joined by a five-atom chain. One of the rings has a boat conformation while the other is disordered, existing as a 4:1 mixture of chair and boat conformations. The S–S distances vary greatly from 187 pm to 239 pm, and S–S–S angles from 91.9° to 127.6°.

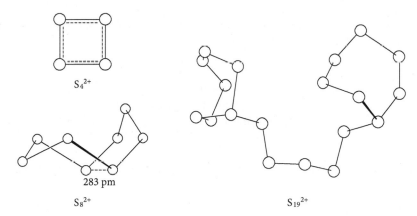

S_4^{2+}

283 pm

S_8^{2+}

S_{19}^{2+}

Fig. 6.4.2 Structures of polyatomic sulfur cations.

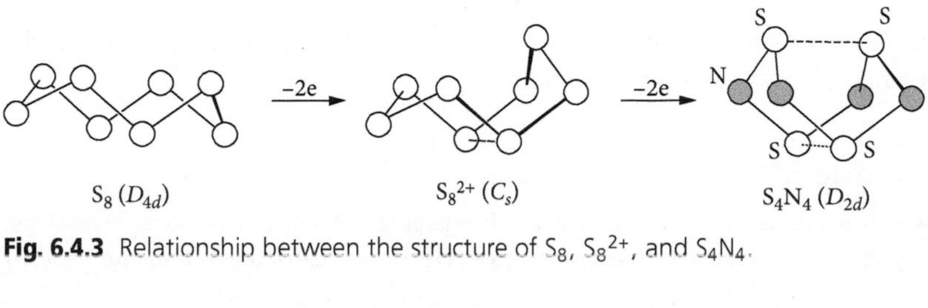

$$S_8\ (D_{4d}) \xrightarrow{-2e} S_8^{2+}\ (C_s) \xrightarrow{-2e} S_4N_4\ (D_{2d})$$

Fig. 6.4.3 Relationship between the structure of S_8, S_8^{2+}, and S_4N_4.

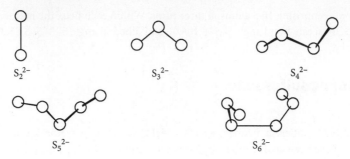

Fig. 6.4.4 Structures of S_n^{2-}.

Table 6.4.3 Structural data of S_n^{2-}

S_n^{2-}	Compound	Bond length/pm	Bond angle (central)	Symmetry
S_2^{2-}	pyrites	208–215	–	$D_{\infty h}$
S_3^{2-}	BaS_3	207.6	144.9°	C_{2v}
S_4^{2-}	Na_2S_4	206.1–207.4	109.8°	C_2
S_5^{2-}	K_2S_5	203.7–207.4	106.4°	C_2
S_6^{2-}	Cs_2S_6	201–211	108.8°	C_2

(2) Anions

The majority of polysulfide anions have acyclic structures. The configurations of S_n^{2-} species are in accord with their bond valence (*b*):

$$b = \tfrac{1}{2}(8n - g)$$
$$= \tfrac{1}{2}[8n - (6n + 2)]$$
$$= n - 1$$

Fig. 6.4.4 shows the puckered chain motifs of some S_n^{2-} species, and their structural data are listed in Table 6.4.3.

6.5 Sulfide Anions as Ligands in Metal Complexes

Sulfur has an extensive coordination chemistry involving the S^{2-} and S_n^{2-} anions, which exhibit an extremely versatile variety of coordination modes.

(1) Monosulfide S^{2-}

Ligands in which S acts as a donor atom are usually classified as soft Lewis bases, in contrast to oxygen donor-atom ligands which tend to be hard Lewis bases. The large size of the S atom and its easily deformed

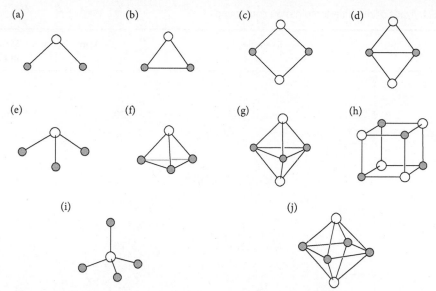

Fig. 6.5.1 Coordination modes of S^{2-} (open circle): (a) $(\mu_2\text{-S})[\text{Au}(\text{Pet}_3)]_2$, (b) $(\mu_2\text{-S})[\text{Pt}(\text{PPh}_3)_2]_2$, (c) $(\mu_2\text{-S})_2\text{Mo}(\text{S})_2\text{Fe}(\text{SPh})_2]^{2-}$, (d) $(\mu_2\text{-S})_2\text{Fe}_2(\text{NO})_4]^{2-}$, (e) $(\mu_3\text{-S})[\text{Au}(\text{PPh}_3)]_3$, (f) $(\mu_3\text{-S})[\text{Co}(\text{CO})_3]_3$, (g) $(\mu_3\text{-S})_2(\text{CoCp})_3$, (h) $(\mu_3\text{-S})_4[\text{Fe}(\text{NO})]_4$, (i) $(\mu_4\text{-S})[\text{Zn}_4(\text{S}_2\text{AsMe}_2)_6]$, (j) $(\mu_4\text{-S})_2\text{Co}_4(\text{CO})_{10}$.

electron cloud account for this difference, and the participation of metal $d\pi$ orbitals in bonding to sulfur is well established. The monosulfide dianion S^{2-} can act either as a terminal or a bridging ligand. As a terminal ligand, S^{2-} donates a pair of electrons to its bonded atom. In the μ_2 bridging mode S is usually regarded as a two-electron donor.

Figs 6.5.1(a)–(d) shows four μ_2 bridging modes. In the μ_3 triply bridging mode S can be regarded as a four-electron donor, using both its two unpaired electrons and one lone pair. The pseudo-cubane structure adopted by some of the μ_3-S compounds is a crucial structural unit in many biologically important systems, for example, the $[(\text{RS})\text{MS}]_4$ (M = Mo, Fe) units which cross-link the polypeptide chains in nitrogenase and ferredoxins. Figs 6.5.1(e)–(h) shows four μ_3 bridging modes. In the μ_4 bridging mode S can be regarded as a four- or six-electron donor, depending on the geometry of the mode. Figs 6.5.1(i)–(j) show two μ_4 bridging modes. As yet there is no molecular compound in which S bridges six or eight metal atoms, but interstitial sulfur is well known.

(2) **Disulfide S_2^{2-}**

The coordination modes of disulfide S_2^{2-} in some representative compounds are listed in Table 6.5.1. Their molecular structures are shown in Fig. 6.5.2.

(3) **Polysulfides S_n^{2-}**

The S_n^{2-} ($n = 3 - 9$) anions generally have a chain structure. The average S–S bond length is smaller in S_n^{2-} ions ($n > 2$) than in S_2^{2-}, and the length of the S–S terminal bond decreases from S_3^{2-} (215 pm) to S_7^{2-} (199.2 pm). These data indicate that the negative charge (filling a π^* antibonding MO) is delocalized over

Table 6.5.1 Types of disulfide coordination compounds

Type		Example	$d(S–S)$/pm	Structure in Fig. 6.5.2
Ia 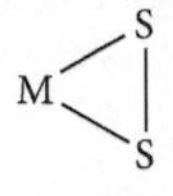		$[Mo_2O_2S_2(S_2)_2]^{2-}$	208	(a)
Ib 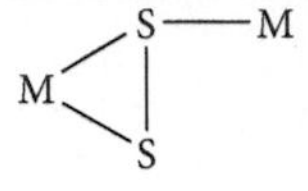		$[Mo_4(NO)_4(S_2)_5(S)_3]^{4-}$	204.8	–
Ic 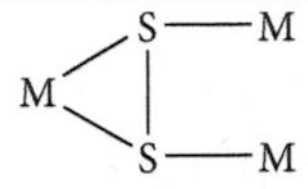		$Mn_4(CO)_{15}(S_2)_2$	207	(b)
Id 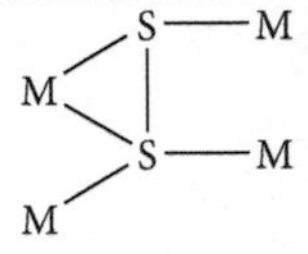		$Mo_4(CO)_{15}(S_2)_2$	209	(b)
IIa 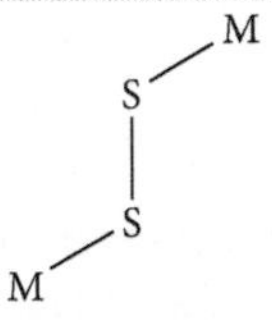		$[Ru_2(NH_3)_{10}S_2]^{4+}$	201.4	(c)
IIb 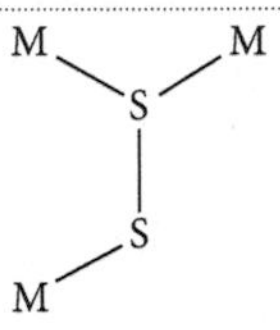		$Co_4Cp_4(\mu_3\text{-}S)_2(\mu_3\text{-}S_2)_2$	201.3	(d)
IIc 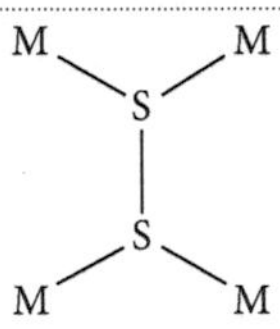		$[SCo_3(CO)_7]_2S_2$	204.2	(e)
III 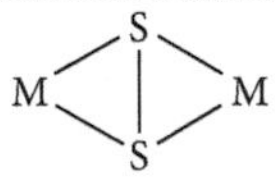		$[Mo_2(S_2)_6]^{2-}$	204.3	(f)

the entire chain, but the delocalization along the chain is less in higher polysulfides. These considerations are important in comparing the S–S bond lengths of the free ions with those of their metal complexes. Fig. 6.5.3 shows the structures of some polysulfide coordination compounds.

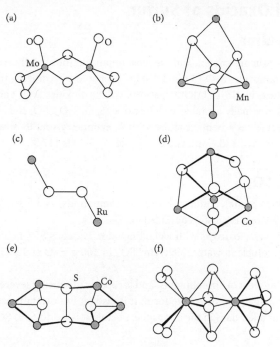

Fig. 6.5.2 Structure of S_2^{2-} coordination compounds: (a) $[Mo_2O_2S_2(S_2)_2]^{2-}$, (b) $Mn_4(CO)_{15}(S_2)_2$, (c) $[Ru_2(NH_3)_{10}S_2]^{4+}$, (d) $Co_4Cp_4(\mu_3\text{-}S)_2(\mu_3\text{-}S_2)_2$, (e) $[S(Co_3(CO)_7)_2S_2$, (f) $[Mo_2(S_2)_6]^{2-}$.

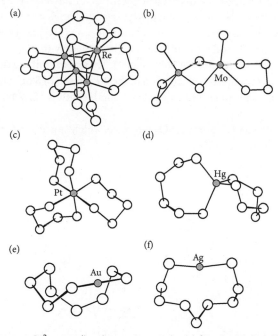

Fig. 6.5.3 Structures of some S_n^{2-} coordination compounds: (a) $[Re_4(\mu_3\text{-}S)_4(S_3)_6]^{4-}$, (b) $[Mo_2(S)_2(\mu\text{-}S)_2(\eta^2\text{-}S_2)(\eta^2\text{-}S_4)]^{2-}$, (c) $[Pt(\eta^2\text{-}S_5)_3]^{2-}$, (d) $[Hg(\eta^2\text{-}S_6)_2]^{2-}$, (e) $[Au(\eta^2\text{-}S_9)]^-$, (f) $[Ag(\eta^2\text{-}S_9)]^-$.

6.6 Oxides and Oxacids of Sulfur

6.6.1 Oxides of sulfur

More than 10 oxides of sulfur are known, and the most important industrially are SO_2 and SO_3. The six homocyclic polysulfur monoxides S_nO ($5 \leq n \leq 10$) are prepared by oxidizing the appropriate cyclo-S_n with trifluoroperoxoacetic acid, $CF_3C(O)OOH$, at $-30\,°C$. The dioxides S_7O_2 and S_6O_2 are also known. In addition, there are the thermally unstable acylclic oxides S_2O, S_2O_2, SO, and the elusive species SOO and SO_4. The S_2O molecule has a bent structure with C_s symmetry, and its structural parameters are: S=O bond length 145.6 pm, S=S 188.4 pm, and bond angle S–S–O is 117.9°.

(1) Sulfur dioxide, SO_2

When sulfide ores (such as pyrite), sulfur-rich organic compounds, and fossil fuels (such as coal) are burned in air, the sulfur therein is mostly converted to SO_2.

Sulfur dioxide is a colorless, toxic gas with a choking odor. Gaseous SO_2 neither burns nor supports combustion. It is readily soluble in water (3927 cm^3 SO_2 in 100 g H_2O at 20 °C) and forms "sulfurous acid."

Sulfur dioxide is a component of air pollutants, and is capable of causing severe damage to human and other animal lungs, particularly in the sulfate form. It is also an important precursor to acid rain. The SO_2 molecule survives for a few days in the atmosphere before it is oxidized to SO_3. The direct reaction:

$$2SO_2(g) + O_2(g) \rightarrow 2SO_3(g)$$

is very slow, but it speeds up markedly in the presence of metal cations such as Fe^{3+}.

Sulfur dioxide is a bent molecule with C_{2v} symmetry; the bond angle is 119°, and the bond length is 143.1 pm, which is shorter than the expected value of a single S–O (176 pm) or S=O double bond (154 pm) calculated from the sum of covalent radii. The S–O bond energy in SO_2 is 548 kJ mol^{-1}, which is larger than that of the unstable molecule SO (524 kJ mol^{-1}). These data indicate that the S–O bond order in SO_2 exceeds 2, a consequence of the participation of the $3d$ orbitals of S in bonding. Sulfur dioxide is homologous with O_3. In the bent molecule O_3 the O–O bond length is 127.8 pm, which is longer than that in O_2 (120.7 pm).

There are lone pairs in the S and O atoms of SO_2 molecule, which as a ligand forms many coordination modes with metal atoms, as shown in Fig. 6.6.1.

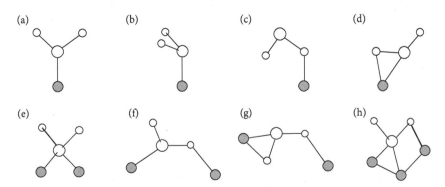

Fig. 6.6.1 Coordination modes of SO_2 to metal atoms (shaded circles).

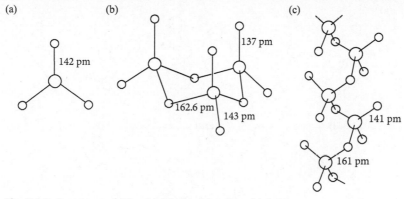

Fig. 6.6.2 Structures of SO_3: (a) $SO_3(g)$, (b) γ–SO_3, (c) β–SO_3.

(2) Sulfur trioxide, SO_3

Sulfur trioxide is made on a huge scale by the catalytic oxidation of SO_2. It is usually not isolated but is immediately converted to H_2SO_4.

In the gas phase, monomeric SO_3 has a D_{3h} planar structure with bond length S–O 142 pm, as shown in Fig. 6.6.2(a). The cyclic trimer $(SO_3)_3$ occurs in colorless orthorhombic γ-SO_3 (mp 16.9 °C), and its structure is shown in Fig. 6.6.2(b). The helical chain structure of β-SO_3 is shown in Fig. 6.6.2(c). A third and still more stable form, α-SO_3 (mp 62°C) involves cross-linkage between the chains to give a complex layer structure. The standard enthalpies of formation of the four forms of SO_3 at 298 K are listed below:

	$SO_3(g)$	α-SO_3	β-SO_3	γ-SO_3
ΔH_f^0/ kJ mol^{-1}	−395.2	−462.4	−449.6	−447.4

Sulfur trioxide reacts vigorously and extremely exothermically with water to give H_2SO_4. Anhydrous H_2SO_4 has an unusually high dielectric constant, and also a very high electrical conductivity which results from the ionic self-dissociation of the compound coupled with a proton-switch mechanism for the rapid conduction of current through the viscous hydrogen-bonded liquid:

$$2H_2SO_4 \rightleftharpoons H_3SO_4^+ + HSO_4^-$$

Photolysis of SO_3 and O_3 mixtures yields monomeric SO_4, which can be isolated by inert-gas matrix techniques at low temperatures (15–78 K). Vibration spectroscopy indicates a sulfuryl group together with either an open SOO branch (C_s structure) or a closed three-membered peroxo ring (C_{2v} structure), the latter being preferred on the basis of the infrared absorption bands. The structures of these two species are illustrated pictorially below.

C_s (planar), or C_1 (non-planar) C_{2v}

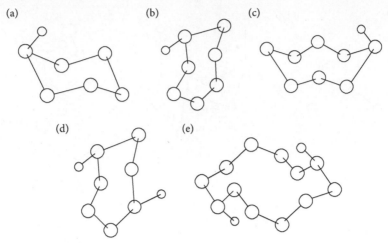

Fig. 6.6.3 Molecular structure of oxides of cyclic poly-sulfur:(a) S_6O, (b) S_7O, (c) S_8O, (d) S_7O_2, (e) $S_{12}O_2$.

(3) **Oxides of cyclic poly-sulfur, S_nO and S_nO_2**

When *cyclo*-S_8, -S_9, and -S_{10} are dissolved in CS_2 and oxidized by freshly prepared $CF_3C(O)O_2H$ at temperatures below $-10\,°C$, modest yields (10–20%) of the corresponding crystalline monoxides S_nO are obtained. Of these compounds, S_5O has not yet been isolated, but it has been prepared in solution by the same method. On the other hand, S_6O has two crystalline modifications: orange α-S_6O with mp $39\,°C$ (dec) and dark orange β-S_6O with mp $34\,°C$ (dec), and its molecular structure is shown in Fig. 6.6.3(a). Also, S_7O is a orange crystal with mp $55\,°C$ (dec), and its molecular structure is shown in Fig. 6.6.3(b). Finally, S_8O is an orange-yellow crystal with mp $78\,°C$ (dec), and its structure is shown in Fig. 6.6.3(c).

In the S_nO_2 compounds, S_7O_2 is a dark orange crystal which decomposes above room temperature. Its molecular structure is shown in Fig. 6.6.3(d). The structure of $S_{12}O_2$ as determined from the orange adduct $S_{12}O_2·2SbCl_5·3CS_2$ is shown in Fig. 6.6.3(e).

6.6.2 **Oxoacids of sulfur**

Sulfur forms many oxoacids, though few of them can be isolated in pure form. Most are prepared in aqueous solution or as crystalline salts of the corresponding oxoacid anions. Table 6.6.1 lists the common oxoacids of sulfur.

(1) **Sulfuric acid and disulfuric acid**

Sulfuric acid is the most important chemical of all sulfur compounds. Anhydrous sulfuric acid is a dense, viscous liquid which is readily miscible with water in all proportions. Sulfuric acid forms hydrogen sulfate (also known as bisulfate, HSO_4^-) and sulfate (SO_4^{2-}) salts with many metals, which are frequently very stable and are important mineral compounds. Fig. 6.6.4(a)–(c) shows the molecular structures of H_2SO_4, HSO_4^-, and SO_4^{2-}.

The crystal structure of sulfuric acid consists of layers of $O_2S(OH)_2$ tetrahedra connected via hydrogen bonds involving the donor OH groups and acceptor O atoms. Fig. 6.6.5 shows a layer of hydrogen-bonded H_2SO_4 molecules. The $O–H···O$ hydrogen bond length is 264.8 pm and the bond angle $O–H···O$ is $170°$.

The hydrogen sulfate (or bisulfate) anion HSO_4^- exists in crystalline salts such as $(H_3O)(HSO_4)$, $K(HSO_4)$, and $Na(HSO_4)$. The bond lengths of HSO_4^- in $(H_3O)(HSO_4)$ are $S–O = 145.6$ pm and $S–OH = 155.8$ pm.

Table 6.6.1 The common oxoacids of sulfur

Formula	Name	Oxidation state of S	Schematic structure	Salt
H_2SO_4	sulfuric	6	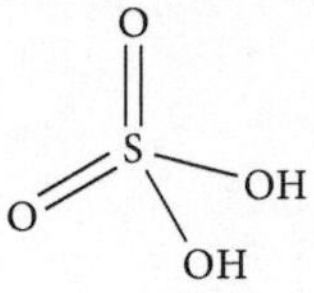	sulfate $SO_4{}^{2-}$
$H_2S_2O_7$	disulfuric	6	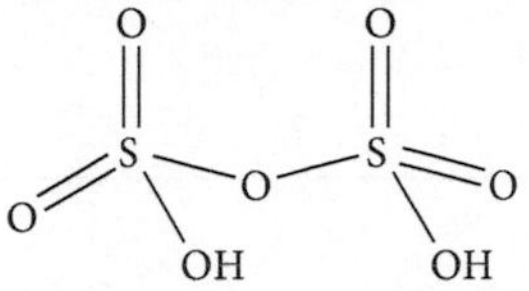	disulfate $O_3SOSO_3{}^{2-}$
$H_2S_2O_3$	thiosulfuric	6, –2	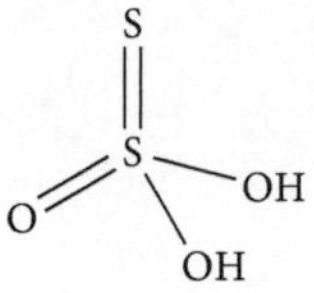	thiosulfate $S_2O_3{}^{2-}$
H_2SO_5	peroxo-monosulfuric	6	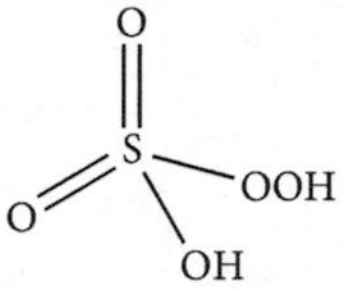	peroxo-monosulfate $OOSO_3{}^{2-}$
$H_2S_2O_8$	peroxo-disulfuric	6	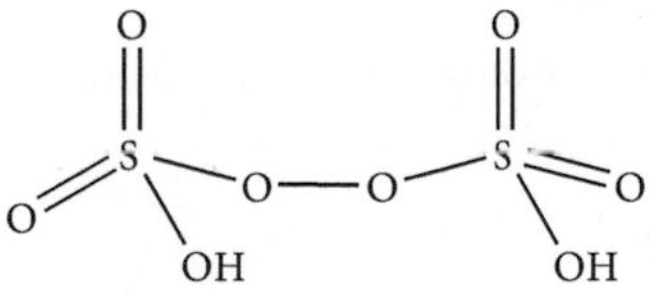	peroxo-disulfate $O_3SOOSO_3{}^{2-}$
$H_2S_2O_6$	dithionic*	5	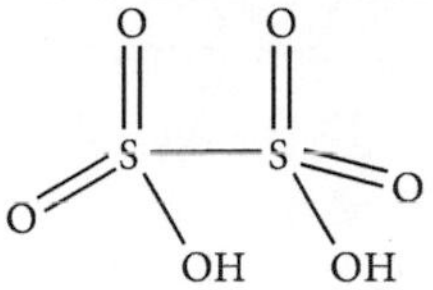	dithionate $O_3SSO_3{}^{2-}$
$H_2S_{n+2}O_6$	polythionic	5, 0	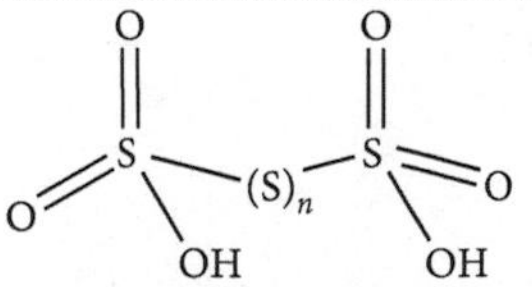	polythionate $O_3S(S)_nSO_3{}^{2-}$
H_2SO_3	sulfurous*	4	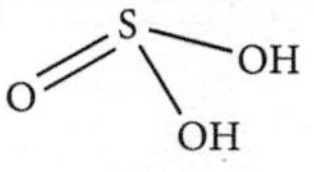	sulfite $SO_3{}^{2-}$

Continued

Table 6.6.1 *Continued*

Formula	Name	Oxidation state of S	Schematic structure	Salt
$H_2S_2O_5$	disulfurous*	5, 3	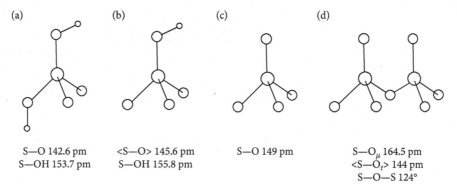	disulfite $O_3SSO_2^{2-}$
$H_2S_2O_4$	dithionous*	3		dithionite O_2SSO_2

* Acids only exist as salts.

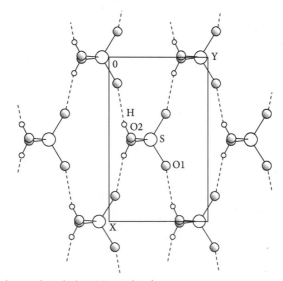

(a)

(b)

(c)

(d)

S—O 142.6 pm
S—OH 153.7 pm

<S—O> 145.6 pm
S—OH 155.8 pm

S—O 149 pm

S—O$_\mu$ 164.5 pm
<S—O$_t$> 144 pm
S—O—S 124°

Fig. 6.6.4 Molecular structure of (a) H_2SO_4, (b) HSO_4^-, (c) SO_4^{2-}, and (d) $S_2O_7^{2-}$.

Fig. 6.6.5 A layer of hydrogen-bonded H_2SO_4 molecules.

Disulfuric acid (also known as pyrosulfuric acid), $H_2S_2O_7$, which is the major constituent of "fuming sulfuric acid," is formed from sulfur trioxide and sulfuric acid:

$$SO_3 + H_2SO_4 \rightarrow H_2S_2O_7$$

Fig. 6.6.4 (d) shows the structure of the $S_2O_7^{2-}$ anion.

(2) Sulfurous acid and disulfurous acid

Sulfurous acid, H_2SO_3, and disulfurous acid, $H_2S_2O_5$, are examples of sulfur oxoacids that do not exist in the free state, although numerous salts derived from them containing the HSO_3^-, SO_3^{2-}, $HS_2O_5^-$, and $S_2O_5^{2-}$ anions are stable solids. An aqueous solution of SO_2, though acidic, contain negligible quantities of the free acid H_2SO_3. The apparent hexahydrate $H_2SO_3 \cdot 6H_2O$ is actually the gas hydrate $6SO_2 \cdot 46H_2O$, in which the SO_2 molecules are enclosed in cages within a host framework constructed from hydrogen-bonded water molecules. Fig. 6.6.6 shows the structures of the anions SO_3^{2-}, $S_2O_5^{2-}$, and $S_2O_4^{2-}$. The hydrogen sulfite (bisulfite) ion HSO_3^- has been found to exist in two isomeric forms: $HO-SO_2^-$ and $H-SO_3^-$.

(3) Thiosulfate, SSO_3^{2-}

In the thiosulfate ion, a terminal S atom replaces an O atom of the sulfate ion. The S–S bond length is 201.3 pm, which indicates essentially single bond character, while the mean S–O bond length is 146.8 pm, which indicates considerable π-bonding between the S and O atoms.

Single-crystal X-ray analysis has shown that the structure of $SeSO_3^{2-}$ is isostructural with the $S_2O_3^{2-}$ ion with a Se–S bond length of 217.5(1) pm.

The thiosulfate ion, in which the terminal S and O atoms can function as ligand sites, is a polyfunctional species in various coordination modes with metal atoms. Fig. 6.6.7 shows the coordination modes of $S_2O_3^{2-}$.

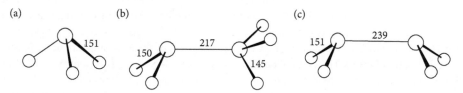

Fig. 6.6.6 Structure of (a) SO_3^{2-}, (b) $S_2O_5^{2-}$, and (c) $S_2O_4^{2-}$ (bond lengths in pm).

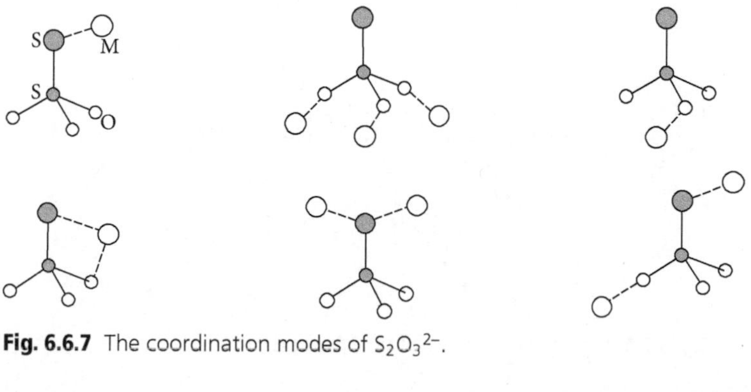

Fig. 6.6.7 The coordination modes of $S_2O_3^{2-}$.

(4) Peroxoacids of sulfur

The peroxoacids of sulfur and their salts all contain the –O–O– group. The salts of $S_2O_8^{2-}$, such as $K_2S_2O_8$, are very convenient and powerful oxidizing agents. Peroxomonosulfuric acid (Caro's acid), H_2SO_5, is a colorless, explosive solid (mp 45 °C), and salts of HSO_5^- are known. In HSO_5^- and $S_2O_8^{2-}$, the S–O (peroxo) and S–O (terminal) bond distances are different. The S–O (peroxo) bond length is about 160 pm, which corresponds to a single bond, and the S–O (terminal) bond length is about 145 pm, which corresponds to a double bond. The structural formulas of HSO_5^- and $S_2O_8^{2-}$ are shown below:

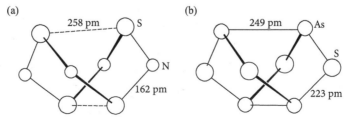

6.7 Sulfur-Nitrogen Compounds

Sulfur and nitrogen are diagonally related elements in the Periodic Table and might therefore be expected to have similar electronic charge densities for similar coordination numbers, and to form cyclic, acyclic, and polycyclic molecules through extensive covalent bonding. Some sulfur-nitrogen compounds exhibit interesting chemical bonding and have unusual properties, as discussed below.

(1) Tetrasulfur tetranitride, S_4N_4

Nitride S_4N_4 is an air-stable compound that can be prepared by passing NH_3 gas into a warm solution of S_2Cl_2 in CCl_4 or benzene. It is a thermochromic crystal: colorless at 83 K, pale yellow at 243 K, orange at room temperature, and deep red above 373 K.

The D_{2d} molecular structure of S_4N_4 is shown in Fig. 6.7.1(a). The atoms of S_4N_4 are arranged so that the electropositive S atoms occupy the vertices of a tetrahedron, while the electronegative N atoms constitute a square that intersects the tetrahedron. All the S–N distances are equal; for gaseous S_4N_4 they are 162.3 pm, which is intermediate between the distances of the S–N single bond (174 pm) and S=N double bond (154 pm). The N–S–N bond angle is 105.3°, S–N–S is 114.2°, and S–S–N is 88.4°. A peculiarity of the S_4N_4 structure is the short distance between the two S atoms connected by a broken line in Fig. 6.7.1(a); at 258 pm, it lies between the S–S single bond length (208 pm) and van der Waals contacting distance S···S 360 pm. This may imply that the S atom uses two electrons for two σ S–N bonds, one electron for delocalized π bonding, and one electron for S···S weak bonding.

Fig. 6.7.1 Molecular structure of (a) S_4N_4 and (b) As_4S_4.

Tetraselenium tetranitride, Se_4N_4, forms red, hygroscopic crystals and is highly explosive. The structure of Se_4N_4 resembles that of S_4N_4 with Se–N bond length of 180 pm and cross-cage Se···Se distance of 276 pm (note that $2r_{cov}(Se) = 2 \times 117$ pm $= 234$ pm).

The homolog realgar, As_4S_4, has an analogous but different structure with the electronegative S atoms at the vertices of a square and the electropositive As atoms at the vertices of a tetrahedron. The As atoms are linked by normal single bonds, as shown by the solid lines between them in Fig. 6.7.1(b). The As–As distance is 249 pm, which is nearly equal to the calculated value of 244 pm (Table 3.4.3).

(2) S_2N_2 and $(SN)_x$

When the heated vapor of S_4N_4 is passed over silver wool at 520 K to 570 K, the unstable cyclic dimer S_2N_2 is obtained. It forms large colorless crystals which are insoluble in water but soluble in many organic solvents.

The molecular structure of S_2N_2, as shown in Fig. 6.7.2(a), is a D_{2h} square-planar ring with S–N edge 165 pm, somewhat analogous to the isoelectronic cation S_4^{2+} (Fig. 6.7.3). The valence-bond representations of the S_2N_2 molecule are as follows:

When colorless S_2N_2 crystals are allowed to stand at room temperature, golden $(SN)_x$ crystals are gradually formed. The $(SN)_x$ chain can conceivably be generated from adjacent square-planar S_2N_2 molecules, and a free radical mechanism has been proposed. Since polymerization can take place with only minor movements of the atoms, the starting material and product are pseudomorphs without alteration of the crystallinity. Fig. 6.7.2 shows the configuration of the $(SN)_x$ chains and the packing of the chains in the crystal.

Polymeric $(SN)_x$ has some unusual properties. For example, it has a bronze color and metallic luster, and its electrical conductivity is about that of mercury metal. Values of the conductivity of $(SN)_x$ depend on the purity and crystallinity of the polymer and on the direction of measurement, being much greater along the fibers than across them. A conjugated single-bond/double-bond system can be formulated, in

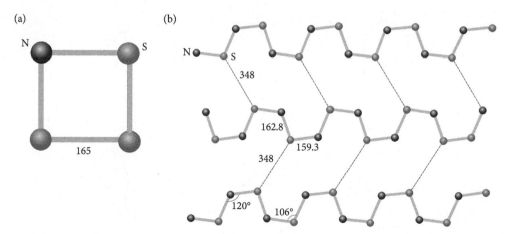

Fig. 6.7.2 Structure of (a) S_2N_2 and (b) polymeric chains in one layer of $(SN)_x$ and the important structural parameters bond lengths in pm.

which every S–N unit has one antibonding π^* electron. The half-filled overlapping π^* orbitals combine to form a half-filled conduction band, in much the same way as the half-filled ns orbitals of alkali metal atoms form a conduction band. However, in $(SN)_x$, the conduction band lies only in the direction of the $(SN)_x$ fibers, so the polymer behaves as an "one-dimensional metal."

(3) Cyclic sulfur-nitrogen compounds

Many cyclic sulfur-nitrogen compounds are known, some of which are shown in Fig. 6.7.3. The general structural features of sulfur-nitrogen compounds are formulated as follows:

(a) The S atom has the ability to form various types of S–S and S–N bonds, some of which contain catenated –S–S– chains that can insert into the cyclic chains as a fragment in molecules. For example, the $S_{11}N_2$ molecule has two S_5 chain fragments, and S_4N_2 and S_4N^- each has one S_3 chain fragment. The S⋯S interactions can vary in strength: 314 pm in S_4N^- as compared to 271–275 pm in $S_4N_5^-$.

(b) The S–N bond distances are in the range of 155 pm to 165 pm, which are shorter than the calculated single bond length, so the S–N bonds have some double bond character. The bond angles vary over a large range: for example, in $S_5N_5^+$ the bond angles are between 138° to 151°. The wide variations of bond distances and angles indicate that the bond types are quite complex.

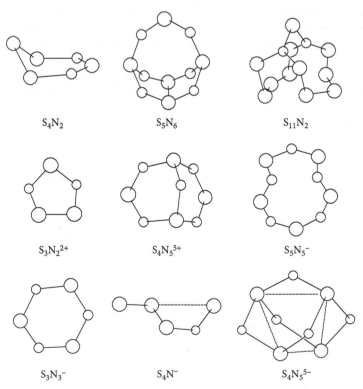

Fig. 6.7.3 Structure of some cyclic sulfur-nitrogen compounds (big circle represents S atom and small circle represents N atom).

(c) Normally the S and N atoms are each bonded to two adjacent atoms, but in some cases they are also three-connected to form polycyclic molecules. Some examples are as S_5N_6, $S_{11}N_2$, $S_4N_5^+$, and $S_4N_5^-$, the structures of which are shown in Fig. 6.7.3.

(d) The conformations of the cyclic S–N molecules exhibit diversity. For example, $S_3N_2^{2+}$ and $S_3N_3^-$ adopt planar conformations which are stabilized by delocalized π bonding, while the majority of these cyclic molecules are non-planar, in which the d orbitals of S atoms also participate in bonding.

6.8 Structural Chemistry of Selenium and Tellurium

6.8.1 Allotropes of selenium and tellurium

Selenium forms several allotropes but tellurium forms only one. The thermodynamically stable form of selenium (α-selenium or gray selenium) and the crystalline form of tellurium are isostructural. In both Te and gray Se, the atoms form infinite, helical chains having three atoms in every turn, the axes of which lie parallel to each other in the crystal, as shown in Fig. 6.8.1.

The distance of two adjacent atoms within the chain are: Se–Se 237 pm and Te–Te 283 pm. Each atom has four adjacent atoms from three different chains at an average distance of Se$\cdots$Se 344 pm, Te$\cdots$Te 350 pm. The inter-chain distance is significantly shorter than expected from the van der Waals separation (380 pm for Se and 412 pm for Te).

Red monoclinic selenium exists in three forms, each containing Se_8 rings with the crown-conformation of S_8 (Fig. 6.8.1). Vitreous black selenium, the ordinary commercial form of the element, comprises an extremely complex and irregular structure of large polymeric rings.

6.8.2 Polyatomic cations and anions of selenium and tellurium

Like their sulfur congener, selenium and tellurium can form polyatomic cations and anions in many compounds.

(1) Polyatomic cations

Fig. 6.8.2 shows the structures of some polyatomic cations of Se and Te that exist in crystalline salts. Fig. 6.8.2(a) shows the square planar geometry of the Se_4^{2+} and Te_4^{2+} cations. In $Se_4(HS_2O_7)_2$ the Se–Se

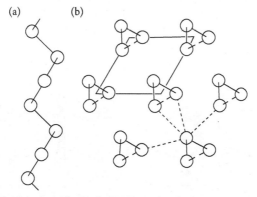

(a)　　　　(b)

Fig. 6.8.1 Structure of α-selenium (or tellurium): (a) side view of Se_x (or Te_x) helical chain, (b) viewed along the helices, the hexagonal unit cell and the coordination environment about one atom is indicated.

distance is 228 pm, and in $Te_4(AsF_6)_2$ the Te–Te distance is 266 pm. These two distances are shorter than those in the respective elemental forms, 237 pm and 284 pm, respectively, being consistent with the effect of some multiple bonding.

The structure of $Te_6{}^{4+}$ is shown in Fig. 6.8.2(b). In this trigonal prismatic cation, the average Te–Te bond length within a triangular face is 268 pm, and the average Te···Te distance between the parallel triangular faces is 313 pm. The $Te_6{}^{4+}$ cation can be considered as a dimer of two $Te_3{}^{2+}$ units that are consolidated by a π^*-π^* 6c-4e bonding interaction. As shown in Fig. 6.8.3, the tellurium $5p_z$ orbitals give rise to six molecular orbitals of $a_1{}'$, e', $a_2{}''$, and e'' symmetry, the first three being used to accommodate eight valence electrons. The e' orbitals are non-bonding within the individual $Te_3{}^{2+}$ units but form bonding interaction between them, and the bonding $a_1{}'$ and antibonding $a_2{}''$ orbitals cancel each other. The formal bond order along each prism edge is therefore 2/3.

Formal addition of two electrons to $Te_6{}^{4+}$ gives $Te_6{}^{2+}$, which takes the shape of a boat-shaped six-membered ring, as shown in Fig. 6.8.2(c). The average length of the pair of weak transannular interactions

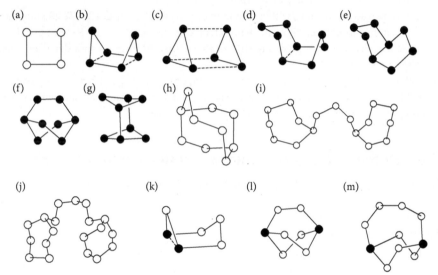

Fig. 6.8.2 Structures of some polyatomic cations of Se and Te. (a) $Se_4{}^{2+}$ in $Se_4(H_2S_2O_7)_2$; $Te_4{}^{2+}$ in $Te_4(AsF_6)_2$. (b) $Te_6{}^{2+}$ in $Te_6(MOCl_4)_2$ (M = Nb, W). (c) $Te_6{}^{4+}$ in $Te_6(AsF_6)_4 \cdot 2AsF_3$. (d) $Te_8{}^{2+}$ in $Te_8(ReCl_6)_2$; $Se_8{}^{2+}$ in $Se_8(AlCl_4)_2$. (e) $Te_8{}^{2+}$ in $Te_8(WCl_6)_2$. (f) $Te_8{}^{2+}$ in $(Te_6)(Te_8)(WCl_6)_4$. (g) $Te_8{}^{4+}$ in $(Te_8)(VOCl_4)_2$. (h) $Se_{10}{}^{2+}$ in $Se_{10}(SbF_6)_2$. (i) $Se_{17}{}^{2+}$ in $Se_{17}(NbCl_6)_2$. (j) $Se_{19}{}^{2+}$ in $Se_{19}(SbF_6)_2$. (k) $(Te_2Se_4)^{2+}$ in $(Te_2Se_4)(SbF_6)_2$. (l) $(Te_2Se_6)^{2+}$ in $(Te_2Se_6)(Te_2Se_8)(AsF_6)_4$. (m) $(Te_2Se_8)^{2+}$ in $(Te_2Se_8)(AsF_6)_2$.

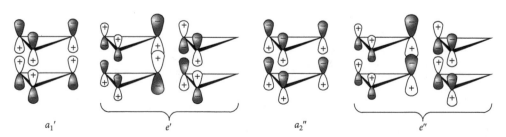

Fig. 6.8.3 Molecular orbitals in $Te_6{}^{4+}$.

is 329 pm, which indicates that the two positive charges are delocalized over all four Te atoms in the rectangular base.

Fig. 6.8.2(d) shows the conformation of and weak central transannular interaction in the Se_8^{2+} and Te_8^{2+} cations, which are isostructural with S_8^{2+}. Their common bicyclic structure can be regarded as being derived from a crown-shaped eight-membered ring by flipping one atom from an *exo* to an *endo* position, with the formally positively charged atoms interacting in transannular linkage. In $Te_8(ReCl_6)_2$, the Te–Te bond length indicates a normal single bond, and the Te$\cdots$Te distance is 315 pm. In $Se_8(AlCl_4)_2$, the Se–Se bond lengths lie in the range 229–236 pm, and Se$\cdots$Se is 284 pm.

Figs 6.8.2(e) and (f) show two other isomeric forms of Te_8^{2+}. In $Te_8(WCl_6)_2$, the Te_8^{2+} cation is composed of two five-membered rings each taking an envelope conformation, with an average Te–Te bond length of 275 pm and a relatively short transannular Te$\cdots$Te bond of 295 pm. In $(Te_6)(Te_8)(WCl_6)_4$, Te_8^{2+} exhibits a bicyclo[2.2.2]octane geometry with two bridge-head Te atoms.

Fig. 6.8.2(g) shows the structure of Te_8^{4+}; this 44 valence electron cluster takes a cube shape with two cleaved edges, the positive charges being located on four three-coordinate Te atoms. The Te_8^{4+} cation can be viewed as two planar Te_4^{2+} ions that have dimerized via the formation of a pair of Te–Te bonds with simultaneous loss of electronic delocalization and distortion from planarity.

Fig. 6.8.2(h), (i), and (j) show the structures of Se_{10}^{2+}, Se_{17}^{2+}, and Se_{19}^{2+}, respectively. They all consist of seven- or eight-membered rings that are connected by short chains. Each homopolyatomic cation has two three-coordinate atoms that formally carry the positive charges. Se_{10}^{2+} has a bicyclo[2.2.4]decane geometry. The Se–Se bond distances vary between 225 pm and 240 pm, and the Se–Se–Se angles range from 97° to 106°. Se_{17}^{2+} and Se_{19}^{2+} comprise a pair of seven-membered rings connected by a three- and four-atom chain, respectively.

Figs 6.8.2(k), (l), and (m) show the structures of $(Te_2Se_4)^{2+}$, $(Te_2Se_6)^{2+}$, and $(Te_2Se_8)^{2+}$, respectively. In these heteropolyatomic cations, the heavier Te atoms generally have a higher coordination number of three and serve as positive charge bearers, which is consistent with the lower electronegativity of Te compared to Se. As expected and confirmed by experiment, a weak transannular Te$\cdots$Te bond exists in the boat-shaped $(Te_2Se_4)^{2+}$ cation. The $(Te_2Se_6)^{2+}$ and $(Te_2Se_8)^{2+}$ cations have bicyclo[2.2.2]octane and bicyclo[2.2.4]decane geometries, respectively, with the Te atoms located at the bridgehead positions.

Fig. 6.8.4 shows the structures of some polymeric cations of Se and Te. In most of these systems, the Te–Te bonds link the Te atoms to form an infinite polymeric chain. The coordination numbers of the Te atoms are normally two or three, but some may attain the value of four in forming hypervalent structures. Various polymeric cations contain four-, five-, or six-membered rings. The four-membered ring is planar, but the larger rings are non-planar. The rings are directly connected or linked by short fragments of one, two, or three atoms. In the heteroatom polymeric cations, the Te atoms invariably occupy the three-coordinate sites.

Figs 6.8.4(a) and (j) show the structures of the coexisting $(Te_4^{2+})_\infty$ and $(Te_{10}^{2+})_\infty$ in $(Te_4)(Te_{10})(Bi_4Cl_{16})$. $(Te_4^{2+})_\infty$ is composed of planar squares of Te atoms that are connected by Te–Te bonds to form an infinite zigzag chain. There are equal numbers of two- and three-coordinate Te atoms, so that the latter carry the positive charges. The bond lengths within each square ring are 275 pm and 281 pm, and the inter-ring bond distance is 297 pm.

Figs 6.8.4(b)–(e) show the structures of $(Te_6^{2+})_\infty$ in $(Te_6)(HfCl_6)$, $(Te_7^{2+})_\infty$ in $(Te_7)(AsF_6)_2$, and $(Te_8^{2+})_\infty$ in $(Te_8)(U_2Br_{10})$ and $(Te_8)(Bi_4Cl_{14})$, respectively. These polymeric zigzag chains are composed of five- or six-membered rings that are linked by one or two atoms.

Figs 6.8.4(f) and (g) show the structures of two related heteroatom polymeric cationic chains composed of Te and Se. In $(Te_{3.15}Se_{4.85})(WOCl_4)_2$, the non-stoichiometric, disordered $(Te_{3.15}Se_{4.85}^{2+})_\infty$ cationic chains are constructed from the linkage of five-membered rings by non-linear three-atom fragments. In $(Te_3Se_4)(WOCl_4)_2$, the $(Te_3Se_4^{2+})_\infty$ chain is composed of planar Te_2Se_2 rings that are connected by non-linear Se–Te–Se fragments.

Figs 6.8.4(h)–(j) show the structures of polymeric cations that contain hypervalent Te atoms. Two kinds of $(Te_7{}^{2+})_\infty$ chains are found separately in $(Te_7)(Be_2Cl_6)$ and $(Te_7)(NbOCl_4)_2$, and $(Te_{10}{}^{2+})_\infty$ exists in $(Te_4)(Te_{10})(Be_4Cl_{16})$. In these polymeric cations, the hypervalent Te atoms each exhibits square-planar coordination to form a $TeTe_4$ unit with Te–Te bond lengths in the range 292–297 pm. In $(Te_7{}^{2+})_\infty$, the $TeTe_4$ unit and a pair of terminal Te atoms constitute an enlarged Te_7 unit composed of two planar squares sharing a common vertex. In $(Te_{10}{}^{2+})_\infty$, the basic Te_{10} structural unit consists of a linear arrangement of three corner-sharing planar squares, and such Te_{10} units are laterally connected by Te–Te bonds to generate a corrugated polymeric ribbon that also contain chair-like six-membered rings.

(2) Polyatomic anions

The chemistry of polyselenides, polytellurides, and their metal complexes is very well established. Typical structures of polyselenide dianions are shown in Fig. 6.8.5. In these species, the Se–Se bond distances vary from 227 pm to 236 pm, and the bond angles from 103° to 110°. The tethered monocyclic structure of $Se_9{}^{2-}$

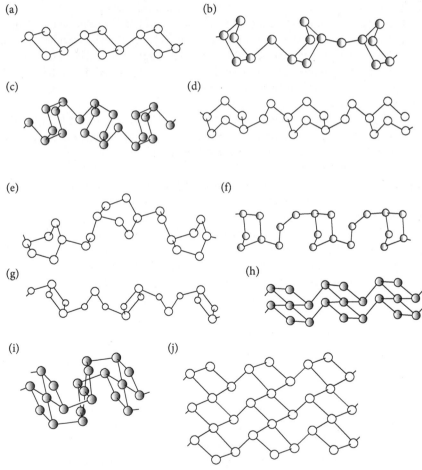

(a)

(b)

(c)

(d)

(e)

(f)

(g)

(h)

(i)

(j)

Fig. 6.8.4 Structures of some polymeric cations of Se and Te. (a) $(Te_4{}^{2+})_\infty$ in $(Te_4)(Te_{10})(Bi_4Cl_{16})$. (b) $(Te_6{}^{2+})_\infty$ in $(Te_6)(HfCl_6)$. (c) $(Te_7{}^{2+})_\infty$ in $(Te_7)(AsF_6)_2$. (d) $(Te_8{}^{2+})_\infty$ in $(Te_8)(U_2Br_{10})$. (e) $(Te_8{}^{2+})_\infty$ in $(Te_8)(Bi_4Cl_{14})$. (f) $(Te_{3.15}Se_{4.85}{}^{2+})_\infty$ in $(Te_{3.15}Se_{4.85})(WOCl_4)_2$. (g) $(Te_3Se_4{}^{2+})_\infty$ in $(Te_3Se_4)(WOCl_4)_2$. (h) $(Te_7{}^{2+})_\infty$ in $(Te_7)(Bi_2Cl_6)$. (i) $(Te_7{}^{2+})_\infty$ in $(Te_7)(NbOCl_4)_2$. (j) $(Te_{10}{}^{2+})_\infty$ in $(Te_4)(Te_{10})(Bi_4Cl_{16})$.

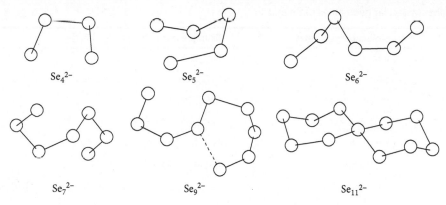

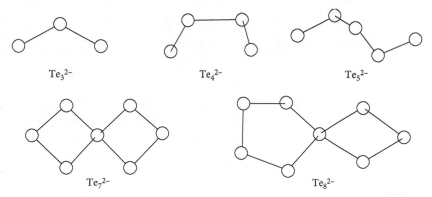

Fig. 6.8.5 Structures of some dianions Se_x^{2-}. For Se_9^{2-}, the longest bond is represented by a broken line.

Fig. 6.8.6 Structures of some dianions Te_x^{2-}.

in the complex $Sr(15\text{-}C5)_2(Se_9)$ has a three-connected Se atom forming two long and one normal Se–Se bonds at 295 pm, 247 pm, and 231 pm (anticlockwise in Fig. 6.8.5, with the longest bond represented by a broken line). The other Se–Se bonds are in the range 227–239 pm.

The dianion Se_{11}^{2-} has a centrosymmetric spiro-bicyclic structure involving a central square-planar Se atom common to the two chair-shaped rings. The shared atom forms four long Se–Se bonds of length 266–268 pm, and the structure may be described as a central Se^{2+} chelated by two $\eta^2\text{-}Se_5^{2-}$ ligands.

Some typical structures of polytelluride dianions, Te_x^{2-}, are shown in Fig. 6.8.6. In these species, the Te–Te bond distances vary from 265 pm to 284 pm. In the bicyclic polytellurides Te_7^{2-} and Te_8^{2-}, the central Te atom each as four long bonds with bond lengths from 292 pm to 311 pm.

The Se_x^{2-} and Te_x^{2-} are effective chelating ligands for both main group and transition metals, giving rise to complexes such as $Sn(\eta^2\text{-}Se_4)_3^{2-}$, $[M(\eta^2\text{-}Se_4)_2]^{2-}$ (M = Zn, Cd, Hg, Ni, Pb), $Ti(\eta^5\text{-}C_5H_5)_2(\eta^2\text{-}Se_5)$, $[Hg(\eta^3\text{-}Te_7)]^{2-}$, and $M_2(\mu_2\text{-}Te_4)(\eta^2\text{-}Te_4)_2$ (M = Cu, Ag).

6.8.3 **Stereochemistry of selenium and tellurium**

Selenium and tellurium exhibit a great variety of molecular geometries as a consequence of the number of stable oxidation states. Various observed structures are summarized in Table 6.8.1, in which A is a central Se or Te atom, X is an atom bonded to A, and E represents a lone pair.

With reference to Table 6.8.1, the stereochemistries of Se and Te compounds are briefly described below.

Table 6.8.1 Molecular geometries of Se and Te

Type	Molecular geometry	Example (structure in Fig. 6.8.7)
AX_2E	**bent**	SeO_2
AX_2E_2	bent	$SeCl_2$, $Se(CH_3)_2$, $TeCl_2$, $Te(CH_3)_2$
AX_3	trigonal planar	SeO_3, TeO_3
AX_3E	trigonal pyramidal	$(SeO_2)_x$ (a), $OSeF_2$ (b)
AX_3E_2	T-shaped	$[SeC(NH_2)_2]_3{}^{2+}$ (c), $C_6H_5TeBr(SC_3N_2H_6)$ (d)
AX_4	tetrahedral	O_2SeF_2, $(SeO_3)_4$
AX_4E	disphenoidal	$Se(C_6H_5)_2Cl_2$, $Se(C_6H_5)_2Br_2$, $Te(CH_3)_2Cl_2$, $Te(C_6H_5)_2Br_2$
AX_5E	square pyramidal	$TeF_5{}^-$ (e), $(TeF_4)_x$ (f)
AX_6	octahedral	SeF_6, TeF_6, $(TeO_3)_x$, $F_5TeOTeF_5$ (g)
AX_6E	octahedral	$SeCl_6{}^{2-}$, $TeCl_6{}^{2-}$
AX_7	pentagonal bipyramidal	$TeF_7{}^-$ (h)

(1) AX_2E type

The SeO_2 molecule in the vapor phase has a bent configuration with Se–O 160.7 pm and O–Se–O 114°. Crystalline SeO_2 is built of infinite chains, in which each Se atom is bonded to three oxygen atoms (AX_3E type) in a trigonal pyramidal configuration. The bond lengths are Se–O_b 178 pm, Se–O_t 173 pm, as shown in Fig. 6.8.7(a).

(2) AX_2E_2 type

Many AX_2E_2 type molecules, such as $Se(CH_3)_2$, $SeCl_2$, $Te(CH_3)_2$, and $TeBr_2$, all exhibit a bent configuration, in which repulsion of lone pairs makes the inter-bond angles smaller than the ideal tetrahedral angle.

	$Se(CH_3)_2$	$SeCl_2$	$Te(CH_3)_2$	$TeCl_2$
A–X (pm)	194.5	215.7	214.2	232.9
X–A–X	96.3°	99.6°	94°	97°

(3) AX_3 type

Monomeric selenium trioxide (SeO_3) and tellurium trioxide (TeO_3) have a trigonal planar structure in the gas phase. In the solid state, SeO_3 forms cyclic tetramers $(SeO_3)_4$, in which each Se atom connects two bridging O atoms and two terminal O atoms, with Se–O_b 177 pm and Se–O_t 155 pm (AX_4 type). The solid-state structure of TeO_3 is a three-dimensional framework, in which Te(VI) forms TeO_6 octahedra (AX_6 type) sharing all vertices.

(4) AX_3E type

Pyramidal molecule $SeOF_2$ has bond lengths Se=O 158 pm and Se–F 173 pm, bond angles F–Se–F 92° and F–Se–O 105°, as shown in Fig. 6.8.7(b). Its dipole moment (2.62 D in benzene) and dielectric constant (46.2 at 20 °C) are both high, and accordingly it is a useful solvent.

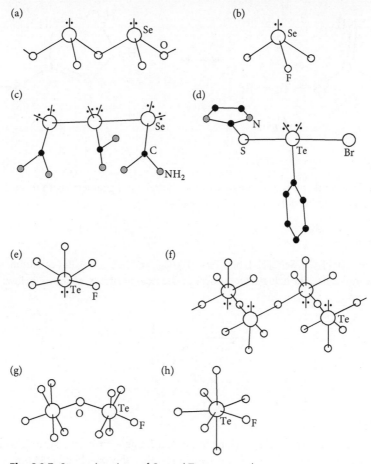

Fig. 6.8.7 Stereochemistry of Se and Te compounds.

(5) **AX₃E₂ type**

The cation $[SeC(NH_2)_2]_3{}^{2+}$ adopts T-shaped geometry, as shown in Fig. 6.8.7(c). In this structure, the central Se atom must bear a formal negative charge to have two lone pairs at the equatorial positions of a trigonal bipyramid. The valence-bond structural formula of this cation is given below:

In the molecule $C_6H_5TeBr(SC_3N_2H_6)$, the Te atom has a similar T-shaped configuration, as shown in Fig. 6.8.7(d). Bonding can be described in terms of resonance between a pair of valence-bond structures:

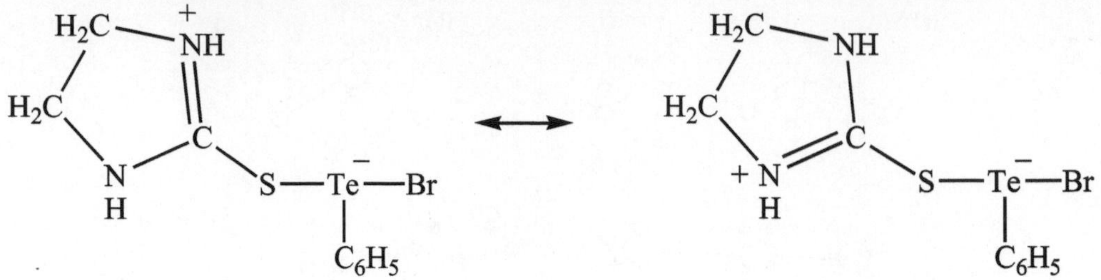

(6) **AX$_4$ type**

The molecule SeO_2F_2 and analogous compounds have tetrahedral geometry with three different bond angles: O–Se–O 126.2°, O–Se–F 108.0°, and F–Se–F 94.1°.

(7) **AX$_4$E type**

The molecules $Se(C_6H_5)_2Cl_2$, $Se(C_6H_5)_2Br_2$, $Te(CH_3)_2Cl_2$, and $Te(C_6H_5)_2Br_2$ constitute this structure type. In all these molecules, the halogen atoms occupy the axial positions, as shown below:

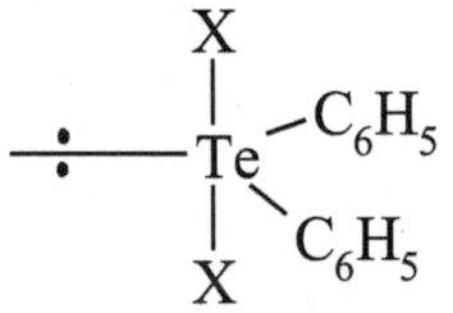

(8) **AX$_5$E type**

The anion TeF_5^- and polymeric $(TeF_4)_x$ belong to the AX$_5$E type with square pyramidal configuration. In TeF_5^-, the bonds in the square base (196 pm) are longer than the axial bond (185 pm), and the bond angles (79°) is smaller than 90°, as shown in Fig. 6.8.7(e). Crystalline $(TeF_4)_x$ has a chain structure, in which TeF_5 groups are linked by bridging F atoms in such a way that alternate pyramids are oriented in opposite directions, as shown in Fig. 6.8.7(f).

(9) **AX$_6$ type**

The hexafluorides SeF_6 and TeF_6 have the expected regular octahedral configuration. In $F_5SeOSeF_5$ and $F_5TeOTeF_5$, the Se and Te atoms take the octahedral configuration, and the four equatorial bonds in each case are bent away from the bridging O atom, as shown in Fig. 6.8.7(g).

(10) **AX$_6$E type**

The anions SeX_6^{2-} and TeX_6^{2-} (X = Cl, Br) adopt regular octahedral geometry, apparently indicating that the lone pair in the valence shell is stereochemically inactive. The observed result can be explained as follows. (a) With increasing size of the central atom the tendency for the lone pair to spread around the core is enhanced. It is drawn inside the valence shell, behaving like an s-type orbital and effectively becoming the outer shell of the core. (b) This tendency is also enhanced by the presence of six bonding pairs in the valence shell, which leaves rather little space for the lone pair. (c) With the addition of the non-bonding electron pair, the core size increases and the core charge decreases from +6 to +4, with the result that the bond pairs move farther from the central nucleus, thus increasing the bond lengths.

In accordance with this, the observed bond lengths of SeX_6^{2-}, TeX_6^{2-}, and SbX_6^{2-} (X = Cl, Br) ions are considerably longer than those expected from the sum of the covalent radii by about 20–25 pm.

(11) AX_7 type

The TeF_7^- ion, an isoelectronic and isostructural analog of IF_7, has a pentagonal bipyramidal structure with Te–F_{ax} 179 pm and Te–F_{eq} 183–190 pm, as shown in Fig. 6.8.5(h). The equatorial F atoms deviate slightly from the mean equatorial plane.

6.8.4 **Chalcogenides of selenium and tellurium**

The meaning of the term chalcogenide (ChB) is described in Section 4.8.1. The number of σ-holes used by a chalcogen donor to bind to negative ions depends on its bonding characteristics. It has been found that there are three main types of donor molecules in the chalcogen bonding range, as shown in Fig. 6.8.8. In Type (a) and Type (b), up to two σ-holes can be present on the chalcogen atom, whereas in Type (c) only one σ-hole is located at the terminus of the chalcogen atom along the extension of the C=Ch bond. In Type (a) the double-coordinating chalcogen atom is part of an electron-deficient aromatic ring; in Type (b) there is divalent exocyclic substituent on (usually aromatic) electron-withdrawing groups, and the chalcogen atom obtains electrons from the divalent outward substituents of an electron-withdrawing group; and in Type (c) only one σ-hole is present at the terminus of the chalcogen atom along the extension of the C=Ch bond (such as from $Se=CF_2$).

In Types (a) and (b), the chalcogen atom serves as an independent bi-coordinated σ-hole ligand and forms a E–Ch···A straight line with the E atom in Fig. 6.8.8. In the case of Type (c), the chalcogen atom extends along the C=Ch direction to combine with a coordinated anion. To date, the most common coordination type found in solids is Type (a).

Example of Type (a): In the structure of $[C_6H_4N_2RSe^+][I_3^-]$ with R = $CH(CMe_3)_2$ shown in Fig. 6.8.9(a), Se···I bonding is due to the action of ChB, and lengthening of the Se–N bond is also observed.

Example of Type (b). The structure of a macrocyclic compound of tellurium is shown in Figs 6.8.9(b) and (c). The ChB interaction between each I⁻ ion with four Te atoms generates an octahedral cage structure.

Example of Type (c). There is as yet no well-characterized example of this type of chalcogen bonding, but quantum-chemical calculation has established that a σ hole at the outer region of the Se atom that forms a C=Se double bond can be paired with an lone pair in nitrogen-containing bases of the type NZ (NZ = N_2, NCH, NH_3, $NHCH_2$, NCLi, and NMe_3) (Fig. 6.8.10).

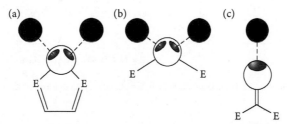

Fig. 6.8.8 Geometry of three types of ChB-anion interactions (represented by dotted lines) in ChB donor atoms with different coordination characteristics (open circle represents a chalcogen atom, ellipse represents a sigma hole, "E" represents an electron-withdrawing atom or group, and "●" represents an external anion A.

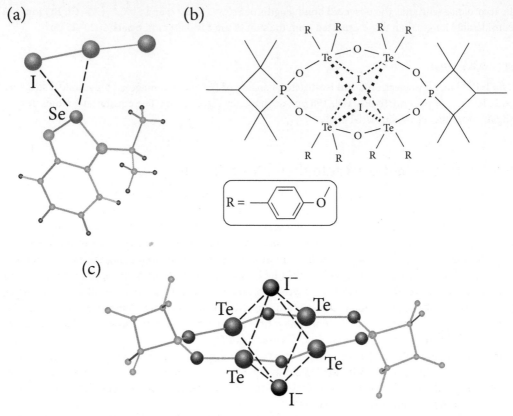

Fig. 6.8.9 (a) ChB interaction of Se···I in $[C_6H_4N_2RSe^+][I_3^-]$, (b) molecular formula of 12-membered macrocyclic tellurium compound $[((p\text{-MeOC}_6H_4)_2Te)_2(\mu\text{-O})(\mu\text{-cycPO}_2)(\mu_4\text{-I})]_2 \cdot 4C_6H_6$ with [cycPO$_2$H = 1,1,2,3,3-pentamethyltrimethylenephosphinic acid] and omission of the solvated benzene molecules, and (c) side view of 12-membered macrocyclic tellurium compound showing ChB interaction between I⁻ ions and Te atoms.

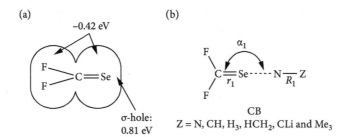

Fig. 6.8.10 (a) Molecular electrostatic potential of $F_2C=Se$. (b) Chalcogen-bonding between $F_2C=Se$ and nitrogen-containing bases.

References

J. Y. C. Lim and P. D. Beer, Sigma-hole interactions in anion recognition, *Chem.* **4**, 731–83 (2018).

X. Guo, X. An, and Q Li, Se···N chalcogen bond and Se···X halogen bond involving $F_2C=Se$: influence of hybridization, substitution, and cooperativity, *J. Phys. Chem. A* **119**, 3518–27 (2015).

V. Chandrasekhar and R. Thirumoorthi, Halide-capped tellurium-containing macrocycles, *Inorg. Chem.* **48**, 10330–7 (2009).

6.8.5 Hypervalent structural features of tellurium chemistry

The reaction of PhTe–TePh with I_2 forms a violet-black "PhTeI" compound; in the solid state it exists as tetrameric molecule $Ph_4Te_4I_4$, in which the individual PhTeI units are linked through weak Te···Te bonds that constitute the sides of a Te_4 square (Fig. 6.8.11(a)). In contrast, the selenium analogue $Ph_4Se_4I_4$ is a centrosymmetric charge-transfer 2:2 complex between diselenide PhSeSePh and I_2 that is stabilized by very weak Se···I contacts (Fig. 6.8.11(b)). The tetrameric structure of $Ph_4Te_4I_4$ is easily disrupted upon addition of PPh_3 to give the 1:1 adduct $Ph_3PTe(Ph)I$ (Fig. 6.8.11(c)).

Binary chalcogen nitrides provide a noteworthy illustration of the hypervalent behavior of tellurium. The tetrachalcogen tetranitrides E_4N_4 (E = S, Se) both adopt an intriguing molecular cage structure stabilized by two weak transannular E···E interactions (Fig. 6.8.12(a)). In marked contrast, the tellurium

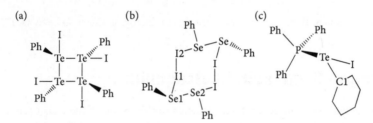

Fig. 6.8.11 (a) Molecular structure of $Ph_4Te_4I_4$. The Te–Te and Te–I bond lengths lie in the range 312.5(2)–317.5(2) pm and 279.9(2)–283.0(2) pm, respectively. (b) Molecular structure of centrosymmetric $Ph_4Se_4I_4$. Bond lengths [pm] and angles [°]: Se1–Se2 234.7, I1–I2 277.5, Se1–I1 234.7, Se2–I2a 358.8, Se1–C6 194.4(7), Se2–C12 192.8(6); Se2–Se1–I1 90.2(1), Se1–Se1–C6 104.3(2), I1–Se1–C6 94.7(2), Se1–Se2–C12a 176.5(1), Se1–Se2–C12 98.4(2), I2a–Se2–C I2 84.0(2), Se1–I1–I2 174.2(1), I1a–I2a–Se2 88.4(1). (c) Molecular structure of $Ph_3PTe(Ph)I$. Selected bond lengths [pm] and angles [°]: Te–P 256.8(2), Te–I 309.3(1), Te–C 1 210.9(7), P–Te–I 179.45(5), P–Te–C1 91.0(2).

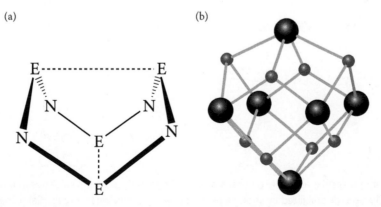

Fig. 6.8.12 Molecular structures of (a) E_4N_4 (E = S, Se) and (b) the Te_6N_8 core unit in $Te_6N_8(TeCl_4)_4$.

analogue of empirical composition Te_3N_4 has a μ^3-nitrido structural motif (Fig. 6.8.12(b)) that was established by structural determination of the Lewis acid adduct $Te_6N_8(TeCl_4)_4$.

The reaction of tellurium tetrachloride with tris(trimethylsilyl)amine in THF gives pale yellow, moisture-sensitive $[Te_6N_8(TeCl_4)_4\cdot4THF]\cdot3.5THF$ crystals in over 80% yield. The rhombic dodecahedral Te_6N_8 core unit, in which the tellurium atoms constitute the corners of a distorted octahedron with the nitrogen atoms positioned above the faces as μ_3 ligands (Fig. 6.8.12(b)). Two opposite corners, Te1 and Te2, are each additionally coordinated by a pair of THF ligands, and the remaining 3.5 THF solvate molecules (one located on a crystallographic two-fold axis) fit into gaps of the crystal packing.

References

S. Kubiniok, W.-W. du Mont, S. Pohl, and W. Saak, The reagent diphenyldiselane/iodine: no phenylselenenyl iodide but a charge transfer complex with cyclic moieties, *Angew. Chem. Int. Ed. Engl.* **27**, 431–2 (1988).

W. Mosa, C. Lau, M. Möhlen, B. Neumöller, and K. Dehnicke, $[Te_6N_8(TeCl_4)_4]$—tellurium nitride stabilized by tellurium tetrachloride, *Angew. Chem., Int. Ed.* **37**, 2840–2 (1998).

P. D. Boyle, W. I. Cross, S. M. Godfrey, C. A. McAuliffe, R. G. Pritchard, S. Sarwar, and J. M. Sheffield, Synthesis and characterization of $Ph_4Te_4I_4$, containing a Te_4 square, and $Ph_3PTe(Ph)I$, *Angew. Chem. Int. Ed.* **39**, 1796–8 (2000).

T. Chivers and R. S. Laitinen, Tellurium: a maverick among the chalcogens, *Chem. Soc. Rev.* **44**, 1725–39 (2015).

6.9 Chalcogen Bonding in Chalcogenides

In general, the strength of a chalcogen bond depends on: i) nature of the chalcogen atom (Te > Se > O > S), ii) Lewis basicity of the interacting partner, iii) polarization of the chalcogen atom (which may be increased by using compounds containing cationic heteroarene or polyfluorinated arene backbones), and iv) R–Ch···Lewis-base interaction angle: strong chalcogen bonds have angles close to 180°. In

(a)   (b)

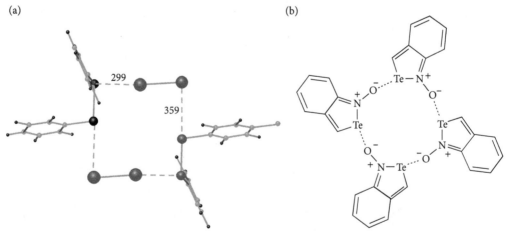

Fig. 6.9.1 Supramolecular assembly via chalcogen bonding. (a) Cyclic complex between molecular I_2 and diphenyldiselenide stabilized by strong linear I···Se halogen and weak angular Se···I chalcogen bonding. (b) Tetracyclic array of iso-tellurazole oxide molecules stabilized by Te···O chalcogen bonding.

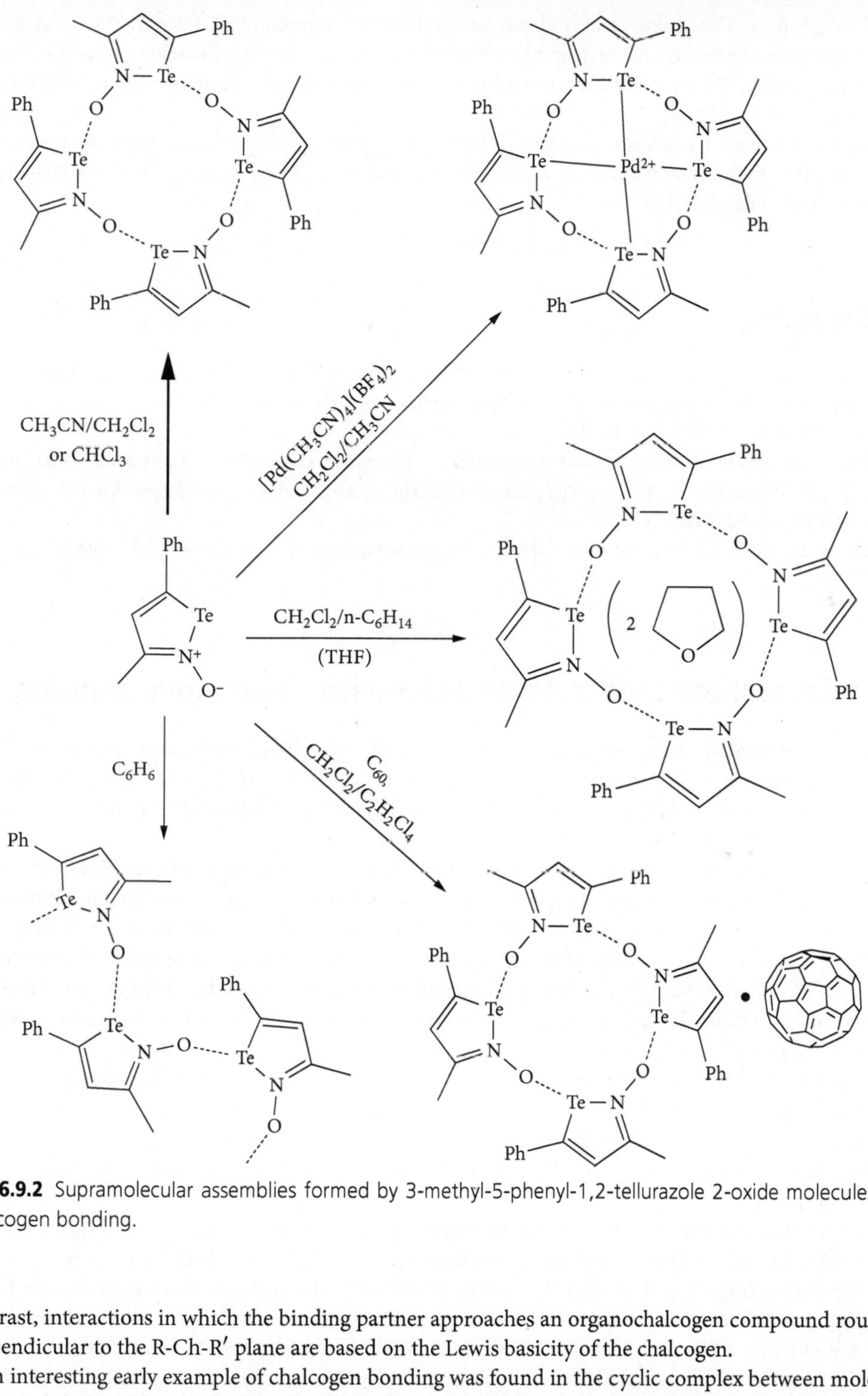

Fig. 6.9.2 Supramolecular assemblies formed by 3-methyl-5-phenyl-1,2-tellurazole 2-oxide molecules via chalcogen bonding.

contrast, interactions in which the binding partner approaches an organochalcogen compound roughly perpendicular to the R-Ch-R′ plane are based on the Lewis basicity of the chalcogen.

An interesting early example of chalcogen bonding was found in the cyclic complex between molecular I_2 and diphenyldiselenide (Fig. 6.9.1(a)). The rectangular coordination pattern is held together by two strong linear halogen bonds (I···Se = 299 pm) and two much weaker angular chalcogen bonds

(Se⋯I = 359 pm). This example also clearly exemplifies the "amphiphilic" (electrophilic and nucle-ophilic) nature of respective halogen and chalcogen substituents. The iso-tellurazole N-oxide molecule shown in Fig. 6.9.1(b) self-assembles into a highly symmetrical tetracyclic array through short Te⋯O chalcogen-bond formation.

3-Methyl-5-phenyl-1,2-tellurazole 2-oxide (**1**) serves as a building block for a variety of supramolecular aggregates that includes zigzag chain, cyclic tetramer and hexamer by forming Te⋯O–N chalcogen-bonded bridges (Fig. 6.9.2).

References

P. C. Ho, P. Szydlowski, J. Sinclair, P. J. W. Elder, J. Kübel, C. Gendy, L. M. Lee, H. Jenkins, J. F. Britten, D. R. Morim, and I. Vargus-Baca, Supramolecular macrocycles reversibly assembled by Te⋯O chalcogen bonding, *Nat. Commun.* **7**, 11299–308 (2016).

L. Vogel, P. Wonner, and S. M. Huber, Chalcogen bonding: an overview. *Angew. Chem. Int. Ed.* **58**, 1880–91 (2019).

P. Scilabra, G. Terraneo, and G. Resnati, The chalcogen bond in crystalline solids: a world parallel to halogen bond, *Acc. Chem. Res.* **52**, 1313–24 (2019).

N. Biot and D. Bonifazi, Chalcogen-bond driven molecular recognition at work, *Coord. Chem. Rev.* **413**, 213243 (2020).

6.10 **Crystalline Solids Assembled with Chalcogen Bonding**

The structural formula and crystal structure of benzo-1,2,5-selenadiazole is shown in Fig. 6.9.1(a). The molecules form trimeric clusters through two distinct chalcogen bonds ($d_{Se⋯N}$ = 315.5 pm). The clusters are further interconnected through weak hydrogen bonds involving both N- and Se-atoms ($d_{N⋯C}$ = 341.7 pm and $d_{Se⋯C}$ = 392.8 pm).

The structural formula and X-ray crystal structure of benzo-1,2,5-selenadiazolium chloride is shown in Fig. 6.9.1(b). Interestingly, the supramolecular ribbon is formed through a combination of multiple chalcogen- and hydrogen-bonds between the Se- and NH-atoms ($d_{Se⋯Cl}$ = 296.1–308.5 pm, $d_{N-H⋯Cl}$ = 308.7 pm). In the crystal structural of N-methylated benzo-1,2,5-selenadiazolium iodide, the molecules are arranged in dimers through the formation of four bifurcated chalcogen bonds ($d_{Se⋯I}$ = 317.6–361.0 pm), in which iodide ions act as bridging cornerstones between a pair of head-to-head methylated selenadiazolium cations (Fig. 6.9.1(c)).

The reaction of N-methylated benzo-1,2,5-selenadiazolium iodide with less than an equivalent of [Me₃O]BF₄ gives a supramolecular trimer in which one benzo-1,2,5-selenadiazole molecule is sand-wiched between two N-methylated benzo-1,2,5-selenadiazolium iodide units (Fig. 6.9.1(d)). This trimer features short Se⋯N contacts ($d_{Se⋯N}$ = 257.3 pm and 293.7 pm) and terminal Se⋯I chalcogen bonds ($d_{Se⋯I}$ = 352.8 pm).

In the crystal structure of benzo-2,1,3-telluradiazole, the molecules are organized into zigzag ribbons through the formation of double chalcogen bonds ($d_{Te⋯N}$ = 268.2–272.0 pm), forming non-covalent four-membered rings (Fig. 6.10.2(a)). The 3,6-di-bromo derivative of benzo-2,1,3-telluradiazole forms chalcogen-bonded dimers ($d_{Te⋯N}$ = 269.6 pm) in the solid state (Fig. 6.10.2(b)). The dimers further link to one another by chalcogen- ($d_{Te⋯Br}$ = 368.3 pm) and H-bonding ($d_{N⋯C}$ = 349.9 pm) interactions. The crystal structure of doubly protonated benzo-2,1,3-telluradiazolium dichloride (Fig. 6.10.2(c)) shows the formation of ribbons each stabilized by a combination of chalcogen- and hydrogen-bonding interactions ($d_{Te⋯Cl}$ = 277.3 pm and $d_{N⋯Cl}$ = 324.5 pm).

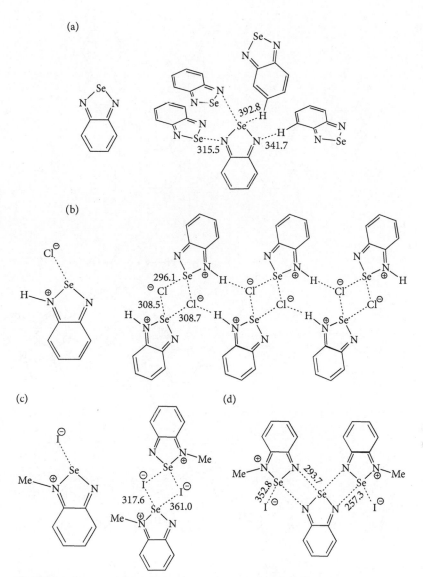

Fig. 6.10.1 Structural formulas and crystal structures of (a) benzo-1,2,5-selenadiazole, (b) benzo-1,2,5-selenadiazole hydrochloride, (c) *N*-methylated benzo-1,2,5-selenadiazolium iodide, and (d) 1:2 adduct of benzo-1,2,5-selenadiazole with *N*-methylated benzo-1,2,5-selenadiazolium iodide. Bond lengths are shown in pm.

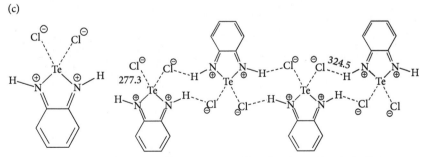

Fig. 6.10.2 Structural formulas of benzo-2,1,3-telluradiazole and its derivatives, together with their zigzag-ribbon structures in the solid state. (a) Top row: benzo-2,1,3-telluradiazole; (b) middle row: 3,6-dibromobenzo-2,1,3-telluradiazole; and (c) bottom row: N,N'-dihydro-benzo-2,1,3-telluradiazolium dichloride salt. Bond lengths are shown in pm.

References

P. C. Ho, P. Szydlowski, J. Sinclair, P. J. W. Elder, J. Kübel, C. Gendy, L. M. Lee, H. Jenkins, J. F. Britten, D. R. Morim, and I. Vargus-Baca, Supramolecular macrocycles reversibly assembled by Te···O chalcogen bonding, *Nat. Commun.* **7**, 11299–308 (2016).

L. Vogel, P. Wonner, and S. M. Huber, Chalcogen bonding: an overview. *Angew. Chem. Int. Ed.* **58**, 1880–91 (2019).

P. Scilabra, G. Terraneo, and G. Resnati, The chalcogen bond in crystalline solids: a world parallel to halogen bond, *Acc. Chem. Res.* **52**, 1313–24 (2019).

N. Biot and D. Bonifazi, Chalcogen-bond driven molecular recognition at work, *Coord. Chem. Rev.* **413**, 213243 (2020).

General References

T. Chivers and R. S. Laitinen, *Chalcogen-Nitrogen Chemistry*, Updated Edition, World Scientific, Singapore, 2021.

J.-X. Lu (ed.), *Some New Aspects of Transitional-Metal Cluster Chemistry*, Science Press, Beijing/New York, 2000.

L. F. Lundegaard, G. Weck, M. I. McMahon, S. Desgreniers, and P. Loubeyre, Observation of an O_8 molecular lattice in the ε phase of solid oxygen. *Nature* **443**, 201–4 (2006).

H. Pernice, M. Berkei, G. Henkel, H. Willner, G. A. Argüello, H. L. McKee and T. R. Webb, Bis(fluoroformyl)trioxide, FC(O)OOOC(O)F. *Angew. Chem. Int. Ed.* **43**, 2843–6 (2004).

T. S. Zwier, The structure of protonated water clusters. *Science* **304**, 1119–20 (2004).

M. Miyazaki, A. Fujii, T. Ebata, and N. Mikami, Infrared spectroscopic evidence for protonated water clusters forming nanoscale cages. *Science*, **304**, 1134–7 (2004).

J.-W. Shin, N. I. Hammer, E. G. Diken, M. A. Johnson, R. S. Walters, T. D. Jaeger, M. A. Duncan, R. A. Christie, and K. D. Jordan, Infrared signature of structures associated with the $H^+(H_2O)_n$ (n = 6 to 27) clusters. *Science*, **304**, 1137–40 (2004).

R. Steudel, O. Schumann, J. Buschmann, and P. Luger, A new allotrope of elemental sulfur: convenient preparation of cyclo-S_{14} from S_8. *Angew. Chem. Int. Ed.* **37**, 2377–8 (1998).

J. Beck, Polycationic clusters of the heavier group 15 and 16 elements, in G. Meyer, D. Naumann, and L. Wesemann (eds), *Inorganic Chemistry in Focus II*, Wiley-VCH, Weinheim, 2005, pp. 35–52.

W. S. Sheldrick, Cages and clusters of the chalcogens, in M. Driess and H. Nöth (eds), *Molecular Clusters of the Main Group Elements*, Wiley-VCH, Weinheim, 2004, pp. 230–45.

Chapter 7

Structural Chemistry of Group 17 and Group 18 Elements

7.1 Elemental Halogens and Polyhalogen Ions

7.1.1 Crystal structures of the elemental halogens

The halogens are diatomic molecules, whose color increases steadily with atomic number. Fluorine (F_2) is a pale yellow gas, bp 85.0 K. Chlorine (Cl_2) is a greenish-yellow gas, bp 239.1 K. Bromine (Br_2) is a dark-red liquid, bp 331.9 K. Iodine (I_2) is a lustrous black crystalline solid, mp 386.7 K, which sublimes and boils readily at 458.3 K. Actually solid iodine has a vapor pressure of 41 Pa at 298 K and 1.2×10^4 Pa at the melting point. In the solid state, the halogen molecules are aligned to give a layer structure. Fluorine exists in two crystalline modifications: a low temperature α-form and a higher temperature β-form, neither of which resembles the orthorhombic layer structure of the isostructural chlorine, bromine, and iodine crystals. Fig. 7.1.1 shows the crystal structure of iodine. Table 7.1.1 gives the interatomic distances in gaseous and crystalline halogens.

The molecules F_2, Cl_2, and Br_2 in the crystalline state have intramolecular distances (X–X) which are nearly the same as those in the gaseous state. In crystalline iodine, the intramolecular I–I bond distance is longer than that in a gaseous molecule, and the lowering of the bond order is offset by the intermolecular bonding within each layer. The closest interatomic distance between neighboring I_2 molecules is 350 pm, which is considerably shorter than twice the van der Waals radius (430 pm). It therefore seems that appreciable secondary bonding interactions occur between the iodine molecules, giving rise to the semiconducting properties and metallic luster; under very high pressure iodine becomes a metallic conductor.

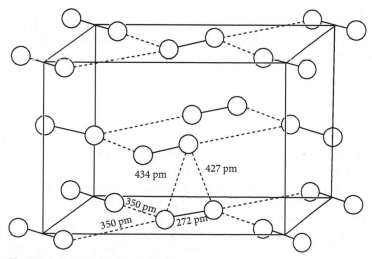

Fig. 7.1.1 Crystal structure of iodine.

Structural Chemistry across the Periodic Table. Thomas Chung Wai Mak, Yu-San Cheung, Gong-Du Zhou, and Ying-Xia Wang. Oxford University Press. © Thomas Chung Wai Mak, Yu-San Cheung, Gong-Du Zhou, and Ying-Xia Wang (2023). DOI: 10.1093/oso/9780198872955.003.0007

Table 7.1.1 Interatomic distances in gaseous and crystalline halogens

X	X–X/pm		X⋯X/pm		Ratio $\frac{\text{X⋯X (Shortest)}}{\text{X–X (Solid)}}$
	Gas	Solid	Within layer	Between layers	
F	143.5	149	324	284	1.91
Cl	198.8	198	332	374	1.68
Br	228.4	227	331	399	1.46
I	266.6	272	350	427	1.29

The distance between layers in the iodine crystal (427 pm and 434 pm) corresponds to the van der Waals distance.

7.1.2 **Homopolyatomic halogen anions**

The homopolyhalogen anions are formed mainly by iodine, which exhibits the highest tendency to form stable catenated anionic species. Numerous examples of small polyiodides, such as I_3^-, I_4^{2-}, I_5^-, and extended discrete oligomeric anionic polyiodides, such as I_7^-, I_8^{2-}, I_9^-, I_{12}^{2-}, I_{16}^{2-}, I_{16}^{4-}, I_{22}^{4-}, I_{29}^{3-}, and polymeric $(I_7^-)_n$ networks have been reported. These polyiodides are all formed by the relatively loose association of several I_2 molecules with several I^- and/or I_3^- anions. In order to assess the association of such species, the following equation between the bond length (d) and bond order (n) has been proposed:

$$d = d_0 - c \log n = 267 \text{ pm} - (85 \text{ pm}) \log n$$

According to this equation, when the distance between two iodine atoms is $d \leq 293$ pm, corresponding to the bond order $n \geq 0.50$, there is a relatively strong bond between them, which is represented by a solid line. The distance d between 293 pm and 352 pm corresponds to bond order n between 0.50 and 0.10, indicating a relatively weak bond, which is represented by a broken line. When the distance is longer than 352 pm, there is only van der Waals interactions between the two molecular/ionic species, and no discrete polyiodide is formed.

Fig. 7.1.2 shows some polyiodides, which have been characterized structurally. All polyiodine anions consist of units of I^-, I_2, and I_3^-. The bond length of the structural components of the polyiodides are often characteristic: 267 pm to 285 pm in I_2 molecular fragments, whereas those of symmetrical triiodide I_3^- are about 292 pm.

The formation and stability of an extended polyiodide species are dependent on the size, shape, and charge of its accompanying cation. In the solid state, the polyiodide species are assembled around a central cation to form discrete or one-, two-, and three-dimensional structure.

Similar to iodine, many advances have been made in the structure and bonding of polybrominated, polychlorinated, and polyfluorinated anions. The existence of polybromide Br_3^-, Br_4^{2-}, Br_5^-, Br_6^{2-}, Br_7^-, Br_8^{2-}, Br_9^-, Br_{10}^{2-}, Br_{11}^-, Br_{20}^{2-}, and Br_{24}^{2-} has been confirmed by experiments.

In $[(\text{ttmgn-Br}_4)(\text{BF}_2)_2](\text{Br}_5)_2$ crystal, [ttmgn represents 1,4,5,8-tetrakis (tetramethylguanidyl)-naphthalene in which all four aromatic hydrogens are replaced by bromine], Br_5^- is a V-shaped species. The Br^- ion at the center connects two Br_2 molecules, as shown in Fig. 7.1.3 (a).

Br_6^{2-} exists in the product $[C_5H_{10}N_2Br]_2[Br_6]$ of the reaction between the urea derivative $C_5H_{10}N_2O$ and oxalyl bromide $(C_2O_2Br_2)$. Its structure is shown in Fig. 7.1.3 (b). It is made up of two nearly linear Br_3^- ions that are almost perpendicular to each other. Each of the Br_3^- ion has a halogen bond $Br^-\cdots Br_2$.

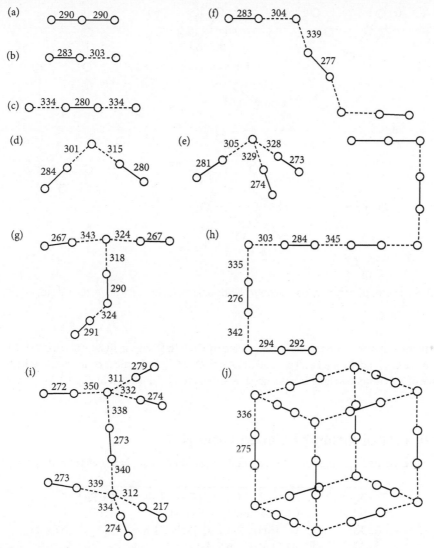

Fig. 7.1.2 Structures of some polyiodides: (a) I_3^- (symmetric) in Ph_4AsI_3, (b) I_3^- (asymmetric) in CsI_3, (c) I_4^{2-} in $Cu(NH_3)_4I_4$, (d) I_5^- in $Fe(S_2CNEt_2)_3I_5$, (e) I_7^- in Ph_4PI_7, (f) I_8^{2-} in $[(CH_2)_6N_4Me]_2I_8$, (g) I_9^- in Me_4NI_9, (h) I_{16}^{4-} in $[(C_7H_8N_4O_2)H]_4I_{16}$, (i) I_{16}^{2-} in $(Cp^*_2Cr_2I_3)_2I_{16}$, (j) $(I_7^-)_n$ network in a unit cell of the $\{Ag[18]aneS_6\}I_7$ complex. Bond lengths are in pm.

Br(3) as the donor is connected to the electrophilic terminal Br(4) of Br(4)–Br(5)–Br(6) and these four Br atoms are approximately linear. The Br(3)⋯Br(4) bond length is 359.3 pm, which is significantly shorter than the sum of van der Waals radii of two Br atoms.

In $[Ph_3PBr][Br_7]$, the Br_7^- anion has a triangular pyramid shape with the Br^- ion at the center bonded to three Br_2 molecules, as shown in Fig. 7.1.3 (c).

The largest polybromide ion Br_{24}^{2-} is known to exist in $(^nBu_4P)_2[Br_{24}]$. This structure is shown in Fig. 7.1.3 (d), the two ends of a central Br_2 molecule are symmetrically connected to a pair of $[Br(Br_2)_3(Br_4)]$ groups.

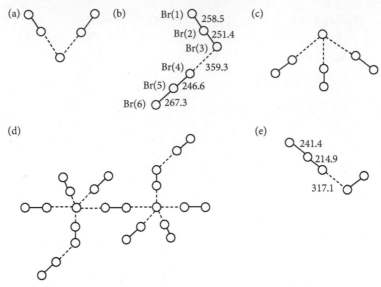

Fig. 7.1.3 Structures of polybromide and polychloride anions: (a) Br_5^-; (b) Br_6^{2-}; (c) Br_7^-; (d) Br_{24}^{2-}; (e) Cl_5^-.

The structure of polychloride ions is exemplified by Cl_5^-. In $[PPh_2Cl_2][Cl_3^-\cdot Cl_2]$, the two Cl–Cl bond lengths in the Cl_3^- moiety are 241.9 pm and 214.4 pm. One Cl_2 molecule is connected at the short end at a distance of 317.1 pm, resulting in a hockeystick shape for Cl_5^- as shown in Fig. 7.1.3 (e).

7.1.3 **Homopolyatomic halogen cations**

The structures of the following homopolyatomic halogen cations have been determined by X-ray analysis:

X_2^+	:	Br_2^+ and I_2^+ (in $Br_2^+[Sb_3F_{16}]^-$ and $I_2^+[Sb_2F_{11}]^-$)
X_3^+	:	Cl_3^+, Br_3^+ and I_3^+ (in X_3AsF_6)
X_4^{2+}	:	I_4^{2+} (in $I_4^{2+}[Sb_3F_{16}]^-[SbF_6]^-$)
X_5^+	:	Br_5^+ and I_5^+ (in X_5AsF_6)
X_{15}^+	:	I_{15}^+ (in $I_{15}AsF_6$)

X_2^+: The bond lengths of Br_2^+ and I_2^+, with a formal bond order of 1½, are 215 pm and 258 pm, respectively, which are shorter than the bond lengths of molecular Br_2 (228 pm) and I_2 (267 pm). This is consistent with the loss of an electron from an antibonding orbital.

X_3^+: These cations have a bent structure (Fig. 7.1.4(a)). The X–X bond lengths are similar to those in gaseous X_2, being consistent with their single-bond character. The bond angles are between the 101° and 104°.

X_4^{2+}: Compound $I_4[Sb_3F_{16}][SbF_6]$ contains an I_4^{2+} cation, which has the shape of a planar rectangle with I–I bond lengths of 258 pm and 326 pm, as shown in Fig. 7.1.4(b).

X_5^+: Cations Br_5^+ and I_5^+ are iso-structural, as shown in Fig. 7.1.4(c).

X_{15}^+: Compound $I_{15}AsF_6$ contains an I_{15}^+ cation, which has the shape of a centrosymmetric zigzag chain. This cation may be considered to be a finite zigzag chain composed of three connected I_5^- units, as shown in Fig. 7.1.4(d).

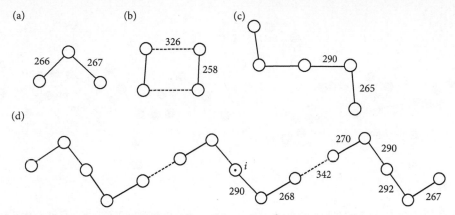

Fig. 7.1.4 Structures of some polyiodine cations: (a) I_3^+, (b) I_4^{2+}, (c) I_5^+, and (d) I_{15}^+. Bond lengths are in pm. Both I_5^+ and I_{15}^+ are centrosymmetric.

7.1.4 **Highest polychloride and polybromide molecular ions**

The square-pyramidal undecachloride molecular ion $[Cl_{11}]^-$ exists in a monomeric crystalline form $[PNP][Cl_{11}]\cdot Cl_2$ (PNP = bis(triphenylphosphoranylidene)iminium) that contains an embedded disordered chlorine molecule (Fig. 7.1.5).

The highest known polychlorine dianion is the dodecachloride dianion $[Cl_{12}]^{2-}$ in $[NMe_3Ph]_2[Cl_{12}]$. The compound crystallizes in the monoclinic space group $P2_1/c$ with two very similar but independent dodecachlorine dianions that occupy inversion centers. In the solid state, each $[Cl_{12}]^{2-}$ dianion is composed of two pentachloride subunits that are interconnected through a Cl_2 moiety (Fig. 7.1.6).

Tetracosabromide $[Br_{24}]^{2-}$ in $[(^nBu)_4P]_2[Br_{24}]$ is the highest polybromine dianion known to date. Its central bromide atom Br1 is coordinated by five Br_2 units [distorted tetragonal pyramid with Br2Br3

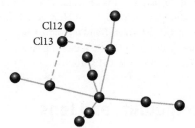

Fig. 7.1.5 Molecular structure of square-pyramidal $[Cl_{11}]^-$ in crystalline $[PNP][Cl_{11}]\cdot Cl_2$. The co-crystallized Cl_2 molecule exhibits disorder, and its major domain is represented by the Cl12 and Cl13 atoms. The broken lines represent intermolecular interactions.

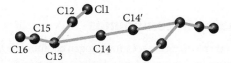

Fig. 7.1.6 Molecular structure of the dodecachloride dianion $[Cl_{12}]^{2-}$ in crystalline $[NMe_3Ph]_2[Cl_{12}]$. Bond lengths for two crystallographically independent $[Cl_{12}]^{2-}$ species are (pm): Cl1–Cl2 203.2(1)/202.8(1), Cl2–Cl3 301.3(1)/295.1(1), Cl3–Cl4 291.5(1)/289.5(1), Cl4–Cl4' 204.1(1)/204.5(1), Cl3–Cl5 273.3(1)/270.3(1), Cl5–Cl6 203.3(1)/205.0(1).

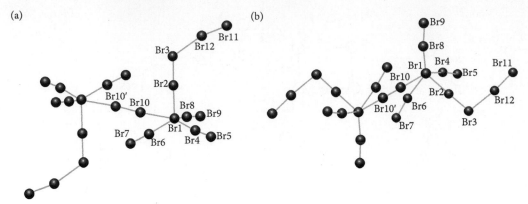

Fig. 7.1.7 The highest known polybromide in $[(^nBu)_4P]_2[Br_{24}]$: (a) centrosymmetric $[Br_{24}]^{2-}$ anion with atom labelling; (b) another view of the $[Br_{24}]^{2-}$ molecular anion showing the tetragonal pyramidal coordination geometry around Br1.

in axial and (Br4Br5, Br6Br7, Br8Br9, Br10Br10′) in basal positions], and a bromine molecule (Br11 and Br12) is end-on coordinated to Br3 (Fig. 7.1.7(a)). Another view of the $[Br_{24}]^{2-}$ species is shown in (Fig. 7.1.7(b)).

References

K. Sonnenberg, P. Pröhm, N. Schwarze, C. Müller, H. Beckers, and S. Riedel, Investigation of large polychloride anions: $[Cl_{11}]^-$, $[Cl_{12}]^{2-}$, and $[Cl_{13}]^-$. *Angew. Chem. Int. Ed.* **57**, 9136–40 (2018).

K. Sonnenberg, L. Mann, F. A. Redeker, B. Schmidt, and S. Riedel, Polyhalogen and Polyinterhalogen Anions from Fluorine to Iodine. *Angew. Chem. Int. Ed.* **59**, 5464–93 (2020).

M. E. Easton, A. J. Ward, T. Hudson, P. Turner, A. F. Masters, and T. Maschmeyer, The Formation of High-order Polybromides in a Room-Temperature Ionic Liquid: From Monoanions ($[Br_5]^-$ to $[Br_{11}]^-$) to the Isolation of $[PC_{16}H_{36}]_2[Br_{24}]$ as Determined by van der Waals Bonding Radii. *Chem. Eur. J.* **21**, 2961–5 (2015).

Z. Mazej, Noble-Gas Chemistry More Than Half a Century after the First Report of the Noble-Gas Compound, *Molecules* **25**, 3014 (2020). doi:10.3390/molecules25133014.

7.2 Interhalogen Compounds and Ions

The halogens form many compounds and ions that are binary or ternary combinations of halogen atoms. There are three basic types: (a) neutral interhalogen compounds, (b) interhalogen cations, and (c) interhalogen anions.

7.2.1 Neutral interhalogen compounds

The halogens react with each other to form binary interhalogen compounds XY, XY_3, XY_5, and XY_7, where X is the heavier halogen. A few ternary compounds are also known, for example, $IFCl_2$ and IF_2Cl. All interhalogen compounds contain an even number of halogen atoms. Table 7.2.1 lists the physical properties of some XY_n compounds.

XY: All six possible diatomic interhalogen compounds between F, Cl, Br, and I are known, but IF is unstable, and BrCl cannot be isolated free from Br_2 and Cl_2. In general, the diatomic interhalogens exhibit

Table 7.2.1 Physical properties of some interhalogen compounds

Compound	Appearance at 298 K	mp/K	bp/K	Bond length*/pm	
ClF	colorless gas	117	173	163	
BrF	pale brown gas	240	293	176	
BrCl	red brown gas	–	–	214	
ICl (α)	ruby red crystal	300	~373	237, 244	
ICl (β)	brownish red crystal	287	–	235, 244	
IBr	black crystal	314	~389	249	
ClF$_3$	colorless gas	197	285	160 (eq)	170 (ax)
BrF$_3$	yellow liquid	282	399	172 (eq)	181 (ax)
IF$_3$	yellow solid	245 (dec)	–	–	
(ICl$_3$)$_2$	orange solid	337 (sub)	–	238 (t)	268 (b)
ClF$_5$	colorless gas	170	260	172 (ba)	162 (ap)
BrF$_5$	colorless liquid	212.5	314	172 (ba)	168 (ap)
IF$_5$	colorless liquid	282.5	378	189 (ba)	186 (ap)
IF$_7$	colorless gas	278 (sub)	–	186 (eq)	179 (ax)

* The XY$_3$ molecule has a T-shaped structure: axial (ax), equatorial (eq); (ICl$_3$)$_2$ is a dimer: bridging (b), terminal (t); XY$_5$ forms a square-based pyramid: apical (ap), basal (ba); XY$_7$ has the shape of a pentagonal bipyramid: equatorial (eq), axial (ax).

properties intermediate between their parent halogens. However, the electronegativities of X and Y differ significantly, so the X–Y bond is stronger than the mean of the X–X and Y–Y bond strengths, and the X–Y bond lengths are shorter than the mean of d(X–X) and d(Y–Y). The dipole moments for polar XY molecules in the gas phase are: ClF 0.88 D, BrF 1.29 D, BrCl 0.57 D, ICl 0.65 D, and IBr 1.21 D.

Iodine monochloride ICl is unusual in forming two modifications: the stable α-form and the unstable β-form, both of which have infinite chain structures and significant I···Cl intermolecular interactions of 294 pm to 308 pm. Fig. 7.2.1(a) shows the chain structure of β-ICl.

XY$_3$: Both ClF$_3$ and BrF$_3$ have a T-shaped structure, being consistent with the presence of 10 electrons in the valence shell of the central atom, as shown in Fig. 7.2.1(b). The relative bond lengths of

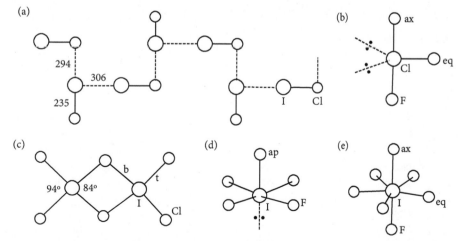

Fig. 7.2.1 Structures of some interhalogen molecules: (a) β-ICl, (b) ClF$_3$, (c) I$_2$Cl$_6$, (d) IF$_5$, and (e) IF$_7$.

$d(X-Y_{ax}) > d(X-Y_{eq})$ and the bond angle of $Y_{ax}-X-Y_{eq} < 90°$ (ClF$_3$ 87.5° and BrF$_3$ 86°) reflect the greater electronic repulsion of the non-bonding pair of electrons in the equatorial plane of the molecule.

Iodine trichloride is a fluffy orange powder that is unstable above room temperature. Its dimer $(ICl_3)_2$ has a planar structure, as shown in Fig. 7.2.1(C), that contains two I–Cl–I bridges (I–Cl distances in the range of 268–272 pm) and four terminal I–Cl bonds (238–239 pm).

XY$_5$: The three fluorides ClF$_5$, BrF$_5$, and IF$_5$ are the only known interhalogens of the XY$_5$ type, and they are extremely vigorous fluorinating reagents. All three compounds occur as a colorless gas or liquid at room temperature. Their structure has been shown to be square pyramidal with the central atom slightly below the plane of the four basal F atoms (Fig. 7.2.1(d)). The bond angles $F_{(ap)}-X-F_{(ba)}$ are: ~90° (ClF$_5$), 85° (BrF$_5$), and 81° (IF$_5$).

XY$_7$: IF$_7$ is the sole representative of this structural type. Its structure, as shown in Fig. 7.2.1(e), exhibits a slight deformation from pentagonal bipyramidal D_{5h} symmetry due to a 7.5° puckering and a 4.5° axial bending displacement. Bond length I–F$_{(eq)}$ is 185.5 pm and I–F$_{(ax)}$ is 178.6 pm.

7.2.2 Interhalogen ions

These ions have the general formulas XY$_n^+$ and XY$_n^-$, where n can be 2, 4, 5, 6, and 8, and the central halogen X is usually heavier than Y. Table 7.2.2 lists many of the known interhalogen ions.

The structures of these ions normally conform to those predicted by the VSEPR theory, as shown in Fig. 7.2.2. Since the anion XY$_n^-$ has two more electrons than the cation XY$_n^+$, they have very different shapes. The anion IF$_5^{2-}$ is planar with lone pairs occupying the axial positions of a pentagonal bipyramid. In [Me$_4$N](IF$_6$), IF$_6^-$ is a distorted octahedron (C_{3v} symmetry) with a sterically active lone pair, whereas both BrF$_6^-$ and ClF$_6^-$ are octahedral. The anion IF$_8^-$ has the expected square antiprismatic structure.

7.2.3 Polyinterhalogen ions

Polyinterhalogen compounds can be subdivided into classical and non-classical polyinterhalides. The former consists of an electropositive central atom surrounded by electronegative halogen atoms, such as in [ICl$_4$]$^-$ and [IF$_6$]$^-$. In non-classical polyinterhalides, the central halide is more electronegative than the coordinating dihalogen or interhalogen molecules (Fig. 7.2.3).

Table 7.2.2 Some interhalogen ions

	XY$_2$		XY$_4$		XY$_5$	XY$_6$	XY$_8$
Cations	ClF$_2^+$	I$_2$Cl$^+$	ClF$_4^+$		–	ClF$_6^+$	–
	Cl$_2$F$^+$	IBr$_2^+$	BrF$_4^+$			BrF$_6^+$	
	BrF$_2^+$	I$_2$Br$^+$	IF$_4^+$			IF$_6^+$	
	IF$_2^+$	IBrCl$^+$	I$_3$Cl$_2^+$				
	ICl$_2^+$						
Anions	BrCl$_2^-$	ClICl$^-$	ClF$_4^-$	I$_2$Cl$_3^-$	IF$_5^{2-}$	ClF$_6^-$	IF$_8^-$
	Br$_2$Cl$^-$	ClIBr$^-$	BrF$_4^-$	I$_2$BrCl$_2^-$		BrF$_6^-$	
	I$_2$Cl$^-$	BrIBr$^-$	IF$_4^-$	I$_2$Br$_2$Cl$^-$		IF$_6^-$	
	FClF$^-$		ICl$_3$F$^-$	I$_2$Br$_3^-$			
	FIBr$^-$		ICl$_4^-$	I$_4$Cl$^-$			
			IBrCl$_3^-$				

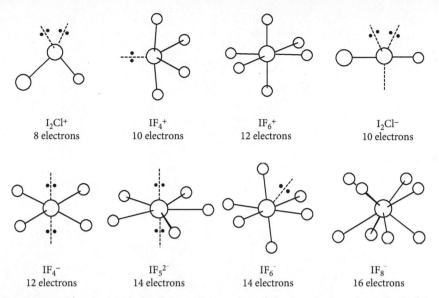

I_2Cl^+
8 electrons

IF_4^+
10 electrons

IF_6^+
12 electrons

I_2Cl^-
10 electrons

IF_4^-
12 electrons

IF_5^{2-}
14 electrons

IF_6^-
14 electrons

IF_8^-
16 electrons

Fig. 7.2.2 Structures of some interhalogen ions. The number of electrons in the valence shell of the central atom is given for each ion.

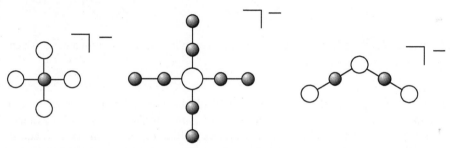

Fig. 7.2.3 Comparison of classical (left) and non-classical polyinterhalide (middle with coordinating dihalogen) and interhalogen (right structure) molecules. Shaded circles represent more electropositive halogen atoms, while the more electronegative atoms are shown as open circles.

The mixed-halogen compound $(Me_4N)^+[I_4Br_5]^-$ crystallized in 1-hexyl-3-methylimidazolium bromide. The structure of the anionic component can be regarded as $[(I_2Br_3)^-·2IBr]$, with its V-shaped $(I_2Br_3)^-$ central unit connected to two external IBr terminals in the *syn* fashion (Fig. 7.2.4).

Synthesis of $(Me_4N)^+[I_4Br_5]^-$ conducted in methylene chloride, instead of 1-hexyl-3-methylimidazolium bromide, also yielded a V-shaped $[I_2Br_3]^-$ central unit, but the pair of outer IBr molecules are connected to it from opposite directions, that is, in the *anti* fashion, to generate a polymeric chain consolidated by halogen–halogen interactions (Fig. 7.2.5).

A series of ternary octanuclear iodine-bromine-chlorine interhalides have been rationally constructed in two steps. Firstly, addition of a di-halogen (ICl or IBr) to triaminocyclopropenium chloride salt $[C_3(NEt_2)_3]Cl$ forms the corresponding trihalide salt ($[ICl_2]^-$ or $[BrICl]^-$). Secondly, addition of a half-equivalent of a second dihalogen, followed by crystallization at low temperature, gives the corresponding $[C_3(NEt_2)_3]_2$ octahalide: addition of Br_2 and IBr to $[ICl_2]^-$ gives $[I_2Br_2Cl_4]^{2-}$ (**1**) and $[I_3BrCl_4]^{2-}$ (**2**), respectively, whereas addition of I_2, Br_2, and IBr to $[BrICl]^-$ gives $[I_4Br_2Cl_2]^{2-}$ (**3**), $[I_2Br_4Cl_2]^{2-}$ (**4**) and $[I_3Br_3Cl_2]^{2-}$ (**5**), respectively. All five $[C_3(NEt_2)_3]_2$ octahalides have been characterized by X-ray crystallography and far–IR spectroscopy.

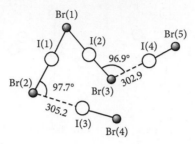

Fig. 7.2.4 Structure of [(I$_2$Br$_3$)$^-$·2IBr] anion.

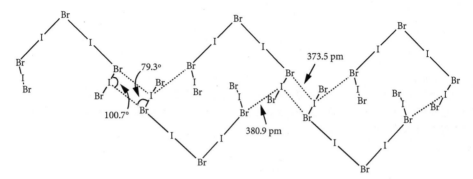

Fig. 7.2.5 Linkage of anionic [I$_4$Br$_5$]$^-$ units via halogen–halogen interactions to form a polymeric chain.

Compounds **1** to **5** crystallize readily from CH$_2$Cl$_2$/diethylether solutions at low temperature in monoclinic space group C2/m with one cation and a half-dianion in the asymmetric unit. In the crystal structure of **1**, the [I$_2$Br$_2$Cl$_4$]$^{2-}$ moiety sits on a special position of site-symmetry 2/m, and its *trans* Z-configuration is typical of octahalides. Each octahalide is sandwiched by two [C$_3$(NEt$_2$)$_3$]$^+$ cations

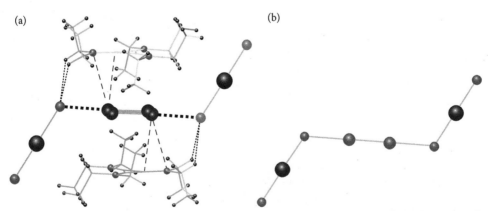

Fig. 7.2.6 (a) X-ray molecular structure of [C$_3$(NEt$_2$)$_3$]$_2$[I$_2$Br$_2$Cl$_4$] **1** showing its orientationally disordered central Br1-Br2 unit and the internal octahalide halogen bonds (• • •), π-hole interactions (– – –), and cation–anion hydrogen bonding (· · ·). (b) *trans* Z-configuration of the [I$_2$Br$_2$Cl$_4$]$^{2-}$ moiety.

and stabilized by weaker interactions with them (Fig. 7.2.6). Very similar structural features are displayed by compounds **2** to **5**.

References

K. Sonnenberg, L. Mann, F. A. Redeker, B. Schmidt, and S. Riedel, Polyhalogen and polyinterhalogen anions from fluorine to iodine, *Angew. Chem. Int. Ed.* **59**, 5464–93 (2020).

L. Mann, P. Voßnacker, C. Müller, and S. Riedel, [NMe$_4$][I$_4$Br$_5$]: a new iodobromide from an ionic liquid with halogen–halogen interactions, *Chem. Eur. J.* **23**, 244–9 (2017).

M. Wolff, A. Okrut, and C. Feldmann, [(Ph)$_3$PBr][Br$_7$], [(Bz)(Ph)$_3$P]$_2$[Br$_8$], [(n–Bu)$_3$MeN]$_2$[Br$_{20}$], [C$_4$MPyr]$_2$[Br$_{20}$], and [(Ph)$_3$PCl]$_2$[Cl$_2$I$_{14}$]: extending the horizon of polyhalides via synthesis in ionic liquids, *Inorg. Chem.* **50**, 11683–94 (2011).

M. S. Abdelbassit and O. J. Curnow, Construction of ternary iodine–bromine–chlorine octahalides, *Chem. Eur. J.* **25**, 13294–8 (2019).

7.3 Charge-Transfer Complexes of Halogens

A charge-transfer (or donor–acceptor) complex is one in which a donor and an acceptor species interact weakly with some net transfer of electronic charge, usually facilitated by the acceptor. The diatomic halogen molecule X_2 has HOMO π^* and LUMO σ^* molecular orbitals, and the σ^* orbital is antibonding and acts as an acceptor. If the X_2 molecule is dissolved in a solvent such as ROH, H_2O, pyridine, or CH_3CN that contains N, O, S, Se, or π electron pairs, the solvent molecule can function as a donor through the interaction of one of its σ or π electron pairs with the σ^* orbital of X_2. This donor–acceptor interaction leads to the formation of a charge-transfer complex between the solvent (donor) and X_2 (acceptor) and alters the optical transition energy of X_2, as shown in Fig. 7.3.1.

Let us take I_2 as an example. The normal violet color of gaseous iodine is attributable to the allowed π^* → σ^* transition. When iodine is dissolved in a solvent, the interaction of I_2 with a donor solvent molecule causes an increase in the energy separation of the π^* to σ^* orbitals from E_1 to E_2, as shown in Fig. 7.3.1. The color of this solution is then changed to brown. (The absorption maximum for the violet solution occurs at 520 nm to 540 nm, and that of a typical brown solution at 460 nm to 480 nm). The electron

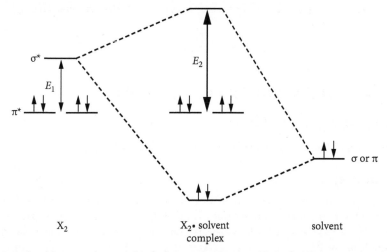

Fig. 7.3.1 Interaction between the σ^* orbital of X_2 and a donor orbital of solvent.

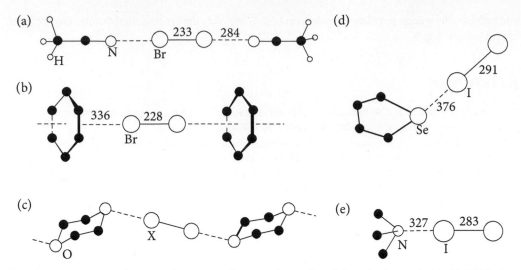

Fig. 7.3.2 Structures of some charge transfer complexes (bond lengths in pm.) (a) $(H_3CCN)_2 \cdot Br_2$, (b) $C_6H_6 \cdot Br_2$, (c) $C_4H_8O_2 \cdot X_2$ (X = Cl, Br, X–X distance: 202 pm for X = Cl and 231 pm for X = Br), (d) $C_4H_8Se \cdot I_2$, and (e) $Me_3N \cdot I_2$.

transition in these $I_2 \cdot$solvent complexes is called a charge transfer transition. However, the most direct evidence for the formation of a charge transfer complex in solution comes from the appearance of an intense new charge transfer band occurring in the near ultraviolet spectrum in the range 230–330 nm.

The structures of many charge-transfer complexes have been determined. All the examples shown in Fig. 7.3.2 share the following common structural characteristics:

(a) The donor atom (D) or π-orbital (π) and X_2 molecule are essentially linear: D···X–X or π···X–X.

(b) The bond lengths of the X–X groups in the complexes are all longer than those in the corresponding free X_2 molecules; the D···X distance is invariably shorter than the sum of their van der Waals radii.

(c) Each X_2 molecule in an infinite chain structure is engaged by donors (D) at both ends, and the D···X–X···D unit is essentially linear, as expected for a σ-type acceptor orbital.

In the extreme case, complete transfer of charge may occur, as in the formation of $[I(py)_2]^+$:

$$2I_2 + 2\left(\text{\large◯}N\right) \longrightarrow \text{\large◯}N \rightarrow \overset{+}{I} \leftarrow N\text{\large◯} + I_3^-$$

7.4 Halogen Oxides and Oxo Compounds

7.4.1 Binary halogen oxides

The structures of some binary halogen oxides are listed in Table 7.4.1.

(1) X_2O molecules

(a) Since fluorine is more electronegative than oxygen, the binary compounds of F_2 and O_2 are named oxygen fluorides, rather than fluorine oxides. Oxide F_2O is a colorless, highly toxic, and explosive gas. The molecule has C_{2v} symmetry, as expected for a molecule with 20 valence electrons and two normal single bonds.

Table 7.4.1 Structure of halogen oxides

Oxide	Molecular structure	Structural formula	Formal oxidation state of X	Bond length/pm and bond angle
X_2O	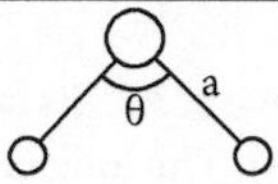	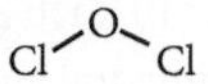	−1 (F) +1	X = F, a = 141, θ = 103° X = Cl, a = 169, θ = 111° X = Br, a = 185, θ = 112°
X_2O_2	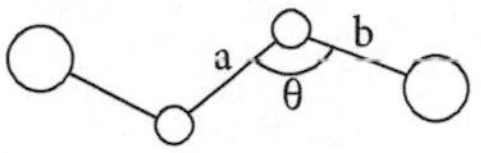	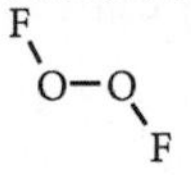	−1	X = F, a = 122, b = 158, θ = 109°
XO_2	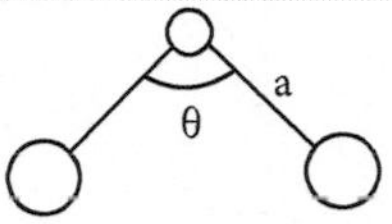	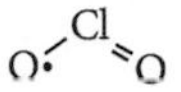	+3	X = Cl, a = 147, θ = 118°
X_2O_3	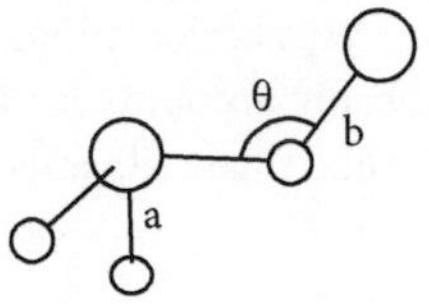	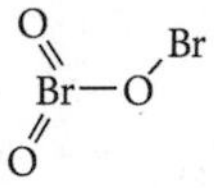	+1, +5	X = Br, a = 161, b = 185, θ = 112°
X_2O_4	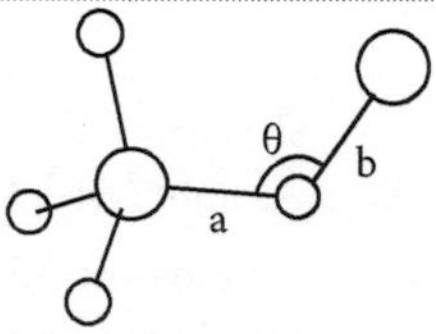	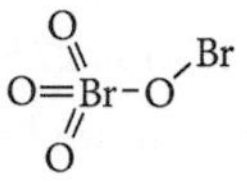	+1, +7	X = Br, a = 161, b = 186, θ = 110°
X_2O_5	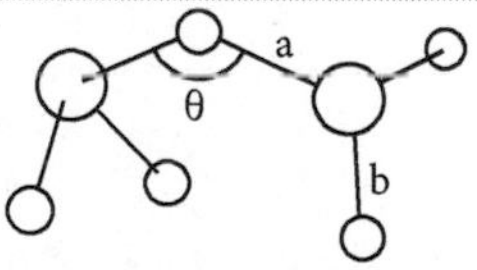	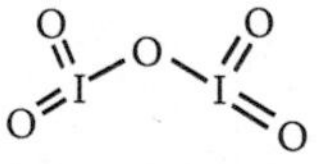	+5	X = Br, a = 188, b = 161, θ = 121° X = I, a = 193(av), b = 179(av), θ = 139°
X_2O_6	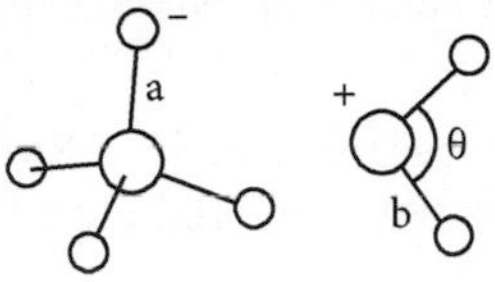		+5, +7	X = Cl, a = 144 (av), b = 141, θ = 119°
X_2O_7	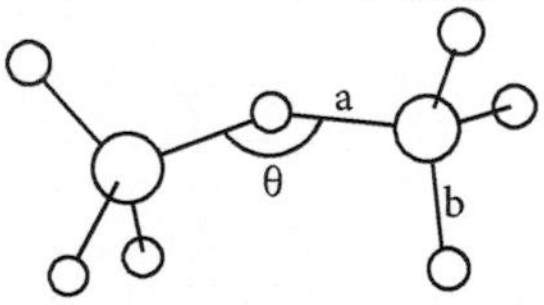	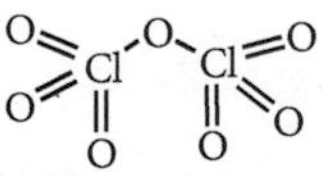	+7	X = Cl, a = 172, b = 142, θ = 119°

(b) Dichlorine monoxide Cl_2O is a yellow-brown gas that is stable at room temperature. There are two linkage isomers, Cl–Cl–O and Cl–O–Cl, but only the latter is stable. The stable form Cl–O–Cl is bent (111°) with Cl–O 169 pm.

(c) Dibromine monoxide Br_2O is a dark-brown crystalline solid that is stable at 213 K (mp 255.6 K with decomposition). The molecule has C_{2v} symmetry in both the solid and vapor phases with Br–O 185 pm and angle Br–O–Br 112°.

(2) X_2O_2 molecules

Only F_2O_2 is known. This is a yellow-orange solid (mp 119 K) that decomposes above 223 K. Its molecular shape resembles that of H_2O_2, although the internal dihedral angel is smaller (87°). The long O–F bond (157.5 pm) and short O–O bond (121.7 pm) in F_2O_2 can be rationalized by resonance involving the following valence bond representations:

(3) XO_2 molecule

Only ClO_2 is known. Chlorine dioxide is an odd-electron molecule. Theoretical calculations suggest that the odd electron is delocalized throughout the molecule, and this probably accounts for the fact that there is no evidence of dimerization in solution, or even in the liquid or solid phase. Its important Lewis structures are shown below:

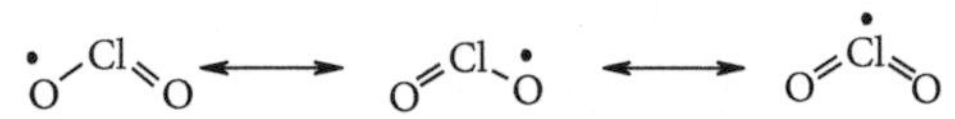

(4) X_2O_3 molecules

(a) Cl_2O_3 is a dark-brown solid which explodes even below 273 K. Its structure has not been determined.

(b) Br_2O_3 is an orange crystalline solid and has been shown by X-ray analysis to be *syn*-BrOBrO_2 with Br^I–O 184.5 pm, Br^V–O 161.3 pm, and angle Br–O–Br 111.6°. It is thus, formally, the anhydride of hypobromous and bromic acid.

(5) X_2O_4 molecules

(a) Chlorine perchlorate, Cl_2O_4, is most likely $ClOClO_3$. Little is known of the structure and properties of this pale-yellow liquid; it is even less stable than ClO_2 and decomposes at room temperature to Cl_2, O_2, and Cl_2O_6.

(b) Br_2O_4 is a pale yellow crystalline solid, whose structure has been shown by EXAFS to be bromine perbromate, $BrOBrO_3$, with Br^I–O 186.2 pm, Br^{VII}–O 160.5 pm, and angle Br–O–Br 110°.

(6) X_2O_5 molecules

(a) Ozonization of Br_2 generates Br_2O_3 and eventually Br_2O_5:

$$Br_2 \xrightarrow{O_3,\ 195\ K} Br_2O_3 \xrightarrow{O_3,\ 195\ K} Br_2O_5$$

$$\text{brown} \qquad\qquad \text{orange} \qquad\qquad \text{colorless}$$

The end product can be crystallized from propionitrile as $Br_2O_5 \cdot EtCN$, and crystal structure analysis has shown that Br_2O_5 is $O_2BrOBrO_2$ with each Br atom pyramidally surrounded by three O atoms, and the terminal O atoms are eclipsed with respect to each other.

(b) I_2O_5 is the most stable oxide of the halogens.

Crystal structure analysis has shown that the molecule consists of two pyramidal IO_3 groups sharing a common oxygen. The terminal O atoms have a staggered conformation, as shown in Table 7.4.1.

(7) X_2O_6 molecule

Cl_2O_6 is actually a mixed-valence ionic compound $ClO_2^+ClO_4^-$, in which the angular ClO_2^+ and tetrahedral ClO_4^- ions are arranged in a distorted CsCl-type crystal structure. Cation ClO_2^+ has Cl–O 141 pm, angle O–Cl–O 119°; ClO_4^- has Cl–O(av) 144 pm.

(8) X_2O_7 molecule

Cl_2O_7 is a colorless liquid at room temperature. The molecule has C_2 symmetry in both gaseous and crystalline states, the ClO_3 groups being twisted from the staggered (C_{2v}) configuration, with Cl–O(bridge) 172.3 pm and Cl–O(terminal) 141.6 pm.

In addition to the compounds mentioned above, other unstable binary halogen oxides are known. The structures of the short-lived gaseous XO radicals have been determined. For ClO, the interatomic distance $d = 156.9$ pm, dipole moment $\mu = 1.24$ D, and bond dissociation energy $D_0 = 264.9$ kJ mol^{-1}. For BrO, $d = 172.1$ pm, $\mu = 1.55$ D, and $D_0 = 125.8$ kJ mol^{-1}. For IO, $d = 186.7$ pm and $D_0 = 175$ kJ mol^{-1}.

The structures of the less stable oxides I_4O_9 and I_2O_4 are still unknown, but I_4O_9 has been formulated as $I^{3+}(I^{5+}O_3)_3$, and I_2O_4 as $(IO)^+(IO_3)^-$.

7.4.2 **Ternary halogen oxides**

The ternary halogen oxides are mainly compounds in which a heavier X atom (Cl, Br, I) is bonded to both O and F. These compounds are called halogen oxide fluorides. The structures of the ternary halogen oxides are summarized in Fig. 7.4.1.

The geometries of these molecules are consistent with the VSEPR model. (a) FClO is bent with C_s symmetry (two lone pairs). (b) FXO_2 is pyramidal with C_s symmetry (one lone pair). (c) FXO_3 has C_{3v}

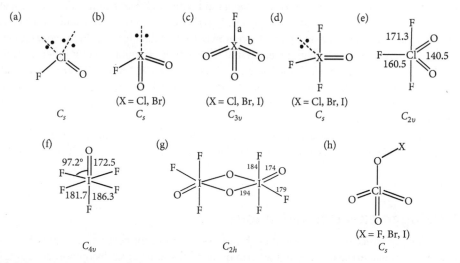

Fig. 7.4.1 Structures of ternary halogen oxides. The bond lengths are in pm. For (c), X = Cl, a = 162 pm, b = 140 pm; X = Br, a = 171 pm, b = 158 pm.

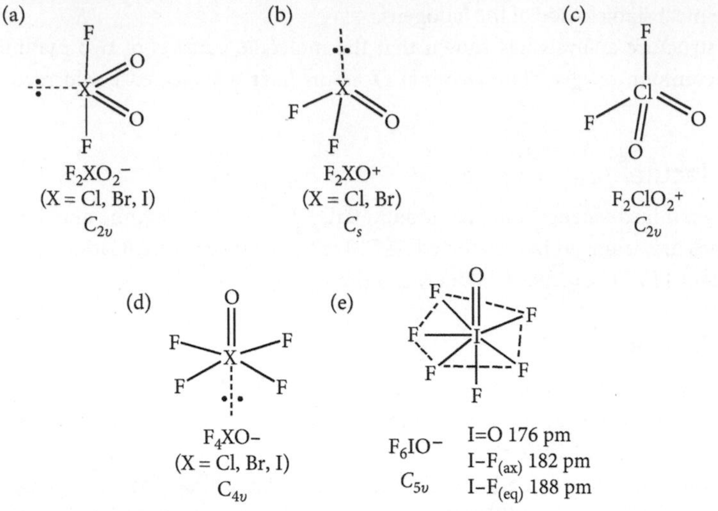

Fig. 7.4.2 Structures of cations and anions of ternary halogen oxides.

Table 7.4.2 Halogen oxoacids

Generic name	Fluorine	Chlorine	Bromine	Iodine
Hypohalous acids	HOF*	HOCl	HOBr	HOI
Halous acids	–	HOClO	–	–
Halic acids	–	HOClO$_2$	HOBrO$_2$	HOIO$_2$*
Perhalic acids	–	HOClO$_3$*	HOBrO$_3$	HOIO$_3$*, (HO)$_5$IO*

* Isolated as pure compounds; others are stable only in aqueous solutions.

symmetry. (d) F$_3$XO is an incomplete trigonal bipyramid with F, O, and a lone pair in the equatorial plane, having C_s symmetry. (e) F$_3$ClO$_2$ is a trigonal bipyramid with one fluorine and both oxygens in the equatorial plane; owing to the strong repulsion from two Cl=O double bonds in the equatorial plane, the axial Cl–F bonds are longer than the equatorial Cl–F bonds, as confirmed by experimental data. (f) F$_5$IO has C_{4v} symmetry, in which the bond lengths are I–F$_{ax}$ 186.3 pm, I–F$_{eq}$ 181.7 pm, and I=O 172.5 pm. Also, the I atom lies above the plane of four equatorial F atoms, and bond angle O–I–F$_{eq}$ is 97.2°. (g) F$_3$IO$_2$ forms oligomeric species, and the dimer (F$_3$IO$_2$)$_2$ has C_{2h} symmetry. (h) O$_3$ClOX are "halogen perchlorate salts."

A variety of cations and anions are derived from some of the above neutral molecules by gaining or losing one F$^-$, as shown in Fig. 7.4.2. Their structures are again in accord with predictions by the VSEPR model.

7.4.3 **Halogen oxoacids and anions**

Numerous halogen oxoacids are known, though most of them cannot be isolated as pure species and are stable only in aqueous solution or in the form of their salts. Anhydrous hypofluorous acid (HOF), perchloric acid (HClO$_4$), iodic acid (HIO$_3$), orthoperiodic acid (H$_3$IO$_6$), and metaperiodic acid (HIO$_4$) have been isolated as pure compounds. Table 7.4.2 lists the halogen oxoacids.

(1) Hypofluorous and other hypohalous acids

At room temperature hypofluorous acid HOF is a gas. Its colorless solid (mp = 156 K) melts to a pale yellow liquid. In the crystal structure, O–F = 144.2 pm, angle H–O–F = 101°, and the HOF molecules are linked by O–H···O hydrogen bonds (bond length 289.5 pm and bond angle O–H···O 163°) to form a planar zigzag chain, as shown below:

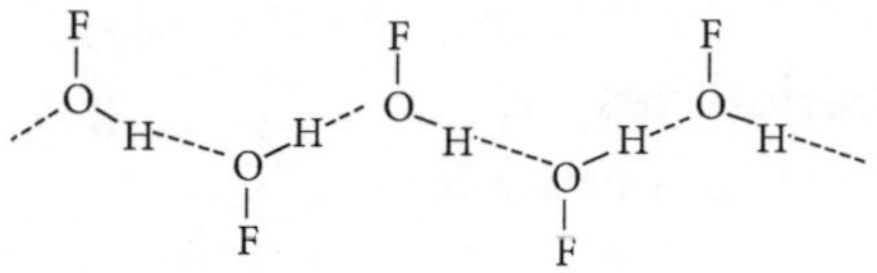

In general, fluorine has a formal oxidation state of –1, but in HOF and other hypohalous acids HOX, the formal oxidation state of F and X is +1.

Hypochlorous acid HOCl is more stable than HOBr and HOI, with Cl–O 169.3 pm and angle H–O–Cl 103° in the gas phase.

(2) Chlorous acid HOClO and chlorite ion ClO₂⁻

Chlorous acid, HOClO, is the least stable of the oxoacids of chlorine. It cannot be isolated in pure form, but exists in dilute aqueous solution. Likewise, HOBrO and HOIO are even less stable, showing only a transient existence in aqueous solution.

The chlorite ion (ClO_2^-) in $NaClO_2$ and other salts has a bent structure (C_{2v} symmetry) with Cl–O 156 pm, O–Cl–O 111°.

(3) Halic acids HOXO₂ and halate ions XO₃⁻

Iodic acid, HIO_3, is a stable white solid at room temperature. In the crystal structure, trigonal HIO_3 molecules are connected by extensive hydrogen bonding with I–O 181 pm, I–OH 189 pm, angle O–I–O 101°, O–I–(OH) 97°.

Halate ions are trigonal pyramidal, with C_{3v} symmetry, as shown below:

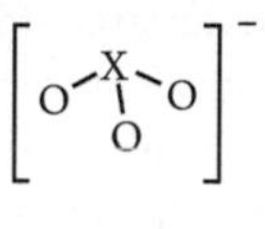

X = Cl, Cl–O 149 pm
X = Br, Br–O 165 pm
X = I, I–O 184 pm
Angles O–X–O 106° to 107°

Note that in the solid state, some metal halates do not consist of discrete ions. For example, in the iodates there are three short I–O distances 177–190 pm and three longer distances 251–300 pm, leading to distorted pseudo-six-fold coordination and piezoelectric properties.

(4) Perchloric acid HClO₄ and perhalates XO₄⁻

Perchloric acid is the only oxoacid of chlorine that can be isolated. The crystal structure of $HClO_4$ at 113 K exhibits three Cl–O distances of 142 pm and a Cl–OH distance of 161 pm. The structure of $HClO_4$, as determined by electron diffraction in the gas phase, shows Cl–O 141 pm, Cl–OH 163.5 pm, and angles O–Cl–(OH) 106°, O–Cl–O 113°.

The hydrates of perchloric acid exist in at least six crystalline forms. The monohydrate is composed of H_3O^+ and ClO_4^- that are connected by hydrogen bonds.

All perhalate ions XO_4^- are tetrahedral, with T_d symmetry, as shown below:

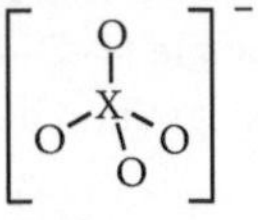

$$\left[\begin{array}{c} O \\ \| \\ O{-}X{-}O \\ | \\ O \end{array} \right]^-$$

X = Cl, Cl–O 144 pm
X = Br, Br–O 161 pm
X = I, I–O 179 pm

(5) Periodic acids and periodates

Several different periodic acids and periodates are known.

(a) Metaperiodic acid HIO_4

Metaperiodic acid HIO_4 consists of one-dimensional infinite chains built up of distorted *cis*-edge-sharing IO_6 octahedra, as shown in Fig. 7.4.3. Until now no discrete HIO_4 molecule has been found.

(b) Orthoperiodic (or paraperiodic) acid $(HO)_5IO$

The crystal structure of orthoperiodic acid, commonly written as H_5IO_6, consists of axially distorted octahedral $(HO)_5IO$ molecules linked into a three-dimensional array by O–H⋯O hydrogen bonds (10 for each molecule, 260–278 pm). The $(HO)_5IO$ molecule has the maximum number of –OH groups surrounding the I(+7) atom and hence it is called orthoperiodic acid.

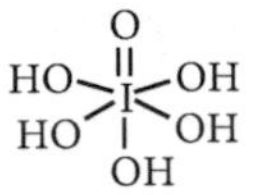

$$\begin{array}{c} O \\ \| \\ HO{-}I{-}OH \\ HO \quad | \quad OH \\ OH \end{array}$$

I=O 178 pm
I–OH 189 pm

The structure of $H_7I_3O_{14}$ does not show any new type of catenation, because this compound exists in the solid state as a stoichiometric phase containing orthoperiodic and metaperiodic acids according to the formula $(HO)_5IO \cdot 2HIO_4$.

The structures of several periodates and hydrogen periodates are shown in Fig. 7.4.4.

7.4.4 Structural features of polycoordinate iodine compounds

Iodine differs in many aspects from the other halogens. Due to the large atomic size and the relatively low ionization energy, it can easily form stable polycoordinate, multivalent compounds. Interest in polyvalent organic iodine compounds arises from several factors: (a) the chemical properties and reactivity of I(III) species are similar to those of Hg(+2), Tl(+3), and Pb(+4), but without the toxic and environmental problems of these heavy metal congeners; (b) the recognition of similarities between organic transition metal complexes and polyvalent main-group compounds such as organoiodine species; and (c) the commercial availability of key precursors, such as $PhI(OAc)_2$.

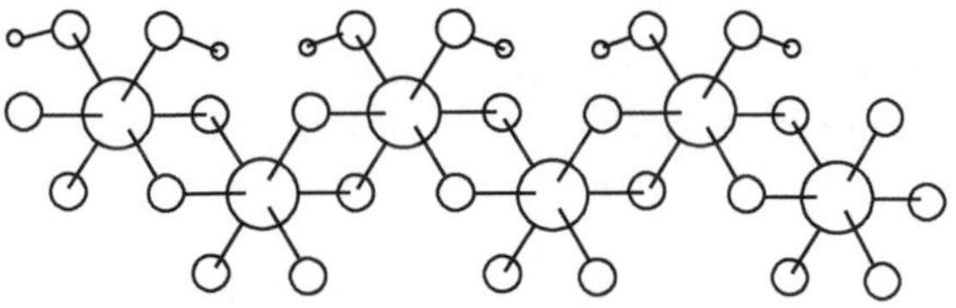

I–(OH) 184 pm
I–O(bridge) 201 pm
I–O(terminal) 191 pm

Fig. 7.4.3 Structure of metaperiodic acid.

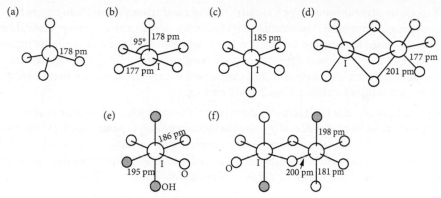

Fig. 7.4.4 Structures of periodates (a)–(d) and hydrogen periodates (e)–(f); small shaded circles represent OH group. (a) IO_4^- in $NaIO_4$, (b) IO_5^{3-} in K_3IO_5, (c) IO_6^{5-} in K_5IO_6, (d) $I_2O_9^{4-}$ in $K_4I_2O_9$, (e) $[IO_3(OH)_3]^{2-}$ in $(NH_4)_2IO_3(OH)_3$, (f) $[I_2O_8(OH)_2]^{4-}$ in $K_4[I_2O_8(OH)_2]\cdot 8H_2O$.

Six structural types of polyvalent iodine species are commonly encountered, as shown below:

The first two types, (a) and (b), called iodanes, are conventionally considered as derivatives of trivalent iodine I(+3). The next two, (c) and (d), periodanes, represent the most typical structural types of pentavalent iodine I(+5). The structural types (e) and (f) are typical of heptavalent iodine I(+7).

The most important structural features of polyvalent iodine compounds may be summarized as follows:

(1) The iodonium ion (type a) generally has a distance of 260–280 pm between iodine and the nearest anion, and may be considered as having pseudotetrahedral geometry about the central iodine atom.

(2) Species of type (b) has an approximately T-shaped structure with a collinear arrangement of the most electronegative ligands. Including the non-bonding electron pairs, the geometry about iodine is a distorted trigonal bipyramid with the most electronegative groups occupying the axial positions and the least electronegative group and both electron pairs residing in equatorial positions.

(3) The I–C bond lengths in both iodonium salts (type a) and iodoso derivatives (type b) are approximately equal to the sum of the covalent radii of I and C atoms, ranging generally from 200 pm to 210 pm.

(4) For type (b) species with two heteroligands of the same electronegativity, both I–L bonds are longer than the sum of the appropriate covalent radii, but shorter than purely ionic bonds. For example, the I–Cl bond lengths in $PhICl_2$ are 245 pm, whereas the sum of the covalent radii of I and Cl is 232 pm. Also, the I–O bond length in $PhI(OAc)_2$ are 215–216 pm, whereas the sum of the covalent radii of I and O is 199 pm.

(5) The geometry of the structural types (c) and (d) can be square pyramidal, pesudo-trigonal bipyrami- dal and pesudo-octahedral. The bonding in I(+5) compounds IL_5 with a square pyramidal structure may be described in terms of a normal covalent bond between iodine and the ligand in the apical position, and two orthogonal, hypervalent 3c-4e bonds accommodating four ligands. The carbon ligand and unshared electron pair in this case should occupy the apical positions, with the most electronegative ligands residing at equatorial positions.

(6) The typical structures of I(+7) involve a distorted octahedral configuration (type e) about iodine in most periodates and oxyfluoride, IOF_5, and the heptacoordinated, pentagonal bipyramidal species (type f) for the IF_7 and IOF_6^- anions. The pentagonal bipyramidal structure can be described as two covalent collinear axial bonds between iodine and ligands in the apical positions and a coplanar, hypervalent 6c-10e bond system for the five equatorial bonds.

7.5 Halogen Bonding

From the structure of the above-mentioned compound between elemental halogen and interhalogen, it can be seen that there is a relatively strong intermolecular force between the halogen molecules or ions, that is, a secondary bond between halogen atoms that is stronger than the van der Waals interaction. This is called halogen bond.

Halogen and hydrogen bonds are similar but not the same. Hydrogen bond can be expressed as R–H···B, where B is a highly electronegative atom. The length of a hydrogen bond refers to the distance between R and B. On the other hand, the halogen bond is expressed as R–X···B, where R can be halogen (Cl, Br, I), or C, N; B is halogen or N, O, S, etc. The halogen bond length refers to the distance between X and B. The halogen bond can also be written as R–X···X', and the bond length refers to the distance between X and X'. In R–X molecules with halogen bonding, the charge distribution of atom X varies with the relative weight of X atom. A heavy halogen atom X_h is more easily polarized than the light halogen atom X_l, so that X_h in R–X_h has a charge distribution as shown in Fig. 7.5.1.

Notably, hydrogen bonds are usually almost linear, that is, the R–H···B angle is close to 180°, while halogen bonds are usually non-linear and the R–H···B angle ranges from 90° to 180°. In halogen bond R–X_h···X_l–R', as in Br(4)–I(3)···Br(2) (or Br(5)–I(4)···Br(3)) in Fig. 7.2.4, Br(2) (or Br(3)) tends to combine with the electrophilic terminal of I(3) (or I(4)), forming a nearly linear configuration. On the other hand, as shown in I(1)–Br(2)···I(3) in Fig. 7.2.4 [and I(2)–Br(3)···I(4)], Br is lighter than I, I(3) tends to bind to the nucleophilic end of Br(2), the bond angle is 97.7° (and I(4) tends to bind with nucleophile of Br(3), and the bond angle is 96.9°).

7.5.1 Halogen-bonded supramolecular capsule

Fig. 7.5.2 shows a supramolecular capsule constructed from four-fold C–I···N halogen-bonding assembly of two resorcin[4]arene cavitand components: [(benzene⊂**1**·4MeOH)]···[(benzene⊂**2**·4MeOH)] formed

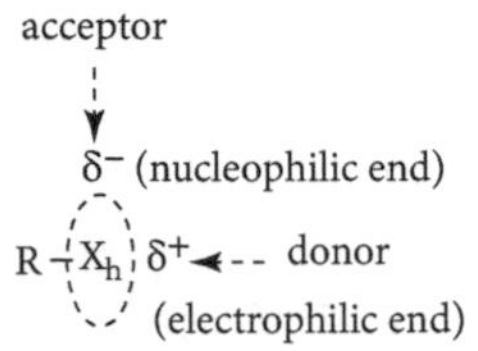

Fig. 7.5.1 Schematic diagram of halogen bonding.

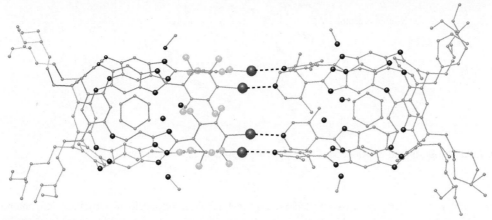

Fig. 7.5.2 Detailed X-ray structure of halogen-bonded supramolecular capsule, which comprises 12 individual molecular components: donor hemisphere on the left and acceptor hemisphere on the right, each encapsulating one benzene molecule, and each cavitand is rigidified by four external MeOH solvate molecules.

by tetra(2,3,5,6-tetrafluoro-4-iodophenyl) as donor and tetra(3,5-dimethylpyridyl) as acceptor. The halogen-bond donor **1** (on the left) and halogen-bond acceptor **2** (on the right) are constructed from resorcin[4]arene cavitands, and R represents a *n*-hexyl group. The crystal structure of the supramolecular capsule determined by X-ray diffraction at 100K showed that the four I···N halogen bonds are all 282 pm in length, and the C–I···N bond angles lie in the range 171–178°. The supramolecular capsule comprises 12 components: donor **1** and acceptor **2** each contains a hemispherical cavity that encapsulates a benzene guest molecule, with all eight co-crystallized methanol solvate molecules located outside it. Fourfold halogen bonding is formed between the two hemi-capsules, and it is the sole stabilizing factor of the molecular capsule.

References

O. Dumele, N. Trapp, and F. Diederich, Halogen bonding supramolecular capsules, *Angew. Chem. Int. Ed.* **54**, 12339–44 (2015).

O. Dumele, B. Schreib, U. Warzok, N. Trapp, C. A. Schalley, and F. Diederich, Halogen-bonded supramolecular capsules in the solid state, in solution, and in the gas phase. *Angew. Chem. Int. Ed.* **56**, 1152–7 (2017).

H. Wang, W. Wang, and W. J. Jin, σ hole bond vs π hole bond: a comparison based on halogen bond, *Chem. Rev.* **116**, 5072–104 (2016).

7.5.2 Organic molecular tessellations assembled by halogen bonding

Tessellation, or tiling on a plane, is the highly ordered arrangement of one or more planar shapes, called prototiles, that can fill the surface without gaps and overlaps to generate attractive geometric patterns. Although molecular tessellations using organic precursors fabricated on a metal surface have been observed with a combination of scanning tunneling microscopy, synchrotron radiation photoelectron

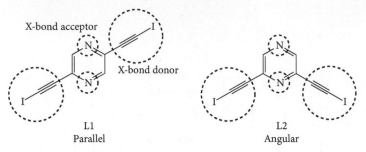

Fig. 7.5.3 Ditopic ligand 2,6-bis(iodoethynyl)pyrazine L1 and its 2,5-regioisomer L2.

spectroscopy, and X-ray spectroscopy techniques, detailed three-dimensional structural information derived from X-ray crystallographic studies of tessellation materials was unavailable until recently. Moreover, exploration in molecular tessellations remains limited to uniform pore patterns in a homogeneous environment owing to the formidable challenge in designing organic molecular building units that readily undergo covalent linkage. Therefore, construction of self-assembled molecular tessellations usually involves two-component systems.

Recently the use of custom-designed multi-functionalized small planar organic molecules interconnected by self-complementary non-covalent interactions succeeded to reveal novel single-component tiling patterns upon crystallization. This study started with a simple pyrazine core possessing two para-nitrogen atoms and two outstretched iodoethynyl arms bearing a coaxial ($\angle = 180°$) or angular relationship ($\angle = 120°$) (Fig. 7.5.3) that facilitate the construction of single-component molecular tessellations.

7.5.2.1 Crystal structure of L1a featuring two-dimensional monohedral rhombic tessellation

Crystal form **L1a** deposited from a solution of 2,5-regioisomer **L1** in acetonitrile belongs to noncentrosymmetric orthorhombic space group $Pca2_1$ (No. 29) with $Z = 16$, so that there are four crystallographically independent **L1** molecules (labelled **A**, **B**, **C**, and **D**) per unit cell. Its crystal structure exhibits stacked sinusoidal layers of monohedral rhombic tessellations formed by intermolecular C–I···N halogen bonding, such that a single independent **L1** molecule is used to construct each successive layer. For instance, two molecules **A** and their glide-related counterparts **A'** are connected to form rhombus **I** (length AA' = 1020 pm, $\angle$A'AA' = 61.8°) through intermolecular C–I···N halogen bonds of lengths 288 pm and 294 pm, respectively (Fig. 7.5.4(a)). Rhombus **I** is replicated along both crystallographic a- and b-axes to a complete tessellation with wallpaper group *cmm* (Fig. 7.5.4(b)). The sinusoidal layers of the rhombic tessellation each composed of independent **L1** molecules **A**, **B**, **C**, and **D** and their 2_1-related counterparts **A''**, **B''**, **C''**, and **D''** constitute a stacked layer structure with an interlayer spacing ranging from 357 pm to 366 pm along the c-axis (Fig. 7.5.4(c)).

7.5.2.2 Crystal structure of Form L2b featuring tacit trihedral snub trihexagonal tessellation

Crystallization of 2,5-bis(iodoethynyl)pyrazine **L2** in benzene or xylene only produced an amorphous solid residue, but evaporation of the solution in toluene gave crystal form **L2b**, which belongs to space group $P1$ (No. 2) with three independent **L2** molecules per unit cell. Intermolecular C–I···N halogen

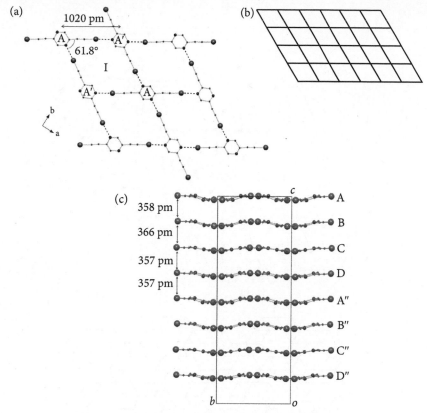

Fig. 7.5.4 (a) Crystal form **L1a** showing the monohedral rhombic tessellation stabilized by Cl⋯N halogen bonding generated from a single independent **L1** molecule. (b) Schematic diagram showing the monohedral rhombic tessellation pattern of a single layer in crystal form **L1a**. (c) Projection diagram along the *a*-axis showing the stacking of wavy tessellated layers. All hydrogen atoms are omitted for clarity.

bonds in the range of 297 pm to 299 pm connecting three independent molecules **A**, **B**, and **C** form triangle **I** with an average edge length of 1030 pm. Furthermore, molecules **B** and **C** and their inversion-related molecules **B′** and **C′** are halogen-bonded in the range 292–299 pm to give rhombus **II**, and three congruent symmetry-related rhombuses **II**, **II′**, and **II″** are each sandwiched between a pair of triangles labelled **I** and **I′** (Fig. 7.5.5 (a)).

A regular halogen-bonded hexagon **III** is formed by sharing six edges **AB′**, **B′C**, **CA′**, **A′B**, **BC′**, and **C′A** of rhombuses **II**, **II′**, and **II″** along with six vertices **A**, **B′**, **C**, **A′**, **B**, and **C′** from triangles **I** and **I′**. Hence, every hexagon **III** is surrounded by triangles and rhombuses in the sequence **I**, **II**, **I′**, **II″**, **I**, **II″**, **I′**, **II**, **I**, **II′**, **I′**, and **II″** running in the clockwise sense. The entire tessellation pattern extends throughout the plane with wallpaper group *p*6 (Fig. 7.5.5(b)). Ordinary snub trihexagonal tessellation is a dihedral tiling pattern composed of two types of prototiles, namely triangles and hexagons. However, in addition to these two types, the trihedral pattern in **L2b** also contains rhombuses, each of which could be bisected by its shorter diagonal to give two triangles to fit the tiling definition. Therefore, the term "tacit trihedral snub trihexagonal tessellation" is coined to designate this subtle difference. The tessellation layers are stacked along the *a*-axis with an interlayer spacing of 356 pm (Fig. 7.5.5(c)).

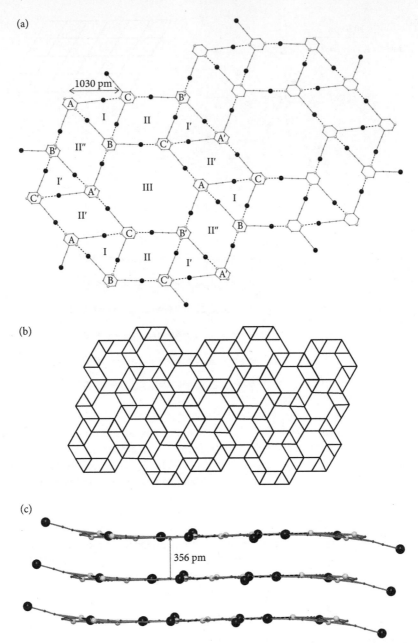

Fig. 7.5.5 (a) Projection diagram showing the tacit trihedral snub trihexagonal tessellation in crystal form **L2b**, which is stabilized by C–I⋯N halogen bonds formed by three independent **L2** molecules labelled **A**, **B**, and **C**. (b) Schematic diagram showing the tessellation pattern of a single layer in. (c) Projection diagram showing the stacking of tiling layers in the crystal structure. All hydrogen atoms are omitted for clarity.

Reference

C.-F. Ng, H.-F. Chow, and T. C. W. Mak, Organic molecular tessellations and intertwined double helices assembled by halogen bonding, *CrystEngComm* **21**, 1130–6 (2019).

7.5.3 **Bifurcated halogen bonding involving two rhodium(I) centers as an integrated σ-hole acceptor**

The $[RhX(COD)]_2$ (X = Cl, Br; COD = 1,5-cyclooctadiene) complexes co-crystallize with various σ–hole iodine donors (I)-based, namely 1,4-diodo-tetrafluorobenzene ($C_6F_4I_2$), nonafluoro-4-iodo-1,1′-biphenyl ($C_{12}F_9I$), tetraiodoethylene (C_2I_4), and iodoform (CHI_3), to give respective co-crystals $[RhCl(COD)]_2 \cdot (C_6F_4I_2)$ (**1**), $[RhBr(COD)]_2 \cdot 0.5(C_6F_4I_2)$ (**2**), $[RhCl(COD)]_2 \cdot (C_{12}F_9I)$ (**3**), $[RhCl(COD)]_2 \cdot 1.5(C_2I_4)$ (**4**), and $[RhCl(COD)]_2 \cdot (CHI_3)$ (**5**).

X-ray diffraction and theoretical studies of complexes **1** to **5** indicate that the d_{z^2} orbitals of two positively charged rhodium(I) centers exhibit sufficient nucleophilicity to form a three-center halogen bond (XB) with the above series of σ_h donating iodine(I)-based organic species. The pair of metal centers functions as an integrated XB acceptor, leading to molecular assembly via formation of a new-type bifurcated

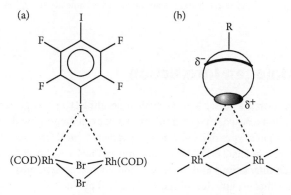

Fig. 7.5.6 (a) $Ar^F(\mu_4\text{-}I)\cdots[Rh_2Br_2]$ and $R^{EWG}(\mu_2\text{-}I)\cdots[Rh, Rh]$ interactions (EWG = electron-withdrawing group) in complex **2**. Hereinafter the contacts shorter than ΣR_{vdW} are drawn as dotted lines, and thermal ellipsoids are shown at 50% probability. (b) New-type bifurcated three-center Rh–X–Rh halogen bond, with X = I.

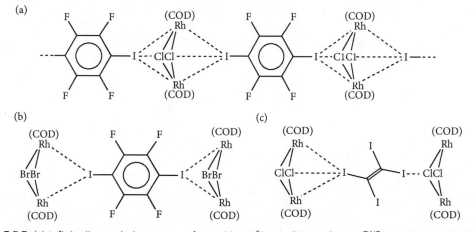

Fig. 7.5.7 (a) Infinite linear-chain structure formed by $Ar^F(\mu_4\text{-}I)\cdots[Rh_2Cl_2]$ and $R^{EWG}(\mu_2\text{-}I)\cdots[Rh, Rh]$ interactions in complex **1**. (b, c) Isolated dinuclear rhodium(I) clusters consolidated by $R^{EWG}(\mu_2\text{-}I)\cdots[Rh, Rh]$ contacts in **2** and **4**, respectively. Complex **3** has two similar conventional two-center contacts I···Rh like those in complex **2**. Complex **5** has two similar conventional two-center contacts I···Rh like those in complex **4**.

three-center Rh–X–Rh halogen bond, such as the one that exists in the crystal structure of complex **2** (Fig. 7.5.6).

In the crystal structure of $[RhCl-(COD)]_2 \cdot (C_6F_4I_2)$ **1**, two iodine centers from two $C_6F_4I_2$ are linked to the Rh_2Cl_2 core of the complex (Fig. 7.5.7(a)), whereas in **2** to **5**, Rh_2X_2 (X = Cl, Br) entities interact with only one I center of an appropriate σ_h donor (Fig. 7.5.7(b) and (c)). In the case of **1**, when the interaction with the iodine atoms occurs above and below the Rh_2Cl_2 plane, the metal core is perfectly planar. In **2** to **5**, the Rh_2X_2 functionality deviates from planarity, with the Rh atoms tilted toward iodine centers of the $REWG^-$ I species; the largest rms deviation of the core was found in **5** (30.2 pm for Br1).

Reference

A. A. Eliseeva, D. M. Ivanov, A. V. Rozhkov, I. V. Ananyev, A. Frontera, and V. Y. Kukushkin, Bifurcated halogen bonding involving two rhodium(i) centers as an integrated σ^- hole acceptor, *J. Am. Chem. Soc.* **1**, 354–61 (2021).

7.6 **Halogen–Halogen Interaction**

Halogen–halogen interaction is a versatile non-covalent bonding type that gives rise to dihalogen (X_2) and trihalogen (X_3) synthons. Based on statistical analysis of crystallographic data from the Cambridge Structural Database, two major types of X_2 interactions, namely Type-I (*trans* and *cis*) and Type-II, have been identified (see top row in Fig. 7.6.1). The bottom row in Fig. 7.6.1 shows the X_3 synthon formed by Type-II halogen–halogen interactions and the X_6 synthon formed by synergistic Type-I *cis* interactions to link six bromine atoms that occupy the corners of a trigonal prism.

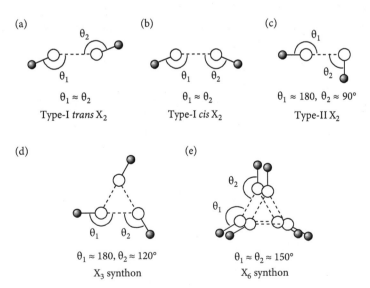

Fig. 7.6.1 Schematic representation of different types of halogen–halogen interactions: (a) Type-I *trans* dihalogen bonding, (b) Type-I *cis* dihalogen bonding, (c) Type-II dihalogen bonding, (d) X_3 synthon formed by Type-II dihalogen interactions, and (e) X_6 synthon formed by Type-I synergistic interaction. Large and small spheres represent halogen and hydrogen atoms, respectively.

The electrostatic model categorized the X_3 synthon having Type-II geometry as stabilizing Coulombic interaction. The first evidence of a stable non-covalent X_6 (Br_6) synthon formed by Type-I synergistic interaction, linking six identical bromine atoms (Fig. 7.6.1(e)), challenges the generally accepted notion of electrostatic nature of halogen bonds. Theoretical evaluations indicate that the lowest-energy geometry of Type-I occurs at $\theta_1 = \theta_2 = 150°$. Experimental realization of the Br_6 synthon, with the predicted minimum energy geometry, has an exchange-correlation component of covalence in stabilizing the synthon. The Br_6 synthon drives the chromophores to form an elegant radial/starburst assembly in the crystalline state, which is further strengthened by intermolecular charge-transfer interactions.

Imidization of 1,8-dibromonaphthalic anhydride yielded 1,8-dibromonaphthalene (2,6-diisopropylphenyl)imide (referred to as $NIBr_2$, chemical formula shown in top left corner of Fig. 7.6.2(a)). Slow evaporation of a 1:3 chloroform/hexane solution of $NIBr_2$ yielded single crystals with $Z = 18$ in centrosymmetric space group $R\bar{3}c$ (No. 167). X-ray crystallographic analysis revealed that the planar naphthalimide moiety of $NIBr_2$ bears an orthogonal 2,6-diisopropylphenyl substituent, and trigonal assembly of three $NIBr_2$ molecules is stabilized by six synergistic intermolecular $Br\cdots Br$ interactions ($d_{Br\cdots Br} = 373$ pm) that constitute the triangular faces of a trigonal prism. This trimeric structural unit (Fig. 7.6.2(a)) is further intercalated by three inverted $NIBr_2$ molecules to form a radial hexameric assembly (Fig. 7.6.2(b)).

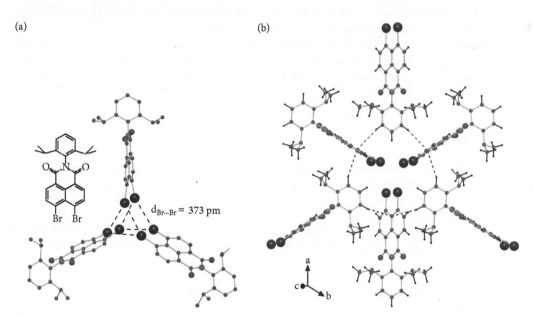

(a) (b)

$d_{Br\cdots Br} = 373$ pm

Fig. 7.6.2 (a) Chemical formula of $NIBr_2$, and assembly of three $NIBr_2$ molecules to form a $(NIBr_2)_3$ trimer via assembly of the Br_6 synthon. (b) Intercalation of $(NIBr_2)_3$ trimeric unit by three inverted $NIBr_2$ molecules to form a radial hexameric assembly stabilized by intermolecular C–H$\cdots$C interactions (dotted lines).

Reference

M. A. Niyas, R. Ramakrishnan, V. Vijay, E. Sebastian, and M. Hariharan, Anomalous halogen–halogen interaction assists radial chromophoric assembly, *J. Am. Chem. Soc.* **141**, 4536–40 (2019).

7.7 Compounds Stabilized by Non-covalent Interactions in the Periodic Table

A σ- or π-hole in a molecule can be regarded as a region of positive electrostatic potential on unpopulated σ* or π (*) orbitals, which is capable of interacting with some electron-dense region. A σ-hole is typically located along the vector of a covalent bond such as X–H or X–Hlg (X = any atom, Hlg = halogen), which is respectively known as a hydrogen or halogen bond donor. A π-hole is typically located perpendicular to the molecular framework of a diatomic π-system such as carbonyl, or a conjugated π-system such as hexafluorobenzene. Anion–π and lone-pair–π interactions are examples of π-hole interactions between conjugated π-systems and anions or lone-pair electrons, respectively.

Over the past decade, it has become clear that σ-holes can be found along covalent bonds formed by chalcogen (X–Ch), pnictogen (X–Pn), and tetrel (X–Tr) atoms. Non-covalent interactions with these synthons are named chalcogen, pnictogen, and tetrel bonds, respectively. This nomenclature indicates the distinct chemical identity of the supramolecular synthon acting as a Lewis acid.

It is obvious that non-covalent interactions exist in many compounds formed by elements belonging to various groups in the Periodic Table. Different types of non-covalent bonding are described in the literature by their specific names: triel, tetrel, pnictogen, chalcogen, halogen, and aerogen for Groups 13, 14, 15, 16, 17, and 18, respectively. Recently, the new nomenclature "spodium bond" is used for Group 12 (Zn, Cd, Hg) compounds, and "regium bond" is proposed for compounds in which Group 10 (Ni, Pd, Pt) or Group 11 (Cu, Ag, Au) elements serve as electron acceptors.

In summary, besides the well-established hydrogen bond, the following specific interactions are employed in the non-covalent bonding description of compounds in the literature: alkali, alkaline-earth, regium, spodium, triel, tetrel, pnictogen, chalcogen, halogen, and aerogen, which cover the majority of elements in the Periodic Table.

References

I. Alkorta, J. Elguero and A. Frontera, Not only hydrogen bonds: other noncovalent interactions, *Crystals*. **10**, 180 (2020). https://doi:10.3390/cryst10030180

A. Bauzá, I. Alkorta, J. Elguero, T. J. Mooibroek, and A. Frontera, Spodium bonds: noncovalent interactions involving group 12 elements, *Angew. Chem. Int. Ed.* **59**, 17482–7 (2020).

J. H. Stenlid, A. J. Johansson, and T. Brinck, σ-Holes and σ-lumps direct the Lewis basic and acidic interactions of noble metal nanoparticles: introducing regium bonds, *Phys. Chem. Chem. Phys.* **20**, 2676–92 (2018).

7.8 Structural Chemistry of Noble Gas Compounds

7.8.1 General survey

All the Group 18 elements (He, Ne, Ar, Kr, Xe, and Rn; rare gases or noble gases) have the very stable electronic configurations ($1s^2$ or ns^2np^6) and are monoatomic gases. The non-polar, spherical nature of the atoms leads to physical properties that vary regularly with atomic number. The only interatomic interactions are weak van der Waals forces, which increase in magnitude as the polarizabilities of the atoms increase and the ionization energies decrease. In other words, interatomic interactions increase with atomic size.

Since its discovery in 1895, helium is known as the only element in the periodic table that forms no chemical compound with any other element. In 2017 an international team of 17 collaborators lead by Chinese and Russian scientists accomplished the synthesis of the first helium-sodium compound $HeNa_2$

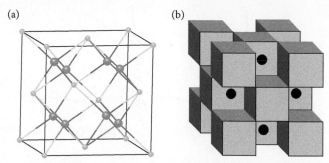

Fig. 7.8.1 Crystal structure of HeNa$_2$ at 300 GPa. Space group *Fm-3m* with *a* = 395 pm, Na atoms occupying Wyckoff position 8*c* (0.25, 0.25, 0.25) and He atoms occupying 4*a* (0,0,0). (a) Ball-and-stick representation: big and small atoms represent Na and He, respectively; (b) Polyhedral representation: half of the Na$_8$ cubes are occupied by He atoms (shown as polyhedra) and the other half by 2e (shown as dots).

at pressures exceeding 115 GPa, about 1 million times that of atmospheric pressure on Earth. X-ray diffraction has shown that it has a fluorite-like crystal structure and further characterized it by other physical methods. The crystal structure is shown in Fig. 7.8.1. It is an electride, which is composed of positively charged ions and electrons that function as anions.

Reference

X. Dong, A. R. Oganov, A. F. Goncharov, E. Stavrou, S. Lobanov, G. Saleh, G. R. Qian, Q. Zhu, C. Gatti, V. L. Deringer, R. Dronskowski, X. F. Zhou, V. B. Prakapenka, Z. Konôpková, I. A. Popov, A. I. Boldyrev, and H. T. Wang, A stable compound of helium and sodium at high pressure. *Nat. Chem.* **9**, 440–5 (2017).

In recent years, the use of matrix-isolation techniques has led to the generation of a wide range of linear triatomic species of the general formula HNgY, where Ng = Kr or Xe, and Y is an electronegative atom or group such as H, halides, pseudohalides, OH, SH, C≡CH, and C≡C–C≡CH. These molecules can be easily detected by the extremely strong intensity of the H–Ng stretching vibration. The observed ν(H–Ng) values for some HXeY molecules are: Y = H, 1166, 1181; Y = Cl, 1648; Y = Br, 1504; Y = I, 1193; Y = CN, 1623.8, Y = NC, 1851.0 cm^{-1}. The electronic structure of HNgY is best described in terms of the ion pair HNg$^+$Y$^-$, in which the HNg fragment is held by a covalent bond, and the interaction between Ng and Y is mostly ionic.

In this section, we turn our attention to the structural chemistry of those noble gas compounds that can be isolated in bulk quantities.

Xenon compounds with direct bonds to the electronegative main-group elements F, O, N, C, and Cl are well established. The first noble-gas compound, a yellow-orange solid formulated as XePtF$_6$, was prepared by Neil Bartlett in 1962. Noting that the first ionization energy of Xe (1170 kJ mol^{-1}) is very similar to that of O$_2$ (1175 kJ mol^{-1}), and that PtF$_6$ and O$_2$ can combine to form O$_2$PtF$_6$, Bartlett replaced O$_2$ by a molar quantity of Xe in the reaction with PtF$_6$ and produced XePtF$_6$. It is now established that the initial product he obtained was a mixture containing diamagnetic Xe^{II}Pt^{IV}F$_6$ as the major product, which is most likely a XeF$^+$ salt of (PtF$_5^-$)$_n$ with a polymeric chain structure, as illustrated below, by analogy with the known crystal structure of XeCrF$_6$.

Xe ○ 193 pm

213 pm

F ○

Pt

If xenon is mixed with a large excess of PtF_6 vapor, further reactions proceed as follows:

$$XePtF_6 + PtF_6 \longrightarrow XeF^+PtF_6^- + PtF_5 \text{ (non–crystalline)}$$

$$XeF^+PtF_6^- + PtF_5 (\text{Warmed} \geq 60°C) \longrightarrow XeF^+Pt_2F_{11}^- \text{ (orange–red solid)}$$

Evidently the Xe(I) oxidation state is not a viable one, and Xe(II) is clearly favored.

Since radon has intense α-radioactivity, information about its chemistry is very limited. Compounds of the other rare gas elements, Ne, Ar, Kr, and Xe, have been reported. For example, experimental investigation of $CUO(Ng)_n$ (Ng = Ar, Kr, Xe; n = 1, 2, 3, 4) complexes in solid neon have provided evidence of their formation. The computed structures of $CUO(Ne)_{4-n}(Ar)_n$ (n = 0, 1, 2, 3, 4) complexes are illustrated in Fig. 7.8.2.

7.8.2 **Stereochemistry of xenon**

Xenon reacts directly with fluorine to form fluorides. Other compounds of Xe can be prepared by reactions using xenon fluorides as starting materials, which fall into four main types:

(1) In combination with F^- acceptors, yielding fluorocations of xenon, as in the formation of $(XeF_5)(AsF_6)$ and $(XeF_5)(PtF_6)$.

(2) In combination with F^- donors, yielding fluoroanions of xenon, as in the formation of $Cs(XeF_7)$ and $(NO_2)(XeF_8)$.

(3) F/H metathesis between XeF_2 and an anhydrous acid, such as

$$XeF_2 + HOClO_3 \longrightarrow F–Xe–OClO_3 + HF.$$

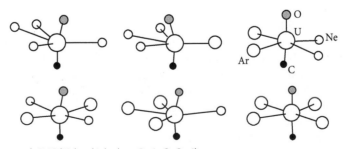

Fig. 7.8.2 Structures of $CUO(Ne)_{4-n}(Ar)_n$ (n = 0, 1, 2, 3, 4).

Table 7.8.1 Structures of some compounds of xenon with fluorine and oxygen

Compound	Geometry/symmetry	Xe–F/pm	Xe–O/pm	Arrangement in the bonded and nonbonded electron pairs in Fig. 7.5.2	
XeF_2	linear, $D_{\infty h}$	200		trigonal bipyramidal	(a)
XeO_3	pyramidal, C_{3v}		176	tetrahedral	(b)
XeF_3^+ in $(XeF_3)(SbF_5)$	T-shaped, C_{2v}	184–191		trigonal bipyramidal	(c)
$XeOF_2$	T-shaped, C_{2v}			trigonal bipyramidal	(d)
XeF_4	square planar, D_{4h}	193		octahedral	(e)
XeO_4	tetrahedral, T_d		174	tetrahedral	(f)
XeO_2F_2	see-saw, C_{2v}	190	171	trigonal bipyramidal	(g)
$XeOF_4$	square pyramidal, C_{4v}	190	170	octahedral	(h)
XeO_3F_2	trigonal bipyramidal, D_{3h}			trigonal bipyramidal	(i)
XeF_5^+ in $(XeF_5)(PtF_6)$	square pyramidal, C_{4v}	179–185		octahedral	(j)
XeF_5^- in $(NMe_4)(XeF_5)$	pentagonal planar, D_{5h}	189–203		pentagonal bipyramidal	(k)
XeF_6	distorted octahedral, C_{3v}	189(av)		capped octahedral	(l)
XeO_6^{4-} in $K_4XeO_6 \cdot 9H_2O$	octahedral, O_h		186	octahedral	(m)
XeF_7^- in $CsXeF_7$	capped octahedral, C_s	193–210		capped octahedral	(n)
XeF_8^{2-} in $(NO)_2XeF_8$	square antiprismatic, D_{4d}	196–208		square antiprismatic	(o)

(4)   Undergo hydrolysis, yielding oxofluorides, oxides, and xenates, such as

$$XeF_6 + H_2O \longrightarrow XeOF_4 + 2HF.$$

Xenon exhibits a rich variety of stereochemistry. Some of the more important compounds of xenon are listed in Table 7.8.1, and their structures are shown in Fig. 7.8.2. The structural description of these compounds depends on whether only nearest neighbor atoms are considered or whether the electron lone pairs are also taken into consideration.

Fig.7.8.3 shows that the known formal oxidation state of xenon ranges from +2 (XeF_2) to +8 (XeO_4, XeO_3F_2, and XeO_6^{4-}), and the structures of the xenon compounds are all consistent with the VSEPR model.

7.8.3 Chemical bonding in xenon fluorides

(1) Xenon difluoride and xenon tetrafluoride

The XeF_2 molecule is linear. A simple bonding description takes the $5p_z$ AO of the xenon atom and the $2p_z$ AO of each fluorine atom to construct the MOs of XeF_2: bonding (σ), non-bonding (σ^n), and antibonding (σ^*), as shown in Fig. 7.8.4. The four valence electrons fill σ and σ^n, forming a 3c-4e σ bond extending over the entire F–Xe–F system. Hence the formal bond order of the Xe–F bond can be taken as 0.5. The remaining $5s$, $5p_x$, and $5p_y$ AOs of the Xe atom are hybridized to form sp^2 hybrid orbitals to accommodate the three lone-pairs, as shown in Fig. 7.8.3(a).

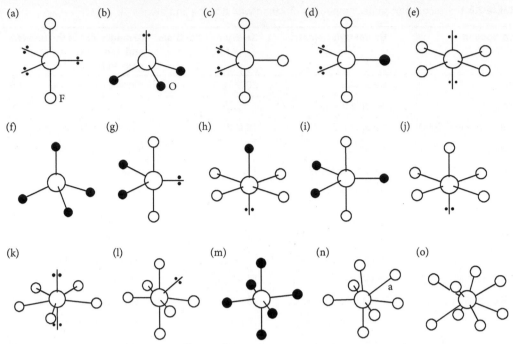

Fig. 7.8.3 Structures of xenon compounds (small shaded circles represent the O atoms): (a) XeF_2, (b) XeO_3, (c) XeF_3^+, (d) $XeOF_2$, (e) XeF_4, (f) XeO_4, (g) XeO_2F_2, (h) $XeOF_4$, (i) XeO_3F_2, (j) XeF_5^+, (k) XeF_5^-, (l) XeF_6, (m) XeO_6^{4-}, (n) XeF_7^-, and (o) XeF_8^{2-}. The electron lone pairs in XeF_7^- and XeF_8^{2-} are not shown in the figure. The XeF_6 molecule (l) has no static structure but is continually interchanging between eight possible C_{3v} structures in which the lone pair caps a triangular face of the octahedron, but in the figure the lone pair is shown in only one possible position. Also, connecting these C_{3v} structures are the transition states with C_{2v} symmetry in which the lone pair pokes out from an edge of the octahedron.

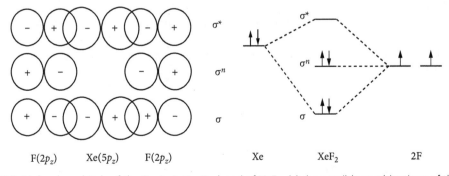

Fig. 7.8.4 Molecular orbitals of the 3c-4e F–Xe–F σ bond of XeF_2: (a) the possible combinations of the AOs of Xe and F atoms, and (b) the schematic energy levels of XeF_2 molecule.

A similar treatment, involving two 3c-4e bonds which are located in X and Y directions, accounts satisfactorily for the planar structure of XeF_4. The remaining $5s$ and $5p_z$ AOs of Xe are hybridized to form sp hybrid orbitals that accommodate the two lone-pair electrons, as shown in Fig. 7.8.3(e). The crystal structure of XeF_2 is analogous to that of α-KrF_2, which is illustrated in Fig. 7.8.13(a).

(2) **Xenon hexafluoride**

There are two possible theoretical models for the isolated XeF_6 molecule: (i) regular octahedral (O_h symmetry), with three 3c-4e bonds and a sterically inactive lone pair of electrons occupying a spherically symmetric s orbital, or (ii) distorted octahedral (C_{3v} symmetry), where the lone pair is sterically active and lies above the center of one face, but the molecule is readily converted into other configurations. All known experimental data are consistent with the following non-rigid model. Starting from the C_{3v} structure, the electron pair can move across an edge between two F atoms (C_{2v} symmetry for the intermediate configuration) to an equivalent position surrounded by three F atoms. This continuous molecular rearrangement, designated as a $C_{3v} \rightarrow C_{2v} \rightarrow C_{3v}$ transformation, involves only modest changes in bond angles and virtually no change in bond lengths.

(3) **Perfluoroxenates XeF_8^{2-} and XeF_7^-**

There are nine electron pairs in XeF_8^{2-}, which are to be accommodated around the Xe atom. From experimental data, the XeF_8^{2-} ion has a square antiprismatic structure (Fig. 7.8.3(o)) showing no distortion that could reveal a possible position for the non-bonding pair of electrons. The lone pair presumably resides in the spherical $5s$ orbital.

There are eight electron pairs to be placed around the Xe atom in XeF_7^-. Six F atoms are arranged octahedrally around the central Xe atom. The approach of the seventh F atom (labelled "a" for this F atom in Fig. 7.8.3(n)), toward the midpoint of one of the faces causes severe distortion of the basic octahedral shape. Such a distortion could easily mask similar effects arising from a non-spherical, sterically active lone pair. The bond between Xe atom and the capping fluorine atom $F_{(a)}$ has the length Xe–$F_{(a)}$ 210 pm, or about 17 pm longer than the other Xe–F bonds, which might suggest some influence of a lone pair in the $F_{(a)}$ direction.

(4) **Complexes of xenon fluorides**

The pentafluoride molecules, such as AsF_5 and RuF_5, act essentially as fluoride ion acceptors, so that their complexes with xenon fluorides can be formulated as salts containing cationic xenon species, e.g., $[XeF][AsF_6]$, $[XeF][RuF_6]$, $[Xe_2F_3][AsF_6]$, $[XeF_3][Sb_2F_{11}]$, and $[XeF_5][AgF_4]$.

In a number of adducts of XeF_2 with metal complexes, the configuration at the Xe atom remains virtually linear, with one F atom lying in the coordination sphere of the metal atom. The ion Xe–F^+ does not ordinarily occur as a discrete ion, but rather is attached covalently to a fluorine atom on the anion.

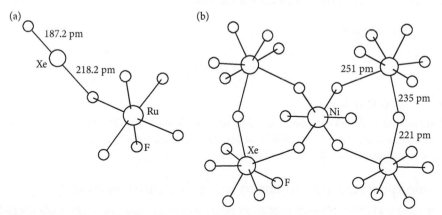

Fig. 7.8.5 Structure of (a) $[XeF][RuF_6]$ and (b) $[Xe_2F_{11}]_2[NiF_6]$.

Table 7.8.2 Xe–F bond lengths in some complexes containing XeF^+

Compound	$Xe-F_{(t)}$/pm	$Xe-F_{(b)}$/pm
$F-Xe-FAsF_5$	187	214
$F-Xe-FRuF_5$	187	218
$F-Xe-FWOF_4$	189	204
$F-Xe-Sb_2F_{10}$	184	235

Fig. 7.8.5(a) shows the structure of $[XeF][RuF_6]$. The terminal $Xe-F_{(t)}$ bond is appreciably shorter than that in XeF_2 (200 pm), while the bridging $Xe-F_{(b)}$ bond is longer. Table 7.8.2 lists the Xe–F bond lengths in some adducts of this type.

The $[Xe_2F_3]^+$ cation in $[Xe_2F_3][AsF_6]$ is V-shaped, as shown below.

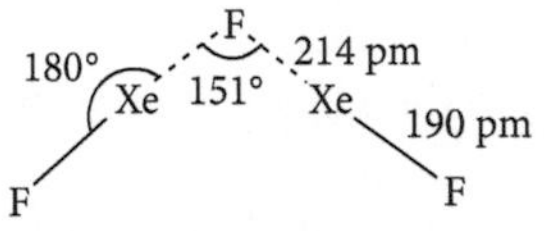

The $[Xe_2F_{11}]^+$ cation can be considered as $[F_5Xe\cdots F\cdots XeF_5]^+$ by analogy to $[Xe_2F_3]^+$. The compounds $[Xe_2F_{11}^+]_2[NiF_6^{2-}]$ and $[Xe_2F_{11}^+][AuF_6^-]$ contain Ni(IV) and Au(V), respectively. Fig. 7.8.5(b) shows the structure of $[Xe_2F_{11}]_2[NiF_6]$.

(5) Metal complexes bearing terminal XeF₂ ligands

XeF_2, as a terminal ligand, can form complexes with a variety of metal ions. Fig. 7.8.6 shows four examples of various coordination modes.

(a) [WOF₄(XeF₂)]

In this structure, the W atom is six-coordinated. One F atom in XeF_2 is directly linked to the W atom, and the other five ligand sites are occupied by O and F atoms as shown in Fig. 7.8.6(a).

(b) [Mg(XeF₂)₄](AsF₆)₂

In this structure, the central Mg atom is coordinated by F atoms of six surrounding ligands: two AsF_6^- anions and four XeF_2 molecules, as shown in Fig. 7.8.6(b).

(c) [Zn(XeF₂)₆](SbF₆)₂

In this structure, the Zn atom is six-coordinated by XeF_2 molecules, as shown in Fig. 7.8.6(c). Notably, the counter anion SbF_6^- does not participate in coordination.

(d) [Mg(XeF₂)(XeF₄)](AsF₆)₂]

In this structure, the Mg atom is six-coordinated by six monodentate ligands: four AsF_6^- anions, one XeF_2 molecule, and one XeF_4 molecule, as shown in Fig. 7.8.6(d). The *trans*-bridged AsF_6^- and *cis*-bridged AsF_6^- units are connected to Mg^{2+} to form layers.

(6) Coordination structure of xenon oxide and fluorine oxide

Some molecules containing covalent bonding of xenon with oxygen can also form weak coordination bonds with other groups.

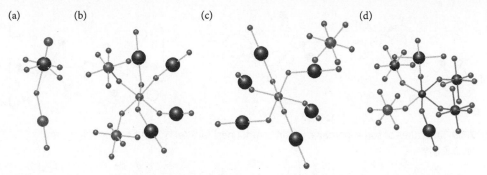

Fig. 7.8.6 Structures of XeF$_2$ complexes. (a) [WOF$_4$(XeF$_2$)], (b) [Mg(XeF$_2$)$_4$](AsF$_6$)$_2$, (c) [Zn(XeF$_2$)$_6$]$^{2+}$ (with only one SbF$_6^-$ ion shown), and (d) [Mg(XeF$_2$)(XeF$_4$)](AsF$_6$)$_2$] (showing coordination mode of the Mg atom in the crystal).

(a) [XeOXeOXe] [μ-F(ReO$_2$F$_3$)$_2$]$_2$

This complex containing Xe(II) oxide is formed by the reaction of ReO$_3$F and XeF$_2$ in anhydrous HF medium at −30 °C. In its crystal structure, the [XeOXeOXe]$^{2+}$ unit has a flat zigzag C_{2h} configuration, with each terminal Xe atom connected to a [(μ-F(ReO$_2$F$_3$)$_2$]$^-$ unit via a Xe⋯F bond, as shown in Fig. 7.8.7.

(b) [NR$_4$]$_3$[Br$_3$(XeO$_3$)$_3$] · 2KrF$_2$ (R = Me, Et)

By reacting XeO$_3$ with (Me$_4$N)Br or (Et$_4$N)Br, (Et$_4$N)$_3$[Br$_3$(XeO$_3$)$_3$] and (Me$_4$N)$_4$[Br$_4$(XeO$_3$)$_4$] can be obtained, respectively. The structures of the cage anions in these two compounds are shown in Fig. 7.8.8. The Xe–Br bond lengths in [Br$_3$(XeO$_3$)$_3$]$^{3-}$ lie in the range 308.4 pm to 331.8 pm. When bromine was replaced by chlorine in the synthesis, the Xe–Cl bond lengths in the isostructural chloroxenonate anion vary from 293.2 pm to 310.1 pm.

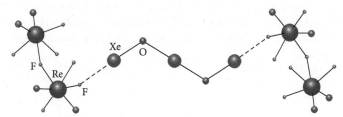

Fig. 7.8.7 Molecular structure of [XeOXeOXe][μ-F(ReO$_2$F$_3$)$_2$]$_2$.

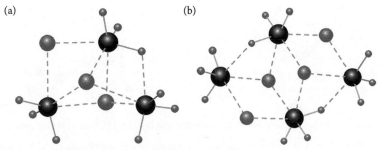

Fig. 7.8.8 Structures of (a) [Br$_3$(XeO$_3$)$_3$]$^{3-}$ and (b) [Br$_4$(XeO$_3$)$_4$]$^{4-}$.

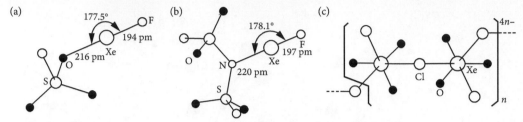

Fig. 7.8.9 Structures of some inorganic xenon compounds: (a) $FXeOSO_2F$, (b) $FXeN(SO_2F)_2$, and (c) $[Xe_2O_6Cl_4]_n^{4n-}$.

(c) $FXeOSO_2F$

When XeF_2 reacts with an anhydrous acid, such as $HOSO_2F$, elimination of HF occurs to yield $FXeOSO_2F$, which contains a linear F–Xe–O group, as shown in Fig. 7.8.9(a).

(d) $FXeN(SO_2F)_2$

This compound is produced by the replacement of a F atom in XeF_2 by a $–N(SO_2F)_2$ group from $HN(SO_2F)_2$. Its molecular structure has a linear F–Xe–N fragment and a planar configuration at the N atom, as shown in Fig. 7.8.9(b).

(e) $Cs_2(XeO_3Cl_2)$

This salt is obtained from the reaction of XeO_3 with CsCl in aqueous HCl solution, and contains an anionic infinite chain. Fig. 7.8.9(c) shows the structure of the chain $[Xe_2O_6Cl_4]_n^{4n-}$.

(7) **Crown-ether complex of xenon oxide**

Recent work showed that xenon trioxide reacts with 15-crown-5 to form a kinetically stable $(CH_2CH_2O)_5XeO_3$ molecular adduct, in which the Xe atom is coordinated by all five O_{crown} atoms (Fig. 7.8.10). The $Xe–O_{crown}$ bonds are predominantly electrostatic and are consistent with σ-hole interaction. X-ray structure analysis showed that the $(CH_2CH_2O)_5XeO_3$ molecule is stabilized by short Xe—O [289.5(1), 293.2(1), 297.0(1) pm] and long Xe⋯O [311.4(1), 312.4(1) pm] bonding interactions. The three

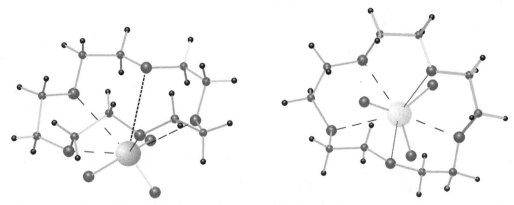

Fig. 7.8.10 Side-on (left) and top-down (right) views of the bonding between molecular components in the crystal structure of $(CH_2CH_2O)_5XeO_3$.

shorter Xe—O contacts are approximately *trans* to the primary polar-covalent Xe–O bonds, with O–Xe⋯O angles of 155.00(4), 159.86(4), and 176.63(4)°, whereas the two longer Xe⋯O contacts subtend significantly smaller O–Xe⋯O angles (129.93(4), 134.79(4)°).

Presently this complex provides the only example of XeO_3 coordinated to a polydentate ligand, and it exhibits the highest coordination number five observed thus far for xenon.

References

P. A. Tucker, P. A. Taylor, J. H. Holloway, and D. R. Russell, The adduct of $XeF_2 \cdot WOF_4$. *Acta Cryst.* **B31**, 906–8 (1975).

M. Tramšek, P. Benkic and B. Žemva, $[Mg(XeF_2)_n](AsF_6)_2$ (n = 4, 2): first compounds of magnesium with XeF_2, *Inorg. Chem.* **43**, 699–703 (2004).

G. Tavcar, E. Goreshnik, and Z. Mazej, Homoleptic $[M(XeF_2)_6]^{2+}$ cations of copper(II) and zinc(II)—syntheses and crystal structures of $[M(XeF_2)_6](SbF_6)_2$ (M = Cu, Zn), *J. Fluorine Chem.* **127**, 1368–73 (2006).

G. Tavcar and B. Žemva, XeF_4 as a ligand for a metal ion, *Angew. Chem. Int. Ed.* **48**, 1432–4 (2009).

M. V. Ivanova, H. P. A. Mercier, and G. J. Schrobilgen, $[XeOXeOXe]^{2+}$, The missing oxide of xenon(II); synthesis, Raman spectrum, and X-ray crystal structure of $[XeOXeOXe][\mu-F(ReO_2F_3)_2]_2$. *J. Am. Chem. Soc.* **137**, 13398–413 (2015).

K. M. Marczenko, H. P. A. Mercier, and G. J. Schrobilgen, A stable crown ether complex with a noble gas compound. *Angew. Chem. Int. Ed.* **57**, 12448–52 (2018).

7.8.4 **Gold-xenon complexes**

Xenon can act as a complex ligand to form M–Xe bonds, especially with gold, which exhibits significant relativistic effects in view of its electronic structure. Some gold-xenon complexes have been prepared and characterized, and their structures are shown in Fig. 7.8.11.

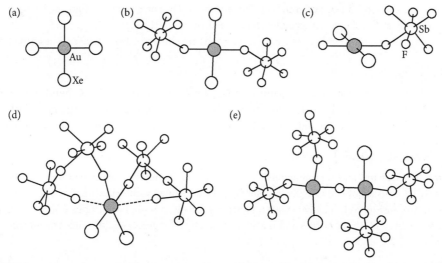

Fig. 7.8.11 Structures of some gold-xenon complexes: (a) $[AuXe_4]^{2+}$, (b) $AuXe_2(SbF_6)_2$, (c) $(AuXe_2F)(SbF_6)$, (d) $(AuXe_2)(Sb_2F_{11})_2$, (e) $(Au_2Xe_2F)^{3+}$ and its immediate anion environment in $[Au_2Xe_2F][SbF_6]_3$.

The compound $[AuXe_4{}^{2+}][Sb_2F_{11}{}^-]_2$ exists in two crystallographically distinct modifications: triclinic and tetragonal. The cation $[AuXe_4]^{2+}$ is square planar with Au–Xe bond lengths ranging from 267.0 pm to 277.8 pm. (Fig. 7.8.11(a)). Around the gold atom, there are three weak Au⋯F contacts of 267.1 pm to 315.3 pm for the triclinic modification, and two contacts of Au⋯F 292.8 pm for the tetragonal modification.

Figs 7.8.11(b) and (c) compare the structures of *trans*-$[AuXe_2][SbF_6]_2$ with the *trans*-$[AuXe_2F][SbF_6]$ moiety in crystalline $[AuXe_2F][SbF_6][Sb_2F_{11}]$. Each Au atom resides in a square planar environment, with two Xe atoms and two F atoms bonded to it. The Au–Xe bond length is 270.9 pm for the former and 259.3 pm to 261.9 pm for the latter.

Fig. 7.8.11(d) shows the structure of $[AuXe_2][Sb_2F_{11}]_2$. The Au–Xe distances of 265.8 pm and 267.1 pm are slightly shorter than those in the $[AuXe_4]^{2+}$ ion.

Fig. 7.8.11(e) shows the structure of the Z-shaped binuclear $[Xe–Au–F–Au–Xe]^{3+}$ ion and its immediate anion environment in $[Au_2Xe_2F][SbF_6]_3$. The Au atom is coordinated by one Xe atom and three F atoms in a square-planar configuration, with Au–Xe bond length 264.7 pm.

The noble gas-noble metal halide complex XeAuF has been characterized in the gas phase using microwave rotational spectroscopy. The Xe–Au bond at 254 pm is short and rigid, and its stretching force constant of 137 Nm^{-1} is the largest of all Ng–M bonds (Ng stands for noble gas, M for noble metal). Its *ab initio* bond energy of 100 kJ mol^{-1} is also the largest of all Ng–M bonds in the NgMX series. The Au–F bond length of 191.8 pm is normal.

Reference

S. A. Cooke and M. C. L. Gerry, XeAuF, *J. Am. Chem. Soc.* **126**, 17000–8 (2004).

7.8.5 **Structures of organoxenon compounds**

More than 10 organoxenon compounds which contain Xe–C bonds have been prepared and characterized. The first structural characterization of a Xe–C bond is performed on $[MeCN–Xe–C_6H_5]^+[(C_6H_5)_2BF_2]^-\cdot MeCN$. The Xe–C bond length is 209.2 pm, and the Xe–N bond length is 268.1 pm, with the Xe atom in a linear environment. This compound has no significant fluorine bridge between cation and anion, as the shortest intermolecular Xe⋯F distance is 313.5 pm.

Other examples of compounds containing Xe–C bonds are shown in Fig. 7.8.12 and described below.

(1) $Xe(C_6F_5)_2$

This is the first reported homoleptic organoxenon(II) compound with two Xe–C bonds of length 239 pm and 235 pm, longer than the corresponding bond in other compounds by about 30 pm. The C–Xe–C unit is almost linear (angle 178°) and the two C_6F_5 rings are twisted by 72.5° with respect to each other.

(2) $[(C_6F_5Xe)_2Cl][AsF_6]$

This is the first isolated and unambiguously characterized xenon(II) chlorine compound. The cation $[(C_6F_5Xe)_2Cl]^+$ consists of two C_6F_5Xe fragments bridged through a chloride ion. Each linear C–Xe–Cl linkage can be considered to involve an asymmetric hypervalent 3c-4e bond. Thus a shorter Xe–C distance (mean value 211.3 pm) occurs and is accompanied by a longer Xe–Cl distance (mean value 281.6 pm). The Xe–Cl–Xe angle is 117°.

(3) $[2,6-F_2H_3C_5N–Xe–C_6F_5][AsF_6]$

In the structure of the cation $(2,6-F_2H_3C_5N–Xe–C_6F_5)^+$, the Xe atom is (exactly) linearly bonded to C and N atoms; the bond lengths are Xe–C 208.7 pm and Xe–N 269.5 pm.

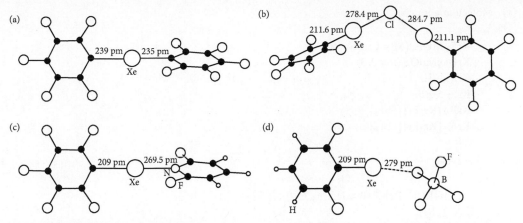

Fig. 7.8.12 Structures of some organoxenon compounds: (a) $Xe(C_6F_5)_2$, (b) the cation $[(C_6F_5Xe)_2Cl]^+$, (c) the cation $(2,6-F_2H_3C_5N-Xe-C_6F_5)^+$, (d) $2,6-F_2H_3C_6-Xe-FBF_3$.

(4) $2,6-F_2H_3C_6-Xe-FBF_3$

In this compound the Xe atom is linearly bonded to C and F atoms; bond lengths are Xe–C 209.0 pm and Xe–F 279.36 pm and the bond angle C–Xe–F is 167.8°.

The structural data of three other organoxenon compounds are listed below:

Compound	Xe–C/pm	Xe–E/pm	C–Xe–E
$(C_6F_5-Xe)(AsF_6)$	(1) 207.9	271.4	170.5°
	(2) 208.2	267.2	174.2°
$(C_6F_5-Xe)(O\overset{O}{\overset{\parallel}{C}}C_6F_5)$	212.2	236.7	178.1°
$(F_2H_3C_6-Xe)(OSO_2CF_3)$	(1) 207.4	268.7	173.0°
	(2) 209.2	282.9	165.1°

In the known organoxenon compounds, the C–Xe–E angle (E = F, O, N, Cl) deviates only slightly from 180°, indicating a hypervalent 3c-4e bond in all cases. The Xe–C bond lengths, except in the compound $Xe(C_6F_5)_2$, vary within the range 208–212 pm. All Xe⋯E contacts are significantly shorter than the sum of the van der Waals radii of Xe and E, indicating at least a weak secondary Xe⋯E interaction.

7.8.6 Krypton compounds

The known compounds of krypton are limited to the +2 oxidation state, and the list includes:

(a) KrF_2

(b) Salts of KrF^+ and $Kr_2F_3^+$:

$[KrF][MF_6]$ (M = P, As, Sb, Bi, Au, Pt, Ta, Ru)
$[KrF][M_2F_{11}]$ (M = Sb, Ta, Nb)
$[Kr_2F_3][MF_6]$ (M = As, Sb, Ta)
$[KrF][AsF_6] \cdot [Kr_2F_3][AsF_6]$

(c) Molecular adducts:

$KrF_2 \cdot MOF_4$ (M = Cr, Mo, W)
$KrF_2 \cdot nMoOF$ (n = 2, 3)
$KrF_2 \cdot VF_5$
$KrF_2 \cdot MnF_4$
$KrF_2 \cdot [Kr_2F_3][SbF_5]_2$
$KrF_2 \cdot [Kr_2F_3][SbF_6]$

(d) Other types

salts of $[RCN-KrF]^+$ (R = H, CF_3, C_2F_5, nC_3F_7)
$Kr(OTeF_5)_2$

(1) Structure of KrF_2

Krypton difluoride KrF_2 exists in two forms in the solid state: α-KrF_2 and β-KrF_2, whose crystal structures are shown in Fig. 7.8.13. Specifically, α-KrF_2 crystallizes in a body-centered tetragonal lattice, with space group $I4/mmm$, and all the KrF_2 molecules are aligned parallel to the c axis. In contrast, β-KrF_2 belongs to tetragonal space group $P4_2/mnm$, in which the KrF_2 molecules located at the corners of the unit cell all lie in the ab-plane and are rotated by 45° with respect to the a axis. The central KrF_2 molecule also lie in the ab-plane, but its molecular axis is orientated perpendicular to those of the corner KrF_2 molecules.

The Kr–F bond length in α-KrF_2 is 189.4 pm and is in excellent agreement with those determined for β-KrF_2, 189 pm, by X-ray diffraction and for gaseous KrF_2 by electron diffraction, 188.9 pm. The interatomic F···F distance between collinearly orientated KrF_2 molecules is 271 pm in both structures.

$Hg(AsF_6)_2$ reacts with a large molar excess of KrF_2 in anhydrous HF to give crystalline $[Hg(KrF_2)_8][AsF_6]_2 \cdot 2HF$, which is the first homoleptic KrF_2 coordination complex of a s-block metal cation. Single-crystal X-ray analysis at –173 °C established that the $[Hg(KrF_2)_8]^{2+}$ cation lies on a crystallographic C_2 axis, with eight KrF_2 ligand molecules constituting a slightly distorted square-antiprismatic

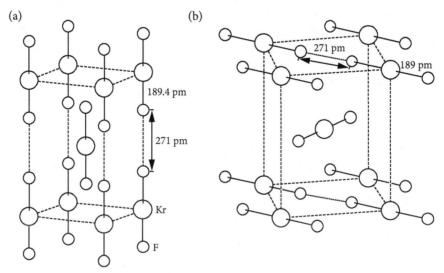

Fig. 7.8.13 Crystal structure of (a) α-KrF_2 and (b) β-KrF_2.

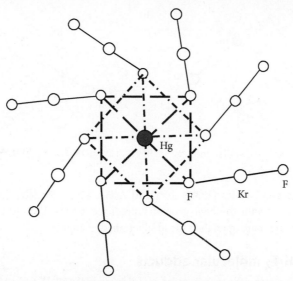

Fig. 7.8.14 Molecular structure of $[Hg(KrF_2)_8]^{2+}$ cation viewed down the crystallographic C_2 axis. The top and bottom faces of the square antiprism are indicated by thin lines. Thick broken lines indicate Hg–F bonds located above the mean molecular plane, and thin broken lines indicate Hg–F bonds located below the mean molecular plane.

coordination sphere around the mercury(II) atom (Fig. 7.8.14). One independent $[AsF_6]^-$ anion exhibits C_2 symmetry with two co-crystallized HF molecules hydrogen-bonded to it, and the other $[AsF_6]^-$ anion occupies an inversion center. The Hg–F bonds in $[Hg(KrF_2)_8]^{2+}$ lying in the range 230.0(1)–241.2(1) pm induce polarization of Kr–F bonds in the peripheral KrF_2 ligands, yielding shorter Kr–F_t terminal bond lengths (range 182.2(1)–185.3(1) pm) and correspondingly longer Kr–F_b bridge bond lengths (range 193.3(1)–195.7(1) pm).

(2) Structures of $[KrF][MF_6]$ (M = As, Sb, Bi)

These three compounds form an isomorphous series, in which the $[KrF]^+$ cation strongly interacts with the anion by forming a fluorine bridge with the pseudo-octahedral anion that is bent about F_b, as shown in Fig. 7.8.15. The terminal Kr–F_t bond lengths in these salts (176.5 pm for $[KrF][AsF_6]$ and $[KrF][SbF_6]$, 177.4 pm for $[KrF][BiF_6]$) are shorter, and the Kr–F_b bridge bond lengths (213.1 pm for $[KrF][AsF_6]$, 214.0 pm for $[KrF][SbF_6]$ and 209.0 pm for $[KrF][BiF_6]$) are longer, than the Kr–F bonds of α-KrF_2 (189.4 pm).

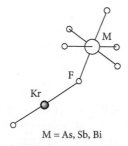

M = As, Sb, Bi

Fig. 7.8.15 Structure of $[KrF][MF_6]$ (M = As, Sb, Bi).

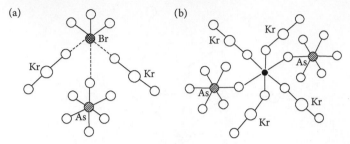

Fig. 7.8.16 Structures of KrF_2 molecular adducts. (a) $[BrOF_2][AsF_6] \cdot 2KrF_2$ and (b) $[Mg(AsF_6)_2] \cdot 4KrF_2$.

The $Kr–F_b–M$ bridge bond angles (133.7° for $[KrF][AsF_6]$, 139.2° for $[KrF][SbF_6]$, and 138.3° for $[KrF][BiF_6]$) are consistent with the bent geometry predicted by the VSEPR arrangements at their respective F_b atoms, but are more open than the ideal tetrahedral angle.

(3) Structures of KrF_2 molecular adducts

$[BrOF_2][AsF_6] \cdot 2KrF_2$ is a rare adduct formed by complex ions $[BrOF_2]^+$ and $[AsF_6]^-$ with two linear KrF_2 molecules. Its structure is determined by X-ray diffraction at –173 °C, as shown in Fig. 7.8.16(a).

The structure of $[Mg(AsF_6)_2] \cdot 4KrF_2$ is characterized by Mg^{2+} ion at the center, and two AsF_6^- ions and four linear KrF_2 as ligands through $Mg \cdots F$ linkage to form an octahedron coordination as shown in Fig. 7.8.16(b).

7.8.7 Mixed noble-gas compounds of krypton(II) and xenon(VI)

The first mixed krypton(II)/xenon(VI) compounds $[F_5Xe(FKrF)AsF_6]$ **1** and $[F_5Xe(FKrF)_2AsF_6]$ **2** were made by low-temperature reaction of $[XeF_5][AsF_6]$ with KrF_2 in anhydrous HF solvent. Subsequent crystallization at low temperature depends on the initial $KrF_2:[XeF_5][AsF_6]$ molar ratio used in carrying out the reaction. Complex **1** was obtained using a $KrF_2:[XeF_5][AsF_6]$ molar ratio of 1.5:1, whereas a stoichiometric excess of KrF_2 (3.5:1 or 2.1:1) resulted in formation of complex **2** as the product.

X-ray crystallography showed that in both complexes the KrF_2 ligands and $[AsF_6]^-$ anions are F-coordinated to the xenon atom of the $[XeF_5]^+$ cation. In complex **1**, the xenon atom of the $[XeF_5]^+$ cation is

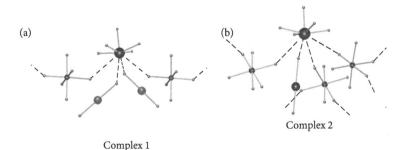

Complex 1

Complex 2

Fig. 7.8.17 X-ray crystal structure of (a) $[F_5Xe(FKrF)AsF_6]$ (**1**) and (b) $[F_5Xe(FKrF)_2AsF_6]$ (**2**). The coordination environment of the Xe atom includes adjacent F atoms generated by space-group symmetry. The Xe, Kr, As, and F atoms are represented by spheres of decreasing sizes.

further coordinated by four terminal F atoms of one KrF_2 and three $[AsF_6]^-$ neighbors [see Fig. 7.8.17(a)]. The coordination geometry of the Xe atom in **1** can be designated as CN_{Xe} = 5 + 4: five primary Xe–F covalent bonds plus four secondary Xe$\cdots$F bonds (one from a KrF_2 ligand and three from three $[AsF_6]^-$ anions). In complex **2**, the $[XeF_5]^+$ cation is also coordinated by four terminal F atoms to give xenon coordination geometry CN_{Xe} = 5 + 4: five primary Xe–F covalent bonds plus four secondary Xe$\cdots$F bonds (from two KrF_2 ligands and two $[AsF_6]^-$ neighbors) [see Fig. 7.8.17(b)]. In both complexes the essentially non-covalent ligand-xenon bonds may be described in terms of σ-hole bonding.

These two complexes significantly extend the XeF_2–KrF_2 analogy and the limited chemistry of krypton by introducing a new class of coordination compound in which KrF_2 functions as a ligand that coordinates to xenon(VI).

Reference

M. Lozinšek, Hélène P. A. Mercier, and G. J. Schrobilgen, Mixed noble-gas compounds of krypton(II) and xenon(VI); $[F_5Xe(FKrF)AsF_6]$ and $[F_5Xe(FKrF)_2AsF_6]$. *Angew. Chem. Int. Ed.* **60**, 8149–56 (2021).

7.8.8 Mixed noble-gas compounds of krypton(II) and xenon(II), and of krypton(II), and xenon(IV)

Reaction of $[XeF][AsF_6]$ with excess KrF_2 at –78°C in anhydrous HF solvent yielded the first mixed Kr^{II}/Xe^{II} noble-gas compound $[FKrFXeF][AsF_6]\cdot0.5KrF_2\cdot2HF$ (**3**). The bent $[FKrFXeF]^+$ cation in **3** is disordered, and one of its two orientations is shown in Fig. 7.8.18(a).

The potent oxidative fluorinating properties of Kr^{II} fluoride species resulted in oxidation of Xe^{II} to Xe^{IV} in anhydrous HF at –60°C to form mixed Kr^{II}/Xe^{IV} cocrystals $([Kr_2F_3][AsF_6])_2\cdot XeF_4$ (**4**) and $XeF_4\cdot KrF_2$ (**5**). Further decomposition at 22 °C resulted in oxidation of Xe^{IV} to Xe^{VI} to give Kr^{II}/Xe^{VI} complexes **1** and **2**, together with a new Kr^{II}/Xe^{VI} complex $[(F_5Xe)_2(\mu\text{-}FKrF)(AsF_6)_2]$ **3**. The $[FKrFXeF][AsF_6]\cdot0.5KrF_2\cdot2HF$ (**3**), $([Kr_2F_3][AsF_6])_2\cdot XeF_4$ (**1**) and $XeF_4\cdot KrF_2$ (**5**) compounds were characterized by low-temperature Raman spectroscopy and single-crystal X-ray diffraction. The structural units of **3** and **4** are shown in Fig. 7.8.18.

As expected, the terminal Ng–F_t bonds of $[FKrFXeF]^+$ in **3** [Kr, 180.6(4) pm; Xe, 188.2(4) pm] are significantly shorter than the $Ng\cdots F_b$ bridge bonds [Kr, 205.5(4) pm; Xe, 217.2(4) pm].

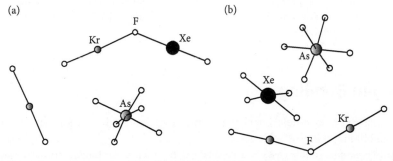

Fig. 7.8.18 (a) X-ray crystal structure of [FKrFXeF] [AsF_6]·0.5KrF$_2$·2HF (**3**), where one of two orientations of the disordered [AsF_6]$^-$ anion is shown. (b) Structural units in the crystal structure of ([Kr$_2$F$_3$][AsF$_6$])$_2$·XeF$_4$ (**4**).

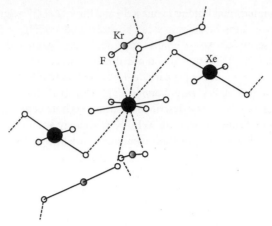

Fig. 7.8.19 Coordination environment around one non-equivalent XeF_4 molecule in the crystal structure of $XeF_4 \cdot KrF_2$ (**5**). The broken lines represent σ-hole type fluorine-bridge interactions.

The low-temperature, high-precision X-ray structure of $XeF_4 \cdot KrF_2$ cocrystal **5** is composed of XeF_4 and KrF_2 molecules that are weakly coordinated through electrostatic, σ-hole type fluorine-bridge interactions, where the shortest $Xe^{IV} \cdots F$ contacts [321.9(1) pm] are close to the sum of the Xe and F van der Waals radii. The XeF_4 and KrF_2 molecules form columns along the c-axis and alternating layers of XeF_4 and KrF_2 molecules along the a and b-axes of their unit cells.

There are two crystallographically non-equivalent XeF_4 molecules in the structural unit of **5**. The Xe atom of one XeF_4 molecule lies on a special position and has four symmetry-related long $Xe^{IV} \cdots F_{Kr}$ contacts of 308.3(2) pm with four neighboring KrF_2 molecules and two symmetry-related long $Xe^{IV} \cdots F_{Xe}$ contacts of 350.8(2) pm with two adjacent XeF_4 molecules (CN_{Xe} = 4 + 6, see Fig. 7.8.19).

The Xe atom of the second independent XeF_4 molecule has four symmetry-related long $Xe^{IV} \cdots F_{Kr}$ contacts [336.7(2) pm] with four neighboring KrF_2 molecules, and four symmetry-related long $Xe^{IV} \cdots F_{Xe}$ contacts [334.1(2) pm] with two neighboring XeF_4 molecules (CN_{Xe} = 4 + 8).

Reference

M. R. Bortolus, H. P. A. Mercier, B. Nguyen, and G. J. Schrobilgen, Syntheses and characterizations of the mixed noble-gas compounds, $[FKr^{II}FXe^{II}F][AsF_6] \cdot 0.5Kr^{II}F_2 \cdot 2HF$, $([Kr^{II}_2F_3][AsF_6])_2 \cdot Xe^{IV}F_4$, and $Xe^{IV}F_4 \cdot Kr^{II}F_2$. *Angew. Chem. Int. Ed.* **60**, 23678–86 (2021).

7.9 **Aerogen Bonding**

Aerogen bonding refers to σ-hole interaction between electron-rich entities that accounts for the supramolecular binding ability of noble gases.

The Group 18 elements (noble gases or aerogens) can use their outer lone electron pairs to generate a new type of supermolecular binding force with adjacent atoms, thereby shortening the bonding distance between them. This effect is called "aerogen bonding interaction," which is illustrated by the following two examples.

Example 1: Supramolecular effects in XeO₃ crystal structure

The XeO_3 molecule has a triangular pyramidal shape, and its valence bond structure is shown in Fig. 7.9.1(a). Within its crystal structure, each XeO_3 molecule interacts with three adjacent XeO_3 neighbors, and the intermolecular Xe⋯O separation lies in the range 280 pm to 290 pm, as shown in Fig. 7.9.1(b). This value is greater than the length of the Xe–O covalent bond (203 pm), but much smaller than 368 pm, the sum of the van der Waals radii of Xe and O. The interaction between Xe and O can be regarded as that between the lone electron pair of Xe atom and the remaining π^* orbitals of three adjacent oxygen atoms, which do not participate in Xe=O bond formation. This secondary interaction shortens the intermolecular Xe⋯O distance.

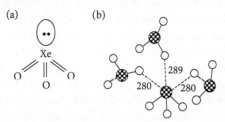

Fig. 7.9.1 Molecular structure of XeO_3 and intermolecular interaction in its crystal structure. (a) Valence bond model of XeO_3 and (b) Xe⋯O distances (in pm) between adjacent XeO_3 molecules.

Example 2: Intermolecular interaction in the crystal structure of XeOF₂·CH₃CN

In the crystal structure of $XeOF_2·CH_3CN$ (Fig. 7.9.2(a)), there are four neighboring CH_3CN molecules around each $XeOF_2$ molecule. The C atoms in three CH_3CN molecules have separate non-covalent interactions with the O atom and two F atoms of the central $XeOF_2$ molecule. The fourth CH_3CN molecule coordinates with the Xe atom through gas-condensation aerogen bonding interaction, involving overlap between the lone electron pair of the Xe atom and the π^* orbital of the N atom (Fig. 7.9.2(b)).

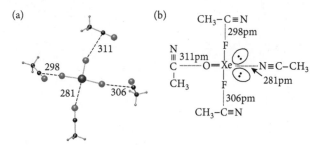

Fig. 7.9.2 (a) Interaction between $XeOF_2$ molecule and four adjacent CH_3CN molecules in the X-ray crystal structure of $XeF_2O·CH_3CN$. (b) Aerogen bonding interaction between the XeF_2O molecule and four neighboring CH_3CN molecules.

References

J. R. DeBackere and G. J. Schrobilgen, A homoleptic KrF_2 complex, $[Hg(KrF_2)_8][AsF_6]_2·2HF$. *Angew. Chem. Int. Ed.* **57**, 13167–71 (2018).

D. H. Templeton, A. Zalkin, J. D. Forrester and S. M. Williamson, Crystal and molecular structure of xenon trioxide, *J. Am. Chem. Soc.* **85**, 817 (1963).

D. S. Brock, V. Bilir, H. P. A. Mercier, and G. J. Schrobilgen, X-ray crystal structure of XeF₂O·CH₃CN, *J. Am. Chem. Soc.* **129**, 3598–611 (2007).

7.10 Directional Bonding Interactions involving Xenon in Crystal Structures

The existence of Xe···F noble-gas (NgB) bonding interaction of 327.9 pm is established in the co-crystal XeF$_2$·2XeF$_4$ (Fig. 7.10.1(a)). In the crystal structure of 2,6-difluorophenyl-xenon trifluoromethanesulfonate, a sulfonate oxygen atom acts as an effective NgB acceptor to form self-assembled dimers each stabilized by a pair of equivalent Xe···O non-covalent bonds (Fig. 7.10.1(b)). In the crystal structure of fluoro-pentafluorophenyl-xenon, two symmetry-equivalent molecules are arranged head to tail in a side-on mode, forming a Xe$_2$F$_2$ parallelogram stabilized by two short 328.8 pm Xe···F NgBs (Fig. 7.10.1(c)). Similar assembly of cyclic structures exist in [C$_6$F$_5$Xe]$^+$[BR$_4$]$^-$ salts (R = C$_6$F$_5$, CN, CF$_3$); the case of R = CN is shown in Fig. 7.10.1(d), where the crystal packing is undoubtedly influenced by intermolecular Xe···N NgB interactions (327.0 pm and 340.2 pm), which are quite short compared with 371 pm (sum of the van der Waals radii of xenon and nitrogen atoms).

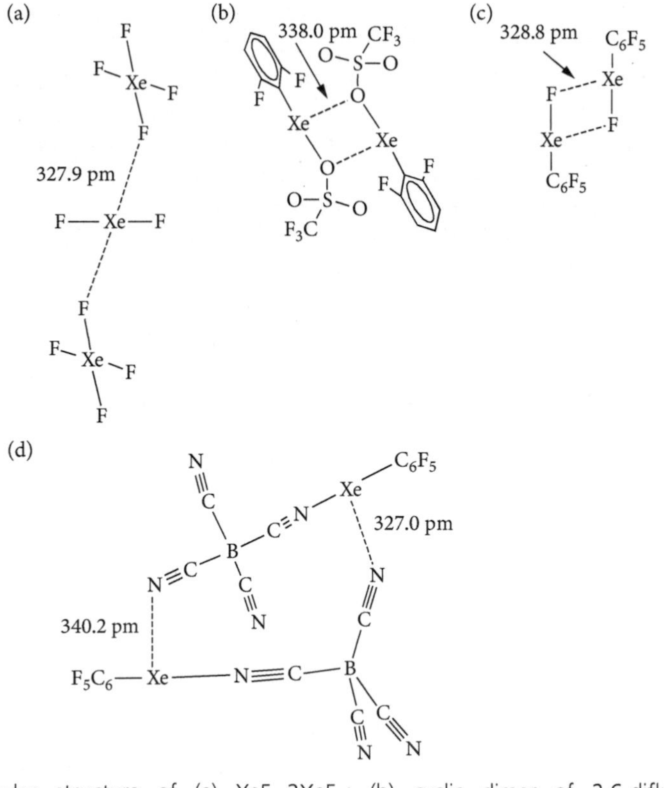

Fig. 7.10.1 Molecular structure of (a) XeF$_2$·2XeF$_4$; (b) cyclic dimer of 2,6-difluorophenyl-xenon trifluoromethanesulfonate; (c) cyclic dimer of fluoro-pentafluorophenyl-xenon; (d) cyclic dimer of [C$_6$F$_5$Xe]$^+$[B(CN)$_4$]$^-$.

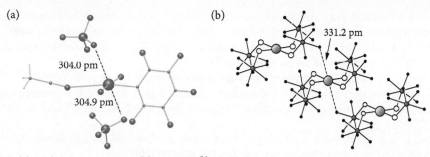

Fig. 7.10.2 (a) Molecular structure of [(MeCN)Xe^{IV}F$_2$(C$_6$F$_5$)] and its stabilization by a pair of BF$_4^-$ ions in the crystalline state. (b) In the solid state, the xenon atom in [Xe^{IV}(OTeF$_5$)$_4$] adopts square-planar coordination geometry, facilitating its π-hole NgB interaction with axial fluorine atoms of neighboring molecules to generate an infinite zigzag ladder.

The X-ray structures of two Xe(IV) derivatives are shown in Fig. 7.10.2. The structure in Part (a) provides the only existing example of a [C$_6$F$_5$XeF$_2$]$^+$ fragment stabilized by a Xe(IV)–C bond. It is stabilized by two NgB contacts with F atoms belonging to two BF$_4^-$ anions that approach the square-planar Xe(IV) center from opposite directions. The Xe···F distances (304.0 pm and 304.9 pm) are significantly shorter than R$_{vdW}$ (363 pm), and they are likely due to electrostatic attraction between the counter-ions (charge-assisted NgB). In the molecular structure of xenon tetrakis(pentafluoroorthotellurate) [Xe(OTeF$_5$)$_4$] shown in Part (b), the xenon atom exhibits square-planar coordination geometry, which facilitates its axial interaction with fluorine atoms (Xe···F distance 331.2 pm) belonging to two adjacent BF$_4^-$ ions. These π-hole NgB interactions govern the formation of an infinite zigzag ladder.

Reference

A. Bauzá and A. Frontera, σ/π-Hole noble gas bonding interactions: insights from theory and experiment, *Coord. Chem. Rev.* **404**, 213112 (2020).

General References

N. Bartlett (ed.), *The Oxidation of Oxygen and Related Chemistry: Selected Papers of Neil Bartlett*, World Scientific, Singapore, 2001.

M. Pettersson, L. Khriachtchev, J. Lundell, and M. Räsänen, Noble gas hydride compounds, in G. Meyer, D. Naumann, and L. Wesemann (eds), *Inorganic Chemistry in Focus II*, Wiley-VCH, Weinheim, 2005, pp. 15–34.

H. Bock, D. Hinz-Hübner, U. Ruschewitz, and D. Naumann, Structure of bis(pentafluorophenyl)xenon, Xe(C$_6$F$_5$)$_2$. *Angew. Chem. Int. Ed.* **41**, 448–50 (2002).

T. Drews, S. Seidal and K. Seppelt, Gold-xenon complexes. *Angew. Chem. Int. Ed.* **41**, 454–6 (2002).

C. Walbaum, M. Richter, U. Sachs, I. Pantenburg, S. Riedel, A.-V. Mudring, and G. Meyer, Iodine-iodine bonding makes tetra(diiodine)chloride, [Cl(I$_2$)$_4$]$^-$, planar, *Angew. Chem. Int Ed.* **52**, 12732–5 (2013).

G. Cavallo, P. Metrangolo, R. Milani, T. Pilati, Arri Priimagi, G. Resnati, and G. Terraneo, The halogen bond, *Chem. Rev.* **116**, 2478–601 (2016).

Mukherjee, S. Tothadi, and G. R. Desiraju, Halogen bonds in crystal engineering: like hydrogen bonds yet different, *Acc. Chem. Res.* **47**, 2514–24 (2014).

L. Khriachtchev, M. Pettersson, N. Runeberg, J. Lundell, and M. Rasanen, A stable argon compound, *Nature* **406**, 874–6 (2000).

D. S. Brock, J. J. Casalis de Pury, H. P. A. Mercier, G. J. Schrobilgen. and B. Silvi, A Rare example of a krypton difluoride coordination compound: $[BrOF_2][AsF_6] \cdot 2KrF_2$, *J. Am. Chem. Soc.* **132**, 3533–42 (2010).

M. Lozinšek, H. P. A. Mercier, D. S. Brock, B. Žemva. and G. J. Schrobilgen, Coordination of KrF_2 to a naked metal cation, Mg^{2+}, *Angew. Chem. Int. Ed.* **56**, 6251–4 (2017).

G. Tavcar and B. Žemva, XeF_4 as a ligand for a metal ion, *Angew. Chem. Int. Ed.* **48**, 1432–4 (2009).

M. V. Ivanova, H. P. A. Mercier, and G. J. Schrobilgen, $[XeOXeOXe]^{2+}$, the missing oxide of Xenon(II); synthesis, Raman spectrum, and X-ray crystal structure of $[XeOXeOXe][\mu\text{-}F(ReO_2F_3)_2]^2$, *J. Am. Chem. Soc.* **137**, 13398–413 (2015).

J. T. Goettel, V. G. Haensch, and G. J. Schrobilgen, Stable chloro- and bromoxenate cage anions; $[X_3(XeO_3)_3]^{3-}$ and $[X_4(XeO_3)_4]^{4-}$ (X = Cl or Br), *J. Am. Chem. Soc.* **139**, 8725–33 (2017).

A. Bauz and A. Frontera, Aerogen bonding interaction: a new supramolecular force? *Angew. Chem. Int. Ed.* **54**, 7340–3 (2015).

Chapter 8

Structural Chemistry of Rare-Earth and Actinide Elements

8.1 Chemistry of Rare-Earth Metals

The lanthanides (Ln) include lanthanum (La) and the following 14 elements (Ce, Pr, Nd, Pm, Sm, Eu, Gd, Tb, Dy, Ho, Er, Tm, Yb, Lu) in which the 4f-orbitals are progressively filled. These 15 elements together with scandium (Sc) and yttrium (Y) are termed the rare-earth metals. The designation of *rare earths* arises from the fact that these elements were first found in rare minerals and were isolated as oxides (called earths in the early literature). In fact, their occurrence in nature is quite abundant. In a broader sense, even the actinides (the 5f elements) are sometimes included in the rare-earth family.

Table 8.1.1 Properties of the rare-earth metals

| Symbol | name | Electronic configuration | | r_M/pm (CN = 12) | $r_{M^{3+}}$/pm (CN = 8) | Crystal structure | Lattice parameters | |
		Metal	M^{3+}				a/pm	c/pm
Sc	scandium	[Ar]$3d^1 4s^2$	[Ar]	164.1	87.1	hcp	330.9	526.8
Y	yttrium	[Kr]$4d^1 5s^2$	[Kr]	180.1	101.9	hcp	364.8	573.2
La	lanthanum	[Xe]$5d^1 6s^2$	[Xe]	187.9	116.0	dhcp	377.4	1217.1
Ce	cerium	[Xe]$4f^1 5d^1 6s^2$	[Xe]$4f^1$	182.5	114.3	ccp	516.1	–
Pr	praseodymium	[Xe]$4f^3 6s^2$	[Xe]$4f^2$	182.8	112.6	dhcp	367.2	1183.3
Nd	neodymium	[Xe]$4f^4 6s^2$	[Xe]$4f^3$	182.1	110.9	dhcp	365.8	1179.7
Pm	promethium	[Xe]$4f^5 6s^2$	[Xe]$4f^4$	(181.0)	(109.5)	dhcp	365	1165
Sm	samarium	[Xe]$4f^6 6s^2$	[Xe]$4f^5$	180.4	107.9	rhom	362.9	2620.7
Eu	Europium	[Xe]$4f^7 6s^2$	[Xe]$4f^6$	204.2	106.6	bcp	458.3	–
Gd	gadolinium	[Xe]$4f^7 5d^1 6s^2$	[Xe]$4f^7$	180.1	105.3	hcp	363.4	578.1
Tb	terbium	[Xe]$4f^9 6s^2$	[Xe]$4f^8$	178.3	104.0	hcp	360.6	569.7
Dy	dysprosium	[Xe]$4f^{10} 6s^2$	[Xe]$4f^9$	177.4	102.7	hcp	359.2	565.0
Ho	holmium	[Xe]$4f^{11} 6s^2$	[Xe]$4f^{10}$	176.6	101.5	hcp	357.8	561.8
Er	erbium	[Xe]$4f^{12} 6s^2$	[Xe]$4f^{11}$	175.7	100.4	hcp	355.9	558.5
Tm	thulium	[Xe]$4f^{13} 6s^2$	[Xe]$4f^{12}$	174.6	99.4	hcp	353.8	555.4
Yb	ytterbium	[Xe]$4f^{14} 6s^2$	[Xe]$4f^{13}$	193.9	98.5	ccp	548.5	–
Lu	lutetium	[Xe]$4f^{14} 5d^1 6s^2$	[Xe]$4f^{14}$	173.5	97.7	hcp	350.5	554.9

Structural Chemistry across the Periodic Table. Thomas Chung Wai Mak, Yu-San Cheung, Gong-Du Zhou, and Ying-Xia Wang. Oxford University Press. © Thomas Chung Wai Mak, Yu-San Cheung, Gong-Du Zhou, and Ying-Xia Wang (2023). DOI: 10.1093/oso/9780198872955.003.0008

The rare-earth metals are of rapidly growing importance, and their availability at quite inexpensive prices facilitates their use in chemistry and other applications. Much recent progress has been achieved in the coordination chemistry of rare-earth metals, in the use of lanthanide-based reagents or catalysts, and in the preparation and study of new materials. Some of the important properties of rare-earth metals are summarized in Table 8.1.1. In this table, r_M is the atomic radius in the metallic state and $r_{M^{3+}}$ is the radius of the lanthanide(III) ion in an eight-coordinate environment.

8.1.1 **Trends in metallic and ionic radii: lanthanide contraction**

The term *lanthanide contraction* refers to the phenomenon of a steady decrease in the radii of the Ln^{3+} ions with increasing atomic number, from La^{3+} to Lu^{3+}, amounting overall to 18 pm. (Table 8.1.1). A similar contraction occurs for the metallic radii and is reflected in many smooth and systematic changes, but there are marked breaks at Eu and Yb. Lanthanide contraction arises because the $4f$ orbitals lying inside the $4d$, $5s$, and $5p$ orbitals provide only incomplete shielding of the outer electrons from the steadily increasing nuclear charge. Therefore the outer electron cloud as a whole steadily shrinks as the $4f$ subshell is filled. In recent years, theoretical work suggests that relativistic effects also play a significant role.

The spectacular irregularity in the metallic radii of Eu and Yb occurs because they have only two valence electrons in the conduction band, whereas the other lanthanide metals have three valence electrons in the $5d/6s$ conduction band. Therefore, lanthanide contraction is not manifested by Eu and Yb in the metallic state or in some compounds such as the hexaborides, where the lower valence of these two elements is evident from the large radii compared with those of neighboring elements.

Lanthanide contraction has important consequences for the chemistry of the third-row transition metals. The reduction in radius caused by the poor shielding ability of the $4f$ electrons means that the third-row transition metals are approximately the same size as their second-row congeners, and consequently exhibit similar chemical behavior. For instance, it has been shown that the covalent radius of gold (125 pm) is less than that of silver (133 pm).

8.1.2 **Crystal structures of the rare-earth metals**

The 17 rare-earth metals are known to adopt five crystalline forms. At room temperature, nine exist in the hexagonal closest packed (hcp) structure, four in the double c-axis hcp (dhcp) structure, two in the cubic closest packed (ccp) structure and one in each of the body-centered cubic packed (bcp) and rhombic (Sm-type) structures, as listed in Table 8.1.1. This distribution changes with temperature and pressure as many of the elements go through a number of structural phase transitions. All of the crystal structures, with the exception of bcp, are closest packed, which can be defined by the stacking sequence of the layers of close packed atoms, and are labeled in Fig. 8.1.1.

hcp:	AB···	dhcp:	ABAC···
ccp:	ABC···	Sm-type:	ABABCBCAC···

If Ce, Eu and Yb are excluded, then the remaining 14 rare-earth metals can be divided into two major sub-groups: (a) The heavy rare-earth metals Gd to Lu, with the exception of Yb and the addition of Sc and Y. These metals adopt the hcp structure. (b) The light rare-earth metals La to Sm, with the exception of Ce and Eu. These metals adopt the dhcp structure. (The Sm-type structure can be viewed as a mixture of one part of ccp and two parts of hcp.) Within each group, the chemical properties of the elements are very similar, so that they invariably occur together in mineral deposits.

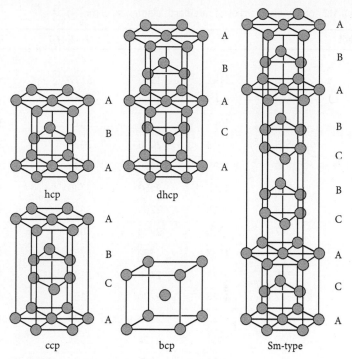

Fig. 8.1.1 Structure types of rare-earth metals.

8.1.3 **Oxidation states**

Due to the stability of the half-filled and filled $4f$ subshell, the electronic configuration of the Ln atom is either $[Xe]4f^n 5d^0 6s^2$ or $[Xe]4f^n 5d^1 6s^2$.

The most stable and common oxidation state of Ln is +3. The principal reason is that the fourth ionization energy I_4 of a rare-earth atom is greater than the sum of the first three ionization energies ($I_1+I_2+I_3$), as listed in Table 8.1.2. The energy required to remove the fourth electron is so great that in most cases it cannot be compensated by chemical bond formation, and thus the +4 oxidation state is largely inaccessible.

Although the +3 oxidation state dominates lanthanide chemistry, other oxidation states are accessible, especially if a $4f^0$, $4f^7$, or $4f^{14}$ configuration is generated. The most common 2+ ions are Eu^{2+} ($4f^7$) and Yb^{2+} ($4f^{14}$), and the most common 4+ ions are Ce^{4+} ($4f^{10}$) and Tb^{4+} ($4f^7$). Of the five lanthanides that exhibit tetravalent chemistry, Nd^{4+} and Dy^{4+} are confined to solid-state fluoride complexes, while Pr^{4+} ($4f^1$) and Tb^{4+} also form the tetrafluoride and dioxide. The most extensive lanthanide(IV) chemistry is that of Ce^{4+}, for which a variety of tetravalent compounds and salts are known (e.g., CeO_2, $CeF_4 \cdot H_2O$). The common occurrence of Ce^{4+} is attributable to the high energy of the $4f$ orbitals at the start of the lanthanide series, such that Ce^{3+} is not sufficiently stable to prevent the loss of an electron.

The crystallographic ionic radii of the rare-earth elements in oxidation states +2 (CN = 6), +3 (CN = 6), and +4 (CN = 6) are presented in Table 8.1.3. The data provide a set of conventional size parameters for the calculation of hydration energies. It should be noted that in most lanthanide(III) complexes the Ln^{3+} center is surrounded by eight or more ligands, and that in aqueous solution the primary coordination sphere has eight and nine aqua ligands for light and heavy Ln^{3+} ions, respectively. The crystal radii of Ln^{3+} ions with CN = 8 are listed in Table 8.1.1.

Table 8.1.2 Ionization energies of rare-earth elements (kJ mol^{-1})

Element	I_1	I_2	I_3	$(I_1+I_2+I_3)$	I_4
Sc	633	1235	2389	4257	7091
Y	616	1181	1980	3777	5963
La	538	1067	1850	3455	4819
Ce	527	1047	1949	3523	3547
Pr	523	1018	2086	3627	3761
Nd	530	1035	2130	3695	3899
Pm	536	1052	2150	3738	3970
Sm	543	1068	2260	3871	3990
Eu	547	1085	2404	4036	4110
Gd	593	1167	1990	3750	4250
Tb	565	1112	2114	3791	3839
Dy	572	1126	2200	3898	4001
Ho	581	1139	2204	3924	4100
Er	589	1151	2194	3934	4115
Tm	597	1163	2285	4045	4119
Yb	603	1176	2415	4194	4220
Lu	524	1340	2022	3886	4360

8.1.4 **Term symbols and electronic spectroscopy**

Atomic and ionic energy levels are characterized by a term symbol of the general form $^{2S+1}L_J$. The values of S, L, and J of lanthanide ions Ln^{3+} in the ground state can be deduced from the arrangement of the electrons in the $4f$ subshell, which are determined by Hund's rules and listed in Table 8.1.4.

Three types of electronic transition can occur for lanthanide compounds. These are: $f \rightarrow f$ transition, $nf \rightarrow (n+1)d$ transition, and ligand → metal f charge transfer transition.

In Ln^{3+} ions, the $4f$ orbitals are radially much more contracted than the d orbitals of transition metals, to the extent that the filled $5s$ and $5p$ orbitals largely shield the $4f$ electrons from the ligands. The result is that vibronic coupling is much weaker in Ln^{3+} compounds than in transition metal compounds, and hence the intensities of electronic transitions are much lower. As many of these electronic transitions lie in the visible region of the electromagnetic spectrum, the colors of Ln^{3+} compounds are typically less intense than those of the transition metals. The colors of the Ln^{3+} ions in hydrated salts are given in Table 8.1.4. The lack of $4f$ orbital and ligand interaction means that the $f \rightarrow f$ transition energies for a given Ln^{3+} change little between compounds, and hence the colors of Ln^{3+} are often characteristic. In viewed of the small interaction of the Ln^{3+} $4f$ orbitals with the surrounding ligands, the $f \rightarrow f$ transition energies in Ln^{3+} compounds are well defined, and thus the bands in their electronic absorption spectra are much sharper.

Since $f \rightarrow d$ transitions are Laporte allowed, they have much higher intensity than $f \rightarrow f$ transitions. Ligand-to-metal charge transfer transitions are also Laporte allowed and also have high intensity. These

Table 8.1.3 Crystallographic ionic radii (pm) and hydration entropies (kJ mol^{-1}) of the rare-earth elements in oxidation states +2, +3, and +4

Element	$r_{M^{2+}}$ (CN = 6)	$-\Delta_{hyd}H°$ (M^{2+})	$r_{M^{3+}}$ (CN = 6)	$-\Delta_{hyd}H°$ (M^{3+})	$r_{M^{4+}}$ (CN = 8)	$-\Delta_{hyd}H°$ (M^{4+})
Sc	—	—	74.5	—	—	—
Y	—	—	90.0	3640	—	—
La	130.4	—	103.2	3372	—	—
Ce	127.8	—	101.0	3420	96.7	6390
Pr	125.3	1438	99.0	3453	94.9	6469
Nd	122.5	1459	98.3	3484	93.6	6528
Pm	120.6	1474	97.0	3520	92.5	6579
Sm	118.3	1493	95.8	3544	91.2	6639
Eu	116.6	1507	94.7	3575	90.3	6682
Gd	114.0	—	93.8	3597	89.4	6726
Tb	111.9	1546	92.1	3631	88.6	6765
Dy	109.6	1566	91.2	3661	87.4	6824
Ho	107.5	1585	90.1	3692	86.4	6875
Er	105.6	1602	89.0	3718	85.4	6926
Tm	103.8	1619	88.0	3742	84.4	6978
Yb	102.6	1631	86.8	3764	83.5	7026
Lu	—	—	86.1	3777	82.7	7069

two types of transitions generally fall in the ultraviolet region, so they do not affect the colors of Ln^{3+} compounds. For easily reduced Ln^{3+} (Eu and Yb), they are at lower energy than the $f \to f$ transitions, and for easily oxidized ligands they may tail into the visible region of the spectrum, giving rise to much more intensely colored complexes.

The Ln^{2+} ions are often highly colored. This arises because the $4f$ orbitals in Ln^{2+} are destabilized with respect to those in Ln^{3+}, and hence lie closer in energy to the $5d$ orbitals. This change in orbital energy separation causes the $f \to d$ transitions to shift from the ultraviolet into the visible region of the spectrum.

The luminescence which arises from $f \to f$ transitions within the Ln^{3+} ion is employed in color television sets, the screens of which contain three phosphor emitters. The red emitter is Eu^{3+} in Y$_2$O$_2$S or Eu^{3+}:Y$_2$O$_3$. The main emissions for Eu^{3+} are between the $^5D_o \to {}^7F_n$ ($n = 4$ to 0) levels. The green emitter is Tb^{3+} in Tb^{3+}:La$_2$O$_2$S. The main emissions for Tb^{3+} are between the $^5D_4$ and ^7F_n ($n = 6$ to 0) levels. The best blue emitter is Ag, Al:ZnS, which has no Ln^{3+} component.

8.1.5 Magnetic properties

The paramagnetism of Ln^{3+} ions arises from their unpaired $4f$ electrons which interact little with the surrounding ligands in Ln^{3+} compounds. The magnetic properties of these compounds are similar to those of the free Ln^{3+} ions. For most Ln^{3+} the magnitude of the spin-orbital interaction in f orbital is sufficiently large, so that the excited levels are thermally inaccessible, and hence the magnetic behavior is

Table 8.1.4 Electron configuration, ground state term symbol, and magnetic properties of Ln^{3+} ions

Ln^{3+}	4f electron configuration	Ground state term symbol	Color of Ln^{3+}	Magnetic moment, μ(298 K)/μ_B	
				Calculated	Observed
La^{3+}	$4f^0$	$^1S_0$	colorless	0	0
Ce^{3+}	$4f^1$	$^2F_{5/2}$	colorless	2.54	2.3–2.5
Pr^{3+}	$4f^2$	$^3H_4$	green	3.58	3.4–3.6
Nd^{3+}	$4f^3$	$^3I_{9/2}$	lilac	3.62	3.5–3.6
Pm^{3+}	$4f^4$	$^5I_4$	pink	2.68	2.7
Sm^{3+}	$4f^5$	$^6H_{5/2}$	pale yellow	0.85	1.5–1.6
Eu^{3+}	$4f^6$	$^7F_0$	colorless	0	3.4–3.6
Gd^{3+}	$4f^7$	$^8S_{7/2}$	colorless	7.94	7.8–8.0
Tb^{3+}	$4f^8$	$^7F_6$	very pale pink	9.72	9.4–9.6
Dy^{3+}	$4f^9$	$^6H_{15/2}$	pale yellow	10.65	10.4–10.5
Ho^{3+}	$4f^{10}$	$^5I_8$	yellow	10.60	10.3–10.5
Er^{3+}	$4f^{11}$	$^4I_{15/2}$	pink	9.58	9.4–9.6
Tm^{3+}	$4f^{12}$	$^3H_6$	pale green	7.56	7.1–7.4
Yb^{3+}	$4f^{13}$	$^2F_{7/2}$	colorless	4.54	4.4–4.9
Lu^{3+}	$4f^{14}$	$^1S_0$	colorless	0	0

determined entirely by the ground level. The effective magnetic moment μ_{eff} of this level is given by the equation:

$$\mu_{eff} = g_J\sqrt{J(J+1)}$$

where

$$g_J = \frac{3}{2} + \frac{S(S+1) - L(L+1)}{2J(J+1)}.$$

The calculated μ_{eff} and observed values from experiments are listed in Table 8.1.4 and shown in Fig. 8.1.2. There is good agreement in all cases except for Sm^{3+} and Eu^{3+}, both of which have low-lying

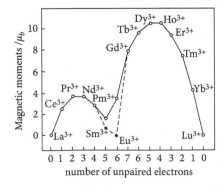

Fig. 8.1.2 Measured and calculated effective magnetic moments (μ_{eff}) of Ln^{3+} ions at 300 K (broken lines represents the calculated values).

excited states ($^6H_{3½}$ for Sm^{3+}, and $^7F_1$ and $^7F_2$ for Eu^{3+}) which are appreciably populated at room temperature.

8.2 Structure of Oxides and Halides of Rare-Earth Elements

The bonding in the oxides and halides of rare-earth elements is essentially ionic. Their structures are determined almost entirely by steric factors and gradual variation across the series, and can be correlated with changes in ionic radii.

8.2.1 Oxides

The sesquioxides M_2O_3 are the stable oxides for all rare-earth elements except Ce, Pr, and Tb, and are the final product of calcination of many salts such as oxalates, carbonates, and nitrates. The sesquioxides M_2O_3 adopt three structural types:

The type-A (hexagonal) structure consists of MO_7 units which approximate to capped octahedral geometry, and is favored by the lightest lanthanides (La, Ce, Pr, and Nd). Fig. 8.2.1(a) shows the structure of La_2O_3.

The type-B (monoclinic) structure is related to type-A, but is more complex as contains three kinds of non-equivalent M atoms, some with octahedral and the remainder with monocapped trigonal prismatic coordination. In the latter type of coordination geometry, the capping O atom is appreciably more distant than those at the vertices of the prism. For example, in Sm_2O_3, the seventh atom is at 273 pm (mean) and the others are at 239 pm (mean). This type is favored by the middle lanthanides (Sm, Eu, Gd, Tb, and Dy).

The type-C (cubic) structure is related to the fluorite structure, but with one-quarter of the anions removed in such a way as to reduce the metal coordination number from eight to six, resulting in two different coordination geometries, as shown in Fig. 8.2.1(b). This type is favored by Sc, Y, and heavy lanthanides from Nd to Lu.

The dioxides CeO_2 and PrO_2 adopt the fluorite structure with cubic unit cell parameters $a = 541.1$ pm and 539.2 pm, respectively, and Tb_4O_7 is closely related to fluorite.

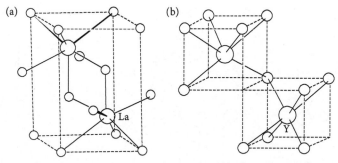

Fig. 8.2.1 (a) Structure of La_2O_3 and (b) the two coordination environments of Y^{3+} in the Y_2O_3 structure.

8.2.2 **Halides**

(1) **Fluorides**

Trifluorides are known for all the rare-earth elements. The structure of ScF_3 is close to the cubic ReO_3 structure, in which Sc^{3+} has octahedral coordination. The YF_3 structure has a nine-coordinate Y^{3+} in a distorted tricapped trigonal prism, eight F^- at approximately 230 pm, and the ninth at 260 pm. The YF_3 structure type is adopted by the $4f$ trifluorides from SmF_3 to LuF_3.

The early light LnF_3 from LaF_3 to HoF_3 adopt the LaF_3 structure, in which the La^{3+} is 11-coordinate with a fully capped, distorted trigonal prismatic coordination geometry. The neighbors of La^{3+} are: seven F^- at 242–248 pm, two F^- at 264 pm, and a further two F^- at 300 pm.

Tetrafluorides LnF_4 are known for Ce, Pr and Tb. In the structure for LnF_4, the Ln^{4+} ion is surrounded by eight F^- forming a slightly distorted square antiprism, which shares its vertices with eight others.

(2) **Chlorides**

Chlorides $ScCl_3$, YCl_3, and later $LnCl_3$ (from $DyCl_3$ to $LuCl_3$) adopt the YCl_3 layer structure, in which the small M^{3+} ion is surrounded by an octahedron of Cl^- neighbors. The early $LnCl_3$ (from $LaCl_3$ to $GdCl_3$) adopt the $LaCl_3$ structure, in which the large Ln^{3+} ion is surrounded by nine approximately equidistant Cl^- neighbors in a tricapped trigonal prismatic arrangement. Fig. 8.2.2 shows the structure of $LaCl_3$.

Chlorides are also known in oxidation state +2 for Nd, Sm, Eu, Dy, and Tm. The structure of the sesquichloride Gd_2Cl_3 is best formulated as $[Gd_4]^{6+}[Cl^-]_6$ and is made up of infinitive chains of Gd_6 octahedra sharing opposite edges, with chlorine atoms capping the triangular faces, as shown in Fig. 8.2.3.

(3) **Bromides and iodides**

The trihalides MBr_3 and MI_3 are known for all the lanthanide elements. The early lanthanide tribromide (La to Pr) adopt the $LaCl_3$ structure, while the later tribromides (from Nd to Lu) and the early triiodides (from La to Nd) form a layer structure with eight-coordinate lanthanide ions.

Ionic dibromides and diiodides are known for Nd, Sm, Eu, Dy, Tm, and Yb. SmI_2, EuI_2, and YbI_2 are useful starting materials for organometallic compounds of these elements in their +2 oxidation states. Iodide SmI_2 is a popular one-electron reducing agent for organic synthesis. The diiodides of La, Ce, Pr, and Gd exhibit metallic properties and are best formulated as $Ln^{3+}(I^-)_2e^-$ with delocalized electrons. The reduction chemistry of lanthanide(II) compounds is discussed in Section 8.5.

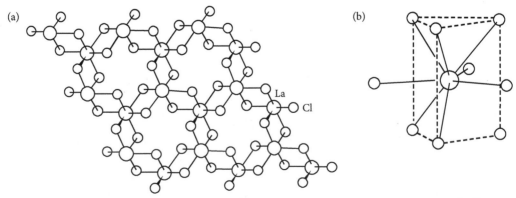

Fig. 8.2.2 Crystal structures of $LaCl_3$: (a) viewed along the c axis, (b) coordination geometry of La^{3+}.

(a) 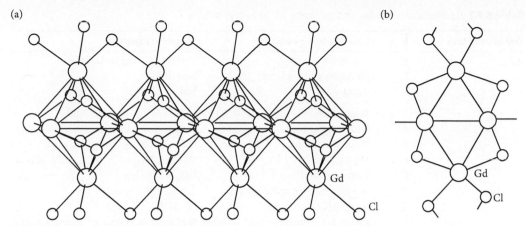 (b)

Fig. 8.2.3 Structure of Gd_2Cl_3: (a) view of a chain aligned along the b axis, (b) view perpendicular to the b axis.

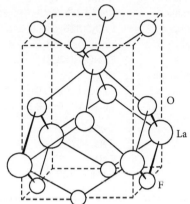

Fig. 8.2.4 Structure of γ-LaOF.

(4) Oxohalides

Many oxohalides of rare-earth elements have been characterized. The crystal of γ-LaOF has a tetragonal unit cell with $a = 409.1$ pm and $c = 583.6$ pm, space group $P4/nmm$. In this structure, each La^{3+} is coordinated by four O^{2-} and four F^- anions, forming a distorted cube, as shown in Fig. 8.2.4. The distance of La–O is 261.3 pm and La–F 242.3 pm.

8.3 Coordination Geometry of Rare-Earth Cations

In lanthanide complexes, the Ln ions are hard Lewis acids, which prefer to coordinate hard bases, such as F, O, N ligands. The f-orbitals are not involved to a significant extent in M–L bonds, so their interaction with ligands is almost electrostatic in nature. Table 8.3.1 lists some examples of the various coordination numbers and geometries of rare-earth cations, which are determined by three factors:

(1) Steric bulk. The ligands are packed around the metal ion in such a way as to minimize inter-ligand repulsion, and the coordination number is determined by the steric bulk of the ligands. The coordination number varies over a wide range from three to twelve, and lower coordination numbers

Table 8.3.1 Coordination number and geometry of rare-earth cations

Oxidation State	CN	Coordination geometry*	Example (structure)
+2	6	octahedral	$Yb(PPh_2)_2(THF)_4$, SmO, EuTe
	7	pentagonal bipyramidal	$SmI_2(THF)_5$
	8	cubic	SmF_2
+3	3	pyramidal	$La[N(SiMe_3)_2]_3$ (Fig. 8.3.1(a))
	4	tetrahedral	$[Lu^tBu_4]^-$
	5	trigonal bipyramidal	$Nd[N(SiHMe_2)_2]_3(THF)_2$
	6	octahedral	$GdCl_4(THF)_2^-$, $ScCl_3$, YCl_3, $LnCl_3$ (Dy-Lu)
	6	trigonal prismatic	$Pr[S_2P(C_6H_{11})_2]_3$
	7	monocapped trigonal prismatic	Gd_2S_3, $Y(acac)_3·H_2O$
	7	monocapped octahedral	La_2O_3 (Fig. 8.3.1(a))
	8	square antiprismatic	$Nd(CH_3CN)(CF_3SO_3)_3L^{**}$ (Fig. 8.3.1(b))
	8	dodecahedral	$Lu(S_2CNEt_2)_4^-$
	8	cubic	$[La(bipyO_2)_4]^{3+}$
	8	bicapped trigonal prismatic	Gd_2S_3
	9	tricapped trigonal prismatic	$Nd(H_2O)_9^{3+}$, $LaCl_3$ (Fig. 8.2.2)
	9	capped square antiprismatic	$LaCl_3(18C6)$ (Fig. 8.3.1(c))
	10	bicapped dodecahedral	$Lu(NO_3)_5^{2-}$ (Fig. 8.3.1(d))
	10	irregular	$Eu(NO_3)_3(12C4)$
	11	fully-capped trigonal prismatic	LaF_3
	11	irregular	$La(NO_3)_3(15C5)$ (Fig. 8.3.1(e))
	12	icosahedral	$La(NO_3)_6^{3-}$ (Fig. 8.3.1(f))
+4	6	octahedral	Cs_2CeCl_6
	8	square antiprismatic	$Ce(acac)_4$
		cubic	CeO_2
	10	irregular	$Ce(NO_3)_4(OPPh_3)_2$
	12	icosahedral	$[Ce(NO_3)_6]^{2-}$

* The polyhedron includes distorted polyhedron.

** L = 1-methyl-1,4,7,10-tetraazacyclododecane

can be achieved with very bulky ligand such as hexamethyldisilylamide. In $La[N(SiMe_3)_2]_3$, the coordination number is only three, as shown in Fig. 8.3.1(a).

(2) Ionic size. The large sizes of lanthanide ions lead to high coordination numbers; eight or nine are very common, and several complexes are known with coordination number 12. For example, the NO_3^- ligand, which has a small bite angle, forms a 12-coordinate La^{3+} complex. Figs 8.3.1(b)–(f) show the structures of Ln^{3+} complexes with coordination numbers ranging from eight to twelve, respectively. The coordination geometries of high coordination complexes are often irregular.

(3) Chelate effect. Higher coordination numbers are usually achieved with chelating ligands, such as crown ethers, EDTA and NO_3^- or CO_3^{2-}. Cations Ln^{3+} form a range of complexes with crown ethers. In the crystal structure of $La(18C6)(NO_3)_3$, the La^{3+} ion is coplanar with the six O-donors of the crown ether. The flexibility of 18C6 allows it to pucker and bind effectively to the smaller later lanthanides. Dibenzo-18C6 is much less flexible than 18C6, and can only form complexes with large early Ln^{3+} ions (La to Nd) in the presence of the strongly coordinating bidentate NO_3^- counterion. Complexes of Ln^{3+} with 15C5 or 12C4 are known for La–Lu with NO_3^- counterions. The larger Ln^{3+} ion cannot fit within the cavity of these ligands, so that it sits above the plane of the macrocycle.

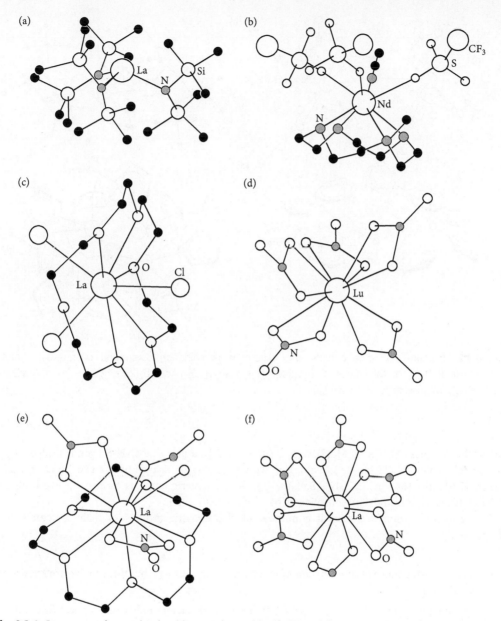

Fig. 8.3.1 Structures of some lanthanide complexes: (a) La[N(SiMe₃)₂]₃, CN = 3; (b) Nd(CH₃CN)(CF₃SO₃)₃ [CH₂- CH₂- N- (CH₂- CH₂- N)₂- CH₂–CH–N- CH₃], CN = 8; (c) LaCl₃(18C6), CN = 9; (d) [Lu(NO₃)₅]²⁻, CN = 10; (e) La(15C5)(NO₃)₃, CN = 11; (f) [La(NO₃)₆]³⁻, CN = 12.

In the coordination chemistry of rare-earth metals, compounds containing M–M bonds are very rare, but the complexes often exist as dimers or oligomers linked by bridging ligands. Fig. 8.3.2 shows the structures of some dimeric and oligomeric coordination compounds: (a) in Rb₅Nd₂(NO₃)₁₁·H₂O, one NO₃⁻ ligand bridges two Nd atoms to form the dimeric anion Nd₂(NO₃)₁₁⁵⁻;

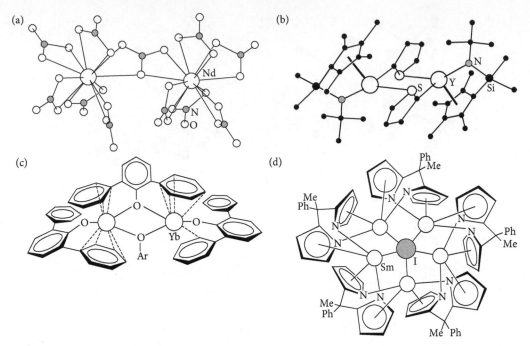

Fig. 8.3.2 Structures of some dimeric and oligomeric coordination compound of rare-earth metals: (a) $[Nd_2(NO_3)_{11}]^{5-}$, (b) $[Y(\eta^5:\eta^1-C_5Me_4SiMe_2N^tBu)(\mu-C_4H_3S)]_2$, (c) $Yb_2(2,6-Ph_2C_6H_3O)_4$, $OAr\equiv(2,6-Ph_2C_6H_3O)$, (d) $\{[MePhC(C_4H_3N)_2]Sm\}_5(\mu_5-I^-)$.

(b) in the complex $[Y(\eta^5:\eta^1-C_5Me_4SiMe_2N^tBu)(\mu-C_4H_3S)]_2$, a pair of $\mu-C_4H_3S$ ligands bridge two Y atoms; (c) in $Yb_2(OC_6H_3Ph_2-2,6)_4(PhMe)_{1.5}$, two $OC_6H_3Ph_2-2,6$ ligands bridge two Yb atoms; and (d) in $K^+(THF)_6\{[MePhC(C_4H_3N)_2]Sm\}_5(\mu_5-I^-)$, the pentameric anion is formed by five bridging $[MePhC(C_4H_3N)_2]$ ligands and consolidated by a central μ_5-I^- anion.

The lanthanide complexes exhibit a number of characteristic features in their structures and properties:

(a) There is a wide range of coordination numbers, generally six to 12, but three to five are known, as listed in Table 8.3.1.

(b) Coordination geometries are determined by ligand steric factors rather than crystal field effects. For examples, the donor oxygen atoms in different ligands coordinate to the La^{3+} ions in different geometries: monocapped octahedral (CN = 7), irregular (CN = 11), and icosahedral (CN = 12).

(c) The lanthanides prefer anionic ligands with hard donor atoms of rather high electronegativity (e.g., oxygen and fluorine) and generally form labile "ionic" complexes that undergo facile ligand exchange.

(d) The lanthanides do not form Ln=O or Ln≡N multiple bonds of the type known for many transition metals and certain actinides.

(e) The $4f$ orbitals in the Ln^{3+} ions do not participate directly in bonding. Their spectroscopic and magnetic properties are thus largely unaffected by the ligands.

8.4 **Organometallic Compounds of Rare-Earth Elements**

In contrast to the extensive carbonyl chemistry of the *d*-transition metals, lanthanide metals do not form complexes with CO under normal conditions. All organolanthanide compounds are highly sensitive to oxygen and moisture, and in some cases they are also thermally unstable.

Organolanthanide chemistry has mainly been developed using the cyclopentadienyl C_5H_5 (Cp) group and its substituted derivatives, such as C_5Me_5 (Cp*), largely because their size allows some steric protection of the large metal center.

8.4.1 **Cyclopentadienyl rare-earth complexes**

The properties of cyclopentadienyl lanthanide compounds are influenced markedly by the relationship between the size of the lanthanide atoms and the steric demand of the Cp group. The former varies from La to Lu according to lanthanide contraction, while the latter varies from the least bulky Cp to highly substituted Cp*, which is appreciably larger. The Sc and Y complexes are very similar to those of lanthanides with proper allowance for the relative atomic sizes.

(1) **Triscyclopentadienyl complexes**

In LaCp$_3$, the coordination requirements of the large La^{3+} ion are satisfied by formation of a polymer, where each La atom is coordinated to three η^5-C_5H_5 ligands with an additional η^2-C_5H_5 interaction, as shown in Fig. 8.4.1(a). The intermediate-size atom Sm forms a simple Sm(η^5-C_5H_5)$_3$ monomeric species, as shown in Fig. 8.4.1(b). The smallest lanthanide, Lu, is unable to accommodate three η^5-C_5H_5 ligands, and LuCp$_3$ adopts a polymeric structure with each Lu coordinated to two η^5-C_5H_5 ligands and two μ_2-Cp ligands as shown in Fig. 8.4.1(c).

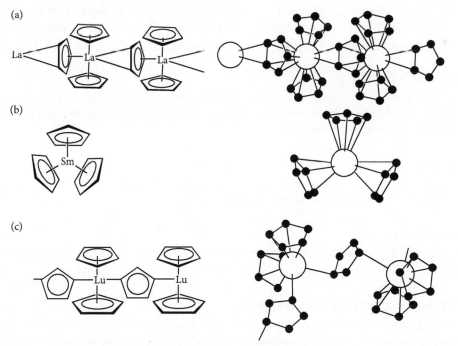

Fig. 8.4.1 Structural formulas and geometries of LnCp$_3$ complexes: (a) LaCp$_3$, (b) SmCp$_3$, and (c) LuCp$_3$.

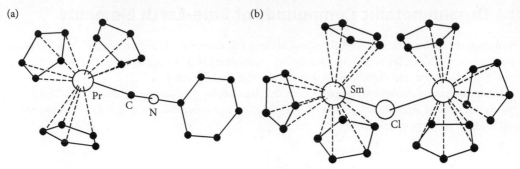

Fig. 8.4.2 Structure of (a) $Cp_3Pr(CNC_6H_{11})$ and (b) $[Cp_3SmClSmCp_3]^-$.

Many adducts of $LnCp_3$ with neutral Lewis bases (such as THF, esters, phosphines, pyridines, and isocyanides) have been prepared and characterized. These complexes usually exhibit pseudo tetrahedral geometry. Fig. 8.4.2(a) shows the structure of $Cp_3Pr(CNC_6H_{11})$ and Fig. 8.4.2(b) shows the structure of the anion of $[Li(DME)_3]^+[Cp_3SmClSmCp_3]^-$, in which two Cp_3Ln fragments are bridged by one anionic ligand.

(2) Biscyclopentadienyl complexes

Many crystal structures of the biscyclopentadienyl rare-earth compounds have been determined; four examples are shown in Fig. 8.4.3.

(a) $(Cp_2LnCl)_2$

The configuration of dimers of this type is dependent on the relative importance of the steric interaction of the Cp rings with the halide bridge versus the interaction between the Cp rings on both metal centers. In $(Cp_2ScCl)_2$, the Cp ring is relatively small in comparison with the halide ligand; accordingly all Cp centroids lie in one plane, and the Sc_2Cl_2 plane is perpendicular to it, as shown in Fig. 8.4.3(a).

(b) $Cp^*_2Yb(MeBeCp^*)$

The Cp^*Yb unit can be coordinated by saturated hydrocarbons, as in $MeBeCp^*$. The positions of the hydrogens on the bridging methyl group show that this is not a methyl bridge having a 3c-2e bond between Yb–C–Be, as shown in Fig. 8.4.3(b).

(c) $Cp_2LuCl(THF)$

This adduct is monomeric with one coordinated THF, forming a pseudo tetrahedral arrangement (each Cp^- ligand is considered to be tridentate), as shown in Fig. 8.4.3(c).

(d) $(Cp^*_2Yb)_2Te_2$

In the chalcogenide complex $Cp^*_2Yb(Te_2)YbCp^*_2$, the Te_2 unit serves as a bridging ligand, as shown in Fig. 8.4.3(d).

(3) Monocyclopentadienyl complexes

Since monocyclopentadienyl rare-earth complexes require four to six σ-donor ligands to reach the stable coordination numbers seven to nine, monomeric complexes must bind several neutral donor molecules. Generally, some of these donor molecules are easily lost, and dinuclear and polynuclear complexes are

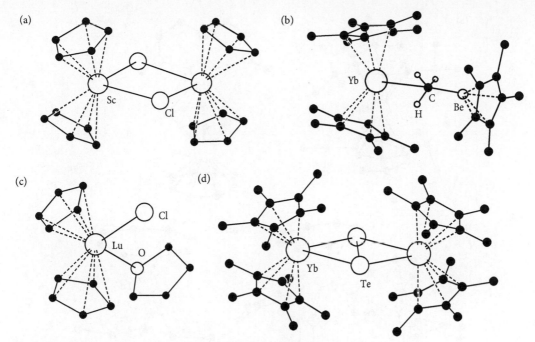

Fig. 8.4.3 Structures of some biscyclopentadienyl rare-earth complexes: (a) $(Cp_2ScCl)_2$, (b) $Cp*_2Yb(MeBeCp*)$, (c) $Cp_2LuCl(THF)$, and (d) $(Cp*_2Yb)_2Te_2$.

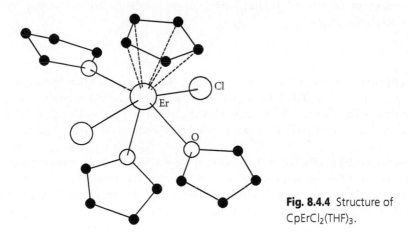

Fig. 8.4.4 Structure of $CpErCl_2(THF)_3$.

formed. This accounts for the fact that monocyclopentadienyl rare-earth complexes exhibit rich structural complexity. Fig. 8.4.4 shows the structure of $CpErCl_2(THF)_3$.

8.4.2 Benzene and cyclooctatetraenyl rare-earth complexes

All the lanthanides and Y are able to bind two substituted benzene rings, such as $Gd(\eta^6\text{-}C_6H_3{}^tBu_3)_2$ and $Y(\eta^6\text{-}C_6H_6)_2$. Fig. 8.4.5(a) shows the sandwich structure of $Gd(\eta^6\text{-}C_6H_3{}^tBu_3)_2$. In this complex, the Gd center is zero-valent.

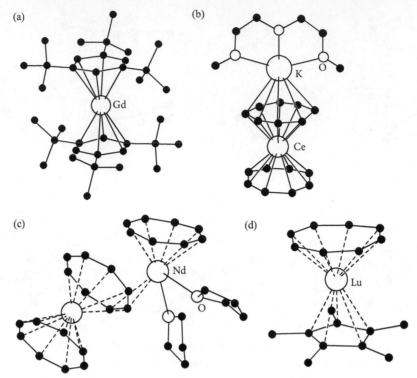

Fig. 8.4.5 Structures of (a) $Gd(\eta^6\text{-}C_6H_3{}^tBu_3)_2$, (b) $[(C_8H_8)_2Ce]K[CH_3O(CH_2CH_2O)_2CH_3]$, (c) $(C_8H_8)_3$ $Nd_2(THF)_2$, and (d) $(C_8H_8)Lu(Cp^*)$.

The large lanthanide ions are able to bind a planar cyclooctatetraene dianion ligand, as shown in Fig. 8.4.5(b)–(d). In $[(C_8H_8)_2Ce]K[CH_3O(CH_2CH_2O)_2CH_3]$, Ce(III) is sandwiched by two $(\eta^8\text{-}C_8H_8)$ dianions, as shown in Fig. 8.4.5(b). The structure of $(C_8H_8)_3Nd_2(THF)_2$ shows a $[(C_8H_8)_2Nd]^-$ anion coordinating to the $[(C_8H_8)Nd(THF)_2]^+$ cation via two carbon atoms of the $C_8H_8{}^{2-}$ ligand, as shown in Fig. 8.4.5(c).

In the sandwich complex $(C_8H_8)Lu(Cp^*)$, as shown in Fig. 8.4.5(d), the molecular structure is slightly bent with a ring centroid-Lu-ring centroid angle of 173°. The methyl groups of Cp^* are bent away from the Lu center by 0.7–2.3° from the idealized planar configuration. The averaged Lu-C bond distances for the $C_8H_8{}^{2-}$ and Cp^{*-} ligands are 243.3 and 253.6 pm, respectively.

8.4.3 Rare-earth complexes with other organic ligands

A variety of "open" π complexes, such as allyl complexes, are known. Generally, the allyl group is η^3-bound to the rare-earth center. Fig. 8.4.6(a) show the structure of the anion in $[Li_2(\mu\text{-}C_3H_5)(THF)_3]^+[Ce(\eta^3\text{-}C_3H_5)_4]^-$.

Using bulky ligands or chelating ligands to saturate the complexes coordinatively or sterically, the rare-earth complexes without π ligands have been obtained and characterized. The La and Sm complexes with bulky alkyl ligands $[CH(SiMe_3)_2]$ are pyramidal, being stabilized by agostic interactions of methyl groups with the highly Lewis acidic metal center, as shown in Fig. 8.4.6(b).

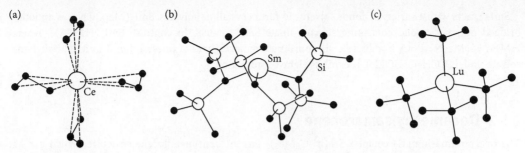

Fig. 8.4.6 Structures of some rare-earth complexes: (a) $[Ce(\eta^3-C_3H_5)_4]^-$, (b) $Sm[CH(SiMe_3)_2]_3$, and (c) $[Lu(^tBu)_4]^-$.

The second half of the lanthanide series react with the more bulky ^tBuLi reagent to form tetrahedrally coordinated complexes, such as $[Li(TMEDA)_2]^+[Lu(^tBu)_4]^-$. Figure 8.4.6(c) shows the structure of the anion $[Lu(^tBu)_4]^-$.

8.5 Reduction Chemistry in Oxidation State +2

Only the elements samarium, europium and ytterbium have significant "normal" chemistry based on true Ln^{2+} ions, although $SmCl_2$ and $EuCl_2$ are well characterized and known for over a century. The compounds of Pr, Nd, Dy, Ho, and Tm in oxidation state +2 are unstable in aqueous solution. Lanthanide(II) compounds are of current interest as reductants and coupling agents in organic and main-group chemistry.

8.5.1 Samarium(II) iodide

SmI_2, which can be prepared conveniently from samarium powder and 1,2-diiodoethane in THF, finds application as a versatile one-electron reducing agent in organic synthesis. Two typical synthetic procedures mediated by SmI_2 are the pinacol coupling of aldehydes and the Barbier reaction, as shown in the following schemes.

pinacol coupling

Barbier reaction

SmI$_2$ reacts with a variety of donor solvents to form crystalline solvates. SmI$_2$(Me$_3$CCN)$_2$ is an iodo-bridged polymeric solid containing six-coordinate Sm^{2+} centers. In contrast, SmI$_2$(HMPA)$_4$, where HMPA is (Me$_3$N)$_3$P=O, is a discrete six-coordinate molecule with a linear I–Sm–I unit. The diglyme solvate SmI$_2$[O(CH$_2$CH$_2$OMe)$_2$]$_2$ exhibits eight-coordination.

8.5.2 **Decamethylsamarocene**

The organosamarium(II) complex Sm(η^5-C$_5$Me$_5$)$_2$ has the bent-metallocene structure, which can be attributed to polarization effects; its THF solvate Sm(η^5-C$_5$Me$_5$)$_2$(THF)$_2$ exhibits the expected pseudo-tetrahedral coordination geometry. [Sm(η^5-C$_5$Me$_5$)$_2$] is a very powerful reductant, and its notable reactions include the reduction of aromatic hydrocarbons such as anthracene and of dinitrogen, as illustrated in the following scheme.

The N–N distance in the dinuclear N$_2$ complex is 129.9 pm, as compared with the bond length of 109.7 pm in dinitrogen, is indicative of a considerable decrease in bond order. An analogous planar system comprising a side-on bonded μ-(bis-η^2)-N$_2$ ligand between two metal centers, with a much longer N–N bond length of 147 pm, occurs in [Cp''$_2$Zr]$_2$(N$_2$), as described in Section 5.1.3.

8.5.3 **Diiodides of thulium, dysprosium, and neodymium**

The molecular structures of LnI$_2$(DME)$_3$ (Ln = Tm, Dy; DME = dimethoxyethane) are shown in Fig. 8.5.1(a) and (b). The larger ionic size of Dy^{2+} compared with Tm^{2+} accounts for the fact that the thulium(II) complex is seven-coordinate whereas the dysprosium(II) complex is eight-coordinate. The

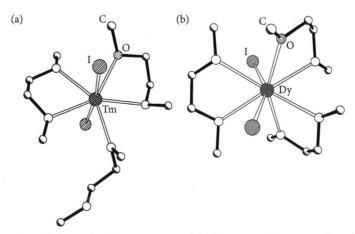

Fig. 8.5.1 Comparison of the molecular structures of (a) hepta-coordinate TmI$_2$(DME)$_3$ and (b) octa-coordinate DyI$_2$(DME)$_3$.

complex $NdI_2(THF)_5$ has pentagonal bipyramidal geometry with the iodo ligands occupying the axial positions.

Thulium(II) complexes are stabilized by phospholyl or arsolyl ligands that can be regarded as derived from the cyclopentadienyl group by replacing one CH group by a P or As atom. Their decreased π-donor capacity relative to the parent cyclopentadienyl system enhances the stability of the Tm(II) center, and stable complexes of the bent-sandwiched type have been isolated.

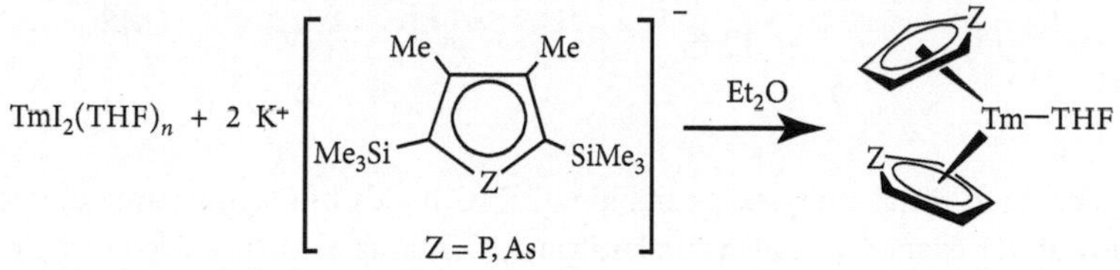

The LnI_2 (Ln = Tm, Dy, Nd) species have been used in a wide range of reactions in organic and organometallic syntheses, including the reduction of aromatic hydrocarbons and the remarkable reductive coupling of MeCN to form a tripodal N_3-ligand, as shown in the following schemes.

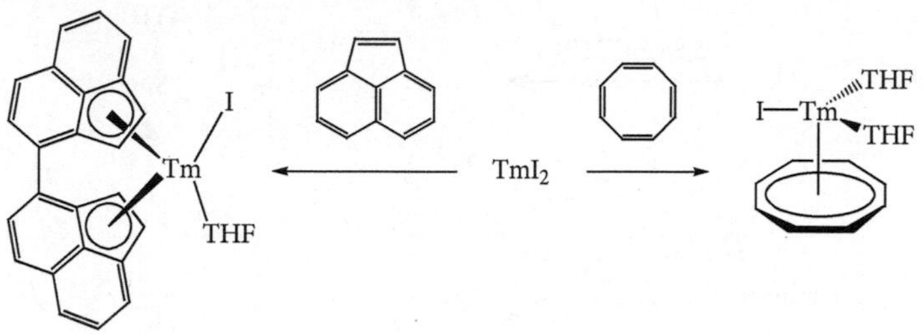

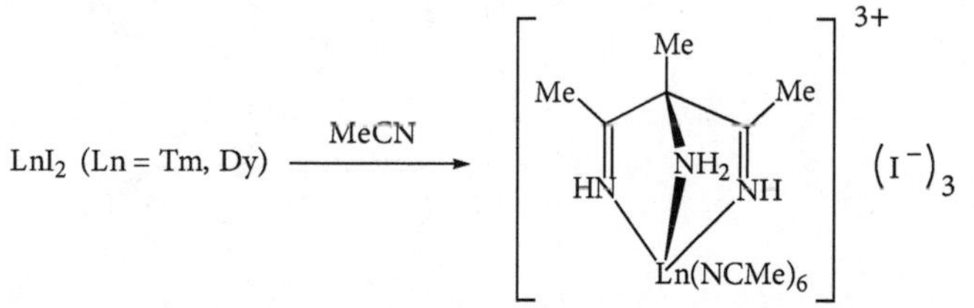

8.6 **RE=C, RE=N, and RE=O Double Bonds in Rare-Earth Metal Complexes**

There are numerous examples of metal-ligand multiple bonds in transition metal complexes, but this kind of multiple bonding is less common in rare-earth metal complexes. Recent progress of this topic has been reviewed by Zhu and colleagues. Some examples are selected and presented in this section.

As early as 2000, the synthesis of samarium carbene complex **1** was reported by Aparna and colleagues. The ligand employed was a dicationic carbene obtained by double deprotonation of bis(iminophosphorano)methane (**2**). The measured Sm–C bond length of 246.7(4) pm was significantly shorter than those of several other neutral rare-earth carbene complexes, and hence it was considered to be a double bond by the investigators.

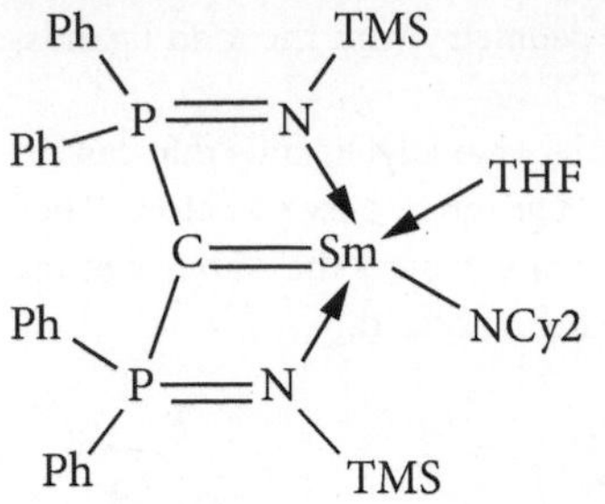

Since then, rare-earth metal complexes exhibiting RE=C bonds but significantly different structural pattern are known. For example, using a similar ligand, Cantat et al. synthesized complex **3**, in which bridging iodides connect two metal-carbene moieties, and **4**, in which two Sm=C bonds are formed at the same metal center. It is interesting to note that these two complexes were obtained with the same reagents but in different stoichiometric ratios.

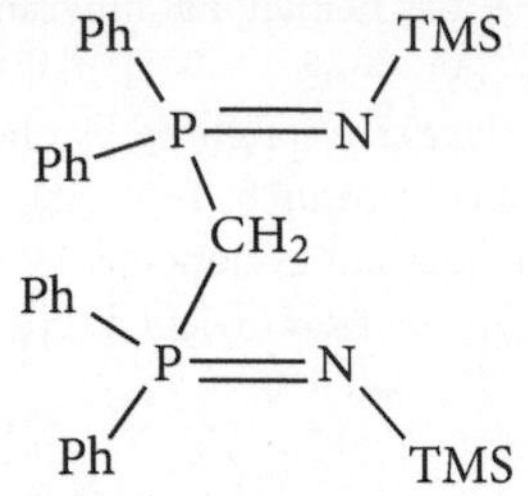

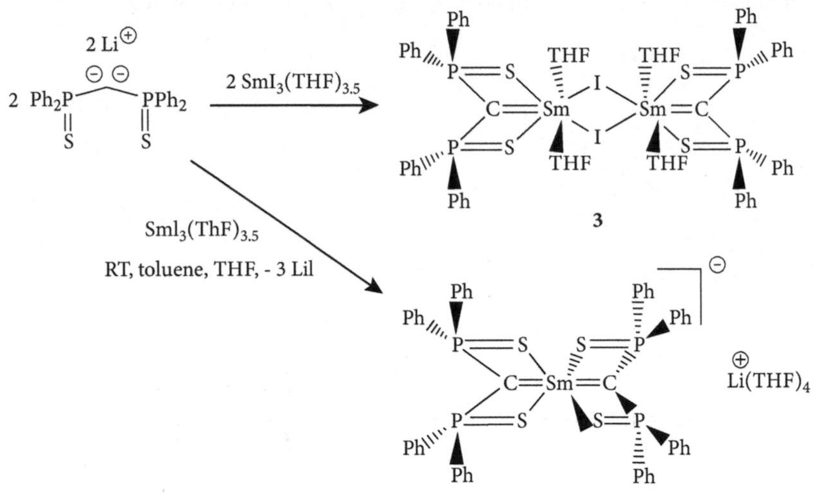

In another strategy employed by Buchard and colleagues, bisalkyl complex **5** was prepared first. In this complex, the Nd atom is coordinated by two ligands through Nd–CH and Nd–N single bonds. The α-proton of one of the two coordinating carbon atom was further removed with a base to form a Nd=C double bond in complex **6**. The bond lengths of the two Nd–C bonds in **6** are 259.2(3) pm and 286.4(3) pm, reflecting the difference in their bond orders.

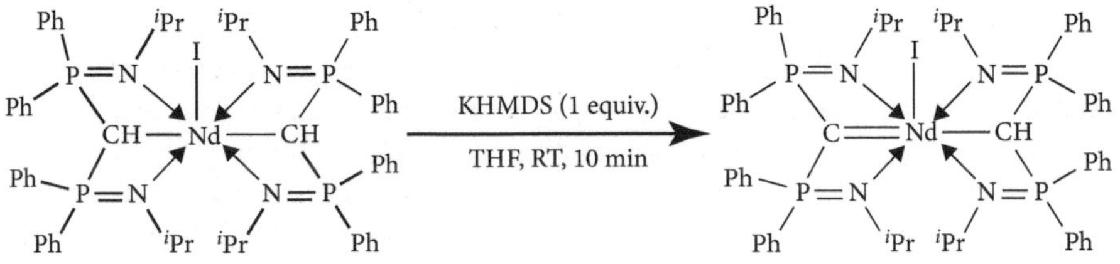

Elimination of the α-proton of coordinating atom of the ligand was also employed in synthesizing complexes **7** featuring both RE–N single and RE=N double bonds.

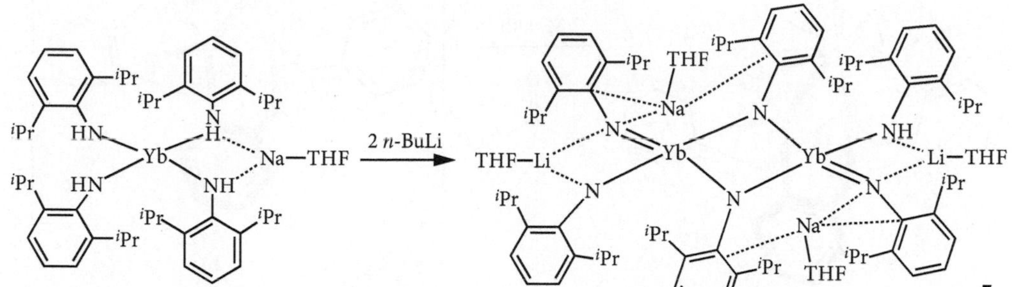

In **7**, the measured Yb=N double bond is longer than the terminal and bridging Yb–N single bonds by about 10 and 20 pm, respectively. It is also notable that the Yb=N–C bond angle comes close to 180°, which further substantiates the manifestation of multiple bonding.

Deprotonation at the coordinating nitrogen atom in complex **8** yielded Ce(IV) complex **9**, in which the Ce=N bond lengths lie in the range 209.8(2) to 212.9(3) pm, depending on the alkali metal counter ion M^+. In another synthesized version of this complex with Cs^+ ion encapsulated by 2.2.2-cryptand so as to eliminate the effect of the M^+ cations, the length of the "pure" Ce=N bond was found to be much shorter at 207.7(3) pm.

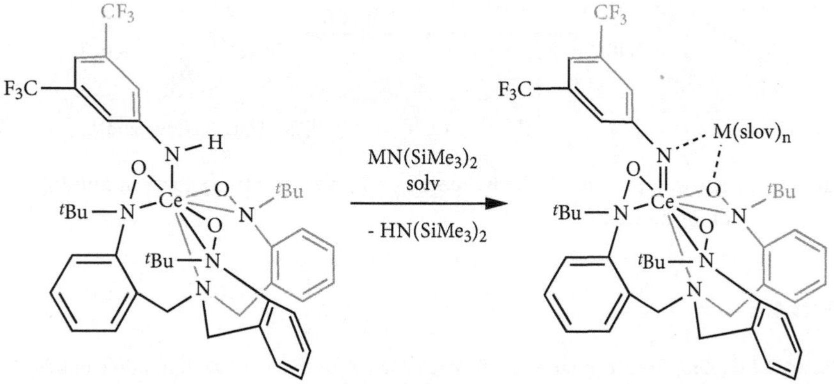

In the same study, oxo-Ce complex **10** was obtained (as a tetramer) by reacting **9** with $Ph_2C=O$:

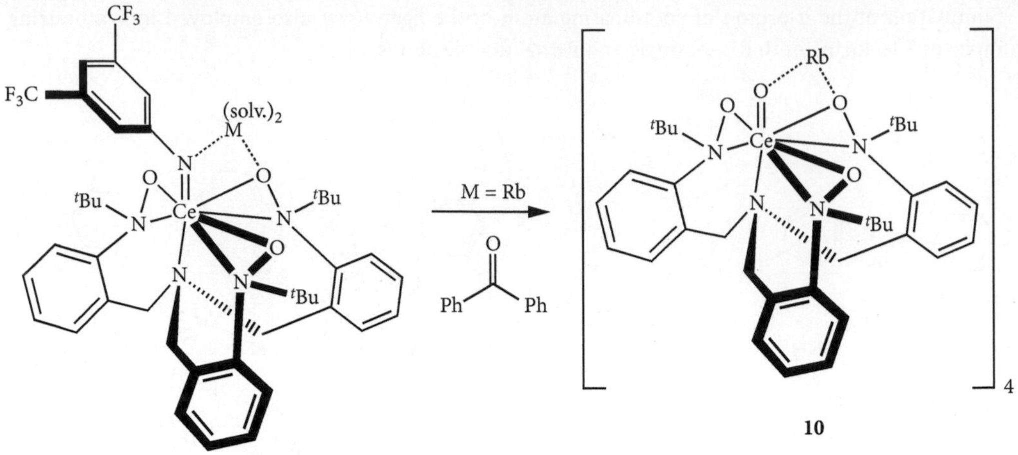

10

The oxo oxygen atom originated from the carbonyl oxygen of $Ph_2C=O$. Inorganic oxygen sources have also been employed to synthesize $Ce^{IV}=O$ complex **11** and **12**:

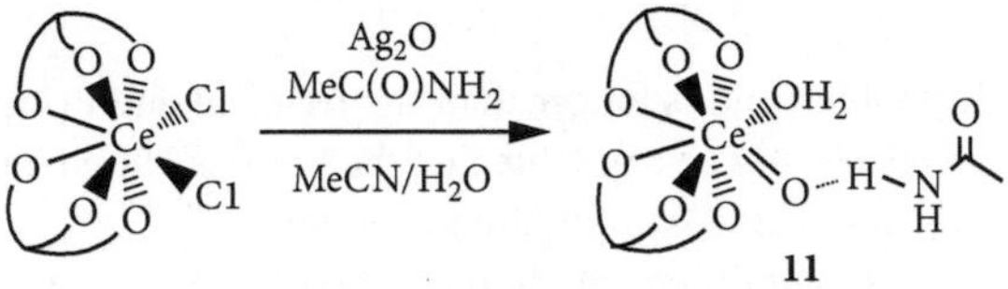

11

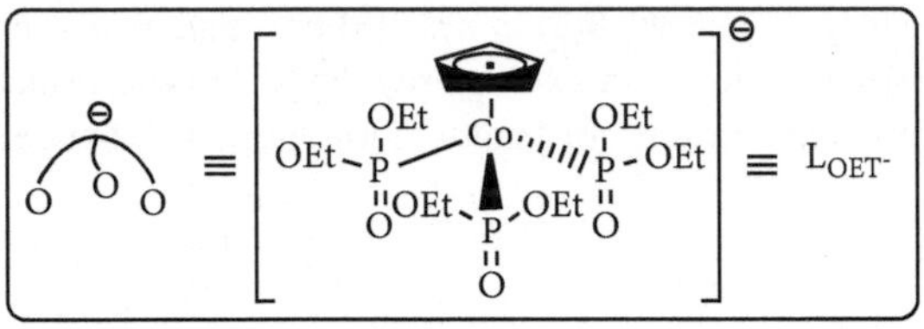

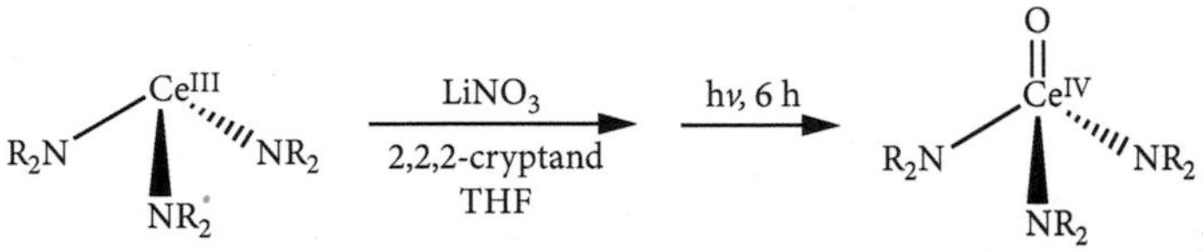

It is notable that the oxo group in tetrahedral complex **12** does not participate in bonding interaction with another metal ion or molecule.

References

Q. Zhu, J. Zhu, and C. Zhu, Recent progress in the chemistry of lanthanide–ligand multiple bonds. *Tetrahedron Lett.* **59**, 514–20 (2018).

K. Aparna, M. Ferguson, and R. G. Cavell, A monomeric samarium bis(iminophosphorano) chelate complex with a Sm=C bond. *J. Am. Chem. Soc.* **122**, 726–7 (2000).

A. J. Arduengo, III, M. Tamm, S. J. McLain, J. C. Calabrese, F. Davidson, and W. J. Marshall, Carbene-lanthanide complexes. *J. Am. Chem. Soc.* **116**, 7927–8 (1994).

H. Schumann, M. Glanz, J. Winterfeld, H. Hemling, N. Kuhn, and T. Kratz, Organolanthanoid-carbene-adducts. *Angew. Chem., Int. Ed.* **33**, 1733–4 (1994).

W. A. Herrmann, F. C. Munck, G. R. J. Artus, O. Runte, and R. Anwander, 1,3-dimethylimidazolin-2-ylidene carbene donor ligation in lanthanide silylamide complexes. *Organometallics* **16**, 682–8 (1997).

T. Cantat, F. Jaroschik, F. Nief, L. Ricard, N. Mézailles, and P. Le Floch, New mono- and bis-carbene samarium complexes: synthesis, X-ray crystal structures and reactivity. *Chem. Commun.* **41**, 5178–80 (2005).

A. Buchard, A. Auffrant, L. Ricard, X. F. Le Goff, R. H. Platel, C. K. Williams, and P. Le Floch, First neodymium(III) alkyl-carbene complex based on bis(iminophosphoranyl) ligands. *Dalton Trans.* **38**, 10219–22 (2009).

H.-S. Chan, H.-W. Li and Z. Xie, Synthesis and structural characterization of imido-lanthanide complexes with a metal-nitrogen multiple bond. *Chem Commun.* **38**, 652–3 (2002).

L. A. Solola, A. V. Zabula, W. L. Dorfner, B. C. Manor, P. J. Carroll, and E. J. Schelter, Cerium(IV) imido complexes: structural, computational, and reactivity studies. *J. Am. Chem. Soc.* **139**, 2435–42 (2017).

Y.-M. So, G.-C. Wang, Y. Li, H. H.-Y. Sung, I. D. Williams, Z. Lin, and W.-H. Leung, A tetravalent cerium complex containing a Ce=O bond. *Angew. Chem., Int. Ed.* **53**, 1626–9 (2014).

M. K. Assefa, G. Wu, and T. W. Hayton, Synthesis of a terminal Ce(IV) oxo complex by photolysis of a Ce(III) nitrate complex. *Chem. Sci.* **8**, 7873–8 (2017).

8.7 **Rare-Earth Polyoxometalates**

POMs are discrete anionic molecular clusters formed by oxo-bridged early transition metal atoms, each being coordinated by six oxygen atoms to form an octahedral MO_6 building block. These building blocks are mainly interconnected by sharing oxygen atoms at the corners or edges. The external O-vertices of such connected blocks can coordinate with d- or f-block metal ions in a way similar to their complexation by multidendate ligands.

The first report of POM dates back to J. Berzelius, who discovered the ammonium salt of $PMo_{12}O_{40}{}^{3-}$ in 1826, although the structural characterization of $H_3PMo_{12}O_{40}\cdot 29H_2O$ by powder X-ray diffraction was carried out much later in 1934. Since then, there are steady increases in global interest of this field and the designed syntheses of new POM species.

Current interest in lanthanide-containing POMs is due to their desirable properties in areas such as luminescence, catalysis, and single-molecule magnets. Selected examples of important lanthanide-containing POMs are described below.

The chemistry of lanthanide-containing POMs is dominated by tungsten-based POMs because there exist a large number of known polytungstate units. Since lanthanide ions have relatively large sizes and high coordination numbers, they tend to be linked by two or more POM units. The structure of the POM $[Ln(W_5O_{18})_2]^{n-}$ (Ln = La, Ce, Pr, Nd, Sm, Ho, Yb, Y for Ln^{3+}, and Ce for Ln^{4+}) is shown below on the left side. The $W_5O_{18}{}^{2-}$ ion consists of five edge-shared octahedral MO_6 units interconnected in a square-pyramidal manner, so that there are four oxygen-containing vertices at its base. Hence a lanthanide ion can be sandwiched by two $W_5O_{18}{}^{2-}$ ions as shown, resulting in square-antiprismatic eight-coordination.

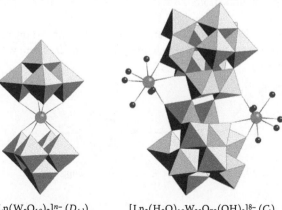

$[Ln(W_5O_{18})_2]^{n-}$ (D_{4d}) $[Ln_2(H_2O)_{10}W_{22}O_{72}(OH)_2]^{8-}$ (C_i)

The family of di-lanthanide coordinated 22-isopolytungstates $[Ln_2(H_2O)_{10}W_{22}O_{72}(OH)_2]^{8-}$ (Ln = La, Ce, Tb, Dy, Ho, Er, Tm, Yb, Lu) illustrates another mode of coordination. In this case two lanthanide(III) ions each coordinates to the polyanion $[H_2W_{22}O_{74}]^{14-}$ laterally through three vertices only, as shown on the right side of the above figure. To compensate for the resulting low coordination number, five water molecules function as terminal aqua ligands for each lanthanide ion.

The use of heteropolyanions provides another approach for synthesizing POMs. Additional heteroatoms X, such as I^{VII} and Te^{VI} in $[XW_6O_{24}]^{n-}$; P^V, Si^{IV}, and B^{III} in $[XW_{12}O_{40}]^{n-}$; and P^V and As^V in $[X_2W_{18}O_{62}]^{n-}$ have been reported in the literature. Examples of lanthanide-containing polyoxotungstates (with and without heteroatom) can be found in a review by Bassil and U. Kortz.

References

J. Berzelius, Beitrag zur näheren Kenntniss des Molybdäns. *Pogg. Ann.*, **6**, 369–92 (1826).

J. F. Keggin, The structure and formula of 12-phosphotungstic acid. *Proc. R. Soc. London, Ser. A*, **144**, 75–100 (1934).

R. D. Peacock and T. J. R. J. Weakley, Heteropolytungstate complexes of the lanthanide elements. Part I. Preparation and reactions. *J. Chem. Soc. A*. 1836–9 (1971).

J. Iball, J. N. Low, and T. J. R. Weakley, Heteropolytungstate complexes of the lanthanoid elements. Part III. Crystal structure of sodium decatungstocerate(IV)–water (1/30). *J. Chem. Soc., Dalton Trans.*, 2021–4 (1974).

M. T. Pope, *Heteropoly and Isopoly Oxometalates*, Springer, Berlin, 1983.

A. H. Ismail, M. H. Dickman, and U. Kortz, 22-Isopolytungstate fragment $[H_2W_{22}O_{74}]^{14-}$ coordinated to lanthanide ions, *Inorg. Chem.* **48**, 1559–65 (2009).

B. S. Bassil and U. Kortz, Recent advances in lanthanide-containing polyoxotungstates, *Z. Anorg. Allg. Chem.* **636**, 2222–31 (2010).

8.8 Imidazolin-2-iminato Complexes with Very Short Metal-Nitrogen Bonds

A number of imidazolin-2-iminato complexes of rare-earth metals have been synthesized (Scheme 8.8.1) and found to exhibit very short metal-nitrogen bonds. The ligand employed, ImDippN$^-$ [1,3-bis(2,6-diisopropylphenyl)imidazolin-2-imide, **1**$^-$], is generated by proton abstraction of **1**-H by LiCH$_2$SiMe$_3$.

Deprotonation of **1**-H results in a formal charge of –1 at the terminal nitrogen atom (**1A**). **1B** was proposed to be a resonance structure of **1**$^-$, which may act as an imido ligand with a formal charge of –2 at the terminal nitrogen atom.

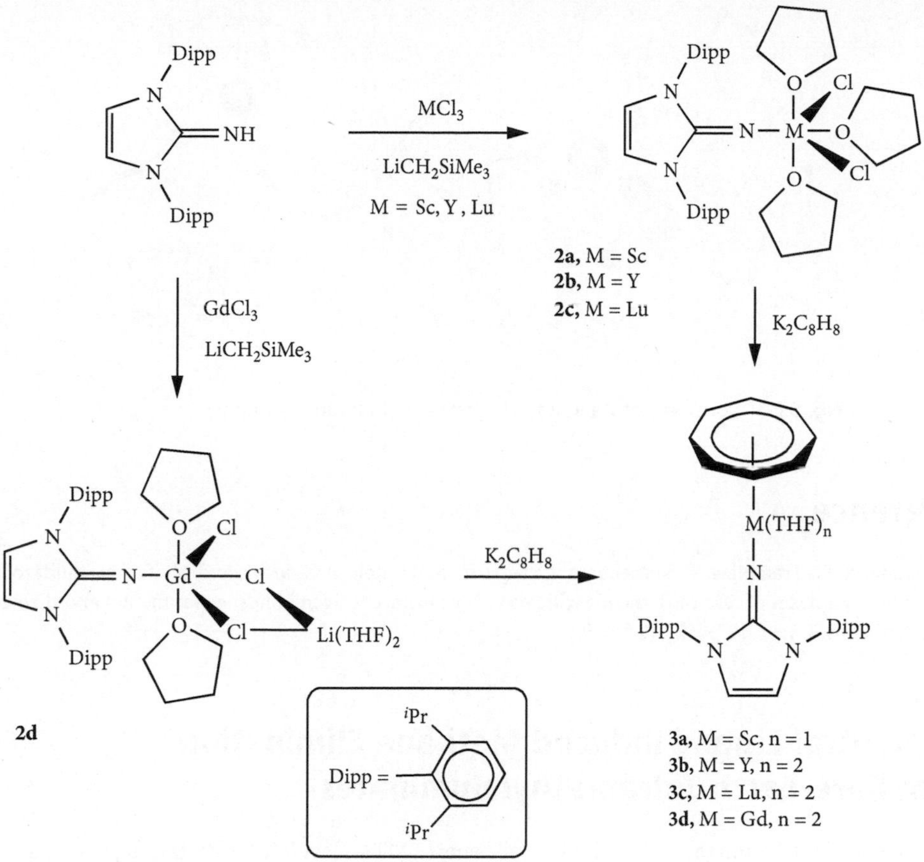

Scheme 8.8.1 Preparation of imidazolin-2-iminato rare-earth metal complexes.

Table 8.8.1 Metal-nitrogen bond lengths (in pm) in **2a** to **2d** and **3a** to **3d**.

M = Sc	M = Y	M = Lu	M = Gd
2a·THF	**2b**·THF	**2c**·THF	**2d**
196.3(2)	212.78(18)	209.1(3)	215.3(3)
3a·THF	**3b**·THF	**3c**·THF	**3d**·THF
198.2(2)/196.5(2)*	216.3(2)	212.2(2)	219.1(2)

* Two independent molecules in the asymmetric unit.

In these complexes, **1**⁻ functions as a linear coordinating ligand to bind M^{III} (M = Sc, Y, Lu and Gd) in forming complexes **2a** to **2d**, which can further react with dipotassium cyclooctatetradienide $[K_2(C_8H_8)]$ to form complexes **3a** to **3d**. In all these complexes, the metal-nitrogen bonds are very short and their values are listed in Table 8.8.1. The structures of **2a** in **2a**·THF and **3a** in **3a**·THF are shown in Fig. 8.10.1. The boron chain in $La_2Re_3B_7$ and B–B bond lengths (in pm).

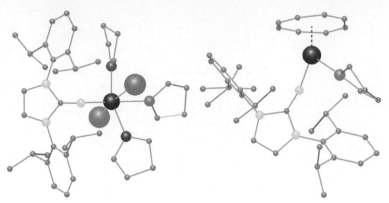

Fig. 8.8.1 Structures of **2a** in **2a**·THF (left) and **3a** in **3a**·THF (right).

Reference

T. K. Panda, A. G. Trambitas, T. Bannenberg, C. G. Hrib, S. Randoll, P. G. Jones, and M. Tamm, Imidazolin-2-iminato complexes of rare earth metals with very short metal-nitrogen bonds: experimental and theoretical studies. *Inorg. Chem.* **48**, 5462–72, (2009).

8.9 Neutral Ligand Induced Methane Elimination from Rare-Earth Tetramethylaluminates

The reaction of 1,3,5-trimethyl-1,3,5-triazacyclohexane (TMTAC) with $[M\{Al(CH_3)_4\}_3]$ (M = La, Y and Sm) yielded rare-earth TMTAC complexes (**1**, **2**, **3**, and **4** in Schemes 8.9.1 to 8.9.3). When TMTAC was added to $[M\{Al(CH_3)_4\}_3]$, the C–H bonds in $[M\{Al(CH_3)_4\}_3]$ were activated with elimination of CH_4 molecules, resulting in methylene groups bridging multi-metal centers. Different pathways have been proposed to describe the methane eliminations.

The structures of **1** and **3** are similar but the latter contains one less CH_4 unit. Both of them have a TMTAC ligand coordinated to the rare-earth metal center on one side, and the carbon atoms of aluminates coordinated from the other side.

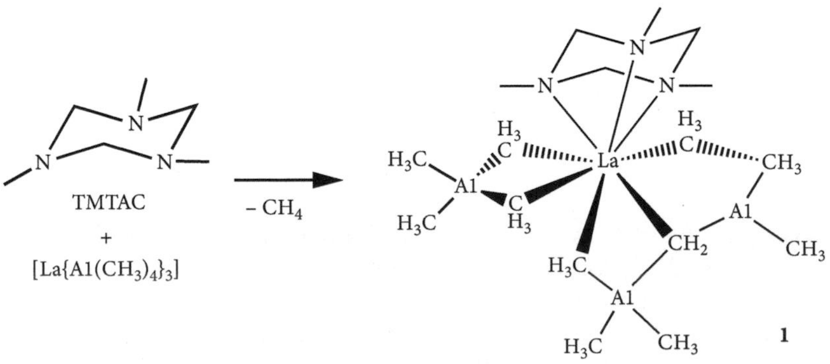

Scheme 8.9.1 Reaction of $[La\{Al(CH_3)_4\}_3]$ with TMTAC.

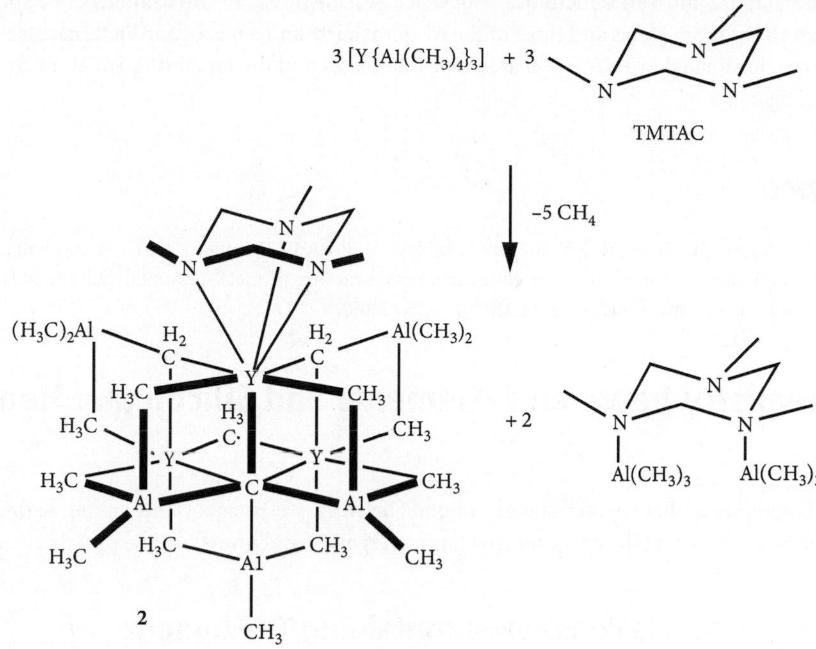

$$3\,[\mathrm{Y}\{\mathrm{Al}(\mathrm{CH}_3)_4\}_3] \;+\; 3\;\; \text{TMTAC}$$

Scheme 8.9.2 Reaction of [Y{Al(CH$_3$)$_4$}$_3$] with TMTAC.

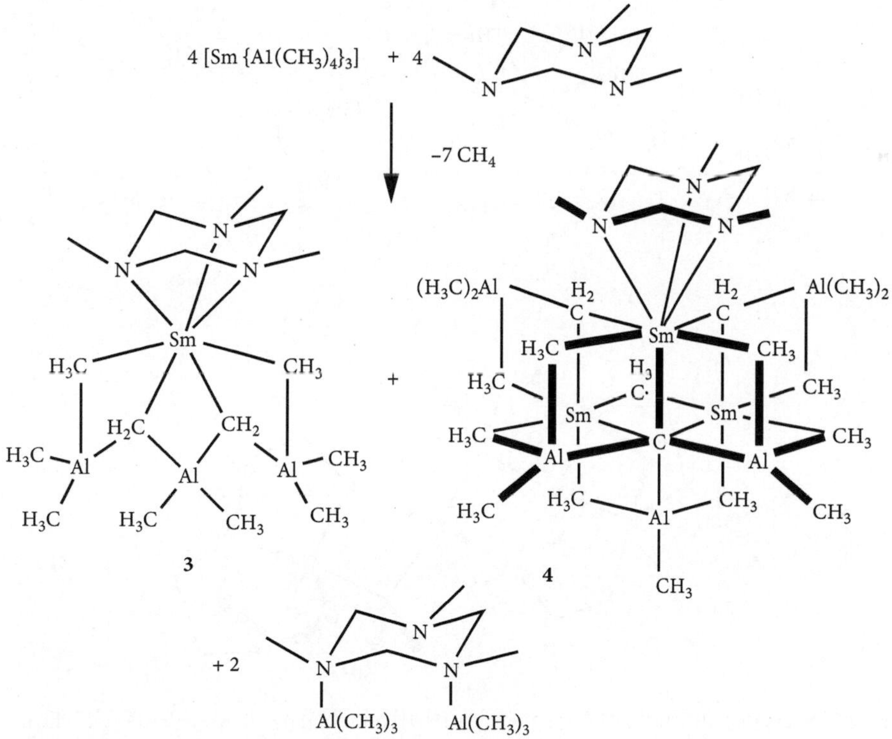

$$4\,[\mathrm{Sm}\{\mathrm{Al}(\mathrm{CH}_3)_4\}_3] \;+\; 4$$

Scheme 8.9.3 Reaction of [Sm{Al(CH$_3$)$_4$}$_3$] with TMTAC.

On the other hand, **2** and **4** are structural analogs each bearing three rare-earth atoms (Y or Sm) and five Al atoms. In **4** all three Sm atoms and three of the Al atoms surround a hexa-coordinate carbide ion, with two Sm atoms coordinated by the carbon atoms of aluminates and the remaining Sm atom coordinated by a TMTAC ligand.

Reference

A. Venugopal, I. Kamps, D. Bojer, R. J. F. Berger, A. Mix, A. Willner, B. Neumann, H.-G. Stammlera, and N. W. Mitzel, Neutral ligand induced methane elimination from rare-earth metal tetramethylaluminates up to the six-coordinate carbide state. *Dalton Trans* **38**. 5755–65 (2009).

8.10 **Chemistry between *f*-Elements and Silicon and Heavy Tetrels**

The study of complexes that contain chemical bonds between *f* elements (lanthanides, actinides) and silicon or the heavier tetrels (Group 14 metals) has undergone rapid growth in the past three decades.

8.10.1 **Ln(II)-silanide complexes containing THF ligands**

THF is the one of the most commonly utilized coordinating solvents in non-aqueous *f*-element chemistry. Some examples are shown in Fig. 8.10.1.

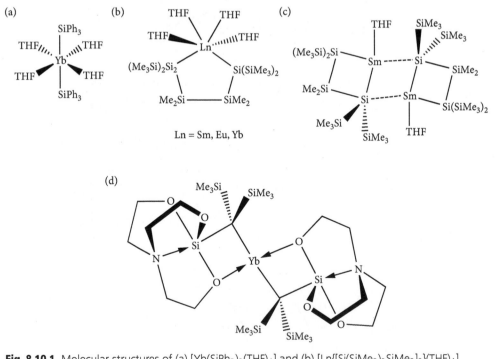

Fig. 8.10.1 Molecular structures of (a) [Yb(SiPh₃)₂(THF)₄] and (b) [Ln{[Si(SiMe₃)₂SiMe₂]₂}(THF)₄], (c) [Sm{Si(SiMe₃)₂SiMe₂}(THF)]₂ and (d) Yb(II) complex [Yb{Si(SiMe₃)₂Si([OCH₂CH₂]₃N)}(THF)₂].

8.10.2 **Ln(II)-silylene complexes**

Silylenes (R_2Si), the silicon analogs of carbenes, are two-electron σ-donor ligands that can form adducts with metals through dative bonding. The first divalent lanthanide silylene complex $(Cp^*)_2Sm\{Si(N^tBuCH)_2\}$ was synthesized by Evans and co-workers in 2003 (Fig. 8.10.2(a)).

Roesky and co-workers have reported the synthesis and reactivity of bis(silylene)-coordinated Eu(II) and Yb(II) [{Ln{SiNSi}-{N(SiMe_3)_2}_2}] complexes, in which SiNSi = {$C_5H_3N[2,6$-{NEt[Si{($N^tBu)_2CPh$}]}, by the addition of one equivalent of SiNSi to [Ln{N(SiMe_3)_2}_2(THF)_2]. The Ln(II) ions in the iso-structural [{Ln{SiNSi}{N(SiMe_3)_2}_2}] complexes adopt distorted tetrahedral arrangements, and the long Ln–Si bonds (Eu, 328.4(2) pm; Yb, 317.5(2) pm) are indicative of weak metal–silicon interactions (Fig. 8.10.2).

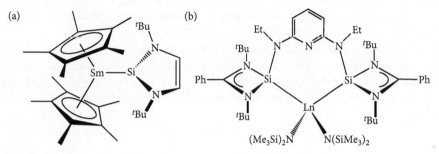

Fig. 8.10.2 Molecular structures of (a) [Sm(Cp*)_2{Si(N^tBuCH)_2}] and (b) iso-structrual bis(silylene)-coordinated Eu(II) and Yb(II) complexes.

8.10.3 **Ln(III)–Si complexes**

In the 1990s Tilley and Rheingold exploited alkane elimination procedure to add a secondary silane $H_2Si(SiMe_3)_2$ at Ln(III) (Ln = Y, Nd, Sm) centers, yielding [Ln(Cp*)_2{SiH(SiMe_3)_2}]. In 2011 Lappert and co-workers reported Ln(III) silylene complexes [Ln(Cp)_3{Si[{N(CH_2-^tBu)}_2C_6H_4-1,2]}] (Ln = Y, Yb). The structures of these two kinds of lanthanide-silane complexes are shown in Fig. 8.10.3.

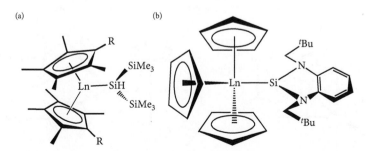

Fig. 8.10.3 Molecular structures of Ln(III)–Si complexes containing (a) secondary silane and (b) silylene ligands.

8.10.4 *f*-Element complexes containing bonds with heavier tetrels (Ge, Sn, and Pb)

The *f*-block germanium complex [Yb(GePh$_3$)$_2$(THF)$_4$] shown in Fig. 8.10.4(a) was synthesized by the redox reaction of Ph$_3$GeCl with excess Yb metal at room temperature for 11 days in THF with concomitant elimination of [YbCl$_2$(THF)$_2$]. Thus far the only published *f*-block complex containing an organogermanium metallacycle is [Yb{(GePh$_2$GePh$_2$)$_2$}(THF)$_4$]. As shown in Fig. 8.10.4(b), the *cis*-arrangement of its ytterbium-bound germanium atoms, with a significantly shorter averaged Yb–Ge bond length of 310.4(2) pm, is in stark contrast with the *trans*- arrangement of GePh$_3$ groups in [Yb(GePh$_3$)$_2$(THF)$_4$] with mean Yb–Ge bond length of 315.6(3) pm.

The reaction of [Yb(C$_{10}$H$_8$)(THF)$_2$] with one equivalent of Ph$_4$Sn for two days formed [Yb(SnPh$_3$)$_2$(THF)$_4$] with a Yb–Sn bond length of 330.5(1) pm, which is isostructural to the lighter [Yb(GePh$_3$)$_2$(THF)$_4$] congener shown in Fig. 8.10.4(a). The reaction of [Yb(C$_{10}$H$_8$)(THF)$_2$] with Ph$_4$Sn yielded a bimetallic ytterbium complex with a triphenyl-tin cap: [(Ph$_3$Sn)Yb(THF)$_2$(μ-η^1:η^6-Ph)$_3$Yb(THF)$_3$], which has a Yb–Sn bond length of 337.9(1) pm [see Fig. 8.10.5 (a)].

Presently there is only one molecular *f*-block lead complex in the literature, which was reported by Zeckert and co-workers in 2013. The reaction of a lithium tris(organo)plumbylene [LiPb(2-py^{6-OtBu})$_3$(THF)] (2-py^{6-OtBu} = 2-C$_5$H$_3$N-6-O^{t}Bu) with Ln(Cp)$_3$ (Ln = Sm, Eu) yielded [Ln(Cp)$_3${Pb(2-py^{6-OtBu})$_3$Li}] adducts (Fig. 8.10.5(b)), which have respective Ln–Pb bond lengths of 326.56(3) pm and 320.38(3) pm.

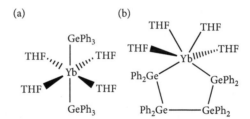

Fig. 8.10.4 Structural formulas of (a) [Yb(GePh$_3$)$_2$(THF)$_4$] and (b) [Yb{(GePh$_2$GePh$_2$)$_2$}(THF)$_4$].

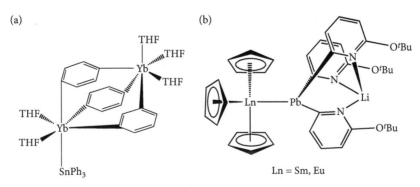

Fig. 8.10.5 Structural formulas of (a) bimetallic ytterbium complex with a triphenyltin cap and (b) f-block lead adduct [Ln(Cp)$_3${Pb(2-py^{6-OtBu})$_3$Li}].

References

B. L. L. Réant, S. T. Liddle, and D. P. Mills, f-element silicon and heavy tetrel chemistry, *Chem. Sci.* **11**, 10871–86 (2020).

K. Zeckert, J. Griebel, R. Kirmse, M. Weiß, and R. Denecke, Versatile reactivity of a lithium tris(aryl)plumbate(ii) towards organolanthanoid compounds: stable lead–lanthanoid–metal bonds or redox processes, *Chem.–Eur. J.* **19**, 7718–22 (2013).

8.11 Introduction to Structural Chemistry of Actinide Compounds

In the Periodic Table, the Actinide Series (or Actinoid Series as recommended by IUPAC) encompasses all 15 elements with atomic numbers from 89 (actinium Ac) to 103 (lawrencium Lr). The informal chemical symbol An is used in general discussions of actinide chemistry.

The actinide series mainly involves filling of the 5f electron shell, although the first two members Ac and 90 (thorium Th) lack any 5f electron, and 96 (curium) and 103 (lawrencium) each has the same 5f number as its preceding element 95 (americium) and 102 (nobelium), respectively. All actinide elements and their compounds are radioactive and require utmost care in their handling.

While ores of thorium [monazite, a mixed-metal phosphate mineral of chemical composition (Ce, La, Nd, Th)(PO_4,SiO_4)] and uranium (pitchblende U_3O_8) occur in nature, actinium, and protoactinium are only obtainable in minute amounts as decay products of ^{235}U and ^{238}U isotopes, and plutonium through neutron capture by uranium. The structural chemistry of actinides is dominated by compounds of uranium, neptunium, and thorium.

[U{N(^tBu)Ar}$_3${Si(SiMe$_3$)$_3$}] with Ar = C_6H_3-3,5-Me$_2$ (Fig. 8.11.1(a)) is the first molecular actinide-silicon complex to have its crystal structure determined by X-ray crystallography in 2001. In 2020, Arnold and co-workers reported the synthesis and solid-state structure of the first U(III)–silylene complexes. Reaction of the amidinate-supported silylenes [Si{PhC(N^tBu)$_2$}(NMe$_2$)] and [Si{PhC(N^iPr)$_2$}$_2$] with [U(Cp')$_3$] (Cp' = Me$_3$SiC$_5$H$_4$) provided [U(Cp')$_3${-Si(R)[PhC(NR')$_2$]}] (R = NMe$_2$ and R' = ^tBu) shown in Fig. 8.11.1(b); the iso-structural complexes (R = PhC(N^iPr)$_2$ and R' = ^iPr) have U–Si bond lengths of 316.37(7) pm and 317.50(6) pm, respectively.

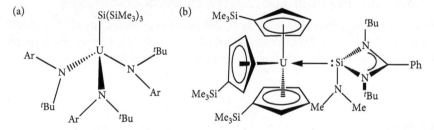

Fig. 8.11.1 Molecular structures of (a) first reported actinide–silicon complex [U{N(^tBu)Ar}$_3${Si(SiMe$_3$)$_3$}] and (b) actinide silylene complex [U(Me$_3$SiC$_5$H$_4$)$_3${-Si(R)[PhC(NR')$_2$]}] (R = NMe$_2$).

8.12 Thorium Coordination Compounds

Thorium is the most abundant radioactive element found in nature, and its common Th(IV) ion can exhibit multiple coordination numbers to bind a wide range of ligands with potential in numerous

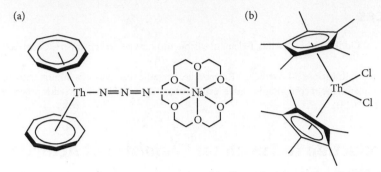

Fig. 8.12.1 Low-coordinate thorium compounds: (a) $[(C_8H_8)_2Th(\mu\text{-}\eta^1{:}\eta^1\text{-}N_3)]Na(18\text{-crown-6})$ and (b) $[(C_5Me_5)_2ThCl_2]$.

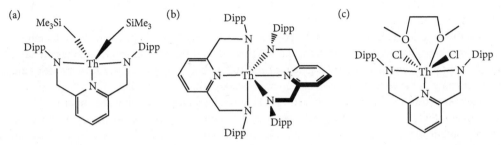

Fig. 8.12.2 (a) Five-coordinate thorium complex $(BDPP)Th(CH_2SiMe_3)_2$, with Dipp = 2,6-diisopropylphenyl, (b) six-coordinate complex $Th(BDPP)_2$, and (c) seven-coordinate pentagonal bipyramidal complex $(BDPP)ThCl_2(dme)$, with dme = dimethoxyethane.

applications. In particular, nine-coordinate Th(IV) has an ionic radius of 109 pm, making it the largest stable tetravalent metal ion.

8.12.1 Low- and medium-coordinate thorium complexes

Fig. 8.12.1 shows representative examples of three- and four-coordinate thorium(IV) complexes.

Examples of thorium(IV) complexes that exhibit medium coordination numbers are those that contain the 2,6-bis(2,6-diisopropylanilidomethyl)pyridine dianion, a *NNN* pincer ligand abbreviated as BDPP (Fig. 8.12.2).

References

A. Hervé, N. Garin, P. Thuéry, M. Ephritikhine, and J.-C. Berthet, Bent thorocene complexes with the cyanide, azide and hydride ligands. *Chem. Commun.* **49**, 6304–6 (2013).

A. Cruz, D. J. H. Emslie, L. E. Harrington, J. F. Britten, and C. M. Robertson, Extremely stable thorium(IV) dialkyl complexes supported by rigid tridentate 4,5-bis(anilido)xanthene and 2,6-bis(anilidomethyl)pyridine ligands. *Organometallics* **26**, 692–701 (2007).

B. Grüner, P. Švec, P. Selucký, and M. Bubeníková, Halogen protected cobalt bis(dicarbollide) ions with covalently bonded CMPO functions as anionic extractants for trivalent lanthanide/actinide partitioning. *Polyhedron* **38**, 103–12 (2012).

8.12.2 **Thorium complexes exhibiting twelve- and ten-coordination**

A 1:1 concentrated methanol mixture of $Th(NO_3)_4$ hydrate and ligand molecule L = *cis*-ethylene bis(diphenylphosphine oxide), upon standing in the refrigerator over three days, deposited compound $[Th(NO_3)_2L_3](NO_3)_2 \cdot 2MeOH$ which crystallized in orthorhombic space group *Pbca* with Z = 8. X-ray structure analysis revealed the presence of both $[Th(NO_3)_3L_2]^+$ and $[Th(NO_3)_5L]^-$ complexes together with two independent methanol solvate molecules in the asymmetric unit. In either Th(IV) molecular ion, all L and nitrate ligands are bound to the metal center in a bidentate manner. The Th–L 1:1 anionic complex is ten-coordinate, and the Th–L 1:2 cationic complex is twelve-coordinate (Fig. 8.12.3).

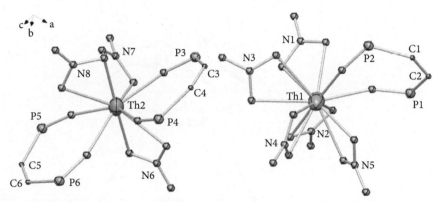

Fig. 8.12.3 Coordination environment of independent molecular ions $[Th(NO_3)_3L_2]^+$ (left) and $[Th(NO_3)_5L]^-$ (right) in the crystal structure of $[Th(NO_3)_2L_3](NO_3)_2 \cdot 2MeOH$, with phenyl rings omitted for clarity.

Reference

P. T. Morse, R. J. Staples, and S. M. Biros, Th(IV) complexes with *cis*-ethylene (diphenylphosphine oxide): X-Ray structures and NMR solution studies. *Polyhedron* **114**, 2–12 (2016).

8.13 *f*-Element Metallocene Complexes

8.13.1 **Cp*₃U complexes**

The discovery of *f*-element tris-(pentamethylcyclopentadienyl) complexes, which were not expected to exist in view of their steric crowding, have led to major advances in the discovery of novel reactions and compounds.

Despite its steric crowding, Cp*₃U reacts with carbon monoxide or *t*BuCN to give the even more crowded adducts Cp*₃UL (L = CO, NC*t*Bu) (Fig. 8.13.1 lower right corner). In the presence of dinitrogen, Cp*₃U is transformed into $Cp*_3U(\eta^1\text{-}N_2)$, demonstrating that end-on nitrogen coordination is possible for *f* elements, but surprisingly Cp*₃U instead reacts with *tert*-butyl isocyanide *t*BuNC to give cyclic trimer $[Cp*_2U(\mu\text{-}CN)(^tBuNC)]_3$ (Fig. 8.13.1 lower left corner).

The sterically crowded Cp*₃M complexes (M = U, Ln) are capable of reacting as reductants via the $(C_5Me_5)^-/C_5Me_5$ redox couple. By combining these reductions with metal-based electron transfers,

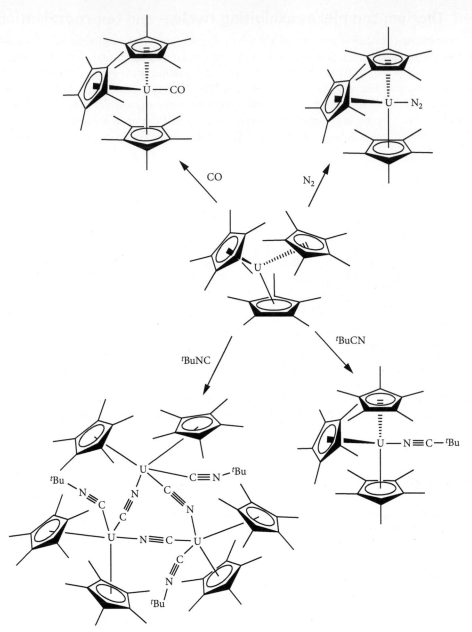

Fig. 8.13.1 Addition reaction and trimerization of over-crowded molecule Cp*₃U.

Cp*₃U can react as a multi-electron reductant, as shown in the reductive coupling of diphenylacety-lene and reductive cleavage of azobenzene, two- and four-electron processes leading to Cp*₂U(C₄Ph₄) and Cp*₂U(=NPh)₂, respectively. The net reaction of Cp*₃U and cyclooctatetraene, giving [(Cp*)-(COT)U]₂(μ-C₈H₈), is accomplished through one U^{III}/U^{IV} and two (C₅Me₅)⁻/C₅Me₅ couples (Fig. 8.13.2).

Addition of pyrazine (pyz) to [Ce^{III}(C₅H₄R)₃] in THF led to the equilibrating formation of [Ce(C₅H₄R)₃(pyz)], while [U^{III}(C₅H₄R)₃] reacted with the Lewis base to give dinuclear uranium(III)

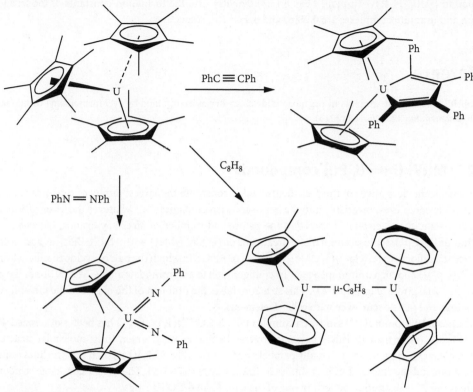

Fig. 8.13.2 Reducing Capacity of Cp*$_3$U.

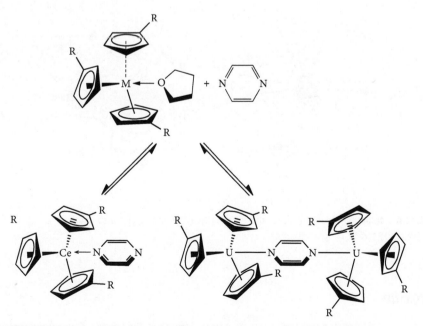

Fig. 8.13.3 Distinct reactions of [M^{III}(C$_5$H$_4$R)$_3$] (M = Ce, U) with pyrazine.

compound [{U(C$_5$H$_4$R)$_3$}$_2$(μ-pyz)] (Fig. 8.13.3). For R = ^{t}Bu, the formation constants of the trivalent cerium and uranium complexes are 0.28(6) and 8(1) × 10^3, respectively.

Reference

M. Ephritikhine, Recent advances in organoactinide chemistry as exemplified by cyclopentadienyl compounds. *Organometallics* **32**, 2464–88 (2013).

8.13.2 **M(IV) (M = U, Pu) compounds**

The metallocene derivatives of the *f*-elements, which constitute the largest family of organolanthanide and actinide complexes, invariably adopt a bent-sandwich configuration, whatever the 4*f* or 5*f* ion, the oxidation state, the electronic charge, and the nature and number of auxiliary ligands. However, dissolution of [U(Cp*)$_2$I$_2$] in acetonitrile or treatment of [U(Cp*)$_2$Me$_2$] with HNEt$_3$BPh$_4$ in acetonitrile produces metallocenes [U(Cp*)$_2$(NCMe)$_5$]X$_2$ (X = I or BPh$_4$), in which the pair of cyclopentadienyl rings of the U(Cp*)$_2$ sandwich unit adopt a parallel configuration to accommodate all five MeCN donor ligands at its equatorial girdle (Fig. 8.13.4(a)). This structure raises the problem of the nature of the metal-ligand interaction and of its occurrence with other *f*-elements.

The unusual plutonium(IV) sandwich complex Pu(1,3-COT″)(1,4-COT″) has been synthesized with 1,4-COT″ = 1,4-bis(trimethylsilyl)cyclooctatetraene dianion as a precursor. In its molecular structure, the two planar C$_5$ rings adopt a virtually parallel arrangement (Fig. 8.13.4(b)), such that the lines joining the Pu atom and the ring centroids is almost co-linear (angle =176.78°). The most remarkable structural feature is the 1,3-substitution pattern in one of the coordinated COT″ rings.

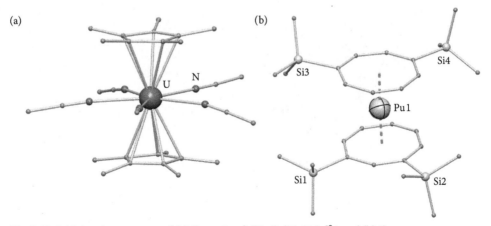

Fig. 8.13.4 Molecular structure of (a) the cation [U(Cp*)$_2$(MeCN)$_5$]$^{2+}$ and (b) the Pu(1,3-COT″)(1,4-COT″) sandwich.

Reference

C. Apostolidis, O. Walter, J. Vogt, P. Liebing, L. Maron, and F. T. Edelmann, A structurally characterized organometallic plutonium(IV) complex. *Angew. Chem. Int. Ed.* **56**, 5066–70 (2017).

8.14 **Organometallic Neptunium Chemistry**

The only neptunium element ($_{93}$Np) available on Earth is man-made. Among its 24 radioisotopes, the most stable are ^{237}Np [$t_{1/2}$ = 2.144(7) × 10^6 y], ^{236}Np [$t_{1/2}$ = 1.54(6) × 10^5 y] and ^{235}Np [$t_{1/2}$ = 396.1(12) d], while all remaining ones have half-lives under 4.5 days.

8.14.1 **Np(IV) complexes**

The most important oxidation state of neptunium is (IV), and anhydrous NpCl$_4$ plays a pivotal role in synthetic chemistry because of three favorable properties: (1) good solubility in polar, aprotic organic solvents; (2) excellent ligand salt metathesis and reductive chemistry; and (3) relative ease of conversion of aqueous Np(IV) chloride solutions into anhydrous solvates such as NpCl$_4$(DME)$_2$, which provides a key entry point for organoneptunium chemistry. Other useful synthetic precursors include neptunates R$_2$[NpCl$_6$] (R = Et, Ph).

The reaction of NpCl$_4$ with excess KCp in toluene produces tetrahedral NpCp$_4$, which is isostructural to its Th and U analogs. In general, Np$^{III/IV}$ redox interconversion is more closely balanced than that of U$^{III/IV}$, which is heavily weighted toward U^{IV}. A fine balance in redox potentials for cyclopentadienyl ligand-supported complexes enables many Np^{III} and Np^{IV} organometallic complexes to be isolated.

Reference

P. L. Arnold, M. S. Dutkiewicz, and O. Walter, Organometallic neptunium chemistry. *Chem. Rev.* **117**, 11460–75 (2017).

8.14.2 **Np(III) cyclopentadiene complexes**

The Np^{IV} complex Np(Cp)$_3$Cl is cleanly reduced by sodium amalgam in diethyl ether, forming the pale green diethyl ether solvate Np(Cp)$_3$(OEt$_2$) that loses solvent readily, affording Np(Cp)$_3$ in excellent yield. Unsolvated Np(Cp)$_3$ crystallizes in the monoclinic system and contains polymeric zigzag chains of {Np^{III}(η^5-Cp)$_2$} units bridged through alternating μ-η^5,η^1-bound cyclopentadienyl groups (Fig. 8.14.1). This is directly comparable to the structures of unsolvated Ln(Cp)$_3$ (Ln = Ce, Ho, Dy, Sm, and La).

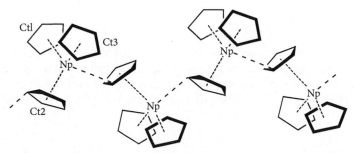

Fig. 8.14.1 Representation of the zigzag chain arrangement in polymeric [Np(Cp)$_3$]$_\infty$. The Np^{III} to C$_5$ ring centroid distances are Np1–Ct1 2.587(5), Np1–Ct2 2.419(6), Np1–Ct3 256.1(10) pm.

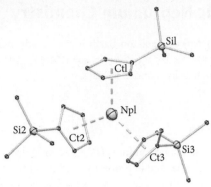

Fig. 8.14.2 Thermal ellipsoid drawing (50% probability for non-H atoms) of NpCp′₃ in the solid state. H atoms omitted. Selected bond lengths [pm] and angles [°]: Np1–Ct1 248.5(2), Np1–Ct2 248.1(2), Np1–Ct3 247.9(2), Np1–Car 273.4(6) to 278.6(4), Si1–(C1–C5) plane –0.382(8), Si2–(C6–C10) plane –0.109(8), Si3–(C11–C15) plane –0.169(8), Ct1–Np1–Ct2 119.86(8), Ct1–Np1–Ct3 120.46(7), Ct2–Np1–Ct3 119.06(8).

Use of sterically more demanding $C_5H_4SiMe_3$ (Cp′) ligand produced olive-green single crystals of NpCp′₃, and X-ray diffraction analysis conducted at –20 °C showed its mononuclear molecular structure with the Np^{III} atom surrounded by three η^5-bound Cp′ rings at average Np–Ct(η^5-Cp′) distance of 248.2(3) pm (Fig. 8.14.2). This means that in NpCp′₃ the C_5 rings bearing the bulky substituents are closer to the metal than in the previously described $[Np(Cp)_3]_\infty$ complex for which Np–Ct distances of 251 pm or 254 pm are found. This means that due to the trigonal planar arrangement of the Cp′ ligands with resulting larger bond angles around the Np^{III} center in NpCp′₃, the metal is able to establish stronger interactions with the more electron-rich Cp′ ligands.

The reaction between Np(Cp)₃Cl and KCp cleanly affords the Np^{III} "ate" product K[Np(Cp)₄], which is the first actinide(III) tetrakis-cyclopentadienyl complex, instead of the anticipated Np(Cp)₄, as Scheme 8.14.1 shown below:

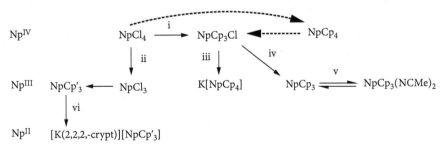

Scheme 8.14.1 NpCp₃ is obtained by reduction from NpCp₃Cl and readily forms the MeCN stabilized solvate. The silylated analog NpCp′₃ is better obtained from the reaction of NpCl₃ with KCp′ and can be reduced to its K salt. Dotted lines indicate literature procedures. Key: (i) excess KCp, PhMe; (ii) Na/Hg, Et₂O, –NaCl; (iii) KCp, –KCl; (iv) Na/Hg, Et₂O, –NaCl; (v) excess MeCN; and (vi) KC₈, 2.2.2-cryptand, THF/Et₂O, –8°C.

The crystal structure of K[Np(Cp)₄] is unique in that it contains two different types of metal-Cp coordination geometries in the same crystal, labeled as Np(A) and Np(B) in Fig. 8.14.3.

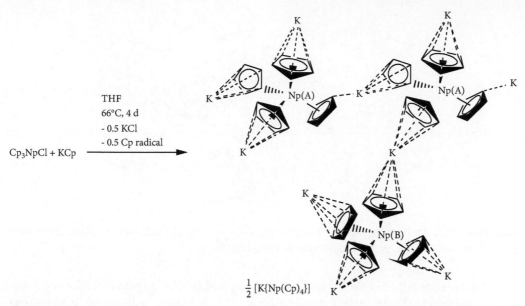

$Cp_3NpCl + KCp \xrightarrow[\substack{- 0.5 \, KCl \\ - 0.5 \, Cp \, radical}]{\substack{THF \\ 66°C, 4 \, d}}$

$\frac{1}{2}\,[K\{Np(Cp)_4\}]$

Fig. 8.14.3 Syntheses of K[Np(Cp)$_4$] showing two types of Np coordination geometries labeled Np(A) and Np(B).

8.14.3 **Np(II) cyclopentadiene complexes**

Reports of the formal oxidation state +2 for uranium, in the form of the U^{II} "ate" complexes [K(2.2.2-cryptand)][U(Cp′)$_3$] and [K(2.2.2-cryptand)][U(Cp″)$_3$] (Cp′ = C$_5$H$_4$SiMe$_3$; Cp″ = C$_5$H$_3$(SiMe$_3$)$_2$) spurred a search for their neptunium(II) analog. Reduction of Np(Cp′)$_3$ by KC$_8$ in the presence of 2.2.2-cryptand afforded black solutions and unstable crystals at –78 °C that were tentatively suggested to be the neptunium(II) complex K(2.2.2-cryptand)[Np(Cp′)$_3$], by analogy with the U^{II} and Th^{II} complexes K(2.2.2-cryptand)[U/Th(Cp′)$_3$], as Scheme 8.14.2 shown below, that are thermally unstable above –10°C.

Unfortunately, radiological concerns precluded mounting of the putative Np^{II} crystals on the diffractometer head at low temperatures, and the diffraction data from the crystals mounted were too poor for a structural determination. However, in line with the presumed increasing stability of lower formal

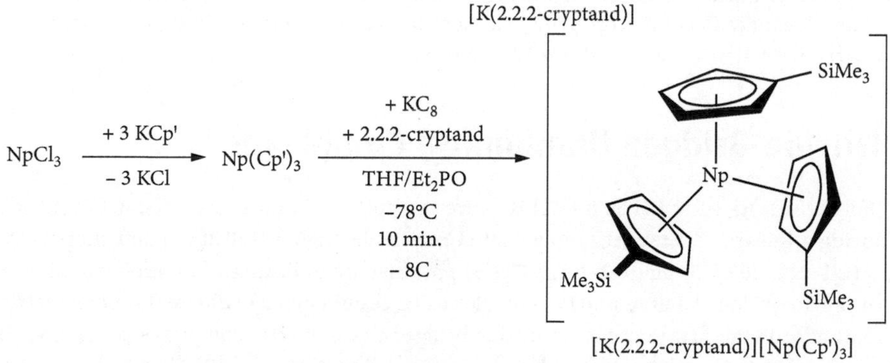

$$NpCl_3 \xrightarrow[\substack{- 3 \, KCl}]{\substack{+ 3 \, KCp'}} Np(Cp')_3 \xrightarrow[\substack{THF/Et_2PO \\ -78°C \\ 10 \, min. \\ - 8C}]{\substack{+ KC_8 \\ + 2.2.2\text{-cryptand}}}$$

[K(2.2.2-cryptand)]

[K(2.2.2-cryptand)][Np(Cp′)$_3$]

Scheme 8.14.2 Synthesis of K(2.2.2-cryptand)[Np(Cp′)$_3$].

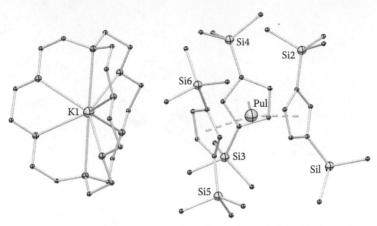

Fig. 8.14.4 Molecular structure of [K(2.2.2-cryptand)]{Pu^{II}[C$_5$H$_3$(SiMe$_3$)$_2$]$_3$} with thermal ellipsoids drawn at the 50% probability level, with hydrogen atoms omitted for clarity.

oxidation states across the row, the Pu^{II} analog K(2.2.2-cryptand)[Pu(Cp″)$_3$] with additional SiMe$_3$ substituents was found to be sufficiently thermally stable at –35 °C to be structurally characterized by X-ray diffraction (Fig. 8.14.4).

8.14.4 Neptunium cyclooctatetraene complexes

Neptunocene Np(COT)$_2$ is prepared by combining two equiv. of K$_2$COT with one equiv. of NpCl$_4$ in a non-polar solvent. It is water-stable but not oxygen-stable, and single-crystal X-ray analysis showed that it is iso-structural with uranocene. Red-purple K[Np(COT)$_2$]·2THF made directly from the reaction of NpBr$_3$ and K$_2$COT in THF is air- and moisture-sensitive. It is readily oxidized to Np^{III}(COT)$_2$, which is assumed to adopt a sandwich structure with either D_{8h} (eclipsed rings) or D_{8d} symmetry (staggered rings).

References

M. S. Dutkiewicz, C. Apostolidis, O. Walter, and P. L. Arnold, Reduction chemistry of neptunium cyclopentadienide complexes: from structure to understanding. *Chem. Sci.* **8**, 2553–61 (2017).

C. J. Windorff, G. P. Chen, J. N. Cross, W. J. Evans, F. Furche, A. J. Gaunt, M. T. Janicke, S. A. Kozimor, and B. L. Scott, Identification of the formal +2 oxidation state of plutonium: synthesis and characterization of {Pu^{II}[C$_5$H$_3$(SiMe$_3$)$_2$]$_3$}$^-$. *J. Am. Chem. Soc.* **139**, 3970–3 (2017).

8.15 Nitride-Bridged Uranium(III) Complexes

Cs[{U(OSi-(O^{t}Bu)$_3$)$_3$}$_2$(μ-N)] (Fig. 8.15.1) is a rare example of a dinuclear uranium(IV) nitride complex featuring a linear U^{IV}≡≡≡N≡≡≡U^{IV} fragment (U≡≡≡N≡≡≡U angle = 170.2(3)°) and short U–N bond distances (U1–N1, 205.8(5) pm; U2–N1, 207.9(5) pm) that are indicative of multiple bonding. Reduction of this nitride-bridged diuranium(IV) complex using cesium metal in different stoichiometric ratios (Fig. 8.15.2) affords the first examples of nitride-bridged uranium(III) complexes containing the heterometallic fragments Cs$_2$[U^{III}≡≡≡N≡≡≡U^{IV}] (**1**) and Cs$_3$[U^{III}≡≡≡N≡≡≡U^{III}] (**2**) (Fig. 8.15.3), which have been characterized by X-ray crystallography.

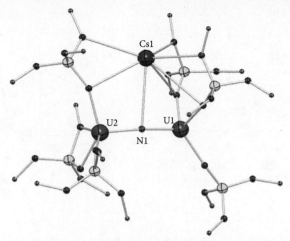

Fig. 8.15.1 Structure of Cs[{U(OSi-(O^tBu)_3)_3}_2(μ-N)] crystallized from a saturated hexane solution.

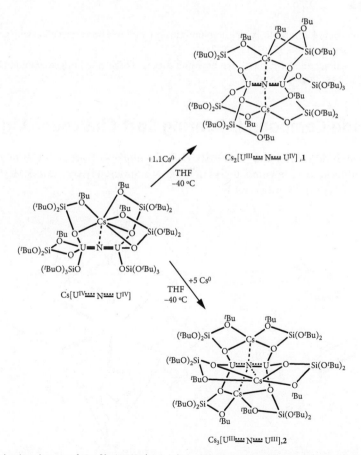

Fig. 8.15.2 Synthetic scheme of Cs_2[{U(OSi(O^tBu)_3)_3}_2(μ-N)] (1) and Cs_3[{U(OSi(O^tBu)_3)_3}_2(μ-N)] (2).

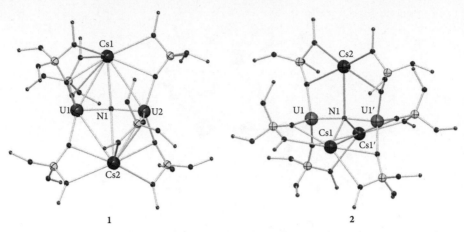

Fig. 8.15.3 Molecular structure of $Cs_2[\{U(OSi(O^tBu)_3)_3\}_2(\mu\text{-}N)]$ (**1**) crystallized from a saturated THF solution and $Cs_3[\{U(OSi(O^tBu)_3)_3\}_2(\mu\text{-}N)]$ (**2**) crystallized from a saturated THF solution. The thermal ellipsoid probability is 50%. Hydrogen atoms, methyl groups, and solvent molecules have been omitted for clarity.

References

L. C. Camp, J. Pécaut, and M. Mazzanti, Tuning uranium-nitrogen multiple bond formation with ancillary siloxide ligands. *J. Am. Chem. Soc.* **135**, 12101–11. (2013).

L. Chatelain, R. Scopelliti, and M. Mazzanti, Synthesis and structure of nitride-bridged uranium(III) complexes, *J. Am. Chem. Soc.* **138**, 1784–7 (2016).

8.16 **Actinide Compounds Bearing Soft Chalcogen Ligands**

The $[U^{IV}(\eta\text{--}C_5Me_4SiMe_2CH_2)_2]$ and $[U^{III}Tp^*_2(CH_2Ph)]$ complexes react with CS_2 to give the mixed tethered alkyl dithiocarboxylate compounds $[U^{IV}(\eta\text{-}C_5Me_4SiMe_2CH_2)(\eta\text{-}C_5Me_4SiMe_2CH_2CS_2)]$ (**3**) and $[U^{III}Tp^*_2(S_2CCH_2Ph)]$ (**4**), respectively.

The trivalent uranium metallocenes [U(MeC$_5$H$_4$)$_3$(THF)] and [U(Me$_3$SiC$_5$H$_4$)$_3$] react with CS$_2$ to form binuclear U(IV) carbon disulfido complexes (R = Me, SiMe$_3$) (**5**) respectively, while the siloxide compound [U(OSi(O^{t}Bu)$_3$)$_2$(μ-OSi(O^{t}Bu)$_3$)]$_2$ is transformed by CS$_2$ into [{U(OSi(O^{t}Bu)$_3$)$_3$}$_2$(μ-η^2(C, S):η^2(S, S)–CS$_2$)] (**6**).

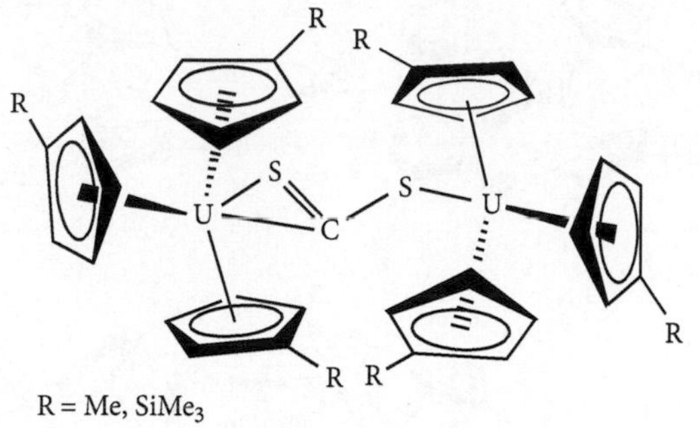

R = Me, SiMe$_3$

5

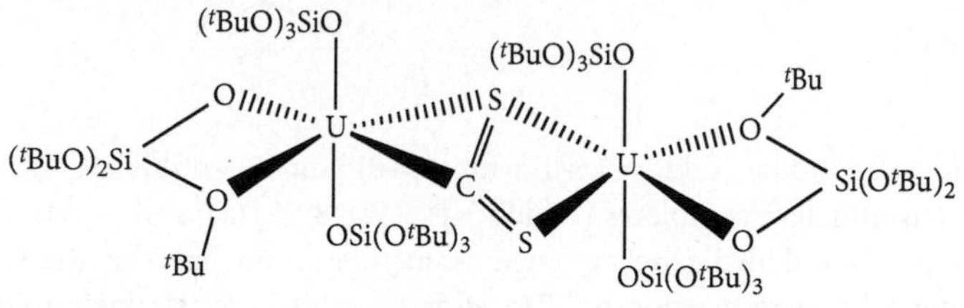

6

The unusual U(IV) hexanuclear complex [{U(COT)}$_4${U(THF)$_3$}$_2$(μ$_3$-S)$_8$] (**7**) was isolated from the reaction of [U(COT)(BH$_4$)$_2$(THF)] with Na[SPS^{OMe}]. Its molecular structure exhibits an octahedron-like skeleton of uranium atoms that are held together by triply-bridging S atoms located above each face of the octahedron. The thioselenophosphinate complexes [An(SSePPh$_2$)$_4$] (An = Th or U) (**8**) were synthesized by salt metathesis reactions between [ThCl$_4$(DME)$_2$] or [UI$_4$(1,4–dioxane)$_2$] and Na(SSePPh$_2$).

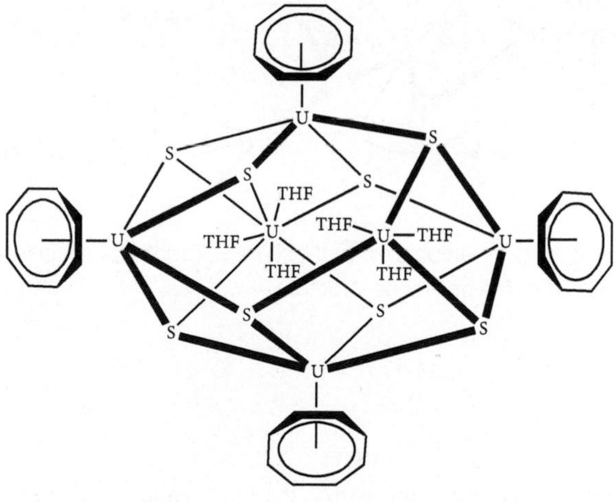

7

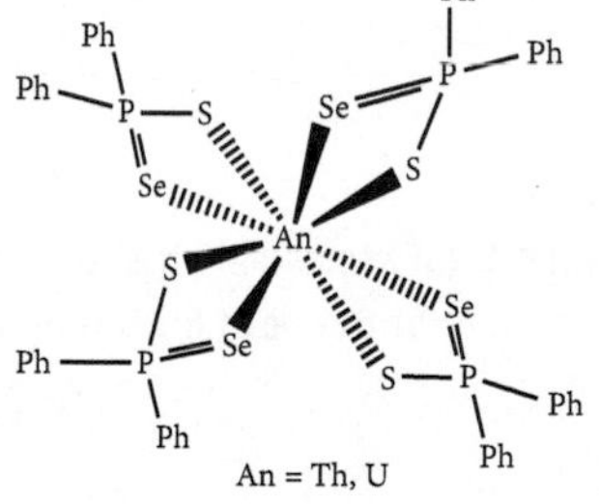

An = Th, U

8

The thorium(IV) complex $[Th(\eta\text{-}1,3\text{-}{}^tBu_2\text{-}C_5H_3)_2(bipy)]$ reacts with CS_2 to give sulfido intermediate $[Th(\eta\text{-}1,3\text{-}{}^tBu_2\text{-}C_5H_3)_2(=S)]$ **(9)**, which is transformed by nucleophilic addition with CS_2 to form the dimeric trithiocarbonate complex $[Th(\eta\text{-}1,3\text{-}{}^tBu_2\text{-}C_5H_3)_2(\mu\text{-}CS_3)]_2$ **(10)**.

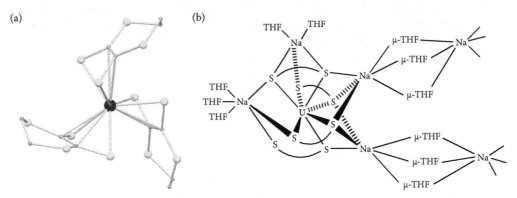

Reaction of UCl_4 and Na_2dddt (dddt = 5,6-dihydro-1,4-dithiine-2,3-dithiolate) in pyridine gives the homoleptic U(IV) trisdithiolene complexes $[U(dddt)_3]^{2-}$ (Fig. 8.16.1(a)) and $[Na\{U(dddt)_3\}_2]^{3-}$, which exhibit large folding of the dddt ligand and significant interaction between the C=C double bond and the metal center. The same reaction in THF afforded crystals of tetrakis(dithiolene) compound $[Na_4(THF)_8U(dddt)_4]_\infty$ composed of zigzag chains, in which $Na_2(\mu\text{-}THF)_3$ fragments serve as bridges between $Na_2(THF)_5U(dddt)_4$ moieties (Fig. 8.16.1(b)).

Fig. 8.16.1 (a) Molecular structure of the $[U(dddt)_3]^{2-}$ anion and (b) $[Na_4(THF)_8U(dddt)_4]_\infty$, for clarity, only the bite of the dddt ligands is represented in the polymeric structure.

References

M. Ephritikhine, The vitality of uranium molecular chemistry at the dawn of the XXIst century. *Dalton Trans.* **35**, 2501–16 (2006).

M. Ephritikhine, Molecular actinide compounds with soft chalcogen ligands. *Coord. Chem. Rev.* **319**, 35–62 (2016).

8.17 **Actinoid Complexes Containing Nitrogen-Rich Ligands**

8.17.1 **Uranium(IV) heptaazide complexes**

There exist two crystalline modifications of the $(Bu_4N)_3[U(N_3)_7]$ salt that feature the binary uranium(IV) heptaazide anion. Dark green Form A recrystallized from $CH_3CN/CFCl_3$ belongs to cubic space group $Pa3$ with $Z = 8$, and yellow-green Form B recrystallized from CH_3CH_2CN belongs to monoclinic space group $P2_1/c$ with $Z = 4$. The structure of in these two crystal forms is shown in Fig. 8.17.1.

In $(Bu_4N)_3[U(N_3)_7]$ **Form A,** the U^{IV} atom and one of three crystallographically independent azide groups in the $[U(N_3)_7]^{3-}$ anion are located on a C_3 symmetry axis. This results in a monocapped octahedron (1:3:3) arrangement of the azide ligands around the central uranium atom. Six azide ligands have their N_α atoms occupying the corners of a trigonal antiprism, whereas the seventh azide group is located along the C_3 axis (Fig. 8.17.1(a)). In contrast, the $[U(N_3)_7]^{3-}$ anion in **Form B** shows a distorted 1:5:1 pentagonal-bipyramidal arrangement of its seven azide ligands around the central uranium(IV) atom (Fig. 8.17.1(b)).

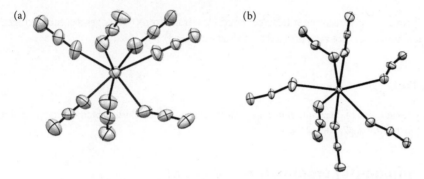

Fig. 8.17.1 Molecular structure of $U(N_3)_7{}^{3-}$ ion in $(Bu_4N)_3[U(N_3)_7]$ crystal (a) **Form A** and (b) **Form B**.

Reference

M.-J. Crawford, A. Ellern, and P. Mayer, $UN_{21}{}^{3-}$: A structurally characterized binary actinide heptaazide anion. *Angew. Chem. Int. Ed.* **44**, 7874–8 (2005).

8.17.2 **Pyridine azido adduct and nitrido-centered uranium azido cluster**

In the crystal structure of $[U(N_3)_4(py)_4]$, the tetravalent uranium center is located on a crystallographic S_4 axis such that the pyridine and terminal azido ligands are arranged around it in distorted dodecahedral geometry (Fig. 8.17.2(a)).

The U^{IV} azido/nitrido complex $\{[(Cs(CH_3CN)_3][U_4(\mu_4\text{-}N)(\mu\text{-}N_3)_8(CH_3CN)_8I_6]\}_\infty$ features a polymeric chain of nitride-centered tetranuclear uranium clusters connected through cesium ions binding the coordinated iodide atoms. The crystal structure contains three independent uranium atoms (U1, U2, and U3) with a symmetry plane passing through U3, U1, and nitride N101 that bisects two equivalent U2 atoms (Fig. 8.17.2(b)). The four uranium cations are connected by eight 1,1-end-on bridging azido

(a) (b)

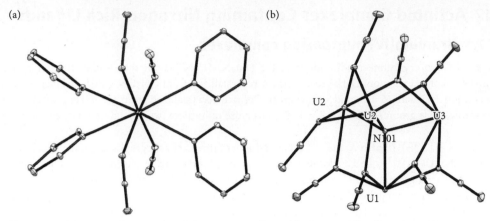

Fig. 8.17.2 (a) Tetranuclear uranium(IV) center coordinated by four pyridine ligands and four terminal azido ligands exhibiting distorted dodecahedral geometry and (b) Azido/nitrido U^{IV} cluster component of the linear polymeric chain $\{[(Cs(CH_3CN)_3][U_4(\mu_4-N)(\mu-N_3)_8(CH_3CN)_8I_6]\}_\infty$.

ligands to form a slightly distorted nitride-centered U_4 tetrahedron (two edges are each bridged by two azido ligands; four edges are each bridged by one azido ligand).

Reference

G. Nocton, J. Pécaut, and M. Mazzanti, A nitrido-centered uranium azido cluster obtained from a uranium azide. *Angew. Chem. Int. Ed.* **47**, 3040–2 (2008).

8.17.3 Uranium(IV) organoazide complexes

X-ray crystallography at −35 °C revealed that the tetraazenido uranium(IV) compound $[CH_2(C_6H_5)NNN(Mes)-\kappa^2N^{1,2}]U[CH_2(C_6H_5)NNN(Mes)-\kappa^2N^{1,3}]_3$ **1** exhibits eight-coordinate, distorted bicapped trigonal prismatic geometry. It is notable that one triazenido ligand is coordinated in a $\kappa^2N^{1,2}$ fashion, while the other three are coordinated in a $\kappa^2N^{1,3}$ fashion (Fig. 8.17.3(a)). In the uranium bis(imido) bis-(triazenido) species $U(NAd)_2[CH_2(C_6H_5)-NNN(Ad)-\kappa^2N^{1,3}]_2(THF)$ complex **2**, where Ad = adamantanyl, the seven-coordinate uranium center adopts distorted pentagonal-bipyramidal geometry, in which two *axial* adamantyl imido ligands serve as the caps and an *equatorial* N_4O set forming the pentagonal plane (Fig. 8.17.3(b)).

Reference

S. J. Kraft, P. E. Fanwick, and S. C. Bart, Exploring the insertion chemistry of tetrabenzyluranium using carbonyls and organoazides. *Organometallics* **32**, 3279–85 (2013).

8.17.4 Thorium 5-methyltetrazolate complexes

$(C_5Me_5)_2Th[\eta^2-(N, N')$-tetrazolate$]_2$ was prepared as an off-white solid from the addition of two molar equivalents of 5-methyl-1H-tetrazole to a solution of $(C_5Me_5)_2Th(CH_3)_2$ in toluene. Its chemical formula and molecular structure determined by single-crystal X-ray diffraction are shown in Fig. 8.17.4.

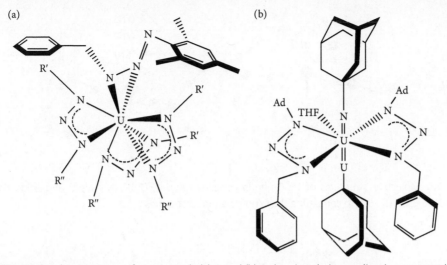

Fig. 8.17.3 Molecular structures of compounds (a) **1** and (b) **2** showing their coordination geometries.

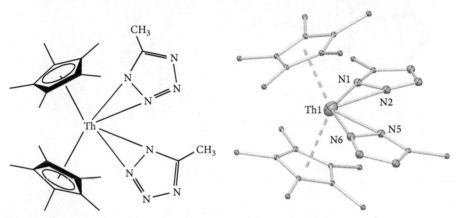

Fig. 8.17.4 Chemical formula and X-ray crystal structure of $(C_5Me_5)_2Th[\eta^2\text{-}(N, N')\text{-tetrazolate}]_2$. Selected bond distances (pm) and angles (°): Th(1)–N(1) 252.4 (2), Th(1)–N(2) 250.1(2), Th(1)–N(5) 249.1 (2),Th(1)–N(6) 254.0 (2), Cg–Th(1) 251.1(2), Cg′–Th(1) 251.4(2); N(1)–Th(1)–N(2) 31.33(6), N(5)–Th(1)–N(6) 31.13(6), Cg–Th(1)–Cg′ 136.65(7).

Reference

K. P. Browne, K. A. Maerzke, N. E. Travia, D. E. Morris, B. L. Scott, N. J. Henson, P. Yang, J. L. Kiplinger, and J. M. Veauthier, Synthesis, characterization, and density functional theory analysis of uranium and thorium complexes containing nitrogen-rich 5-methyltetrazolate ligands. *Inorg. Chem.* **55**, 4941–50 (2016).

8.17.5 **Uranium(VI) nitride complexes**

In 2020 the terminal nitride $[NBu_4][U(OSi(O^tBu)_3)_4(N)]$ was prepared upon photolysis with UV light of the sterically demanding terminal U(IV) azide analog $[NBu_4][U(OSi(O^tBu)_3)_4(N_3)]$. X-ray crystallographic analysis showed that the uranium(VI) atom U1 in $[U(OSi(O^tBu)_3)_4(N)]^-$ anion exhibits slightly

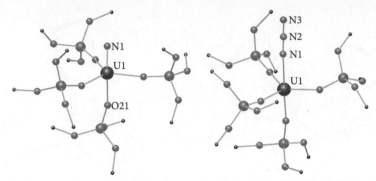

Fig. 8.17.5 Molecular structures of U(IV) nitrido complex [U(OSi(O^{t}Bu)$_3$)$_4$(N)]$^-$ (left) and azido complex [U{OSi(O^{t}Bu)$_3$}$_4$(N$_3$)]$^-$ (right) comparing their coordination geometries.

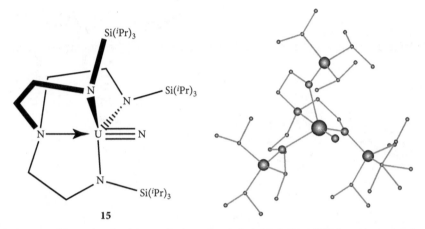

Fig. 8.17.6 Structural formula of stable terminal uranium(VI) nitride [U(Tren^{TIPS})N] and its molecular structure determined by X-ray crystallography.

distorted trigonal-bipyramidal geometry (Fig. 8.17.5, left figure), with three siloxide ligands lying on the equatorial plane, and the axial positions occupied by the nitride ligand and a fourth siloxide ligand. The U1–N1 bond distance (176.9(2) pm) is slightly shorter than the U–nitride bond distance (179.9(7) pm) reported for tris(triisopropylsilylamidoethyl)amine uranium(VI) nitride [U(Tren^{TIPS})N] (Fig. 8.17.6) and compares well with the value calculated for the matrix-isolated terminal nitride complex UNF$_3$ (176 pm). The U1–O21 bond distance at 207.3(1) pm is approximately 6% smaller compared to the mean value 220(3) pm of the equatorial U–O$_{siloxide}$ bond distances. This is indicative of an inverse *trans* influence, which is often manifested in high-valent f-element complexes. The U1–N1 bond distance is 176.9 (2) pm. In the [U{OSi(O^{t}Bu)$_3$}$_4$(N$_3$)]$^-$ anion (Fig. 8.17.5, right figure), the U1–N1, N1–N2, and N2–N3 bond distances determined by X-ray crystallography are 237.5(4) pm, 118.7(5) pm, and 116.5(6) pm, respectively.

References

L. Barluzzi, R. Scopelliti, and M. Mazzanti, Photochemical synthesis of a stable terminal uranium(VI) nitride, *J. Am. Chem. Soc.* **142**, 19047–51 (2020).

D. M. King, F. Tuna, E. J. L. McInnes, J. McMaster, W. Lewis, A. J. Blake, and S. T. Liddle, Isolation and characterization of a uranium(VI)–nitride triple bond, Nat. Chem. **5**, 482–8 (2013).

8.18 **Bent Anionic Thorocenes**

Reaction of linear thorocene $(COT)_2Th$ $(COT = \eta\text{-}C_8H_8)$ with sodium salts Na(18-crown-6)X {X = CN, N_3, H} yielded bent thorocene derivatives $[(COT)_2Th(X)]^-$ (X = CN, N_3) and bimetallic $[\{(COT)_2Th\}_2(\mu\text{-}H)]^-$, whereas only $[(COT)_2U(CN)]^-$ can be formed from $(COT)_2U$ (Scheme 8.18.1).

The molecular structures of the cyano- and azido-bridged complexes, **13** and **14** respectively, are shown in Fig. 8.18.1. The CN^- ligand in **13** is best refined with a Th–C rather than a Th–N linkage, which is the case with all known uranium cyanide compounds.

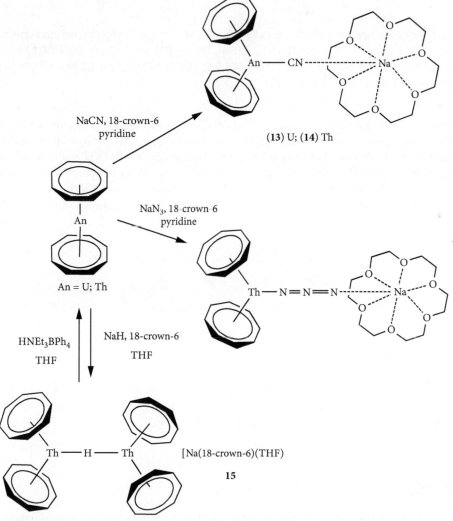

Scheme 8.18.1 Synthesis of bent actinocene complexes.

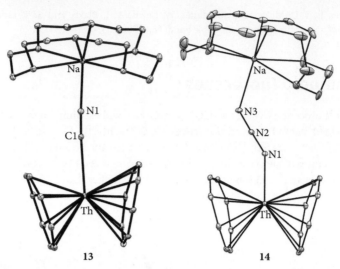

Fig. 8.18.1 Molecular structures of complexes **13** (left) and **14** (right) with all hydrogen atoms omitted. In **13**, Th–C1 2.648(4), C1–N1 1.157(6), Na–N1 2.365(4), Th···Cg 209 and 210 pm, Cg = COT ring center; Cg···Th···Cg 150°. In **14**, Th–N1 251.8(4), N1–N2 117.9(5), N2–N3 115.9(5), Na–N3 248.4(4), Th···Cg 210 pm; Cg···Th···Cg 149, N1–N2–N3 178.3(5)°.

In the [Na(18-crown-6)(THF)$_2$][{(COT)$_2$Th}$_2$(μ-H)] salt, both complex cation and anion occupy crystallographic C_2 sites. In the [{COT$_2$Th}$_2$(μ-H)]$^-$ anion **15** (Fig. 8.18.2), the hydride H$^-$ bridges two bent (COT)$_2$Th units that interlock in an almost perpendicular fashion. The COT ring centroids around the Th cores are positioned at the apices of a distorted tetrahedron, and the two planes containing the Th and H1 atoms and the two ring centroids in each (COT)$_2$Th fragment intersect with a dihedral angle of 83.71°. In the isomorphous [K(18-crown-6)(THF)$_2$]$^+$ complex cation, the corresponding dihedral angle is 89.81°.

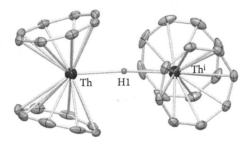

Fig. 8.18.2 Perspective view of the [{(COT)$_2$Th}$_2$(μ-H)]$^-$ anion (**15**) with all hydrogen atoms omitted except for bridging hydride H1. Symmetry code: i = 1−x, −y, z. Selected bond lengths and angles: Th–H1 2.323(7), Th···Cg 209 and 210 pm, Cg···Th···Cg 148°; Cg = COT ring centroid.

Reference

A. Hervé, N. Garin, P. Thuéry, M. Ephritikhine, and J.-C. Berthet, Bent thorocene complexes with the cyanide, azide and hydride ligands. *Chem. Commun.* **49**, 6304–6 (2013).

8.19 **Linear Trans-bis(imido) Actinyl(V) Complexes**

The reaction of $UCl_4(^tBu_2bpy)_2$ ($^tBu_2bipy$ = 4,4'-di-*tert*-butyl-2,2'-bipyridyl) with 4 equiv. of LiN-HDipp (Dipp = 2,6-$^iPr_2C_6H_3$) in THF, followed by addition of 0.5 equiv. of CH_2Cl_2, generated $U(NDipp)_2(^tBu_2bpy)_2Cl$. This U̲(V) complex exhibits distorted pentagonal-bipyramidal coordination geometry (Fig. 8.19.1(a)), and it is the first monomeric bis(imido) uranium(V) species.

Treatment of $NpCl_4(DME)_2$ with 2 equiv. of $^tBu_2bipy$, 4 equiv. of LiNHDipp and excess CH_2Cl_2 resulted in isolation of $Np(NDipp)_2(^tBu_2bipy)_2Cl$. This naptunium(V) complex crystallizes in space group *Pbcn* with Z = 4 as the $2(H_2NDipp)\cdot O(C_4H_8)$ solvate, with its Np-Cl bond lying on a crystallographic C_2 axis, such that the Np(V) center exhibits seven-coordinate distorted pentagonal-bipyramidal coordination geometry (Fig. 8.19.1(b)). In fact, this study provided the first example of a transuranic imido group, as well as the first molecular Np-ligand multiple-bond beyond the dioxo neptunyl cation.

(a)   (b)

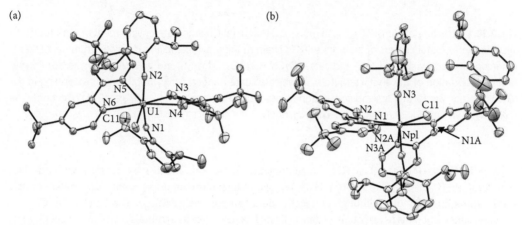

Fig. 8.19.1 Iso-structural pentagonal bipyramidal *trans*-bis(imido) actinyl(V) complexes: (a) $U(NDipp)_2(^tBu_2bpy)_2Cl$, U1–N1 = 197.7(4), U1–N2 = 198.0(4), U1–N3 = 268.3(4), U1–N4 = 259.4(4), U1–N5 = 259.5(4), U1–N6 = 265.9(4), U1–Cl1 = 272.9(2) pm; (b) $Np(NDipp)_2(^tBu_2bipy)_2Cl$, Np1–N3 = 196.0(3), Np1–N3A = 196.1(3), Np1–N1 = 260.2(3), Np1–N2 = 262.7(3), Np1–Cl1 = 227.6(1) pm.

References

R. E. Jilek, L. P. Spencer, R. A. Lewis, B. L. Scott, T. W. Hayton, and J. M. Boncella, A direct route to bis(imido)uranium(V) halides via methesis of uranium tetrachloride. *J. Am. Chem. Soc.* **134**, 9876–8 (2012).

J. L. Brown, E. R. Batista and J. M. Boncella, A. J. Gaunt, S. D. Reilly, B. L. Scott, and N. C. Tomson, A linear *trans*-bis(imido) naptunium(V) actinyl analog: $Np^V(NDipp)_2(^tBu_2bipy)_2Cl$ (Dipp = 2,6-$^iPr_2C_6H_3$). *J. Am. Chem. Soc.* **137**, 9583–6 (2015).

8.20 **First Well-Characterized MOF of a Transplutonium Element**

The transplutonium element americium is a by-product in nuclear reactors that continues to emit radiation for thousands of years. In a recent study of its fundamental coordination chemistry, radioactive trivalent ^{243}Am ions separated from spent nuclear fuels were incorporated into

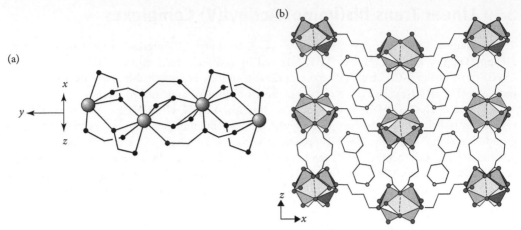

Fig. 8.20.1 Structure of $[Am_2(C_6H_8O_4)_3(H_2O)_2]\cdot(C_{10}H_8N_2)$. (a) Coordination environment around four consecutive Am^{3+} metal centers of an infinite [010] chain running parallel to the *b*-axis. Each metal center is nine-coordinated by six $C_6H_8O_4^{2-}$ (adipate, $^-O_2C(CH_2)_4CO_2^-$) ligands and a water molecule. (b) Polyhedral representation of the crystal structure, in which adipate-bridged Am^{3+} metal centers extending in the [010] direction are further assembled through several adipate ligands in the [100] and [001] directions. The $C_{10}H_8N_2$ (4,4′-bipyridine) molecules occupy channels in the crystal structure.

a $[Am_2(C_6H_8O_4)_3(H_2O)_2]\cdot(C_{10}H_8N_2)$ metal-organic framework (MOF) by analogy with the isostructural $[Pr_2(C_6H_8O_4)_3(H_2O)_2]\cdot(C_{10}H_8N_2)$ crystal architecture containing chemically similar but very stable lanthanide(III) ions. Although radioactive decay progressively damaged the Am(III) MOF, least-squares refinement of its X-ray intensity data collected 26 days after its synthesis revealed the details of its crystal structure (Fig. 8.20.1). After the course of three months managed, its unit cell parameters increased by 2%, and the diffraction data became worse.

References

D. T. de Lill, N. S. Gunning, and C. L. Cahill, Toward templated metal-organic frameworks: synthesis, structures, thermal properties, and luminescence of three novel lanthanide-adipate frameworks. *Inorg, Chem.* **44**, 258–66 (2005)].

J. A. Ridenour, R. G. Surbella III, A. V. Gelis, D. Koury, F. Poineau, K. R. Czerwinski, and C. L. Cahill, An americium-containing metal-organic framework: a platform for studying transplutonium elements, *Angew. Chem. Int. Ed.* **58**, 16508–11 (2019).

8.21 New Types of Metal-ligand δ and φ Bonding in Actinide Complexes

Delta (δ) and phi (φ) bonds are formed by metal-ligand (M–L) orbital overlap that involve four and six orbital lobes, respectively. In the vast majority of known metal–ligand (M–L) δ and φ bonds, the metal orbitals are aligned to the ligand orbitals in a "head-to-head" or "side-to-head" fashion.

8.21.1 δ and φ back-donation in An^{IV} metallacycles

Two fundamentally new types of M–L bonding interactions, namely "head-to-side" δ and "side-to-side" φ back-donation, are found in complexes of metallacyclopropenes and metallacyclocumulenes of actinides (Pa–Pu), making them distinct from their corresponding Group 4 analogs. In addition to the known $(C_8H_8)_2$Th and $(C_8H_8)_2$U complexes, recent theoretical calculations have been performed on complexes of Pa, Np, and Pu. In contrast with conventional expectation of decreasing An–C bond length due to actinide contraction, the An–C bond distance increases from Pa to Pu. It is now established that direct L–An σ and π donations combined with An–L δ or φ back-donations are crucial in explaining this non-classical trend of An–L bond lengths in both series, underscoring the significance of δ/φ back-donation interactions, particularly in complexes of Pa and U.

Due to the availability of d-electrons and f-electrons, chemical bonding in transition-metal, lanthanide, and actinide compounds may exhibit exotic δ and φ bonding modes. Common examples of compounds exhibiting these bonds include various dimetals M_2, either bare or surrounded by stabilizing ligands, with multiple M–M bonds. A typical δ bond reported in such systems features two nodal planes passing through the M–M axis arising from face-to-face overlap of two d_{xy} or $d_{x^2-y^2}$ atomic orbitals (Fig. 8.21.1(a)).

For more than two metal atoms (M_n, $n > 2$), another type of δ bonding can be achieved through interaction of d_{z^2} AOs (Fig. 8.21.1(b)). In such case the δ bond features two parallel nodal planes located above and below the plane of the metal atoms, thus giving rise to δ aromaticity. In contrast to the M–M δ bonds, metal–ligand (M–L) interactions involving formation of δ bonds are less common. These bonds differ qualitatively from the M–M δ bonds in different aspects: first, they are formed by covalent overlap between occupied d or f metal orbitals and unoccupied ligand orbital(s), and hence are called δ back bonds; second, such M–L interactions occur with "head-to-head" orbital overlap (Fig. 8.21.1(c)). The

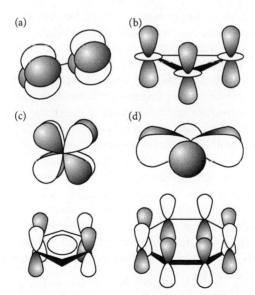

Fig. 8.21.1 Schematic representation of AOs comprising δ and φ bonds. (a) M–M δ interaction between two d_{xy} (or $d_{x^2-y^2}$) AOs. (b) M–M–M δ interaction among three d_{z^2} AOs. (c) "head-to-head" M–L δ interaction between f_{xyz} (or $f_{y(x^2-z^2)}$) AO of metal (top) and unoccupied ligand orbital (bottom). (d) "side-to-head" M–L φ back-donation interaction between $f_{x(x^2-3y^2)}$ (or $f_{y(3x^2-y^2)}$) AO of metal (top) and unoccupied ligand orbital (bottom).

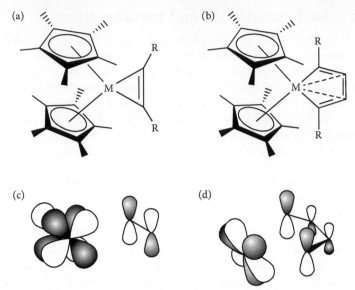

Fig. 8.21.2 Structures of actinide and transition-metal metallacycles and their respective M–L δ and φ interactions. (a) Metallacyclopropene (η^5-C$_5$Me$_5$)$_2$An[η^2-C$_2$R$_2$]. (b) Metallacyclocumulene (η^5-C$_5$Me$_5$)$_2$An[η^4-C$_4$R$_2$] (R = trimethylsilyl, phenyl; M = Ti, Zr, Th, Pa, U, Np, Pu). (c) "head-to-side" M–L δ interaction between f_{xyz} (or $f_{y(x^2-z^2)}$) AO of metal and unoccupied orbital of the cyclopropene ligand in (η^5-C$_5$Me$_5$)$_2$An[η^2-C$_2$R$_2$]. (d) "side-to-side" M–L φ interaction between f_{yz^2} (or f_{xz^2}) AO of metal and unoccupied orbital of the cyclocumulene ligand in (η^5-C$_5$Me$_5$)$_2$An[η^4-C$_4$R$_2$], An = Pa–Pu.

even rarer M–L φ back-bonds are formed by covalent overlap between occupied f metal orbital(s) and unoccupied ligand orbital(s) in a "side-to-head" fashion (Fig. 8.21.1(d)).

Both carbon K-edge X-ray absorption spectra (XAS) and DFT calculations on thorocene and uranocene have shown that the 5f orbitals engage in significant d-type mixing with the C$_8$H$_8^{2-}$ ligands, which increases as the 5f orbitals drop in energy on moving from Th^{4+} to U^{4+}.

There are two novel types of M–L back-bonding in actinide metallacyclopropenes (η^5-C$_5$Me$_5$)$_2$An[η^2-C$_2$(SiMe$_3$)$_2$] (Fig. 8.21.2(a)) and metallacyclocumulenes (η^5-C$_5$Me$_5$)$_2$An[η^4-C$_4$(SiMe$_3$)$_2$] (Fig. 8.21.2(b)) for the actinide series from Th to Pu, that is, "head-to-side" δ (Fig. 8.21.2(c)) and "side-to-side" φ (Fig. 8.21.2(d)) M–L interactions. These two unique bonding modes are made possible by a number of factors. First, there is a smaller number of carbon atoms in the ligand interacting with the metal center (two in metallacyclopropenes and four in metallacyclocumulenes) than in previously reported complexes featuring M–L δ or φ back-bonding, where this number ranged from six to 16. Second, availability of f electrons for the formation of such bonds, unlike transition metals and actinides without f electrons. Third, co-planar positioning of the propene and cumulene ligands with the metal center facilitates efficient overlap of carbon 2p orbitals with metal 5f orbitals (Figs 8.21.2(c) and (d)), making them superior to other systems featuring M–L δ/φ back bonding where the M center is located out of the plane of ligand carbon atoms (Figs 8.21.1(c) and (d)).

References

L. J. Clouston, R. B. Siedschlag, P. A. Rudd, N. Planas, S. Hu, A. D. Miller, L. Gagliardi, and C. C. Lu, Systematic variation of metal-metal bond order in metal-chromium complexes. *J. Am. Chem. Soc.* **135**, 13142–8 (2013).

B. O. Roos, P.-Å. Malmquist, and L. Gagliardi, Exploring the actinide-actinide bond: theoretical studies of the chemical bond in Ac_2, Th_2, Pa_2, and U_2, *J. Am. Chem. Soc.* **128**, 17000–6 (2006).

S. G. Minasian, J. M. Keith, E. R. Batista, K. S. Boland, D. L. Clark, S. A. Kozimor, R. L. Martin, David K. Shuh, and T. Tyliszczak, New evidence for 5f covalency in actinocenes determined from carbon K-edge XAS and electronic structure theory. *Chem. Sci.* **5**, 351–9 (2014).

M. P. Kelley, I. A. Popov, J. Jung, E. R. Batista, and P. Yang, δ and φ back-donation in An^{IV} metallacycles, *Nat. Commun.* 11, 1558 (2020). <https://doi.org/10.1038/s41467-020-15197-w>

General References

H. C. Aspinall, *Chemistry of the f-Block Elements*, Gordon and Breach, Amsterdam, 2001.

S. Cotton, *Lanthanide and Actinide Chemistry*, Wiley, Chichester, 2006.

N. Kaltsoyannis and P. Scott, *The f Elements*, Oxford University Press, Oxford, 1999.

S. D. Barrett and S. S. Dhesi, *The Structure of the Rare-Earth Metal Surfaces*, Imperial College Press, London, 2001.

C. H. Huang, *Coordination Chemistry of Rare-Earth Elements* (in Chinese), Science Press, Beijing, 1997.

K. A. Gschneidner, Jr., L. Eyring, G. R. Choppin, and G. H. Lander (eds), *Handbook on the Physics and Chemistry of Rare Earths*, Vol. 18: *Lanthanides/Actinides: Chemistry*, North-Holland, Amsterdam, 1994.

Thematic issue on "Frontiers in Lathanide Chemistry", *Chem. Rev.* **102**, 6 (2002).

S. T. Liddle, D. P. Mills, and L. S. Natrajan (eds), *The Lanthanides and Actinides: Synthesis, Reactivity, Properties and Applications*, World Scientific, Singapore, 2021.

Chapter 9

Metal-Metal Bonds and Transition-Metal Clusters

9.1 Bond Valence and Bond Number of Transition-Metal Clusters

A dinuclear transition metal complex contains two transition metal atoms each surrounded by a number of ligands. A transition-metal cluster has a core of three or more metal atoms that are directly bonded with each other to form a discrete molecule containing metal-metal (M–M) bonds. The first reported organometallic complex that possesses a M–M bond is $Fe_2(CO)_9$. Since this Fe–Fe bond is supported by three bridging CO ligands, as shown in Fig. 9.1.1(a), it cannot be taken as definitive proof of direct metal–metal interaction. In contrast, the complexes $Re_2(CO)_{10}$, $Mn_2(CO)_{10}$, and $[MoCp(CO)_3]_2$, with no bridging ligands, provide unequivocal examples of unsupported M–M bonding. In these species the metal atoms are considered to be bonded through a single bond of the 2c-2e type. Fig. 9.1.1(b) shows the molecular structure of $Mn_2(CO)_{10}$.

Since polynuclear complexes and cluster compounds are in general rather complicated species, the application of quantitative methods for describing bonding is not only difficult but also impractical. Qualitative approaches and empirical rules often play an important role in treating such cases. We have used the octet rule and bond valence to describe the structure and bonding of boranes and their derivatives (Sections 3.3 and 3.4). Now we use the 18-electron rule and bond valence to discuss the bonding and structure of polynuclear transition metal complexes and clusters.

The metal atoms in most transition metal complexes and clusters obey the 18-electron rule. The bond valence, b, of the skeleton of complex $[M_nL_p]^{q-}$ can be calculated from the formula:

$$b = \tfrac{1}{2}(18n - g)$$

where g is the total number of valence electrons in the skeleton of the complex, which is the sum of the following three parts:

(a) n times the number of valence electrons of metal atom M;

(b) p times the number of valence electrons donated to the metal atoms by ligand L;

(c) q electrons from the net charge of the complex.

The bond valence b of the skeleton of a complex or cluster corresponds to the sum of the bond numbers of the metal-metal bonds. For an M–M single bond, the bond number is equal to 1; similarly, the bond number is 2 for an M=M double bond, 3 for an M≡M triple bond, 4 for an M≣M quadruple bond, and 2 for a 3c-2e MMM bond.

The majority of cluster compounds have carbonyl ligands coordinated to the metal atoms. Some clusters bear NO, CNR, PR_3, and H ligands, while others contain interstitial C, N, and H atoms. The compounds can be either neutral or anionic, and the common structural metal building blocks are triangles, tetrahedra, octahedra, and condensed clusters derived from them. The carbonyl ligand has two special features.

Structural Chemistry across the Periodic Table. Thomas Chung Wai Mak, Yu-San Cheung, Gong-Du Zhou, and Ying-Xia Wang. Oxford University Press. © Thomas Chung Wai Mak, Yu-San Cheung, Gong-Du Zhou, and Ying-Xia Wang (2023). DOI: 10.1093/oso/9780198872955.003.0009

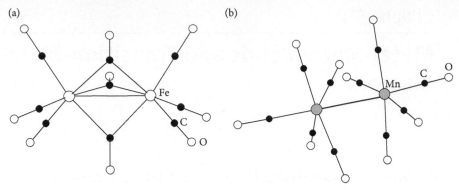

Fig. 9.1.1 Structure of dinuclear transition metal complexes: (a) $Fe_2(CO)_9$ and (b) $Mn_2(CO)_{10}$.

First, the carbonyl ligand functions as a two-electron donor in the terminal-, edge-, or face-bridging mode, and neither changes the valence of the metal skeleton nor requires the involvement of additional ligand electrons. In contrast, the Cl ligand may change from a one- (terminal) to a three- (edge-bridging) or five- (face-bridging) electron donor. Second, the synergic bonding effects which operate in the alternative bridging modes of the carbonyl ligand lead to comparable stabilization energies and therefore readily favor the formation of transition metal cluster compounds.

As mentioned in Sections 3.3 and 3.4, the structures of most known boranes and carboranes are based on the regular deltahedron and divided into three types: *closo*, *nido*, and *arachno*. In a *closo*-structure the skeletal B and C atoms occupy all the vertices of the polyhedron. In the cases of *nido*- and *arachno*-structures, one and two of the vertices of the appropriate polyhedron remain unoccupied, respectively. On each skeletal B and C atom, there is always a H atom (or some other simple terminal ligand such as halide) that points away from the center of the polyhedron and is also linked to the skeleton by a radial 2c-2e single bond. In a polyhedron of n vertices, there are $4n$ atomic orbitals. Among them, n orbitals are used up for n terminal B–H bonds, the remaining $3n$ atomic orbitals being available for skeletal bonding. In the skeleton of *closo*-borane $B_nH_n^{2-}$, there are $4n+2$ valence electrons (n B atoms contribute $3n$ electrons, n H atoms donate n electrons, and the charge accounts for two electrons), and hence $2n+2$ electrons or $n+1$ electron pairs are available for skeletal bonding. These numbers are definite. For the transition metal cluster compounds, however, the electron counts vary with different bonding types. For example, octahedral M_6 clusters of different compounds form various structures and bond types. Some examples are discussed below and shown in Fig. 9.1.2.

Fig. 9.1.2(a) shows the structure of $[Mo_6(\mu_3\text{-}Cl)_8Cl_6]^{2-}$. In this structure, eight μ_3-Cl and six terminal Cl atoms are coordinated to the Mo_6 cluster. Each μ_3-Cl donates five electrons and each terminal Cl donates one electron. Thus the g value is:

$$g = 6 \times 6 + (8 \times 5 + 6 \times 1) + 2 = 84$$

The bond valence of the Mo_6 cluster is:

$$b = \tfrac{1}{2}(6 \times 18 - 84) = 12$$

The bond valence b precisely matches 12 2c-2e Mo–Mo bonds, as shown in the front and back views of the Mo_6 cluster shown in the right side of Fig. 9.1.2(a).

Fig. 9.1.2(b) shows the structure of $[Nb_6(\mu_2\text{-}Cl)_{12}Cl_6]^{4-}$. In this structure, there are 12 μ_2-Cl atoms each donating three electrons, plus six terminal μ_1-Cl atoms each donating one electron, to the Nb cluster. The g value is:

$$g = 6 \times 5 + (12 \times 3 + 6 \times 1) + 4 = 76$$

The bond valence of the Nb_6 cluster is:

$$b = \tfrac{1}{2}(6 \times 18 - 76) = 16$$

In this Nb_6 cluster, each edge is involved in bonding with a μ_2-Cl ligand, so the edges do not correspond to 2c-2e Nb–Nb bonds. Each face of the Nb_6 cluster forms a 3c-2e Nb–Nb–Nb bond, and has bond number 2. The sum of bond number, 16, is just equal to the bond valence of 16, as shown in the front and back views of the Nb cluster in Fig. 9.1.2(b).

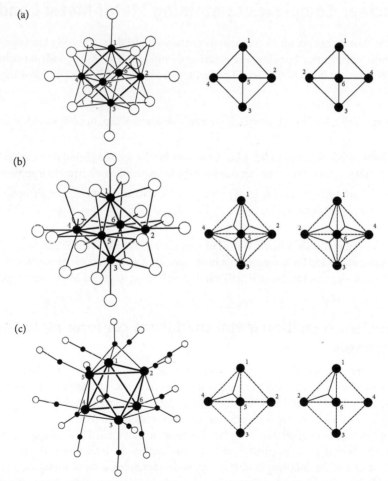

Fig. 9.1.2 Structure and bonding of three octahedral clusters: (a) $[Mo_6(\mu_3\text{-}Cl)_8Cl_6]^{2-}$; (b) $[Nb_6(\mu_2\text{-}Cl)_{12}Cl_6]^{4-}$; (c) $Rh_6(\mu_3\text{-}CO)_4(CO)_{12}$.

Fig. 9.1.2(c) shows the structure of $Rh_6(\mu_3\text{-}CO)_4(CO)_{12}$. The CO group as a ligand always donates two electrons to the M_n cluster. The g value and bond valence of the Rh_6 cluster are:

$$g = 6 \times 9 + (4 \times 2 + 12 \times 2) = 86$$
$$b = \tfrac{1}{2}(6 \times 18 - 86) = 11$$

The b value of Rh_6 is 11, which just matches the b value of $B_6H_6^{2-}$. The Rh_6 forms four 3c-2e Rh–Rh–Rh bonds and three 2c-2e Rh–Rh bonds. Various bond formulas can be written for the Rh_6 cluster, one of which is shown in the right side of Fig. 9.1.2(c). These formulas are equivalent by resonance and the octahedron has overall O_h symmetry.

9.2 Dinuclear Complexes Containing Metal-Metal Bonds

Studies on dinuclear transition metal compounds containing metal-metal bonds have deeply enriched our understanding of chemical bonding. The nature and implication of M–M bonds are richer and more varied than the covalent bond of the second period representative elements. The following points may be noted.

(a) The d atomic orbitals of transition metals are available for bonding, in addition to the s and p atomic orbitals.

(b) The number of valence orbitals increases from four for the second period elements to nine for the transition metal atoms. Thus the octet rule suits the former and 18-electron rule applies to the latter.

(c) The use of d-orbitals for bonding leads to the possible formation of a quadruple bond (and even a quintuple bond) between two metal atoms, in addition to the familiar single-, double-, and triple-bonds.

(d) The transition metal atoms interact with a large number of ligands in coordination compounds. The various geometries and electronic factors have significant effects on the properties of metal-metal bonds. The 18-electron rule is not strictly valid, thus leading to varied M–M bond types.

9.2.1 Dinuclear transition metal complexes conforming to the 18-electron rule

When the structure of a dinuclear complex has been determined, the structural data may be used to enumerate the number of valence electrons available in the complex, to determine its bond valence, and to understand the properties of the metal-metal bond. For a dinuclear complex, the bond valence $b = \tfrac{1}{2}(18 \times 2 - g)$. Table 9.2.1 lists the data for some dinuclear complexes.

In the complex $Ni_2(Cp)_2(\mu_2\text{-}PPh_2)_2$, with two bridging ligands ($\mu_2\text{-}PPh_2$) linking the Ni atoms, the bond valence equals zero, indicating that there is no bonding interaction between them.

$Mn_2(CO)_{10}$ is one of the simplest dimeric compounds containing a metal-metal single bond unsupported by bridging ligands, and its diamagnetic behavior is accounted for in compliance with the 18-electron rule. The structures of $Tc_2(CO)_{10}$, $Re_2(CO)_{10}$, and $MnRe(CO)_{10}$, which are isomorphous with $Mn_2(CO)_{10}$, have also been determined. The measured M–M bond lengths (pm) are: Tc–Tc 303.6, Re–Re 304.1, and Mn–Re 290.9. The structure of $Fe_2(CO)_9$, shown in Fig. 9.1.1(a), has an Fe–Fe single bond of 252.3 pm, which is further stabilized by the bridging carbonyl ligands.

Table 9.2.1 Dinuclear complexes.

Complex	g	b	M–M/pm	Bond properties
$Ni_2(Cp)_2(\mu_2\text{-PPh}_2)_2$	36	0	336	Ni⋯Ni, no bonding
$(CO)_5Mn_2(CO)_5$	34	1	289.5	Mn–Mn, single bond
$Co_2(\mu_2\text{-CH}_2)(\mu_2\text{-CO})(Cp^*)_2$	32	2	232.0	Co=Co, double bond
$Cr_2(CO)_4(Cp)_2$	30	3	222	Cr≡Cr, triple bond
$[Mo_2(\mu_2\text{-O}_2CMe)_2(MeCN)_6]^{2+}$	28	4	213.6	Mo≣Mo, quadruple bond

Many dinuclear complexes have a M=M double bond. Some examples are:

$$Co_2(CO)_2Cp_2{}^*, \text{ Co=Co 233.8 pm;}$$
$$Fe_2(NO)_2Cp_2, \text{ Fe=Fe 232.6 pm;}$$
$$Re_2(\mu - Cl)_2Cl_4(dppm)_2, \text{ Re=Re 261.6pm;}$$
$$Mo_2(OR)_8 \left(R={}^iPr, {}^tBu\right), \text{ Mo=Mo 252.3 pm.}$$

Most compounds with triple and quadruple bonds are formed by Re, Cr, Mo, and W. The ligands in such compounds are in general relatively hard Lewis bases such as halides, carboxylic acids, and amines. Nevertheless, in some cases π-acceptor ligands such as carbonyl, phosphines, and nitrile are also present.

9.2.2 **Quadruple bonds**

The recognition and understanding of the quadruple bond is one of the most important highlights in modern inorganic chemistry. The overlap of d atomic orbitals can generate three types of molecular orbitals: σ, π, and δ. These molecular orbitals can be used to form a quadruple bond between two transition metal atoms under appropriate conditions. In a given compound, however, not all of these orbitals are always available for multiple metal-metal bonding. Thus the 18-electron rule does not always hold for the dinuclear complexes and needs to be modified according to their structures.

The most interesting aspect of the crystal structure of $K_2[Re_2Cl_8]\cdot 2H_2O$ is the presence of the dianion $Re_2Cl_8{}^{2-}$ (Fig. 9.2.1), which possesses an extremely short Re–Re bond distance of 224.1 pm, as compared with an average Re–Re distance of 275 pm in rhenium metal. Another unusual feature is the eclipsed configuration of the Cl atoms with a Cl⋯Cl separation of 332 pm. As the sum of van der Waals radii of two Cl atoms is 360 pm, the staggered configuration would normally be expected for $Re_2Cl_8{}^{2-}$. These two features are both attributable to the formation of a Re≣Re quadruple bond.

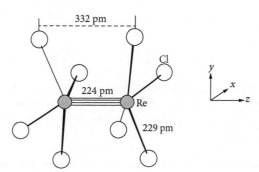

Fig. 9.2.1 Structure of $[Re_2Cl_8]^{2-}$.

The bonding in the skeleton of the $[Re_2Cl_8]^{2-}$ ion can be formulated as follows: each Re atom uses its square planar set of dsp^2 ($d_{x^2-y^2}$, s, p_x, p_y) hybrid orbitals to overlap with ligand Cl p-orbitals to form Re–Cl bonds. The p_z atomic orbital of Re is not available for bonding. The remaining d_{z^2}, d_{xz}, d_{yz}, and d_{xy} atomic orbitals on each Re atom overlap with the corresponding orbitals on the other Re atom to generate the following MOs:

$d_{z^2} \pm d_{z^2}$	$\rightarrow$	σ and σ^* MO
$d_{xz} \pm d_{xz}$	$\rightarrow$	π and π^* MO
$d_{yz} \pm d_{yz}$	$\rightarrow$	π and π^* MO
$d_{xy} \pm d_{xy}$	$\rightarrow$	δ and δ^* MO

Fig. 9.2.2 shows the pairing up of d AOs of two Re atoms to form MOs and the ordering of the energy levels. In the $[Re_2Cl_8]^{2-}$ ions, the two Re atoms have 16 valence electrons including two from the negative charge. Eight valence electrons are utilized to form eight Re–Cl bonds, and the remaining eight occupy four metal-metal bonding orbitals to form a quadruple bond: one σ bond, two π bonds, and one δ bond, leading to the $\sigma^2\pi^4\delta^2$ configuration.

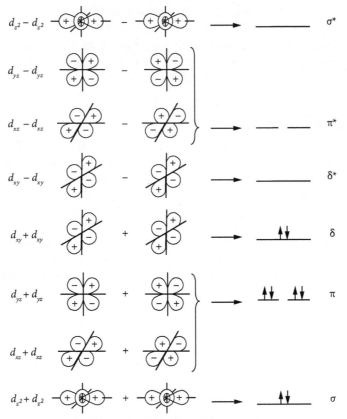

Fig. 9.2.2 Overlap of d orbitals leading to the formation of a quadruple bond between two metal atoms. Note that the z-axis of each metal atom is taken to point toward the other, such that if a right-handed coordinate system is used for the atom on the left, a left-handed coordinate system must be used for the atom on the right.

The structures of two transition metal complexes that contain a Mo≣Mo quadruple bond are shown in Fig. 9.2.3. In the compound $[Mo_2(O_2CMe)_2(NCMe)_6](BF_4)_2$, each Mo atom is six-coordinate,

(a)   (b)

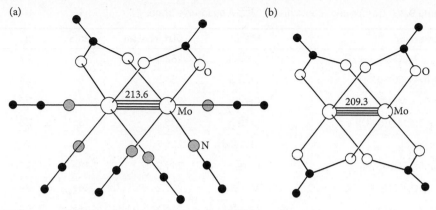

Fig. 9.2.3 Structure of (a) $[Mo_2(O_2CMe)_2(NCMe)_6]^{2+}$ and (b) $Mo_2(O_2CMe)_4$.

and all nine valence AOs are used for bonding according to the 18-electron rule. In the cation $[Mo_2(O_2CMe)_2(NCMe)_6]^{2+}$,

$$g = 2 \times 6 + 2 \times 3 + 6 \times 2 - 2 = 28, \text{and}$$
$$b = \frac{1}{2}(2 \times 18 - 28) = 4.$$

The Mo–Mo bond distance of 213.6 pm. is in accord with the quadruple Mo≣Mo bond. In the molecule $Mo_2(O_2CMe)_4$, each Mo atom is five-coordinate and has one empty AO(p_z) which is not used for bonding. It obeys the 16-electron rule with $g = 2 \times 6 + 4 \times 3 = 24$ and $b = \frac{1}{2}(2 \times 16 - 24) = 4$. The observed Mo–Mo distance of 209.3 pm is consistent with that expected of a quadruple bond.

A large number of dinuclear complexes containing a M≣M quadruple bond formed by the Group 16 and 17 metals Cr, Mo, W, Re, and Tc have been reported in the literature. Selected examples are listed in Table 9.2.2.

The quadruple bond can undergo a variety of interesting reactions, as outlined in Fig. 9.2.4. (1) There is a rich chemistry in which ligands are exchangable, and virtually every type of ligand can be used except the strong π-acceptors. (2) Addition of a mononuclear species to an M≣M bond can yield a trinuclear cluster. (3) Two quadruple bonds can combine to form a metallacyclobutadiyne ring. (4) Oxidative addition of acids to generate a M≡M bond (particularly W≡W) is a key part of molybdenum and tungsten chemistry. (5) Phosphines can act as reducing agents as well as ligands to give products with triple bonds of the $\sigma^2\pi^4\delta^2\delta^{*2}$ type, as in $Re_2Cl_4(PEt_3)_4$. (6) Photo excitation by the $\delta \rightarrow \delta^*$ transition can lead to reactive species which are potentially useful in photosensitizing various reactions, including the splitting of water. (7) Electrochemical oxidation or reduction reduces the bond order and generates reactive intermediates. (8) With π-acceptor ligands, the M≣M bonds are usually cleaved to give mononuclear products, which are sometimes inaccessible by any other synthetic route.

9.2.3 Bond valence of metal-metal bond

In dinuclear complexes, the bond valence of the metal-metal (M–M) bond can be calculated from its number of bonding electrons, g_M. A simple procedure for counting the metal-metal bond valence is as follows:

(a) Calculate the g value in the usual manner.

(b) Calculate the number of valence electron used for M-L bonds, g_L.

Table 9.2.2 Compounds containing a M≣M quadruple bond

Compound	M≣M	Distance/pm	Rule	g
$Cr_2(2\text{-MeO-5-Me-}C_6H_3)_4$	Cr≣Cr	182.8	16e	24
$Cr_2[MeNC(Ph)NMe]_4$	Cr≣Cr	184.3	16e	24
$Cr_2(O_2CMe)_4$	Cr≣Cr	228.8	16e	24
$Cr_2(O_2CMe)_4(H_2O)_2$	Cr≣Cr	236.2	18e	28
$Mo_2(hpp)_4$	Mo≣Mo	206.7	16e	24
$K_4[Mo_2(SO_4)_4]\cdot 2H_2O$	Mo≣Mo	211.0	18e	28
$[Mo_2(O_2CCH_2NH_3)_4]Cl_4\cdot 3H_2O$	Mo≣Mo	211.2	16e	24
$Mo_2[O_2P(OPh)_2)]_4$	Mo≣Mo	214.1	16e	24
$W_2(hpp)_4\cdot 2NaHBEt_3$	W≣W	216.1	16e	24
$W_2(O_2CPh)_4(THF)_2$	W≣W	219.6	18e	28
$W_2(O_2CCF_3)_4$	W≣W	222.1	16e	24
$W_2Cl_4(P^nBu_3)_4\cdot C_7H_8$	W≣W	226.7	16e	24
$(Bu_4N)_2Tc_2Cl_8$	Tc≣Tc	214.7	16e	24
$K_2[Tc_2(SO_4)_4]\cdot 2H_2O$	Tc≣Tc	215.5	16e	24
$Tc_2(O_2CCMe_3)_4Cl_2$	Tc≣Tc	219.2	18e	28
$(Bu_4N)_2Re_2F_8\cdot 2Et_2O$	Re≣Re	218.8	16e	24
$Na_2[Re_2(SO_4)_4(H_2O)_2]\cdot 6H_2O$	Re≣Re	221.4	18e	28
$[Re_2(O_2CMe)_2Cl_4(\mu\text{-pyz})]_n$	Re≣Re	223.6	18e	28

Note: hpp is the anion of 1,3,4,6,7,8-hexahydro-2*H*-pyrimido-[1,2-*a*]-pyrimidine (Hhpp); pyz is pyrazine.

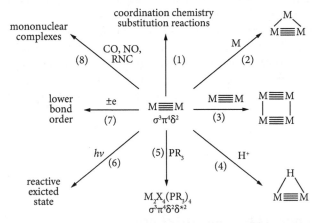

Fig. 9.2.4 Some reaction types of dimetal compounds containing an M≣M quadruple bond.

(c) Calculate the number of valence electron used for M-M bonds, $g_M = g - g_L$.

(d) Assign the g_M electrons to the following orbitals according to the energy sequence: σ, (π_x, π_y), δ, δ^*, (π_x^*, π_y^*) and σ^*, as shown in Fig. 9.2.2.

Some examples are presented below:

(1) $Mo_2(O_2CMe)_4$

$$g = 2 \times 6 + 4 \times 3 = 24, \quad g_L = 8 \times 2 = 16, \quad g_M = g - g_L = 24 - 16 = 8.$$

The electron configuration is $\sigma^2\pi^4\delta^2$, and the bond order is 4, that is, the bond valence is 4. For this Mo≣Mo bond, the bond length is 209.3 pm.

(2) $[Mo_2(O_2CMe)_2(MeCN)_6]^{2+}$

$$g = 2 \times 6 + 2 \times 3 + 6 \times 2 - 2 = 28, \quad g_L = 2 \times 5 \times 2 = 20, \quad g_M = 28 - 20 = 8.$$

The electron configuration is $\sigma^2\pi^4\delta^2$, and the bond valence is 4. So again this is a Mo≣Mo bond, and the bond length is 213.6 pm.

(3) $Re_2Cl_4(PEt_3)_4$

$$g = 2 \times 7 + 4 \times 1 + 4 \times 2 = 26, \quad g_L = 8 \times 2 = 16, \quad g_M = 26 - 16 = 10.$$

The structure of the molecule is shown in Fig. 9.2.5(a). The electron configuration is $\sigma^2\pi^4\delta^2\delta^{*2}$. The bond valence is 3, indicating a Re≡Re triple bond, and the bond length is 223.2 pm.

(4) $Mo_2(CH_2SiMe_3)_6$

The molecule has D_{3d} symmetry as shown in Fig. 9.2.5(b).

$$g = 2 \times 6 + 6 \times 1 = 18, \quad g_L = 6 \times 2 = 12, \quad g_M = 18 - 12 = 6.$$

The electron configuration is $\sigma^2\pi^4$. The length of the Mo≡Mo triple bond is 216.7 pm.

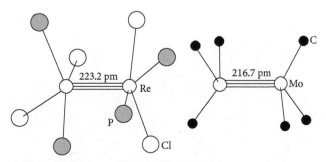

Fig. 9.2.5 Structure of (a) $Re_2Cl_4(PEt_3)_4$ and (b) $Mo_2(CH_2SiMe_3)_6$.

(5) $Mo_2(OR)_8$, (R = ^{i}Pr, ^{t}Bu)

$$g = 2 \times 6 + 8 \times 1 = 20, \qquad g_L = 8 \times 2 = 16, \qquad g_M = 20 - 16 = 4.$$

The electron configuration is $\sigma^2 \pi_x^1 \pi_y^1$, indicating a Mo=Mo double bond. The experimental bond distance is 252.3 pm, and the molecule is a paramagnetic species.

9.2.4 Quintuple bonding in a dimetal complex

Recently, structural evidence for the first quintuple bond between two metal atoms was found in the dichromium(I) complex Ar′CrCrAr′ (where Ar′ is the sterically encumbering monovalent 2,6-bis[(2,6-diisopropyl)phenyl]phenyl ligand). This complex exists as air- and moisture-sensitive dark-red crystals that remain stable up to 200 °C. X-ray diffraction revealed a centrosymmetric molecule with a planar *trans*-bent C–Cr–Cr–C backbone with measured structural parameters Cr–Cr 183.51(4) pm, Cr–C 213.1(1) pm and C–Cr–Cr 102.78(3)°. Characterization of the compound is further substantiated by magnetic and spectroscopic data, as well as theoretical computations. Fig. 9.2.6 shows the molecular geometry and structural formula of Ar′CrCrAr′.

In principle, a homodinuclear transition metal species can form up to six bonds using the ns and five $(n-1)d$ valence orbitals. In a simplified bonding description of Ar′CrCrAr′, the planar C–Cr–Cr–C skeleton has idealized molecular symmetry C_{2h} with reference to a conventional z axis lying perpendicular to it. For each chromium atom, a *local z* axis is chosen to be directed toward the other chromium atom, and the *local x* axis to lie in the skeletal plane. Each chromium atom (electronic configuration $3d^5 4s^1$) uses it $4s$ orbital to overlap with a sp^2 hybrid orbital on the *ipso* carbon atom of the terphenyl ligand to form a Cr–C σ bond. That leaves five d orbitals at each Cr(I) center for the formation of a five-fold (i.e., quintuple) metal-metal bond, which has one σ ($d_{z^2} + d_{z^2}$; symmetry species A_g), two π ($d_{yz} + d_{yz}$, $d_{xz} + d_{xz}$; A_u, B_u), and two δ ($d_{x^2-y^2} + d_{x^2-y^2}$, $d_{xy} + d_{xy}$; A_g, B_g) components. The situation is actually more complex as mixing of the chromium $4s$, $3d_{z^2}$, and $3d_{x^2-y^2}$ orbitals (all belonging to A_g) can occur. In an alternative bonding scheme, each chromium atom may be considered to be $d_{z^2}s$ hybridized. The outward-extended $(s - d_{z^2})$ hybrid orbital is used to form a Cr–C σ bond. Note that this type of overlap results in a C–Cr–Cr angle of 90° and allows free rotation of the C–Cr bond about the Cr–Cr axis, and it is the steric repulsion between the pair of bulky Ar′ ligands that accounts for the obtuse C–Cr–Cr bond angle and the *trans*-bent geometry of the central C–Cr–Cr–C core. The pair of chromium $(s + d_{z^2})$ hybrid orbitals aligned along the common local z axis overlap to form the metal-metal σ bond, while the π and δ bonds are formed in the manner described above. In either bonding description, there is a formal bond order of five between the chromium(I) centers.

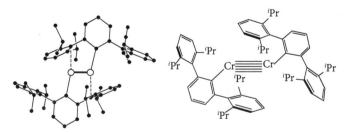

Fig. 9.2.6 Molecular geometry and structural formula of the dinuclear complex Ar′CrCrAr′ (Ar′ = C_6H_3-2,6(C_6H_3-2,6-^{i}Pr$_2$)$_2$.

9.2.5 **Quintuple-bonded homodinuclear chromium complexes with lateral coordination**

The quintuple bond was first observed in Ar'CrCrAr' (**1**, as shown in Fig. 9.2.6). Since then, a number of new dichromium complexes stabilized by quintuple bonding have been synthesized.

Typical examples are those with bidentate *N*-ligands coordinating the metal pair laterally, such as **2** and **3**:

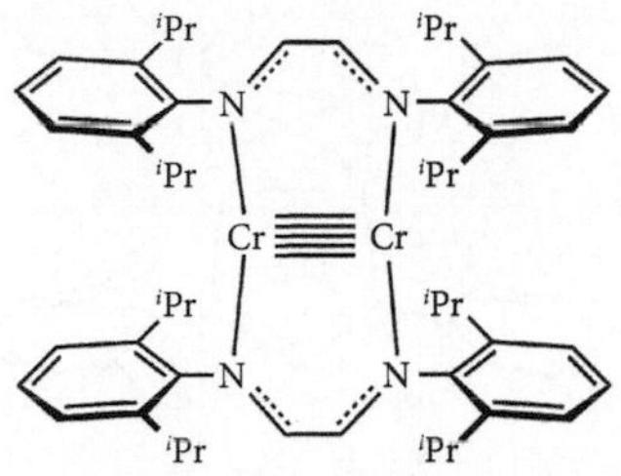

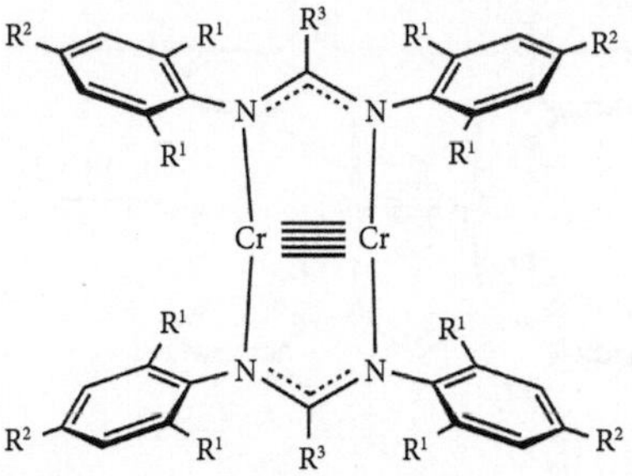

Since the two N atoms are close to each other in the bidentate ligand, it is expected that the Cr–Cr quadruple bond distance is shorter in **2** (~ 180 pm) and **3** (~ 174 pm) than in **1** (181–184 pm). The Cr–Cr bond distance can be further shortened by increasing the bulkiness of the substituents of *N*-bidentate ligands. For example, a very short distance of 170.56(12) pm was found for **4**:

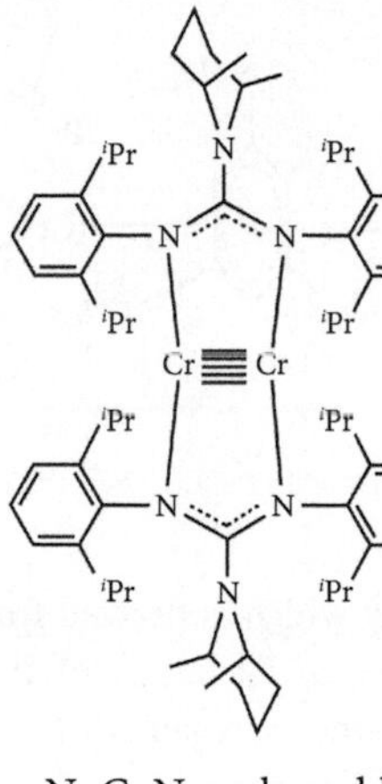

Steric repulsion accounts for a smaller N–C–N angle and hence a shorter N⋯N distance to bring the two Cr atoms closer.

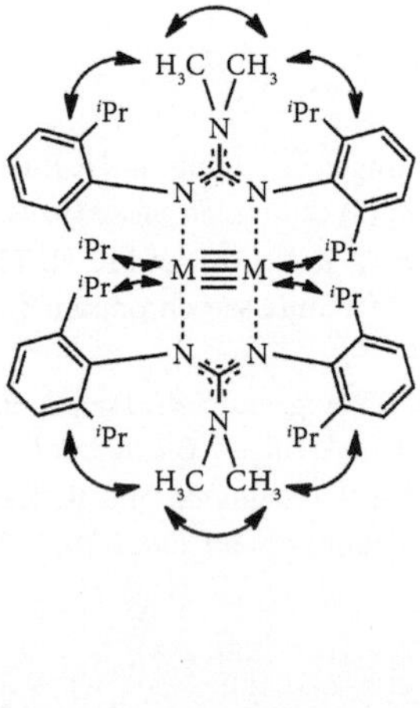

The quintuple Cr–Cr bond undergoes a variety of reactions with simple inorganic and organic molecules. An interesting case is the reaction between the Cr–Cr quintuple bond and an alkyne. When one molar equivalent of $Me_3SiC\equiv CH$ molecule reacts with the Cr–Cr quintuple bond in **5**, a [2+2] cycloaddition product **6** is obtained with the two Cr atoms and the alkyne linking to each other to form a pyramidal moiety.

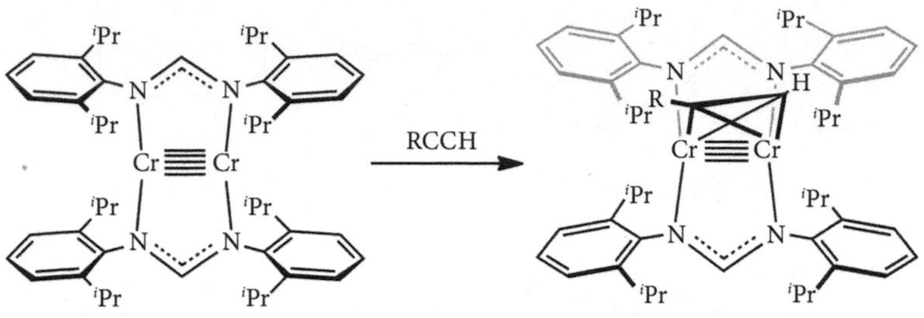

When **6** is allowed to react with two $Me_3SiC\equiv CH$ molecules, a benzene ring is formed between the two Cr atoms in product **7**.

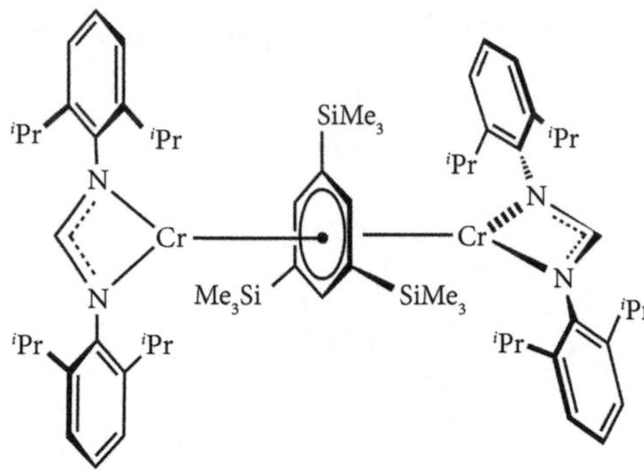

In compound **7** the central benzene ring, which is derived from cyclic-trimerization of Me_3SiCCH, is sandwiched by two Cr-amidinate moieties. Exclusion of the central $1,3,5\text{-}(Me_3Si)_3C_6H_3$ moiety regenerates **5**, and the Cr–Cr quintuple bond is regained. In this sense, **5** acts as a catalyst for the cyclotrimerization of alkyne $Me_3SiC\equiv CH$.

References

T. Nguyen, A. D. Sutton, S. Brynda, J. C. Fettinger, G. J. Long, and P. P. Power, Synthesis of a stable compound with fivefold bonding between two chromium(I) centers. *Science* **310**, 844–7 (2005).

K. A. Kreisel, G. P. A. Yap, P. O. Dmitrenko, C. R. Landis, and K. H. Theopold, The Shortest metal–metal bond yet: molecular and electronic structure of a dinuclear chromium diazadiene complex. *J. Am. Chem. Soc.* **129**, 14162–3 (2007).

C.-W. Hsu, J.-S. K. Yu, C.-H. Yen, G.-H. Lee, Y. Wang, and Y.-C. Tsai, Quintuply-bonded dichromium(I) complexes featuring metal–metal bond lengths of 1.74 Å. *Angew. Chem. Int. Ed.* **47**, 9933–6 (2008).

A. Noor, T. Bauer, T.K. Todorova, B. Weber, L. Gagliardi, and R. Kempe, The ligand-based quintuple bond-shortening concept and some of its limitations. *Chem. Eur. J.* **19**, 9825–32 (2013).

A. K. Nair, N. V. S. Harisomayajula, and Y.-C. Tsai, The lengths of the metal-to-metal quintuple bonds and reactivity thereof. *Inorganica Chim. Acta.* **424**, 51–62 (2015).

Y.-S. Huang, G.-T. Huang, Y.-L. Liu, J.-S. K. Yu, and Y.-C. Tsai, Reversible cleavage/formation of the chromium–chromium quintuple bond in the highly regioselective alkyne cyclotrimerization. *Angew. Chem. Int. Ed.* **56**, 15427–31 (2017).

9.2.6 **Theoretical investigation on sextuple bonding**

Beyond the quintuple bond, the next challenge to chemists is sextuple bonding. To our knowledge, no stable chemical species containing a sextuple bond has been synthesized, except for diatomic Mo_2 in an inert matrix at low temperature. In terms of a conventional bonding scheme, a sextuple bond between two transition-metal atoms may be formed by symmetry matching between pairs of five nd orbitals and one $(n+1)s$ orbital. When the resulting bonding orbitals of $2\sigma_g + 2\pi_u + 2\delta_g$ are fully occupied, a sextuple bond is obtained with a formal bond order of 6. However, a theoretical study by Roos and colleagues showed that the effective bond order (EBO) of the metal-metal bond in Mo_2 and its congeners, Cr_2 and W_2, are 5.2, 3.5, and 5.2, respectively. Hence the metal-metal bond in Mo_2 and W_2 are considered sextuple but not that in Cr_2.

Through theoretical calculations, Chen *et. al.* proposed the use of electron-donating ligands to obtain a complex with a M_2 core held by a sextuple bond. Complexation changes the energy, and possibly the order too, of the orbitals contributing to the M–M bond. In their illustration based on $C_6H_6(W_2)C_6H_6$, the ground state configuration of the W–W bond changes from $(\pi_{nd})^4(\sigma_{nd})^2(\sigma_{(n+1)s})^2(\delta_{nd})^4$ to $(\sigma_{nd})^2(\delta_{nd})^4(\pi_{nd})^4(\delta_{nd}^*)^2$ upon complexation, so that the formal bond order changes from 6 to 4. In contrast, the order of the MOs of the W–W bond remains unchanged when electron-donating 12-crown-4 ether (**8**) is used as a ligand. Their results also showed that the complexation of Mo_2, W_2, Re_2^{2+} and Tc_2^{2+} with 1,7-diaza-12-crown-4 ether (**9**) and 1,4,7,10-tetraza-12-crown-4 ether (**10**) may increase the M–M bond significantly (by as much as 0.88 for the complex between Re_2^{2+} and **10**).

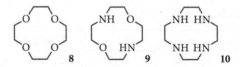

On the other hand, Joy and Jemmis investigated the possibility of shortening the metal–metal bond in Cr_2 and Mo_2 by halogen bonding. In their theoretical results, the metal–metal bonds are shortened by 1.5 pm and 1.7 pm, respectively, through the coordination between a metal atom and the iodine atom of CF_3I. The rationale is that the $\sigma_{(n+1)s}$ orbital in Cr_2 and Mo_2 is repulsive and the destabilizing nature is reduced when the electrons in this orbital are extracted into the strong σ-hole on the iodine atom.

References

B. O. Roos, A. C. Borin, and L. Gagliardi, Reaching the maximum multiplicity of the covalent chemical bond. *Angew. Chem. Int. Ed.* **46**, 1469–72 (2007).

Y. Chen, J.-y. Hasegawa, K. Yamaguchi, and S. Sakaki, A coordination strategy to realize a sextuply-bonded complex. *Phys. Chem. Chem. Phys.* **19**, 14947–54 (2017).

J. Joy and E. D. Jemmis, A halogen bond route to shorten the ultrashort sextuple bonds in Cr_2 and Mo_2. *Chem. Commun.* **53**, 8168–71 (2017).

9.3 Clusters with Three or Four Transition Metal Atoms

9.3.1 Trinuclear clusters

Table 9.3.1 lists the structural data and bond valences of some trinuclear clusters. In $Os_3(CO)_9(\mu_3\text{-}S)_2$, the Os_3 unit is in a bent configuration with two Os–Os bonds of average length 281.3 pm, and the other Os···Os distance (366.2 pm) is significantly longer. The cluster $(CO)_5Mn\text{-}Fe(CO)_4\text{-}Mn(CO)_5$ adopts a linear configuration. The other clusters are all triangular. The Fe_3 skeleton of $Fe_3(CO)_{12}$ has 48 valence electrons, and the Fe_3 unit contains three Fe–Fe single bonds. The Os_3 skeleton of $Os_3H_2(CO)_{10}$, a 46-electron triangular cluster, has one Os=Os double bond and two Os–Os single bonds. The length of the Os=Os double bond is 268.0 pm, and the Os–Os single bonds are 281.8 pm and 281.2 pm.

The remaining three clusters $[Mo_3(\mu_3\text{-}S)_2(\mu_2\text{-}Cl)_3Cl_6]^{3-}$, $[Mo_3(\mu_3\text{-}O)(\mu_2\text{-}O)_3F_9]^{5-}$, and $Re_3(\mu_2\text{-}Cl)_3(CH_2SiMe_3)_6$ all have nearly equilateral M_3 skeletons. According to the calculated bond valences of 5, 6, and 9 for the metal-metal bonds in these compounds, the bond types are Mo---Mo (bond order 1⅔), Mo=Mo, and Re≡Re, respectively.

9.3.2 Tetranuclear clusters

Table 9.3.2 lists the bond valence and structural data of some tetranuclear clusters.

The bond valence of $Re_4(\mu_3\text{-}H)_4(CO)_{12}$ is 8, and the tetrahedral Re_4 skeleton can be described in two ways: (1) resonance between valence-bond structures, leading to a formal bond order of 1⅓, and (2) four 3c-2e ReReRe bonds. Since there are already four μ_3-H capping the faces, description (2) is not as good as (1).

Table 9.3.1 Some trinuclear clusters

Cluster	g	b	M–M/pm	Figure
$Os_3(CO)_9(\mu_3\text{-}S)_2$	50	2	Os–Os, 281.3	(a)
$Mn_2Fe(CO)_{14}$	50	2	Mn–Fe, 281.5	(b)
$Fe_3(CO)_{12}$	48	3	Fe–Fe, 281.5	(c)
$Os_3H_2(CO)_{10}$	46	4	two Os–Os, 281.5	(d)
			Os=Os, 268.0	
$[Mo_3(\mu_3\text{-}S)_2(\mu_2\text{-}Cl)_3Cl_6]^{3-}$	44	5	Mo--- Mo, 261.7*	(e)
$[Mo_3(\mu_3\text{-}O)(\mu_2\text{-}O)_3F_9]^{5-}$	42	6	Mo=Mo, 250.2	(f)
$Re_3(\mu_2\text{-}Cl)_3(CH_2SiMe_3)_6$	36	9	Re≡Re, 238.7	(g)

* Bond order of 1⅔.

Table 9.3.2 Some tetranuclear clusters

Cluster	g	b	M–M/pm	Figure
$Re_4(\mu_3\text{-}H)_4(CO)_{12}$	56	8	6Re≡≡Re, 291*	(a)
$Ir_4(CO)_{12}$	60	6	6 Ir–Ir, 268	(b)
$Re_4(CO)_{16}^{2-}$	62	5	5 Re–Re, 299	(c)
$Fe_4(CO)_{13}C$	62	5	5 Fe–Fe, 263	(d)
$Co_4(CO)_{10}(\mu_4\text{-}S)_2$	64	4	4 Co–Co, 254	(e)
$Re_4H_4(CO)_{15}^{2-}$	64	4	4 Re–Re, 302	(f)
$Co_4(\mu_4\text{-}Te)_2(CO)_{11}$	66	3	3 Co–Co, 262	(g)
$Co_4(CO)_4(\mu\text{-}SEt)_8$	68	2	2 Co–Co, 250	(h)

(a) (b) (c) (d) (e) (f) (g) (h)

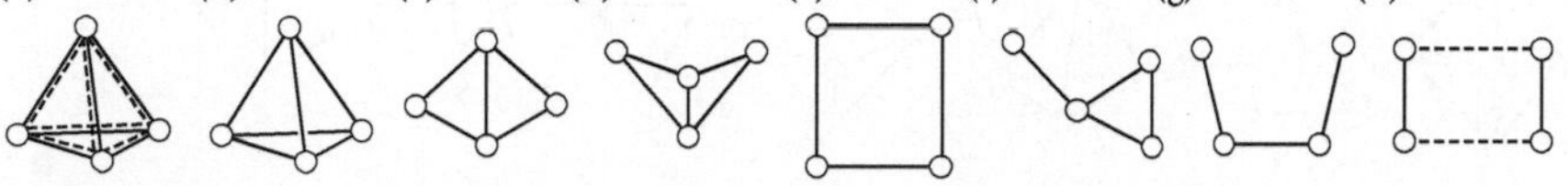

* Bond order of 1⅓.

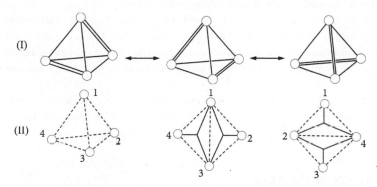

(I)

(II)

In other examples listed in Table 9.3.2, the calculated bond valence *b* is just equal to the number of edges of the corresponding metal skeleton, signifying 2c-2e M–M single bonds.

9.4 **Clusters with More Than Four Transition Metal Atoms**

9.4.1 **Pentanuclear clusters**

Selected examples of pentanuclear metal clusters are listed in Table 9.4.1. In these examples, the calculated bond valence *b* is exactly the same as the number of edges of the metal skeleton, indicating 2c-2e M–M bonds. In general, electron-rich species have lower bond valence and more open structures than the electron-deficient ones.

The structures and skeletal bond valences of $Os_5(CO)_{16}$ and $B_5H_5^{2-}$ are similar as a pair, as are also $Fe_5C(CO)_{15}$ and B_5H_9. But the bonding types in the boranes and the metal clusters are not the same. Since every B atom in a polyhedral borane has three AOs for bonding of the B_n skeleton, any vertex that

Table 9.4.1 Some pentanuclear clusters

Cluster	g	b	No. of edges	Figure
$Os_5(CO)_{16}$	72	9	9	(a)
$Fe_5C(CO)_{15}$	74	8	8	(b)
$Os_5H_2(CO)_{16}$	74	8	8	(c)
$Ru_5C(CO)_{15}H_2$	76	7	7	(d)
$Os_5(CO)_{18}$	76	7	7	(e)
$Os_5(CO)_{19}$	78	6	6	(f)
$Re_2Os_3H_2(CO)_{20}$	80	5	5	(g)

(a)　(b)　(c)　(d)　(e)　(f)　(g)

is more than three-connected must involve multi-center bonds. In the transition metal skeleton, the M_n atoms form either 2c-2e single bonds or 3c-2e multi-center bonds.

Some clusters have 76 valence electrons based on a trigonal bipyramidal skeleton, such as $[Ni_5(CO)_{12}]^{2-}$, $[Ni_3Mo_2(CO)_{16}]^{2-}$, $Co_5(CO)_{11}(PMe_2)_3$, and $[FeRh_4(CO)_{15}]^{2-}$, which are not shown in Table 9.4.1. The additional four valence electrons compared to $Os_5(CO)_{16}$ have a significant effect on its geometry, and the bond lengths to the apical metal atoms are increased, and the bond valence b is decreased.

The cluster $[Os_5C(CO)_{14}(O_2CMe)I]$ has 78 valence electrons; it is not shown in Table 9.4.1, and its bond valence is equal to 6: $[b = \frac{1}{2}(5 \times 18 - 78) = 6]$. This cluster has a deformed trigonal bipyramidal geometry with no bonding in the equatorial plane of the bipyramid.

9.4.2 **Hexanuclear clusters**

Table 9.4.2 lists the bond valence and structural data of some selected examples of hexanuclear clusters. The first three clusters have different b values, yet they are all octahedral with 12 edges. There are three stable types of bonding schemes for an octahedron, as shown in Fig. 9.1.2: (a) $Mo_6Cl_{14}^{2-}$, $g = 84$, $b = 12$; in this cluster there are 12 2c-2e bonds at the 12 edges. (b) $Nb_6Cl_{18}^{4-}$, $g = 76$, $b = 16$; in this cluster there are eight 3c-2e bonds on the eight faces of an octahedron. (c) $Rh_6(CO)_{16}$, $g = 86$, $b = 11$; in this cluster there are four 3c-2e bonds on four faces and three 2c-2e bonds at three edges, as in the case of $B_6H_6^{2-}$.

Other clusters listed in Table 9.4.2 have the property that their b value equals the number of edges of their M_n skeletons.

9.4.3 **Clusters with seven or more transition metal atoms**

Table 9.4.3 listed the bond valence and structural data of some selected examples of high-nuclearity clusters each consisting of seven or more transition metal atoms. The skeletal structures of these clusters are shown in Fig. 9.4.1.

When the number of metal atoms in a cluster increases, the geometries of the clusters become more complex, and some are often structurally better described in terms of capped or decapped polyhedra and

Table 9.4.2 Some hexanuclear clusters

Cluster	g	b	No. of edges	Figure
$Mo_6(\mu_3\text{-Cl})_8Cl_6{}^{2-}$	84	12	12	(a)
$Nb_6(\mu_2\text{-Cl})_{12}Cl_6{}^{4-}$	76	16	12	(a)
$Rh_6(CO)_{16}$	86	11	12	(a)
$Os_6(CO)_{18}$	84	12	12	(b)
$Os_6(CO)_{18}H_2$	86	11	11	(c)
$Os_6C(CO)_{16}(MeC\equiv CMe)$	88	10	10	(d)
$Ru_6C(CO)_{15}{}^{2-}$	90	9	9	(e)
$Os_6(CO)_{20}[P(OMe)_3]$	90	9	9	(f)
$Co_6(\mu_2\text{-}C_2)(\mu_4\text{-S})(CO)_{14}$	92	8	8	(g)

(a) (b) (c) (d) (e) (f) (g)

Table 9.4.3 Clusters with more than six transition metal atoms

Cluster	g	b	Structure (Fig. 9.4.1)	Remark
$Os_7(CO)_{21}$	98	14	(a)	capped octahedron
$[Os_8(CO)_{22}]^{2-}$	110	17	(b)	para-bicapped octahedron
$[Rh_9P(CO)_{21}]^{2-}$	130	16	(c)	capped square antiprism; iso-bond valence with B_9H_{13}
$[Rh_{10}P(CO)_{22}]^-$	142	19	(d)	bicapped square antiprism; iso-bond valence with $B_{10}H_{10}{}^{2-}$
$[Rh_{11}(CO)_{23}]^{3-}$	148	25	(e)	three face-sharing octahedra
$[Rh_{12}Sb(CO)_{27}]^{3-}$	170	23	(f)	icosahedron; iso-bond valence with $B_{12}H_{12}{}^{2-}$

condensed polyhedra. For example, the first and second clusters listed in Table 9.4.3 are a capped octahedron and a bicapped octahedron, respectively. Consequently, capping or decapping with a transition metal fragment to a deltapolyhedral cluster leads to an increase or decrease in the cluster valence electron count of 12. When a transition metal atom caps a triangular face of the cluster, it forms three M–M bonds with the vertex atoms, so according to the 18-electron rule, the cluster needs an additional $18 - 6 = 12$ electrons. The parent octahedron of $[Os_6(CO)_{18}]^{2-}$ has $g = 86$, the monocapped octahedron $Os_7(CO)_{21}$ has $g = 98$, and the bicapped octahedron $[Os_8(CO)_{22}]^{2-}$ has $g = 110$.

The metal cluster of $[Rh_{10}P(CO)_{22}]^-$ forms a deltapolyhedron, which has $g = 142$, as shown in Fig. 9.4.1(d). The skeleton of $[Rh_9P(CO)_{21}]^{2-}$ is obtained by removal of a vertex transition metal fragment. The skeletal valence electron count of $[Rh_9P(CO)_{21}]^{2-}$ gives $g = 142 - 12 = 130$.

The metal cluster of $[Rh_{11}(CO)_{23}]^{3-}$ is composed of three face-sharing octahedra, as shown in Fig. 9.4.1(e). The metal cluster of $[Rh_{12}Sb(CO)_{27}]^{3-}$ consists of an icosahedron with an encapsulated Sb atom at its center.

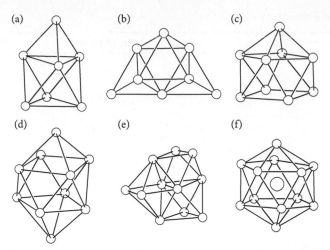

Fig. 9.4.1 Structures of some transition metal clusters: (a) $Os_7(CO)_{21}$, (b) $[Os_8(CO)_{22}]^{2-}$, (c) $[Rh_9P(CO)_{21}]^{2-}$, (d) $[Rh_{10}P(CO)_{22}]^-$, (e) $[Rh_{11}(CO)_{23}]^{3-}$, (f) $[Rh_{12}Sb(CO)_{27}]^{3-}$.

Generally, capped or decapped deltapolyhedral clusters are characterized by the number of skeletal valence electrons g:

$$g = (14n + 2) \pm 12m$$

where n is the number of M atoms in the parent deltapolyhedron and m is the number of capped (+) or decapped (–) metal fragments.

A classical example of correlation of structure with valence electron count of transition metal clusters is shown in Fig. 9.4.2. There the structures of a series of osmium clusters are systematized by applying the capping and decapping procedures.

9.4.4 **Anionic carbonyl clusters with interstitial main-group atoms**

There is much interest in transition metal carbonyl clusters containing interstitial (or semi-interstitial) atoms in view of the fact that insertion of the encapsulated atom inside the metallic cage increases the number of valence electrons but leaves the molecular geometry essentially unperturbed. The clusters are generally anionic, and the most common interstitial heteroatoms are carbon, nitrogen, and phosphorus. Some representative examples are displayed in Fig. 9.4.3.

The core of the anionic carbonyl cluster $[Co_6Ni_2(C)_2(CO)_{16}]^{2-}$ consists of two trigonal prisms sharing a rectangular face (Fig. 9.4.3(a)). All four vertical edges and two horizontal edges, one on the top face and the other on the bottom face, are each bridged by a carbonyl group. The two Co* atoms each has two terminal carbonyl groups, and the Ni and Co atoms each has one.

The $[Os_{18}Hg_3(C)_2(CO)_{42}]^{2-}$ cluster is composed of two tri-capped octahedral $Os_9(C)(CO)_{21}$ units sandwiching a Hg_3 triangle (Fig. 9.4.3(b)). Each corner Os atom in the top and bottom faces has three terminal carbonyl groups, and the remaining Os atoms each has two.

The core of the $[Fe_6Ni_6(N)_2(CO)_{24}]^{2-}$ cluster comprises a central Ni_6 octahedron that shares a pair of opposite faces with two Ni_3Fe_3 octahedra, as shown in Fig. 9.4.3(c). The interstitial N atoms occupy the centers of the Ni_3Fe_3 octahedra. Each Fe atom has two terminal carbonyl groups, and each Ni atom has one.

The metallic core of the $[Rh_{28}(N)_4(CO)_{41}H_x]^{4-}$ cluster is composed of three layers of Rh atoms in a close-packed ABC sequence, as shown in Fig. 9.4.3(d). Four N and an unknown number of H atoms occupy the octahedral holes.

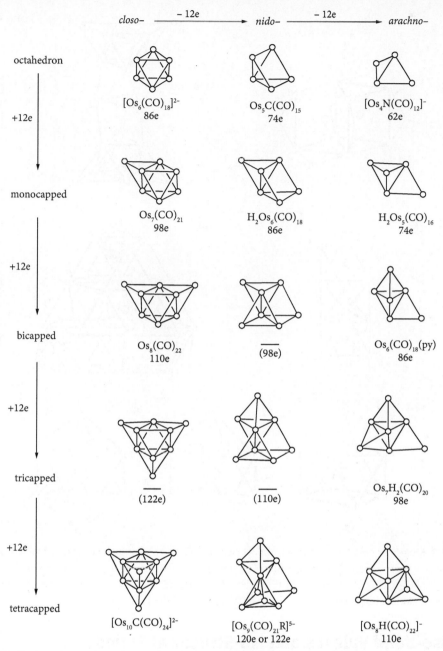

Fig. 9.4.2 The structures of osmium carbonyl compounds vary with an increase or decrease of the valence electrons.

The metal atoms in $[Ru_8(P)(CO)_{22}]^-$ constitute a square antiprismatic assembly (Fig. 9.4.3(e)). Two opposite slant edges are each bridged by a carbonyl group. Each Ru bridged atoms has two terminal carbonyl groups, and the remaining four each has three. Comparison of this cluster core with those of $[Rh_9(P)(CO)_{21}]^{2-}$ (Fig. 9.4.3(f)) and $[Rh_{10}(S)(CO)_{22}]^{2-}$ (Fig. 9.4.3(g)) shows that the latter two are derived from successive capping of the rectangular faces of the square antiprism.

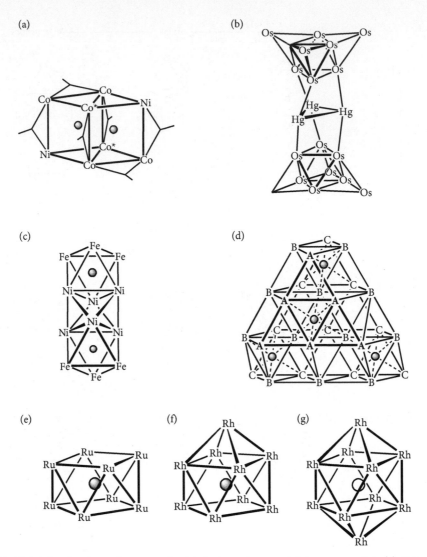

Fig. 9.4.3 Molecular structure of some carbonyl cluster anions containing encapsulated heteroatoms; all terminal CO groups are omitted for clarity. (a) $[Co_6Ni_2(C)_2(CO)_{16}]^{2-}$; only the bridging CO groups are shown. (b) $[Os_{18}Hg_3(C)_2(CO)_{42}]^{2-}$. (c) $[Fe_6Ni_6(N)_2(CO)_{24}]^{2-}$. (d) $[Rh_{28}(N)_4(CO)_{41}H_x]^{4-}$; to avoid clutter, not all Rh–Ph bonds are included. (e) $[Ru_8(P)(CO)_{22}]^-$. (f) $[Rh_9(P)(CO)_{21}]^{2-}$. (g) $[Rh_{10}(S)(CO)_{22}]^{2-}$.

9.5 **Iso-Bond Valence and Iso-Structural Series**

For a cluster consisting of n_1 transition metal atoms and n_2 main-group atoms, the bond valence b is evaluated as follows:

$$b = \tfrac{1}{2}(18n_1 + 8n_2 - g),$$

where g is the number of valence electrons of the skeleton formed by the n_1 transition metals and n_2 main-group atoms.

When a BH group of the octahedral cluster $(BH)_6^{2-}$ is replaced by a CH^+ group, both g and b retain their values and the structure of the cluster anion $(BH)_5CH^-$ remains octahedral. On the other hand, when a BH group of $(BH)_6^{2-}$ is replaced by a $Ru(CO)_3$ group, the b value still remains the same. But g increases its value by 10, as a BH group contributes four electrons to the skeleton, while a $Ru(CO)_3$ group contributes 14 (eight from Ru and two from each CO). Therefore, replacement of one or more BH groups in $(BH)_6^{2-}$ by either CH^+ or $Ru(CO)_3$ groups results in a series of iso-bond valence and iso-structural clusters. The structures of some members of this series, $(BH)_6^{2-}$, $(BH)_4(CH)_2$, $[Ru(CO)_3]_4(CH)_2$, and $[Ru(CO)_3]_6^{2-}$ are shown in Fig. 9.5.1. Similar substitutions by either transition-metal or representative-element groups give rise to a variety of cluster compounds, and five iso-structural series of clusters containing both transition metal and main-group element components are displayed in Fig. 9.5.2.

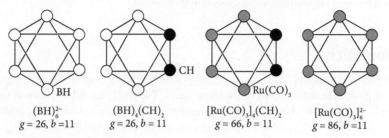

$(BH)_6^{2-}$
$g = 26, b = 11$

$(BH)_4(CH)_2$
$g = 26, b = 11$

$[Ru(CO)_3]_4(CH)_2$
$g = 66, b = 11$

$[Ru(CO)_3]_6^{2-}$
$g = 86, b = 11$

Fig. 9.5.1 Iso-bond valence and iso-structural series of $B_6H_6^{2-}$.

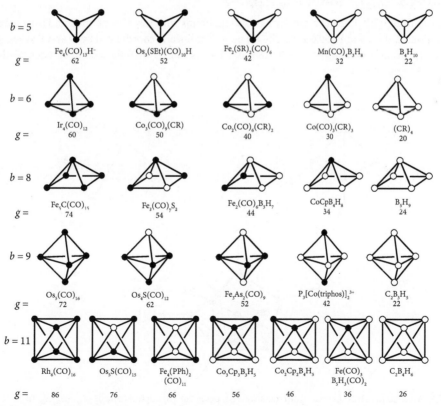

Fig. 9.5.2 Iso-structural series of transition metal and main-group clusters.

Clearly the structure of a given cluster depends on electronic and geometric factors, among others. Hence the structure of a compound cannot be predicted until it has been determined experimentally. Still, based on the bond valence and structural principle illustrated above, an educated guess on the structure of a cluster becomes feasible. In addition, the bond valence concept provides a useful link between apparently dissimilar clusters such as $(BH)_6^{2-}$ and $[Ru(CO)_3]_4(CH)_2$.

9.6 Selected Topics in Metal-Metal Interactions

Since the 1980s, studies on metal clusters and metal string complexes have revealed unusual interactions between metal atoms, some of which are discussed in this section.

9.6.1 Aurophilicity

The term aurophilicity (or aurophilic attraction) refers to the formally non-bonding but attractive interaction between gold(I) atoms in gold cluster compounds. The Au(I) atom has a closed shell electron configuration: $[Xe] 4f^{14} 5d^{10} 6s^0$. Normally, repulsion exists between the non-bonding homo-atoms. However, there is extensive crystallographic evidence of attractions between gold(I) cations. Fig. 9.6.1 shows the structures of three Au(I) compounds.

(1) $O[AuP(o\text{-}tol)_3]_3^+$ (o-tol = $C_6H_4Me\text{-}2$)

In $O[AuP(o\text{-}tol)_3]_3(BF_4)$, the O atom forms covalent bonds with three Au atoms in a OAu_3 pyramidal configuration, as shown in Fig. 9.6.1(a). The Au atoms are linearly coordinated by O and P atoms. In this structure, the mean Au⋯Au distance is 308.6 pm, which is shorter than the sum of van der Waals radii, $2 \times 166 = 332$ pm.

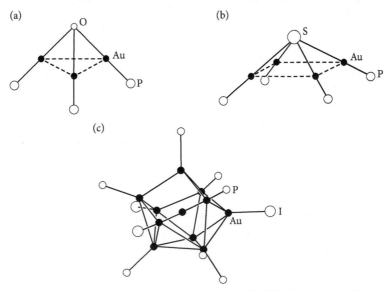

Fig. 9.6.1 Molecular structure of gold cluster compounds: (a) $O[AuP(o\text{-}tol)_3]_3^+$, (b) $S[AuP(o\text{-}tol)_3]_4^{2+}$, (c) $Au_{11}I_3[P(p\text{-}C_6FH_4)_3]_7$.

(2) S[AuP(*o*-tol)₃]₄²⁺

In S[AuP(*o*-tol)₃]₄(ClO₄)₂, the SAu₄ unit takes a square-pyramidal configuration, as shown in Fig. 9.6.1(b). The Au atoms are near linearly coordinated by S and P atoms. The Au···Au distances lie in the range of 288.3 pm to 293.8 pm, and the mean distance is 293.0 pm. The distance of the S atom to the center of the Au₄ basal plane is 130 pm.

(3) Au₁₁I₃[P(*p*-C₆FH₄)₃]₇

The structure of the molecule is shown in Fig. 9.6.1(c). In the Au₁₁ cluster, the central Au atom is surrounded by 10 Au atoms, which form an incomplete icosahedron (lacking two vertices) with each vertex carrying one terminal iodo ligand or P(*p*-C₆FH₄)₃ group. The mean Au···Au distance from the central atom to the surrounding atoms is 268 pm, and the mean distance between the ten Au vertices is 298 pm.

In other gold(I) cluster compounds, such as tetrahedral $[(AuL)_4(\mu_4\text{-N})]$ and $[(AuL)_4(\mu_4\text{-O})]^{2+}$, trigonal-bipyramidal $[(AuL)_5(\mu_5\text{-C})]^+$, $[(AuL)_5(\mu_5\text{-N})]^{2+}$, and $[(AuL)_5(\mu_5\text{-P})]^{2+}$, and octahedral $[(AuL)_6(\mu_6\text{-C})]^{2+}$ and $[(AuL)_6(\mu_6\text{-N})]^{3+}$ (L = PPh₃ or PR₃), the Au···Au distances lie in the range 270 pm to 330 pm. These data substantiate that aurophilicity is a common phenomenon among gold cluster complexes.

Aurophilicity presumably arises from relativistic modification of the gold valence AOs energies, which brings the 5*d* and 6*s* orbitals into close proximity in the energy-level diagram. In more recent theoretical studies, the effect is primarily attributed to electron correlation, which takes precedence over 6*s*/5*d* hybridization. To date, the origin of the aurophilicity has not yet been unambiguously established.

Making use of the concept of aurophilicity, simple gold compounds can be combined to yield complicated oligomeric aggregates in designed synthesis. Fig. 9.6.2 shows the structures of three oligomeric molecules.

9.6.2 **Argentophilicity and mixed metal complexes**

By analogy to aurophilicity, argentophilicity has been demonstrated to exist in silver cluster complexes. In the crystal structures of a variety of silver(I) double and multiple salts containing a fully encapsulated acetylide dianion C₂²⁻ (IUPAC name acetylenediide) in different polyhedral silver cages (see Fig. 14.3.11), there exist many Ag···Ag contacts that are shorter than twice the van der Waals radii of silver (2 × 170 pm = 340 pm). Further details are given in Chapter 10.

Taking advantage of both aurophilicity and argentophilicity, tetranuclear mixed-metal complexes which contain pairs of Au and Ag atoms have been prepared, as shown in Fig. 9.6.3.

In the preparation of mixed gold/silver polynuclear complexes, aurophilicity and argentophilicity have been utilized to promote cluster formation. Fig. 9.6.4 shows the cores of several mixed gold-silver clusters.

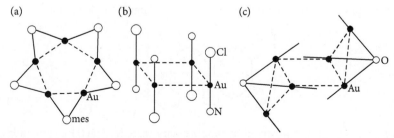

Fig. 9.6.2 Structure of some gold oligomeric aggregate molecules: (a) Au₅(mes)₅ (mes = C₆H₃Me₃-2,4,6); (b) [LAuCl]₄, L = N(CH₂)₄CH₂; (c) [O(AuPPh₃)₃]₂.

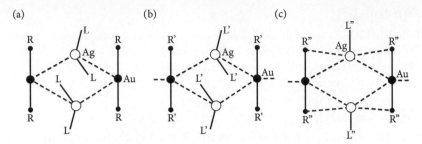

Fig. 9.6.3 Some mixed Au and Ag metal complexes (filled circle, Au; open circle, Ag): (a) $[Au(CH_2PPh_2)_2]_2[Ag(OClO_3)_2]_2$, (b) $[Au(C_6F_5)]_2[Ag(C_6H_6)]_2$, and (c) $[Au(C_6F_5)]_2[Ag(COMe_2)]_2$.

The $[Au_{13}Ag_{12}]$ molecular skeletons (a) and (b) of $[(Ph_3P)_{10}Au_{13}Ag_{12}Br_8]SbF_6$ and $[(p\text{-}tol_3P)_{10}Au_{13}Ag_{12}Br_8]Br$, respectively, can each be considered as two centered icosahedra sharing a common vertex. In an alternative description, the 25 metal atoms constitute three fused icosahedra, with a common pentagon shared between an adjacent pair of fused icosahedra. Structure (a) adopts the **ses** (staggered-eclipsed-staggered) configuration, and the sequence of relative positioning of atoms is ABBA. Structure (b) adopts the **sss** (staggered-staggered-staggered) configuration, and the sequence is ABAB. The structure of (c) consists of the three centered icosahedra, each of which uses one edge to form a central triangle, with an additional Ag atom lying above and below it. The structure of (d) consists of six Au atoms that form a planar six-membered ring, with one Ag atom located at the center; each edge of the Au_6 hexagon is bridged by a bridging C atom, with three C atoms lying above the plane and three below it.

9.6.3 **Mercurophilicity**

Metallophilicity refers to the weak interaction between metal atoms with closed shell electronic configurations, such as Au(I), Ag(I), and Hg(II). Metallophilicity for these Group 11 and 12 species (termed "aurophilicity," "argentophilicity," and "mercurophilicity," respectively) have been reviewed. Aurophilicity and argentophilicity are introduced in Sections 9.6.1 and 9.6.2, respectively. In comparison, mercurophilicity is much less investigated, and some examples are presented in this section.

Metallophilicity can be briefly classified as "unsupported" and "supported." Unsupported metallophilicity refers to the interaction between molecules, and the few examples of Hg(II) compounds include the relatively simple $Hg(CH_2Ph)_2$ and a trinuclear metallacyclic complex cation with Cu^+ at its center (**11**). The intermolecular Hg···Hg distances in these two compounds are 354 pm and 320.3(4) pm, respectively. An even shorter distance of 314.63(6) pm was observed in $Hg(SiMe_3)_2$. The mercurophilic interaction in $(HgX_2)_2$ with different Xs (e.g., X = H, halogen and $SiMe_3$) has been investigated theoretically, and the value ranges from about 2 kJ mol^{-1} to 150 kJ mol^{-1}.

$$\left[\begin{array}{c} \text{Hg-----Cu-------Hg} \end{array}\right]^+ \quad \textbf{11}$$

Supported mercurophilicity is observed in simple binary mercury Hg(II) compounds such as the crystalline form of HgO (**12**), which features zigzag chains each composed of almost linear O–Hg–O segments. This feature brings adjacent Hg atoms close enough (330 ± 2 pm) to each other for

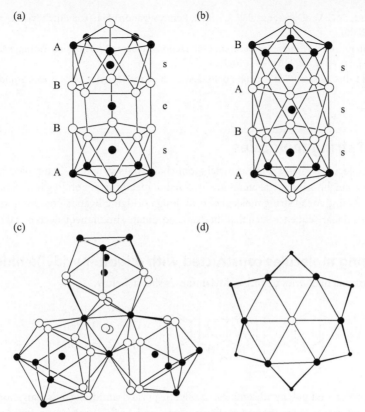

Fig. 9.6.4 Skeletal structure of some mixed Au/Ag clusters (filled circle, Au; open circle, Ag): (a) $[Au_{13}Ag_{12}]$ in $[(Ph_3P)_{10}Au_{13}Ag_{12}Br_8]SbF_6$, (b) $[Au_{13}Ag_{12}]$ in $[(p\text{-tol}_3P)_{10}Au_{13}Ag_{12}Br_8]Br$, (c) $[Au_{18}Ag_{20}]$ in $[(p\text{-tol}_3P)_{12}Au_{18}Ag_{20}Cl_{14}]$, and (d) $[AgAu_6C_6]$ in $[Ag(AuC_6H_2(CHMe_2)_3)_6CF_3SO_3$.

mercurophilic interaction. Examples of supported mercurophilicity in other types of compounds, for example, metallo-complexes and mercuroborands, can be found in References.

330 pm

12

References

H. Schmidbaur and A. Schier, Aurophilic interactions as a subject of current research: an up-date. *Chem. Soc. Rev.* **41**, 370–412 (2012).

H. Schmidbaur and A. Schier, Argentophilic interactions. *Angew. Chem. Int. Ed.* **54**, 746–84 (2015).

H. Schmidbaur and A. Schier, Mercurophilic interactions. *Organometallics.* **34**, 2048–66 (2015).

S. J. Faville, W. Henderson, T. J. Mathieson, and B. K. Nicholson, Solid-state aggregation of mercury bis-acetylides, $Hg(C{\equiv}CR)_2$, R=Ph, SiMe₃. *J. Organomet. Chem.* **580**, 363–9 (1999).

U. Patel, H. B. Singh, and G. Wolmershäuser, Synthesis of a metallophilic metallamacrocycle: A $Hg^{II}{\cdots}Cu^{I}{\cdots}Hg^{II}{\cdots}Hg^{II}{\cdots}Cu^{I}{\cdots}Hg^{II}$ interaction. *Angew. Chem. Int. Ed.* **44**, 1715–7 (2005).

N. L. Pickett, O. Just, D. G. VanDerveer, and W. S. Rees Jr, Reinvestigation of bis(trimethylsilyl)mercury. *Acta Cryst.* **C56**, 412–3 (2000).

J. Echeverría, J. Cirera, and S. Alvarez, Mercurophilic interactions: a theoretical study on the importance of ligands. *Phys. Chem. Chem. Phys.* **19**, 11645–54 (2017).

K. Aurivillius and I.-B. Carlsson, The structure of hexagonal mercury(II) oxide. *Acta Chem. Scand.* **12**, 1297–304 (1958).

9.6.4 **Metal string molecules**

A metal string molecule contains a linear metal-atom chain in its structure. In a molecule of this type, all or a part of the neighboring metal atoms are involved in metal-metal bonding interactions. A general strategy of synthesizing metal string molecules is to design bridging ligands possessing multiple donor sites arranged in a linear sequence such that they can coordinate simultaneously to metal centers.

(1) **Metal string molecules constructed with oligo(α-pyridyl)amido ligands**

The common formula of a series of polypyridylamines is shown below:

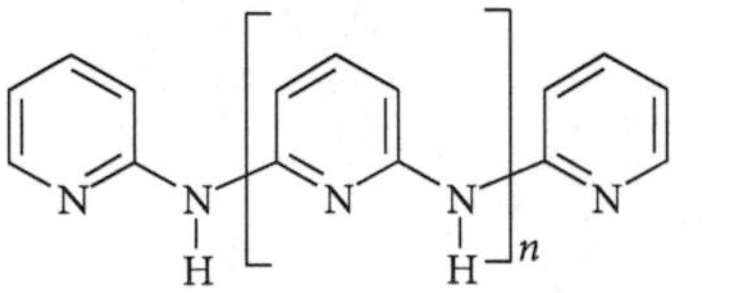

$n = 0$, Hdpa
$n = 1$, H_2tdpa
$n = 2$, H_3teptra
$n = 3$, H_4peptea

Four fully deprotonated polypyridylamines, or oligo(α-pyridyl)amido ligands, can coordinate simultaneously to metal atoms from the upper, lower, front, and back directions to form a metal string molecule:

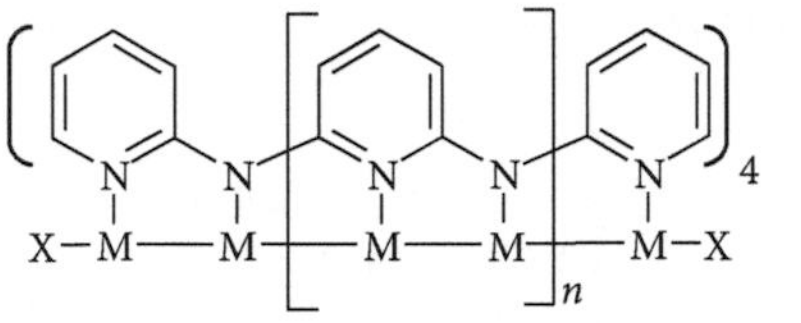

n	M (II)
0	Cr, Ru, Co, Rh, Ni, Cu
1	Cr, Co, Ni
2	Cr, Ni
3	Cr, Ni

As an oligo(α-pyridyl)amido ligand has an odd number of donor sites, the corresponding metal string molecule has the same number of metal atoms.

From the reactions of Ni(II) salts with polypyridylamines, a series of metal string molecules, in which the number of nickel atom varies from 3 (i.e., $n = 0$) to 9 (i.e., $n = 3$), have been prepared. The structures of these molecules are very similar. Fig. 9.6.5 shows the structure of $[Ni_9(\mu_9\text{-peptea})_4Cl_2]$, ($H_4$peptea = pentapyridyltetramine). The nine Ni atoms are in a straight line, and the distances between atoms are approximately equal. Every N atom in the four (α-pyridyl)amido ligands coordinates to one Ni atom. Each Ni atom is coordinated by four N atoms to form a square that is oriented perpendicular to the metal string. Due to steric repulsion between H atoms of the neighboring pyridine rings, the (α-pyridyl)amido ligands are helically distributed around the string axis, as shown in Fig. 9.6.5.

The Ni–Ni and Ni–N distances in a family of related metal string molecules, in which the numbers of Ni atoms are 3, 5, 7, and 9, are shown in Fig. 9.6.6. The Ni–Ni distances increase from the center toward each terminal, and the Ni–N distances are nearly equal except for the outermost ones. These effects are attributable to the fact that, unlike the inner metal atoms, each terminal metal atom has square-pyramidal coordination.

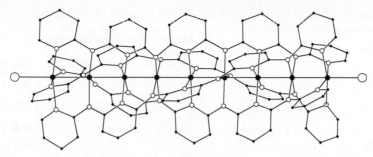

Fig. 9.6.5 Structure of [Ni$_9$(μ_9-peptea)$_4$Cl$_2$] (filled circle, Ni; large open circle, Cl; small open circle, N; and small filled circle, C).

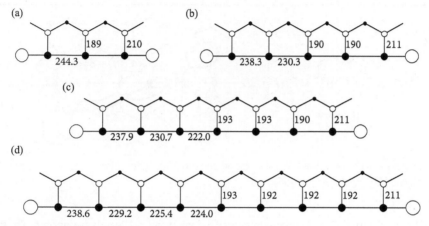

Fig. 9.6.6 Bond lengths in metal string molecules (in pm): (a) Ni$_3$(μ_3-dpa)$_4$Cl$_2$, (b) Ni$_5$(μ_5-tpda)$_4$Cl$_2$, (c) Ni$_7$(μ_7-teptra)$_4$Cl$_2$, (d) Ni$_9$(μ_9-peptea)$_4$Cl$_2$.

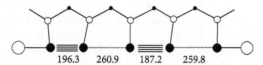

Fig. 9.6.7 Structure of Cr$_5$(tpda)$_4$Cl$_2$ (bond length in pm) (large filled circle, Cr; small filled circle, C; large open circle, Cl; small open circle, N).

The metal string molecule [Cr$_5$(tpda)$_4$Cl$_2$]·2Et$_2$O·4CHCl$_3$ contains alternately long and short metal-metal distances, as shown in Fig. 9.6.7. The short Cr–Cr distances are 187.2 pm and 196.3 pm, which correspond to the quadruple bond Cr≣Cr. The long Cr⋯Cr distances are 259.8 pm and 260.9 pm, which are indicative of non-bonding interaction.

(2) **Hexanuclear metal string cationic complexes**

Modification of the H$_2$tpda ligand by substitution of the central pyridyl group with a naphthyridyl group gives rise to the new ligand 2,7-bis(α-pyridylamino)-1,8-naphthyridine (H$_2$bpyany), which has been used to generate a series of hexanuclear metal string complexes of the general formula [M$_6$(μ_6-bpyany)$_4$X$_2$]Y$_n$ (M = Co, Ni; X$^-$ = terminal monoanionic ligand; Y$^-$ = counter monoanion; n = 1, 2). All compounds contain a linear hexanuclear cation that is helically supported by four bpyany^{2-} ligands and conforms

Table 9.6.1 Hexanuclear metal string complexes and average M–M bond distances

M	X	Y	n	Bond distance (pm) averaged for D_4 symmetry		
				outermost	mid-way	innermost
Co	NCS	PF$_6$	2	231.3(1)	225.5(1)	224.5(1)
Co	NCS	PF$_6$	1	231.3(1)	227.3(1)	225.6(1)
Co	CF$_3$SO$_3$	CF$_3$SO$_3$	2	228.3(1)	224.3(1)	226.7(1)
Co	CF$_3$SO$_3$	CF$_3$SO$_3$	1	228.4(1)	224.9(1)	225.1(1)
Ni	NCS	BPh$_4$	2	240.3(1)	231.4(1)	229.6(1)
Ni	Cl	PF$_6$	1	241.1(3)	228.5(3)	220.2(3)

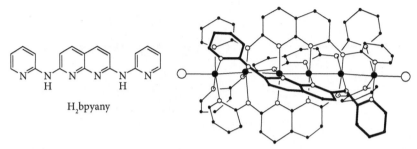

H$_2$bpyany

Fig. 9.6.8 Structural formula of the H$_2$bpyany ligand and molecular geometry of the hexanuclear monocation in crystalline [Co$_6$(μ_6-bpyany)$_4$Cl$_2$]PF$_6$.

approximately to idealized D_4 molecular symmetry if the axial terminal ligands are ignored. The structure of a representative example is shown in Fig. 9.6.8.

The averaged metal-metal bond lengths in the series of linear M$_6{}^{12+}$ (M = Co, Ni) complexes and their M$_6{}^{11+}$ one-electron reduction products are tabulated in Table 9.6.1. In all complexes the outermost M–M distance is in general slightly longer than the inner bond distances, but neither the nature of the axial ligands nor the addition of one electron to the Co$_6{}^{12+}$ system result in significant structural changes. In contrast, the innermost Ni–Ni bond shows a substantial decrease of 9.4(3) pm upon one-electron reduction of the Ni$_6{}^{12+}$ system. The crystallographic data are consistent with the proposed model of a delocalized electronic structure for the Co$_6{}^{n+}$ (n = 11, 12) complexes, whereas the extra electron in the Ni$_6{}^{12+}$ system partakes in a δ bond constructed from $d_{x^2-y^2}$ orbitals of the naphthyridyl-coordinated nickel atoms.

(3) **Mixed-valence metal string complexes of gold**

The structures of two gold metal string molecules are shown in Fig. 9.6.9. The formula of molecule (a) in this figure is:

$$\left[\begin{array}{c} \overset{\displaystyle \overset{Ph_2}{P}}{|} \qquad R \qquad \overset{\displaystyle \overset{Ph_2}{P}}{|} \\ R\!-\!Au\!-\!Au\!-\!Au\!-\!Au\!-\!Au\!-\!R \\ |\qquad\quad |\qquad\quad | \\ \underset{Ph_2}{P} \qquad R \qquad \underset{Ph_2}{P} \end{array} \right]^{+} (AuR_4)^{-} \qquad R = C_6F_5$$

(a)

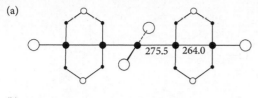

(b)

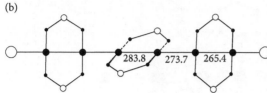

Fig. 9.6.9 Structures of two mixed-valence metal string molecules (large filled circle, Au; large open circle, R group; small filled circle, P; small open circle, C; bond lengths in pm).

From the Au–Au bond lengths shown and related theoretical calculations, the valence states of the Au atoms are in the sequence of Au(III)–Au(I)–Au(I)–Au(I)–Au(III). The molecular cation is composed of the central unit $[Au(C_6F_5)_2]^-$ and two outer dinuclear gold cations $[Au_2(PPh_2)_2C_6F_5]^+$, with the central unit donating electrons to the outer units.

The formula of molecule (b) in Fig. 9.6.9 is:

$$\left[R\!-\!Au\!-\!Au\!-\!Au\!-\!Au\!-\!Au\!-\!Au\!-\!R \right]^{2+} (ClO_4^-)_2 \quad R = C_6F_3H_2$$

From the Au–Au bond lengths shown and theoretical calculations, the valence states of the Au atoms are identified as Au(III)–Au(I)–Au(I)–Au(I)–Au(I)–Au(III).

(4) **Heteronuclear metal-string complexes (HMSCs)**

The design and synthesis of HMSCs is an important topic in metal-string complexes. Since the report of the first HMSC in 2007, more HMSCs have been synthesized. In terms of metal arrangement, there are three general classes of HMSCs: (i) M_A–M_B–M_A, (ii) M_A–M_A–M_B, and (iii) M_A–M_B–M_C. Representative examples of M_A–M_B–M_A and M_A–M_A–M_B types are described in this part. More details of these two types and examples of M_A–M_B–M_C can be found in the review by Hua and colleagues.

A number of M_A–M_B–M_A HMSCs have been obtained by the reactions between dpa$^-$ and combinations of M_A and M_B ions:

$$2\,M_A{}^{2+} + M_B{}^{2+} + 4\,dpa^- + 2\,Cl^- \rightarrow \quad Cl\text{-}M_A - M_B - M_A\text{-}Cl$$

$$\left(M_A = Mn^{II}, Co^{II}, Cu^{II};\ M_B = Pd^{II}, Pt^{II};\ M_A = Fe^{II}, M_B = Pd^{II}. \right)$$

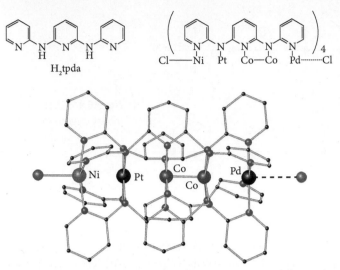

Fig. 9.6.10 Structural formula and molecular structure of [NiPtCo₂Pd(tpda)₄Cl₂]. H₂tpda = tripyridyldiamine.

To account for the regioselectivity of the metal framework (M_A–M_B–M_A instead of M_A–M_A–M_B), it should be noted that the coordination environment is different for the terminal and central metal centers. Ignoring the relatively weak metal-metal interaction, the coordination environment for the terminal metal centers is square-pyramidal and that for the central metal center is square-planer. The latter one is more favorable for d^8 metal ions and hence Pd^{II} and Pt^{II} preferably take up the central position.

HMSCs of M_A-M_A-M_B type are synthesized with a different strategy. Ions of M_B have been used to react with the binuclear complex $(M_A)_2(dpa)_4$, in which M_A = Cr, Mo, or W. In $(M_A)_2(dpa)_4$, the two M_A centers form a $(M_A)_2$ moiety stabilized by M_A-M_A quadrupole bonding. The quadrupole bond remains intact during the reaction, resulting in regioselectivity of the M_A-M_A-M_B metal framework.

Stepwise synthesis has successfully produced a series of heterometallic pentanuclear metal strings containing two to four kinds of divalent transition metal ions aligned in one chain. The X-ray molecular structure of the heterometallic pentanuclear complex [NiPtCo₂Pd(tpda)₄Cl₂] is shown in Fig. 9.6.10. This complex has the shortest reported Co(II)–Co(II) single bond (210.5(9) pm) reported in the literature. The synthesis and X-ray crystallographic characterization of a novel heteroheptanuclear metal-string complex [Ni₃Ru₂Ni₂(μ₇-teptra)₄(NCS)₂](PF₆) supported by tetra-pyridyl-tri-amine (H₃teptra) ligands (Fig. 9.6.11) yielded a remarkably short Ru–Ru bond distance of 224.99(3) pm, which is indicative of a unique metal-metal interaction in the mixed-valence [Ru₂]⁵⁺ (S = 3/2) unit. The complex exhibits a relatively high magnetic moment value of 4.55 Bohr magnetons (B.M.) at 4 K, which increases rapidly to 6.00 B.M. at 30 K and remains at 6.11 B.M. from 50 K to 300 K as shown by SQUID measurements, indicating a high-spin (S ≥ 3/2) system which is further supported by the analyses of EPR spectra at low temperatures.

(5) Long-chain defective metal-string complex

Through the use of a pyrimidine and naphthyridine-containing triamine ligand, N²-(pyrimidin-2-yl)-N⁷-(2-(pyrimidin-2-ylamino)-1,8-naphthyridin-7-yl)-1,8-naphthyridine-2,7-diamine (H₃N₉-2pm), a defective metal string complex [Ni₈.₃₃(N₉-2pm)₄(NCS)₂](PF₆)₂ (**1**) was successfully synthesized and structurally characterized. X-ray crystallographic studies revealed that complex **1** crystallizes in monoclinic

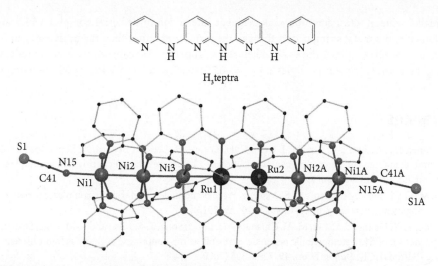

H_3teptra

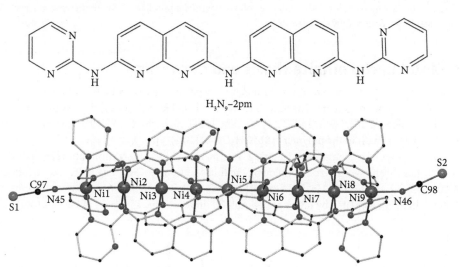

Fig. 9.6.11 Molecular structure of heteroheptanuclear metal-string complex cation $[Ni_3Ru_2Ni_2(\mu_7\text{-}teptra)_4(NCS)_2]^+$ in its PF_6^- salt. Bond lengths (pm) and bond angles (°): Ni1–Ni2 237.71(6), Ni2–Ni3 230.22(5), Ni3–Ru1 224.99(3), Ni1–$N_{av.}$ 209.5(3), Ni2–$N_{av.}$ 190.2(3), Ni3–$N_{av.}$ 198.7(3), Ru1–$N_{av.}$ 202.3(3), Ni1–N15 204.8(3); Ni1–Ni2–Ni3 178.79(2), Ni2–Ni3–Ru1 179.82(2), Ni3–Ru1–Ru2 179.39(2), Ni2–Ni1–N15 178.49(9), Ni1–N15–C41 160.2(3), N15–C41–S1 178.3(4).

H_3N_9–2pm

Fig. 9.6.12 X-ray molecular structure of the dicationic complex in $[Ni_{8.33}(N_9\text{-}2pm)_4 (NCS)_2](PF_6)_2$ (**1**). Thermal ellipsoids are drawn at the 50% probability level. The hydrogen atoms have been omitted for clarity. There are 3.5 ether and 2 acetonitrile solvent molecules in the crystal lattice, which are not shown in this figure.

space group $P2_1/c$. As shown in Fig. 9.6.12, the complex consists of a linear metal-string and four deprotonated supporting N_9-2pm^{3-} ligands which are helically wrapped around it. Nickel(II) atoms Ni(1)–Ni(3) and Ni(5)–Ni(9) have site-occupancy factors of 1, while that of Ni(4) is 0.33.

The small J value of –2.90 cm^{-1} for weak magnetic interaction in defective complex **1** indicates that spin exchange in the metal string occurs through the metal core rather than the bridging ligands. Electrochemical study on complex **1** shows abundant redox properties and capacity to carry out reduction by displaying three reversible redox couples at $E_{1/2}$ = –0.35, –0.69, and –0.88 V in its cyclic voltammogram.

References

M.-M. Rohmer, I. P.-C. Liu, J.-C. Lin, M.-J. Chiu, C.-H. Lee, G.-H. Lee, M. Bénard, X. Lopez, and S.-M. Peng, Structural, magnetic, and theoretical characterization of a heterometallic polypyridylamide complex. *Angew. Chem. Int. Ed.* **46**, 3533–6 (2007).

S.-A. Hua, M.-C. Cheng, C.-H. Chen, and S.-M. Peng, From homonuclear metal string complexes to heteronuclear metal string complexes. *Eur. J. Inorg. Chem.* 2510–23 (2015).

M.-C. Cheng, R.-X. Huang, Y.-C. Liu, M.-H. Chiang, G.-H. Lee, Y. Song, T.-S. Lina, and S.-M. Peng, Structures and paramagnetism of five heterometallic pentanuclear metal strings containing as many as four different metals: NiPtCo$_2$Pd(tpda)$_4$Cl$_2$, *Dalton Trans.* **49**, 7299–303 (2020).

C.-C. Chiu, M.-C. Cheng, S.-H. Lin, C.-W. Yan, G.-H. Lee, M.-C. Chang, T.-S. Lin, and S.-M. Peng, Structure and magnetic properties of a novel heteroheptanuclear metal string complex [Ni$_3$Ru$_2$Ni$_2$(μ_7-teptra)$_4$(NCS)$_2$](PF$_6$), *Dalton Trans.* **49**, 6635–43 (2020).

J. A. Chipman and J. F. Berry, Paramagnetic metal–metal bonded heterometallic complexes, *Chem. Rev.* **120**, 2409–47 (2020).

R. H. Ismayilov, F. F. Valiyev, N. V. Israfilov, W.-Z. Wang, G.-H. Lee, S.-M. Peng, and B A. Suleimanov, Long chain defective metal string complex with modulated oligo-α-pyridylamino ligand: Synthesis, crystal structure and properties, *J. Mol. Struct.* 1200 (2020) 126998.

9.6.5 **Metal-based infinite chains and networks**

Infinite metal-based chains are expected to be much more promising as conducting inorganic "molecular wires" than short-chain oligomers. The infinite rhodium chain, [Rh(CH$_3$CN)$_4^{1.5+}$]$_\infty$ consists of alternating Rh–Rh distances of 284.42 pm and 292.77 pm, and is a semiconductor. Fig. 9.6.13 shows a section of the infinite cationic chain in the polymer [{Rh(CH$_3$CN)$_4$}(BF$_4$)$_{1.5}$]$_\infty$.

A series of polymeric complexes featuring the metallophilic interaction between gold(I) and thallium(I) has been synthesized employing acid-base strategy. For example, the treatment of Bu$_4$N[Au(C$_6$Cl$_5$)$_2$] with TlPF$_6$ in THF gave [AuTl(C$_6$Cl$_5$)$_2$]$_n$, which consists of an infinite linear (Tl···Au···)$_\infty$ chain consolidated by unsupported Au^I···Tl^I interactions, as shown in Fig. 9.6.14(a).

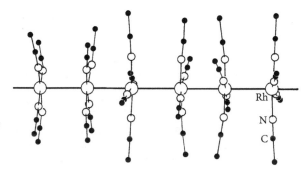

Fig. 9.6.13 Structure of a section of the infinite rhodium chain, [Rh(CH$_3$CN)$_4^{1.5+}$]$_\infty$.

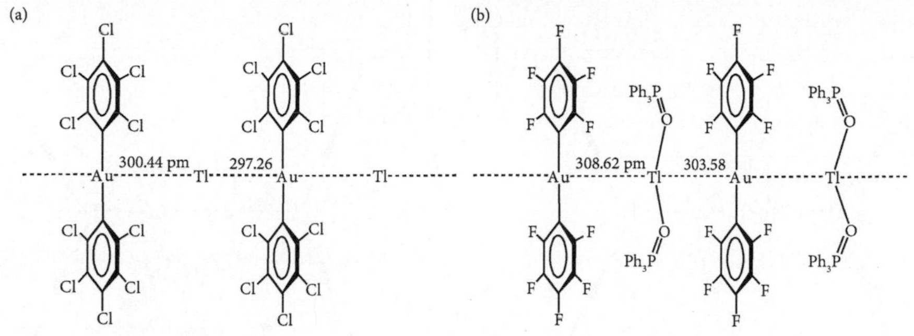

Fig. 9.6.14 Structure of (a) $[AuTl(C_6Cl_5)_2]_n$ and (b) $[AuTl(C_6F_5)_2(Ph_3P=O)_2]_n$.

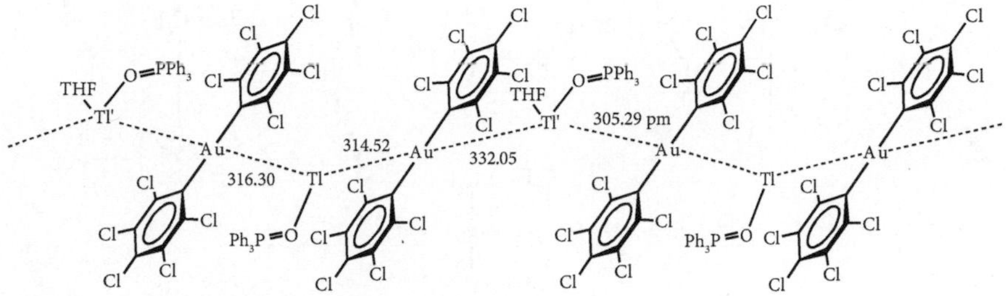

Fig. 9.6.15 Structure of $[AuTl(C_6Cl_5)_2(Ph_3P=O)_2(THF)]_n$.

In the presence of triphenylphosphine oxide, and also depending on the nature of the pentahalophenyl group employed, similar synthetic reactions afforded $[AuTl(C_6F_5)_2(Ph_3P=O)_2]_n$ and $[AuTl(C_6Cl_5)_2(Ph_3P=O)_2(THF)]_n$ with the phosphine oxide incorporated into their polymeric structures. In the pentafluorophenyl complex, thallium(I) adopts a distorted trigonal-bipyramidal geometry with one equatorial coordination site filled by a stereochemically active lone pair, forming a linear $(Tl\cdots Au\cdots)_\infty$ chain, as shown in Fig. 9.6.14(b). In contrast, the pentachlorophenyl complex comprises an infinite zigzag $(Tl\cdots Au\cdots Tl'\cdots Au\cdots)_\infty$ chain constructed from two kinds of thallium(I) centers with distorted trigonal-pyramidal and pseudo-tetrahedral geometries for Tl and Tl′ (each possessing a stereochemically active lone pair), respectively, as shown in Fig. 9.6.15.

Similar acid-base reactions with the introduction of the exo-bidentate bridging ligand 4,4′-bipyridine led to the formation of $[AuTl(C_6F_5)_2(bipy)]_n$ and $[Au_2Tl_2(C_6Cl_5)_4(bipy)_{1.5}(THF)]_n$, which exhibit higher-dimensional polymeric structures. In the pentafluorophenyl complex, linear tetranuclear $Tl\cdots Au\cdots Au\cdots Tl$ units are linked through bipyridine bridges to form a honeycomb-like network, as shown in Fig. 9.6.16.

The pentachlorophenyl complex contains two kinds of thallium(I) centers: atom Tl is coordinated by two bipyridyl N atoms, whereas atom Tl′ is bound to a THF ligand and a bipyridyl N atom, as shown in Fig. 9.6.17. The asymmetric unit contains two non-equivalent bridging 4,4′-bypyridine ligands, one of which occupies a 1 site in the unit cell and necessarily exists in the planar configuration. Heterometallophilic interaction generates infinite $(Tl\cdots Au\cdots Tl'\cdots Au\cdots)_\infty$ zigzag chains that are linked by the first kind of bridging bipyridine ligand (located in a general position and non-planar) across Tl and Tl′ to form a brick-like layer. The centrosymmetric bipyridine, which is bound to Tl and represented by a dangling rod in the figure, connects the Tl atoms in adjacent layers to form a double layer.

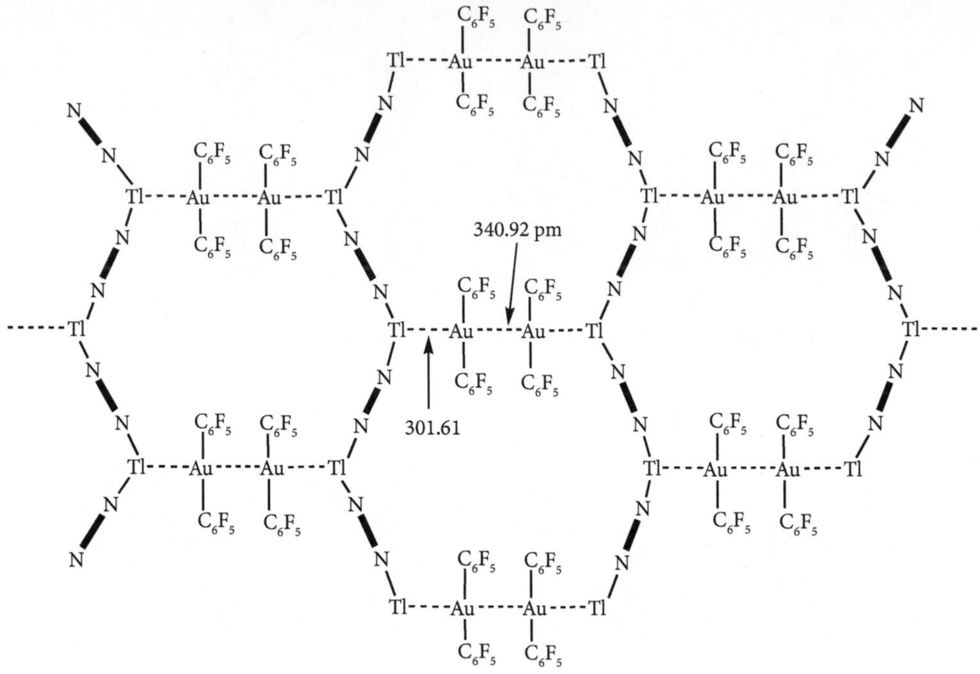

Fig. 9.6.16 Layer structure of $[AuTl(C_6F_5)_2(bipy)]_n$. The bridging 4,4'-bypyridine ligand is represented by a thick rod joining its two terminal N atoms.

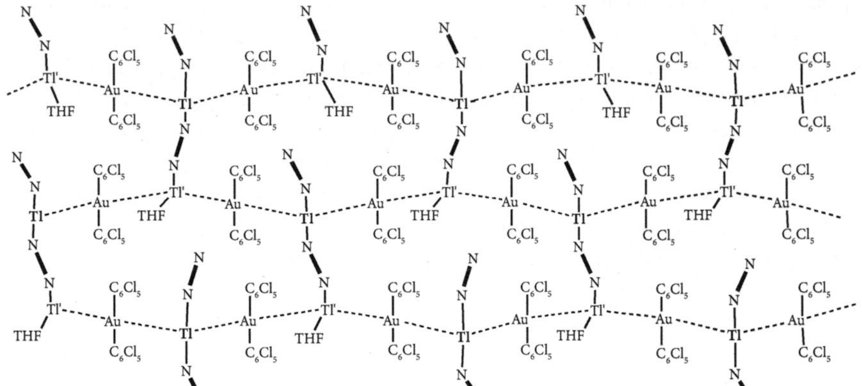

Fig. 9.6.17 Layer structure of $[Au_2Tl_2(C_6Cl_5)_4(bipy)_{1.5}(THF)]_n$. Atom Tl is shown in bold-face type to distinguish it from atom Tl'. Each bridging 4,4'-bypyridine ligand is represented by a thick rod joining the terminal N atoms.

9.6.6 Heterometallic macrocycles and cages

Research in supramolecular coordination complexes has emerged as a promising new research area in the past several decades. A wide variety of molecular species of this kind with different shapes and sizes have been synthesized, and their potential applications have been proposed. For example, molecular square

13 is an ideal host for electron-poor guests such as nitro-aromatics, which are common components in many explosives. A nitro-aromatic guest acts as a quencher for the fluorescence of **13**, which serves as an effective detector of explosives.

13

The [6Pd+8Ru] cage **14** has been shown to be able to accommodate photosensitive 2,2-dimethoxy-2-phenylacetophenone (DMPA) molecule as a guest and provide photoprotection for DMPA. The cage is therefore a potential container of photosensitizers and also found application in drug delivery and photodynamic therapy.

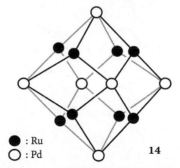

● : Ru
○ : Pd   **14**

One typical approach to form metal macrocycles is similar to polymer condensation, such as that for nylon, in the sense that two types of metal complexes with identical head and tail (for each type) can join in an alternating and head-to-end manner. Since the two types of metal complexes can be separately prepared in advance, different metals can be introduced to achieve heterometallic property in the final product. Quite often, the joining point is at the metal centre of one of the metal complexes which carries a good leaving group (such as OTf⁻) or relatively weak ligand (such as Cl⁻). The joining is carried out by substitution via the attack of a nucleophile (e.g., a nitrogen lone pair), as is illustrated by the following scheme:

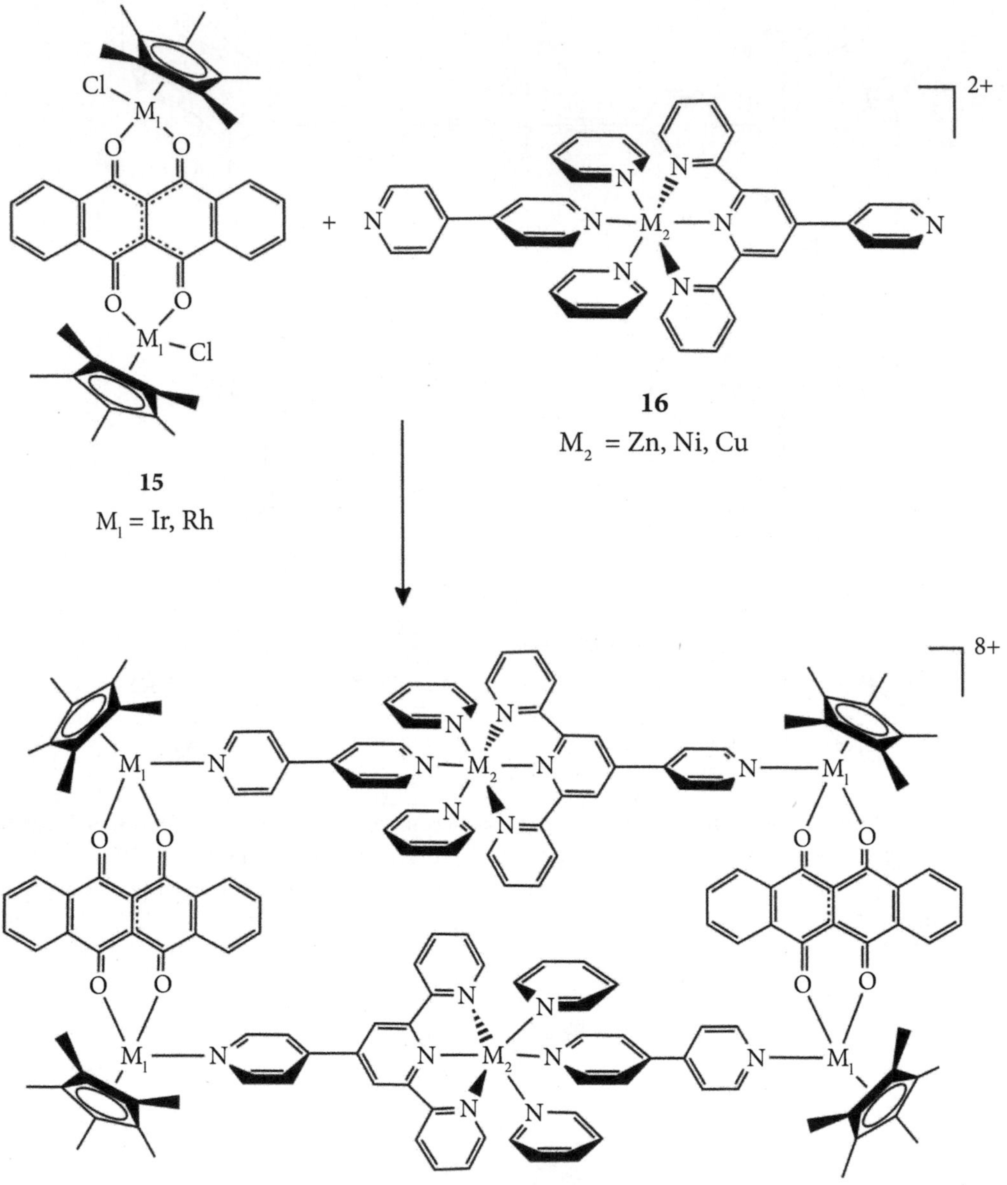

15

M_1 = Ir, Rh

16

M_2 = Zn, Ni, Cu

17

In this reaction, both **15** and **16** are metal complexes bearing different metal centres. Complex **16** can act as a metallo-bidentate ligand to bridge two M_1 centers of separate **15** molecules by replacing two chloride ligands in a 2:2 cyclic manner to generate rectangular macrocycle **17**.

Starting with a homologous series of dinuclear Ir(III) metal complexes, Zhang and colleagues were able to produce a homologous series of rectangular bimetallic macrocycles via cyclization reaction with a 4-pyridinylboron-capped iron(II) clathrochelate:

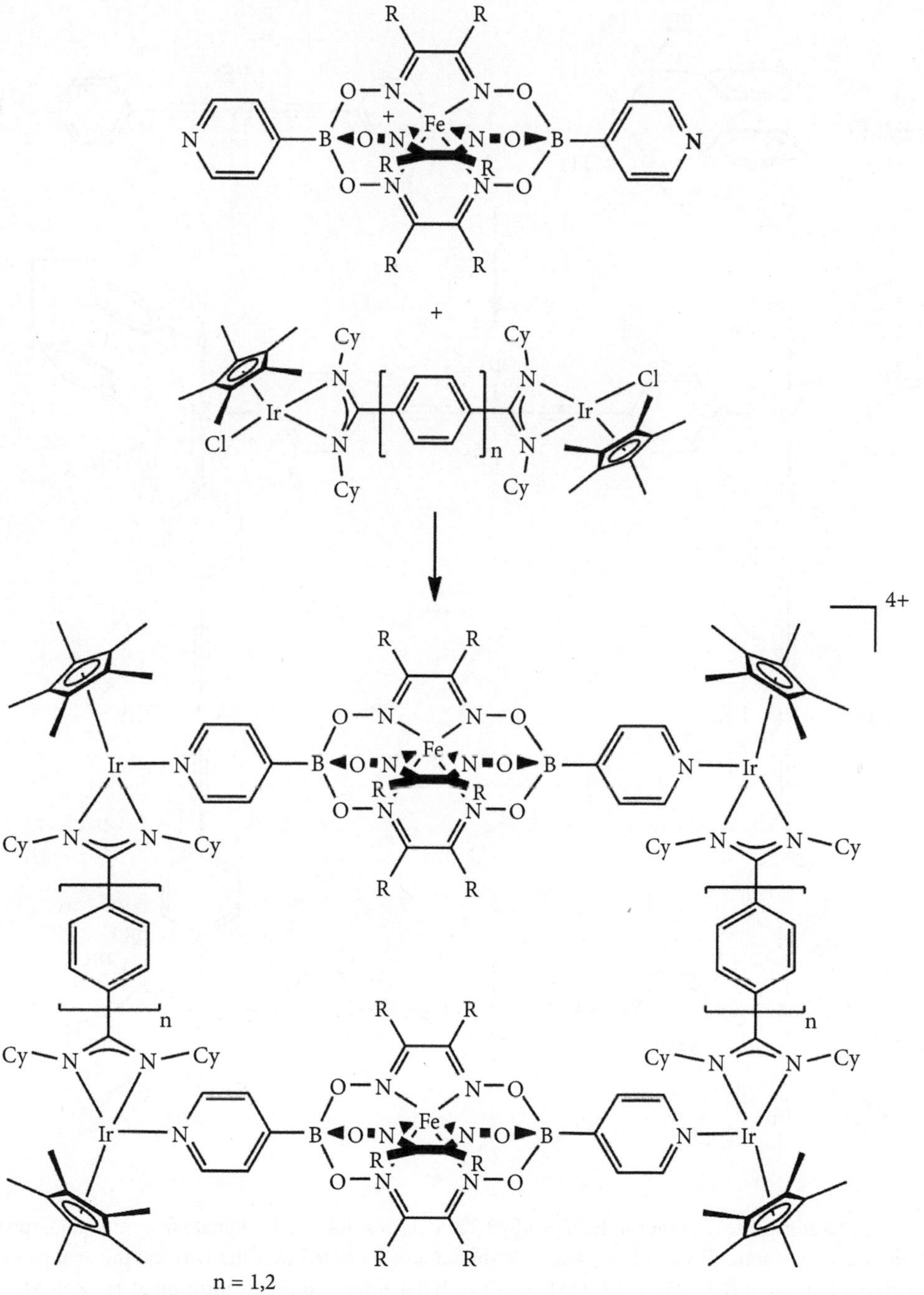

n = 1,2

On the other hand, at the extreme when both leaving groups (such as OTf) bond to the same metal atom, the latter atom is attacked twice by nucleophiles. The result is the formation of a square macrocycle, such as **20**:

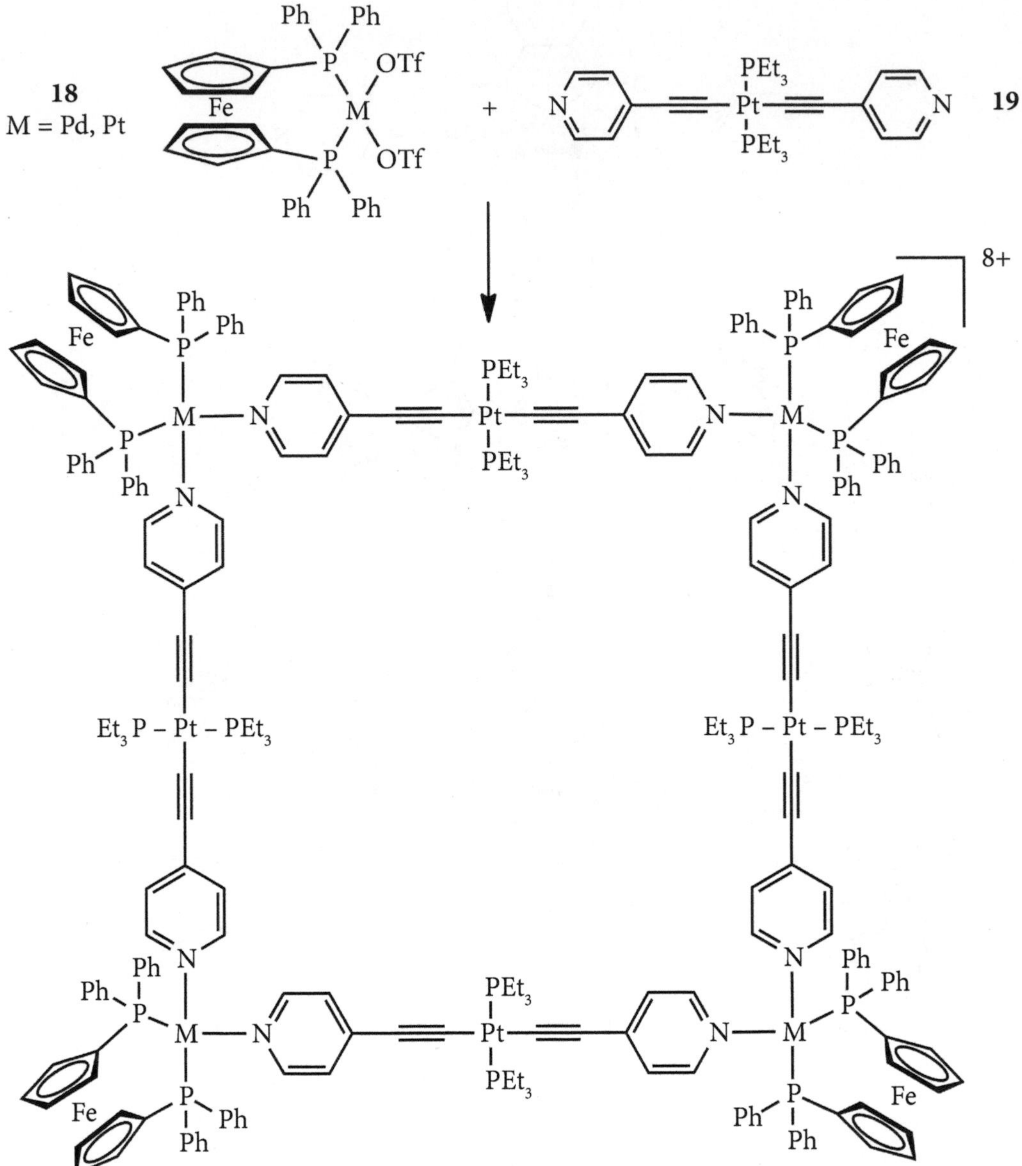

Complex **18** constitutes a corner in final product **20**. All four sides of **20** originate from the same species (**19**) and hence a square product is obtained. It should also be noted that the two leaving groups in **18** are in *cis* configuration (i.e., ∠TfO–M–OTf ≈ 90°) such that after double substitution at **18**, ∠N–M–N is also close to a right angle.

A less intuitive approach is to attach two different types of metal to a ligand before coordination to the metals. This approach proved to be successful in the following synthetic scheme:

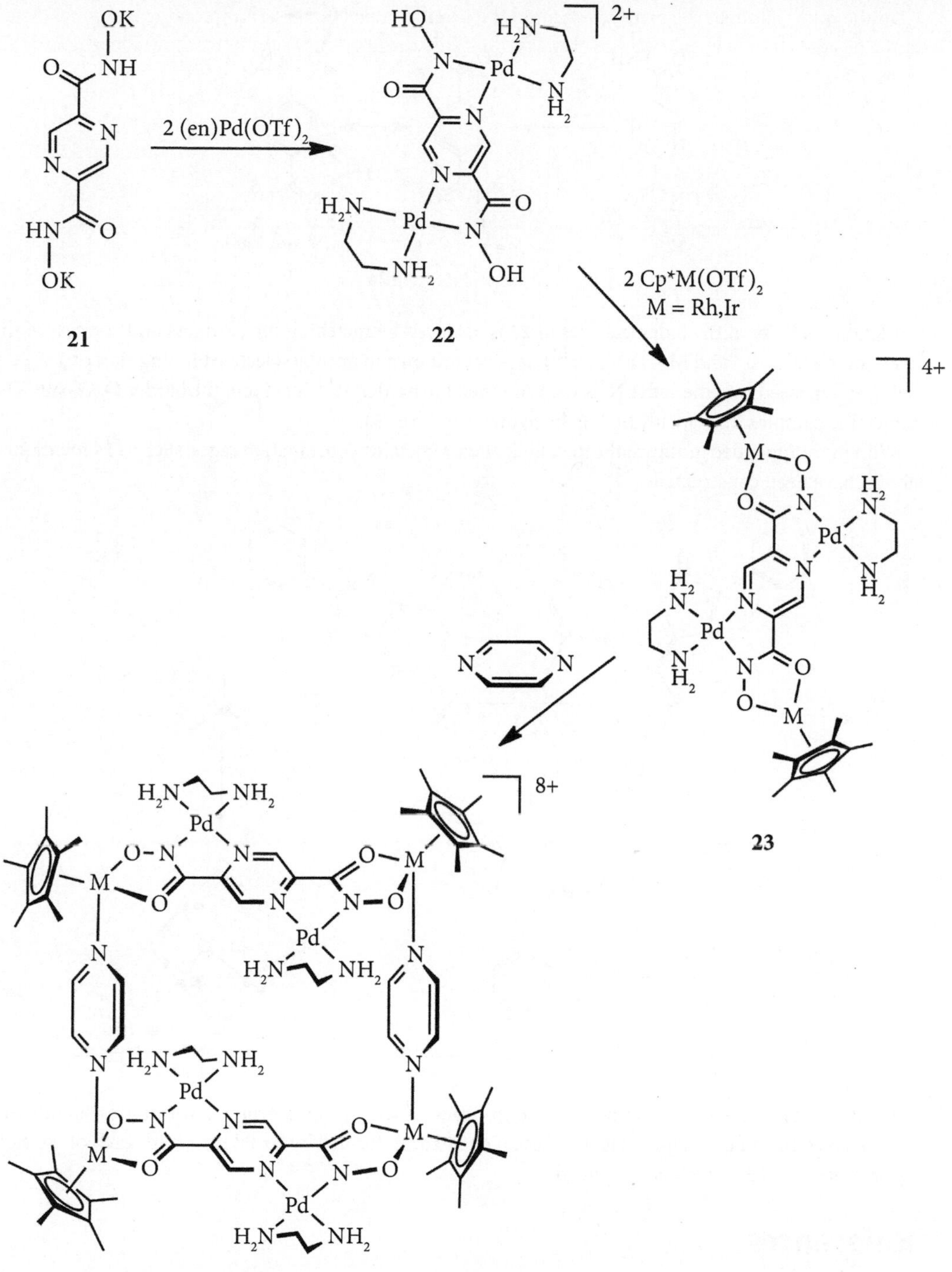

The synthesis starts with non-metallic species **21**, which contains two N, N'- and two O, O'-bidentate sites upon double deprotonation:

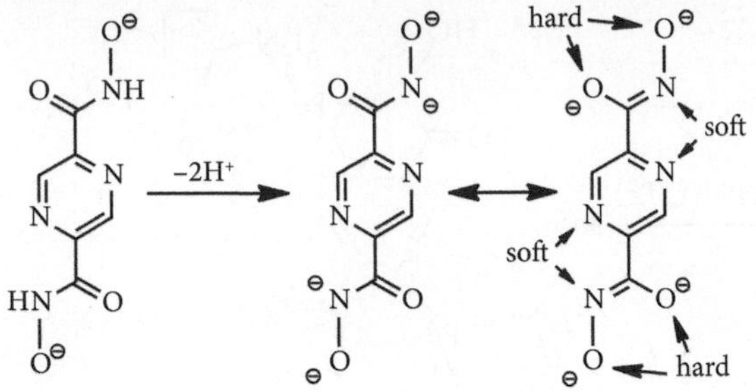

24

Metal selectivity of the bidentate sites in **24** is achieved by matching the hardness and softness of the sites and metals. Pd^{II} and M^{III} (Rh^{III}, Ir^{III}) have been shown to occupy selective binding sites of **24**: Pd^{II}, being softer, is fed into the softer N, N' site and then the harder M^{III} is fed into the harder O, O' site. The bimetallic complex **22** is finally tied up by pyrazine to form **23**.

With more appended groups at the metalloligands, bimetallic coordination cages, such as **14** mentioned above, have been constructed:

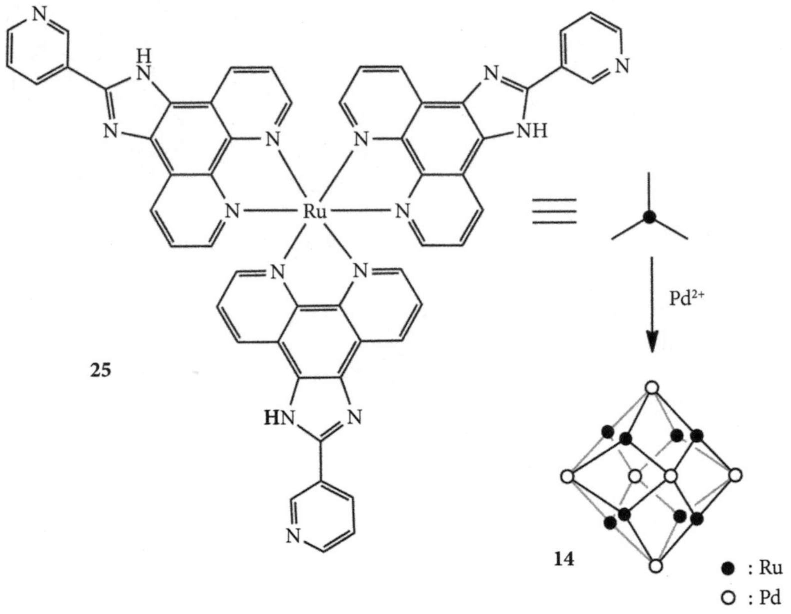

25

14

● : Ru
○ : Pd

Complex **25** contains three appended groups, each of which can coordinate with a transition metal center. Cage **14** can be roughly visualized as an octahedron formed by six Pd ions, with each of the eight faces capped by **25** as a tridentate ligand.

References

V. Vajpayee, H. Kim, A. Mishra, P. S. Mukherjee, P. J. Stang, M. H. Lee, H. K. Kim, and K. W. Chi, Self-assembled molecular squares containing metal-based donor: synthesis and application in the sensing of nitro-aromatics. *Dalton Trans.* **40**, 3112–5 (2011).

H. Sohn, R. M. Calhoun, M. J. Sailor, and W. C. Trogler, Detection of TNT and picric acid on surfaces and in seawater by using photoluminescent polysiloles. *Angew. Chem. Int. Ed.* **40**, 2104–5 (2001).

K. Li, L.-Y. Zhang, C. Yan, S.-C. Wei, M. Pan, L. Zhang, and C.-Y. Su, Stepwise assembly of $Pd_6(RuL_3)_8$ nanoscale rhombododecahedral metal–organic cages via metalloligand strategy for guest trapping and protection. *J. Am. Chem. Soc.* **136**, 4456–9 (2014).

Y.-Y. Zhang, W.-X. Gao, L. Lin, and G.-X. Jin, Recent advances in the construction and applications of heterometallic macrocycles and cages. *Coord. Chem. Rev.* **344**, 323–44 (2017).

J.-J. Liu, Y.-J. Lin, and G.-X. Jin, Box-like heterometallic macrocycles derived from bis-terpyridine metalloligands. *Organometallics.* **33**, 1283–90 (2014).

Y.-Y. Zhang, Y.-J. Lin, and G.-X. Jin, Nano-sized heterometallic macrocycles based on 4-pyridinylboron-capped iron(II) clathrochelates: syntheses, structures and properties. *Chem. Commun.* **50**, 2327–9 (2014).

L. Zhang, Y.-J. Lin, Z.-H. Li, and G.-X. Jin, Rational design of polynuclear organometallic assemblies from a simple heteromultifunctional ligand. *J. Am. Chem. Soc.* **137**, 13670–8 (2015).

9.7 **Comproportionation Reactions**

The term "comproportionation reaction" is used to designate a chemical reaction in which two reactants, each containing the same element but in different oxidation states, gives a single product in which the element has an intermediate oxidation number. For example, an aqueous solution of a copper(II) salt reacts with metallic copper to produce copper(I) ions.

$$Cu^{2+} (aq, blue) + Cu^0(s) \downarrow \longrightarrow 2Cu^+ (aq, colorless\ to\ red)$$

Similarly, the reaction of $Hg(NO_3)_2$ with $K[B(CN)_4]$ forms $Hg[B(CN)_4]_2$, which undergoes comproportionation reaction with elemental mercury to give $Hg_2[B(CN)_4]_2$.

$$Hg^{2+}\big[B(CN)_4{}^-\big]_2 + Hg^0\ (liquid) \longrightarrow Hg_2{}^{2+}\big[B(CN)_4{}^-\big]_2$$

The field of organometallic compounds is particularly rich in examples, two of which are presented in Eqs. (1a) and (1b).

$$2Nb(\eta^6\text{-}1,3,5\text{-}Me_3C_6H_3)_2 + 7CO$$
$$\longrightarrow \big[Nb(\eta^6\text{-}1,3,5\text{-}Me_3C_6H_3)_2(CO)\big]Nb(CO)_6 + 2\,(1,3,5\text{-}Me_3C_6H_3) \tag{1a}$$

$$3Co_2(CO)_8 + 12py \rightleftharpoons 2\big[Co(py)_6\big]\big[Co(CO)_4\big]_2 + 8CO \tag{1b}$$

Through reaction (1a), the 18-electron niobium(+1): niobium(–1) product is obtained from the reaction of the 17-elecron niobium(0) bis-mesitylene derivative with carbon monoxide. In reaction (1b), the cobalt(0) reactant is in equilibrium with the cobalt(+2): two cobalt(–1) product in 3:2 ratio; this reaction is reversible and largely shifted to the right depending on the partial carbon monoxide pressure and temperature.

9.7.1 **Synthesis of $[Cu_{17}({}^tBuC\equiv C)_{16}(MeOH)]^+$ cationic cluster**

In a more recent example, the reaction of $Cu(BF_4)_2$ and copper metallic powder in the presence of ${}^tBuC\equiv CH$ in methanol at room temperature provided cluster compound $[Cu_{17}({}^tBuC\equiv C)_{16}(MeOH)](BF_4)$ in high yield. The complex crystallizes as discrete cation-anion pairs, and in the cationic cluster the sterically bulky ${}^tBuC\equiv C^-$ ligands cover the surface of the assembled

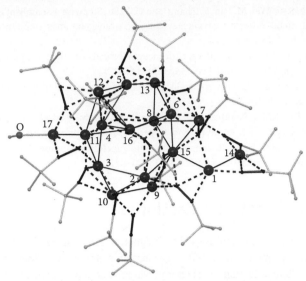

Fig. 9.7.1 Molecular structure of the cationic component of [Cu$_{17}$(^{t}BuC≡C)$_{16}$(MeOH)](BF$_4$). The ethynide groups are shown as thick black rods. The carbon atoms of the ethynide group are represented as small black spheres, and their bonds to Cu(I) atoms are indicated by broken lines. Cu(I)···Cu(I) contacts (< 280 pm) are represented by thin black lines. H atoms have been omitted for clarity.

Cu(I) kernel (Fig. 9.7.1). The pie-shaped [Cu$_{17}$(^{t}BuC≡C)$_{16}$(MeOH)]$^+$ cationic cluster can be roughly described as a Cu$_6$/Cu$_4$/Cu$_7$ triangular double-layered structure, which is consolidated by 16 ^{t}BuC≡C$^-$ ligands in various coordinating modes: μ_2-η^1,η^1; μ_2-η^1,η^2; μ_3-η^1,η^1,η^1; μ_3-η^1,η^1,η^2; and μ_4-η^1,η^1,η^1,η^2. The corrugated Cu$_6$ and Cu$_7$ layers are defined by Cu1–Cu6 and Cu7–Cu13, respectively. Cu16 at the central region of the pie is coordinated by two butyl alkynide ligands with end-on η^1-interaction to form a linear Cu(^{t}BuC≡C)$_2$ unit. Cu15 is located at the rim of the pie and bonded to three η^1-alkynide ligands. Two pairs of alkynide ligands radiating outward act like "arms" to hold peripheral Cu14 and Cu17. In comparison, Cu14 is ligated by two alkynide ligands with side-on η^2-interaction apart from η^1- and η^2-bonding, whereas Cu17 is additionally coordinated by a MeOH molecule. Most Cu(I)···Cu(I) distances in the range 245.8(2)–279.7(2) pm are shorter than twice the van der Waals radius of Cu atoms (2 × 140 pm), indicating the presence of closed-shell cuprophilic interaction.

Reference

L.-M. Zhang and T. C. W. Mak, Comproportionation synthesis of copper(I) alkynyl complexes encapsulating polyoxomolybdate templates: bowl-shaped Cu$_{33}$ and peanut-shaped Cu$_{62}$ nanoclusters. *J. Am. Chem. Soc.* **138**, 2909–12 (2016).

9.7.2 Isolation of high oxidation state Mn$_x$ clusters

The reaction of Mn(O$_2$C^{t}Bu)$_2$ and N^{n}Bu$_4$MnO$_4$ with an excess of pivalic acid in the presence of Mn(ClO$_4$)$_2$ and N^{n}Bu$_4$Cl in hot MeCN led to the isolation of [Mn$_8$O$_6$(OH)(O$_2$C^{t}Bu)$_9$Cl$_3$(^{t}BuCO$_2$H)$_{0.5}$(MeCN)$_{0.5}$] (**1**). In contrast, the reaction of Mn(NO$_3$)$_2$ and N^{n}Bu$_4$MnO$_4$ in hot MeCN with an excess of pivalic acid gave a different octanuclear complex, [Mn$_8$O$_9$(O$_2$C^{t}Bu)$_{12}$] (**2**). The latter reaction but with Mn(O$_2$C^{t}Bu)$_2$ in place of Mn(NO$_3$)$_2$, and in a MeCN/THF solvent medium, gave

$[Mn_9O_7(O_2C^tBu)_{13}(THF)_2]$ (3). Complexes 1 to 3 exhibit rare or unprecedented Mn_x topologies: 1 possesses a $[Mn^{III}_7Mn^{IV}(\mu_3\text{-}O)_4(\mu_4\text{-}O)_2(\mu_3\text{-}OH)(\mu_4\text{-}Cl)(\mu_2\text{-}Cl)]^{8+}$ core consisting of two body-fused Mn_4 butterfly units attached to the remaining Mn atoms via bridging O^{2-}, OH^-, and Cl^- ions. In contrast, 2 possesses a $[Mn_6^{IV}Mn_2^{III}(\mu_3\text{-}O)_6(\mu\text{-}O)_3]^{12+}$ core consisting of two $[Mn_3O_4]$ incomplete cubanes linked by their O^{2-} ions to two Mn^{III} atoms. The cores of 1 and 2 are unprecedented in Mn chemistry. The $[Mn^{III}_9(\mu_3\text{-}O)_7]^{13+}$ core of 3 also contains two body-fused Mn_4 butterfly units, but they are linked to the remaining Mn atoms in a different manner than in 1. This work thus demonstrates the continuing potential of comproportionation reactions for isolating high oxidation state Mn_x clusters, and the sensitivity of the product identity to minor changes in the reaction conditions.

Single crystals of 1·3MeCN, 2·MeCN, and 3·1/3THF·2/3MeCN were used for X-ray data collection at 173 K. Compound 1·3MeCN contains a mixed-valent $[Mn_8(\mu_3\text{-}O)_4(\mu_4\text{-}O)_2(\mu_3\text{-}OH)(\mu_4\text{-}Cl)(\mu_2\text{-}Cl)]^{10+}$ core: $Mn^{III}_7Mn^{IV}$, with Mn6 identified as the Mn^{IV} atom (Fig. 9.7.2). The Mn oxidation states and the protonation levels of O^{2-}, OH^-, and carboxylate O atoms were confirmed by bond-valence sum (BVS) calculations. The core can be described as comprising two $[Mn_4(\mu_3\text{-}O)_2]$ butterfly units: Mn1/Mn2/Mn5/Mn8 and Mn3/Mn5/Mn6/Mn7 that are fused together by sharing "body" atom Mn5. The remaining Mn4 atom is connected to four "wing-tip" atoms (Mn1/Mn3/Mn7/Mn8) via two μ_4-O^{2-} ions (O3 and O7) that also bridges Mn5, which is formally seven-coordinate. Finally, there are three additional monatomic bridges: at one end of the core, μ_3-OH^- ion (O5) bridges Mn3/Mn6/Mn7; at the other end, chloride ion μ_4-Cl1 bridges Mn1/Mn2/Mn8/Mn4; on one side, μ_2-Cl2 bridges Mn1/Mn2. Apart from seven-coordinate Mn5, all other Mn centers exhibit near-octahedral geometry.

The structure of the mixed-valent complex $[Mn_8O_9(O_2C^tBu)_{12}]$ (2) is shown in Fig. 9.7.3. It contains a $[Mn_8(\mu_3\text{-}O)_6(\mu\text{-}O)_3]^{12+}$ core at the $Mn^{III}_2Mn^{IV}_6$ level. Its core consists of two $Mn^{IV}_3O_4$ partial cubane units bridged by μ-O^{2-} ion O9 and two Mn^{III} atoms Mn6 and Mn8. All Mn atoms exhibit near-octahedral coordination, and the Mn^{III} Jahn–Teller elongation axes involve only carboxylate groups (O23–Mn6–O28 and O32–Mn8–O10) and are nearly parallel. Peripheral ligation is provided by 12 η^1:η^1:μ-pivalate groups. The molecule has virtual two-fold rotational symmetry with the C_2 axis passing through the μ-O^{2-} ion O9.

The complex $[Mn_9O_7(O_2C^tBu)_{13}(THF)_2]$ (3) contains a homovalent $[Mn^{III}_9(\mu_3\text{-}O)_7]^{13+}$ core, as shown in Fig. 9.7.4. Like 1, complex 2 contains two Mn_4 butterfly units, Mn9/Mn3/Mn2/Mn6 and Mn8/Mn2/Mn1/Mn5, fused at body atom Mn2. This unit is connected to two Mn atoms, Mn4 and Mn7, by three additional μ_3-O^{2-} ions, O8/O18/O29. As a result, the Mn1–Mn2–Mn3 angle deviates

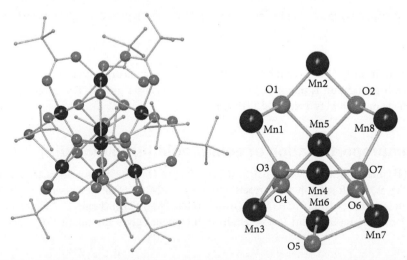

Fig. 9.7.2 Structure of complex 1 (left) and the labeled core (right). H atoms have been omitted for clarity.

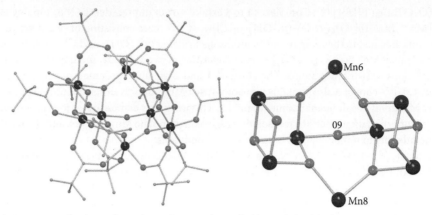

Fig. 9.7.3 Structure of octanuclear cluster in complex **2** (left), and the labeled core emphasizing its two open-faced cubane units (right). H atoms are omitted for clarity.

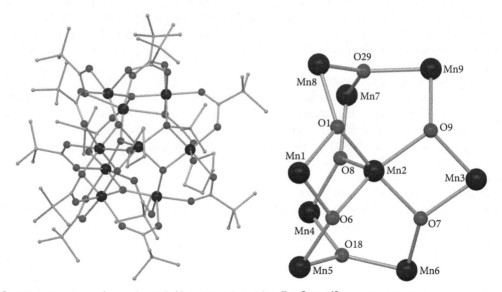

Fig. 9.7.4 Structure of complex **3** (left) and its labeled $[Mn^{III}_9(\mu^3\text{-}O)_7]^{13+}$ core (right). H atoms are omitted for clarity.

markedly from linearity (143°). All the Mn^{III} atoms are near-octahedral and display Jahn–Teller elongations. Peripheral ligation is provided by eleven $\eta^1:\eta^1:\mu$-pivalate groups, two pivalates in the rarer $\eta^1:\eta^2:\mu$-pivalate mode, and two terminal THF groups.

9.7.3 Comproportionation of gold(I)/gold(III) to digold(II)

The digold(II) dichloride complex $[Au_2Cl_2(\mu\text{-}\{CH_2\}_2PPh_2)_2]$ (**1**), which was synthesized and fully characterized (Fig. 9.7.5, left side) in 1986, reacts with alkyne PhC≡CH in the presence of KOH to form the pale yellow complex $[Au^I(i\text{-}\{CH_2\}_2PPh_2)_2Au^{III}$ (C≡CPh)$_2]$ (**2**) (Scheme 1). Treatment of **2** with two equivalents of $[Ag(ClO_4)tht]$ (tht = tetrahydrothiophene) gives the cationic digold(II) complex

$[Au_2(tht)_2-(\mu-\{CH_2\}_2PPh_2)_2](ClO_4)_2$ (**3**). This reaction is reversible, as treatment of **3** with PhC≡CH and KOH regenerates the original mixed-valence complex **2**. This facile and reversible comproportionation between gold(I)/gold(III) and digold(II) ylide complexes is remarkable in transition metal chemistry. Complex **3** reacts with KCl to give the starting digold(II)dichloride dimer **1**, thus completing a cycle from digold(II) to gold(I)/gold(III) and back to the original digold(II) complex.

The molecular structure of **2** (Fig. 9.7.5, right side) consists of two $(CH_2)_2PPh_2$ units bridging a linearly coordinated gold(I) and a square-planar gold(III) atom, forming an eight-membered organometallic ring. In addition, two phenylacetylide ligands coordinate to the gold(III) center in a *trans* arrangement. With reference to the formal gold-gold bond length of 260.0(1) pm in the starting digold(II) dichloride complex **1**, the considerably longer Au⋯Au distance of 296.87(2) pm in **2** is indicative of attractive aurophilic interaction.

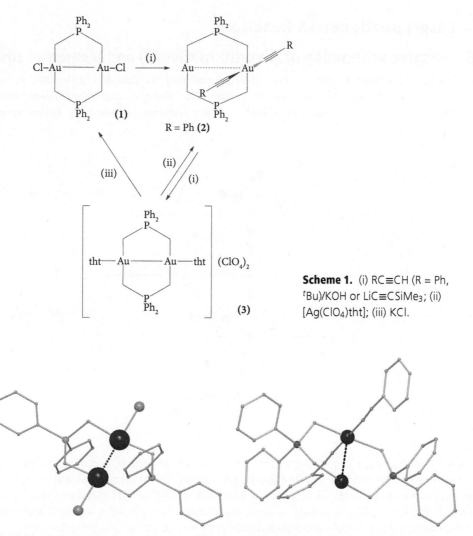

Scheme 1. (i) RC≡CH (R = Ph, ^tBu)/KOH or LiC≡CSiMe_3; (ii) [Ag(ClO_4)tht]; (iii) KCl.

Fig. 9.7.5 Molecular structure of complex **1** (left) and **2** (right). H atoms are omitted for clarity.

References

S. Mukherjee, K. A. Abboud, W. Wernsdorfer, and G. Christou, Comproportionation reactions to manganese(III/IV) pivalate clusters: a new half-integer spin single-molecule magnet, *Inorg. Chem.* **52**, 873–84 (2013).

H. H. Murray III, J. P. Fackler, Jr, L. C. Porter, and A. M. Mazany, The reactivity of [Au(CH$_2$)$_2$PPh$_2$]$_2$ with CCl$_4$. The oxidative addition of CCl$_4$ to a dimeric gold ylide complex to give Au^{II} and Au^{III} CCl$_3$ adducts. The X-ray crystal structure of [Au(CH$_2$)$_2$PPh$_2$]$_2$Cl$_2$, [Au(CH$_2$)$_2$PPh$_2$]$_2$(CCl$_3$)Cl, and [Au(CH$_2$)$_2$PPh$_2$]$_2$(CCl$_3$)Cl$_3$, *J. Chem. Soc. Chem. Commun.* 321–2 (1986).

L. A. Méndez, J. Jiménez, E. Cerrada, F. Mohr, and M. Laguna, A family of alkynylgold(III) complexes [Au^I((μ-{CH$_2$}$_2$PPh$_2$)$_2$Au^{III}(C≡CR)$_2$] (R = Ph, ^{t}Bu, Me$_3$Si): facile and reversible comproportionation of gold(I)/gold(III) to digold(II), *J. Am. Chem. Soc.* **127**, 852–3 (2005).

9.8 Disproportionation Reactions

9.8.1 Disproportionation of silver(II) to silver(I) and elemental silver

The unusual +2 oxidation state of silver can be stabilized by macrocyclic ligands, especially aza-crowns and nitrogen heterocycles. Following this strategy, the mixed-valent complex [Ag^{II}(tmc)(BF$_4$)][Ag$^I_6$(C$_2$)(CF$_3$CO$_2$)$_5$(H$_2$O)]·H$_2$O **1** was prepared by dissolving freshly prepared

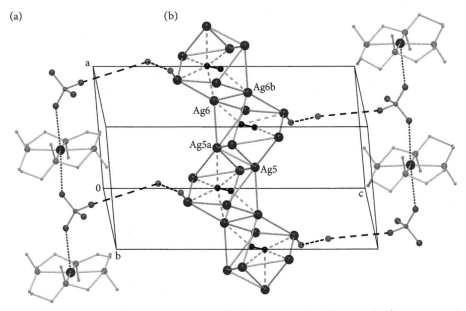

Fig. 9.8.1 Crystal structure of **1**. Hydrogen bonds are shown as dashed lines, and trifluoroacetate ligands have been omitted for clarity. (a) [Ag^{II}(tmc)(BF$_4$)]$^+_\infty$ cationic column, in which the disordered [Ag^{II}(tmc)]$^{2+}$ cation is represented by its major configuration, and the weak axial Ag^{II}···F interactions are shown by dashed open lines. (b) [Ag$^I_6$(C$_2$)(CF$_3$CO$_2$)$_5$(H$_2$O)]$^-_\infty$ zigzag chain formed from edge-sharing silver(I) dodecahedra. The mid-points of the Ag5–Ag5a and Ag6–Ag6b bonds are located at alternating inversion centers.

Ag_2C_2 (silver acetylide) in an aqueous solution of CF_3CO_2Ag and $AgBF_4$, to which 1,4,8,11-tetramethyl-1,4,8,11-tetraazacyclotetradecane (tmc) was then added. The colorless solution turned rapidly to dark red along with the precipitation of black metallic silver, which was removed by filtration. The red filtrate was allowed to stand without disturbance, and dark red block-like crystals of **1** were obtained in ca. 40% yield after several days. The crystal structure of **1** is shown in Fig. 9.8.1.

$$Ag_2C_2 + 5CF_3CO_2Ag + AgBF_4 + tmc \xrightarrow{H_2O} \left[Ag^{II}(tmc)(BF_4)\right]\left[Ag^I_6(C_2)(CF_3CO_2)_5(H_2O)\right] + Ag \downarrow$$

9.8.2 Disproportionation of organotin(I) compound to diorganostannylene and elemental tin

The intramolecularly coordinated organotin(I) compound $\{4\text{-}^tBu\text{-}2,6\text{-}[P(O)(O\text{-}^iPr)_2]_2C_6H_2Sn\}_2$ (**1**) readily undergoes disproportionation reaction to give the corresponding diorganostannylene $\{4\text{-}^tBu\text{-}2,6\text{-}[P(O)(O\text{-}^iPr)_2]_2C_6H_2\}_2Sn$ (**2**) and elemental tin. Their molecular structures are elucidated by X-ray crystallographic analyses of single crystals of **1** and the toluene solvate $\mathbf{2}\cdot C_7H_8$ (Fig. 9.8.2).

In the crystal structure of **1**, the Sn(1) and Sn(2) atoms are each four-coordinate and exhibit distorted pseudo-trigonal-bipyramidal configuration, with the O(1)/O(2) and O(3)/O(4) atoms occupying the axial positions at Sn(1) and Sn(2), respectively. The equatorial positions are occupied by the C(1)/Sn(2)/lone pair (at Sn1) and the C(31)/Sn(1)/lone pair (at Sn2). In the $\mathbf{2}\cdot C_7H_8$ solvate, the four-coordinate Sn atom exhibits distorted pseudo-square-pyramidal configuration, with the C(1), C(1A), O(1), and O(1A) atoms occupying four equatorial positions with an electron lone pair occupying the apical position.

$$RSnSnR \xrightarrow{\Delta T} R_2Sn + Sn$$

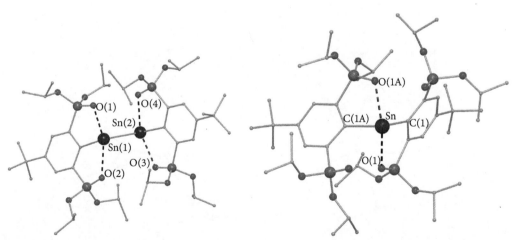

Fig. 9.8.2 Disproportionation reaction scheme and molecular structures of **1** and **2** (the co-crystallized toluene molecule in $\mathbf{2}\cdot C_7H_8$ is not shown). H atoms are omitted for clarity.

9.8.3 **Disproportionation of tin monobromide**

The reaction of a metastable Sn^IBr solution with $LiSi(SiMe_3)_3$ leads to the metalloid cluster compound $Sn_{10}[Si(SiMe_3)_3]_6$ (**1**) in a moderate yield of *ca.* 17%. In its molecular structure established by X-ray crystallography, the 10 tin atoms are arranged in the form of a Sn_{12} centaur polyhedron, which comprises an empty Sn_8 cube with two adjacent square faces topped by a Sn_2 bond to generate two adjacent pentagonal pyramidal extensions (Fig. 9.8.3, left). The observed tin–tin bond distances lie within the normal range of 285–314 pm for metalloid tin cluster compounds. The shortest Sn1–Sn2 bond can be regarded as the side of a cubic face, whereas the longest Sn8–Sn9 bond corresponds to the edge of a triangular face in an icosahedron.

As tin atoms Sn3, Sn6, Sn8, and Sn9 in metalloid cluster **1** are not involved in coordination by the $Si(SiMe_3)_3$ ligand, the average oxidation state of the tin atoms is 0.6. Consequently **1** is a reduction product of the disproportionation reaction that yields elemental tin. As the reaction starts with the monohalide Sn^IBr, oxidized species with an average oxidation state of the tin atoms larger than **1** must also be present in the reaction solution.

Compound **1** forms a co-crystallized 1:1 complex with the cyclotristannene $Sn_3[Si(SiMe_3)_3]_4$ **2** in the form of black, diamond shaped crystals. X-ray crystallographic analysis showed that **2** (Fig. 9.8.3, right) consists of a central three-membered Sn_3 ring, where two tin atoms (Sn12, Sn13) are each bound to one $Si(SiMe_3)_3$ ligand while the third tin atom (Sn11) is bound to two $Si(SiMe_3)_3$ ligands, so that **2** can be regarded as an unsaturated cyclic cyclotristannene.

Of most significance are the dimensions of the Sn_3 ring in **2**. While the Sn11–Sn12 and Sn11–Sn13 bond distances of 284 pm lie in the normal range for a single bond, the Sn12–Sn13 bond distance of 258 pm is the shortest among all distannenes structurally characterized thus far.

Synthesis of metalloid cluster **1** yielded the dark red lithium salt $Li(THF)\{Sn[Si(SiMe_3)_3]_3\}$ **3** as a byproduct via reaction of a metastable Sn^IBr solution with $LiSi(SiMe_3)_3$ in THF. X-Ray crystal structure analysis revealed that the lithium cation is linearly coordinated by the tin atom at a Sn···Li distance of 274 pm and the THF molecule at a O···Li distance of 186 pm (Fig. 9.8.4). These distances are quite short,

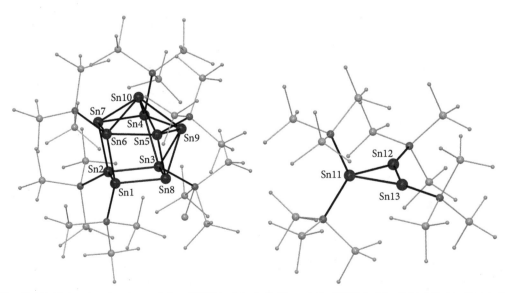

Fig. 9.8.3 Molecular structure of $Sn_{10}[Si(SiMe_3)_3]_6$ **1** (left) and $Sn_3[Si(SiMe_3)_3]_4$ **2** (right). H atoms are omitted for clarity.

Fig. 9.8.4 Molecular structure of $[Li(THF)]^+\{Sn[Si(SiMe_3)_3]_3\}^-$ **3**. H atoms are omitted for clarity.

being normally in the range of 286 pm and 192 pm, respectively, and the shortening effect is attributable to the low coordination number of two for the lithium cation.

9.8.4 Disproportionation of tin sulphonate

Disproportionation reactions between equimolar quantities of $R_2Sn(X)OSO_2Me$ [X = OMe or OH] and ethylmalonic/maleic acid in acetonitrile under mild conditions afford new diorganotin dicarboxylates $R_2Sn(O_2CBu^nR'COOH)_2$ along with disulphonates $R_2Sn(OSO_2Me)_2$. Similar reactions of the tin precursors with pyridine-2-carboxylic acid provide an access to novel trinuclear tin complexes $R_6Sn_3(O_2CC_5H_4N\text{-}2)_3(OSO_2Me)_3$. The structures of representative complexes $^nBu_2Sn[O_2CCH(Et)COOH]_2$ (**4**), $^nBu_2Sn(OSO_2Me)_2$ (**5**) and $R_6Sn_3(O_2CC_5H_4N\text{-}2)_3(OSO_2Me)_3$ (**6**) have been determined by X-ray crystallography.

Compound **4** is monomeric with bicapped tetrahedron geometry by virtue of anisobidentate coordination of one carboxylate group of each ligand, while the other carboxylic acid group remains free (Fig. 9.8.5, left). The polymeric structure of **5** features centrosymmetric eight-membered rings comprising bridging methanesulfonate groups and near-perfect octahedral geometry around each tin atom (Fig. 9.8.5, right).

Compound **6** crystallizes as the solvate **6**·2H$_2$O·Et$_2$O. The molecular structure of **6** reveals self-assembly of the mixed ligand tin ester $^nBu_2Sn(O_2CC_5H_4N\text{-}2)OSO_2Me$ and its disproportionated products $^nBu_2Sn(O_2CC_5H_4N\text{-}2)_2$ and $^nBu_2Sn(OSO_2Me)_2$, which are coordinatively associated by varying bonding modes of pyridine-2-carboxylate groups. The O(1) and N(1) atoms of the ligand is bonded to

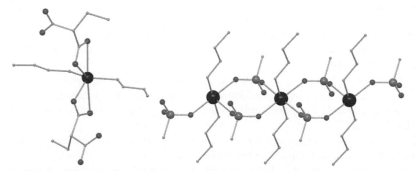

Fig.9.8.5 Molecular structure of $^nBu_2Sn[O_2CCH(Et)COOH]_2$ **4** (left) and polymeric structure of $^nBu_2Sn(OSO_2Me)_2$ **5** (right). H atoms are omitted for clarity.

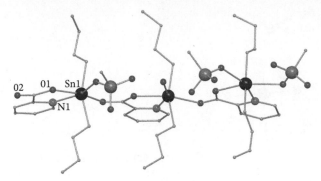

Fig.9.8.6 Polymeric structure of $^nBu_6Sn_3(O_2CC_5H_4N-2)_3(OSO_2Me)_3$ **6** in the solvate **6**·$2H_2O$·Et_2O.

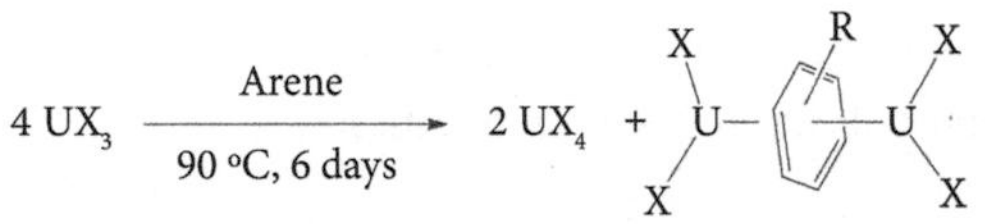

$$4\ UX_3 \xrightarrow[\text{90 °C, 6 days}]{\text{Arene}} 2\ UX_4\ +\ \text{inverse-sandwich complex}$$

Fig. 9.8.7 Formation of inverse-sandwich arene complex UX_3 (X = aryloxide, amide), which comprises a pair of bent UX_2 moieties and two-electron reductive activation of the central arene component.

Sn(1) in a chelating fashion [Sn(1)–O(1) 210.9(6) pm, Sn(1)–N(1) = 230.5(7) pm], while the other carboxylic oxygen atom O(2) remains free (Fig. 9.8.6). Compound **6** exhibits a polymeric structure with centrosymmetric eight-membered rings each composed of pairwise bridging methanesulfonate groups between consecutive tin atoms.

9.8.5 **Uranium(III) disproportionation**

The trivalent uranium complex UX_3 (X = aryloxide, amide) dissolved in an arene solvent readily undergoes spontaneous disproportionation, which involves transference of an electron and a X-ligand, enabling the resulting bent divalent UX_2 moieties to bind an arene solvent molecule to form an inverse-sandwich $[X_2U(\mu\text{-}\eta^6:\eta^6\text{-arene})UX_2]$ molecule and a tetravalent UX_4 byproduct (Fig. 9.8.7).

References

Q.-M. Wang and T. C. W. Mak, Induced assembly of a catenated chain of edge-sharing silver(I) dodecahedra with embedded acetylide by silver(II)-tmc (tmc = 1,4,8,11-tetramethyl-1,4,8,11-tetraazacyclotetradecane), *Chem. Commun.* **37**, 807–8 (2001).

M. Wagner, C. Dietz, S. Krabbe, S. G. Koller, C. Strohmann, and K. Jurkschat, {4-^{t}Bu-2,6-[P(O)(O-^{i}Pr)$_2$]$_2C_6H_2$Sn}$_2$: an intramolecularly coordinated organotin(I) compound with a Sn–Sn single bond, its disproportionation toward a diorganostannylene and elemental tin, and its oxidation with PhI(OAc)$_2$, *Inorg. Chem.* **51**, 6851–9 (2012).

C. Schrenk, I. Schellenberg, R. Pöttgen, and A. Schnepf, The formation of a metalloid Sn_{10}[Si(SiMe$_3$)$_3$]$_6$ cluster compound and its relation to the α↔β tin phase transition, *Dalton Trans.* **39**, 1872–6 (2010).

C. Schrenk and A. Schnepf, Sn_3[Si(SiMe$_3$)$_3$]$_3^-$ and Sn_3[Si(SiMe$_3$)$_3$]$_4$: first insight into the mechanism of the disproportionation of a tin monohalide gives access to the shortest double bond of tin, *Chem. Commun.* **46**, 6756–8 (2010).

R. Shankar, M. Kumar, S. P. Narula, and R. K. Chadha, Disproportionation reactions of (methoxy/hydroxy)diorganotin(IV) methanesulfonates with carboxylic acids: synthesis and structure of new diorganotin(IV) carboxylates, *J. Organomet. Chem.* **671**, 35–42 (2003).

P. L. Arnold, S. M. Mansell, L Maron, and D. McKay, Spontaneous reduction and C–H borylation of arenes mediated by uranium(III) disproportionation. *Nature Chem.* **4**, 668–74 (2012).

General References

F. A. Cotton, C. A. Murillo, and R. A. Walton (eds.), *Multiple Bonds between Metal atoms*, 3rd ed., Springer, New York, 2005.

M. Gielen, R. Willem, and B. Wrackmeyer (eds), *Unusual Structures and Physical Properties in Organometallic Chemistry*, Wiley, West Sussex, 2002.

J.-X. Lu (ed.), *Some New Aspects of Transition-Metal Cluster Chemistry*, Science Press, Beijing/New York, 2000.

P. Braunstein, L. A. Oro, and P. R. Raithby (eds), *Metal Clusters in Chemistry: Vol. 1 Molecular Metal Cluster; Vol. 2 Catalysis and Dynamics and Physical Properties of Metal Clusters; Vol. 3 Nanomaterials and Solid-state Cluster Chemistry*, Wiley-VCH, Weinheim, 1999.

J. P. Collman, R. Boulatov, and G. B. Jameson, The first quadruple bond between elements of different groups. *Angew. Chem. Int. Ed.* **40**, 1271–4 (2001).

T. Nguyen, A. D. Sutton, M. Brynda, J. C. Fettinger, G. J. Long, and P. P. Power, Synthesis of a stable compound with fivefold bonding between two chromium(I) centers. *Science.* **310**, 844–7 (2005).

P. Pyykkö, Strong closed-shell interaction in inorganic chemistry. *Chem. Rev.* **97**, 579–636 (1997).

N. Kaltsoyannis, Relativistic effects in inorganic and organometallic chemistry. *J. Chem. Soc. Dalton Trans.* 1–11 (1997).

C.-H. Chien, J.-C. Chang, C.-Y. Yeh, G.-H. Lee, J.-M. Fang, Y. Song, and S.-M. Peng, *Dalton Trans.* **35**, 3249–56 (2006).

Chapter 10

Supramolecular Structural Chemistry

10.1 Introduction

Supramolecular chemistry is a highly interdisciplinary field of science covering the chemical, physical, and biological features of molecular assemblies that are organized and held together by intermolecular interactions. The basic concepts and terminology were introduced by J.-M. Lehn, who together with D. J. Cram and C. J. Pedersen were awarded the 1987 Nobel Prize in Chemistry. In the words of Lehn, supramolecular chemistry may be defined as chemistry beyond the molecule; that is, the study of organized entities of higher complexity (supermolecule) resulting from the association of two or more chemical species consolidated by intermolecular forces. The relationship of supermolecules to molecules and intermolecular binding is analogous to that of molecules to atoms and covalent bonds (Fig. 10.1.1).

A clarification about vocabulary in the chemical literature: the prefix in the word supermolecule (a noun) is derived from the Latin *super*, meaning "more than" or "above"; it should not be used interchangeably with the prefix *supra* in the word supramolecular (an adjective), which means "beyond" or "at a higher level than."

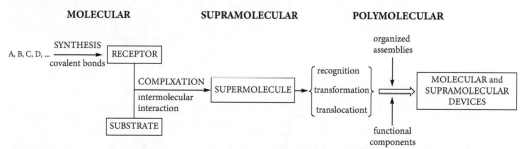

Fig. 10.1.1 Conceptual development from molecular to supramolecular chemistry: molecules, supermolecules, molecular devices, and supramolecular devices.

10.1.1 Intermolecular interactions

Intermolecular interactions constitute the core of supramolecular chemistry. The design of supermolecules requires a clear understanding of the nature, strength, and spatial attributes of intermolecular bonding, which is a generic term that includes ion pairing (Coulombic), hydrophobic and hydrophilic interactions, hydrogen bonding, host-guest complementarity, π–π stacking, and van der Waals interactions. For inorganic systems, coordination bonding is included in this list if the metal acts as an attachment template. Intermolecular interactions in organic compounds can be classified as (a) isotropic, medium-range forces that define molecular shape, size, and close packing, and (b) anisotropic, long-range forces, which are electrostatic and involve heteroatom interactions. In general, isotropic forces (van der Waals interactions) usually mean dispersive and repulsive forces, including $C\cdots C$, $C\cdots H$, and $H\cdots H$ interactions, while most interactions involving heteroatoms (N, O, Cl, Br,

Structural Chemistry across the Periodic Table. Thomas Chung Wai Mak, Yu-San Cheung, Gong-Du Zhou, and Ying-Xia Wang. Oxford University Press. © Thomas Chung Wai Mak, Yu-San Cheung, Gong-Du Zhou, and Ying-Xia Wang (2023). DOI: 10.1093/oso/9780198872955.003.0010

I, P, S, and Se, etc.) with one another or with carbon and hydrogen are anisotropic in character, including ionic forces, strongly directional hydrogen bonds (O–H···O, N–H···O), weakly directional hydrogen bonds (C–H···O, C–H···N, C–H···X, where X is a halogen, and O–H···π), and other weak forces such as halogen···halogen, nitrogen···nitrogen, sulfur···halogen, and so on. In a crystal, various strong and weak intermolecular interactions coexist (sometimes in delicate balance, as is demonstrated by the phenomenon of polymorphism) and consolidate the three-dimensional (3-D) scaffolding of the molecules.

10.1.2 Molecular recognition

The concept of molecular recognition has its origin in effective and selective biological functions, such as substrate binding to a receptor protein, enzyme reactions, assembly of protein-DNA complexes, immunological antigen-antibody association, reading of the genetic code, signal induction by neuro-transmitters, and cellular recognition. Many of these functions can be performed by artificial receptors, whose design requires an optimal match of the steric and electronic features of the non-covalent inter-molecular forces between substrate and receptor. Some of the recognition processes that have been well studied by chemists include spherical recognition of metal cations by cryptates, tetrahedral recognition by macrotricyclic cryptands, recognition of specific anions, and the binding and recognition of neutral molecules through Coulombic, donor–acceptor, and in particular hydrogen-bonding interactions (see Fig. 10.1.2).

Molecular recognition studies are typically carried out in solution, and the effects of intermolecular interactions are often probed by spectroscopic methods.

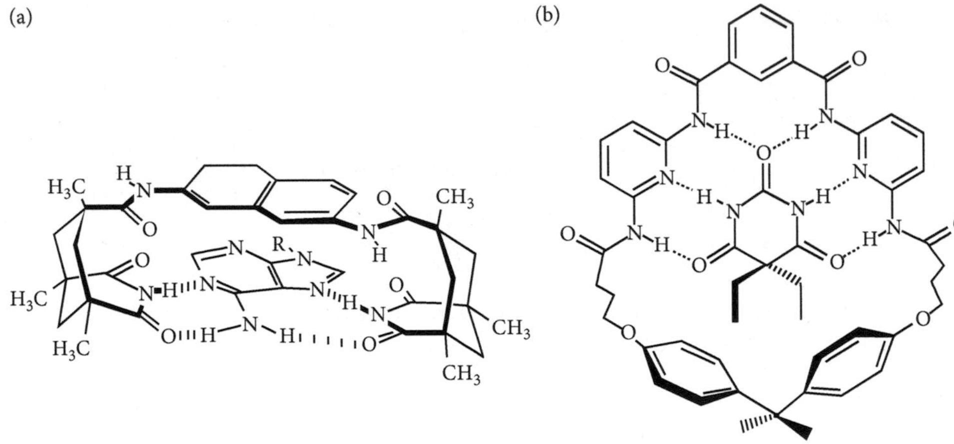

Fig. 10.1.2 Molecular recognition through hydrogen bonding of (a) adenine in a cleft, and (b) barbituric acid in a macrocyclic receptor (right).

10.1.3 Self-assembly

The term "self-assembly" is used to designate the evolution toward spatial confinement through spontaneous connection of molecular components, resulting in the formation of discrete or extended entities at either the molecular or the supramolecular level. Molecular self-assembly yields covalent structures, while in supramolecular self-assembly several molecules spontaneously associate into a single, highly structured supramolecular aggregate. In practice, self-assembly can be achieved if the molecular components are loaded with recognition features that are mutually complementary; that is, they contain two or more interaction sites for establishing multiple connections. Thus well-defined molecular and supramolecular architectures can be spontaneously generated from specifically "engineered" building blocks. For example, self-assembly occurs with interlocking of molecular components using π–π interactions (Fig. 10.1.3) and the formation of capsules with some curved molecules bearing complementary hydrogen bonding sites (Fig. 10.1.4).

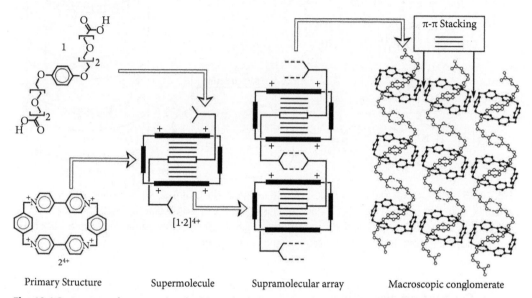

Primary Structure   Supermolecule   Supramolecular array   Macroscopic conglomerate

Fig. 10.1.3 Aspects of supramolecular hierarchy in increasing superstructural complexity.

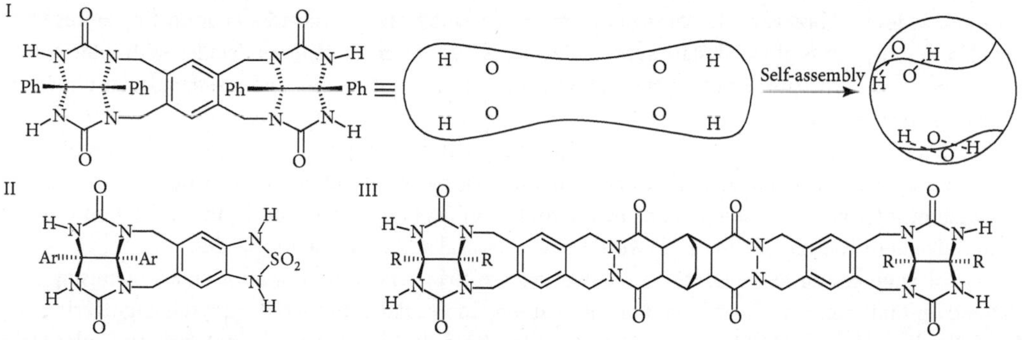

Fig. 10.1.4 A tennis ball-shaped molecular aggregate can be constructed by the self-assembly of curved molecule I. Tetrameric assembly of II generates a pseudo-spherical capsule. Dimeric assembly of III can be induced by the introduction of smaller molecules of appropriate size and shape into a spherical complex.

The cyclic octapeptide *cyclo*-[–(D-Ala–L-Glu–D-Ala–L-Gln)$_2$–] has been designed by Ghadiri and co-workers to generate a hydrogen-bonded organic nanotube having an internal diameter *ca.* 0.7–0.8 nm (Fig. 10.1.5).

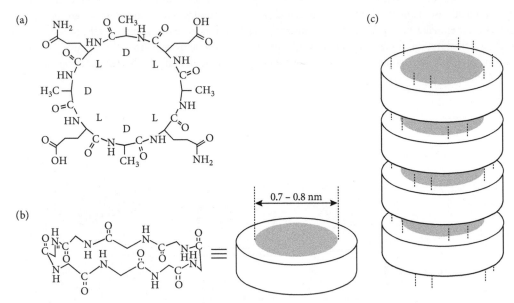

Fig. 10.1.5 (a) Structural formula of *cyclo*-[–(D-Ala–L-Glu–D-Ala–L-Gln)$_2$–]; Ala = alanine, Glu = glutamic acid, Gln = glutamine, D or L indicates chirality at the carbon atom. (b) Perspective view of the backbone of the flat, ring-shaped octapeptide. (c) Tubular architecture generated from a stack of octapeptide molecules held by intermolecular hydrogen bonding.

10.1.4 **Crystal engineering**

Structural chemists and crystallographers rightfully regard an organic crystal as the "supermolecule *par excellence*," being composed of Avogadro's number of molecules self-assembled by mutual recognition at an amazing level of precision. In contrast to a molecule, which is constructed by connecting atoms with covalent bonds, a crystal (solid-state supermolecule) is built by connecting molecules with intermolecular interactions. The process of crystallization is one of the most precise and spectacular examples of molecular recognition.

The determination of crystal structures by X-ray crystallography provides precise and unambiguous data on intermolecular interactions. Crystal engineering has been defined by Desiraju as "the understanding of intermolecular interactions in the context of crystal packing and in the utilization of such knowledge in the design of new solids with desired physical and chemical properties."

Crystal engineering and molecular recognition are twin tenets of supramolecular chemistry that depend on multiple matching of functionalities among molecular components. Crystal engineering has been developed by structural and physical chemists with a view to design new materials and solid-state reactions, whereas molecular recognition has been developed by physical organic chemists interested in mimicking biological processes. The methodologies and goals of these two related fields are summarized in Table 10.1.1.

Table 10.1.1 Comparison of crystal engineering and molecular recognition

	Crystal engineering	Molecular recognition
(1)	Concerned with the solid state	Concerned mainly with solution phase
(2)	Considers both convergent and divergent binding of molecules	Most cases only focus on convergent binding of molecules
(3)	Intermolecular interaction are examined directly in terms of their geometrical features obtained from X-ray crystallography	Intermolecular interactions are studied indirectly in terms of association constants obtained from various spectroscopic (NMR, UV, etc.) methods
(4)	Design strategies involve the control of the 3-D arrangement of molecules in the crystal; such an arrangement ideally results in desired chemical and physical properties	Design strategies are confined to the mutual recognition of generally two species: the substrate and the receptor; such recognition is expected to mimic some biological functionality
(5)	Both strong and weak interactions are considered independently or jointly in the design strategy	Only strong interactions such as hydrogen bonding are generally used for the recognition event
(6)	The design may involve either single-component species or multi-component species; a single-component molecular crystal is a prime example of self-recognition	The design usually involves two distinct species: the substrate and the receptor; ideas concerning self-recognition are poorly developed
(7)	In host-guest complexes, the host cavity is composed of several molecules whose synthesis may be fairly simple; the geometry and functionality of the guest molecules are often of significance in the complexation	In host-guest complexes, the host cavity is often a single macrocyclic molecule whose synthesis is generally tedious; the host framework rather than the guest molecule plays a critical role in the complexation
(8)	Systematic retrosynthetic pathways may be deduced with the Cambridge Structural Database (CSD) to design new recognition patterns using both strong and weak interactions	There is no systematic set of protocols for the identification of new recognition patterns; much depends on individual style and preferences

10.1.5 Supramolecular synthon

In the context of organic synthesis, the term "*synthon*" was introduced by Corey in 1967 to refer to "structural units within molecules which can be formed and/or assembled by known or conceivable synthetic operations." This general definition was modified by Desiraju for supramolecular chemistry as: "*Supramolecular synthons* are structural units within supermolcules which can be formed and/or assembled by known or conceivable synthetic operations involving intermolecular interactions." The goal of crystal engineering is to recognize and design synthons that are sufficiently robust to be carried over from one network structure to another, which ensures generality and predictability. Some common examples of supramolecular synthons are shown in Fig. 10.1.6.

It should be emphasized that supramolecular synthons are derived from designed combinations of interactions and are not identical to the interactions. A supramolecular synthon incorporates both chemical and geometrical recognition features of two or more molecular fragments; that is, both explicit and implicit involvement of intermolecular interactions. However, in the simplest cases, a

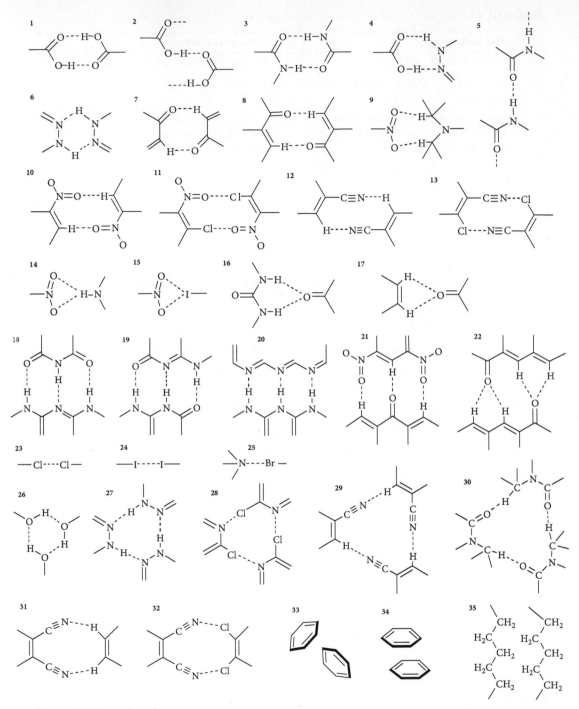

Figure 10.1.6 *continued*

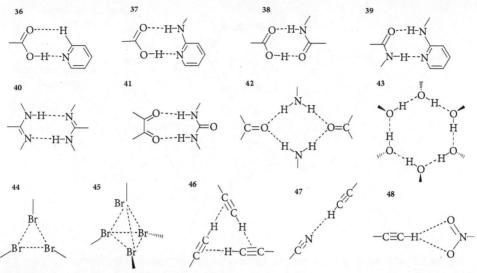

Fig. 10.1.6 Representative supramolecular synthons. Synthon No. **1-35** are taken from G. R. Desiraju, *Angew. Chem. Int. Ed.* **34**, 2311 (1995). No. **33**: known as EF (edge-to-face) and **34**: OFF (offset face-to-face) phenyl–phenyl interactions; M. L. Scudder and I. G. Dance, *Chem. Eur. J.* **8**, 5456 (2002); I. Dance, Supramolecular inorganic chemistry, in G. R. Desiraju (ed.), *The Crystal as a Supramolecular Entity, Perspectives in Supramolecular Chemistry*, Vol. 2, Wiley, New York, 1996, pp. 137–233; **36**: T. Steiner, *Angew. Chem. Int. Ed.* **41**, 48 (2002); **37**: A. Nangia, *CrystEngComm* **17**, 1 (2002); **38**: P. Vishweshwar, A. Nangia and V. M. Lynch, *CrystEngComm* **5**, 164 (2003); **39**: F. H. Allen, W. D. S. Motherwell, P. R. Raithby, G. P. Shields, and R. Taylor, *New J. Chem.* 25 (1999); **40**: R. K. Castellano, V. Gramlich, and F. Diederich, *Chem. Eur. J.* **8**, 118 (2002); **41**: C.-K. Lam and T. C. W. Mak, *Angew. Chem. Int. Ed.* **40**, 3453 (2001); **42**: M. D. Hollingsworth, M. L. Peterson, K. L. Pate, B. D. Dinkelmeyer, and M. E. Brown, *J. Am. Chem. Soc.* **124**, 2094 (2002); **43**: observed in classical hydroquinone clathrates and phenolic compounds, T. C. W. Mak and B. R. F. Bracke, Hydroquinone clathrates and diamondoid host lattices, in D. D. MacNicol, F. Toda, and R. Bishop (eds.), *Comprehensive Supramolecular Chemistry*, Vol. 6, Pergamon Press, New York, 1996, pp. 23–60; **44**. C. K. Broder, J. A. K. Howard, D. A. Keen, C. C. Wilson, F. H Allen, R. K. R. Jetti, A. Nangia and G. R. Desiraju, *Acta Crystallogr.* **B56**, 1080 (2000); **45**: D. S. Reddy, D. C. Craig, and G. R. Desiraju, *J. Am. Chem. Soc.* **118**, 4090 (1996); **46**: B. Goldfuss, P. v. R. Schleyer, and F. Hampel, *J. Am. Chem. Soc.* **119**, 1072 (1997); **47**: P. J. Langley, J. Hulliger, R. Thaimattam, and G. R. Desiraju, *New J. Chem.* 307 (1998); **48**. B. Moulton and M. J. Zaworotko, *Chem. Rev.* **101**, 1629 (2001).

single interaction may be regarded as a synthon, for instance Cl···Cl, I···I, or N···Br (Figs 10.1.6 (**23–25**)). Besides the strong hydrogen bonds (N–H···O and O–H···O), which are expected to be frequently involved in supramolecular synthons (Figs 10.1.6 (**1–5**)), weak hydrogen bonds of the C–H···X variety and π–π interactions may also be significant. Although such weaker interactions have low energies in the range of 2 to 20 kJ mol^{-1}, their cumulative effects on molecular association and crystal structure and packing are just about as predictable as the effects of conventional hydrogen bonding.

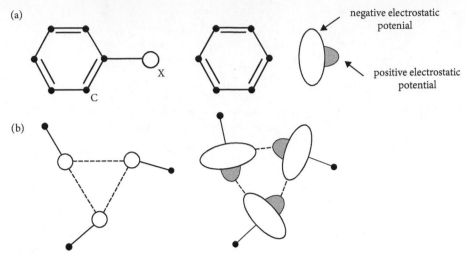

Fig. 10.1.7 (a) Areas of positive and negative electrostatic potentials at the halogen substituent of a phenyl ring and (b) stabilization of the X···X trimer supramolecular synthon **44**.

The nature of the X···X interaction in trimer synthon **44** is illustrated in Fig. 10.1.7. The C–X bond in a halo-substituted phenyl ring is polarized, so that there are regions of positive and negative electrostatic potentials around the X atom. The cyclic interaction of three C–X groups optimizes electrostatic potential overlap in the halogen trimer system.

10.2 **Hydrogen-Bond Directed Assembly**

Hydrogen bonding is an indispensable tool for designing molecular aggregates within the fields of supramolecular chemistry, molecular recognition, and crystal engineering. It is well recognized that in organic crystals certain building blocks or supramolecular synthons have a clear pattern preference, and molecules that contain these building blocks tend to crystallize in specific arrangements with efficient close packing. As already mentioned in Section 1.2.1, three important rules are generally applicable to the formation of hydrogen bonds between functional groups in neutral organic molecules: (a) all strong donor and acceptor sites are fully utilized, (b) intramolecular hydrogen bonds giving rise to a six-membered ring will form in preference to intermolecular hydrogen bonds, and (c) the remaining proton donors and acceptors not used in (b) will form intermolecular hydrogen bonds to one another. For additional rules and a full discussion the reader is referred to the papers of Etter (*Acc. Chem. Res.* **23**, 120–6 (1990); *J. Phys. Chem.* **95**, 4601–10 (1991)). Fig. 10.2.1 shows a hydrogen-bonded polar sheet where 3,5-dinitrobenzoic acid and 4-aminobenzoic acid are co-crystallized.

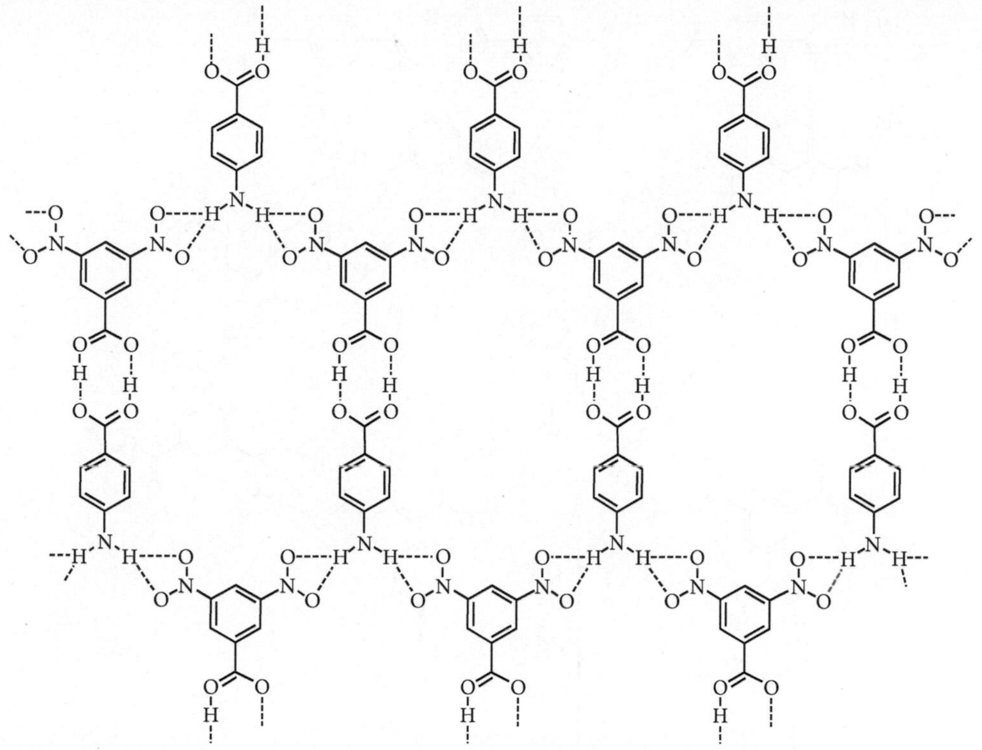

Fig. 10.2.1 Polar sheet formed by 3,5-dinitrobenzoic acid and 4-aminobenzoic acid using the nitro-amine and carboxylic acid dimer motif.

10.2.1 Supramolecular architectures based on the carboxylic acid dimer synthon

Carboxylic acids are commonly used as pattern controlling functional groups for the purpose of crystal engineering. The most prevalent hydrogen bonding patterns formed by carboxylic acids are the dimer and the catemer. Carboxylic acids containing small substituent groups (formic acid, acetic acid) form the catemer motif, while most others (especially aromatic carboxylic acids) form dimers, although not exclusively. In the case of di- and poly-carboxylic acids, terephthalic acid and isophthalic acid form linear and zigzag ribbons (or tapes), respectively, trimesic acid (1,3,5-benzenetricarboxylic acid) with its three-fold molecular symmetry forms a hydrogen-bonded sheet, and adamantane-1,3,5,7-tetracarboxylic acid forms a diamondoid network (Fig. 10.2.2).

If a bulky hydrophobic group is introduced at the 5-position of isophthalic acid, a cyclic hexamer (rosette) is generated. In the crystal structure of trimesic acid, the voids are filled through interpenetration of two honeycomb networks with additional stabilization by π–π stacking interactions. In the host lattice of adamantane-1,3,5,7-tetracarboxylic acid, the large voids are filled by interpenetration and small guest molecules.

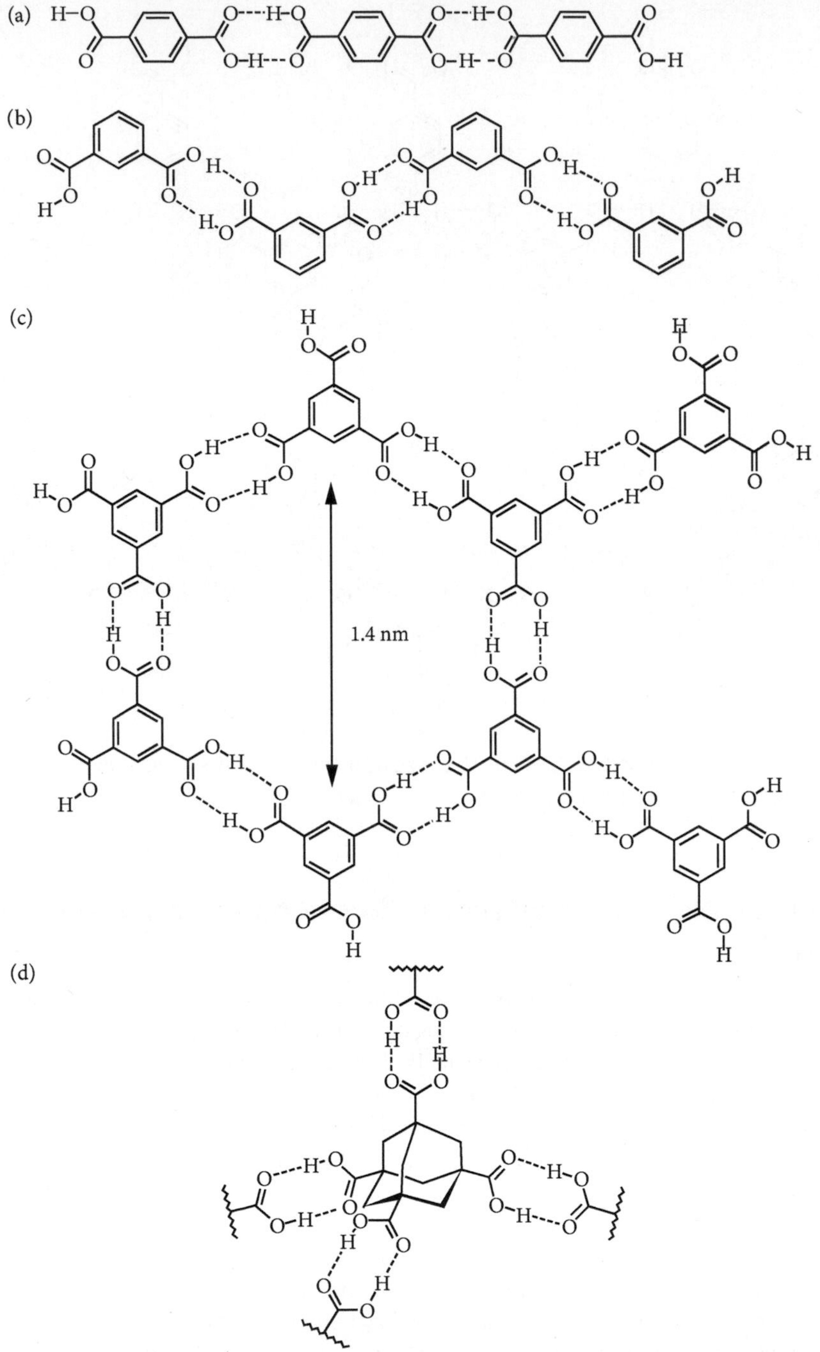

Fig. 10.2.2 One-dimensional linear tape (a) and crinkled (or zigzag) tape (b), two-dimensional sheet (or layer) (c), and three-dimensional network (or framework) (d) held together by the carboxylic acid dimer synthon.

10.2.2 **Graph-set encoding of hydrogen-bonding pattern**

The robust intermolecular motifs found in organic systems can be used to direct the synthesis of supramolecular complexes in crystal engineering. In the interest of adopting a systematic notation for the topology of hydrogen-bonded motifs and networks, a graph-set approach has been suggested by Etter and Ward. This provides a description of hydrogen-bonding schemes in terms of four pattern designators (**G**), that is, infinite chain (**C**), ring (**R**), discrete complex (**D**), and intramolecular (self-associating) ring (**S**), which together with the degree of the pattern (n, the number of atoms comprising the pattern), the number of donors (d), and the number of acceptors (a) are combined to form the quantitative graph-set descriptor $G_d^a(n)$. Examples of the use of these quantitative descriptors are given in Fig. 10.2.3.

Preferable hydrogen-bonding patterns of a series of related compounds containing a particular type of functional group can be obtained by graph-set analysis. For example, most primary amides prefer forming cyclic dimers and chains with a common hydrogen-bonding pattern $C(4)[R_2^1(6)]$. Furthermore, important insights may be gained by graph-set analysis of two seemingly unrelated organic crystals, which may lead to similar hydrogen-bonding patterns involving different functional groups. Some of the most common supramolecular synthons found in the Cambridge Structure Database are illustrated below.

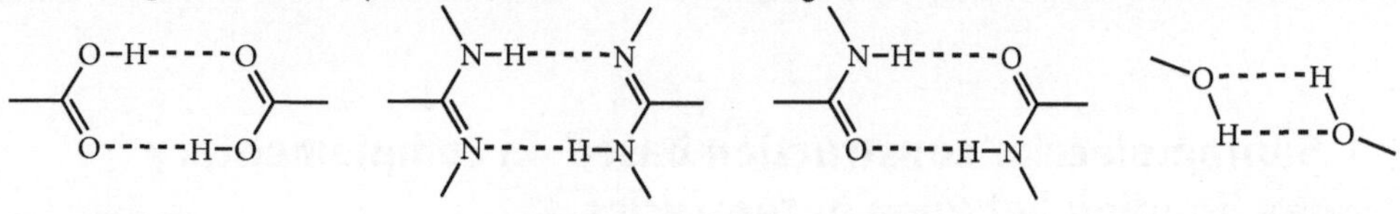

$$R_2^2(8) \qquad\qquad R_2^2(8) \qquad\qquad R_2^2(8) \qquad\qquad R_2^2(4)$$

The same set of hydrogen bond donors and acceptors may be connected in alternate ways to generate distinguishable motifs, giving rise to different polymorphic forms. A good example is 5,5-diethylbarbituric acid, for which three crystalline polymorphs that exhibit polymeric ribbon structures are shown in Fig. 10.2.4.

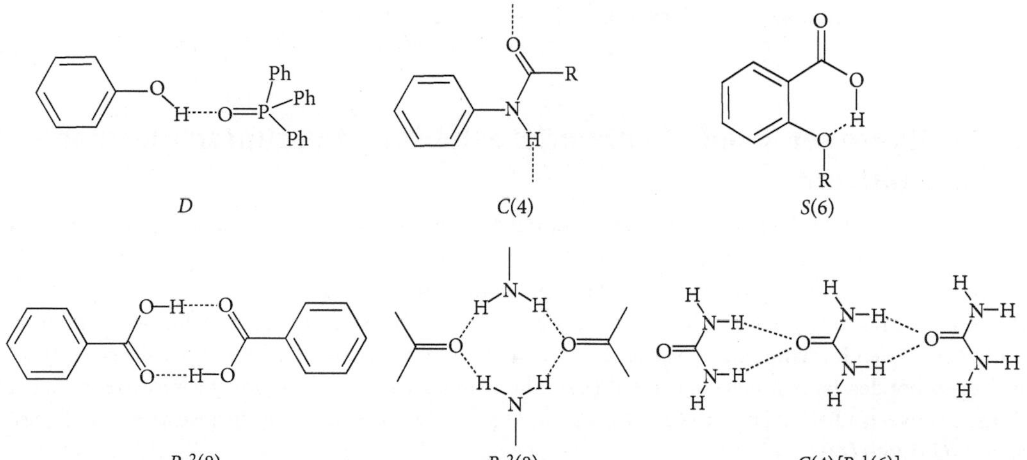

$$D \qquad\qquad C(4) \qquad\qquad S(6)$$

$$R_2^2(8) \qquad\qquad R_4^2(8) \qquad\qquad C(4)[R_2^1(6)]$$

Fig. 10.2.3 Some examples of graph-set descriptors of hydrogen-bonded structural motifs.

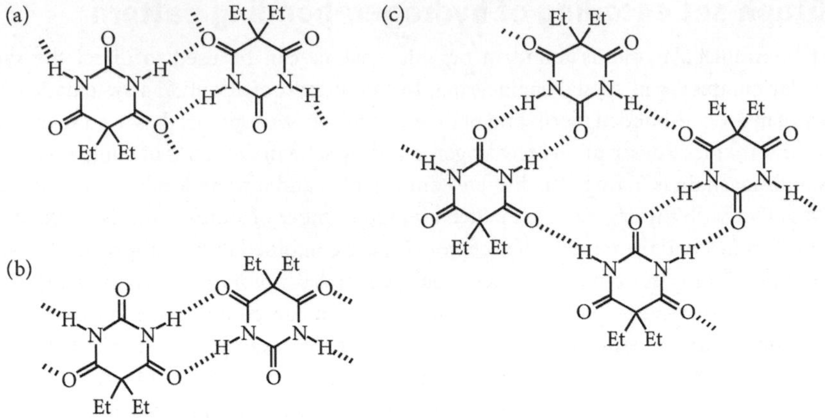

Fig. 10.2.4 Patterns of hydrogen bonding found in three polymorphic forms of 5,5-diethylbarbituric acid. The graph-set descriptors are $C(6)[R_2^2(8)R_4^2(12)]$ for (a), $C(10)[R_2^2(8)]$ for (b), and $C(6)[R_2^2(8)R_4^4(16)]$ for (c).

10.2.3 Supramolecular construction based on complementary hydrogen bonding between heterocycles

The simple heterocyclic compounds melamine and cyanuric acid possess perfectly matched sets of donor–acceptor sites, 6/3 and 3/6 respectively, for complementary hydrogen bonding to form a planar hexagonal network (Fig. 10.2.5). Three distinct structural motifs can be recognized in this extended array: (a) linear tape, (b) crinkled tape, and (c) cyclic hexameric aggregate (rosette). Using barbituric acid derivatives and 2,4,6-triaminopyrimidine derivatives, the group of Whitesides synthesized all three preconceived systems. The conceptual design involves disruption of N–H···O and N–H···N hydrogen bonding in specific directions by introducing suitable hydrophobic bulky groups (Fig. 10.2.6).

The group of Reinhoudt has reported the construction of a D_3 hydrogen-bonded assembly of three calix[4]arene bismelamine and six barbituric acid derivatives (Fig. 10.2.7).

10.2.4 Hydrogen-bonded networks exhibiting the supramolecular rosette pattern

Ward has shown that the self-assembly of cations and anions in a guanidinium sulfonate salt, through precise matching of donor and acceptor sites, gives rise to a layer structure displaying the rosette motif, which is not planar but corrugated since the configuration at the S atom is tetrahedral (Fig. 10.2.8). If the sulfonate R group is small, a bilayer structure with interdigitated substituents is formed (structural motif (a)). When R is large, a single layer structure with substituents alternating on opposite sides of the hydrogen-bonded layer is obtained (motif (b)). Alternatively, the use of disulfonates provides covalent linkage between adjacent layers to generate a pillared 3-D network, which may enclose a variety of guest species G (motif (c)).

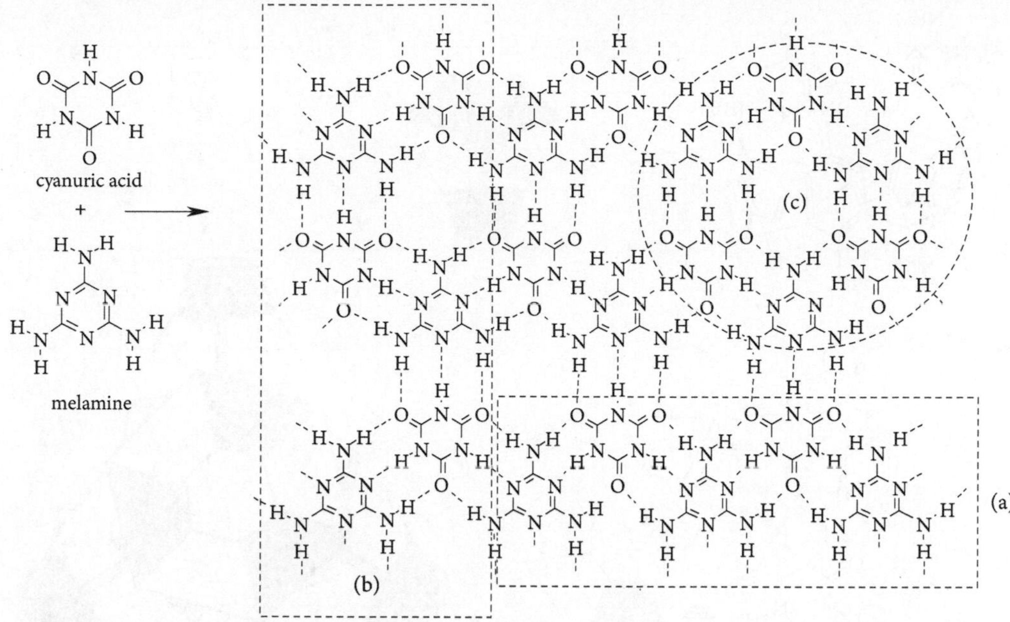

Fig. 10.2.5 Hexagonal layer structure of the 1:1 complex of melamine and cyanuric acid. Three kinds of assembly in lower dimensions are possible: (a) linear tape, (b) crinkled tape, and (c) rosette.

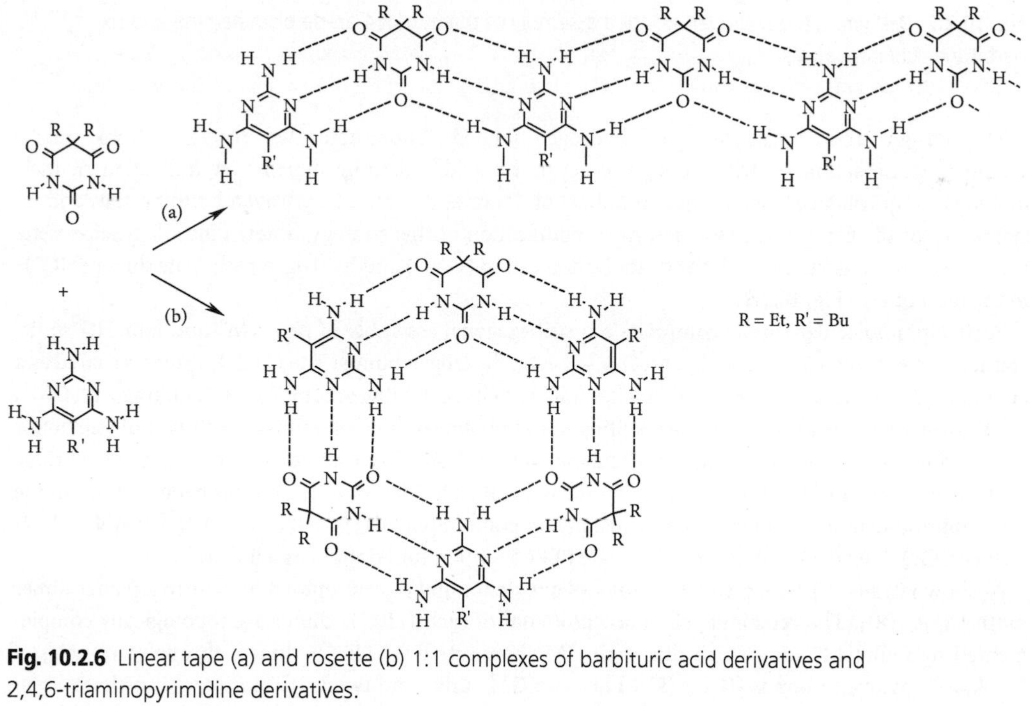

Fig. 10.2.6 Linear tape (a) and rosette (b) 1:1 complexes of barbituric acid derivatives and 2,4,6-triaminopyrimidine derivatives.

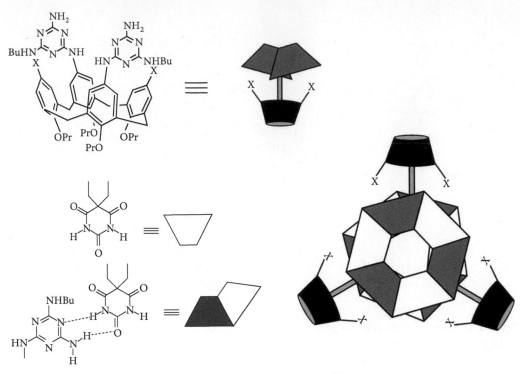

Fig. 10.2.7 D_3-Symmetric hydrogen-bonded assembly of three calix[4]arene bismelamine and six barbituric acid derivatives.

The design and construction of hydrogen-bonded "supramolecular rosettes" from guanidinium/organic sulfonate, trimesic acid, or cyanuric acid/melamine depend on utilization of their topological equivalence; that is, equal numbers of donor and acceptor hydrogen bonding sites and C_3 symmetry of the component moieties. As a modification of this strategy, a new kind of "fused-rosette ribbon" can be constructed with the guanidinium cation (GM$^+$) and hydrogen carbonate dimer (HC$^-$)$_2$ in the ratio of 1:1 (Fig. 10.2.9).

Each supramolecular rosette comprises a quasihexagonal assembly of two GM$^+$ and four HC$^-$ units connected by strong N_{GM}–H$\cdots$O$_{HC}$ and O$_{HC}$–H$\cdots$O$_{HC}$ hydrogen bonds. The (HC$^-$)$_2$ dimer is shared as a common edge of adjacent rosettes and makes full use of its remaining acceptor sites in linking with GM$^+$. On the other hand, each GM$^+$ in the resulting linear ribbon (or tape) still possesses a pair of free donor sites, and it is anticipated that some "molecular linker" with suitable acceptor sites may be used to bridge an array of parallel ribbons to form a sheet-like network. This design objective has been realized in the synthesis and characterization of the inclusion compound 5[C(NH$_2$)$_3$$^+$]·4(HCO$_3$$^-$)·3[(^{n}Bu)$_4N^+$]·2[1,4-C$_6H_4$-(COO$^-$)$_2$]·2H$_2$O with the terephthalate (TPA^{2-}) anion functioning as a linker.

As shown in Fig. 10.2.10, each HC$^-$ provides one donor and one acceptor site to form a planar dimer motif [**A**, R$_2$^{2}(8)]. The remaining eight acceptor sites of each (HC$^-$)$_2$ dimer are topologically complemented by four GM$^+$ units, such that each GM$^+$ connects two (HC$^-$)$_2$ dimers through two pairs of N–H$_{syn}$$\cdots$O hydrogen bonds [**B**, R$_2$^{2}(8)]. Thus two GM$^+$ units and two (HC$^-$)$_2$ dimers constitute a planar, pseudocentrosymmetric, quasihexagonal supramolecular rosette [**C**, R$_6$^{4}(12)] with inner and outer diameter of ca. 0.55 and 0.95 nm, respectively. In the resulting fused-rosette ribbon, the remaining two

(a)

(b)

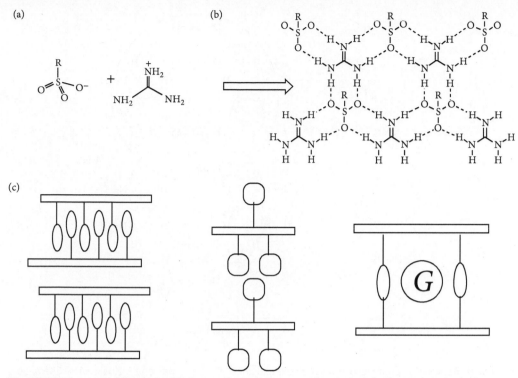

(c)

Fig. 10.2.8 Top: assembly of guanidinium and sulfonate groups by N–H⋯O hydrogen bonds to give a corrugated rosette layer. Bottom: structural motifs (a), (b), and (c), which are formed depending on the size of R and the nature of the sulfonate group used.

exo-orientated donor sites of each GM⁺ unit form a pair of N–H$_{anti}$⋯O hydrogen bonds [**D**, $R_2^2(8)$] with a TPA²⁻ carboxylate group. Thus two types of ladders are developed: type (I) [TPA²⁻ composed of C(10) to C(17) and O(13) to O(16)] is consolidated by two independent water molecules that alternately bridge carboxylate oxygen atoms of neighboring steps by pairs of donor O_w–H⋯O hydrogen bonds, generating a centrosymmetric ring motif [**E**, $R_4^4(22)$] and pentagon pattern [**F**, $R_6^3(12)$]; in the type (II) ladder [TPA²⁻ composed of C(18) to C(25) and O(17) to O (20)], carboxylate oxygen atoms O(18)' and O(19) belonging to adjacent steps are connected by the remaining GM⁺ ions, which are not involved in rosette formation, via two pairs of donor hydrogen bonds [**G**, $R_2^1(6)$] (Fig. 10.2.11(a)). The remaining two pairs of donor sites of each free GM⁺ ion are linked to a carboxylate oxygen atom and a water molecule of TPA²⁻ column (I) of an adjacent layer to form a pentagon motif [**H**, $R_3^2(8)$], thus yielding a 3-D pillared layer structure (Fig. 10.2.11(b)). The large voids in the pillar region generate nanoscale channels extending along the [100] direction. The dimensions of the cross section of each channel are ca. 0.8 × 2.2 nm, within which three independent [(ⁿBu)₄N]⁺ cations are aligned in separate columns in a well-ordered manner.

Drawing upon the above successful design of a linear "fused-rosette ribbon" assembled from the (HC⁻)₂ dimer and GM⁺ in 1:1 molar ratio, it would be challenging to attempt the hydrogen-bond mediated construction of two premeditated anionic rosette-layer architectures using guanidinium and ubiquitous C_3-symmetric oxo-anions that carry *unequal* charges, namely guanidinium-carbonate **I** and guanidinium-trimesate **II**, as illustrated in Fig. 10.2.12.

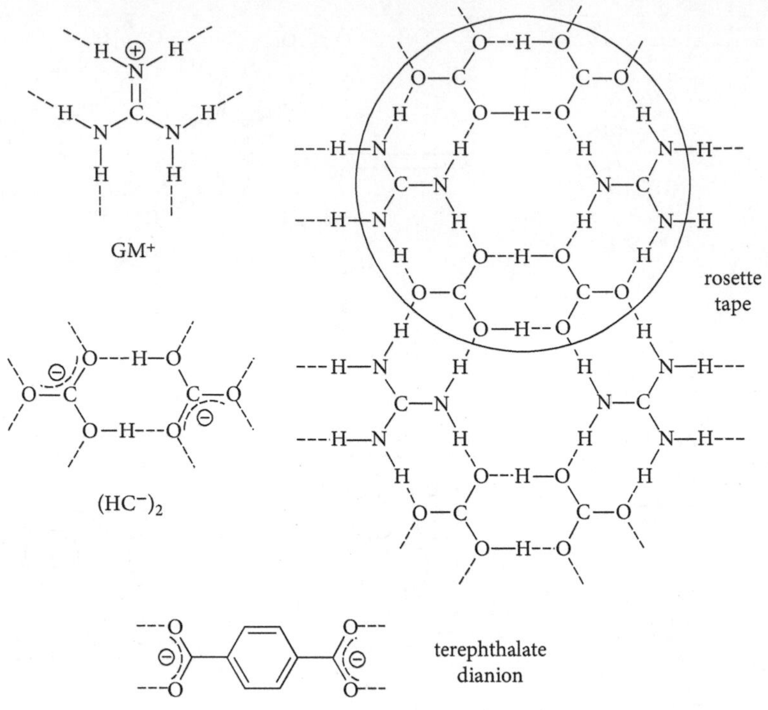

Fig. 10.2.9 Design of supramolecular rosette tape and linker. (T. C. W. Mak and F. Xue, *J. Am. Chem. Soc.* **122**, 9860–1 (2000).)

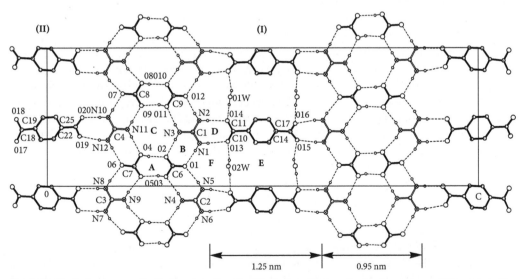

Fig. 10.2.10 Projection on (010) showing the two-dimensional network. Atom types are differentiated by size and shading, and hydrogen bonds are represented by broken lines.

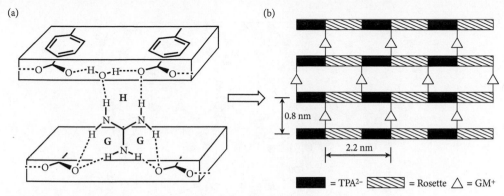

Fig. 10.2.11 (a) Hydrogen-bonding motifs involving linkage of the free guanidinium ion to neighboring rosette ribbon-terephthalate layers. (b) Schematic presentation of the pillared layer structure.

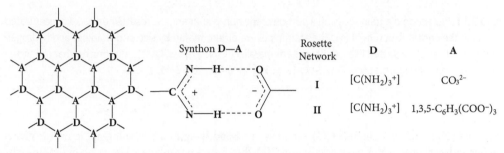

Fig. 10.2.12 Design of supramolecular rosette layers. (C. -K. Lam and T. C. W. Mak, *J. Am. Chem. Soc.* **127**, 11,536–7 (2005).)

In principle, the negatively charged, presumably planar network **I** can be combined with one molar equivalent of tetraalkylammonium ion R_4N^+ of the right size as interlayer template to yield a crystalline inclusion compound of stoichiometric formula $(R_4N^+)[C(NH_2)_3^+]CO_3^{2-}$ that is reminiscent of the graphite intercalates. Anionic network **II**, on the other hand, needs twice as many monovalent cations for charge balance, and furthermore possesses honeycomb-like host cavities of diameter ~700 pm that have to be filled by suitable guest species. The expected formula of the corresponding inclusion compound is $(R_4N^+)_2[C(NH_2)_3^+] [1,3,5-C_6H_3(COO^-)_3] \cdot G$, where G is an entrapped guest moiety with multiple hydrogen-bond donor sites to match the nearly planar set of six carboxylate oxygens that line the inner rim of each cavity.

Crystallization of $(R_4N^+)[C(NH_2)_3^+]CO_3^{2-}$ by variation of R, based on conceptual network **I**, was not successful. Taking into account the fact that the guanidinium ion can function as a pillar between layers and the carbonate ion is capable of forming up to twelve acceptor hydrogen bonds, as observed in crystalline bis(guanidinium) carbonate and $[(C_2H_5)_4N^+]_2 \cdot CO_3^{2-} \cdot 7(NH_2)_2CS$, the synthetic strategy was modified by incorporating a second guanidinium salt $[C(NH_2)_3]X$ as an extra component. After much experimentation with various combinations of R and X, the targeted construction of network **I**, albeit in undulating form, was realized through the isolation of crystalline $4[(C_2H_5)_4N^+] \cdot 8[C(NH_2)_3^+] \cdot 3(CO_3)^{2-} \cdot 3(C_2O_4)^{2-} \cdot 2H_2O$ (**1**).

In the asymmetric unit of (**1**), there are two independent carbonate anions and five independent guanidinium cations, which are henceforth conveniently referred to by their carbon atom labels in bold type.

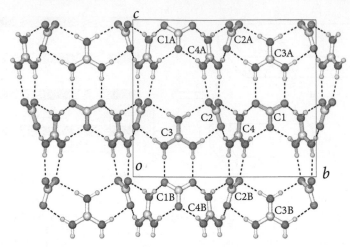

Fig. 10.2.13 Projection diagram showing a portion of the non-planar anionic rosette network **I** concentrated at $a = \frac{1}{4}$ in the crystal structure of (**1**). The atom types are differentiated by size and shading, and hydrogen bonds are indicated by dotted lines. Adjacent anti-parallel $\{[C(NH_2)_3]^+ \cdot CO_3{}^{2-}\}_\infty$ ribbons run parallel to the b-axis. Symmetry transformations: A ($\frac{1}{2} - x$, $1 - y$, $\frac{1}{2} + z$), B ($\frac{1}{2} - x$, $y - \frac{1}{2}$, $z - \frac{1}{2}$).

Carbonate **C(1)** and guanidinium **C(3)** each has one bond lying in a crystallographic mirror plane; together with carbonate **C(2)** and guanidinium **C(4)**, they form a non-planar zigzag ribbon running parallel to the b-axis, neighboring units being connected by a pair of strong N^+—$H\cdots O^-$ hydrogen bonds (Fig. 10.2.13). Adjacent anti-parallel $\{[C(NH_2)_3]^+ \cdot CO_3{}^{2-}\}_\infty$ ribbons are further cross-linked by strong N^+—$H\cdots O^-$ hydrogen bonds to generate a highly corrugated rosette layer, which is folded into a plane-wave pattern by guanidinium **C(5)** (Fig. 10.2.4). Guanidinium **C(6)** and **C(7)** protruding away from carbonate **C(1)** and **C(2)** are hydrogen-bonded to the two-fold disordered oxalate ion containing **C(8)** and **C(9)** (Fig. 10.2.14), forming a pouch that cradles the disordered $(C_2H_5)_4N^+$ ion. The carbonate ions **C(1)** and **C(2)** each form eleven acceptor hydrogen bonds, only one fewer than the maximum number. The resulting composite hydrogen-bonded layers at $a = \frac{1}{4}$ and $\frac{3}{4}$ are interconnected by $[(C_2O_4{}^{2-} \cdot (H_2O)_2]_\infty$ chains derived from centrosymmetric oxalate **C(10)** and water molecules O1w and O2w via strong ^+N—$H\cdots O^-$ hydrogen bonds to generate a complex three-dimensonal host framework, within which the second kind of disordered $(C_2H_5)_4N^+$ ions are accommodated in a zigzag fashion within channels extending along the [010] direction.

The predicted assembly of guanidinium-trimesate network **II** was achieved through the crystallization of $[(C_2H_5)_4N^+]_2 \cdot [C(NH_2)_3{}^+] \cdot [1,3,5\text{-}C_3H_3(COO^-)_3] \cdot 6H_2O$ (**2**). The guanidinium and trimesate ions are connected together by pairs of strong charge-assisted ^+N—$H\cdots O^-$ hydrogen bonds to generate an essentially planar rosette layer with large honeycomb cavities (Fig. 10.2.15 (left)). Three independent water molecules constitute a cyclic $(H_2O)_6$ cluster of symmetry **2**, which is tightly fitted into each host cavity by adopting a flattened-chair configuration in an out-of-plane orientaion, with $O\cdots O$ distances comparable to 275.9 pm in deuterated ice I_h. Each water molecule has its ordered hydrogen atom pointing outward to form a strong O—$H\cdots O^-$ hydrogen bond with a carboxylate oxygen on the inner rim of the cavity. The well-ordered $(C_2H_5)_4N^+$ guests, represented by large yellow spheres, are sandwiched between anionic rosette host layers with an interlayer spacing of ~750 pm (Fig. 10.2.15 (right)).

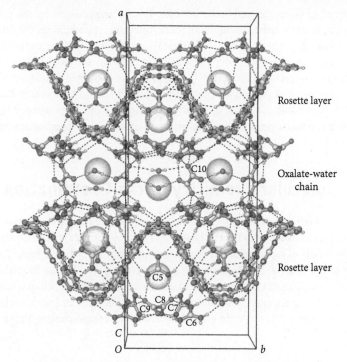

Fig. 10.2.14 Perspective view of the crystal structure of (**1**) along [001]. The undulating guanidinium-carbonate rosette network **I** appears as a sinusoidal cross-section. Adjacent composite hydrogen-bonded layers are interconnected by a $[C_2O_4^{2-} \cdot (H_2O)_2]_\infty$ chain. The disordered oxalate is shown in one possible orientation, and the two different types of disordered Et_4N^+ ions (represented by large semi-transparent spheres) are included in the pouches and the zigzag channels running parallel to the [010] direction, respectively.

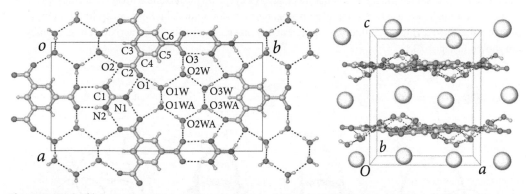

Fig. 10.2.15 (Left) Projection diagram showing the hydrogen-bonding scheme in the infinite rosette layer **II** of (**2**). Only one of the two cyclic arrangements of disordered H atoms lying on the edges of each $(H_2O)_6$ ring is displayed. Symmetry transformation: A $(1 - x, y, \frac{1}{2} - z)$. (Right) Sandwich-like crystal structure of (**2**) viewed along the *b*-axis.

The malleability of guanidinium-carbonate network **I**, rendered possible by the prolific hydrogen-bond accepting capacity of its carbonate building block, opens up opportunities for further exploration of supramolecular assembly. The flattened-chair $(H_2O)_6$ guest species, filling the cavity within robust guanidinium-trimesate layer **II** and being comparable to that in the host lattice of bimesityl-3,3'-dicarboxylic acid, may conceivably be replaced by appropriate hydrogen-bond donor molecules. The present anionic rosette networks are unlike previously reported neutral honeycomb lattices of the same (6,3) topology, thus expanding the scope of de novo engineering of *charge-assisted* hydrogen-bonded networks using ionic modular components, from which discrete molecular aggregates bearing the rosette motif may be derived.

10.3 Supramolecular Chemistry of the Coordination Bond

The predictable coordination geometry of transition metals and the directional characteristics of interacting sites in a designed ligand provide the blueprint (or programmed instructions) for the rational synthesis of a wide variety of supramolecular inorganic and organometallic systems. Current research is concentrated in two major areas: (a) the construction of novel supermolecules from the intermolecular association of a few components and (b) the spontaneous organization (or self-assembly) of molecular units into 1-, 2-, and 3-D arrays. In the solid state, the supermolecules and supramolecular arrays can further associate with one another to yield gigantic macroscopic conglomerates, that is supramolecular structures of higher order.

10.3.1 Principal types of supermolecules

The supermolecules that have been synthesized include large metallocyclic rings, helices, and host-guest complexes, as well as interlocked structures such as catenanes, rotaxanes, and knots.

The simplest catenane is [2]catenane which contains two interlocked rings. A polycatenane has three or more rings interlocked in a one-to-one linear fashion. A rotaxane has a ring component (or a bead) threaded by a linear component (or a string) with a stopper at each end. A polyrotaxane has several rings threaded onto the same string. A molecular necklace is a cyclic oligorotaxane with several rings threaded by a closed loop. There are two topological isomers for a trefoil knot. The Borromean link is composed of three interlocked rings such that the scission of any one ring unlocks the other two. These topologies are shown in Fig. 10.3.1.

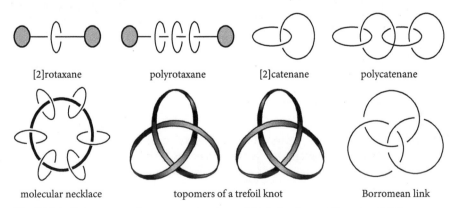

| [2]rotaxane | polyrotaxane | [2]catenane | polycatenane |

| molecular necklace | topomers of a trefoil knot | Borromean link |

Fig. 10.3.1 Catenanes, rotaxanes, molecular necklace, knots, and Borromean link.

10.3.2 **Some examples of inorganic supermolecules**

(1) **Ferric wheel**

The best known example of a large metallocyclic ring is the "ferric wheel" $[Fe(OMe)_2(O_2CCH_2Cl)]_{10}$ prepared from the reaction of oxo-centered trinuclear $[Fe_3O(O_2CCH_2Cl)_6(H_2O)_3](NO_3)$ and $Fe(NO_3)_3 \cdot 9H_2O$ in methanol. An X-ray analysis showed that it is a decameric wheel having a diameter of about 1.2 nm with a small and unoccupied hole in the middle (Fig. 10.3.2). Each pair of iron(III) centers are bridged by two methoxides and one O, O'-chloroacetate group.

(2) **Hemicarceplex**

A carcerand is a closed-surface, globular host molecule with a hollow interior that can enclose guest species such as small organic molecules and inorganic ions to form a carceplex. A hemicarcerand is a carcerand that contains portals large enough for the imprisoned guest molecule to escape at high temperatures, but otherwise remains stable under normal laboratory conditions. The hemicarcerand host molecule designed by Cram shown in Fig. 10.3.3 has idealized D_{4h} symmetry with its four-fold axis roughly coincident with the long axis of the ferrocene guest, which lies at an inversion center and hence adopts a fully staggered D_{5d} conformation. The 1,3-diiminobenzene groups connecting the northern and southern hemispheres of the hemicarceplex are arranged like paddles in a paddle wheel around the circumference of the central cavity.

(3) **[2]Catenane Pt(II) complex**

The coordination bond between a pyridyl N atom and Pt(II) is normally quite stable, but it becomes labile in a highly polar medium at high concentration. Fig 10.3.4 shows the overall one-way transformation of a

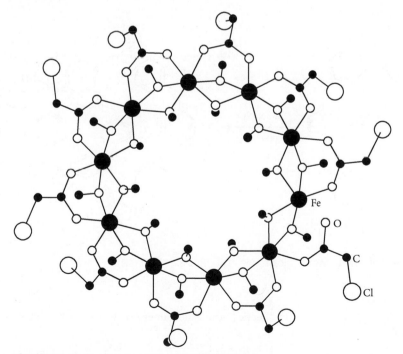

Fig. 10.3.2 Molecular structure of the ferric wheel.

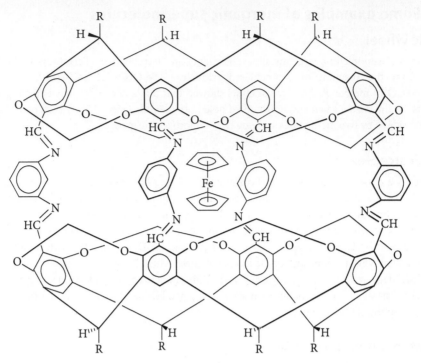

Fig. 10.3.3 A hemicarceplex consisting of a hemicarcerand host molecule enclosing a ferrocene guest molecule.

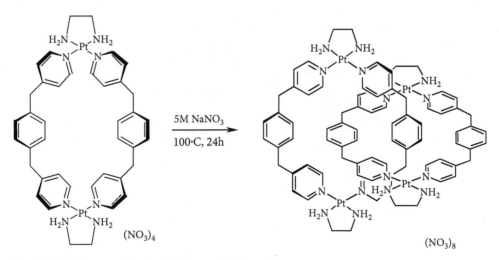

Fig. 10.3.4 Self-assembly of a [2]catenane Pt(II) complex.

binuclear cyclic Pt(II) complex into a dimeric [2]catenane framework, which can be isolated as its nitrate salt upon cooling.

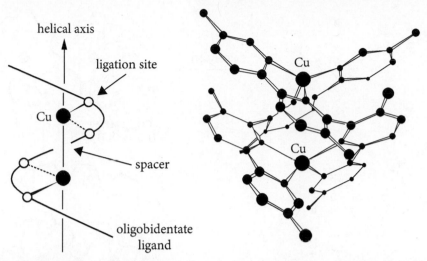

Fig. 10.3.5 (a) Structural features of a *M*-type helicate (only one strand is shown). (b) A Cu(I) double helicate containing an oligopyridine ligand.

(4) Helicate

In a helicate the polytopic ligand winds around metal ions lying on the helical axis, so that its ligation sites match the coordination requirements of the metal centers. The two enantiomers of a helicate are designated *P*- for a right-handed helix and *M*- for a left-handed helix (Fig. 10.3.5(a)). An example of a *M*-type Cu(I) double helicate is shown in Fig. 10.3.5(b). Double helicates containing up to five Cu(I) centers and triple helicates involving the wrapping of oligobidentate ligands around octahedral Co(II) and Ni(II) centers have been synthesized.

(5) Molecular trefoil knot

The first molecular trefoil knot was successfully synthesized according to the scheme shown in Fig. 10.3.6. A specially designed ligand consisting of two diphenolic 1,10-phenanthroline units tied together by a tetramethylene tether was reacted with $[Cu(MeCN)_4]BF_4$ to give a dinuclear double helix. This precursor was then treated under high dilution conditions with two equivalents of the diiodo derivative of hexaethylene glycol in the presence of Cs_2CO_3 to form the cyclized complex in low yield. Finally, demetallation of this helical dicopper complex yielded the desired free trefoil knot with retention of topological chirality. It should be noted that in the crucial cyclization step, the double helical precursor is in equilibrium with a non-helical species, which leads to an unknotted dicopper complex consisting of two 43-membered rings arranged around two Cu(I) centers in a face-to-face manner (Fig. 10.3.7).

(6) Molecular Borromean link

The connection between chemistry and topology is exemplified by the elegant construction of a molecular Borromean link that consists of three identical interlocked rings. In the designed synthesis, each component ring comprises two short and two long segments. The programmed, one-step self-assembly process is indicated by the scheme illustrated in Fig. 10.3.8.

Self-assembly of the molecular Borromean link (a dodecacation **BR**$^{12+}$ is achieved by a template-directed cooperative process that results in over 90% yield. Each of the three component rings (L)

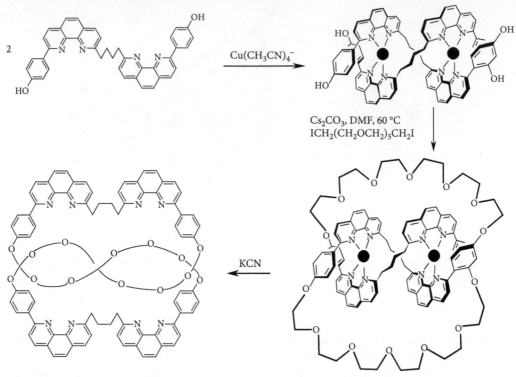

Fig. 10.3.6 Synthetic scheme for dicopper and free trefoil knots. (C. O. Dietrich-Buchecker, J. Guilhem, C. Pascard, and J.-P. Sauvage, *Angew. Chem. Int. Ed.* **29**, 1154–6 (1990).)

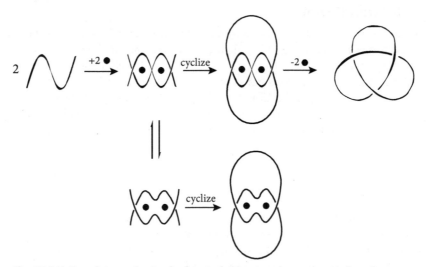

Fig. 10.3.7 Template synthesis of a free trefoil knot and an unknotted product.

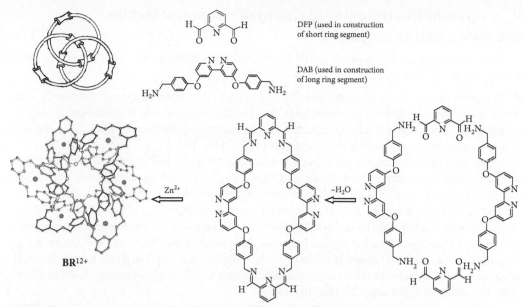

Fig. 10.3.8 Synthetic scheme for one-step supramolecular assembly of the molecular Borromean link $[L_3Zn_6]^{12+} \equiv \mathbf{BR}^{12+}$; its three component interlocked rings are differentiated by different degrees of shading. Each octahedral zinc(II) ion is coordinated by an *endo*-N$_3$ ligand set and a chelating bipyridyl group belonging to a different ring, the monodentate acetate ligand being omitted for clarity. (K. S. Chichak, S. J. Cantrill, A. R. Pease, S.-H Chiu, G. W. V. Cave, J. L. Atwood, and J. F. Stoddart, *Science (Washington)* **304**, 1308–12 (2004).)

in $\mathbf{BR}^{12+}$ is constructed from [2+2] macrocyclization involving two DFP (2,6-diformylpyridine) and two DAB (diamine containing a 2,2′-bipyridyl group) molecules. Kinetically labile zinc(II) ions serve efficiently as templates in multiple molecular recognition that results in precise interlocking of three macrocyclic ligands.

$$6\,\text{DFB} + 6\,[\text{DABH}_4](\text{CF}_3\text{CO}_2)_4 + 9\,\text{Zn}(\text{CF}_3\text{CO}_2)_2 \xrightarrow{\text{CH}_3\text{OH}} [\text{L}_3(\text{ZnCF}_3\text{CO}_2)_6]\cdot 3\,[\text{Zn}(\text{CF}_3\text{CO}_2)_4] + 12\text{H}_2\text{O}$$
$$\|\|$$
$$[\mathbf{BR}(\text{CF}_3\text{CO}_2)_6]\cdot 3\,[\text{Zn}(\text{CF}_3\text{CO}_2)_4]$$

An X-ray analysis of $[\mathbf{BR}(\text{CF}_3\text{CO}_2)_6]\cdot3[\text{Zn}(\text{CF}_3\text{CO}_2)_4]$ revealed that the hexacation $[\mathbf{BR}(\text{CF}_3\text{CO}_2)_6]^{6+}$ has S_6 symmetry with each macrocyclic ligand L adopting a chair-like conformation (Fig. 10.3.8). The interlocked rings are consolidated by six Zn(II) ions, each being coordinated in a slightly distorted octahedral geometry by the *endo*-tridentate diiminopyridyl group of one ring, the *exo*-bidentate bipyridyl group of another ring, and an oxygen atom of a CF_3CO_2^- ligand. Each bipyridyl group is sandwiched unsymmetrically between a pair of phenolic rings at π–π stacking distance of 361 and 366 pm in different directions.

In the crystal packing, the $[\mathbf{BR}(\text{CF}_3\text{CO}_2)_6]^{6+}$ ions are arranged in hexagonal arrays with intermolecular π–π stacking interactions of 331 pm and C–H⋯O=C hydrogen bonds (H⋯O 252 pm), generating columns along c that accommodate the $[\text{Zn}(\text{CF}_3\text{CO}_2)_4]^{2-}$ counterions.

10.3.3 Synthetic strategies for inorganic supermolecules and coordination polymers

Two basic approaches have been developed for the synthesis of inorganic supermolecules and 1-, 2-, and 3-D coordination polymers:

(1) Transition metal ions are employed as nodes and bifunctional ligands as spacers. Commonly used spacer ligands are pseudohlides such as cyanide, thiocyanate, and azide, and N-donor ligands such as pyrazine, 4,4′-bipyridine, and 2,2′-bipyrimidine. Besides discrete supermolecules, some 1-, 2-, and 3-D architectural motifs generated from this strategy are shown in Fig. 10.3.9.

If all nodes at the boundary of a portion of a motif are bound by terminal ligands, a discrete molecule will be formed. An example is the square grid shown in Fig. 10.3.10. Reaction of the tritopic ligand 6,6′-bis[2-(6-methylpyridyl)]-3,3′-bipyridazine (Me$_2$bpbpz) with silver triflate in 2:3 molar ratio in nitromethane results in self-assembly of a complex of the formula [Ag$_9$(Me$_2$bpbpz)$_6$](CF$_3$SO$_3$)$_9$. X-ray analysis showed that the [Ag$_9$(Me$_2$bpbpz)$_6$]$^{9+}$ cation is in the form of a 3 × 3 square grid, with two sets of Me$_2$bpbpz ligands positioned above and below the mean plane of the silver centers, as shown in Fig. 10.3.10(a), so that each Ag(I) atom is in a distorted tetrahedral environment. The grid is actually distorted into a diamond-like shape due to the curved nature of the ligand, and the angle between the mean planes of the two sets of ligands is about 72° (Fig. 10.3.10(b)).

(2) Exodentate multitopic ligands are used to link transition metal ions into building blocks. Some examples of such ligands are 2,4,6-tris(4-pyridyl)-1,3,5-triazine, oligopyridines, and 3- and 4-pyridyl-substituted porphyrins.

In the following sections, selected examples from the recent literature are used to illustrate the creative research activities in transition-metal supramolecular chemistry.

10.3.4 Molecular polygons and tubes

(1) Nickel wheel

The reaction of hydrated nickel acetate with excess 6-chloro-2-pyridone (HCHP) produces in 60% yield a dodecanuclear nickel complex, which can be recrystallized from tetrahydrofuran as a solvate of stoichiometry [Ni$_{12}$(O$_2$CMe)$_{12}$(chp)$_{12}$(H$_2$O)$_6$(THF)$_6$]. X-ray structure analysis revealed that this wheel-like molecule (Fig. 10.3.11) lies on a crystallographic three-fold axis. There are two kinds of nickel atoms in distorted octahedral coordination: Ni(1) is bound to three O atoms from acetate groups, two O atoms from chp ligands, and an aqua ligand, whereas Ni(2) is surrounded by two acetate O atoms, two chp O atoms, an aqua ligand, and the terminal THF ligand. All ligands, other than THF, are involved in bridging pairs of adjacent nickel atoms. The structure of this nickel metallocycle resembles that of the decanuclear "ferric wheel" [Fe(OMe)$_2$(O$_2$CCH$_2$Cl)]$_{10}$ (see Fig. 10.3.2). Both complexes feature a closed chain of intersecting M$_2$O$_2$ rings, with each ring additionally bridged by an acetate ligand. They differ in that in the ferric wheel the carboxylate ligands are all exterior to the ring, whereas in the nickel wheel half of the acetate ligands lie within the central cavity.

(2) Nano-sized tubular section

The ligand 2,4,6-tris[(4-pyridyl)methylsulfanyl]-1,3,5-triazine (tpst) possesses nine possible binding sites to transition metals for the assembly of supramolecular systems. Its reaction with AgNO$_3$ in a 1:2 molar ratio in DMF/MeOH followed by addition of AgClO$_4$ produces Ag$_7$(tpst)$_4$(ClO$_4$)$_2$(NO$_3$)$_5$(DMF)$_2$. In the crystal structure, two TPST ligands coordinate to three silver(I) ions to form a bicyclic ring. Two

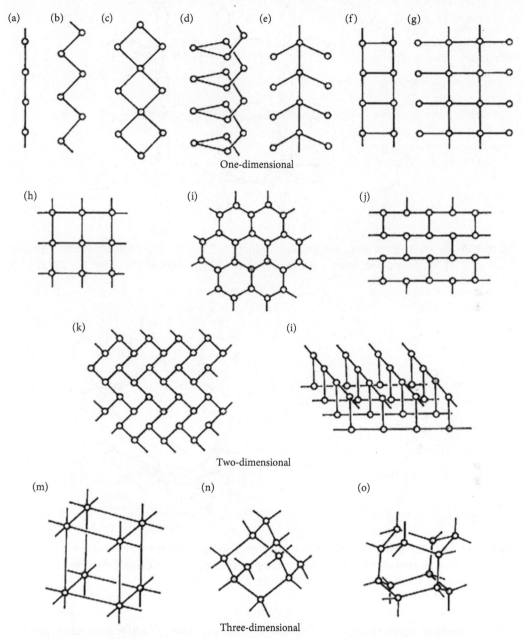

Fig. 10.3.9 Schematic representation of the motifs generated from the connection of transition metals by bifunctional spacer ligands. One-dimensional: (a) linear chain, (b) zigzag chain, (c) double chain, (d) helical chain, (e) fishbone, (f) ladder, and (g) railroad. Two dimensional: (h) square grid, (i) honeycomb, (j) brick wall, (k) herringbone, and (l) bilayer. Three-dimensional: (m) six-connected network, (n) four-connected diamondoid network, and (o) four-connected ice-I_h network.

(a)   (b)

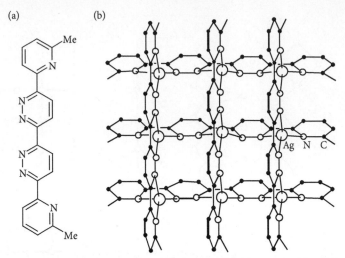

Fig. 10.3.10 The 3 × 3 square molecular grid $[Ag_9(Me_2bpbpz)_6]^{9+}$: (a) structural formula and (b) molecular structure. (P. N. W. Baxter, J-M. Lehn, J. Fischer and M-T. Youinou. *Angew. Chem. Int. Ed.* **33**, 2284–7 (1994).)

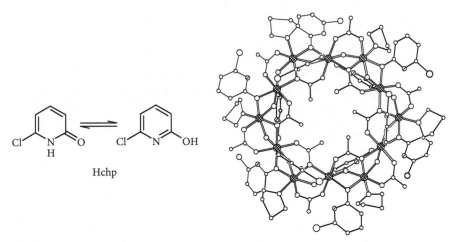

Hchp

Fig. 10.3.11 Molecular structure of the nickel wheel. The bridging pyridone and terminal THF ligands point alternately above and below the mean plane of the twelve nickel atoms. (A. J. Blake, C. M. Grant, S. Parsons, J. M. Rawson, and R. E. P. Winpenny. *Chem. Commun.* 2363–4 (1994).)

such rings are fitted together by Ag–N and Ag–S bonds involving a pair of bridging silver ions to generate a nano-sized tubular section $[Ag_7(tpst)_4]$ with dimensions of 1.34 × 0.96 × 0.89 nm, which encloses two perchlorate ions and two DMF molecules (Fig. 10.3.12). The tubular sections are further linked by additional Ag–N and Ag–S bonds to form an infinite chain. The nitrate ions are located near the silver ions and imbedded in between the linear polymers.

The two independent TPST ligands are each bound to four silver atoms but in different coordination modes: three pyridyl N plus one thioether S, or three pyridyl N plus one triazine N. The silver atoms exhibit two kinds of coordination modes: normal linear AgN_2 and a very unusual AgN_2S_2 mode of distorted square-planar geometry.

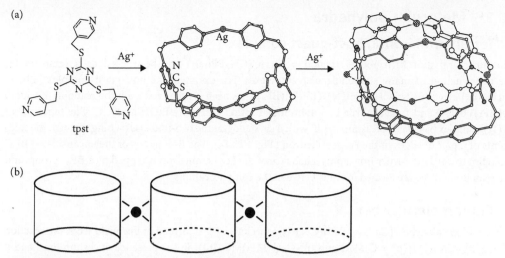

Fig. 10.3.12 (a) Formation and structure of the nano-sized tubular section [Ag$_7$(tpst)$_4$]. Bonds formed by bridging silver ions are indicated by broken lines. (b) Linkage of tubular sections to form a linear polymer. (M. Hong, Y. Zhao, W. Su, R. Cao, M. Fujita, Z. Zhou, and A. S. C. Chan. *Angew. Chem. Int. Ed.* **39**, 2468–70 (2000).)

(3) Infinite square tube

The reaction of 2-aminopyrimidine (apym), Na[N(CN)$_2$] and M(NO$_3$)$_2$·6H$_2$O (M = Co, Ni) gives M[N(CN)$_2$]$_2$(apym), which consists of a packing of infinite molecular tubes of square cross-section. In each tube, the metal atoms constitute the edges and three connecting N(CN)$_2^-$ (dicyanamide) ligands form the sides (Fig. 10.3.13). The octahedral coordination of the metal atoms is completed by chains of two-connecting ligands (with the amide nitrogen uncoordinated) which occupy the outside of each edge, as well as monodentate apym ligands. The length of a side of the molecular tube is 486.4 pm for M = Co and 482.1 pm for M = Ni. In the crystal structure extensive hydrogen bonding between the tubes occurs via the terminal apym ligands.

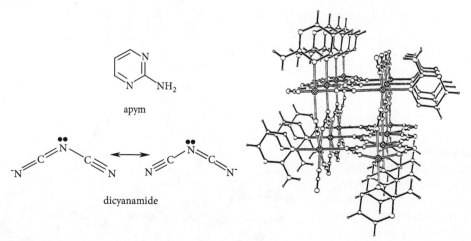

Fig. 10.3.13 Structure of a square molecular tube in M[N(CN)$_2$]$_2$(apym) (M = Co, Ni). (P. Jensen, S. R. Batten, B. Moubaraki, and K. S. Murray, *Chem. Commun.* 793–4 (2000).)

10.3.5 **Molecular polyhedra**

(1) **Tetrahedral iron(II) host-guest complex**

The structural unit $M(tripod)^{n+}$ (tripod = $CH_3C(CH_2PPh_3)_3$) has been used as a template for the construction of many supramolecular complexes. For example, the host-guest complex $[BF_4 \subset \{(tripod)_3Fe\}_4(trans\text{-}NCCH=CHCN)_6(BF_4)_4](BF_4)_3$ has been synthesized from the reaction of tripod, $Fe(BF_4)_3 \cdot 6H_2O$ and fumaronitrile in a 4:4:6 molar ratio in $CH_2Cl_2/EtOH$ at 20 °C. The tetranuclear Fe(II) complex has idealized symmetry T, with a crystallographic two-fold axis passing through the midpoints of a pair of edges of the Fe_4 tetrahedron (Fig. 10.3.14). The B–F bonds of the encapsulated BF_4^- ion point toward the corner iron atoms. Each face of the Fe_4 tetrahedron is capped by a BF_4^- group, and the remaining three are located in voids between the complex cations.

(2) **Cubic molecular box**

The cyanometalate box $\{Cs \subset [Cp^*Rh(CN)_3]_4[Mo(CO)_3]_4\}^{3-}$ is formed in low yield from the reaction of $[Cp^*Rh(CN)_3]^-$ (Cp^* = C_5Me_5) and $(\eta^6\text{-}C_6H_3Me_3)Mo(CO)_3$ in the presence of cesium ions, and it can be crystallized as a Et_4N^+ salt. The Cs^+ ion serves as a template in the self assembly of the anionic molecular box, which has a cubic $Rh_4Mo_4(\mu\text{-}CN)_{12}$ core with three exterior carbonyl ligands attached to each Mo and a Cp^* group to each Rh. The encapsulated Cs^+ ion has a formal coordination number of 12 if interaction with the centers of cyano groups is considered (Fig. 10.3.15).

(3) **Lanthanum square antiprism**

A tris-bidentate pyrazolone ligand, 4-(1,3,5-benzenetricarbonyl)-tris(3-methyl-1-phenyl-2-pyrazoline-5-one (H_3L), has been designed and synthesized. When this rigid C_3-symmetric ligand was reacted with $La(acac)_3$ in dimethylsufoxide (DMSO), the complex $La_8L_8(DMSO)_{24}$ was obtained in 81% yield. An X-ray analysis revealed a square-antiprismatic structure with idealized D_{4d} symmetry, with each L^{3-} ligand occupying one of the eight triangular faces (Fig. 10.3.16). In the coordination sphere of the La^{3+} ion, six

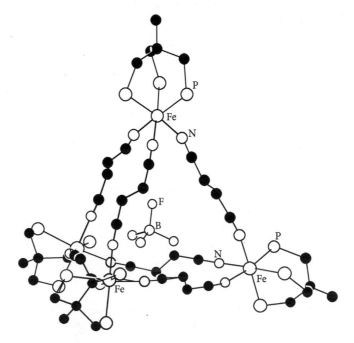

Fig. 10.3.14 Molecular structure of the tetrahedral Fe(II) host-guest complex cation. The capping BF_{4-} groups and the phenyl rings of the tripod ligands have been omitted for clarity. (S. Mann, G. Huttner, L. Zsolnai, and K. Heinze, *Angew. Chem. Int. Ed.* **35**, 2808–9 (1996).)

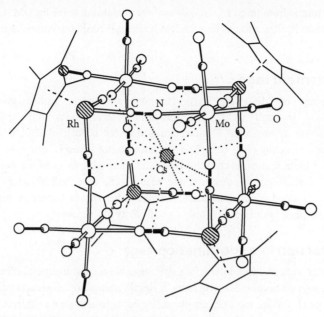

Fig. 10.3.15 Structure of the molecular cube that encloses a Cs⁺ ion. (K. K. Klausmeyer, S. R. Wilson, and T. B. Rauchfuss. *J. Am. Chem. Soc.* **121**, 2705–11 (1999).)

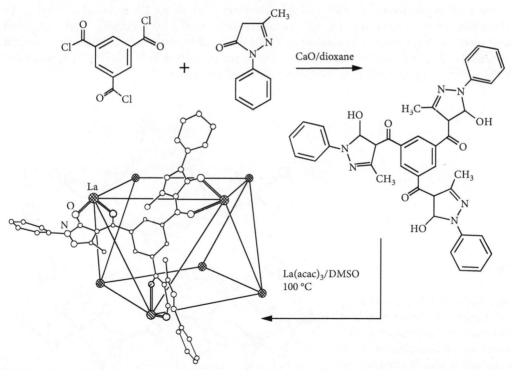

Fig. 10.3.16 Synthesis and structure of the La₈L₈ cluster. Only one L³⁻ ligand is shown, and the coordinating DMSO molecules have been omitted for clarity. (K. N. Raymond and J. Xu. *Angew. Chem. Int. Ed.* **39**, 2745–7 (2000).)

sites are filled by O atoms from three L^{3-} ligands and the remaining three by DMSO molecules which point into the central cavity. It was found that the DMSO ligands could be replaced partially by methanol in recrystallization.

(4) Super-adamantoid cage

A 2:3 molar mixture of $MeC(CH_2PPh_2)_3$ (triphos) and silver triflate gives $[Ag_6(triphos)_4(CF_3SO_3)_4](CF_3SO_3)_2$ in high yield. The inorganic "super-adamantoid" cage $[Ag_6(triphos)_4(CF_3SO_3)_4]^{2+}$ exhibiting approximate T molecular symmetry is formed with the $CF_3SO_3^-$ ion as a template. In the cage structure (Fig. 10.3.17), an octahedron of silver(I) ions is bound by two sets of tritopic triphos and triflate ligands, each occupying four alternating faces of the octahedron. Thus six silver ions and four triphos ligands constitute one adamantane core, and likewise the silver ions and triflate ligands form a second adamantane core. A novel feature in this structure is the "endo-methyl" conformation of the triphos ligand, leaving only a small cavity at the center.

(5) Chemical reaction in a coordination cage

Flat panel-like ligands with multiple interacting sites have been used for metal-directed self-assembly of many fascinating supramolecular 3D structures. A simple triangular "molecular panel" is 2,4,6-tris(4-pyridyl)-1,3,5-triazine (L), which has been employed by Fujita to assemble a discrete $[\{Pt(bipy)\}_6L_4]^{12+}$ (bipy = 2,2′-bipyridine) coordination cage in quantitative yield by treating $Pt(bipy)(NO_3)_2$ with L in a 3:2 molar ratio. In this complex cation, the Pt(II) atoms constitute an octahedron, and the triangular panels (L ligands) are located at four of the eight faces (Fig. 10.3.18). The $Pt(bipy)^{2+}$ fragment thus serves as a *cis*-protected coordination block each linking a pair of molecular panels. The nano-sized central cavity with a diameter of ~1 nm is large enough to accommodate several guest molecules. Different types of guest species such as adamantane, adamantane carboxylate, *o*-carborane, and anisole have been used. In particular, C-shaped molecules such as *cis*-azobenzene and *cis*-stilbene derivatives can be encapsulated in the cavity as a dimer stabilized by the "phenyl-embrace" interaction.

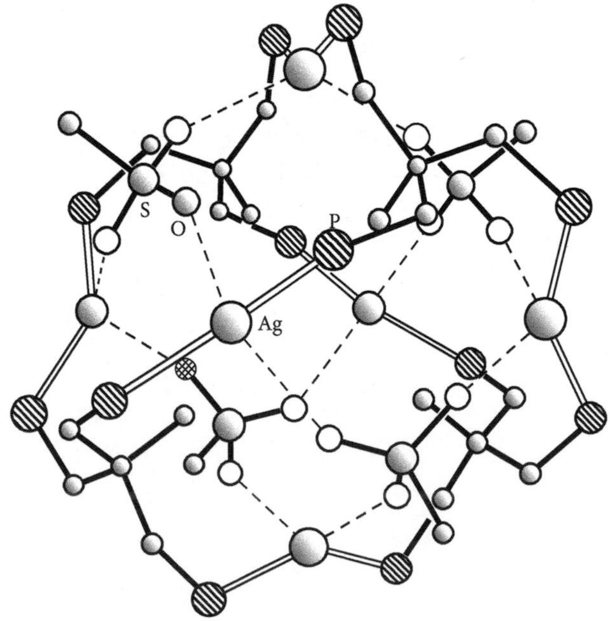

Fig. 10.3.17 The super-adamantoid core of the $[Ag_6(triphos)_4(CF_3SO_3^-)_4]^{2+}$ cage. The broken lines indicate one of the two adamantane cores formed by silver ions and triflate ligands, whose F atoms have been omitted for clarity. (S. L. James, D. M. P. Mingos, A. J. P. White, and D. Williams. *Chem Commun.* 2323–4 (2000).)

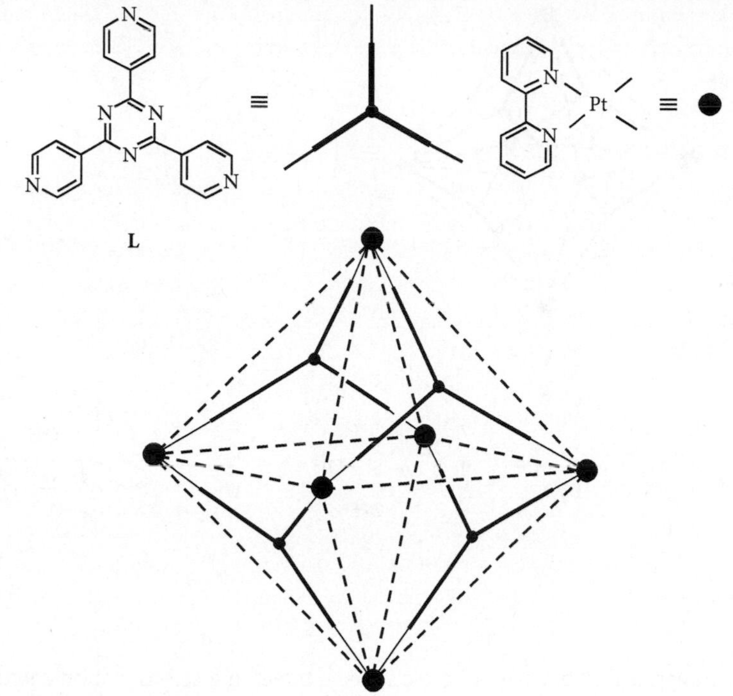

Fig. 10.3.18 Self-assembly of [{Pt(bipy)}$_6$L$_4$]$^{12+}$ cage.

Useful chemical reactions have been carried out in the nano-sized cavity, as illustrated by the in situ isolation of a labile cyclic siloxane trimer (Fig. 10.3.19). In the first step, three to four molecules of phenyltrimethoxysilane enter the cage and are hydrolyzed to siloxane molecules. Next, condensation takes place in the confined environment to generate the cyclic trimer {SiPh(OH)O–}$_3$, which is trapped and stabilized in a pure form. The overall reaction yields an inclusion complex [{SiPh(OH)O–}$_3$ ⊂ {Pt(bipy)}$_6$L$_4$](NO$_3$)$_{12}$·7H$_2$O, which can be crystallized from aqueous solution in 92% yield. The all-*cis* configuration of the cyclic siloxane trimer and the structure of the inclusion complex have been determined by NMR and ESI-MS.

(6) Nanoscale dodecahedron

The dodecahedron is a Platonic solid that contains 12 fused pentagons formed from 20 vertices and 30 edges. An organic molecule of this exceptionally high icosahedral (I_h) symmetry is the hydrocarbon dodecahedrane C$_{20}$H$_{20}$, which was first synthesized by Paquette in 1982. Recently an inorganic analog has been obtained from edge-directed self-assembly of a metallocyclic structure (Fig. 10.3.20) in a remarkably high 99% yield. The tridentate ligand at each vertex is tris(4′-pyridyl)methanol, and the linear bidentate subunit at each edge is bis[4,4′-(*trans*-Pt(PEt$_3$)$_2$(CF$_3$SO$_3$))]benzene. The dodecahedral molecule carries 60 positive charges and encloses 60 CF$_3$SO$_3^-$ anions, and its estimated diameter d along the three-fold axis is about 5.5 nm. Using bis[4,4′-(*trans*-Pt(PPh$_3$)$_2$(CF$_3$SO$_3$))]biphenyl as a longer linear linker, d for the resulting enlarged dodecahedron increases to about 7.5 nm.

(7) High-spin rhombic dodecahedron

The cluster [Mn^{II}{Mn^{II}(MeOH)$_3$}$_8$(μ-CN)$_{30}${Mo(CN)$_3$}$_6$]·5MeOH.2H$_2$O has a pentadecanuclear core of idealized O_h symmetry, as shown in Fig. 10.3.21. The nine Mn(II) ions constitute a body-centered

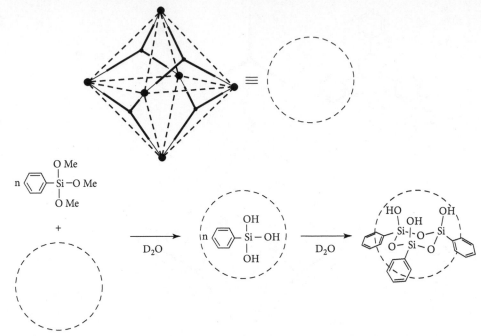

Fig. 10.3.19 Generation and stabilization of cyclic siloxane trimer in a self-assembled coordination cage. (M. Yoshizawa, T. Kusukawa, M. Fujita, and K. Yamaguchi. *J. Am. Chem. Soc.* **122**, 6311–2, (2000).)

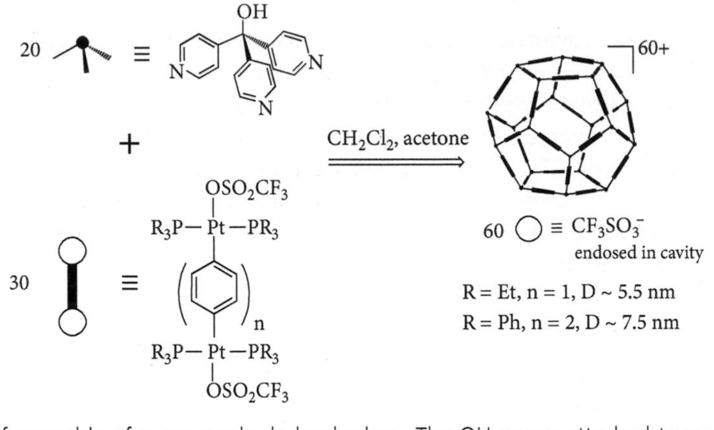

Fig. 10.3.20 Self-assembly of a nanoscale dodecahedron. The OH group attached to each quaternary C atom of the tris(4'-pyridyl)methanol molecule has been omitted for clarity. (B. Olenyuk, M. D. Levin, J. A. Whiteford, J. E. Shield, and P. J. Stang. *J. Am. Chem. Soc.* **121**, 10434–5 (2000).)

cube, the six Mo(V) ions define an octahedron, and the two polyhedra interpenetrate each other so that the peripheral atoms exhibit the geometry of a rhombic dodecahedron. Each pair of adjacent metal centers is linked by a μ-cyano ligand with Mo bonded to C and Mn bonded to N. Each outer Mn(II) atom is surrounded by three methanol ligands, leading to octahedral coordination. Similarly, three terminal cyano ligands are bound to each Mo(V) to establish an eight-coordination environment.

Actually the cluster has a lower symmetry with a crystallographic C_2 axis passing through the central Mn atom and the midpoints of two opposite Mo⋯Mo edges. The resulting neutral cluster has the highest

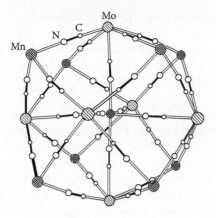

Fig. 10.3.21 The $[Mn_9(\mu\text{-}CN)_{30}Mo_6]$ core of the high-spin cluster. (J. Larionova, M. Gross, M. Pilkington, H. Andres, H. Stoeckli-Evans, H. U. Güdel, and S. Decurtins. *Angew. Chem. Int. Ed.* **39**, 1605–9 (2000).)

spin ground state observed to date with $S = 25\frac{1}{2}$. [Note that $S = 2\frac{1}{2}$ for Mn(II) and $\frac{1}{2}$ for Mo(V); thus the total spin of the system is $9 \times 2\frac{1}{2} + 6 \times \frac{1}{2} = 25\frac{1}{2}$.]

10.4 **Selected Examples in Crystal Engineering**

Examples from the recent literature that illustrate various approaches in the rational design of novel crystalline materials are given in this section.

10.4.1 **Diamondoid networks**

The 3-D network structure of diamond can be considered as constructed from the linkage of nodes (C atoms) with rods (C–C bonds) in a tetrahedral pattern. From the viewpoint of crystal engineering, in a diamondoid network the node can be any group with tetrahedral connectivity, and the linking rods (or linker) can be all kinds of bonding interactions (ionic, covalent, coordination, hydrogen-bond, and weak interactions) or molecular fragment.

The molecular skeletons of adamantane, $(CH_2)_6(CH)_4$, and hexamethylenetetramine, $(CH_2)_6N_4$ (Fig. 10.4.1) constitute the characteristic structural units of diamondoid networks containing one kind and two kinds of four-connected nodes, respectively. If the rod is long, the resulting diamondoid network becomes quite porous, and stability can only be achieved by interpenetration. If two diamondoid networks interpenetrate to form the crystal structure, the degree of interpenetration ρ is equal to two. Examples in which ρ ranges from two to nine are known.

(a)　　　　　　　　　(b)

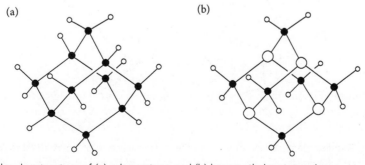

Fig. 10.4.1 Molecular structure of (a) adamantane and (b) hexamethylenetetramine.

Some crystalline compounds that exhibit diamondoid structures are listed in Table 10.4.1. The rod linking a pair of nodes can be either linear or non-linear.

In the crystal structure of Cu_2O, each O atom is surrounded tetrahedrally by four Cu atoms, and each Cu atom is connected to two O atoms in a linear fashion. Hence the node is the O atom, and the rod is O–Cu–O. Fig. 10.4.2 shows a single Cu_2O diamondoid network, and the crystal structure is composed of two interpenetrating networks.

In the crystal structure of ice-VII, which is formed under high pressure, the node is the O atom, and the rod is a hydrogen bond. Since the H atoms are disordered, the hydrogen bond is written as O···H···O in Table 10.4.1, indicating equal population of O–H···O and O···H–O. The degree of interpenetration is two, as shown in Fig. 10.4.3.

Figs 10.4.4(a) and 10.4.4(b) illustrate the crystal structure of the 1:1 complex of tetraphenylmethane and carbon tetrabromide. The nodes comprise $C(C_6H_5)_4$ and CBr_4 molecules, and the each linking rod is the weak interaction between a Br atom and a phenyl group. The hexamethylenetetramine-like structural unit is outlined by broken lines. Figs 10.4.4(c) and 10.4.4(d) show the crystal structure of tetrakis(4-bromophenyl)methane, which has a distorted diamondoid network based on the hexamethylenetetramine building unit. If the synthon composed of the aggregation of four Br atoms is

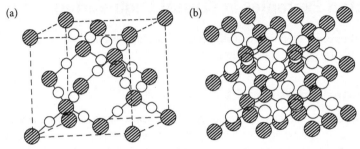

Fig. 10.4.2 Crystal structure of Cu_2O: (a) single diamondoid network and (b) two interpenetrating networks.

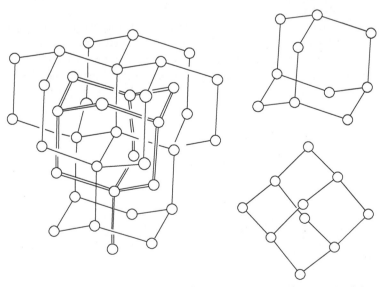

Fig. 10.4.3 (Left) Two interpenetrating networks in the crystal structure of ice-VII. (Right) Two views of an adamantane-like structural unit.

Table 10.4.1 Diamondoid networks

Compound	Node	Linking rod	Bond type between nodes	Degree of inter-penetration ρ	Remarks
Diamond	C	C–C	covalent	none	Fig. 10.4.2
M_2O (M = Cu, Ag, Pb)	O	O–M–O	ionic/covalent	two	H atom disordered
Ice-VII	O	O···H···O	H-bond	two	Fig. 10.4.3
KH_2PO_4	$H_2PO_4^-$	O–H···O	H-bond	two	K^+ in channel
methane- tetraacetic acid	$C(CH_2COOH)_4$	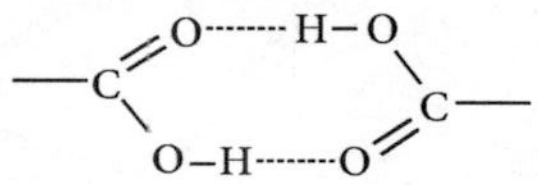	cyclic dimeric H-bond	three	
$(CH_2)_6N_4 \cdot CBr_4$	$(CH_2)_6N_4$, CBr_4	N···Br	N···Br interaction	two	Fig. 10.4.4(a); Fig. 10.4.4(b)
$C(C_6H_5)_4 \cdot CBr_4$	$C(C_6H_5)_4$, CBr_4	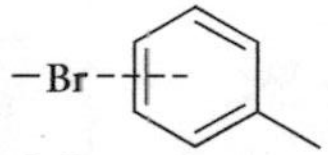	Br···phenyl interaction	none	
$C(4\text{-}C_6H_4Br)_4$	$C(C_6H_4)_4$ unit, Br_4 synthon	C–Br	covalent	three	tetrahedral Br_4 synthon consolidated by weak Br···Br interaction Fig. 10.4.4(c); Fig. 10.4.4(d)
$M(CN)_2$ (M = Zn, Cd)	M	M←C≡N→M	coordination	two	
$[Cu(L)_2]BF_4$ (L = $p\text{-}C_6H_4(CN)_2$)	Cu	Ag←NCC_6H_4CN→Ag	coordination	five	BF_4^- in channel Fig. 10.4.5
$C(C_6H_4C_2C_5NH_4O)_4 \cdot 8CH_3CH_2COOH$	$C(C_6H_4C_2C_5NH_4O)_4$	double N–H···O	cyclic dimeric H-bond	seven	Fig. 10.4.6
$[Ag(L_2]XF_6$ (L = 4,4'-$NCC_6H_4\text{-}C_6H_4CN$, X = P, As, Sb)	Ag	Ag←$NCC_6H_4C_6H_4CN$→Ag	coordination	nine	XF_6^- in channel
$[Ag(L)]BF_4 \cdot xPhNO_2$ (L = $C(4\text{-}C_6H_4CN)_4$)	Ag, $C(4\text{-}C_6H_4CN)_4$	CN→Ag	coordination	none	BF_4^- and $PhNO_2$ guest species in cavity
$[Mn(CO)_3(\mu\text{-}OH)]_4 \cdot (H_2NCH_2CH_2NH_2)$	$[Mn(CO)_3\text{-}(\mu\text{-}OH)]_4$	O–H···NH_2CH_2–CH_2NH_2···H–O	H-bond	three	

considered as a node, then two kinds of nodes (Br_4 synthon and quatenary C atom) are connected by rods consisting of p-phenylene moeities.

Fig. 10.4.5 shows the crystal structure of $[Cu\{1,4\text{-}C_6H_4(CN)_2\}_2]BF_4$. The nodes are the four-connected Cu atoms, each being coordinated tetrahedrally by $N{\equiv}C\text{-}C_6H_4\text{-}C{\equiv}N$ ligands as rods. The adamantane-like structure unit is shown in Fig. 10.4.5(a), and repetition of such units along a two-fold axis leads to five-fold interpenetration. The remaining space is filled by the BF_4^- ions.

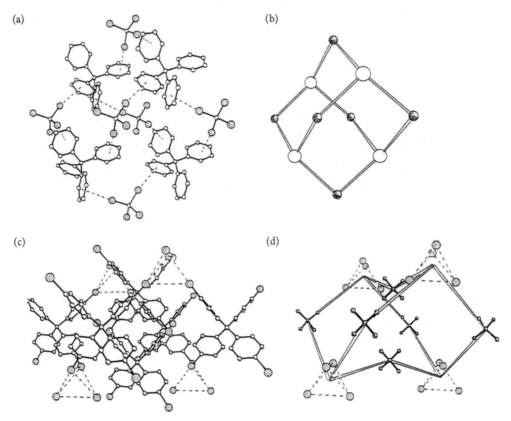

Fig. 10.4.4 (a) Crystal structure and (b) diamondoid network of $C(C_6H_5)_4{\cdot}CBr_4$. (c) Crystal structure and (d) distorted diamondoid network of $C(4\text{-}C_6H_4Br)$.

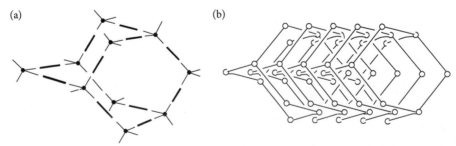

Fig. 10.4.5 (a) Single diamond network of $[Cu\{1,4\text{-}C_6H_4(CN)_2\}_2]$ and (b) fivefold interpenetration in the crystal structure.

Fig. 10.4.6 shows the crystal structure of $C(C_6H_4C_2C_5NH_4O)\cdot8C_2H_5COOH$. The whole $C(C_6H_4C_2C_5NH_4O)_4$ molecule serves as a node, and such nodes are connected by rods each comprising a pair of N–H···O hydrogen bonds between pyridine groups. As the resulting diamondoid network is quite open, seven-fold interpenetration occurs, and the remaining space is used to accommodate the ethanol guest molecules.

In the crystal structure of $[Ag\{C(4\text{-}C_6H_4CN)_4\}]BF_4\cdot x PhNO_2$, the structural unit is of the hexamethylenetetramine type (Fig. 10.4.7). The nodes are Ag atoms and $C(4\text{-}C_6H_4CN)_4$ molecules, and the

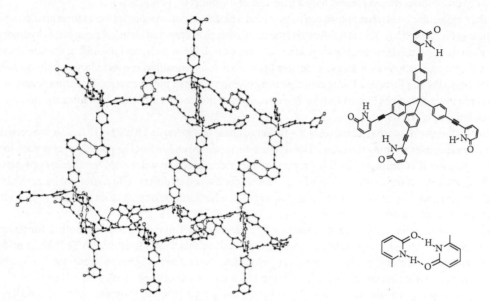

Fig. 10.4.6 Crystal structure of $C(C_6H_4C_2C_5NH_4O)_4\cdot8CH_3CH_2COOH$. The structure of the host molecule is shown in the upper right and the pair-wise linkage of pyridone groups is shown at the lower right.

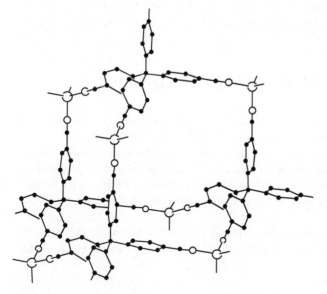

Fig. 10.4.7 Hexamethylenetetramine-like structure unit in the diamondoid network of $[Ag\{C(4\text{-}C_6H_4CN)_4\}]BF_4$.

rods are CN→Ag coordination bonds. The void space is filled by the nitrobenzene guest molecules and BF_4^- ions.

10.4.2 Interlocked structures constructed from cucurbituril

Cucurbituril is a hexameric macrocyclic compound with the formula $(C_6H_6N_4O_2)_6$ that is shaped like a pumpkin which belongs to the botanical family *Cucurbitaceae*. This macrocyclic cavitand has idealized symmetry D_{6h} with a hydrophobic internal cavity of about 0.55 nm. The two portals, which are each laced by six hydrophilic carbonyl groups, have a diameter of 0.4 nm (Fig. 10.4.8(a)).

Like a molecular bead, cucurbituril can be threaded with a linear diammonium ion to form an inclusion complex (Fig. 10.4.8(b)). This is stabilized by the fact that each protonated amino N atom forms hydrogen bonds to three of the six carbonyl groups at its adjacent portal. Since each rotaxane unit has two terminal pyridyl groups, it can serve as an *exo*-bidentate ligand to link up transition metals to form a polyrotaxane which may take the form of a linear coordination polymer, a zigzag polymer, a molecular necklace, or a puckered layer network (Fig. 10.4.8(c)). The structures of the linear and zigzag polyrotaxane polymers are shown in Fig. 10.4.9.

The 2-D network is constructed from the fusion of chair-like hexagons with Ag(I) ions at the corners and rotaxane units forming the edges. The nitrate ions lie above and below the puckered layer such that each Ag(I) ion is coordinated by three rotaxanes and a nitrate ion in a distorted tetrahedral geometry. In the crystal structure, two sets of parallel 2-D networks stacked in different directions make a dihedral angle of 69°, and they interpenetrate in such a way that a hexagon belonging to one set interlocks with four hexagons of the other set, and vice versa.

The rotaxane building unit can be modified by replacing the 4-pyridyl group by another functional group such as 3-cyanobenzyl. When a rotaxane unit built in this way is treated with $Tb(NO_3)_3$ under hydrothermal conditions, the cyano group is converted to the carboxylate group to generate a 3-D coordination polymeric network. The basic building block of the framework consists of a binuclear Tb^{3+} center and two types of rotaxane units: Type I having bridging 3-phenylcarboxylate terminals and Type II having chelating carboxylate terminals (Fig. 10.4.10).

The binuclear terbium centers and Type I rotaxanes form a 2-D layer. Stacked layers are further interconnected via Type II rotaxanes to form a 3-D polyrotaxane network, which has an inclined α-polonium topology with the binuclear terbium centers behaving as six-connected nodes (Fig. 10.4.10). The void space in the crystal packing is filled by a free rotaxane unit, NO_3^- and OH^- counter ions, and water molecules.

10.4.3 Inorganic crystal engineering using hydrogen bonds

The utilization of hydrogen bonding in inorganic crystal design has gained prominence in recent years. A conceptual framework for understanding supramolecular chemistry involving metals and metal complexes is provided by the domain model according to Dance and Brammer (Fig. 10.4.11).

The central *Metal Domain* of a metal complex consists of the metal atom M, a metal hydride group, or a number of metal atoms if a metal cluster complex is considered. The *Ligand Domain* is composed of ligand atoms L that are directly bonded with the metal center(s). The *Periphery Domain* is the outmost part of the complex, consisting of those parts of the ligand that are not strongly influenced by electronic interaction with the metal center.

Hydrogen bonding arising from donor groups (M)O–H and (M)N–H is commonly observed. The σ-type coordinated ligands are hydroxy (OH), aqua (OH$_2$), alcohol (ROH), and amines (NH$_3$, NRH$_2$,

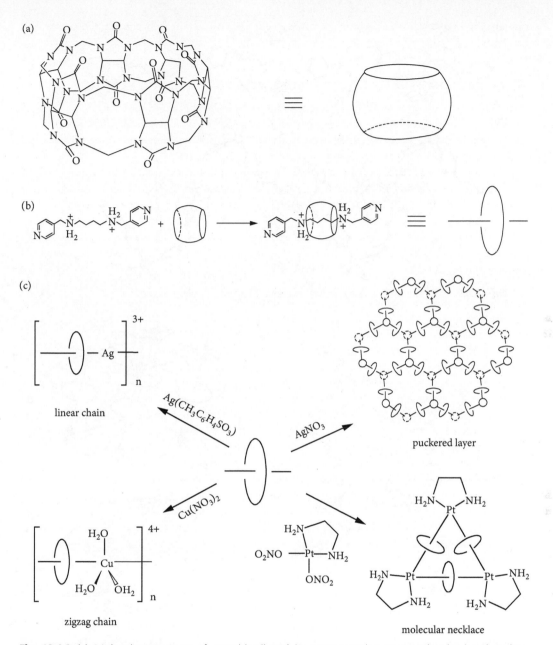

Fig. 10.4.8 (a) Molecular structure of cucurbituril and its representation as a molecular bead or barrel, (b) cucurbituril threaded with a linear diammonium ion to form an inclusion complex, and (c) rotanane building unit obtained by threading cucurbituril with diprotonated N, N′-bis(4-pyridylmethyl)-1,4-diaminobutane, and its subsequent reactions to yield linear polymers, a puckered layer and a molecular triangle.

(a)

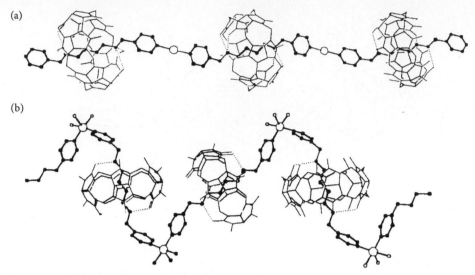

(b)

Fig. 10.4.9 Structure of (a) a linear polytaxane and (b) a zigzag polyrotaxane.

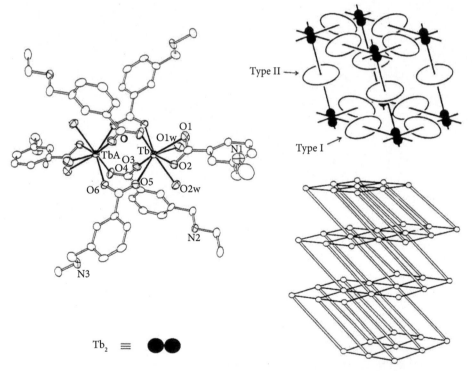

Fig. 10.4.10 (a) Coordination geometry around the centrosymmetric binuclear terbium center. (b) Unit cell (top) and schematic representation (bottom) of the α-polonium-type network. Two contacting black circles stand for a binuclear terbium center, and the solid and open rods representation Type I and Type II rotaxane, respectively. (E. Lee, J. Heo and K. Kim, *Angew. Chem. Int. Ed.* **39**, 2699–701, (2000).)

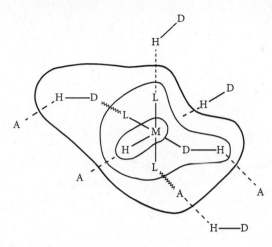

Fig. 10.4.11 Domain model for hydrogen bonding involving metal complexes. Here D and A represent donor and acceptor atom, respectively. Note that both M–H and D–H units remain intact.

NR_2H). Acceptors include halide-type (M–X with X = F, Cl, Br, I), hydride-type (D–H···H–M and D–H···H–E with E = B, Al, Ga), and carbonyl-type (D–H···OC–M).

Some examples of the coordination polymers consolidated by hydrogen bonding are discussed below.

(1) **Chains**

In the crystal structure of $Fe[\eta^5\text{-CpCOOH}]_2$, a hydrogen-bonded chain with carboxyl groups interacting via the $R_2^2(8)$ dimer synthon is formed (Fig. 10.4.12(a)). The complex $[Ag(nicotinamide)_2]CF_3SO_3$ has a ladder structure propagated via the N–H···O catemer with rungs comprising $R_2^2(8)$ amide dimer interactions (Fig. 10.4.12(b)). Two $CF_3SO_3^-$ ions (not shown) lie inside each centrosymmetric macrocyclic ring, and their O atoms form N–H···O–S–O···H–N hydrogen-bonded and weak Ag···O···Ag bridges. The crystal structure of $[Ru(\eta^5\text{-Cp})(\eta^5\text{-1-}p\text{-tolyl-2-hydroxyindenyl})]$ features a zigzag chain linked by O–H···π(Cp) hydrogen bonds (Fig. 10.4.12(c)).

The crystal structures of [Pt(NCN–OH)Cl] and [Pt(SO$_2$)(NCN–OH)Cl], where NCN is the tridentate pincer ligand $\{C_6H_2\text{-4-(OH)-2,6-}(CH_2NMe_2)_2\}^-$, are compared in Fig. 10.4.13. The colorless complex [Pt(NCN–OH)Cl] consists of a sheet-like array of parallel zigzag chains connected via O–H···Cl(Pt) hydrogen bonds. Reversible uptake of SO_2 is accompanied by a color change, resulting in an orange complex [Pt(SO$_2$)(NCN–OH)Cl] in which the coordinated SO_2 ligand is involved in a donor–acceptor S···Cl interaction.

(2) **Two-dimensional networks**

The O–H···O$^-$ hydrogen-bonded square grid found in $[Pt(L_2)(HL)_2]\cdot 2H_2O$ (HL = isonicotinic acid) is shown in Fig. 10.4.14(a). Water molecules (not shown) occupy channels in the three-fold interpenetrated network. The crystal structure of $[Zn(SC(NH_2)NHNH_2)_2(OH)_2][1,4\text{-}O_2CC_6H_4CO_2]\cdot 2H_2O$ has a brick-wall sheet structure (Fig. 10.4.14(b)). Each *N*, *S*-chelating thiosemicarbazide ligand forms two donor hydrogen bonds with one carboxylate group of a terephthalate ion and one donor hydrogen bond with another terephthalate ion. The layers are linked via N–H···O and O–H···O hydrogen bonds to the water molecules (not shown). The complex $[Ag(nicotinamide)_2]PF_6$ has a cationic herringbone layer constructed from amide N–H···O hydrogen bonds (Fig. 10.4.14(c)). Such layers are further cross-linked by N–H···F and C–H···F hydrogen bonds involving the PF_6^- ions (not shown).

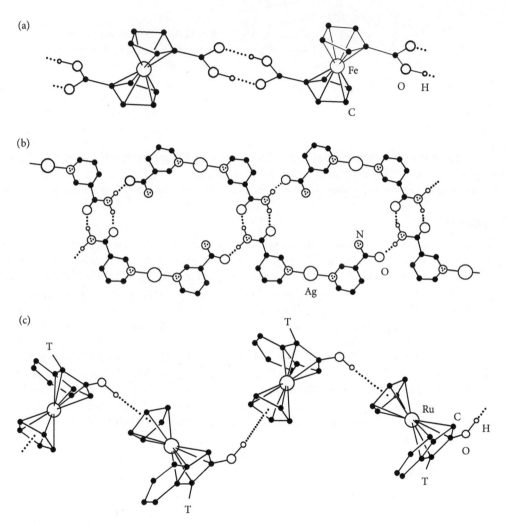

Fig. 10.4.12 Hydrogen-bonded chain in (a) Fe[η^5-CpCOOH]$_2$, (b) [Ag(nicotinamide)$_2$]$^+$, and (c) [Ru(η^5-Cp)(η^5-1-p-tolyl-2-hydroxyindenyl)] (T stands for p-tolyl group).

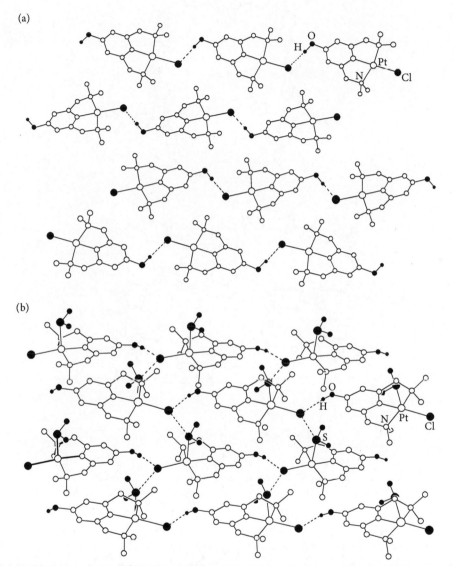

Fig. 10.4.13 (a) Zigzag chain of [Pt(NCN–OH)Cl] held by O–H⋯Cl hydrogen bonds. (b) Two-dimensional network of [Pt(SO$_2$)(NCN–OH)Cl] consolidated by additional S⋯Cl donor–acceptor interactions.

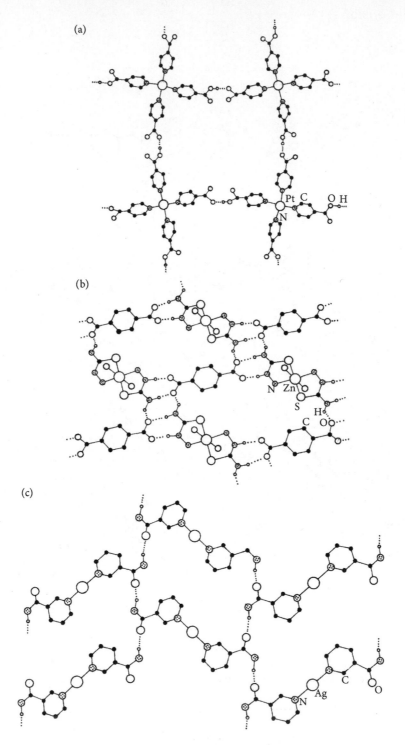

Fig. 10.4.14 (a) Square grid in [Pt(L₂)(HL)₂]·2H₂O (HL = isonicotinic acid). (b) Brick-wall sheet in [Zn(thiosemicarbazide)(OH)₂](terephthate)·2H₂O. (c) Cationic layer structure of [Ag(nicotinamide)₂]PF₆.

(3) **Three-dimensional networks**

In the complex $[Cu(L)_4]PF_6$ (L = 3-cyano-6-methylpyrid-2(1H)-one), the hydrogen-bonded linkage involves the amido $R_2^2(8)$ supramolecular synthon (Fig. 10.4.15(a)). Each Cu(I) center serves as a tetra-hedral node in a four-fold interpenetrated cationic diamondoid network. The tetrahedral metal cluster $[Mn(\mu_3\text{-}OH)(CO)_3]_4$ can be used to construct a diamondoid network with the linear 4,4'-bipyridine spacers (Fig. 10.4.15(b)).

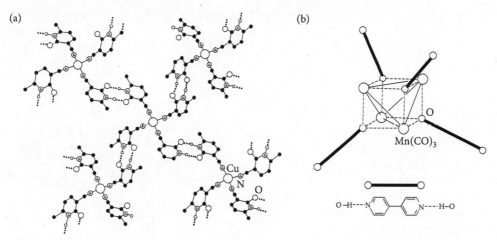

Fig. 10.4.15 (a) Part of the cationic diamondoid network in $[Cu(L)_4]PF_6$ (L = pyridone ligand). (b) Molecular components and linkage mode of the diamondoid network in $[Mn(\mu_3\text{-}OH)(CO)_3]_4 \cdot 2(bipy) \cdot 2CH_3CN$.

10.4.4 **Generation and stabilization of unstable inorganic/organic anions in urea/thiourea complexes**

Urea, thiourea, or their derivatives are often employed as useful building blocks for supramolecular architectures because they contain amido functional group which can form moderately strong N—H···X hydrogen bonds with rather well-defined and predictable hydrogen-bonding patterns. Furthermore, in the presence of anions, the hydrogen bond is strengthened by two to three times (40 to 190 kJ mol^{-1}) com-pared with the bond strength involving uncharged molecular species (10 to 65 kJ mol^{-1}). Hence, making use of this kind of charge-assisted N—H···X$^-$ hydrogen bonding interactions and bulky tetraalkylammo-nium cations as guest templates, some unstable organic anions A$^-$ can be generated in situ and stabilized in urea/thiourea–anion host frameworks. A series of inclusion compounds of the type $R_4N^+A^- \cdot m(NH_2)_2CX$ (where X = O or S) with novel topological features have been characterized.

(1) **Dihydrogen borate**

As mentioned in Section 3.5.1, the transient species $[BO(OH)_2]^-$ has been stabilized by hydrogen-bonding interactions with the nearest urea molecules in the host framework of the inclusion compound $[(CH_3)_4N]^+[BO(OH)_2]^- \cdot 2(NH_2)_2CO \cdot H_2O$. A perspective view of the crystal structure along the [010] direction is presented in Fig. 10.4.16. The host lattice consists of a parallel arrangement of unidirectional channels whose cross-section has the shape of a peanut. The diameter of each spheroidal half is about

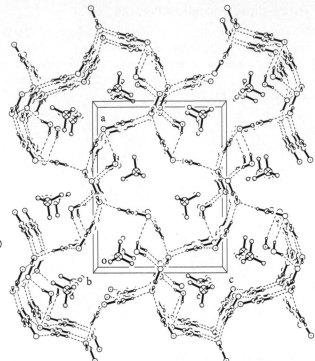

Fig. 10.4.16 Crystal structure of $[(CH_3)_4N]^+[BO(OH)_2]^-\cdot2(NH_2)_2CO\cdot H_2O$ showing the channels extending parallel to the *b* axis and the enclosed cations. Broken lines represent hydrogen bonds, and the atoms are shown as points for clarity. (Q. Li, F. Xue, and T. C. W. Mak, *Inorg. Chem.* **38**, 4142–5 (1999).)

704 pm, and the separation between two opposite walls at the waist of the channel is about 585 pm. The well-ordered tetramethylammonium cations are accommodated in double columns within each channel.

(2) **Allophanate and 3-thioallophanate**

Allophanate esters $H_2NCONHCOOR$ are among the oldest organic compounds recorded in the literature. The parent allophanic acid, $H_2NCONHCOOH$, is not known in the free state, whereas inorganic allophanate salts are unstable and readily hydrolyzed by water to carbon dioxide, urea and carbonate. However, the elusive allophanate anion can be generated in situ and stabilized in the following three inclusion compounds:

(a) $[(CH_3)_4N]^+[NH_2CONHCO_2]^-\cdot5(NH_2)_2CO$

(b) $[(^nC_3H_7)_4N]^+[NH_2CONHCO_2]^-\cdot3(NH_2)_2CO$

(c) $[(CH_3)_3N^+CH_2CH_2OH][NH_2CONHCO_2]^-\cdot(NH_2)_2CO$

A part of the host framework in $[(CH_3)_4N]^+[NH_2CONHCO_2]^-\cdot5(NH_2)_2CO$ is shown in Fig. 10.4.17. Two neighboring allophanate anions are arranged in a head-to-tail fashion and connected by a urea molecule with N–H···O and charge-assisted N–H···O⁻ hydrogen bonds to generate a zigzag ribbon. This ribbon is further joined to another ribbon related to it by an inversion center via pairs of N–H···O⁻ hydrogen bonds to form a double-ribbon.

The hitherto unknown 3-thioallophanate anion has been trapped in the host lattice of $[(^nC_4H_9)_4N]^+[H_2NCSNHCO_2]^-\cdot(NH_2)_2CS$. The cyclic structure and molecular dimensions of the allophanate and 3-thioallophanate ions are compared in Fig. 10.4.18. As expected, the C–O bond

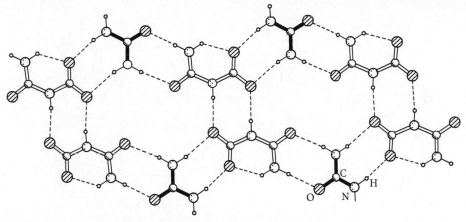

Fig. 10.4.17 Part of the host framework in $[(CH_3)_4N]^+[H_2NCONHCO_2]^-\cdot5(NH_2)_2CO$ showing the double-ribbon constructed by allophanate anions and one of the independent urea molecules. (T. C. W. Mak, W.-H. Yip, and Q. Li, *J. Am. Chem. Soc.* **117**, 11995–6 (1995).)

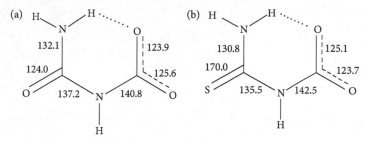

Fig. 10.4.18 Bond lengths (pm) of the (a) allophanate and (b) 3-thioallophanate anion. (C.-K. Lam, T.-L. Chan, and T. C. W. Mak, *CrystEngComm.* **6**, 290–2 (2004).)

involved in intramolecular hydrogen bonding in the 3-thioallophanate anion is longer than that in the allophanate anion.

(3) Valence tautomers of the rhodizonate dianion

The rhodizonate dianion $C_6O_6^{2-}$ (Fig. 10.4.19) is a member of a series of planar monocyclic oxo-carbon dianions $C_nO_n^{2-}$ ($n = 3$, deltate; $n = 4$, squarate; $n = 5$, croconate; $n = 6$; rhodizonate) which have been recognized as non-benzenoid aromatic compounds. However, this six-membered ring species is not stable in aqueous solution as it readily undergoes oxidative ring contraction reaction to the croconate dianion, and the decomposition is catalyzed by alkalis. Recently, this relatively unstable species has been generated in situ and stabilized by hydrogen-bonding in two novel inclusion compounds $[(^nC_4H_9)_4N^+]_2C_6O_6^{2-}\cdot2(m\text{-}OHC_6H_4NHCONH_2)\cdot2H_2O$ (Fig. 10.4.20) and $[(^nC_4H_9)_4N^+]_2C_6O_6^{2-}\cdot2(NH_2CONHCH_2CH_2NHCONH_2)\cdot3H_2O$ (Fig. 10.4.21), respectively.

The measured dimensions of the $C_6O_6^{2-}$ species in $[(^nC_4H_9)_4N^+]_2C_6O_6^{2-}\cdot2(m\text{-}OHC_6H_4$ $NHCONH_2)\cdot2H_2O$ and $[(^nC_4H_9)_4N^+]_2C_6O_6^{2-}\cdot2(NH_2CONHCH_2CH_2NHCONH_2)\cdot3H_2O$ nearly conform to idealized D_{6h} and C_{2v} molecular symmetry, corresponding to distinct valence tautomeric structures that manifest nonbenzenoid aromatic and enediolate character, respectively (Fig. 10.4.22).

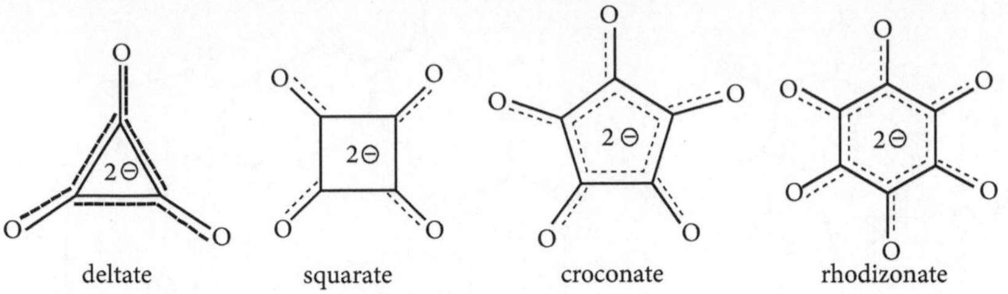

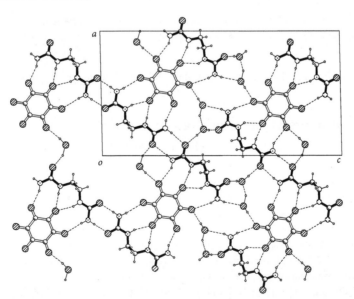

deltate squarate croconate rhodizonate

Fig. 10.4.19 Structural formulas of cyclic oxocarbon dianions.

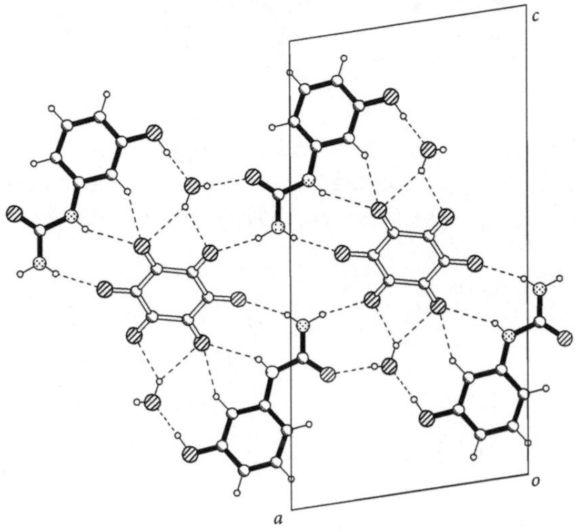

Fig. 10.4.20 Projection along the *b*-axis showing extensive hydrogen bonding interactions around the centrosymmetric rhodizonate dianion in the host lattice of $[(^nC_4H_9)_4N^+]_2C_6O_6^{2-} \cdot 2(m\text{-}OHC_6H_4NHCONH_2)\cdot 2H_2O$. (C.-K. Lam and T. C. W. Mak, *Angew. Chem. Int. Ed.* **40**, 3453–5 (2001).)

Fig. 10.4.21 Projection along the *b*-axis showing the hydrogen bonding interactions within the puckered rhodizonate-bisurea-water layer of $[(^nC_4H_9)_4N^+]_2C_6O_6^{2-} \cdot 2(NH_2CONHCH_2CH_2NHCONH_2)\cdot 3H_2O$. (C.-K. Lam and T. C. W. Mak, *Angew. Chem. Int. Ed.* **40**, 3453–5 (2001).)

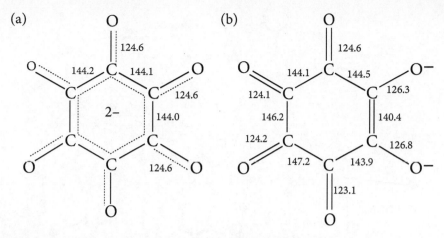

Fig. 10.4.22 Bond lengths (pm) of (a) D_{6h} (note that in this case the dianion is located at a site of symmetry) and (b) C_{2v} valence tautomers of the rhodizonate dianion. (C.-K. Lam and T. C. W. Mak, *Angew. Chem. Int. Ed.* **40**, 3453–5 (2001).)

Occurrence of the charge-localized structure of $C_6O_6^{2-}$ in $[(^nC_4H_9)_4N^+]_2C_6O_6^{2-}\cdot 2(NH_2CONHCH_2CH_2NHCONH_2)\cdot 3H_2O$, as well as its noticeable deviation from idealized C_{2v} molecular symmetry, can be attributed to unequal hydrogen bonding interaction with its two neighboring bisurea donors and a pair of water molecules (Fig. 10.4.21).

(4) Valence tautomers of the croconate dianion

In the host lattice of $[(^nC_3H_7)_4N^+]_2C_5O_5^{2-}\cdot 3(NH_2)_2CO\cdot 8H_2O$, the croconate anion resides in a rather symmetrical hydrogen bonding environment (Fig. 10.4.23) and thus its measured dimensions are consistent with its expected charge-delocalized D_{5h} structure in the ground state as shown in Fig. 10.4.24(a). In the host lattice of $[(C_2H_5)_4N^+]_2C_5O_5^{2-}\cdot 3(CH_3NH)_2CO$, the croconate dianion resides on a two-fold axis, being directly linked to three 1,3-dimethylurea molecules through pairs of N–H···O hydrogen bonds to form a semi-circular structural unit (Fig. 10.4.25). In particular, each type O1 oxygen atom forms two strong acceptor hydrogen bonds with the N–H donors from a pair of 1,3-dimethylurea molecules, while each type O2 oxygen atom forms only one N–H···O hydrogen bond. In contrast, the solitary type O3 oxygen atom is stabilized by two weak C–H···O hydrogen bonds with neighboring tetra-*n*-butylammonium cations. Such a highly unsymmetrical environment engenders a sharp gradient of hydrogen-bonding donor strength around the croconate ion, which is conducive to stabilization of its C_{2v} valence tautomer with significantly different C–C and C–O bond lengths around the cyclic system as shown in Fig. 10.4.24(b).

The examples given in this section show that crystal engineering provides a viable route to breaking the degeneracy of canonical forms of a molecular species, and an elusive anion can be generated in situ and stabilized in a crystalline inclusion compound through hydrogen-bonding interactions with neighboring hydrogen-bond donors, such as urea/thiourea or their derivatives, which can function as supramolecular stabilizing agents.

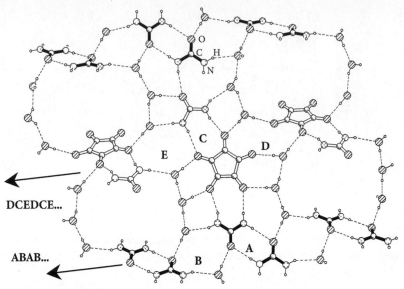

Fig. 10.4.23 Projection diagram on the (10) plane showing the hydrogen-bonding scheme for a portion of the host lattice in $[(^nC_3H_7)_4N^+]_2C_5O_5{}^{2-}\cdot3(NH_2)_2CO\cdot8H_2O$. The slanted vertical (urea dimer-croconate-urea)$_\infty$ chain constitutes a side wall of the [110] channel system. The parallel [urea dimer–(H$_2$O)$_2$]$_\infty$ **ABAB...** and [croconate–urea–(H$_2$O)$_4$]$_\infty$ **DCEDCE...** ribbons define the channel system in the c direction. (C.-K. Lam, M.-F. Cheng, C.-L. Li, J.-P. Zhang, X.-M. Chen, W.-K. Li, and T. C. W. Mak, *Chem. Commun.* 448–9 (2004).)

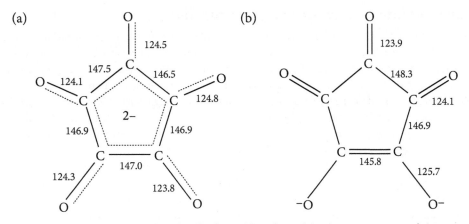

Fig. 10.4.24 Bond lengths (pm) and angles (°) of D_{5h} (a) and C_{2v} (b) valence tautomers of the croconate dianion. (C.-K. Lam, M -F. Cheng, C.-L. Li, J.-P. Zhang, X.-M. Chen, W.-K. Li, and T. C. W. Mak, *Chem. Commun.* 448–9 (2004).)

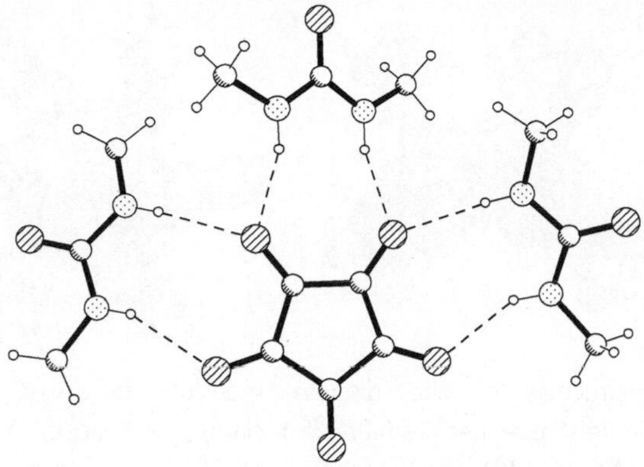

Fig. 10.4.25 Hydrogen-bonding environment of the croconate dianion in the crystal structure of $[(C_2H_5)_4N^+]_2C_5O_5{}^{2-}\cdot3(CH_3NH)_2CO$. (C.-K. Lam, M.-F. Cheng, C.-L. Li, J.-P. Zhang, X.-M. Chen, W.-K. Li, and T. C. W. Mak, *Chem. Commun.* 448–9 (2004).)

10.4.5 Supramolecular assembly of silver(I) polyhedra with embedded acetylenediide dianion

Since 1998, a wide range of double, triple, and quadruple salts containing Ag_2C_2 (IUPAC name silver acetylenediide, commonly known as silver acetylide or silver carbide in the older literature) as a component have been synthesized and characterized (see Section 4.3.10 and the comprehensive review by Bruce and Low cited at the end of this chapter). Such Ag_2C_2-containing double, triple and quadruple salts can be formulated as $Ag_2C_2\cdot nAgX$, $Ag_2C_2\cdot mAgX\cdot nAgY$, and $Ag_2C_2\cdot lAgX\cdot mAgY\cdot nAgZ$, respectively. The accumulated experimental data indicate that the acetylenediide dianion C_2^{2-} preferentially resides inside a silver(I) polyhedron of the type $C_2@Ag_n$ ($n = 6$–10), which is jointly stabilized by ionic, covalent (σ, π, and mixed) and argentophilic interactions (see Section 9.6.2). However, the $C_2@Ag_n$ cage is quite labile and its formation can be influenced by various factors such as solvent, reaction temperature, as well as the co-existence of anions, crown ethers, tetraaza macrocycles, organic cations, neutral ancillary ligands, and exo-bidentate bridging ligands, so that cage size and geometry are in general unpredictable.

(1) Discrete molecules

To obtain discrete molecules, one effective strategy is to install protective cordons around the $C_2@Ag_n$ moiety with neutral, multidentate ligands that can function as blocking groups or terminal stoppers.

Small crown ethers have been introduced as structure-directing agents into the Ag_2C_2-containing system to prevent catenation and inter-linkage of silver polyhedra. Judging from the rather poor host-guest complementarity of Ag(I) (soft cation) with a crown ether (hard O ligand sites), the latter is expected not to affect the formation of $C_2@Ag_n$, but to act as a capping ligand to an apex of the polyhedral silver cage.

$[Ag_2C_2\cdot5CF_3CO_2Ag\cdot2(15C5)\cdot H_2O]\cdot3H_2O$ (15C5 = [15]crown-5) can be obtained from an aqueous solution containing silver acetylenediide, silver trifluoroacetate, and 15C5. The discrete $C_2@Ag_7$ moiety in $[Ag_2C_2\cdot5CF_3CO_2Ag\cdot2(15C5)\cdot H_2O]\cdot3H_2O$ is a pentagonal bipyramid, with four equatorial edges bridged by $CF_3CO_2{}^-$ groups while the two apical Ag atoms are each attached to a 15C5, as shown in Fig. 10.4.26(a). In this discrete molecule, one equatorial Ag atom is coordinated by a monodentate $CF_3CO_2{}^-$ and the other by an aqua ligand.

In $[Ag_{14}(C_2)_2(CF_3CO_2)_{14}(dabcoH)_4(H_2O)_{1.5}]\cdot H_2O$, the core is a Ag_{14} double cage constructed from edge-sharing of two triangulated dodecahedra. Apart from trifluoroacetate and aqua ligands, there are four monoprotonated dabco ligands surrounding the core unit, each being terminally coordinated to a silver(I) vertex, as shown in Fig. 10.4.26(b).

For the discrete molecule $[Ag_8(C_2)(CF_3CO_2)_6(L^1)_6]$ (L^1 = 4-hydroxyquinoline) of $\bar{3}$ symmetry displayed in Fig. 10.4.26(c), the encapsulated acetylenediide dianion in the rhombohedral Ag_8 core is

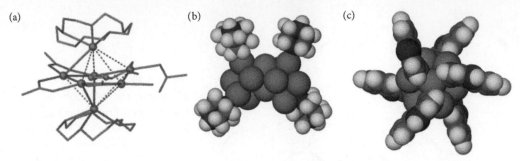

Fig. 10.4.26 (a) Crown-sandwiched structure of the discrete molecule in $[Ag_2C_2 \cdot 5CF_3CO_2 Ag \cdot 2(15C5) \cdot H_2O] \cdot 3H_2O$. The F and H atoms have been omitted for clarity. (b) Space-filling drawing of the discrete supermolecule in $[Ag_{14}(C_2)_2(CF_3CO_2)_{14}(dabcoH)_4(H_2O)_{1.5}] \cdot H_2O$. (dabco = 1,4-diazabicyclo[2.2.2]octane). The trifluoroacetate and aqua ligands have been omitted for clarity. (c) Space-filling drawing of the discrete supermolecule $[Ag_8(C_2)(CF_3CO_2)_6(L^1)_6]$ viewed along its $\bar{3}$ symmetry axis (L^1 = 4-hydroxyquinoline). The trifluoroacetate ligands have been omitted for clarity. (Q.-M. Wang and T. C. W. Mak, *Angew. Chem. Int. Ed.* **40**, 1130–3 (2001); Q.-M. Wang, T. C. W. Mak, *Inorg. Chem.* **42**, 1637–43 (2003); X.-L. Zhao, Q.-M. Wang, and T. C. W. Mak, *Inorg. Chem.* **42**, 7872–6 (2003).)

disordered about a crystallographic three-fold axis that bisects the C≡C bond and passes through two opposite corners of the rhombohedron. Apart from the μ_2-O, O' trifluoroacetate liagnds, there are six keto-L^1 ligands surrounding the polynuclear core, each bridging an edge in the μ-O mode.

In the structures shown in Fig. 10.4.26, the hydrophobic tails of perfluorocarboxylates, together with the bulky ancillary ligands, successfully prevent linkage between adjacent silver polyhedra, thus leading to discrete supermolecules.

(2) Chains and columns

Free betaines are zwitterions bearing a carboxylate group and a quaternary ammonium group, and the prototype of this series is the trimethylammonio derivative commonly called betaine ($Me_3N^+CH_2COO^-$, IUPAC name trimethylammonioacetate, hereafter abbreviated as Me_3bet). Owing to their permanent bipolarity and overall charge neutrality, betaine and its derivatives (considered as carboxylate-type ligands) have distinct advantages over most carboxylates as ligands in the formation of coordination polymers: (a) synthetic access to water-soluble metal carboxylates; (b) generation of new structural varieties, such as complexes with metal centers bearing additional anionic ligands, and those with variable metal to carboxylate molar ratios; (c) easy synthetic modification of ligand property by varying the substituents on the quaternary nitrogen atom or the backbone between the two polar terminals. The introduction of such ligands into the Ag_2C_2-containing system has led to isolation of supramolecular complexes showing chain-like or columnar structures.

$[(Ag_2C_2)_2(AgCF_3CO_2)_9(L^2)_3]$ has a columnar structure composed of fused silver(I) double cages: a triangulated dodecahedron and a bicapped trigonal prism, each encapsulating an acetylenediide dianion. Such a neutral column is coated by a hydrophobic sheath composed of trifluoroacetate and L^2 ligands, as shown in Fig. 10.4.27(a).

The core in $[(Ag_2C_2)_2(AgCF_3CO_2)_{10}(L^3)_3] \cdot H_2O$ is a double cage generated from edge-sharing of a square-antiprism and a distorted bicapped trigonal-prism, with each single cage encapsulating an acetylenediide dianion. Double cages of this type are fused together to form a helical column, which is surrounded by a hydrophobic sheath composed of trifluoroacetate and L^3 ligands, as shown in Fig. 10.4.27(b).

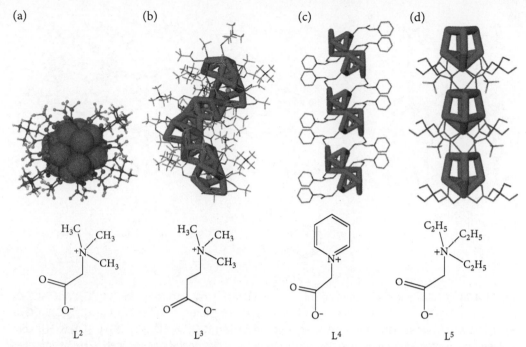

Fig. 10.4.27 (a) Projection along an infinite silver(I) column with enclosed C_2^{2-} species and hydrophobic sheath in $[(Ag_2C_2)_2(AgCF_3CO_2)_9(L^2)_3]$. (b) Perspective view of the infinite silver(I) helical column with C_2^{2-} species embedded in its inner core and an exterior coat comprising anionic and zwitterionic carboxylates in $[(Ag_2C_2)_2(AgCF_3CO_2)_{10}(L^3)_3]\cdot H_2O$. (c) Infinite chain generated from the linkage of $(C_2)_2@Ag_{16}$ double cages by μ_2-O,O' L^4 ligands in $[(Ag_2C_2)(AgC_2F_5CO_2)_6(L^4)_2]$. (d) Infinite chain constructed from $C_2@Ag_9$ polyhedra connected by L^5 and trifluoroacetate bridges in $[(Ag_2C_2)(AgCF_3CO_2)_7(L^5)_2(H_2O)]$. (X-L. Zhao, Q-M, Wang, and T. C. W. Mak, *Chem. Eur. J.* **11**, 2094–102 (2005).)

The building unit in $[(Ag_2C_2)(AgC_2F_5CO_2)_6(L^4)_2]$ is a centrosymmetric double cage, in which each half encapsulates an acetylenediide dianion. Each single cage is an irregular monocapped trigonal prism with one appended atom. The L^4 ligand acting in the μ_2-O, O' coordination mode links a pair of double cages to form an infinite chain, as shown in Fig. 10.4.27(c).

In $[(Ag_2C_2)(AgCF_3CO_2)_7(L^5)_2(H_2O)]$, the basic building unit is a distorted monocapped cube. The trifluoroacetate and L^5 ligands act as μ_3-bridges across adjacent single cage blocks to form a bead-like chain (Fig. 10.4.27(d)).

(3) Two-dimensional structures

The incorporation of ancillary *N,N'*- and *N,O*-donor ligands into the Ag_2C_2-containing system has led to a series of 2-D structures.

In the synthesis of $[(Ag_2C_2)(AgCF_3CO_2)_4(L^6)(H_2O)]\cdot H_2O$ under hydrothermal reaction condition, the starting ligand 4-cyanopyridine undergoes hydrolysis to form 4-pyridine-carboxamide (L^6). The basic building unit is a $C_2@Ag_8$ single cage in the shape of a distorted triangulated dodecahedron. Such dodecahedra share edges to form a zigzag composite chain, which are further linked via L^6 to generate a 2-D network (Fig. 10.4.28(a)).

The L^1 ligand, H_3O^+ species, and the anionic polymeric system $\{[Ag_{11}(C_2)_2(C_2F_5CO_2)_9(H_2O)_2]^{2-}\}_\infty$ comprise $(L^1\cdot H_3O)_2[Ag_{11}(C_2)_2(C_2F_5CO_2)_9(H_2O)_2]\cdot H_2O$. The basic building unit in the latter is a Ag_{12}

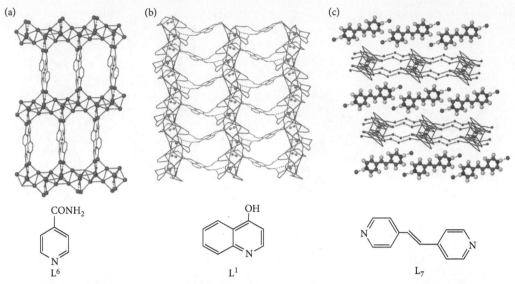

Fig. 10.4.28 (a) Ball-and-stick drawing of the two-dimensional structure in $[(Ag_2C_2)(AgCF_3CO_2)_4$ $(L^6)(H_2O)]\cdot H_2O$. (b) Schematic showing of the layer structure in $(L^1\cdot H_3O)_2[Ag_{11}(C_2)_2(C_2F_5CO_2)_9$ $(H_2O)_2]\cdot H_2O$. (c) Two-dimensional host layer structure of $[Ag_8(C_2)(CF_3CO_2)_8(H_2O)_2]\cdot(H_2O)_4\cdot(L^7H_2)$ constructed from hydrogen bonds linking the silver(I) chains, with hydrogen-bonded $(bpeH_2\cdot 2H_2O)^{2+}$ moieties are accommodated between the host layers. (X.-L. Zhao, and T. C. W. Mak, *Dalton Trans.* 3212–7 (2004); X.-L. Zhao, Q.-M. Wang, and T. C. W. Mak, *Inorg. Chem.* **42,** 7872–6 (2003). X.-L. Zhao, and T. C. W. Mak, *Polyhedron* **24**, 940–8 (2005).)

double cage composed of two irregular monocapped trigonal antiprisms sharing an edge. The double cages are fused together to generate an infinite, sinuous anionic column. The oxygen atoms of L^1 and the water molecule are bridged by a proton to give the cationic aggregate $L^1\cdot H_3O^+$, which link the columns into a layer structure via hydrogen bonds with the pentafluoropropionate groups (Fig. 10.4.28(b)).

$[Ag_8(C_2)(CF_3CO_2)_8(H_2O)_2]\cdot(H_2O)_4\cdot(L^7H_2)$ represents a rare example of a hydrogen-bonded layer-type host structure containing $C_2@Ag_n$ that features the inclusion of organic guest species. The core is a centrosymmetric $C_2@Ag_8$ single cage in the shape of a slightly distorted cube. The trifluoroacetate ligands functioning in the μ_3-coordination mode further interlink the single cages into an infinite zigzag silver(I) chain. Of the three independent water molecules, one forms an acceptor hydrogen bond with a terminal of the L^7H_2 dication, the other serves as an aqua ligand bonded to a silver atom, and the third lying on a two-fold axis functions as a bridge between aqua ligands belonging to adjacent silver(I) chains. The centrosymmetric L^7H_2 ions each hydrogen-bonded to a pair of terminal water molecules are accommodated between adjacent layers, forming an inclusion complex, as shown in Fig. 10.4.28(c).

(4) Three-dimensional structures

The strategy of using $C_2@Ag_n$ polyhedra as building blocks for the assembly of new coordination frameworks via introduction of potentially exo-bidentate nitrogen/oxygen-donor bridging ligands between agglomerated components has led to the isolation of 3-D supramolecular complexes exhibiting interesting crystal structures.

In $(Ag_2C_2)(AgCF_3CO_2)_8(L^8)_2(H_2O)_4$, square-antiprismatic $C_2@Ag_8$ cores are linked by trifluoroacetate groups to generate a columnar structure. Hydrogen bonds with the amino group of L^8 and aqua ligands serving as donors and the oxygen atoms of the trifluoroacetate group as acceptors further connect the columns into a 3-D scaffold (Fig. 10.4.29(a)).

The building block in $(L^9H)_3 \cdot [Ag_8(C_2)(CF_3CO_2)_9] \cdot H_2O$ is a $C_2@Ag_8$ single cage in the shape of triangulated dodecahedron located on a two-fold axis. Silver cages of such type are connected by μ_3-O, O, O' trifluoroacetate ligands to form a zigzag anionic silver(I) column along the a direction. All three independent L^9 molecules are protonated to satisfy the overall charge balance in the crystal structure. Notably, the resulting L^9H cations play a key role in the construction of the 3-D architecture. As shown in Fig. 10.4.29(b), the silver(I) columns are interconnected by hydrogen bonding with the protonated L^9 serving as donors and O, F atoms of trifluoroacetate ligands as acceptors to form the 3-D network.

In $[Ag_7(C_2)(CF_3CO_2)_2(L^{10})_3]$, the basic building block is a centrosymmetric $(C_2)_2@Ag_{14}$ double cage, with each half taking the shape of a distorted bicapped trigonal prism. Such double cages are fused together to form an infinite column. Each silver(I) column is linked to six other radiative silver(I) columns via L^{10}, and every three neighboring silver columns encircle a triangular hole, thus resulting in a (3,6) (or 3^6) topology, as displayed in Fig. 10.4.29(c).

(5) Mixed-valent silver(I, II) compounds containing Ag$_2$C$_2$

To investigate the effect of coexisting metal ions on the assembly of polyhedral silver(I) cages, macrocyclic N-donor ligand 1,4,8,11-tetramethyl-1,4,8,11-tetraazacyclotetradecane (tmc) has been used for in situ generation of $[Ag^{II}(tmc)]$. Mixed-valent silver complexes $[Ag^{II}(tmc)(BF_4)]$ $[Ag^I_6(C_2)(CF_3CO_2)_5(H_2O)] \cdot H_2O$ and $[Ag^{II}(tmc)][Ag^{II}(tmc)(H_2O)]_2[Ag^I_{11}(C_2)(CF_3CO_2)_{12}(H_2O)_4]_2$ have been isolated and structurally characterized.

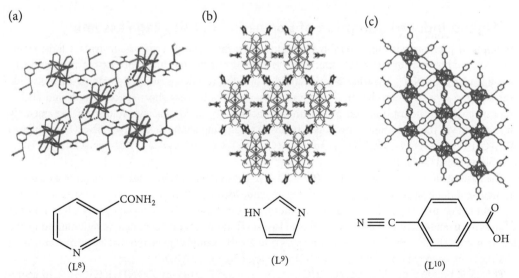

Fig. 10.4.29 (a) Three-dimensional architecture of $(Ag_2C_2)(AgCF_3CO_2)_8(L^8)_2(H_2O)_4$ resulting from the linkage of silver columns via hydrogen bonds. (b) Three-dimensional architecture in $(L^9H)_3 \cdot [Ag_8(C_2)(CF_3CO_2)_9] \cdot H_2O$ generated from covalent silver chains linked by hydrogen bonds. (c) The (3,6) covalent network in $[Ag_7(C_2)(CF_3CO_2)_2(L^{10})_3]$ constructed the linkage of L^{10} with silver(I) columns. (X.-L. Zhao, and T. C. W. Mak, *Dalton Trans.* 3212–7 (2004); X.-L. Zhao, and T. C. W. Mak, *Polyhedron*, **25**, 975–82 (2006).)

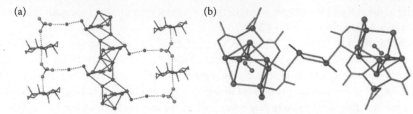

Fig. 10.4.30 (a) Crystal structure of $[Ag^{II}(tmc)(BF_4)][Ag^{I}_6(C_2)(CF_3CO_2)_5(H_2O)]\cdot H_2O$. (b) Perspective view of the structure of the dimeric supramolecular anion in $[Ag^{II}(tmc)][Ag^{II}(tmc)(H_2O)]_2[Ag^{I}_{11}(C_2)(CF_3CO_2)_{12}(H_2O)_4]_2$. The F atoms of the $CF_3CO_2^-$ ligands and some $CF_3CO_2^-$ are omitted for clarity. (Q.-M. Wang, and T. C. W. Mak, *Chem. Commun.* 807–8 (2001); Q.-M. Wang, H. K. Lee, and T. C. W. Mak, *New J. Chem.* **26**, 513–5 (2002).)

In $[Ag^{II}(tmc)(BF_4)][Ag^{I}_6(C_2)(CF_3CO_2)_5(H_2O)]\cdot H_2O$, the addition of tmc leads to disproportionation of silver(I) to give elemental silver and complexed silver(II), the latter being stabilized by tmc to form $[Ag^{II}(tmc)]^{2+}$. Weak axial interactions of the d^9 silver(II) center with adjacent BF_4^- serve to link the complexed Ag(II) cations into a $\left[Ag^{II}(tmc)(BF_4)\right]^+_\infty$ column, which further induces the assembly of a novel anionic zigzag chain constructed from edge-sharing of silver(I) triangulated dodecahedra each enclosing a C_2^{2-} species (Fig. 10.4.30(a)).

In $[Ag^{II}(tmc)][Ag^{II}(tmc)(H_2O)]_2[Ag^{I}_{11}(C_2)(CF_3CO_2)_{12}(H_2O)_4]_2$, which lacks the participation of BF_4^- ions, the cations do not line up in a 1-D array and instead a dimeric supramolecular cluster anion is generated (Fig. 10.4.30(b)).

(6) Ligand-induced disruption of polyhedral $C_2@Ag_n$ cage assembly

Attempts to interfere with the assembly process to open the $C_2@Ag_n$ cage or construct a large single cage for holding two or more C_2^{2-} species were carried out via the incorporation of the multidentate ligand pyzCONH$_2$ (pyrazine-2-carboxamide) into the reaction system. Pyrazine-2-carboxamide was selected as a structure-directing component by virtue of its very short spacer length and chelating capacity, and the introduction of the amide functionality could conceivably disrupt the assembly of $C_2@Ag_n$ via the formation of hydrogen bonds. The ensuing study yielded two silver(I) complexes $Ag_{12}(C_2)_2(CF_3CO_2)_8(2\text{-pyzCONH}_2)_3$ and $Ag_{20}(C_2)_4(C_2F_5CO_2)_8(2\text{-pyzCOO})_4(2\text{-pyzCONH}_2)(H_2O)_2$ exhibiting novel $C_2@Ag_n$ motifs.

The basic structural unit of $Ag_{12}(C_2)_2(CF_3CO_2)_8(2\text{-pyzCONH}_2)_3$ comprises the fusion of a distorted triangulated dodecahedral Ag$_8$ cage containing an embedded C_2^{2-} dianion and an open fish-like $Ag_6(\mu_6\text{-}C_2)$ motif (Fig. 10.4.31(a)). In the $Ag_6(\mu_6\text{-}C_2)$ motif, one carbon atom is embraced by four silver atoms in a butterfly arrangement and the other bonds to two silver atoms. Its existence can be rationalized by the fact that it is stabilized by four surrounding pyrazine-2-carboxamide ligands so that steric overcrowding obstructs the aggregation of silver(I) into a closed cage (Fig. 10.4.31(b)).

The basic building block in $Ag_{20}(C_2)_4(C_2F_5CO_2)_8(2\text{-pyzCOO})_4(2\text{-pyzCONH}_2)(H_2O)_2$ is an aggregate composed of three polyhedral units: an unprecedented partially opened cage $(C_2)_2@Ag_{13}$ (Fig. 10.4.31(c)) and two similar distorted $C_2@Ag_6$ trigonal prisms. A pair of C_2^{2-} dianions are completely encapsulated in the Ag$_{13}$ cage. For simplicity, this single cage can be visualized as composed of two distorted cubes sharing a common face, with cleavage of four of the edges and capping of a lateral face. The two embedded C_2^{2-} dianions retain their triple bond character with similar C–C bond lengths of 118(2) pm.

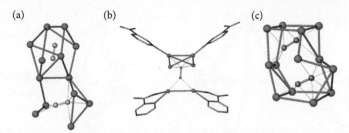

Fig. 10.4.31 (a) Basic building unit in $Ag_{12}(C_2)_2(CF_3CO_2)_8(2\text{-pyzCONH}_2)_3$. (b) Open fish-like $Ag_6(\mu_6\text{-}C_2)$ motif coordinated by four pyrazine-2-carboxamide ligands in $Ag_{12}(C_2)_2(CF_3CO_2)_8(2\text{-pyzCONH}_2)_3$. (c) The $(C_2)_2@Ag_{13}$ in $Ag_{20}(C_2)_4(C_2F_5CO_2)_8(2\text{-pyzCOO})_4(2\text{-pyzCONH}_2)(H_2O)_2$. (X.-L. Zhao, and T. C. W. Mak, *Organometallics* **24**, 4497–9 (2005).)

10.4.6 **Supramolecular assembly with the silver(I)-ethynide synthon**

In 2004, the silver carbide (Ag_2C_4) was synthesized as a light gray powder, which behaves like its lower homologue Ag_2C_2, being insoluble in most solvents and highly explosive in the dry state when subjected to heating or mechanical shock. Using Ag_2C_4 and the crude polymeric silver ethynide complexes [R–$(C{\equiv}CAg)_m]_\infty$ (R = aryl; m = 1 or 2) as starting materials, a variety of double and triple silver(I) salts containing 1,3-butadiynediide and related carbon-rich ethynide ligands have been synthesized. Investigation of the coordination modes of the ethynide moiety in these compounds led to the recognition of a new class of supramolecular synthons R–$C{\equiv}C{\supset}Ag_n$ (n = 4, 5), which can be utilized to assemble a series of 1-D, 2-D, and 3-D networks together with argentophilic interactions, π–π stacking, silver-aromatic interactions, and hydrogen bonding.

(1) **Silver(I) complexes containing the C_4^{2-} dianion**

In all of its silver(I) complexes, the linear $^{-}C{\equiv}C{-}C{\equiv}C^{-}$ dianion exhibits an unprecedented μ_8-coordination mode, each terminal being capped by four silver(I) atoms (Figure 10.4.32). However, the σ-type and π-type silver-ethynide interactions play different roles in symmetrical and unsymmetrical μ_8-coordination. Furthermore, coexisting ancillary anionic ligands, nitrile groups, and aqua molecules also influence the coordination environment around each terminal ethynide, which takes the form of a butterfly-shaped, barb-like, or planar Ag_4 basket. The carbon-carbon triple and single bond lengths in C_4^{2-} are in good agreement with those observed in transition-metal 1,3-butadiyne-1,4-diyl complexes. The Ag⋯Ag distances within the Ag_4 baskets are all shorter than 340 pm, suggesting the existence of significant Ag⋯Ag interactions.

The [$Ag_4C_4Ag_4$] aggregates *vide supra* can be further linked by other anionic ligands such as nitrate and perfluorocarboxylate groups, and/or water molecules, to produce various 2-D or 3-D coordination networks. In the structure of $Ag_2C_4 \cdot 6AgNO_3 \cdot 2H_2O$, the [$Ag_4C_4Ag_4$] aggregates arranged in a pseudo-hexagonal array are connected by one nitrate group acting in the $\mu_3\text{-}O, O', O''$ plus O, O'-chelating mode to form a thick layer normal to [100] (Fig. 10.4.33(a)). Linkage of adjacent layers by the remaining two independent nitrate groups, abetted by O-H⋯O(nitrate) hydrogen bonding involving the aqua ligand, then generates a 3-D network.

In the crystal structure of $Ag_2C_4 \cdot 16AgC_2F_5CO_2 \cdot 24H_2O$, each [$Ag_4C_4Ag_4$] unit connects with eight such units by eight [$Ag_2(\mu\text{-}O_2CC_2F_5)_4$] bridging ligands to form a (4,4) coordination network (Fig. 10.4.33(b)). Through the linkage of the C_4 carbon chains perpendicular to this network, an infinite channel is aligned along the [001] direction, and each accommodates a large number of pentafluoroethyl groups.

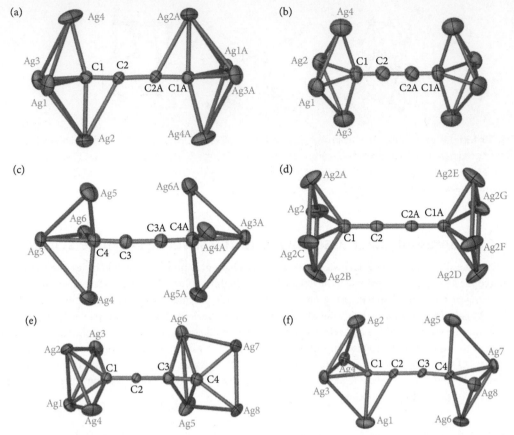

Fig. 10.4.32 Observed μ_8-coordination modes of C_4^{2-} dianion. (a) Symmetrical mode with two butterfly-shaped baskets in $Ag_2C_4 \cdot 6AgNO_3 \cdot nH_2O$ ($n = 2, 3$). (L. Zhao, and T. C. W. Mak, *J. Am. Chem. Soc.* 126, 6852–3 (2004).) (b) Symmetrical mode only with Ag-C σ bonds in $Ag_2C_4 \cdot 16AgC_2F_5CO_2 \cdot 6CH_3CN \cdot 8H_2O$. (c) Barb-like μ_8-coordination mode with two linearly coordinated C≡C-Ag bonds in triple salt $Ag_2C_4 \cdot AgF \cdot 3AgNO_3 \cdot 0.5H_2O$. (d) Symmetrical mode with two parallel planar Ag_4 aggregates in $Ag_2C_4 \cdot 16AgC_2F_5CO_2 \cdot 24H_2O$. (e) Unsymmetrical μ_8-coordination mode with one butterfly-shaped Ag_4 basket and one planar Ag_4 aggregate in $Ag_2C_4 \cdot 6AgCF_3CO_2 \cdot 7H_2O$. (f) Unsymmetrical μ_8-coordination mode with two butterfly-shaped Ag_4 baskets in triple salt $Ag_2C_4 \cdot 4AgNO_3 \cdot 2Ag_2FPO_3$.

In the crystal structure of $Ag_2C_4 \cdot AgF \cdot 3AgNO_3 \cdot 0.5H_2O$, $[Ag_4C_4Ag_4]$ aggregates are mutually connected through the linkage of nitrate groups and sharing of some silver atoms to form a silver column. The fluoride ions bridge these silver columns in the μ_3-mode to produce a structurally robust 3-D coordination network (Fig. 10.4.33(c)). With an external silver atom and two carboxylato oxygen atoms as bridging groups, the $[Ag_4C_4Ag_4]$ aggregates in $Ag_2C_4 \cdot 10AgCF_3CO_2 \cdot 2[(Et_4N)CF_3CO_2] \cdot 4(CH_3)_3CCN$ are linked to form a 2-D rosette layer, in which the C_4^{2-} dianion acts as the shared border of two metallacycles (Fig. 10.4.33(d)).

(2) Silver(I) complexes of isomeric phenylenediethynides with the supramolecular synthons $Ag_n \subset C_2$-x-C_6H_4-$C_2 \supset Ag_n$ ($x = p, m, o$; $n = 4, 5$)

The above study of silver(I) 1,3-butadiynediide complexes suggests that the $Ag_4 \subset C_2$-R-$C_2 \supset Ag_4$ moiety may be conceived as a synthon for the assembly of coordination networks, by analogy to the plethora

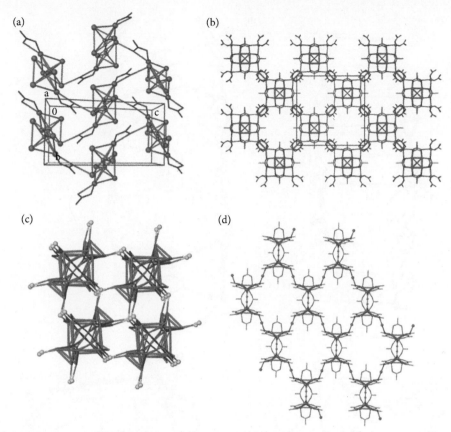

Fig. 10.4.33 Some examples of 2-D and 3-D coordination networks of Ag_2C_4. (a) Pseudohexagonal array of $[Ag_4C_4Ag_4]$ aggregates linked by an independent nitrate group in $Ag_2C_4 \cdot 6AgNO_3 \cdot 2H_2O$. (b) (4,4) network in *ab* plane of $Ag_2C_4 \cdot 16AgC_2F_5CO_2 \cdot 24H_2O$, which is composed of $[Ag_4C_4Ag_4]$ aggregates connected by bridging $[Ag_2(\mu\text{-}O_2CC_2F_5)_4]$ groups. (c) 3-D coordination network in $Ag_2C_4 \cdot AgF \cdot 3AgNO_3 \cdot 0.5H_2O$ through the coordination of μ_3-fluoride ligands. (d) Rosette layer in $Ag_2C_4 \cdot 10AgCF_3CO_2 \cdot 2[(Et_4N)CF_3CO_2] \cdot 4(CH_3)_3CCN$ composed of $[Ag_4C_4Ag_4]$ aggregates linked by one external silver atoms and two trifluoroacetate groups.

of well-known supramolecular synthons that involve hydrogen bonding and other weak intermolecular interactions (see Section 10.1.5). With reference to 1,3-butadiynediide as a standard, the *p*-phenylene ring was introduced as the bridging R group in the supramolecular synthon $Ag_n \subset C_2\text{-}R\text{-}C_2 \supset Ag_n$ with a lengthened linear π-conjugated backbone, and the aromatic ring of the resulting *p*-phenylenediethynide dianion could presumably partake in π–π stacking and silver-aromatic interaction. The isomeric *m*- and *o*-phenylenediethynides were also investigated in order to probe the influence of varying the relative orientation of the pair of terminal ethynide groups.

In $2[Ag_2(p\text{-}C{\equiv}CC_6H_4C{\equiv}C)] \cdot 11AgCF_3CO_2 \cdot 4CH_3CN \cdot 2CH_3CH_2CN$, the *p*-phenylenediethynide ligand exhibits the highest ligation number reported to date for the ethynide moiety by adopting an unprecedented $\mu_5\text{-}\eta^1$ mode. The Ag_{14} aggregate, being constructed essentially from two independent $Ag_n \subset C_2\text{-}(p\text{-}C_6H_4)\text{-}C_2 \supset Ag_n$ ($n = 4, 5$) synthons through argentophilic interaction and continuous π–π stacking, is connected to its symmetry equivalents to form a broken silver(I) double chain along the *a*-axis. Adjacent Ag_{14} segments within a single chain are bridged by the oxygen atoms of two independent trifluoroacetate groups, and the pair of single chains are arranged in interdigitated fashion (Fig. 10.4.34(a)).

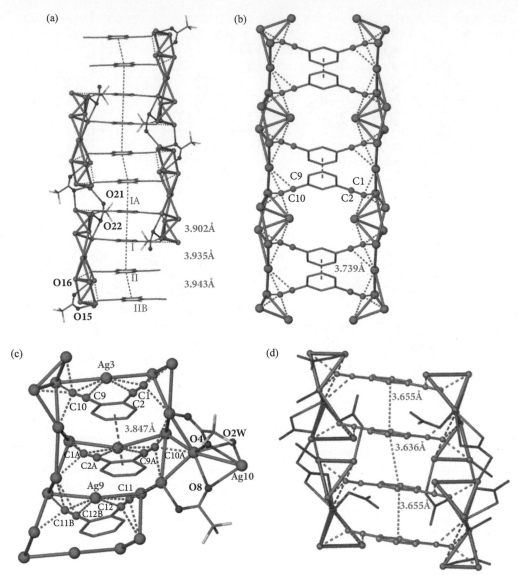

Fig. 10.4.34 (a) Broken silver(I) double chain in $2[Ag_2(p-C{\equiv}CC_6H_4C{\equiv}C)]{\cdot}11AgCF_3CO_2{\cdot}4CH_3CN{\cdot}$ $2CH_3CH_2CN$ stabilized by continuous π-π stacking between parallel p-phenylene rings. (b) Silver double chain in $Ag_2(m-C{\equiv}CC_6H_4C{\equiv}C){\cdot}6AgCF_3CO_2{\cdot}3CH_3CN{\cdot}2.5H_2O$ assembled by argentophilic interaction and π-π interaction between adjacent pairs of m-phenylene rings. (c) Widened silver chain in $3[Ag_2(o-C{\equiv}CC_6H_4C{\equiv}C)]{\cdot}14AgCF_3CO_2{\cdot}2CH_3CN{\cdot}9H_2O$ constructed through the cross-linkage of two narrow silver chains by bridging silver atoms with pair-wise π-π interaction between the o-phenylene rings. (d) Broken silver(I) double chain in $Ag_2(m-C{\equiv}CC_6H_4C{\equiv}C)]{\cdot}5AgNO_3{\cdot}3H_2O$ stabilized by continuous π-π stacking between parallel m-phenylene rings. (L. Zhao, and T. C. W. Mak, *J. Am. Chem. Soc.* **127**, 14966–7 (2005).)

In contrast to the popular μ_1,μ_1-coordination mode of m-phenylenediethynide in most transition metal complexes, this ligand exhibits two different terminal ethynide bonding modes, namely, μ_4-$\eta^1,\eta^1,\eta^1,\eta^1$ and μ_4-$\eta^1,\eta^1,\eta^1,\eta^2$, in the crystal structure of $Ag_2(m$-$C\equiv CC_6H_4C\equiv C)\cdot6AgCF_3CO_2\cdot3CH_3CN\cdot2.5H_2O$. Through inversion centers located between successive pairs of m-phenylene rings, the Ag_nCC_2-(m-C_6H_4)-C_2⊃Ag_n ($n = 4$) synthon is extended along the a direction by Ag···Ag interactions to form a silver double chain (Fig. 10.4.34(b)), which are further consolidated by π–π interaction between adjacent m-phenylene rings protruding alternatively on either side.

In the crystal structure of $3[Ag_2(o$-$C\equiv CC_6H_4 C\equiv C)]\cdot14AgCF_3CO_2\cdot2CH_3CN\cdot9H_2O$, the ethynide moieties bond to silver atoms via three different coordination modes: μ_4-$\eta^1,\eta^1,\eta^1,\eta^2$, μ_4-$\eta^1,\eta^1,\eta^2,\eta^2$ and μ_5-$\eta^1,\eta^1,\eta^1,\eta^1,\eta^2$ (Fig. 10.4.34(c)). The two independent Ag_nCC_2-(o-C_6H_4)-C_2⊃Ag_n ($n = 4, 5$) synthons mutually associate to generate an undulating silver chain consolidated by argentophilic interaction, pairwise π–π interaction between the o-phenylene rings, and linkage by silver atoms Ag3 and Ag9 across two silver single chains. Bridged by silver atom Ag10 through Ag···Ag interaction and three oxygen atoms (O2W, O4, O8), the silver columns are linked to form a wave-like layer with o-phenylene groups protruding on both sides.

The decreasing separation of the pair of ethynide groups in the above three complexes is accompanied by strengthened argentophilic interaction at the expense of weakened π–π stacking, yielding a broken double chain, a double chain, and a silver layer, respectively. When the pair of ethynide vectors make an angle of 60°, sharing of a common silver atom for the Ag_n caps occurs in the o-phenylenediethynide complex.

The crystal structure of $Ag_2(m$-$C\equiv CC_6H_4C\equiv C)]\cdot5AgNO_3\cdot3H_2O$ features a broken silver(I) double chain analogous to that of the p-phenylenediethynide complex (see Fig. 10.4.34(a)). However, the single chains containing the Ag_{14} aggregates are arranged in parallel fashion, each matching Ag_{14} pair being connected by two m-phenylenediethynide ligands (Fig. 10.4.34(d)). Linkage of a series of Ag_nCC_2-(m-C_6H_4)-C_2⊃Ag_n ($n = 4$) fragments by intra-chain bridging nitrate groups with continuous π–π interaction between adjacent m-phenylene rings engenders the broken silver(I) double chain.

The structural correlation between various silver-ethynide supramolecular synthons affords a rationale for the preponderant existence of C_2@Ag_n ($n = 6$-10) polyhedra in Ag_2C_2 complexes (see Section 10.4.5). If the linear $^-C\equiv C$–$C\equiv C^-$ chain were contracted to a C_2^{2-} dumbbell, further overlap between atoms of the terminal Ag_n caps would conceivably yield a closed cage with six to ten vertices (Fig. 10.4.35).

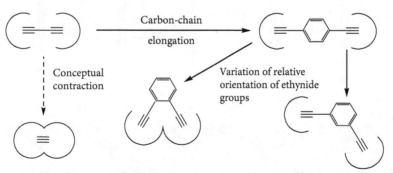

Fig. 10.4.35 Schematic diagram showing the structural relationship between the supramolecular synthons $Ag_4 \subset C_2$—$C_2 \supset Ag_4$, $Ag_n \subset C_2$—R—$C_2 \supset Ag_n$ (R = p-, m-, o-C_6H_4; $n = 4, 5$) and C_2@Ag_n ($n = 6$–10). The circular arc represents a Ag_n ($n = 4, 5$) basket.

(3) Silver(I) arylethynide complexes containing R–C$_2$⊃Ag$_n$ (R = C$_6$H$_5$, C$_6$H$_4$Me-4, C$_6$H$_4$Me-3, C$_6$H$_4$Me-2, C$_6$H$_4^t$Bu-4; n = 4, 5)

π–π stacking or π–π interactions are important non-covalent intermolecular interactions, which contribute much to self-assembly when extended structures are formed from building blocks with aromatic moieties. In relation to the rich variety of π–π stacking in the crystal structure of silver(I) complexes of phenylenediethynide, related silver complexes of phenylethynide and its homologues with different substituents (–CH$_3$, –C(CH$_3$)$_3$) or the –CH$_3$ group in different positions (o-, m-, p-) are investigated.

In the crystal structure of 2AgC≡CC$_6$H$_5$·6AgC$_2$F$_5$CO$_2$·5CH$_3$CN, the ethynide group composed of C1 and C2 is capped by a square-pyramidal Ag$_5$ basket in an unprecedented μ_5-η^1,η^1,η^1,η^1,η^2 coordination mode and the other one comprising C9 and C10 by a butterfly-shaped Ag$_4$ basket in a μ_4-η^1,η^1,η^1,η^2 coordination mode, as shown in Fig. 10.4.36(a). With an inversion center located at the center of the Ag1···Ag1A bond, two Ag$_5$ baskets share an edge to engender a Ag$_8$ aggregate, whereas another Ag$_8$ aggregate results from fusion of a pair of inversion-related Ag$_4$ baskets. Two adjacent Ag$_8$ aggregates are linked

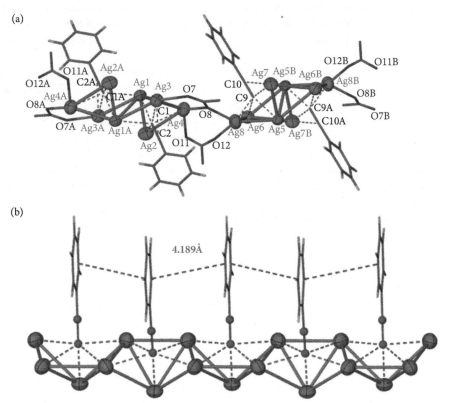

Fig. 10.4.36 (a) Coordination modes of the independent phenylethynide ligands in 2AgC≡CC$_6$H$_5$·6AgC$_2$F$_5$CO$_2$·5CH$_3$CN. (b) Coordination mode of the C$_6$H$_5$C≡C$^-$ ligand in AgC≡CC$_6$H$_5$·3AgCF$_3$CO$_2$·CH$_3$CN. The silver column is connected by edge sharing between adjacent square-pyramidal Ag$_5$ aggregates, and continuous π–π stacking of phenyl rings occurs on one side of the column. (L. Zhao, W.-Y. Wong, and T. C. W. Mak, *Chem. Eur. J.* **12**, 4865–72 (2006).)

by two pentafluoropropionate groups via μ_3-O, O',O', and μ_2-O, O' coordination modes, respectively, to generate an infinite column along the [111] direction. No π–π interaction is observed in this complex.

An infinite array of parallel phenyl rings stabilized by π–π stacking (center-to-center distance 418.9 pm) occurs in the complex AgC≡CC$_6$H$_5$·3AgCF$_3$CO$_2$·CH$_3$CN (Fig. 10.4.36(b)). The capping square-planar Ag$_5$ baskets are fused through argentophilic interactions via edge-sharing to form an infinite coordination column along the [100] direction, with continuous π–π stacking of phenyl rings lying on the same side of the column.

When substituents are introduced into the phenyl group, the π–π stacking between consecutive aromatic rings is affected by the size of the substituted groups and their positions. In the crystal structure of 2AgC≡CC$_6$H$_4$Me-4·6AgCF$_3$CO$_2$·1.5CH$_3$CN (Fig. 10.4.37(a)), the methyl group has a little influence on the formation of a silver column stabilized by π–π stacking, which is almost totally identical with the structure of AgC≡CC$_6$H$_5$·3AgCF$_3$CO$_2$·CH$_3$CN (see Fig. 10.4.36(b)). However, when a more bulky *tert*-butyl group is employed, the π–π stacking system is interrupted despite the formation of a similar silver chain (Fig. 10.4.37(b)). The entire C$_6$H$_4$^{t}Bu-4 moiety rotates around the C(^{t}Bu)–C(phenyl) single bond to generate a highly disordered structure. On the other hand, putting a *meta*-methyl group on the phenyl ring can sustain π–π stacking, but the constitution of the silver chain is changed from edge-sharing to vertex-sharing (Fig. 10.4.37(c)). Finally, use of the 2-methyl-substituted phenyl ligand entirely destroys the π–π stacking and even breaks the Ag···Ag interactions between Ag$_n$ caps to form a silver chain connected by trifluoroacetate groups (Fig. 10.4.37(d)).

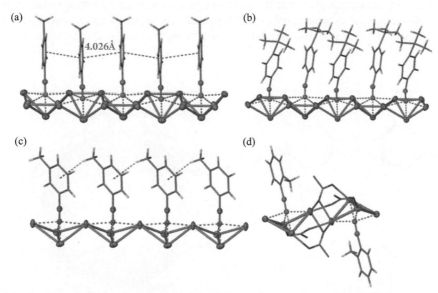

Fig. 10.4.37 (a) Silver column in 2AgC≡CC$_6$H$_4$Me-4·6AgCF$_3$CO$_2$·1.5CH$_3$CN connected by the fusion of square-pyramidal Ag$_5$ baskets and stabilized by continuous π–π stacking of phenyl rings. (b) Similar silver column in AgC≡CC$_6$H$_4$^{t}Bu-4·3AgCF$_3$CO$_2$·CH$_3$CN connected only by argentophilic interaction. (c) Silver chain in AgC≡CC$_6$H$_4$Me-3·2AgCF$_3$SO$_3$ through atom sharing. (d) Silver chain in AgC≡CC$_6$H$_4$Me-2·4AgCF$_3$CO$_2$·H$_2$O through the connection of trifluoroacetate groups.

10.4.7 Self-assembly of nanocapsules with pyrogallol[4]arene macrocycles

Recent studies have shown that the bowl-shaped C-alkyl substituted pyrogallol[4]arene macrocycles readily self-assemble to form a gobular hexameric cage, which is structurally robust and remains stable even in aqueous media (Fig. 10.4.38). Slow evaporation of a solution of C-heptylpyrogallol[4]arene in ethyl acetate gives crystalline [(C-heptylpyrogallol[4]arene)$_6$(EtOAc)$_6$(H$_2$O)]·6EtOAc, and X-ray analysis revealed that the large spheroidal supermolecule is stabilized by a total of 72 O–H···O hydrogen bonds (four intramolecular and eight intermolecular per macrocycle building block). The nanosized molecular capsule, having an internal cavity volume of about 1.2 nm^3, contains six ethyl acetate molecules and one water molecule; the methyl terminal of each encapsulated ethyl acetate guest molecule is orientated toward a bulge on the surface, and the single guest water molecule resides at the center of the capsule. In the crystal structure, the external ethyl acetate solvate molecules are embedded within the lower-rim alkyl legs at the base of each of the macrocycles, and the nanocapsules are arranged in hexagonal closest packing.

With reference to the unique architecture of this hydrogen-bonded hexameric capsule **I** (Fig. 10.4.39(a)), it was noted that the pyrogallol[4]arene building block has the potential of serving as a multidentate ligand through deprotonation of some of the upper-rim phenolic groups. As envisaged, treatment of C-propan-3-ol pyrogallol[4]arene with four equivalents of Cu(NO$_3$)$_2$·3H$_2$O in a mixture of acetone and water yielded a large neutral coordination capsule **II** [Cu$_{24}$(H$_2$O)$_x$(C$_{40}$H$_{40}$O$_{16}$)$_6$ ⊂ (acetone)$_n$] where $x \geq 24$ and $n = 1$–6. Single-crystal X-ray analysis established that retro-insertion of 24 Cu(II) meter centers into the hexameric framework results in substitution of 48 of the 72 phenolic protons, leaving the remaining 24 intact for intramolecular hydrogen bonding (O···O 0.2400–0.2488 nm). As shown in Fig. 10.4.39(b), the large coordination capsule **II** may be viewed as an octahedron with the six 16-membered macrocylic rings located at its corners, and each of its eight faces is capped by a planar cyclic [Cu$_3$O$_3$] unit of dimensions Cu–O 0.1911 to 0.1980 nm, O–Cu–O 85.67 to 98.23°, and Cu–O–Cu 140.96 to 144.78°.

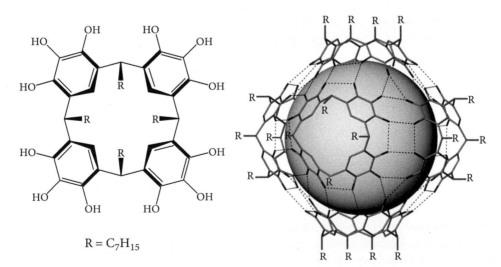

R = C$_7$H$_{15}$

Fig. 10.4.38 Assembly of six C-heptylpyrogallol[4]arene molecules by intra- and inter-molecular hydrogen bonds to form a globular supermolecule with a host cavity of volume ~ 1.2 nm^3. H atoms are omitted for clarity, and hydrogen bonds are represented by dotted lines. (G. V. C. Cave, J. Antesberger, L. J. Barbour, R. M. McKinley, and J. L. Atwood, *Angew. Chem. Int. Ed.* **43**, 5263–6 (2004).)

(a)   (b)

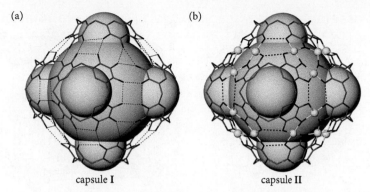

capsule **I**   capsule **II**

Fig. 10.4.39 Hydrogen-bonded supramolecular capsule **I** compared with coordination capsule **II**. The external aliphatic groups are omitted for clarity. Note that a pair of intermolecular O–H⋯O hydrogen bonds is replaced by four square-planar Cu–O coordination bonds in the metal-ion insertion process that generates the isostructural inorganic analog. (R. M. McKinley, G. V. C. Cave, and J. L. Atwood, *Proc. Nat. Acad. Sci.* **102**, 5944–8 (2005).)

Definitive location of all guest molecules inside the cavity is somewhat ambiguous owing to inexact stoichiometry and disorder. However, the (+)-MALDI mass spectra of **II** indicate that each individual capsule encloses different mixtures of water and acetone. In particular, two peaks implicated the presence of 24 entrapped water molecules that occupy axial coordination sites orientated toward the center of the cavity.

The pair of supramolecular capsules **I** and **II** represents a landmark in the construction of large coordination cages using multi-component ligands. This elegant blueprint approach is facilitated by the robustness of the hydrogen-bonded assembly **I**, which serves as a template for metal-ion insertion at specific sites with conservation of structural integrity. Notably, capsule size is virtually unchanged as the center-to-corner distance of the four phenolic O atoms belonging to the eight-membered intermolecular hydrogen-bonded ring (0.1883–0.1976 nm) in **I** closely matches the Cu–O bond distance of 0.1913–0.1978 nm in **II**.

10.4.8 Reticular design and synthesis of porous metal-organic frameworks

The design and synthesis of metal-organic frameworks (MOFs) has yielded a large number of solids that possess useful gas and liquid adsorption properties. In particular, highly porous structures constructures constructed from discrete metal-carboxylate clusters and organic links have been demonstrated to be amenable to systematic variation in pore size and functionality.

Consider the structure of the discrete tetranuclear $Zn_4O(CH_3CO_2)_6$ molecule, which is isostructural with $Be_4O(CH_3CO_2)_6$, the structure of which is shown in Fig. 9.5.2. Replacement of each acetate ligand by one half of a linear dicarboxylate produces a molecular entity that can conceivably be interconnected to identical entities to generate an infinite coordination network. An illustrative example is $Zn_4O(BDC)_6$, referred to as MOF-5, which is prepared from Zn(II) and benzene-1,4-dicarboxylic acid (H_2BDC) under solvothermal conditions. In the crystal structure, a $Zn_4O(CO_2)_6$ fragment comprising four fused ZnO_4 tetrahedra sharing a common vertex and six carboxylate C atoms constitute a "secondary building unit" (SBU). Connection of such octahedral SBUs by mutually perpendicular *p*-phenylene ($-C_6H_4-$) links leads to an infinite primitive cubic network, as shown in Fig. 10.4.40. Alternatively, the smaller Zn_4O

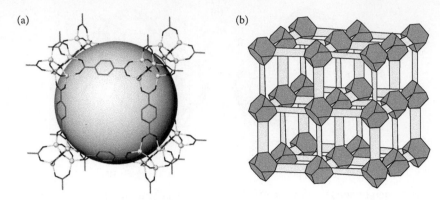

Fig. 10.4.40 Crystal structure of MOF-5. (a) Zn_4O tetrahedra joined by benzenedicarboxylate linkers. H atoms are omitted for clarity. (b) The topology of the framework (primitive cubic net) shown as an assembly of $(Zn_4O)O_{12}$ clusters (represented as truncated tetrahedra) and *p*-phenylene (—C_6H_4—) links (represented by rods). (O. M. Yaghi, M. O'Keeffe, N. W. Ockwig, H. K. Chae, M. Eddaoudi, and J. Kim, *Nature* **423**, 705–14 (2003).)

fragment (an oxo-centered Zn_4 tetrahedron) can be regarded as the SBU and the corresponding organic linker is the whole benzene-1,4-dicarboxylate dianion.

The resulting MOF-5 structure has exceptional stability and porosity as both the SBU and organic link are relatively large and inherently rigid. This reticular (which means "having the form of a (usually periodic) net") design strategy, based on the concept of discrete SBUs of different shapes (triangles, squares, tetrahedra, octahedra, etc.) considered as "joints" and organic links considered as "struts", has been applied to the synthesis and utilization of a vast number of MOF structures exhibiting varying geometries and network topologies. Based on the $Zn_4O(CO_2)_6$ SBU in the prototype MOF-5 (also designated as IRMOF-1), a family of isoreticular and isostructural cubic frameworks with diverse pore sizes and functionalities has been constructed, including IRMOF-6, IRMOF-8, IRMOF-11, and IRMOF-16, which are illustrated in Fig. 10.4.41.

An example of a porous framework that is isoreticular, but not isostructural, with MOF-5 is MOF-177, which incorporates the extended organic linker 1,3,5-benzenetribenzoate (BTB). The framework of crystalline MOF-177, $Zn_4O(BTB)_2\cdot(DEF)_{15}(H_2O)_3$ where DEF = diethyl formamide, has an ordered structure with an estimated surface area of 4,500 m^2 g^{-1}, which greatly exceeds those of zeolite Y (904 m^2g^{-1}) and carbon (2,030 m^2 g^{-1}).

As shown in Fig. 10.4.42, the underlying topology of MOF-177 is a (6,3)-net with the center of the octahedral $Zn_4O(CO_2)_6$ cluster as the six-connected node and the center of the BTB unit as the thee-connected node. Its exceptionally large pores are capable of accommodating polycyclic organic guest molecules such as bromobenzene, 1-bromonaphthalene, 2-bromonaphthalene, 9-bromoanthracene, C60, and the polycyclic dyes Astrazon Orange R and Nile Red.

Furthermore, the ability to prepare these kinds of MOFs in high yield and with adjustable pore size, shape, and functionality has led to their exploration as gas storage materials. Thermal gravimetric and gas sorption experiments have shown that IRMOF-6, bearing a fused hydrophobic unit C_2H_4 in its organic link, has the optimal pore aperture and rigidity requisite for maximum uptake of methane. Activation of the porous framework was achieved by exchanging the included guest molecules with chloroform, which was then removed by gradual heating to 800 °C under an inert atmosphere. The evacuated framework

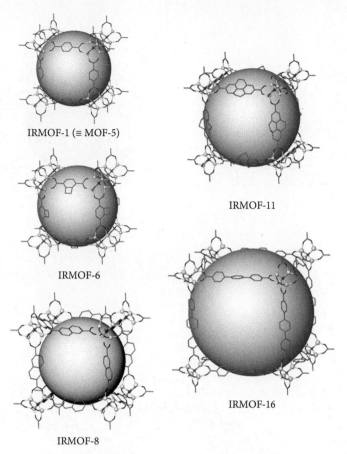

IRMOF-1 (≡ MOF-5)

IRMOF-11

IRMOF-6

IRMOF-16

IRMOF-8

Fig. 10.4.41 Comparison of cubic fragments in the respective three-dimensional extended structures of IRMOF-1 (≡ MOF-5), IRMOF-6, IRMOF-8, IRMOF-11, and IRMOF-16. (M. Eddaoudi, J. Kim, N. Rosi, D. Vodak, J. Wachter, M. O'Keeffe, and O. M. Yaghi, *Science* **295**, 469–72 (2004).)

has a stability range of 100–400 °C, and the methane sorption isotherm measured in the range 0–40 atm at room temperature has an uptake of 240 cm^3(STP)/g [155 cm^3(STP)/cm^3] at 298K and 36 atm. On a volume-to-volume basis, the amount of methane sorbed by IRMOF-6 at 36 atm amounts to 70% of that stored in compressed methane cylinders at ~205 atm. As compared to IRMOF-6, IRMOF-1 under the same conditions has a smaller methane uptake of 135 cm^3(STP)/g.

The isoreticular MOFs based on the $Zn_4O(CO_2)_6$ SBU also possesses favorable sorption properties for the storage of molecular hydrogen. At 77 K, microgravimetric sorption measurements gave H_2 (mg/g) values of 13.2, 15.0, 16.0, and 12.5 for IRMOF-1, IRMOF-8, IRMOF-11, and MOF-177, respectively. At the highest pressures attained in the measurements, the maximum uptake values for these frameworks are 5.0, 6.9, 9.3, and 7.1 molecules of H_2 per Zn_4OL_x unit where L stands for a linear dicarboxylate.

MOF-177 has been demonstrated to act like a super sponge in capturing vast quantities of carbon dioxide at room temperature. At moderate pressure (about 35 bar), its voluminous pores result in a gravimetric CO_2 uptake capacity of 33.5 mmol/g, which far exceeds those of the benchmark adsorbents zeolite 13X (7.4 mmol/g at 32 bar) and activated carbon MAXSORB (25 mmol/g at 35 bar). In terms of volume

(a)   (b)

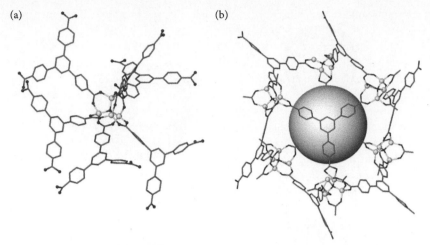

Fig. 10.4.42 Crystal structure of MOF-177. (a) A central Zn_4O unit coordinated by six BTB ligands. (b) The structure viewed down [001]. (H. K. Chae, D. Y. Siberio-Pérez, J. Kim, Y. B. Go, M. Eddaoudi, A. J. Matzger, M. O'Keeffe, and O. M. Yaghi, *Nature* **427**, 523–7 (2004).)

(a)   (b)

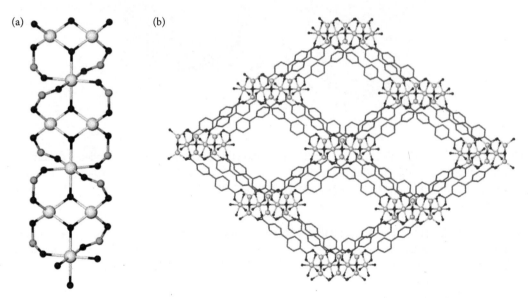

Fig. 10.4.43 Channel structure of MOF-69A: (a) ball-and-stick representation of inorganic SBU; (b) SBUs connected by biphenyl links. The DEF and water molecules have been omitted for clarity. (N. L. Rosi, J. Kim, M. Eddaoudi, B. Chen, M. O'Keeffe, and O. M. Yaghi, *J. Am. Chem. Soc.* **127**, 1504–18 (2005).)

capacity, a container filled with MOF-177 can hold about twice the amount of CO_2 versus the benchmark materials, and 9 times the amount of CO_2 stored in an empty container under the same conditions of temperature and pressure.

The previous structural and sorption studies of MOFs are all based on the discrete $Zn_4O(CO_2)_6$ SBU. Recent development has demonstrated that rod-shaped metal-carboxylate SBUs can also give rise to a variety of stable solid-state architectures and permanent porosity. Three illustrative examples are presented below.

In the crystal structure of $Zn_3(OH)_2(BPDC)_2 \cdot (DEF)_4(H_2O)_2$ where BPDC = 4,4′-biphenyldicarboxylate (MOF-69A), there are tetrahedral and octahedral Zn(II) centers coordinated by four and two carboxylate groups, respectively, in the *syn, syn* mode, with each μ_3-hydroxide ion bridging three metal centers (Fig. 10.4.43(a)). The infinite Zn-O-C rods are aligned in parallel fashion and laterally connected to give a 3-D network (Fig. 10.4.43(b)), forming rhombic channels of edge 1.22 nm and 1.66 nm along the longer diagonal, into which the DMF and water guest molecules are fitted.

The Mn–O–C rods in MOF-73, $Mn_3(BDC)_3 \cdot (DEF)_2$, are constructed from a pair of linked six-coordinate Mn(II) centers (Fig. 10.4.44(a)). One metal center is bound by two carboxylate groups acting in the *syn, syn* mode, one in the bidentate chelating mode, and a fourth one in the *syn, anti* mode. The other metal center has four carboxylates bound in the *syn, syn* mode and two in the *syn, anti* mode. Each rod is built of corner-linked and edge-linked MnO_6 octahedra, and is connected to four neighboring rods by *p*-phenylene links. The resulting 3-D host framework (Fig. 10.4.44(b)) has rhombic channels of dimensions $1.12 \times 0.59 \ nm^2$ that are filled by the DEF guest molecules.

The structure of MOF-75, $Tb(TDC) \cdot (NO_3)(DMF)_2$ where TDC = 2,5-thiophenedicarboxylate, contains eight-coordinate Tb(III) that is bound by four carboxylate groups all acting in the *syn, syn* mode, one bidentate nitrate ligand, and two terminal DMF ligands (Fig. 10.4.45(a)). The Tb–O–C rod orientated in the *a* direction consists of linked TbO_8 bisdisphenoids with the carboxyl carbon atoms forming a twisted ladder. Lateral linkage of rods in the *b* and *c* directions by the thiophene units generate rhombic

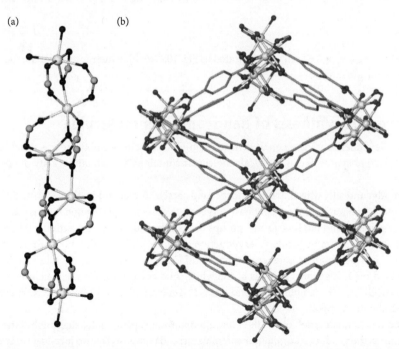

(a)   (b)

Fig. 10.4.44 Channel structure of MOF-73: (a) ball-and-stick representation of inorganic SBU; (b) SBUs connected by *p*-phenylene links. The DEF molecules have been omitted for clarity. (N. L. Rosi, J. Kim, M. Eddaoudi, B. Chen, M. O'Keeffe, and O. M. Yaghi, *J. Am. Chem. Soc.* **127**, 1504–18 (2005).)

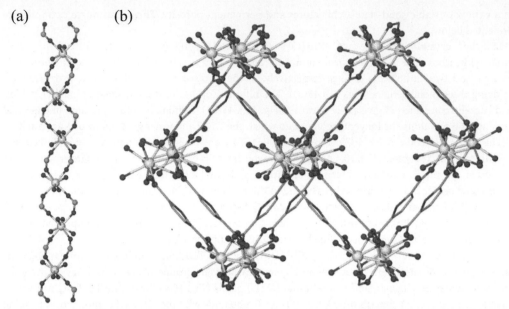

Fig. 10.4.45 Channel structure of MOF-75: (a) ball-and-stick representation of inorganic SBU; (b) SBUs connected by thiophene links. The DMF molecules and nitrate ions have been omitted for clarity. (N. L. Rosi, J. Kim, M. Eddaoudi, B. Chen, M. O'Keeffe, and O. M. Yaghi, *J. Am. Chem. Soc.* **127**, 1504–18 (2005).)

channels measuring 0.97×0.67 nm^2, as illustrated in Fig. 10.4.45(b), which accommodate the DMF guest molecules and nitrate ions.

10.4.9 One-pot synthesis of nanocontainer molecule

Dynamic covalent chemistry has been used in an atom-efficient self-assembly process to achieve a nearly quantitative one-pot synthesis of a nanoscale molecular container with an inner cavity of approximately 1.7 nm^3.

In a thermodynamically driven, trifluoroacetic acid catalyzed reaction in chloroform, six cavitands **1** and twelve ethylenediamine linkers condense to generate an octahedral nanocontainer **2**, as shown in Fig. 10.4.46. Each cavitand has four formyl groups on its rim, and these react with the 24 amino groups of the linkers to form 24 imine bonds. After reduction of all the imine bonds with NaBH$_4$, the hexameric nanocontainer can be isolated via reversed-phase HPLC as the trifluoroacetate salt **2**·24CF$_3$COOH in 63% yield based on **1**. Elemental analysis of the white solid corresponds to the stoichiometric formula **2**·24CF$_3$COOH·9H$_2$O. The simplified ^{1}H and ^{13}C NMR spectra of **2** and **2**·24CF$_3$COOH are consistent with their octahedral symmetry.

If the same reaction is carried out with either 1,3-diaminopropane or 1,4-diaminobutane in place of ethylenediamine, the product is an octaimino hemicarcerand composed of two face-to-face cavitands that are connected by four diamino bridging units.

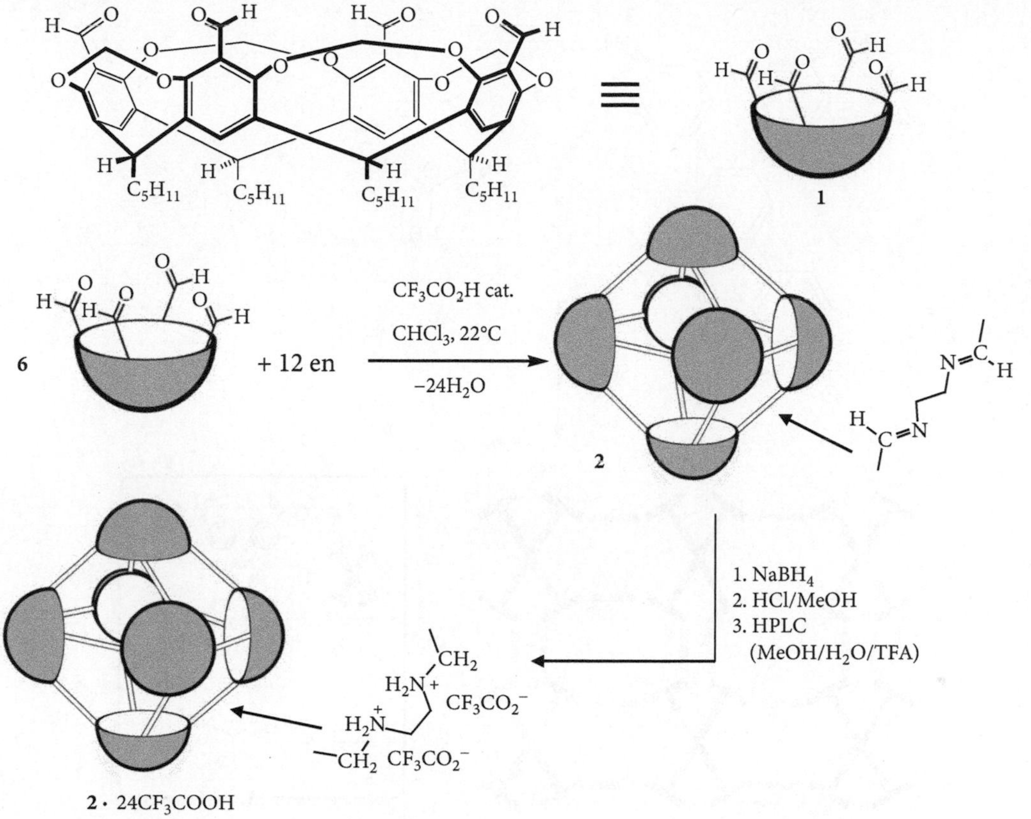

Fig. 10.4.46 Reaction scheme showing the thermodynamically controlled condensation of tetraformylcavitand **1** with ethylenediamine to form an octahedral nanocontainer **2**, which undergoes reduction to yield the trifluoroacetate salt **2**·24CF$_3$COOH.

10.5 Hydrogen-Bonded Organic Frameworks

Hydrogen-bonded organic frameworks (HOFs) are in general self-assembled through comparably weak hydrogen bonding interactions among molecular organic linkers, and hence they are mostly fragile and difficult to stabilize.

Over the years, the designed construction of lamellar to cylindrical GS hydrogen-bonded host frameworks from the linkage of guanidinium (G) and sulfonate (S) groups of organomonosulfonates and disulfonates has yielded over 450 host-guest compounds.

A representative example is the guanidinium–sulfonate framework built from 1,2,4,5-tetra(4-sulfonatophenyl)benzene (TSPB), an aromatic tetrasulfonate which can potentially generate three distinct types of host architectures: lamellar, a zeolite-like framework constructed from truncated octahedra, or stacks of 2-D grids, as illustrated schematically in Fig. 10.5.1. The guanidinium cations form hydrogen-bonded bridges between adjacent TSPB molecules along the *a* and *c* axes to generate square-like GS

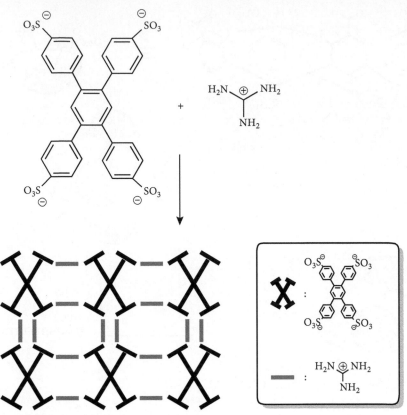

Fig. 10.5.1 Schematic representation of a two-dimensional GS hydrogen-bonded grid formed by guanidinium cations and TSPB anions. Linkage of a stack of such grids by GS hydrogen bonds in the third dimension generates the GS channel-type host framework.

channels (Type I) along the *b* axis with edge lengths of approximately 680 pm. Channel I is surrounded by two larger hexagonal channels created by the cleft of TSBP, one with a cross section of 900 pm × 1550 pm. (Type II) and the other 1170 pm × 1160 pm. (Type III). All three types of channels are occupied by dioxane guest molecules in the ratio 1:2:2.

Further investigation has shown that the crystal architecture (i.e. lamellar versus cylindrical) and the shape of GS cylinders can be regulated in a predictable way by the molecular symmetries and conformational constraints of the organopolysulfonates building blocks.

Two porous hydrogen-bonded organic frameworks (HOFs) based on 4,4′-biphenyldisulfonic acid and 1,5-napthalenedisulfonic acid that are non-covalently bonded to guanidinium ions to form infinite pillar-brick type arrangements are reported to exhibit ultrahigh proton conduction values (σ) 0.75×10^{-2} S cm^{-1} and 1.8×10^{-2} S cm^{-1} respectively under humidified conditions.

Two types of hydrogen-bonded hexagonal network (HexNet) structures (CPSM-1 and CPSM-2) have been assembled based on the C_3-symmetric buckybowl derivative hexakis(carboxyphenyl)sumanene CPSM. CPSM-1 has a wavy HexNet structure with an alternate alignment of upward and downward

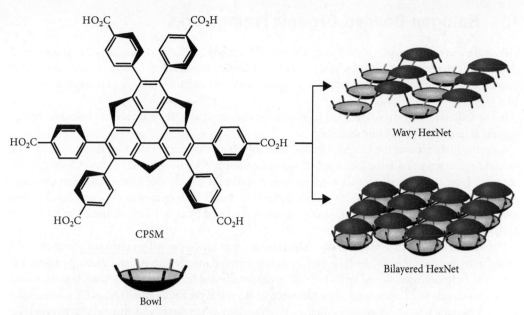

Fig. 10.5.2 Two kinds of hexagonal networks (HexNets) constructed with C_3-symmetric bowl-shaped molecule CPSM possessing six peripheral hydrogen-bonding groups: non-planar wavy CPSM-1 and bilayered CPSM-2 HexNets.

facing bowls, whereas CPSM-2 has a bi-layered HexNet structure composed of hamburger-shaped bowl dimers (Fig. 10.5.2). This work demonstrates that non-planar π-systems can be networked two-dimensionally by an appropriate supramolecular synthon to achieve structurally well-defined unique bumpy π-sheets.

References

W. Xiao, C. Hu, and M. D. Ward, Guest exchange through single crystal–single crystal transformations in a flexible hydrogen-bonded framework, *J. Am. Chem. Soc.* **136**, 14200–6 (2014).

Y. Liu, W. Xiao, J. J. Yi, C. Hu, S.-J. Park, and M. D. Ward, Regulating the architectures of hydrogen-bonded frameworks through topological enforcement, *J. Am. Chem. Soc.* **137**, 3386–92, (2014).

A. Karmakar, R. Illathvalappil, B. Anothumakkool, A. Sen, P. Samanta, A, V. Desai, S. Kurungot, and S. K. Ghosh, Hydrogen-bonded organic frameworks (hofs): a new class of porous crystalline proton-conducting materials, *Angew. Chem. Int. Ed.* **55**, 10667–71 (2016).

I. Hisaki, H. Toda, H. Sato, N. Tohnai, and H. Sakurai, A hydrogen-bonded hexagonal buckybowl framework, *Angew. Chem. Int. Ed.* **56**, 15294–8 (2017).

I. Hisaki, C. Xin, K. Takahashi, and T. Nakamura, Designing hydrogen-bonded organic frameworks (HOFs) with permanent porosity, *Angew. Chem. Int. Ed.* **58**, 11160–70 (2019).

I. Hisaki, Y. Suzuki, E. Gomez, Q. Ji, N. Tohnai, T. Nakamura, and A. Douhal, Acid responsive hydrogen-bonded organic frameworks, *J. Am. Chem. Soc.* **141**, 2111–21 (2019).

10.6 **Halogen-Bonded Organic Frameworks**

The origin and characteristics of the halogen bond are described in Chapter 7. In general, a halogen bond XB can be represented as D⋯X—Y, where X is an electrophilic halogen atom (Lewis acid, XB donor), D is a donor of electron density (Lewis base, XB acceptor), and Y is a carbon, nitrogen, or halogen atom (Fig. 10.6.1).

In the following paragraphs, selected highlights on the supramolecular assembly of halogen-bonded organic frameworks (XOFs) are described.

A recent study reports the halogen-bonded assembly of a prolate speroidal supramolecular capsule involving matching of a tetra(2,3,5,6-tetrafluoro-4-iodophenyl) XB donor with a tetra(3,5-lutidyl) XB acceptor. (Fig. 10.6.2) Single-crystal X-ray analysis revealed the presence of 12 molecular components: XB donor and acceptor resorcin[4]arene hemispheric cavitands each encapsulating one benzene guest molecule, and the resulting supramolecular capsule is stabilized by eight co-crystallized MeOH solvate molecules.

The reactions between a tetrahedrally-shaped tecton, tetrakis-(4-(iodoethynyl)phenyl)methane, and tetraphenylphosphonium halides (Fig. 10.6.3) readily afforded interpenetrated and densely packed diamondoid architectures sustained by C–I⋯X⁻ (X⁻ = chloride, bromide, iodide) interactions. In this series of robust iso-structural halogen-bonded supramolecular networks, the halide anions act as four-connecting nodes, while the tetraphenylphosphonium cations function as templates and structural supports (Fig. 10.6.4).

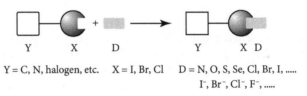

Y = C, N, halogen, etc.   X = I, Br, Cl   D = N, O, S, Se, Cl, Br, I, ...
I⁻, Br⁻, Cl⁻, F⁻, ...

Fig. 10.6.1 Scheme showing formation of the halogen bond.

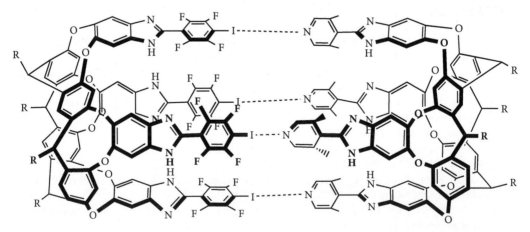

Fig. 10.6.2 Halogen-bonded supramolecular capsule (R = *n*-hexyl) structurally characterized by X-ray crystallography at 100 K; the measured I⋯N halogen bond distances and C–I...N bond angles are 4 × 282 pm and 171–178°.

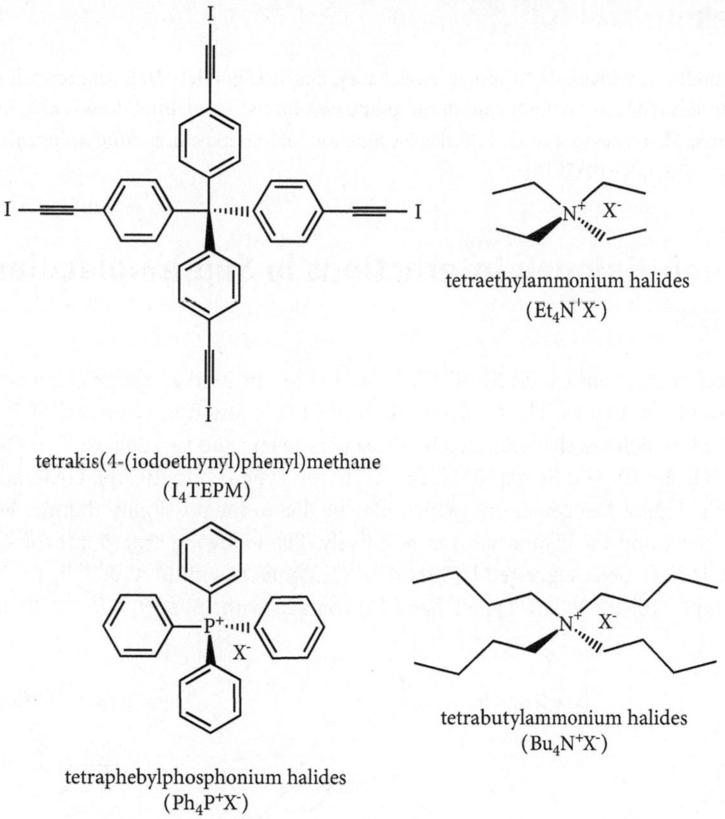

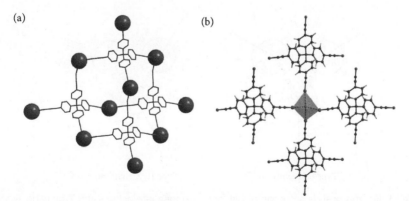

tetrakis(4-(iodoethynyl)phenyl)methane
(I_4TEPM)

tetraethylammonium halides
($Et_4N^+X^-$)

tetraphebylphosphonium halides
($Ph_4P^+X^-$)

tetrabutylammonium halides
($Bu_4N^+X^-$)

Fig. 10.6.3 Structural formulas of tetrakis{4-(iodoethynyl)phenyl}methane (left) and a series of quaternary ammonium and phosphonium halide salts (right) in this study ($X^- = Cl^-$, Br^-, I^-).

(a)   (b)

Fig. 10.6.4 (a) Crystal structure of I_4TEPM·$Ph_4P^+I^-$ emphasizing its diamondoid nature (hydrogen atoms and tetraphenylphosphonium cations have been omitted for clarity). (b) Tetrahedral coordination environment around an iodide ion.

References

O. Dumele, B. Schreib, U. Warzok, N. Trapp, C. A. Schalley, and F. Diederich, Halogen-bonded supramolecular capsules in the solid state, in solution, and in the gas phase, *Angew. Chem. Int. Ed.* **56**, 1152–7 (2017).

C. A. Gunawardana, M. Đaković, and C. B. Aakeröy, Diamondoid architectures from halogen-bonded halides, *Chem. Commun.* **54**, 607–10 (2018).

10.7 Halogen-Halogen Interactions in Supramolecular Frameworks

Halogen⋯halogen interactions (X⋯X; X = F, Cl, Br, and I) can be used as design elements in the assembly of supramolecular networks. The $C-X_1\cdots X_2-C$ contacts are found in symmetrical Type I and bent Type II geometries, which are characterized by three parameters and two angles: $\theta_1 = C-X_1\cdots X_2$ and $\theta_2 = X_1\cdots X_2-C$ ($\theta_1 = \theta_2$ for Type I; $\theta_1 \approx 180°$ and $\theta_2 \approx 90°$ for Type II, Fig. 10.7.1). Generally, the θ_2 angle often occurs at the lighter halogen atoms, which may be due to the possibility that the heavier halogen is polarized positivity and the lighter halogen negatively. The following criteria for the classification of Type I and type II have been suggested by Desiraju: (1) contacts with $0° \leq |\theta_1 - \theta_2| \leq 15°$ are Type I; (2) contacts with $|\theta_1 - \theta_2| \geq 30°$ are Type II; and (3) contacts with $15° \leq |\theta_1 - \theta_2| \leq 30°$ are quasi-Type

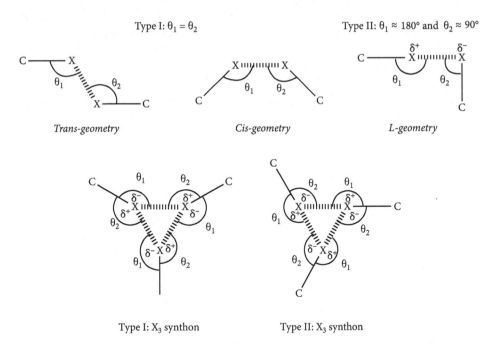

Fig. 10.7.1 Geometrical classification of halogen⋯halogen interactions and Type I and Type II X_3 supramolecular synthons.

I/Type II. Owing to the electrostatic nature of X⋯X contacts, the pure Type II is mainly in the region of $|\theta_1 - \theta_2| = 45$–$50°$ for Cl⋯Cl, 55–60° for Br⋯Br, and 65–70° for I⋯I interactions, which indicated that the angular attributes of the Type II interaction become more sharply demarcated when the polarizability increases. For the distances, Type I predominates at shortest distances, while Type II is more frequent closer to the sum of the van der Waals radii owing to the electrostatic nature of Type II contacts. However, when the distance is greater than the sum of the van der Waals radii, Type I is again favored. A significant result of these studies is that Type I is a geometry-based contact that arises from close packing and is found for all halogens, whereas type II arises from an electrophile–nucleophile pairing, and only Type II exhibits true XB. In addition, organic fluorine prefers Type I–F⋯F contacts, whereas Cl, Br, and I prefer Type II contacts. In contrast, heterohalogen⋯halogen interactions are predominately of Type II geometry because of the greater polarizability of the electron density associated with the heavier halogen. In addition, the X_3 (Cl_3, Br_3, I_3) supramolecular synthons are often used as design elements in crystal engineering.

Recent years have witnessed an increasing number of fascinating metal-containing supramolecular networks assembled by halogen-halogen bonding. Metal-bound halogens and halide ions are strongly nucleophilic and serve as halogen bonding acceptors, whereas carbon-bound halogens are strongly electrophilic and act as halogen bond donors. Hence XB plays a major role in the control of intermolecular recognition and self-assembly processes. As shown in Fig. 10.7.2, supramolecular networks based on XB can be formed with halometallates.

Presently the types of dihalogen interactions shown in Fig. 10.7.3 have been identified in the assembly of supramolecular architectures.

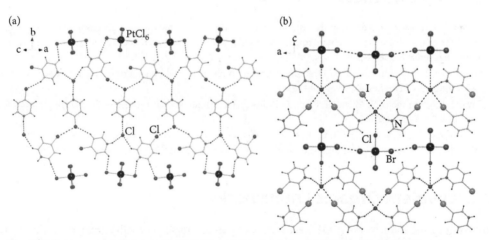

Fig. 10.7.2 Non-covalent interactions in the crystal structure of (a) $(4\text{-ClpyH})_3[PtCl_6]Cl$ and (b) $(3\text{-IpyH})_2[AuBr_{3.35}Cl_{0.65}]Br_{0.30}Cl_{0.70}$.

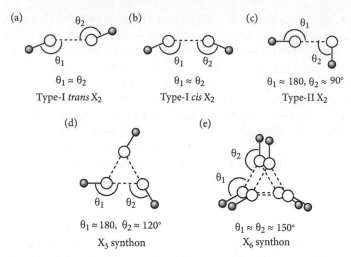

Fig. 10.7.3 Schematic representation of different types of dihalogen interactions: (a) Type-I *trans* dihalogen bonding, (b) Type-I *cis* dihalogen bonding, (c) Type-II dihalogen bonding, (d) X_3 Synthon formed by Type-II dihalogen interactions, and (e) X_6 Synthon.

Reference

M. A. Niyas, R. Ramakrishnan, V. Vijay, E. Sebastian, and M. Hariharan, Anomalous halogen–halogen interaction assists radial chromophoric assembly, *J. Am. Chem. Soc.* **141**, 4536–40 (2019).

General References

F. Zordan, S. L. Purver, H. Adams, and L. Brammer, Halometallate and halide ions: nucleophiles in competition for hydrogen bond and halogen bond formation in halopyridinium salts of mixed halide-halometallate anions, *CrystEngComm* **7**, 350–4 (2005).

A. Mukherjee, S. Tothadi, and G. R. Desiraju, Halogen bonds in crystal engineering: like hydrogen bonds yet different, *Acc. Chem. Res.* **47**, 2514–24 (2014).

G. Cavallo, P. Metrangolo, R. Milani, T. Pilati, A. Priimagi, G. Resnati, and G. Terraneo, The halogen bond, *Chem. Rev.* **116**, 2478–601 (2016).

B. Li, S.-Q. Zang, L.-Y. Wang, and T. C. W. Mak, Halogen bonding: A powerful, emerging tool for constructing high-dimensional metal-containing supramolecular networks, *Coord. Chem. Rev.* **308**, 1–21 (2016).

10.8 Covalent Organic Frameworks

Covalent organic frameworks (COFs) are a class of porous organic crystalline materials whose backbone is composed entirely of light elements (B, C, N, O, Si, S) connected by conventional covalent bonds. A breakthrough in their assembly was made in 2005, when high-yielding crystalline COF-1 $[(C_3H_2BO)_6 \cdot (C_9H_{12})_1]$ was efficiently synthesized from a simple "one-pot" condensation reaction in which three phenyldiboronic acid $\{C_6H_4[B(OH)_2]_2\}$ molecules converge to form a planar B_3O_3 (boroxine) ring with the elimination of three water molecules (Fig. 10.8.1).

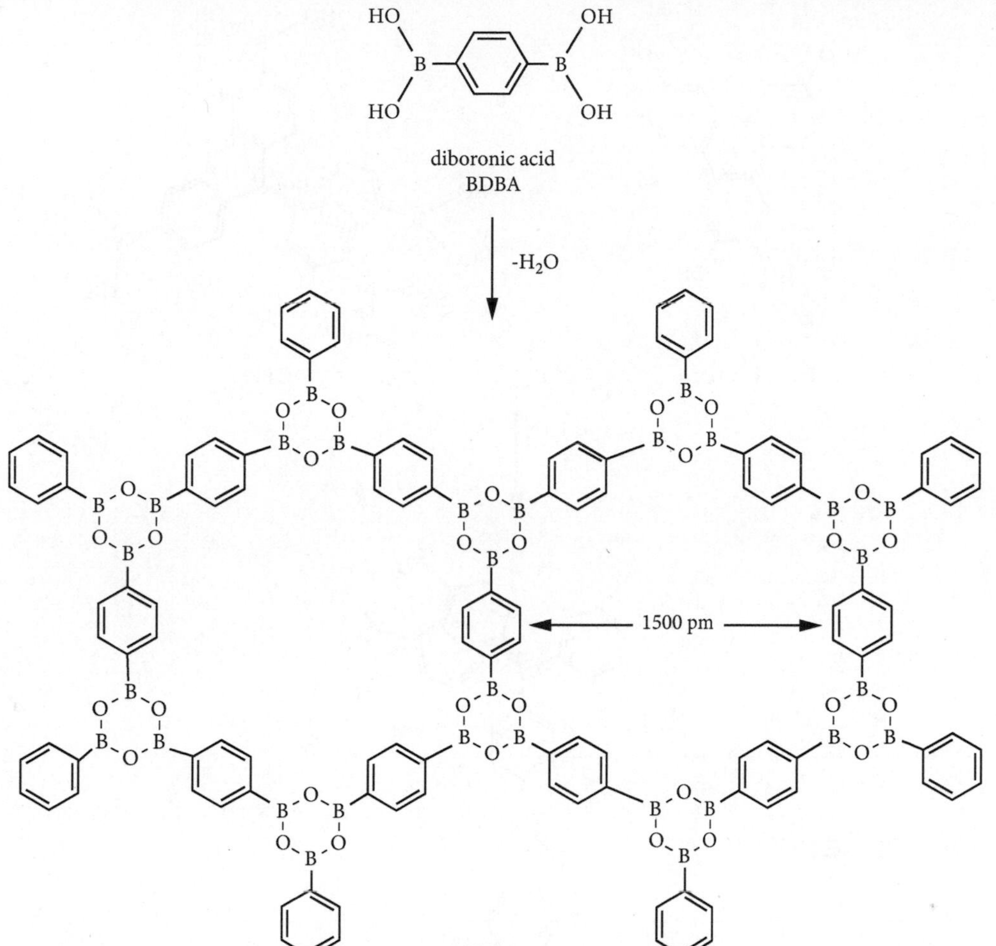

Fig. 10.8.1 Condensation reaction of phenyldiboronic acid produces extended COF-1, which has a pore size of 700 pm.

In a landmark case, COF-108 was prepared through the reaction between triangular hexahydroxytriphenylene (HHTP) and tetrahedral tetra(4-dihydroxyborylphenyl)methane (TBPM) to produce a 3-D network material with 3-D pores and an extremely low density of $0.17\ g\ cm^{-3}$ (Fig. 10.8.2). This synthetic route is remarkably general in scope and has been applied to a large array of boronic acids. For example, isoreticular series of layered COFs with hexagonal or square topologies have been reported. This resulted in families of related materials with systematically increased pore sizes, reaching pore apertures as large as 4.7 nm. In addition, COFs have also been synthesized through the formation of C–N bonds.

Click chemistry has been applied to generate porous organic networks. A Husigen 1,3-dipolar cycloaddition reaction employing tetrakis(4-ethynylphenyl)methane and tetrakis(4-azidophenyl)methane yielded the porous organic network shown in Fig. 10.8.3.

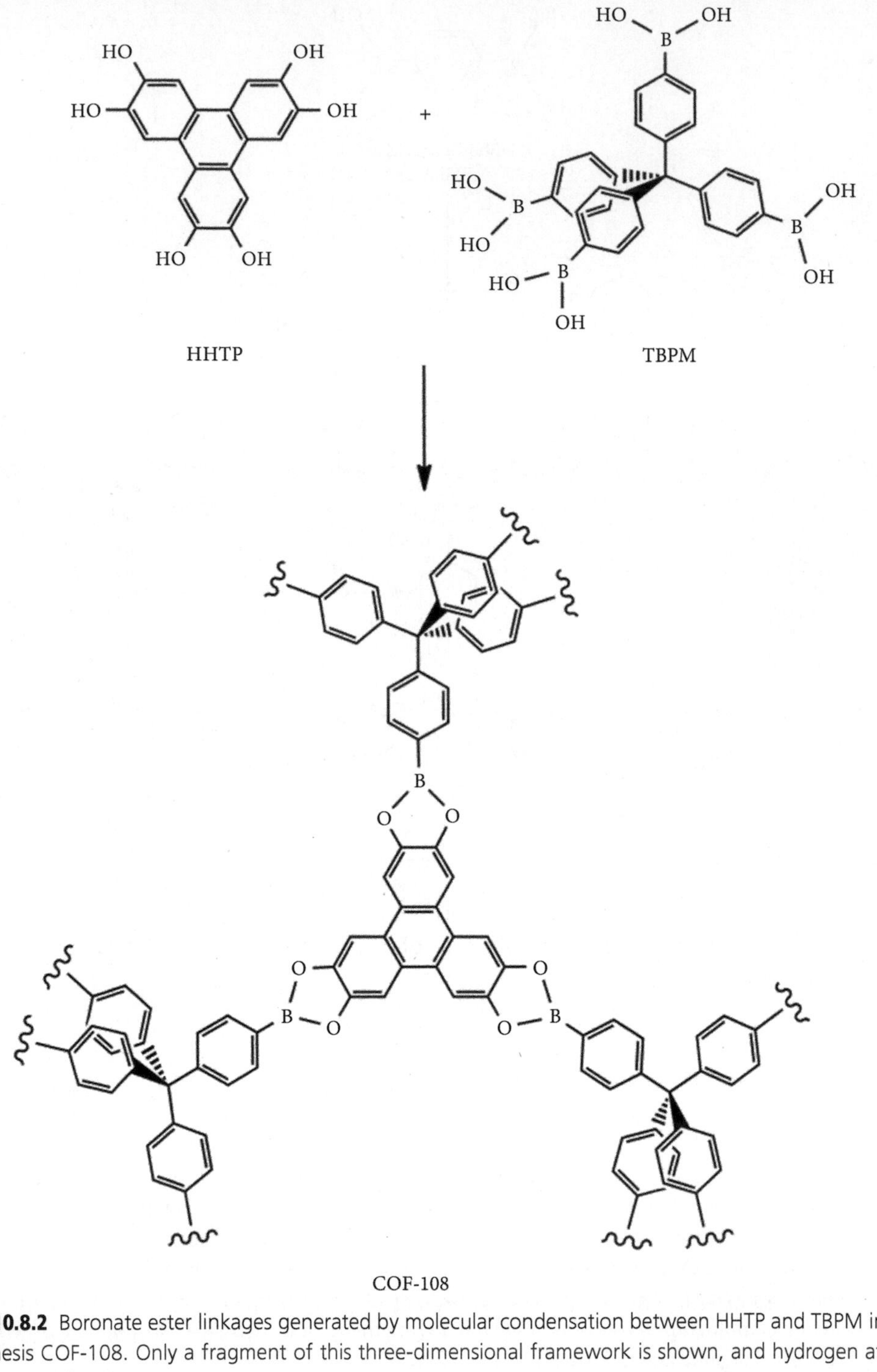

Fig. 10.8.2 Boronate ester linkages generated by molecular condensation between HHTP and TBPM in the synthesis COF-108. Only a fragment of this three-dimensional framework is shown, and hydrogen atoms are omitted for clarity.

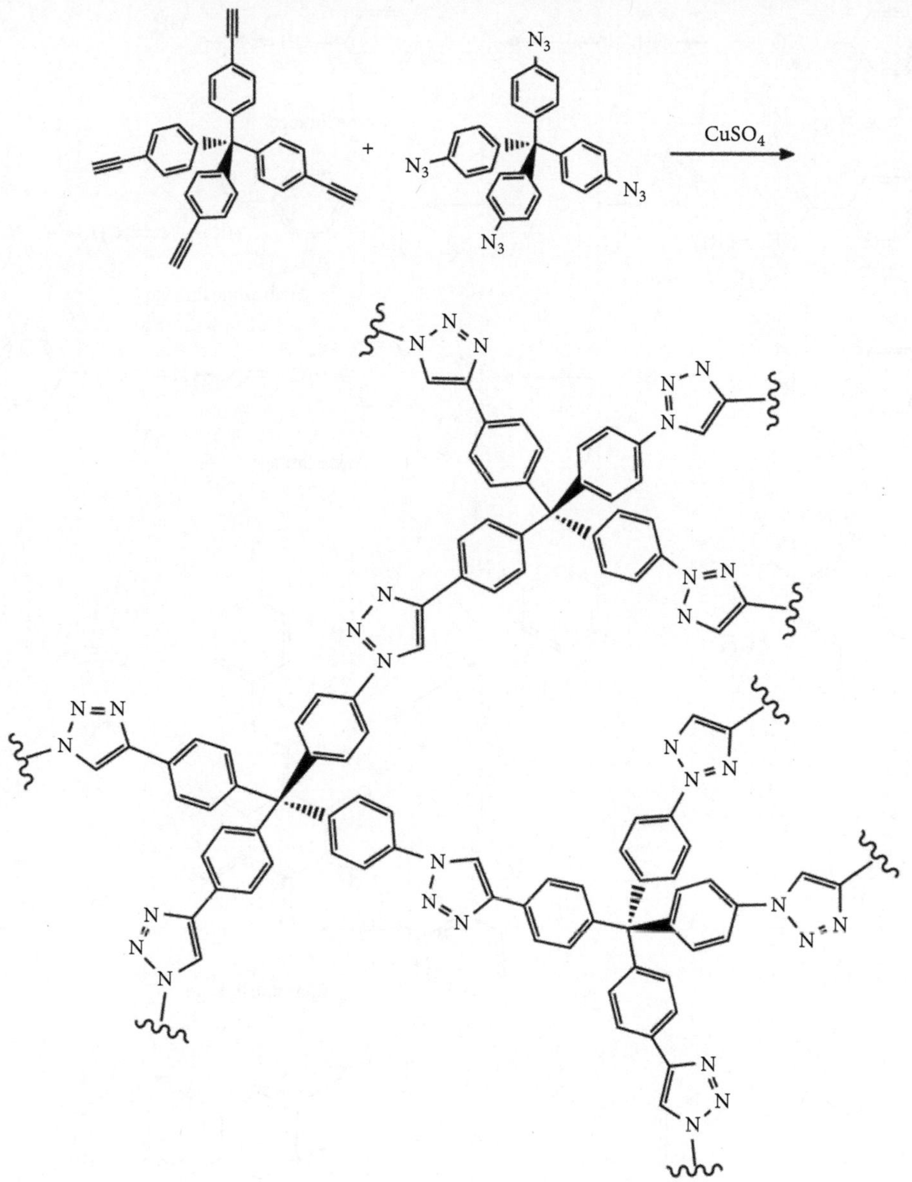

Fig. 10.8.3 Crystalline three-dimensional COF generated through click chemistry.

Recently the use of Schiff-base chemistry or dynamic imine-chemistry has been developed for the synthesis of covalent organic frameworks (COFs). The chemical reactions employed for obtaining this specific class of COFs are summarized in Fig. 10.8.4.

To date, a good number of stable 2-D COFs and a few 3-D COFs with different topologies have been synthesized. Fig. 10.8.5 shows the synthetic scheme and structure of a dual-core 2-D COF obtained by combining D_{2h} symmetric and C_2 symmetric monomers.

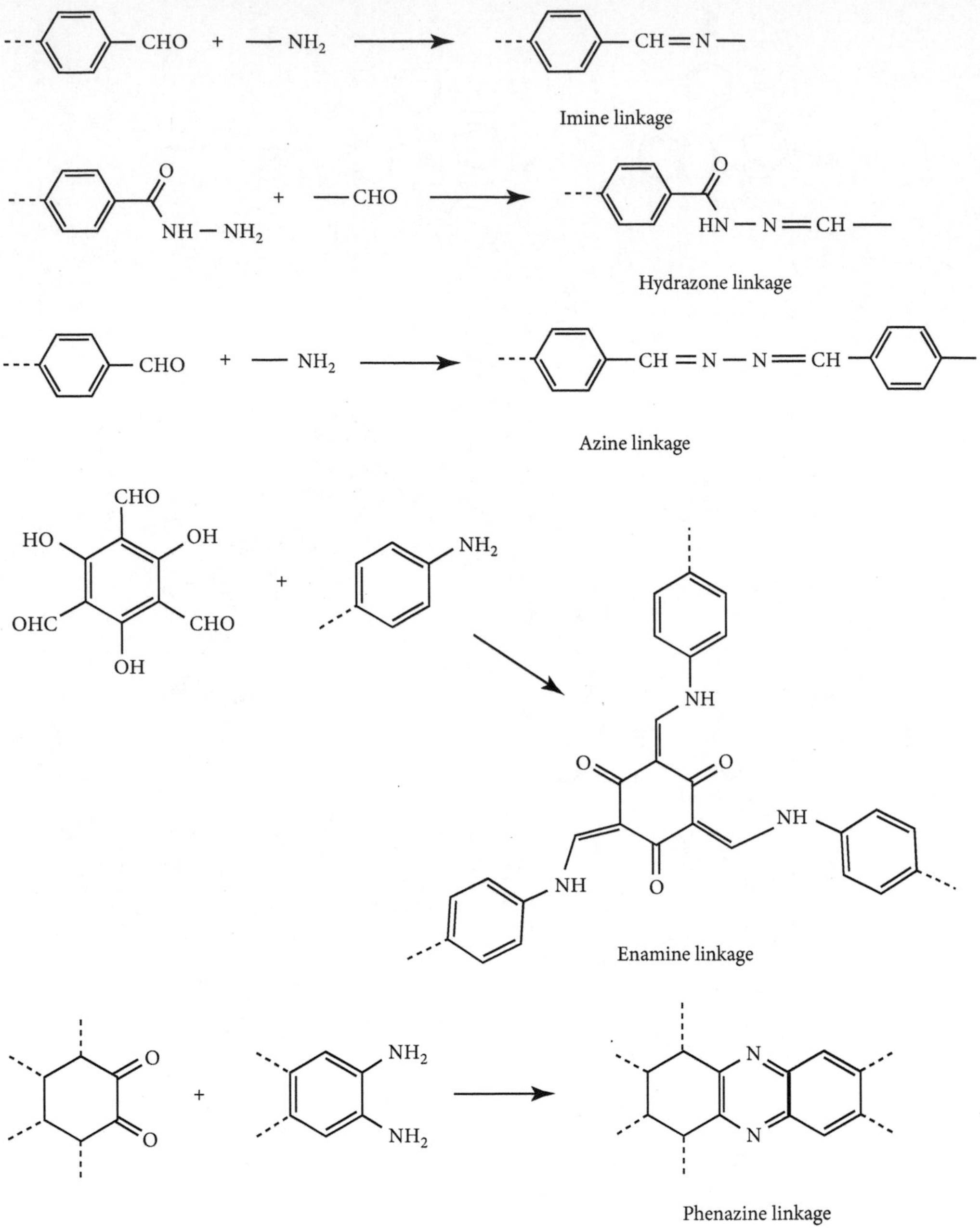

Fig. 10.8.4 Dynamic chemical reactions used for the preparation of COFs via Schiff-base chemistry.

The structure of a crystalline COF featuring azine linkage assembled by condensation between hydrazine and 1,3,6,8-tetrakis(4-formylphenyl)pyrene TFPPY under hydrothermal condition is shown in Fig. 10.8.6. The pyrene units occupy the vertices and the bent diazabutadiene (C=N–N=C–) linkers

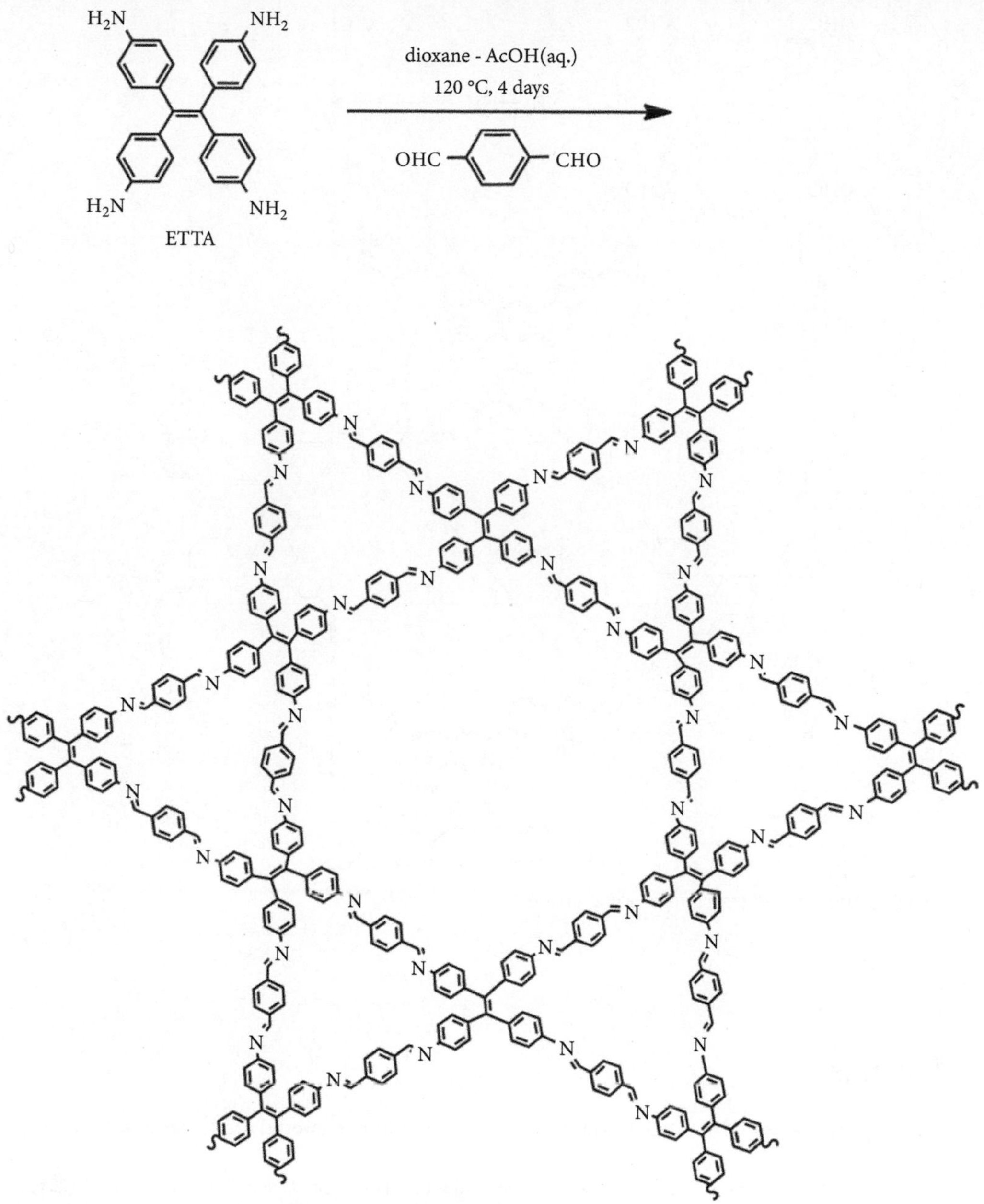

Fig. 10.8.5 Synthesis and structure of a dual-pore 2D COF.

are located at the edges of rhombic polygon sheets, which adopt the AA-stacking mode to form ordered pyrene columns and microporous rhombic channels.

Further information on the assembly of supramolecular architectures by covalent bonding are provided in the following reviews.

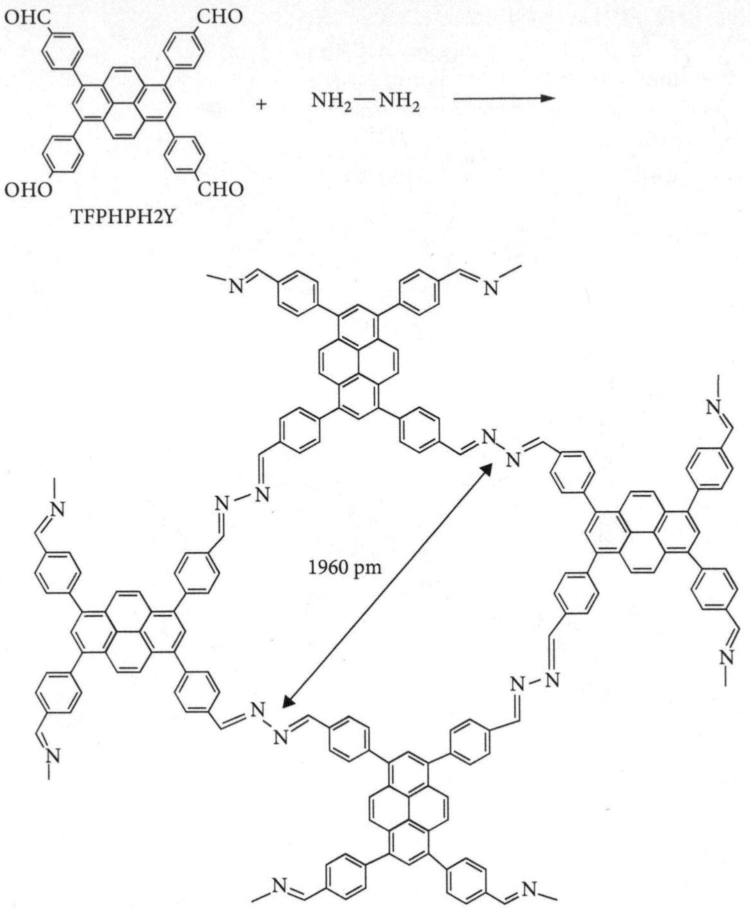

Fig. 10.8.6 Synthesis and structure of azine-linked COF.

References

P. J. Waller, F Gándara, and O. M. Yaghi, Chemistry of covalent organic frameworks, *Acc. Chem. Res.* **48**, 3053–63 (2015).

J. Jiang, Y. Zhao, and O. M. Yaghi, Covalent chemistry beyond molecules, *J. Am. Chem. Soc.* **138**, 3255–65 (2016).

C. S. Diercks and O. M. Yaghi, The atom, the molecule, and the covalent organic framework, *Science* **355**, eaal1585 (2017).

S. Kandambeth, K. Dey, and R. Banerjee, Covalent organic frameworks: chemistry beyond the structure, *J. Am. Chem. Soc.* **141**, 1807–22 (2019).

C. Gropp, T. Ma, N. Hanikel, and O. M. Yaghi, Design of higher valency in covalent organic frameworks, *Science* **370**, 6515, eabd6406 (2020). DOI:10.1126/science.abd6406

K. Geng, T. He, R. Liu, S. Dalapati, K. T. Tan, Z. Li, S. Tao, Y. Gong, Q. Jiang, and D. Jiang, Covalent organic frameworks: design, synthesis, and functions, *Chem. Rev.* **120**, 8814–933 (2020).

G. A. Leith, A. A. Berseneva, A. Mathur, K. C. Park, and N. B. Shustova, A multivariate toolbox for donor–acceptor alignment: MOFs and COFs, *Trends Chem.* **2**, 4, 367–82 (2020).

10.9 **Metal-Organic Frameworks**

Metal-organic frameworks (MOFs) now constitute a major research topic in materials chemistry in view of their immense scope in actual and potential applications. As numerous papers covering this field are published annually, it would be prudent to list several pioneering papers and authoritative reviews as recommended reading to students who aspire to pursue postgraduate studies in this area.

References

B. F. Hoskins and R. Robson, Design and construction of a new class of scaffolding-like materials comprising infinite polymeric frameworks of 3D-linked molecular rods. a reappraisal of the $Zn(CN)_2$ and $Cd(CN)_2$ structures and the synthesis and structure of the diamond-related frameworks $[N(CH_3)_4][Cu^IZn^{II}(CN)_4]$ and $Cu^I[4,4',4'',4'''$- tetracyanotetraphenylmethane]$BF_4 \cdot xC_6H_5NO_2$, *J. Am. Chem. Soc.* **112**, 1546–54 (1990). This pioneering paper laid the foundation of blossoming research on metal-organic networks.

H. Furukawa, K. E. Cordova, M. O'Keeffe, and O. M. Yaghi, The chemistry and applications of metal-organic frameworks, *Science* **341**, 1230444 (2013).

T. R. Cook, Y.-R. Zheng, and P. J. Stang, Metal–organic frameworks and self-assembled supramolecular coordination complexes: comparing and contrasting the design, synthesis, and functionality of metal–organic materials, *Chem. Rev.* **113**, 734–77 (2013).

J.-P. Zhang, P.-Q. Liao, H.-L.g Zhou, R.-B. Lin, and X.-M. Chen, Single-crystal X-ray diffraction studies on structural transformations of porous coordination polymers, *Chem. Soc. Rev.* **43**, 5789–814 (2014).

W.-X. Zhang, P-Q Liao, R.-B. Lin, Y.-S. Wei, M.-H. Zeng, and X.-M. Chen, Metal cluster-based functional porous coordination polymers, *Coord. Chem. Rev.* **293–4**, 263–78 (2015).

A. Schoedel, M. Li, D. Li, M. O'Keeffe, and O. M. Yaghi, Structures of metal–organic frameworks with rod secondary building units, *Chem. Rev.* **116**, 12466–535 (2016).

J.-P. Zhang, H.-L. Zhou, D.-D. Zhou, P.-Q. Liao, and X.-M. Chen, Controlling flexibility of metal–organic frameworks, *Nat. Sci. Rev.* **5**, 907–19, (2018). https://doi.org/10.1093/nsr/nwx127

M. J. Kalmutzki, N. Hanikel, and O. M. Yaghi, Secondary building units as the turning point in the development of the reticular chemistry of MOFs. *Sci. Adv.* **4**, eaat9180 (2018).

M. Ding, R. W. Flaig, H.-L. Jiang, and O. M. Yaghi, Carbon capture and conversion using metal-organic frameworks and MOF-based materials, *Chem. Soc. Rev.* **48**, 2783–828 (2019).

Q. Wang and D. Astruc, State of the art and prospects in metal–organic framework (MOF)-based and MOF-derived nanocatalysis, *Chem. Rev.* **120**, 1438–511 (2020).

M. Kalaj, K. C. Bentz, S. Ayala, Jr., J. M. Palomba, K. S. Barcus, Y. Katayama, and S. M. Cohen, MOF-polymer hybrid materials: from simple composites to tailored architectures, *Chem. Rev.* **120**, 8267–302 (2020).

X. Wu, L. K. Macreadie, and P. A. Gale, Anion binding in metal-organic frameworks, *Coord. Chem. Rev.* **432** 213708 (2021).

10.10 **Synthetic Molecular Machines**

Over the past decades, chemists have continued to achieve impressive success in major research areas such as the total synthesis of natural products, the design of enantioselective catalysts, the assembly of functional materials, and dynamic molecular systems involving control and utility of motion at the nano scale.

The metal-organic framework MOF-5 having composition $Zn_4O(BDC)_3$ (BDC = 1,4-benzenedicarboxylate) has a 3-D porous structure in which robust inorganic $[OZn_4]^{6+}$ cluster units are joined to an array of octahedral $[O_2C\text{-}C_6H_4\text{-}CO_2]^{2-}$ (1,4-benzenedicarboxylate, BDC) groups to form a robust and highly porous cubic framework, which can adsorb hydrogen gas up to 4.5 weight percent (17.2 H_2 molecules per formula unit) at 78 K and 1.0 weight percent at room temperature and 20 bar pressure.

A recent study on a topologically similar iso-reticular metal-organic framework BODCA-MOF (BODCA = 1,4-bicyclo[2.2.2]octane dicarboxylic acid) constructed from molecular components that resemble gyroscopes housed within solid frames (Fig. 10.10.1). Using spin-lattice relaxation ^{1}H solid-state NMR at 29.49 and 13.87 MHz in the temperature range of 2.3–80 K, it was shown that internal rotation of the alicyclic cage linker occurs with a potential energy barrier of 0.185 kcal mol^{-1}. Since each cage has an exterior case surrounding a rotating axis, the crystal has a solid exterior but contains moving internal parts, thereby providing the first demonstration that a single material can be both static and moving, or amphidynamic. Computer simulations of the crystal demonstrated that the "BODCA spheres" spin constantly, each either clockwise or counterclockwise, at up to 50 billion rotations per second, as fast as they would behave in empty space.

Notably, the Nobel Prize in Chemistry 2016 was awarded jointly to Jean-Pierre Sauvage, Sir J. Fraser Stoddart, and Bernard L. Feringa "for the design and synthesis of molecular machines." Students using this book are recommended to read their Nobel Lectures that feature each winner's personal account on developing his scientific pursuits and achievements.

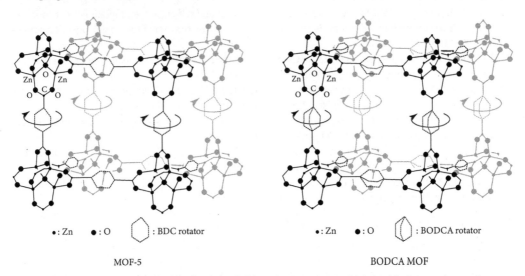

•: Zn ●: O ⬡ : BDC rotator

MOF-5

•: Zn ●: O ⬡ : BODCA rotator

BODCA MOF

Fig. 10.10.1 Isostructural networks of MOF-5 (left) and BODCA-MOF (right) with the corresponding 1,4-benzenedicarboxylate (BDC) and 1,4-bicyclo[2.2.2]octane dicarboxylic acid (BODCA) rotators and the static carboxylate and Zn_4O clusters (hydrogen atoms omitted for clarity).

References

C. S. Vogelsberg, F. J. Uribe-Romo, A. S. Lipton, S. Yanga, K. N. Houk, S. Brown, and M. A. Garcia-Garibay, Ultrafast rotation in an amphidynamic crystalline metal organic framework, *PNAS.* **114**, 13613–8 (2017).

J.-P. Sauvage, From chemical topology to molecular machines (Nobel Lecture), *Angew. Chem. Int. Ed.* **56**, 11080–93 (2017).

J. F. Stoddart, Mechanically interlocked molecules (MIMS)—molecular shuttles, switches, and machines (Nobel Lecture), *Angew. Chem. Int. Ed.* **56**, 11094–125 (2017).

B. L. Feringa, The art of building small: from molecular switches to motors (Nobel Lecture), *Angew. Chem. Int. Ed.* **56**, 11060–78 (2017).

V. García-López, D. Liu, and J. M. Tour, Light-activated organic molecular motors and their applications, *Chem. Rev.* **120**, 79–124 (2020).

I. Aprahamian, The future of molecular machines, *ACS Cent. Sci.* **6**, 347–58 (2020).

10.11 **Supramolecular Chemistry in Materials Science Research**

The chemical and physical properties of molecular material depend upon the contacts between its component parts and their arrangement. In the past decade, supramolecular chemistry has played a dominant role in the development of principles, properties, and applications in molecular materials research. Readers are referred to the following exemplary presentations on some of the most exciting and inspiring recent achievements.

References

E. Busseron, Y. Ruff, E. Moulin, and N. Giuseppone, Supramolecular self-assemblies as functional nanomaterials, *Nanoscale* **5**, 7098–140 (2013).

H. Yang, B. Yuan, and X. Zhang, Supramolecular chemistry at interfaces: host–guest interactions for fabricating multifunctional biointerfaces, *Acc. Chem. Res.* **47**, 2106–15 (2014).

D. Prochowicz, A. Kornowicz, and J. Lewiński, Interactions of native cyclodextrins with metal ions and inorganic nanoparticles: fertile landscape for chemistry and materials science, *Chem. Rev.* **117**, 13461–501 (2017).

D. B. Amabilino, D. K. Smith, and J. W. Steed, Supramolecular materials, *Chem. Soc. Rev.* **46**, 2404–20 (2017).

A. K. Nangia and G. R. Desiraju, Crystal engineering: an outlook for the future, *Angew. Chem. Int. Ed.* **58**, 4100–7 (2019).

General References

Books

J.-M. Lehn, *Supramolecular Chemistry: Concepts and Perspectives*, VCH, Weinheim, 1995.

G. R. Desiraju and T. Steiner, *The Weak Hydrogen Bond in Structural Chemistry and Biology*, Oxford University Press, New York, 1999.

J. W. Steed and J. L. Atwood, *Supramolecular Chemistry*, Wiley, Chichester, 2000.

J.-P. Sauvage (ed.), *Transition Metals in Supramolecular Chemistry*, Wiley, Chichester, 1999.

M. Fujita (ed.), *Molecular Self-assembly: Organic versus Inorganic Approaches* (Structure and Bonding, Vol. 96), Springer, Berlin, 2000.

G. R. Desiraju (ed.), *Crystal Design: Structure and Function*, Wiley, Chichester, 2003.

F. Toda and R. Bishop (eds), *Separations and Reactions in Organic Supramoilecular Chemistry*, Wiley, Chichester, 2004.

J. W. Steed and. L. Atwood, *Supramolecular Chemistry*, 3rd ed., Chichester, 2021.

Supramolecular Assembly

N. W. Ockwig, O. Delgardo-Friedrichs, M. O'Keeffe, and O. M. Yaghi, Reticular chemistry: occurrence and taxonomy of nets and grammar for the design of frameworks. *Acc. Chem. Res.* **38**, 176–82, (2005).

D. Zhang, T. K. Ronson, and J. R. Nitschke, Functional capsules via subcomponent self-assembly, *Acc. Chem. Res.* **51**, 2423–36 (2018).

S. Datta, M. L. Saha, and P. J. Stang, Hierarchical assemblies of supramolecular coordination complexes, *Acc. Chem. Res.* **51**, 2047–63 (2018).

M. K. Corpinot and D.-K. Bučar, A Practical Guide to the design of molecular crystals, *Cryst. Growth Des.* **19**, 1426–53 (2019).

P. Chakraborty, A. Nag, A. Chakraborty, and T. Pradeep, Approaching materials with atomic precision using supramolecular cluster assemblies, *Acc. Chem. Res.* **52**, 2–11(2019).

A. K. Nangia and G. R. Desiraju, Crystal engineering: an outlook for the future, *Angew. Chem. Int. Ed.* **58**, 4100–7 (2019).

J. D. Wuest, Atoms and the void: modular construction of ordered porous solids, Nature Communications **11**, Article number: 4652 (2020).

Metal-Organic Frameworks and Covalent-Organic Frameworks

M. J. Kalmutzki, N. Hanikel, and O. M. Yaghi, Secondary building units as the turning point in the development of the reticular chemistry of MOFs. *Sci. Adv.* **4**, eaat9180 (2018).

M. Ding, R. W. Flaig, H.-L. Jiang, and O. M. Yaghi, Carbon capture and conversion using metal-organic frameworks and MOF-based materials, *Chem. Soc. Rev.* **48**, 2783–828 (2019).

G. A. Leith, A. A. Berseneva, A. Mathur, K. C. Park, and N. B. Shustova, A multivariate toolbox for donor–acceptor alignment: MOFs and COFs, *Trends Chem.*, **2**, 4, 367–82 (2020).

X. Chen, K. Geng, R. Liu, K. T. Tan, Y. Gong, Z. Li, S. Tao, Q. Jiang, and D. Jiang, Covalent organic frameworks: chemical approaches to designer structures and built-in functions, *Angew. Chem. Int. Ed.* **59**, 5050 –91 (2020).

Metal Nanoclusters

L. Liu and A. Corma, Metal catalysts for heterogeneous catalysis: from single atoms to nanoclusters and nanoparticles, *Chem. Rev.* **118**, 4981–5079 (2018).

Z. Lei, X.-K. Wan, S.-F. Yuan, Z.-J. Guan, and Q.-M. Wang, Alkynyl approach toward the protection of metal nanoclusters, *Acc. Chem. Res.* **51**, 2465–74 (2018).

X. Kang and M. Zhu, Tailoring the photoluminescence of atomically precise nanoclusters, *Chem. Soc. Rev.* **48**, 2422–57 (2019).

L. L.-M. Zhang, G. Zhou, G. Zhou, H.-K. Lee, N. Zhao, O. V. Prezhdo, and T. C. W. Mak: Core-dependent properties of copper nanoclusters: valence-pure nanoclusters as NIR TADF emitters and mixed-valence ones as semiconductors. *Chem. Sci.* **10**, 10122–8 (2019).

Y. Du, H. Sheng, D. Astruc, and M. Zhu, Atomically precise noble metal nanoclusters as efficient catalysts: a bridge between structure and properties, *Chem. Rev.* **120**, 526–622 (2020).

X.-Y. Dong, Y. Si, J.-S. Yang, C. Zhang, Z. Han, P. Luo, Z.-Y. Wang, S.-Q. Zang, and T. C. W. Mak, Ligand engineering to achieve ratiometric oxygen sensing in a silver cluster-based metal-organic framework. *Nat. Commun.* (2020) 11, 3678 <https://doi.org/10.1038/s41467-020-17200-w>

Y. Jin, C. Zhang, X.-Y. Dong, S.-Q. Zang, and T. C. W. Mak, Shell engineering to achieve modification and assembly of atomically-precise silver clusters, *Chem. Soc. Rev.* (2021). DOI:10.1039/d0cs01393e

Appendix I

New Types of Chemical Bonding Described in This Book

Recommended Books and Monographs

The titles in this appendix booklist, including many time-honored and treasured classics that have stood the test of time, are broadly divided into six categories for readers who wish to delve deeper into specific topics.

Inorganic Chemistry

1. J. Emsley, *The Elements*, 3rd edn, Clarendon Press, Oxford, 1998.

2. J. Emsley, *Nature's Building Blocks: An A-Z Guide to the Elements*, Oxford University Press, 2011.

3. G. Rayner-Canham, *The Periodic Table: Past, Present, and Future*, World Scientific, Singapore, 2020.

4. J. E. Huheey, E. A. Keiter, and R. L. Keiter, *Inorganic Chemistry: Principles of Structure and Reactivity*, 4th edn, Harper Collins, New York, 1993.

5. B. E. Douglas, D. H. McDaniel, and J. J. Alexander, *Concepts and Models of Inorganic Chemistry*, 3rd edn, Wiley, New York, 1993.

6. P. W. Atkins, T. L. Overton, J. P. Rourke, M. T. Weller, and F. A. Armstrong, *Inorganic Chemistry*, 5th edn, W. H. Freeman, New York, 2010.

7. G. L. Miessler, P. J. Fisher, and D. A. Tarr, *Inorganic Chemistry*, 5th edn, Pearson, Upper Saddle River, New Jersey, 2013.

8. C. E. Housecroft and A. G. Sharpe, *Inorganic Chemistry*, 5th edn, Pearson Education, Essex, 2018.

9. J. E. House, *Inorganic Chemistry*, 3rd edn, Elsevier/Academic Press, London, 2020.

10. N. S. Hosmane, *Advanced Inorganic Chemistry: Applications in Everyday Life*, Elsevier/Academic Press, London, 2017.

11. A. F. Wells, *Structural Inorganic Chemistry*, 5th edn, Oxford University Press, Oxford, 1984.

12. U. Müller, *Inorganic Structural Chemistry*, 2nd edn, Wiley, Chichester, 2006.

13. W.-K. Li, G.-D. Zhou, and T. C. W. Mak, *Advanced Structural Inorganic Chemistry*, Oxford University Press, Oxford, 2008.

14. D. M. P. Mingos, *Essential Trends in Inorganic Chemistry*, Oxford University Press, Oxford, 1998.

15. D. M. P. Mingos and D. J. Wales, *Introduction to Cluster Chemistry*, Prentice-Hall, Englewood Cliffs, New Jersey, 1990.

16. F. A. Cotton, G. Wilkinson, C. A. Murillo, and M. Bochmann, *Advanced Inorganic Chemistry*, 6th edn, Wiley, New York, 1999.

17. N. N. Greenwood and A. Earnshaw, *Chemistry of the Elements*, 2nd edn, Butterworth-Heinemann, Oxford, 1997.

18. R. Xu, W. Pang, J. Yu, Q. Huo, and J. Chen (eds.), *Chemistry of Zeolites and Related Porous Materials: Synthesis and Structure*, Wiley (Asia), Singapore, 2007.

19. S. T. Liddle, D. P. Mills, and L. S. Natrajan (eds.), *The Lanthanides and Actinides: Synthesis, Reactivity, Properties and Applications*, World Scientific, Singapore, 2022.

20. R. Xu and Y. Xu (eds.), *Modern Inorganic Synthetic Chemistry*, 2nd edn, Elsevier, Amsterdam, 2017.

21. R, Cao (ed.), *Advanced Structural Chemistry: Tailoring Properties of Inorganic Materials and their Applications*, Vols 1–3, Wiley-VCH, Hoboken, New Jersey, 2021.

Chemical Bonding and Molecular Structure

22. R. McWeeny, *Coulson's Valence*, 3rd edn, Oxford University Press, Oxford, 1979.

23. N. W. Alcock, *Bonding and Structure: Structural Principles in Inorganic and Organic Chemistry*, Ellis Horwood, New York, 1990.

24. A. Haaland, *Molecules and Models: The Molecular Structures of Main Group Element Compounds*, Oxford University Press, Oxford, 2008.

25. R. J. Gillespie and P. L. A. Popelier, *Chemical Bonding and Molecular Geometry from Lewis to Electron Densities*, Oxford University Press, New York, 2001.

26. A. Rauk, *Orbital Interaction Theory of Organic Chemistry*, 2nd edn, Wiley, New York, 2001.

27. I. Fleming, *Molecular Orbitals and Organic Chemical Reactions: Reference Edition*, Wiley, London, 2010.

Molecular Symmetry and Group Theory

28. Y. Öhrn, *Elements of Molecular Symmetry*, Wiley, New York, 2000.

29. A. M. Lesk, *Introduction to Symmetry and Group Theory for Chemists*, Kluwer, Dordrecht, 2004.

30. K. C. Molloy, *Group Theory for Chemists: Fundamental Theory and Applications*, 2nd edn, Woodhead Publishing, Cambridge, 2011.

31. M. Ladd, *Symmetry of Crystals and Molecules*, Oxford University Press, New York, 2014.

32. P. R. Bunker and P. Jensen, *Fundamentals of Molecular Symmetry*, Institute of Physics Publishing, Bristol, 2005.

33. S. F. A. Kettle, *Symmetry and Structure: Readable Group Theory for Chemists*, 3rd edn, Wiley, Chichester, 2007.

34. F. A. Cotton, *Chemical Applications of Group Theory*, 3rd edn, Wiley, New York, 1990.

Symmetry and Structure of Crystalline Solids

35. G. Burns and A. M. Glazer, *Space Groups for Solid State Scientists*, 2nd edn, Academic Press, New York, 1990.

36. T. Hahn (ed.), *International Tables for Crystallography, Volume A: Space-group symmetry*, 5th edn (corrected reprint), Kluwer Academic, Dordrecht, 2005.

37. E. Prince (ed.), *International Tables for Crystallography, Volume C: Mathematical, physical and chemical tables*, 3rd edn, Kluwer Academic, Dordrecht, 2004.

38. M. O'Keeffe and B. G. Hyde, *Crystal Structures. I. Patterns and Symmetry*, Mineralogical Society of America, Washington, DC, 1996; paperback, Dover Publications, 2020.

39. B. Douglas and S.-M. Ho, *Structure and Chemistry of Crystalline Solids*, Springer, New York, 2006.

40. B. G. Hyde and S. Andersson, *Inorganic Crystal Structures*, Wiley, New York, 1989.

41. Á. Vegas. *Structural Models of Inorganic Crystals. From the Elements to the Compounds*, Editorial Universitat Politècnica de València, 2018.

42. R. C. Buchanan and T. Park, *Materials Crystal Chemistry*, Marcel Dekker, New York, 1997.

43. I. D. Brown, *The Chemical Bond in Inorganic Chemistry: The Valence Bond Model*, Oxford University Press, New York, 2002.

44. J. M. Robertson, *Organic Crystals and Molecules*, Cornell University Press, New York, 1953.

45. G. A. Jeffrey, *An Introduction to Hydrogen Bonding*, Oxford University Press, New York, 1997.

46. G. R. Desiraju, J. J. Vittal and A. Ramanan, *Crystal Engineering: A Textbook*, World Scientific, Singapore, 2011.

47. Bernstein, *Polymorphism in Molecular Crystals*, 2nd edn, Oxford University Press, New York, 2020.

48. J. G. S. Rohrer, *Structure and Bonding in Crystalline Materials*, Cambridge University Press, Cambridge, 2001.

49. E. W. T. Tiekink and J. J. Vittel (eds.), *Frontiers in Crystal Engineering*, Wiley, Chichester, 2006.

50. E. W. T. Tiekink, J. J. Vittel, and M. J. Zaworotko (eds.), *Organic Crystal Engineering* (Frontiers in Crystal Engineering II), Wiley, Chichester, 2010.

51. E. W. T. Tiekink and J. Zukerman-Schpector (eds.), *The Importance of Pi-Interactions in Crystal Engineering* (Frontiers in Crystal Engineering III), Wiley, Chichester, 2012.

X-Ray Crystallography

52. C. Hammond, *The Basics of Crystallography and Diffraction*, 4th edn, Oxford University Press, New York, 2018 (reprinted).

53. P. G. Radaelli, *Symmetry in Crystallography: Understanding the International Tables*, Oxford University Press, New York, 2011.

54. J. P. Glusker and K. N. Trueblood, *Crystal Structure Analysis: A Primer*, 3rd edn, Oxford University Press, New York, 2010.

55. M. Ladd and R. Palmer, *Structure Determination by X-Ray Crystallography: Analysis by X-rays and Neutrons*, 5th edn, Springer, New York, 2013.

56. W. Massa, *Crystal Structure Determination*, 2nd edn, Springer-Verlag, Berlin, 2004, corrected 5th printing, 2010.

57. W. Clegg (ed.), A. J. Blake, W. Clegg, J. M. Cole, J. S. O. Evans, P. Main, S. Parsons, and D. J. Watkin, *Crystal Structure Analysis: Principles and Practice*, 2nd edn, Oxford University Press, New York, 2009.

58. J. D. Dunitz, *X-Ray Analysis and the Structure of Organic Molecules*, 2nd Corrected Reprint, VCH Publishers, New York, 1995.

59. J. P. Glusker, M. Lewis, and M. Rossi, *Crystal Structure Analysis for Chemists and Biologists*, VCH Publishers, New York, 1994.

60. C. Giacovazzo (ed.), C. Giacovazzo, H. L. Monaco, G. Artioli, D. Viterbo, M. Milanesio, G. Ferraris, G. Gilli, P. Gilli, G. Zanotti, and M. Catti, *Fundamentals of Crystallography*, 3rd edn, Oxford University Press, 2011.

61. U. Müller, *Symmetry Relationships between Crystal Structures*, Oxford University Press, Oxford, 2013.

62. I. Hargittai and B. Hargittai (eds.), *Science of Crystal Structures: Highlights in Crystallography*, Springer International Publishing Switzerland, 2015.

63. T. C. W. Mak and G.-D. Zhou, *Crystallography in Modern Chemistry: A Resource Book of Crystal Structures*, Wiley-Interscience, New York, 1992; Wiley Professional Paperback Edition, 1997.

Chemical Bonding and Spectroscopic Study

64. I. Pauling, *The Nature of the Chemical Bond and the Structure of Molecules and Crystals: An Introduction to Modern Structural Chemistry*, 3rd edn, Cornell University Press, Ithaca, New York, 1960.

65. W. J. Hehre, L. Radom, P. v. R. Schleyer, and J. A. Pople, *Ab initio Molecular Orbital Theory*, Wiley, New York, 1986.

66. F. Weinhold and C. R. Landis, *Valency and Bonding: A Natural Bond Orbital Donor-Acceptor Perspective*, Cambridge University Press, Cambridge, 2005.

67. P. W. M. Jacobs, *Group Theory with Applications in Chemical Physics*, Cambridge University Press, Cambridge. 2005.

68. S. K. Kim, *Group Theoretical Methods and Applications to Molecules and Crystals*, Cambridge University Press, Cambridge, 1999.

69. B. S. Tsukerblat, *Group Theory in Chemistry and Spectroscopy: A Simple Guide to Advanced Usage,* Academic Press, London, 1994.

70. J. Cioslowski (ed.), *Quantum-Mechanical Prediction of Thermochemical Data*, Kluwer, Dordrecht, 2001.

71. M. Reiher and A. Wolf, *Relativistic Quantum Chemistry: The Fundamental Theory of Molecular Science*, Wiley-VCH, Weinheim, 2009.

72. K. Balasubramanian (ed.), *Relativistic Effects in Chemistry, Part A and Part B*, Wiley, New York, 1997.

73. J. B. Foresman and A. Frisch, *Exploring Chemistry with Electronic Structure Methods*, 3rd ed., Gaussian, Inc.: Wallingford, CT, 2015.

74. I. B. Bersuker, *Electronic Structure and Properties of Transition Metal Compounds: Introduction to the Theory*, 2nd edn, Wiley, New Jersey, 2010.

75. R. G. Parr and W. Yang, *Density-Functional Theory of Atoms and Molecules*, Oxford University Press, New York, 1989.

76. D. S. Sholl and J. A. Steckel, *Density Functional Theory: A Practical Introduction*, Wiley, Hoboken, NJ, 2009.

77. P. F. Bernath, *Spectra of Atoms and Molecules*, 2nd edn, Oxford University Press, New York, 2005.

78. K. Nakamoto, *Infrared and Raman Spectra of Inorganic and Coordination Compounds (Part A: Theory and Applications in Inorganic Chemistry; Part B: Applications in Coordination, Organometallic, and Bioinorganic Chemistry)*, 6th edn, Wiley, Hoboken, New Jersey, 2009.

79. R. A. Nyquist, *Interpreting Infrared, Raman, and Nuclear Magnetic Resonance Spectra, Vols. 1 & 2*, Academic Press, San Diego, CA, 2001.

80. J.R. Ferraro and K. Nakamoto, *Introductory Raman Spectroscopy*, 2nd edn, Academic Press, San Diego, California, 2003.

81. J. P. Fackler, Jr., and L. R. Falvello (eds.), *Techniques in Inorganic Chemistry*, CRC Press, Taylor & Francis, Boca Raton, Florida, 2011.

82. J. Demaison, J. E. Boggs, and A. G. Csaszar (eds.), *Equilibrium Molecular Structures: From Spectroscopy to Quantum Chemistry*, CRC Press, Boca Raton, Florida, 2011.

83. D. W. H. Rankin, N. Mitzel, and C. Morrison, *Structural Methods in Molecular Inorganic Chemistry*, Wiley, Chichester, 2013.